DYNAMICS OF ATMOSPHERIC MOTION

(FORMERLY THE CEASELESS WIND)

JOHN A. DUTTON

Professor of Meteorology and Dean
College of Earth and Mineral Sciences
The Pennsylvania State University

DOVER PUBLICATIONS, INC.
New York

About the frontispiece

This striking satellite photo illustrates many of the phenomena discussed in the book, and the patterns in the clouds reveal the wanderings of the wind.

A cyclone off the east coast of North America has a cold front trailing south toward Florida and a warm front extending to the east across the North Atlantic, joining the fronts trailing westward from a storm system in the eastern North Atlantic. Here, then, is a picture of a polar front and the associated wave cyclones.

A tropical cyclone (hurricane Alma) appears just north of Guiana on the South American coast, and the intertropical convergence zone can be seen extending east-west across the Atlantic just north of the equator.

The high clouds strung out over the central part of South America suggest the subtropical jet stream, while the arcing cirrus patterns off the east coast of South America at the bottom of the picture are typical of jet-stream structure and suggest the polar-front jet.

Finally, many of the oceanic cloudy regions demonstrate the regular arrangements typical of cellular convection. (*National Environmental Satellite Service photograph, courtesy of Vincent J. Oliver.*)

Published in Canada by General Publishing Company, Ltd., 30 Lesmill Road, Don Mills, Toronto, Ontario.
Published in the United Kingdom by Constable and Company, Ltd., 3 The Lanchesters, 162–164 Fulham Palace Road, London W6 9ER.

This Dover edition, first published in 1995, is an unabridged and unaltered republication of the Dover edition of 1986 which was published under the book's original title, *The Ceaseless Wind: An Introduction to the Theory of Atmospheric Motion*. The 1986 Dover edition was a corrected and enlarged republication of the work first published by the McGraw-Hill Book Company, New York, in 1976. A new preface and a new chapter (Chapter 15) were written for the 1986 Dover edition by the author who also made corrections in the text. Material that originally appeared on the end papers was relocated to a new appendix (Appendix 3).

Library of Congress Cataloging-in-Publication Data

Dutton, John A.
[Ceaseless wind]
Dynamics of atmospheric motion / John A. Dutton.
p. cm.
Previously published: The ceaseless wind. New York: Dover Publications, 1986.
Includes bibliographical references and index.
ISBN 0-486-68486-5
1. Atmospheric circulation. 2. Atmospheric sciences. 3. Dynamic meteorology. I. Title.
QC880.4.A8D88 1995
551.5′17—dc20 94-25265
CIP

Manufactured in the United States of America
Dover Publications, Inc., 31 East 2nd Street, Mineola, N.Y. 11501

To those from whom I've learned
And to their teachers too;
To those whom I have taught,
And to their students too.

CONTENTS

PREFACE TO THE DOVER EDITION

Atmospheric science has achieved, in the decade since the first edition of this book, a deeper understanding of atmospheric processes. Improved methods of using both theoretical and numerical models to simulate and predict the evolution of atmospheric flow patterns have stimulated a clarification of the motivations for modeling and a more precise assessment of the potential rewards. Advances both in theory and in observational and computational technology offer new opportunities for progress in dynamic meteorology.

The major challenges cited in the first edition—understanding interactions between scales, understanding the sensitivity of solutions, developing a theory of climate, and understanding moist convection—remain formidable despite a decade of progress. An increasing awareness of the importance of interactions between the atmosphere, ocean, land, and biosphere in controlling the evolution of the planet gives these challenges new breadth and prominence. Multidisciplinary efforts to understand the earth as an integrated system are converging toward the development of an earth-system science that will be concerned with planetary-scale versions of the same challenges.

Atmospheric science, because of its maturity in modeling and in the mathematical representation of physical systems, will have the opportunity to be a leader in the development of this new science. The progress being made in understanding nonlinear aspects of atmospheric flow, in seeking topological representations of flow evolution, and in developing a formal structure within which to examine the process of modeling, will stimulate both atmospheric science and associated disciplines. These topics are addressed in a new chapter (Chap. 15) prepared for this edition.

As atmospheric science takes advantage of new technological opportunities and attacks new problems, the fundamental concepts will be increasingly important for organizing comprehension of complex interactions and for developing advanced modes of thought. Thus, with this book now being made available to the broad audience reached by Dover Publications, it is my hope that it will stimulate interest in atmospheric science and will continue to contribute to an understanding of the theory of atmospheric motion.

JOHN A. DUTTON

PREFACE TO THE FIRST EDITION

The aim of this book is to make the theory of atmospheric motion easier to learn and easier to teach. That is not an easy task, for most of the enthusiastic students who begin to study meteorology are interested in aspects of the atmosphere that are more exciting to them than pages filled with equations.

But the phenomena that have captured their interest are all part of the complex chain of processes in an atmosphere that is forced into a ceaseless motion by thermal effects on a rotating planet. To study and understand these processes, to predict them and control them, it is necessary to have a theory that explains them. And this requires concepts and techniques from mathematics, thermodynamics, mechanics, and fluid dynamics. These ideas, arising in separate disciplines, must be combined and unified in order to comprehend atmospheric events.

Most students arrive in a dynamics classroom for the first time with their mathematical training untried by applications and their physics unassailed by the complexities of moving fluids. Teachers of dynamics know well that they cannot proceed with the meteorological derivations without carefully integrating the necessary preliminary mathematics and physics.

That is what I have tried to do in this book, and it is one of the reasons for its length. Another reason is that I have tried to be thorough, both in the topics covered and in the explanations of the topics under discussion. I have tried also to be moderately rigorous, for I am convinced that students understand rigorous arguments better and more thoroughly than careless logic or explanations in which the difficult parts are left fuzzy.

I envision the book as being useful as a text in either undergraduate or graduate dynamics courses. It would be most useful to the students if their introductory courses were composed of the usual range of material selected from the various places where it occurs in the book, and if their more advanced courses went on from these points, covering additional aspects of the theory that are presented here. This would provide both student and teacher with a consistency of notation and the advantage of having the earlier material readily available for reference or review. The last few chapters can either be used to introduce the topics covered there, or individually as texts for graduate courses with supplementary material provided by the lecturer.

The book is also intended to be useful as a reference for researchers in meteorology—another reason for the attempt at completeness. I have often been frustrated in my own work by having to embark on lengthy searches for a theorem or a result. In compiling the material and choosing topics, I have paid particular attention to this problem in an attempt to ensure that most concepts that are needed by working dynamicists are included. The index has been constructed with the aim of making sure that things can be found quickly when they are needed.

The text is augmented by a liberal sprinkling of some 400 problems of various levels of difficulty. Those of particular meteorological significance or those that give results often used in later developments are marked with an asterisk. The problems are not graded by difficulty, on the theory that working meteorologists do not encounter problems specified in advance as easy or hard. A word of caution: A correct solution to most of the problems can be obtained quickly and economically if the correct technique is used. Only a few of them need to involve lengthy calculations.

The same concept motivates the organization of the book. Dynamic meteorology can be done quite quickly and economically once the basic ideas from mathematics, thermodynamics, and fluid dynamics are well in hand. Hence I have collected these concepts in what I hope is a logical grouping. Some students and teachers may prefer to proceed sequentially through the book; others may prefer to take up topics of interest and refer back to the preliminary material when necessary. The first approach is probably the more comfortable one for the expert; the second one is perhaps better for the student.

Textbooks can be written from various viewpoints. Most dynamics texts proceed in about the same sequence, isolating the thermodynamic ideas at the beginning from the discussion of motion that follows. I have tried to integrate these concepts and, in fact, to emphasize the thermodynamic aspects of atmospheric motion. The rotation of the earth, the shallowness of the atmosphere, and the curvature of the earth, all profoundly affect atmospheric motions—but those motions are created, driven, and maintained by thermodynamic forcing and thermodynamic processes. The thermodynamics creates a response, the mechanics of fluid flow and the geometry of the planet guide the response into the atmospheric patterns that we observe.

I am confident that such an integrated approach to atmospheric dynamics, along with an increasing sophistication in mathematical methods, will provide the solutions to the challenging problems that face meteorology today. As pointed out in the first chapter, these problems are departures from the traditional ones of dynamics, and some of the most useful approximations and techniques will not be valid. But however the successful approaches are developed, they will begin from the present theory of atmospheric motion. In this book, I have tried to organize and present that theory in such a manner that today's students will be well prepared to use it to solve these problems.

If I have succeeded in accomplishing only part of the goals I have set for this book, I shall be well rewarded. The success of any teacher, and of textbooks too, must be judged by the success of the students who learn from them.

In developing the material and writing the book, I have had the assistance of several friends and colleagues, some of whom tried out portions of the book on their own students. Most of all, I have had the benefit of using preprints of the material with my own students for more than 5 years, discovering on their faces and on their examination papers both the strong points of this approach and the weak parts that needed improvement.

The bibliographic notes at the ends of chapters contain the references I used in preparing the material and some of the classic works on the various topics. I have certainly not tried to be exhaustive; that is a job for research monographs, not textbooks. If students need to know a special result or concept to read a text, then it should be in the text; they have troubles enough without being sent off in search of references to prepare their lesson. Additional reading is most effectively assigned by a well-informed instructor, who can take advantage of recent literature and developments in making such assignments. Some of the problems derive from articles or books I have read; credit is given (by last name) for those for which I know the source.

Finally, a word about errors. Writing a book is a constant war against errors, and the longer the book, the more opportunity the errors have to win now and then. Despite my best efforts, I fear that some will have crept in. There can be typographical errors, mathematical mistakes, and errors of omission, in addition to logical errors, and those that stem from ignorance. I will be grateful to readers who might write about any they find.

GUIDE TO USE OF THE BOOK

This book is intended to serve as the text for several undergraduate and graduate courses in dynamic meteorology. The plan adopted in writing it was to present each of the main topics over a range from the basic versions introduced in a first course through the more advanced material covered in graduate courses.

There are many different possible paths through the book, and each instructor will surely want to choose his own, depending in part on his own interests and the abilities of his students. The suggestions given here are presented in part as an indication of my own choices and the usual practice, and are aimed in part at the independent reader.

Undergraduate Dynamics. The topics usually covered in undergraduate courses include:

Basic concepts of dynamics and modeling	(Chap. 1, to be woven in as the course proceeds)
Physical and mathematical fundamentals, including numerical weather prediction	Chap. 2
Atmospheric thermodynamics	Appendix 1, Secs. 3.2 and 3.3
Hydrostatics and model atmospheres	Secs. 4.1 to 4.3
Parcel motion in the vertical	Sec. 4.4 through Sec. 4.4.1
Atmospheric structure	Sec. 4.6
Vector analysis and integration theorems	Chap. 5 through Sec. 5.3.1
Fluid motion in inertial coordinates	Chap. 6 through Sec. 6.4 or through Sec. 6.7
Meteorological equations of motion	Chap. 7 through Sec. 7.3
Models of the wind: geostrophic, thermal, and gradient winds	Secs. 9.1, 9.2, 9.3, 9.5, 9.6
The vorticity equation	Secs. 10.1, 10.2
Applications of vorticity to diagnosis and prediction	Secs. 10.4, 10.6
Determination of vertical motion	Secs. 9.8, 10.5
Moist thermodynamics	Chap. 8

Graduate Dynamics. A reasonable goal for a first course in graduate dynamics is to ensure that the students have a firm and integrated understanding of the logic and motivations, as well as methodology, of the material listed above. From this foundation, an attempt can be made to develop some creativity in using the methods of dynamic meteorology to tackle more advanced material, special cases, and applications.

In my own graduate dynamics course, I put particular emphasis on the implications of nonlinearity (beginning with Sec. 2.5) and basic concepts of numerical integration of the equations of motion, thermodynamics of atmospheric motion including the material of Chap. 3 and Sec. 6.7, the nonlinear theory of parcel motion and stability in Secs. 4.4 and 4.5, integral theorems and cartesian tensor analysis in Chap. 5, the mathematical model of fluid flow in Sec. 5.7, special forms of the equations of motion in Sec. 6.5, the axiomatic approach of Sec. 6.8, boundary conditions in Sec. 6.9, coordinate transformations in Chap. 7, and wind models in Chap. 9.

Advanced Courses. The remaining material of the book can be used as texts for a variety of advanced courses in dynamics. Among the possibilities are a course in atmospheric energetics, using Chap. 11, and one or more courses in atmospheric wave motion using Chaps. 12 to 14, a course in quasi-geostrophic theory using Chap. 14, a course in tensor methods using Secs. 5.4.2 and 5.5, and with applications from Secs. 5.6.4 and 5.7, generalized coordinates in Sec. 7.4, advanced vorticity analysis in Secs. 10.11 and 10.12, and a course in vorticity and circulation using Chap. 10.

ACKNOWLEDGMENTS

Several colleagues and students have provided me with enthusiastic assistance in the preparation of this book, and I am grateful to them. Dr. Robert E. Livezey aided me throughout the many years of manuscript preparation with constructive criticism, discovery of errors, and careful proofreading. Hampton N. Shirer created the computer codes and graphic routines for figures in Chaps. 5, 10, and 12; B. Parhami did the codes for those in Chap. 8. The computer analyses necessary for the figures in Chap. 11 were done with routines developed by Dr. Dennis Deaven. The material for Sec. 4.6 was assembled by Thomas T. Warner. The lecture notes of Prof. John B. Hovermale were an invaluable aid in developing Chap. 8. Profs. Donald R. Johnson and Richard A. Anthes used portions of the book in their dynamics courses and provided me with careful accounts of both their reactions and those of their students.

The development of some of the new results that appear in the book was part of research performed under National Science Foundation Grants 1595X and 40754X. The penultimate draft was prepared in the quiet of the Danish countryside during a year's visit with the Meteorology Group of the Research Establishment Risø of the Danish Atomic Energy Commission.

Finally, the whole project would have been impossible without the devoted efforts of Mrs. Judy Rippel, Mrs. Kandie Baluch, Mrs. Joyce Sabol, and Mrs. Tanya Sharer, who carefully converted manuscript into typescript and then cheerfully retyped the almost endless revisions.

JOHN A. DUTTON

THE CEASELESS WIND

The creature who atoned the beasts he hunted
By painting timeless portraits on the wall
Surely watched the aspens tremble
Before an angry sky with rumbling voice
And jagged, blinding tongue.

His symbols on the stone became
A wheel that changed the face of Earth.
They showed him how to sail the void,
To walk the soil of other worlds
And look back upon the patterns of the wind.

It brings the water from the sea to shore,
Transports the seed to rooting place,
And lifts the falcon to her meal and mate.
But the rhythmic chorus of the wind
Is etched in mind and not upon a scroll.

Each vast current, each small eddy
Is conceived upon a balance
And is born in dynamic labor.
Each passes on and leaves the monument
Of thermal calm, of thermal peace.

The ocher marks have changed
To integrals and bits in core.
Still there is no picture of the artist,
And his restless mind has yet to touch
The secret of the wind, the ceaseless wind.

J. A. D.

PART ONE

Foundations of Atmospheric Dynamics

Atmospheric dynamics combines the knowledge of thermodynamics about heat and energy transfer and that of mechanics about motion to produce a theory of thermally forced motion on a rotating planet. In this first part, we consider the basic methods for describing the structure and behavior of the atmosphere, develop a thermodynamic theory of sufficient scope for our later applications, analyze the vertical structure of the atmosphere, and collect and motivate the mathematical and kinematical methods that will be used to describe the motion fields of the atmosphere in the later parts of the book.

1
TO TOUCH THE WIND

The central goal of atmospheric science is to learn about the winds, to understand why they come to life, why they take the forms and patterns they do, why they change and evolve and finally die in the birth of new winds.

The motions of the atmosphere are complex and can be studied and described in many ways—on charts, with graphs, with equations, or with computers. But all knowledge about the wind contributes to the theory of atmospheric motion, and although this theory is a human creation, we believe that it mirrors nature's plan for the winds. With it we can reach for, and perhaps even touch, some of the wind's own truth.

1.1 THE GOALS OF DYNAMIC METEOROLOGY

The theoretical study of atmospheric motion is the province of dynamic meteorology. The subject involves a variety of techniques and concepts, and it has two major goals. The first is to provide understanding of the many facets involved in the phenomenon of atmospheric motion; the second is to provide a rational basis for prediction of future atmospheric events.

Considerable progress has been made toward these goals. We understand the consistency in atmospheric structure, and we can model most of the mechanisms at work in the large-scale atmospheric flows. We have a rather complete picture of the relations between the mass, thermal, and velocity fields, and the processes involved in the atmosphere's energy cycle are fairly well known. We can predict the sequence of events in the evolution of a cyclone, and we can predict how average quantities measured in the atmosphere's turbulent surface layer will be related.

But there are several aspects of atmospheric motion that are not yet clearly comprehended. We have little theoretical knowledge about the reasons the atmosphere has the structure it does, or why its patterns, storms, and winds evolve in the fairly regular sequence that we observe. The atmosphere presumably could respond to the solar forcing in many conceivable ways, but it always makes its own choice from the many possibilities, and the reasons for these choices are not yet clear.

The relations between features of vastly different size are not yet well understood, either. Motion in the atmosphere ranges from the globe-girdling jet stream to small, turbulent eddies (Fig. 1.1). These motion systems of different scale affect each other both in direct and subtle ways, but dynamics is just beginning to investigate those relations.

Finally, there is little knowledge about the controls on climate. We have a much firmer grasp on the dynamics of a developing cyclone than we have on a theory of climate that will reveal the essential factors determining the long-term average results of the circulation.

Predictions of either tomorrow's weather or the climate of the next decade could be, and are, made by many techniques. The contribution of dynamic meteorology to prediction is to provide a logical basis for using information about present conditions to make predictions about the future. Such predictions are presumably, but not yet always, better than predictions based on intuition or subjective techniques.

Both in studying atmospheric phenomenology and in providing prediction techniques, dynamic meteorology uses a model of the atmosphere. This model involves a number of features that are replicas of the actual atmosphere—an ideal gas replaces air, it is affected by a planetary gravitational force and forces that arise from planetary rotation, the gas is described by specific heats and other constants that have the same values as those of air. But the motion of the actual atmosphere is replaced by equations that govern the flow of the imaginary model atmosphere. These partial differential equations that prescribe the evolution of the model atmosphere are often called the laws of atmospheric motion.

That these laws are accurate seems beyond doubt. But their complexity foils their application in many ways, and serious problems develop with limited observational data, with inadequate speed or size of computers, with phenomena that are modeled correctly by the laws but are of limited interest in weather prediction. Still they are essential to the comprehension of atmospheric processes and those of any fluid, and the basic laws of fluid dynamics are one of man's great intellectual achievements. For they summarize with but a handful of symbols processes so diverse

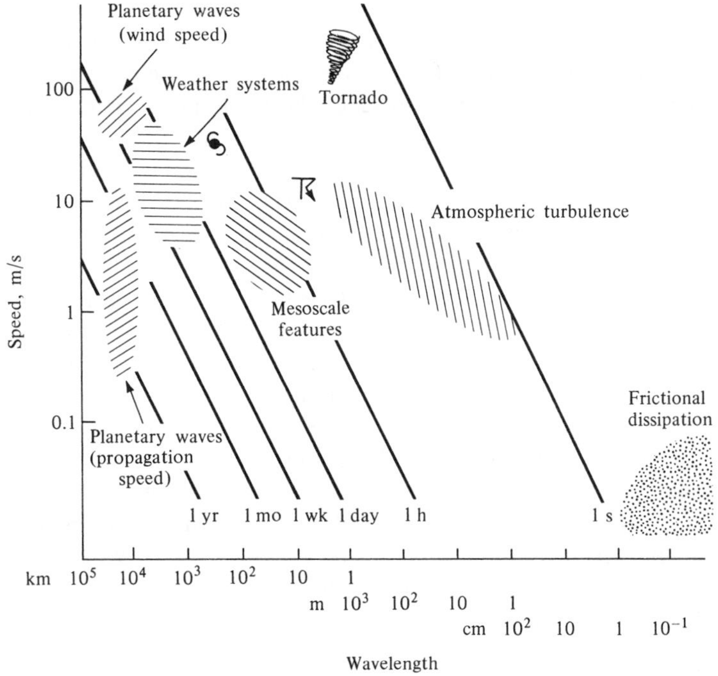

FIGURE 1.1
Scales of atmospheric phenomena. For many of these phenomena, it is possible to choose different length or speed scales, thus giving quite different time scales.

and complex that the effective application of the summary to most actual cases is still beyond our capability.

The present theory of atmospheric motion is incomplete in an important way, however. For although we have the laws of motion, we do not know the essential reason why they take precisely the form they do. And although the laws of motion are in a sense derived, they are basically empirical statements about facts we have observed to be true of fluid motion. This is not to imply that they are trivial or simple, for as all students of fluid dynamics or atmospheric motion come to appreciate, the laws of motion are an amazing creation. Given the exact initial state of a fluid, given the exact conditions on the boundary, and given a powerful enough way to compute, the equations will predict the successive states of the fluid for time unending.

Even though this is true, there are simple questions we cannot yet answer. Why does vorticity tend to be conserved? Why does the largest eddy have the size it does? Why does a wake tend to be wavelike? There is an answer to such questions: "Because the equations of motion say so." But now we ask: "Why do the equations of motion say so—what basic principle is thus satisfied? Why do the equations and nature choose this alternative from the infinite number of possibilities that are present?"

The same sort of questions arise in atmospheric science. Why are there subtropical anticyclones? Why does the total energy of the atmosphere remain nearly constant? Why does the polar-front jet meander?

Thus the basic problems of fluid dynamics and atmospheric dynamics revolve around the question of why the observed responses are those that are chosen. We can, with the equations of motion, reproduce a mathematical or numerical model of the same response nature does for a specific set of stimuli. But where we are guided by laws formed from experience, nature makes the choice for reasons we do not know. These choices must be motivated by some attempt at economy or efficiency. For some reason, it must be more acceptable to take one route toward the goal of thermodynamic equilibrium than another.

Should we ever discover the reason and be able to express it compactly in one mathematical statement, we shall have achieved the ultimate in parsimony. Such an economy in assumptions has long been sought in science and is embodied in the principle of parsimony: That theory is best which proceeds from the fewest assumptions and still gives empirically correct results. We may, of course, find that the statement "X must be true" will yield the equations of motion and answer all questions about atmospheric motion save one: "Why must X be true?"

1.2 THE LANGUAGE OF SCIENCE

Theories start from axioms or assumptions. The assumptions may be based upon empirical evidence, upon analogy with those of successful theories in other fields, upon inspiration or intuition, or upon the results of critical and creative thought. Once the assumptions are stated, the theory develops according to logical inference. The assumptions contain the implications of the theory, and in combination and restatement they produce a chain of conclusions and predictions that can be tested against the facts observed in nature. The validity of the assumptions, then, is tested by comparison of their implications and predictions with reality. Invalid theories are usually useless because the assumptions or axioms do not reflect reality accurately. Theories may also falter in the crucial test because errors of logic are made in proceeding from assumptions to predictions.

Theories are usually constructed in a halting way. The observations that need explanation accumulate, and by trial and error assumptions that might yield explanation of these data are formulated, modified, and compared once again to reality. The crucial test of the theory lies in its ability to predict correct results in cases different from those included in the initial data—to predict new facts which can be verified by observational programs based upon the suggestions of the theory of what should be true.

Theories, then, involve assumptions and logical inference based upon the assumptions. Both parts of a theory are equally important, and errors both in formulating the assumptions and in developing the inference are equally disastrous. But if we are to find errors in stating the assumptions, then we must be sure that the chains of logical inference and deduction are free from error.

The implications can be deduced with careful statements shorn of the ambiguity of casual language. They can be found with graphical aids or with drawings and models, or they can be deduced by manipulation of symbols according to the rules of logic. The more complicated the situation and the larger the possibility for error or ambiguity, the more efficient it becomes to use formal methods of logic. Thus the implications and assumptions in most physical theories are connected by mathematical arguments. The reasons are simple: We usually want quantitative predictions from the assumptions, we usually find it easier to work with the concepts and symbols of mathematics than to invent our own personalized systems of verbalized concepts, we usually find it easier to proceed mathematically without making errors of logic than to proceed with alternative methods.

So mathematics is the universal language of science. As we transact our daily business with other humans in our written and spoken language, we carry on the day's affairs with our theories and other scientists with mathematics. The fact is that correct meteorological reasoning must embody the laws of motion, either as mathematical equations and implications or as an understanding gained from long experience and empirical insight. It is also true that most beginners will understand intuitive arguments and graphical presentations more readily than equations and mathematical arguments. But these forms of inference are generally limited; they are often oversimplifications. In the final analysis, they limit both the depth with which a problem can be studied and the application of the results. Pedagogy in the atmospheric sciences must somehow combine motivation, development of intuitive understanding of the atmosphere in motion, and progress toward the mathematical maturity required for effective use of the laws of atmospheric motion.

Thus the issue of mathematics in dynamic meteorology must be faced. The only alternative is to leave the mathematics out, to leave it for another day. This has great pedagogical disadvantages and is a disservice to any serious student. The greater fault is to use the mathematics wrongly, to try to make it easy rather than do it correctly.

And the issue of mathematics must be faced in the training of weather forecasters, too. There is always the student with the lament: "I just want to be a good forecaster–I don't want to learn equations." But with few exceptions, good forecasts represent the careful application of dynamic principles. And dynamics without mathematics is tortuous at best and prone to error at worst. The best dynamicists are certainly not necessarily the best forecasters, for prediction also requires an ability to commune with the atmosphere that is the gift of few. But the best forecasters usually have good backgrounds in dynamics, and nearly all will sincerely say they wished they had learned more mathematics, more dynamics, and more about computer science.

In the end, the conclusion must be that mathematics is the most efficient way to conceptualize and reason about atmospheric processes. As an example, the complete equations of motion as written in this book involve the following meager list of symbols:

Variables: $\mathbf{v}, p, \rho, T, \mathbf{R}, r$

Operators: $\dfrac{\partial}{\partial t}, \nabla$

Constants: $g, \Omega, c_V, R, k, \mu$

No chain of words, however long, no series of drawings, however revealing, can ever be the combination of truth about the motion of the atmosphere that evolves when these few symbols are arranged to form the equations of motion.

But beware of any implication that the use of symbols and operators in an argument makes it infallible. Errors can be made in mathematics, too, and will lead to nonsense just as rapidly as any other kind of error. There are rules of operation and theorems about validity of the results that can and should be consulted in the process of developing, or studying, a mathematical argument. Errors usually arise from lack of mathematical experience, from attempts to reduce actual complexity to a false simplicity, and, of course, from the nemesis that plagues us all—carelessness.

1.3 TOWARD THERMODYNAMIC PEACE

Any observer of the atmosphere must be struck by the fact that it is always in motion—motion that ranges from great jet streams high in the sky, through the vortices called cyclones and hurricanes, through thunderstorms and tornadoes, through clear air and boundary layer turbulence with dominant wavelengths of a kilometer or so, through the small eddies a few meters in size we can see ruffling a wheat field, and down to the eddies a millimeter in size where friction takes its toll.

Natural philosophers have recognized for a long time that the sun provides the energy for all atmospheric motion. The earth intercepts a small fraction of the energy that is continually radiated away from the sun. And because the earth is spherical in shape, the amount of solar energy falling on the earth decreases with increasing latitude. More heat is received at the equator than in middle latitudes, more in mid-latitudes than at the poles. This uneven distribution of incoming radiation thus leads to differential heating of the atmosphere, to the creation of a global temperature gradient.

As we shall see, temperature gradients cause motion in fluids. The differential heating thus forces the atmosphere into motion, and its flow patterns and those of the oceans labor unceasingly at the task of carrying heat from the tropics to the polar regions.

The air and the earth lose energy by radiation to space, and because of the heat transported by the atmosphere and the oceans, the outgoing radiation is almost constant with latitude. The two radiation currents—incoming from the sun and outgoing from the earth—result in a net heating in equatorial regions and a net cooling in the polar regions. In between, the atmosphere and oceans transport the heat, trying to eliminate the temperature gradients, trying to reach an equilibrium.

This is a process that has gone on as long as air and water have covered the earth, but it is hardly a straightforward process—we do not see the simple, direct convection of warm air poleward matched by a companion flow of cold air toward the equator. Instead the most pronounced feature of the atmosphere's circulation is a broad, deep belt of westerly winds with a high-speed core that forms the great winding polar-front jet stream. This giant river of air, flowing from west to east, cannot by itself carry heat from south to north.

But the heat is indeed transported, and so there must be subtle aspects of the girdle of westerly winds that accomplish the task. There are disturbances in this zonal flow, the cyclones and anticyclones we see near the surface and the associated wavelike patterns higher in the atmosphere. In these disturbances, the winds have northerly and southerly components that can carry heat in the proper direction. These eddies provide the warm, moist poleward currents and the equatorward flows of cold, dry air that accomplish the transport.

We can study the averages of this process in many ways, some more revealing than others. When we use a particular meteorological coordinate system and average around the globe, we shall find that there is indeed a direct thermodynamic circulation evident in the average motion. Warm air rises in equatorial regions, flows poleward, sinks upon cooling, and returns toward the equator. Thus the net result of the disturbances is a flow of the kind we might have expected to find in the beginning.

But why is the heat transport not direct? What does the atmosphere accomplish or gain by such a convoluted, complicated way of doing its thermodynamic business?

As we shall see, we know a great deal about atmospheric motion. We know enough to see the wondrous consistency of its form and structure, of its alternations and its changes. We know how it begins, we know why it must be an unceasing motion, we know enough to use the theory and modern computers to predict the wind patterns of tomorrow.

But we know more about the details than we do about the grand plan, we know more about the winds than we do about the constraints of global thermodynamics, just as we know more about men than we do about mankind.

1.4 THE MAJOR SIMPLIFICATIONS

As described briefly in the previous section, the atmosphere's response to solar heating is a complex one including motions that have wavelengths ranging from the circumference of the planet to millimeters or less. The total phenomenon of atmospheric motion involves heating and cooling by radiation, other transfers of thermodynamic energy, phase changes of water, expansion and contraction of moving masses of air, conversions of energy from one form to another, and accelerations caused by deviations from equilibrium that are induced by the thermal forcing.

Although much of this complete process can be described with sufficiently powerful computers, not all of it can, and the progress that is possible with analytical methods depends upon our being able to study simplified models. Relying on observations of the atmosphere to suggest approximations that are useful and not too inimical to accuracy, dynamic meteorologists have found some crucial features of large-scale flow that permit simplifications to be made.

Meteorological theory begins by formulating suitable versions of Newton's laws, the thermodynamic energy equation, and the law of mass conservation so that they apply to a moving parcel (or small mass) of air. This is the basic step in the derivation of the equations of atmospheric motion.

The next step is to move from a description based on the moving parcels to a mathematical description based on a fixed coordinate system. This change from parcels to points is made with a relation that gives the rate of change in a moving parcel as the sum of rate of change at a point and changes brought by the wind field over the point. This is the basic step in making the equations useful. As we shall see, it also makes them nonlinear, and nonlinearity is perhaps the most serious difficulty in the theory of atmospheric motion.

Having taken these first two steps, we arrive at the partial differential equations that form the mathematical model atmosphere. In their most complete form, the set involves seven equations in the seven variables: pressure, temperature, density, three orthogonal components of the wind velocity, and a suitable moisture variable.

These equations are useful in this original form for studying the energetics of atmospheric motion and for forming numerical models that could be used on a computer either to make weather predictions or to study various properties of atmospheric flow. They apply to all scales of atmospheric motion.

At this point, dynamic meteorologists have historically divided into two groups: those interested in the features of the flow involving weather systems and other large-scale patterns, and those interested in small-scale flow, in convection near the earth's surface, or in atmospheric turbulence. In both groups, the assumption is made that the effect of widely different scales of motion can be ignored. Thus large-scale problems are generally treated without considering turbulence; turbulence studies use a model without large-scale flows or in which the large-scale flow forces, but does not respond, to the turbulence.

The explicit assumption of independence of flow in one range of scales from flows in other ranges has been a useful one in meteorology, and most of the present theory of atmospheric motion is strongly dependent on it.

The next major step involves using observations of the atmosphere to find ways to eliminate variables, and thus equations. In essence, the most important of these kinds of assumptions are related to the fact that the atmosphere is very nearly in configurations that would be special solutions to the equations of motion. The *hydrostatic approximation* provides a relation between pressure and density—the same relation that would prevail if the atmosphere were motionless. The *isentropic approximation* provides a relation between pressure and temperature—the same relation that would prevail if the atmosphere were not being heated and if it were frictionless and unable to conduct heat. The *geostrophic approximation* provides the same relation between the velocity and the mass field that would prevail if the flow were not accelerated.

These may seem like severe approximations, but they are remarkably accurate ones in much of the atmosphere. The three listed above, for example, will give a solution to the equations of motion that describes the basic balances involved in the westerly current that circles the globe in mid-latitudes. With slight modifications, they can be used to provide a more complicated model, but one that gives a reasonably accurate picture of cyclone development and could include some of the thermal forcing that drives the flow.

The final approximation that we shall discuss at this point concerns the shape of the earth itself. A mathematician starting out to study the atmosphere would undoubtedly begin in spherical coordinates, recognizing that the shape of the earth must surely be taken into account. But for many problems, it is actually sufficient to imagine that the flow occurs in cartesian coordinates, which is a great simplification over the equations in spherical coordinates. As the scales of the flows in question become larger, the sphericity of the earth does become significant, but meteorologists have developed an approximation that sidesteps this problem too. The actual effect of curvature on the force arising from the earth's rotation is replaced by a linear term that provides an adequate approximation to the dynamics without introducing any serious mathematical complexity.

Together these approximations provide an almost incredible simplification of the original set of equations. In fact, we shall end up with one partial differential equation in one variable that describes the motions of cyclonic scale. Much of present meteorological thought is oriented toward this quasi-geostrophic equation, its assumptions, and its implications.

Most lacking in this equation, and the model it represents, has been dependence of atmospheric flow on its thermodynamic aspects. It is in the thermodynamics that we find the origin of atmospheric motion, and it is surely in thermodynamics that we shall find the answers to the "why" questions that still elude answers even though we understand the "how" of atmospheric phenomena in many cases.

Despite the success with which simplifications have been found for the large-scale mathematical model, the situation is quite different at smaller scales. There most of these approximations do not work, or do not work all the time. The basic problem is that the flow is not near a steady solution, and in essence everything is out of balance and changing rapidly in both time and space. The only technique that has been generally successful is to simplify the problem by considering averages over many cases, and even this introduces a new problem that can be solved only by finding suitable approximations for relations between statistical quantities.

1.5 CHALLENGES FACING DYNAMIC METEOROLOGY

In the nineteenth century, meteorologists groped toward the solution of two problems. The first was finding the equations of motion and discovering their basic implications. The second was describing the mid-latitude cyclone (do the winds blow toward the

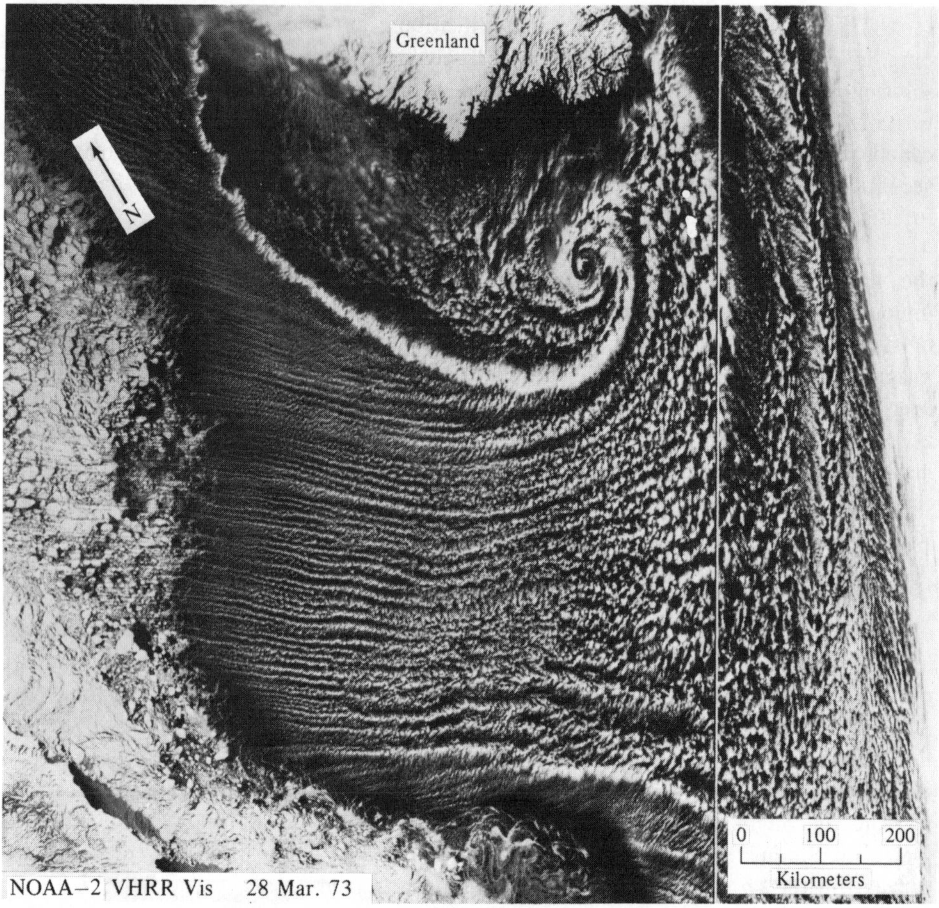

FIGURE 1.2
A mesoscale vortex in the North Atlantic. This vortex has formed in cold, northwest flow from the North American continent. Pack ice is evident along the coastline on the left of the photo (note the more liquid appearance where it is apparently melting just offshore of the date, 28 Mar. 73). As the cold air is warmed by the sea, convective rolls or cloud streets are formed, and the decreasing stability due to the upward heat flux is evidenced by the increasing size of the cloud elements south of the vortex. The cloud streaks extend back over the pack ice, indicating that there is some heat flux through the broken ice field. The pack ice has pulled away from the shore, forming a lead on the Labrador coast at about 56°N, as can be seen just north of the H in VHRR. Two dry tongues are apparent in the southwest flow south of the vortex, one along the interference line in the photo, the other just south of the cloud band trailing out of the vortex.

This cloud band can be interpreted as a small front associated with the vortex by virtue of the obvious change in wind velocity across the band. Note center or around the cyclone?) and constructing the first theoretical explanation of the evolution of these weather-producing systems.

So far in the twentieth century, considerable progress in large-scale meteorology is apparent. We have a detailed knowledge of the structure of the atmosphere, and we have consistent theoretical explanations for most of the processes and phenomena we observe. Even some of the formidable problems posed by the nonlinearity of the equations are being tackled successfully.

This progress has resulted in interest turning from the traditional problems of dynamic meteorology to new ones in which some of the difficulties avoided by the classical approximations are being faced. We shall divide these challenges of modern dynamics into four rather broad groups:

1 Interactions between scales. Every scale of atmospheric motion affects every other scale because of the nonlinearity in the equations. This fact can no longer be avoided by simply assuming that the motions can be separated into independent ranges in scale. An obvious example arises as attention turns to the so-called mesoscale phenomena that lie between cyclones and turbulence, thus including motions with wavelengths of tens or hundreds of kilometers. The mesoscale features are significant in local weather, and understanding them is a key to greatly improved forecasts (Fig. 1.2). They are also significant in dispersion of atmospheric pollutants.

These mesoscale motions lie in the range which students of both large-scale flow and atmospheric turbulence have ignored, but they can be ignored no longer.

Other relations between motions of different scales are also important. Small eddies and turbulence provide the conduit for frictional dissipation in all large-scale flows, but they have been ignored in both theoretical and numerical studies. Clouds and their circulation patterns are important in cyclones, in the climate, and in the hydrological cycle, but nearly all theoretical models of the atmosphere are assumed to be dry because cloud processes are so complicated. At the largest scales of atmospheric motion, the nonlinear processes create strong gradients, and small-scale instabilities produce turbulence that destroys them. This process is not considered in studying large-scale motion, and it is not generally considered in studying turbulence—a process that destroys the forcing that created it.

that lateral waves are forming on the front as it trails through the Davis Strait into the west coast of Greenland.

As indicated by the approximate distance scale on the photo, the closed circulation in the vortex itself is about 100 km across, and the minifront trails another 1000 km to the west and northwest. The vortex occurs in a flow of much larger scale typical of the northwest flow behind a major cold front, which in this case is over the horizon to the northeast. The amount of cloudiness, wind speed and direction, and possibly the precipitation at a station would all be affected by the location of the station relative to the vortex and its minifront. (*Photograph courtesy of Vincent J. Oliver, National Environmental Satellite Service.*)

All these interactions will have to be studied and included in a comprehensive model of the atmosphere, but it is not at all clear what mathematical techniques or physical approximations will permit us to deal explicitly with the fact that all the motions are linked together.

2 Sensitivity of solutions. A fundamental assumption made in producing simplified models is that small causes have small effects—that only dominant terms in the equations need to be retained. If this assumption were valid, then it should also be true that flows starting from slightly different initial conditions should remain similar. Both of these assumptions are not generally valid, and the modern studies of the predictability of atmospheric flow are devoted to studying the resulting problem. Thus dynamics is faced with determining the sensitivity of atmospheric flows to slight changes in the initial conditions, the boundary conditions, or the approximations used to simplify the equations.

In a qualitative sense, the answer to the sensitivity questions is that the effects of small causes may take a long time to affect the flow, but eventually they will become significant. Thus weather predictions based on limited initial observations or approximate equations will eventually fail.

Sensitivity of flows to small causes is also involved in the study of interrelations of phenomena at different scales. Turbulence may be a small cause for changes in the jet stream, but its effects must be modeled if a truly accurate theory of large-scale dynamics is to be developed.

3 Climate and modes of response. The need for a theory of climate has become apparent as mankind begins to be seriously threatened by shortages of food and energy and by environmental degradation. The paleoclimatological records provide ample evidence that there have been sensibly large changes in the climate, but meteorological theory has almost no information about how such changes occurred and even less about why.

For example, it is not known whether the observed changes in climate represent a natural, long-period oscillation of the atmosphere and ocean system, or whether they were caused by perturbations in the amount of energy received due to changes in the sun itself or because the sky was obscured by volcanic dust clouds.

In approaching the theory of climate, in determining what controls the long-term averages of atmospheric variables and processes, we must develop a thorough understanding of the controls on the modes by which the atmosphere responds to the thermal forcing. To create a successful theory of climate, we must find the reason that the atmosphere makes the choice it does, and we shall probably find, as stated earlier, that the choices are basically thermodynamic ones.

4 Moist convection. Most of modern meteorological theory ignores the fact that the atmosphere contains water in its three phases. The theories of motion cannot deal with moist air because of the complexity of the phase change processes, and most of the theories of the phase changes themselves ignore the complexity introduced by atmospheric motion.

But clouds and precipitation are obviously an important part of weather and climate, and their presence and processes also have significant effects on other scales of atmospheric motion.

The difficulty is that the processes involved in phase changes take place on a very much smaller scale than other meteorological events. Since water evaporates and condenses on the molecular scale, the phase change process can be treated only in a mathematical or numerical model of larger-scale features as a random one. Thousands of clouds can fall between the grid points of a numerical model, and some way must be found to model their influence on the numerical solutions.

Here again, the nearby, simple steady solution that affords us useful approximations is lacking. Even if we were to study a steady model of a towering thunderstorm, some way of providing a suitable source of inflow moisture and some way of carrying out the mathematical equivalent of the resulting condensation would have to be effected in the model.

So the models of atmospheric motion that are used to study dynamics, large-scale flow, and climate shall have to incorporate vapor, water, and ice to represent atmospheric processes truly.

The form the solutions to problems in these four areas will take is not necessarily clear. But we can be confident that they will be built upon the foundation of present knowledge, that today's theories of atmospheric motion will evolve into the new ones that meet these challenges.

2

SOME BASIC PHYSICAL AND MATHEMATICAL CONCEPTS OF ATMOSPHERIC SCIENCE

Atmospheric science is devoted to the determination of quantitative relations between variables that describe the state and motion of the atmosphere. The description involves specifying both what is occurring and how the events vary from place to place and from time to time. We must develop a suitable system of quantities that describe the mass, thermal, and motion fields of the atmosphere, and we must find ways to utilize the mathematical operations available for determining rates of changes and average properties.

If the atmosphere were still, it would be quite easy to describe its properties. But the majority of phenomena of interest in meteorology are connected with the motions of the atmosphere. The winds are at the core of atmospheric science, for they affect every process that occurs, and it is the wind that is most difficult to predict, but must be predicted in order to predict anything else. So understanding the motion of the atmosphere is the key to understanding many of its other facets. In this chapter we begin to assemble the concepts with which we shall study atmospheric motion.

2.1 FUNDAMENTAL PHYSICAL CONCEPTS

The concepts of the distance between objects and the time interval between events are key ideas in the development of physical theories. They serve atmospheric science both as the basic scale with which we perceive the occurrence of physical events and as the means for describing the place and time of occurrence of those events.

Thus we begin with the coordinates (x_1,x_2,x_3) that describe location in a three-dimensional space and a coordinate t that represents the time. The choice of the origin of the spatial coordinate system will become important later, but for now we may select it at will. We might choose it to be the center of the earth, and use the variables r, representing radial distance from the earth's center, ϕ, representing latitude, and λ, representing longitude, to describe position in the atmosphere.

Although this is a natural system for atmospheric studies, its use leads to certain complications in the manipulation of equations that are avoided by the use of a rectangular cartesian coordinate system (x,y,z). For many meteorological problems, the curvature of the earth may be ignored and we instead analyze the problem that would exist if the atmosphere were placed on an infinite, rotating plane; in doing this the general convention is that x increases toward the east, y increases toward the north, and z is zero at the surface and increases upward. We shall see later that the curvature effect important in large-scale flow can be introduced as a simple approximation in this coordinate system.

In most of this book, the mathematical analysis will apply to this imaginary atmosphere over a plane surface; it is always possible to reanalyze the problem or to convert the results directly to apply to a spherical earth.

The three spatial coordinates and one in time are the *independent variables*; they will be defined here simply and intuitively, thereby avoiding philosophical dissension. The spatial coordinates are measures of distance or length, and length is what we measure with a ruler. The temporal coordinate, time, is what we measure with a clock. The tip of a balance arm changes its position as we put different objects on one of the pans. Here then is evidence of a property of objects that causes a balance arm to change position. We shall define this property as *mass*, and hence mass is proportional to what we compare with a balance.

The final fundamental concept is temperature; for now, we shall simply define temperature as the quantity we measure with a thermometer. It is a property of substances that determines the direction of the flow of heat, and hence our definition is consistent with the physics of a thermometer. We shall see later that temperature is a measure of molecular energy, but it is convenient in meteorology to consider it a fundamental quantity.

These four quantities along with data about original standard values in the metric system are shown in Table 2.1. These *fundamental physical quantities* cannot be expressed as functions of each other without introduction of other quantities.

The fundamental quantities of which a physical variable is composed are referred to as its *dimensions*. For example, pressure will be defined as a force per unit area, so its dimensions are M/LT^2. If p denotes pressure, this relation can be written as

$$[p] = \frac{M}{LT^2} \tag{1}$$

The square brackets denote the operation of taking the dimensions of p.

Physical equations must be dimensionally homogeneous. That is, the dimensions of every term must be the same—force and energy cannot be summed.† Thus, in the ordinary differential equation for the position x of a harmonic oscillator,

$$\frac{d^2x}{dt^2} + b\frac{dx}{dt} + kx = 0 \tag{2}$$

we must have $[b] = T^{-1}$ and $[k] = T^{-2}$.

The fundamental quantities or dimensions are often expressed in different *units*. The metric units are in general use in science today, and meteorologists have most often employed the centimeter-gram-second (cgs) system, but there is now a trend toward meter-kilogram-second (mks) units (see Appendix 2). However, units are sometimes changed to produce numerical values as close as possible to unity. Instead

†Any equation worth writing deserves to have its dimensions checked, and the person writing it has a duty to do so. Although the fact that the dimensions are the same in each term does not guarantee that the equation is correct, if they do not check, the equation is surely wrong. The habit of always checking dimensions in equations will serve the student well and be an invaluable guide through the thicket of inevitable errors.

Table 2.1 PHYSICAL QUANTITIES

Fundamental quantities

Quantity	Symbol	Operational means of definition	Original metric standard
Length	L	Ruler	$1\ \text{m} \sim 10^{-7}D$ D = earth's circumferential distance from equator to pole
Time	T	Clock	86,400 s = 1 solar day 365.25 solar days = 1 solar year
Mass	M	Balance	1 g ~ mass of 1 cm^3 of H_2O at 4°C
Temperature	K	Thermometer	Water boils at 100°C Ice melts at 0°C

Derived quantities

Velocity	$V = L/T$	Energy	$E = ML^2/T^2$
Acceleration	$A = L/T^2$	Power	$P = E/T$
Force	$F = ML/T^2$	Entropy	$S = E/K$

Note: The solar day is the interval required for one earth rotation with respect to the sun; the solar year is the interval required for one revolution around the sun. The sidereal day is the interval required for one rotation with respect to a distant star. Because of the earth's revolution, there is one more sidereal day than there are solar days per year.

of 10^4 m we would write 10 km, and instead of 10^3 erg/cm^2 · s we would choose 1 J/m^2 · s = 1 W/m^2. It is obvious that changes in units do not change dimensions.

PROBLEM 2.1.1 Suppose that you were endowed with a ruler to measure length, a clock to measure time, and an energy meter to measure energy. Develop an appropriate new system of fundamental and derived quantities.

2.2 FUNDAMENTAL MATHEMATICAL CONCEPTS AND OPERATIONS

The variables we use to describe atmospheric processes, such as wind speed, temperature, or pressure, are functions of the independent variables. If we denote one of these atmospheric variables by f, then we express the functional dependence of f on position and time in the form

$$f = f(x,y,z,t) \tag{1}$$

or if the coordinates are not yet specified, as

$$f = f(x_1,x_2,x_3,t) \tag{2}$$

This notion of a function that produces a numerical value for each combination of its arguments is an important one in constructing an efficient description of atmospheric phenomena. It is essential to note that we are assuming that we can pass from air as a collection of discrete molecules to a representation of atmospheric properties as functions of space and time. The method by which this transformation is accomplished will be discussed later in the chapter.

We shall always assume that the functions representing atmospheric variables are well-behaved: thus they are always assumed to be continuous functions of each of their arguments, and we shall see that they must be differentiable to various orders if the equations of motion are to make sense. Thus theorems that require continuity of a function or its first derivative may always be applied to meteorological variables.

The two basic mathematical operations we shall use are differentiation and integration. The notations $\partial f/\partial t$, $\partial f/\partial x$, $\partial f/\partial y$, $\partial f/\partial z$ denote the partial derivatives with respect to time and the three spatial coordinates. These derivatives are sometimes denoted by subscripts: $\partial f/\partial x = f_x$. If a meteorological variable were a function of only one variable, say x, then we might write $df/dx = f_x$. The measurement of derivatives on weather charts is generally done with finite difference versions of the usual limit definition.

Nearly all the integrals in this book will be definite integrals over spatial domains, although a few integrations over a time domain will occur. Integration over the volume of the entire atmosphere will also be used frequently, and these three-dimensional integrals will always be denoted by

$$\int f(x,y,z,t)\,dV = \underset{\text{atm}}{\iiint} f(x,y,z,t)\,dx\,dy\,dz \tag{3}$$

Here we have given the representation of dV in cartesian coordinates. In other coordinates, it would take the form appropriate to the system in use.

It is a useful habit always to note what variables are left whenever an integral is encountered. In Eq. (3), the integration is with respect to x, y, and z and the result is a function of time. If Eq. (3) were now integrated between t_1 and t_2, *the result would be a number.*

The definition of integration as the inverse of differentiation is convenient when simple functions are considered. But more advanced concepts, such as the Riemann or Lebesgue integral, are sometimes needed in meteorological theory to permit efficient calculations to be made with definite integrals. We cannot go into these definitions here, but we shall have to use some theorems about integration. These theorems thus assume that a suitable definition of the integral has been stated, and when we specify that a function f is integrable, it means that it must be integrable with respect to the definition in use.

The main reason for using integrals in atmospheric science is that they provide quantitative summaries of masses of information whose details we cannot hope to comprehend. But many of the integrals that occur cannot be evaluated directly; we often obtain estimates with:

Theorem Mean-value theorem for integrals Let

$$I = \int_a^b f(x)g(x)\,dx \tag{4}$$

where f and g are both integrable. Then if $g(x)$ is always of the same sign,

$$I = K\int_a^b g(x)\,dx \tag{5}$$

where min $[f] \leqslant K \leqslant$ max $[f]$. If f is continuous, then there exists a point x^* such that

$$I = f(x^*)\int_a^b g(x)\,dx \tag{6}$$

PROOF Without a loss of generality, let $g(x) \geqslant 0$. Then

$$\min\,[f]\int_a^b g(x)\,dx \leqslant I \leqslant \max\,[f]\int_a^b g(x)\,dx \tag{7}$$

This proves the first assertion. If f is continuous, then K must be some actual value $f(x^*)$ of f between the maximum and minimum of f. Hence we have

$$f(x^*) = K = \frac{\int_a^b f(x)g(x)\,dx}{\int_a^b g(x)\,dx} \tag{8}$$

which proves the second assertion. ////

PROBLEM 2.2.1 Prove that the finite difference approximate derivative $[f(x+L) - f(x-L)]/2L$ is the average value of the exact derivative $\partial f/\partial x$ over the domain $(x - L, x + L)$.

*PROBLEM 2.2.2 Let $f = f(u,v)$ where $u = u(x,y)$ and $v = v(x,y)$. Prove the *chain rule* of partial differentiation that

$$\frac{\partial f}{\partial x} = \frac{\partial f}{\partial u}\frac{\partial u}{\partial x} + \frac{\partial f}{\partial v}\frac{\partial v}{\partial x}$$

What then is

$$df = \frac{\partial f}{\partial x}dx + \frac{\partial f}{\partial y}dy$$

PROBLEM 2.2.3 Why was the condition that $g(x)$ be of one sign necessary in the proof of the mean-value theorem? Prove the theorem for $g(x) \leqslant 0$.

PROBLEM 2.2.4 Analyze the expression

$$\int [af(x) + bg(x)]^2\,dx$$

in order to prove the Schwarz inequality

$$\left|\int f(x)g(x)\,dx\right| \leqslant \left[\int f^2(x)\,dx \int g^2(x)\,dx\right]^{1/2}$$

When is the equal sign valid?

2.3 VARIABLES USED TO DESCRIBE ATMOSPHERIC PROCESSES

It clearly would be impossible to develop any sort of description or theory of atmospheric motion that took account of the behavior of each of the molecules in the air. The variables we consider could all be defined as suitable averages that describe the molecular motion, and the equations of motion could be derived as they are in statistical mechanics. But in atmospheric science, we actually ignore the fact that our definitions are based upon molecular concepts, and once we have described the macroscopic variables, we treat the air as though it were a continuum substance, not a collection of discrete molecules.

In the following discussion of basic meteorological variables, we wish to emphasize that we use both local and global concepts. The local variables are the ones we use when we study the details of atmospheric processes. The global variables are integrals of the local ones over the entire atmosphere and are useful in studying the dynamics of the atmosphere in summary form.

We have the following basic variables:

Density Meteorology and fluid mechanics do not use mass in the same way as ordinary mechanics does because we are concerned with millions of molecules at a time. The relevant concept is *density*, defined to be the total mass of all the molecules in a unit volume. Thus if we have N molecules, each of mass m, in a volume with sides of length L, we have

$$\rho = \frac{Nm}{L^3} \tag{1}$$

and this relation is frequently written as

$$\rho = \frac{M}{V}$$

where M is the total mass in a volume V. The reciprocal of the density, defined as the *specific volume*,

$$\alpha = \frac{1}{\rho} \tag{2}$$

will often appear.

The associated global quantity is the total mass of the atmosphere, given by the integral

$$M = \int \rho \, dV \tag{3}$$

The density also appears in another global quantity, the potential energy

$$P = \int g\rho z \, dV \tag{4}$$

where g is the acceleration of gravity. Thus the potential energy depends on the distribution of density with altitude z.

Pressure Pressure, defined to be the normal force per unit area exerted by the air, can be explained as the normal force arising from molecular impacts on a surface placed in the air. In this case, it is shown in statistical mechanics that the pressure is

$$p = \frac{\rho \overline{c^2}}{3} \tag{5}$$

where $\overline{c^2}$ denotes the average of the square of the molecular velocities.

Pressure can also be explained as representing the cumulative weight of all the air above a given level. We shall investigate this idea more fully in Chap. 4.

Finally, we can revert to the continuum description. Two adjacent masses of air can exert both normal and tangential forces on each other. The normal force is the pressure; the tangential forces are part of the frictional forces. So long as there is no macroscopic motion, all three definitions of pressure are equivalent. We shall develop the refinements necessary for air in motion later in the book.

Pressure as a force per unit area has dimensions of M/LT^2; in cgs units it is expressed as dynes per square centimeter. The average pressure at the surface of the earth is about 1013.3×10^3 dyn/cm^2, and 10^6 dyn/cm^2 is called 1 *bar* of pressure. Meteorologists, however, use millibars and the average pressure is expressed as 1013.3 mbar. When pressure occurs in a mathematical expression, it usually must be converted from millibars into the cgs or mks units in which the other variables of the equation are expressed.   1 bar = 10^5 Pascal = 100 KPA

Temperature and the equation of state The introduction of the concept of temperature is a main feature that distinguishes thermodynamics from ordinary mechanics. The temperature of a body or substance is a property of its state that is not considered in the analyses of motion performed in classical mechanics but is important in studies of the behavior of gases or other fluids.

Continuing for the moment with our definition of temperature as that property which is measured by a thermometer, we cite two empirical laws about the behavior of gases. Boyle's law states that at constant temperature the pressure and volume are related by

$$pV = \text{const} \tag{6}$$

Charles' law says that at constant pressure

$$V = T\,\text{const} \tag{7}$$

A gas for which these two laws hold exactly is called an *ideal* gas. Real gases obey these laws only approximately and under certain conditions, but the great simplification that results from assuming that air is an ideal gas has led to the adoption of this assumption in the vast majority of meteorological calculations.

Because Eq. (6) is a relation between pressure and volume and Eq. (7) is a relation between volume and temperature, we can infer that there must be some general relation between these three variables in the form

$$F(p,V,T) = 0 \tag{8}$$

Such an expression is called an *equation of state*. On this basis we conclude that the constant in Eq. (6) can be at most a function of temperature and the constant in Eq.

(7) can only be a function of pressure. We shall, therefore, rewrite these equations as

$$pV = f(T) \quad \text{and} \quad V = g(p)T \tag{9}$$

where f and g are functions to be determined. The ratio of these two equations gives

$$g(p)p = \frac{f(T)}{T} \tag{10}$$

The left side is a function of pressure only, and the right side is a function of T only, so both sides must be equal to a constant. Hence we have

$$g(p) = \frac{K}{p} \tag{11}$$

and

$$f(T) = KT \tag{12}$$

where K is the necessary constant, as yet undetermined. Now both of Eqs. (9) may be written as

$$pV = KT \tag{13}$$

The temperature defined in this way is called the *gas temperature.*

Avogadro's law assures us that at equal pressure and temperature all ideal gases contain equal numbers of molecules in equal volumes. If we then define a mole to be an amount of mass of the gas equal in grams to its molecular weight, we can define the *molar volume* to be the volume occupied by a mole of the gas at a given pressure and temperature. But now Avogadro's law is equivalent to the statement that at equal pressure and temperature, the molar volume of any ideal gas is identical.

Application of Eq. (13) to a molar volume V_{mol} therefore results in the universal form

$$pV_{\text{mol}} = R^*T \tag{14}$$

which is valid for the molar volume of any ideal gas. We have replaced K with R^*, which is a universal constant because it applies to 1 mol of gas. If the mass of 1 mol of gas is μ, then we have for the specific volume

$$\alpha = \frac{V_{\text{mol}}}{\mu} \tag{15}$$

The equation of state for an ideal gas thus becomes, upon division of both sides by μ,

$$p\alpha = \frac{R^*}{\mu}T = RT \tag{16}$$

and this can be applied to any gas whose molar mass μ is known. The value of R^* is 8.31 J/deg · mol and the molar mass for air is 28.97 g/mol so that the gas constant for air is $R = 0.287$ J/g · deg. The equation of state is depicted graphically in Fig. 2.1.

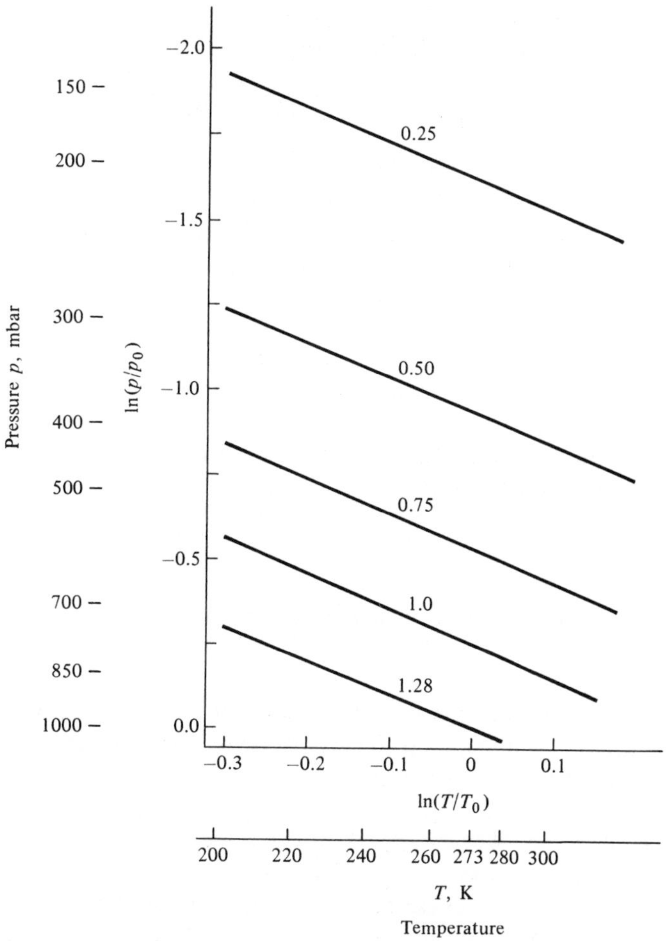

FIGURE 2.1
The density of air in kilograms per cubic meter as given by the equation of state for various values of pressure and temperature. The values $p_0 = 1000$ mbar and $T_0 = 273$ K were used.

It is a nearly universal practice in theoretical meteorology (and a necessity in equations like the equation of state) to measure temperature on an absolute scale. The most common one is the Kelvin scale. At standard pressure, ice melts at 273.16 K and water boils at 373.16 K. Celsius (or centigrade) temperatures are obtained from Kelvin temperatures by subtracting 273.16 from the Kelvin value of temperature.

A global quantity associated with temperature is the internal energy, defined by

$$I = \text{const} \int \rho T \, dV \tag{17}$$

The constant here will be specified in the next chapter. Note that by virtue of the equation of state (16) we may write this (with a new constant) as

$$I = \text{const} \int p \, dV \tag{18}$$

The wind velocity Finally, we come to the description of motion, a concept which implies that an object has moved from one place to another. With an ideal tracer that moved exactly with the air, we could determine the trajectory $(x(t),y(t),z(t))$ of a certain part of the air. From the trajectory we can find the velocity components $u = dx/dt$, $v = dy/dt$, and $w = dz/dt$ (see Fig. 2.2). The air velocity $\mathbf{v}$ can be expressed in components ($\mathbf{v} = (u,v,w)$) or as the combination of a total speed $|\mathbf{v}| = (u^2 + v^2 + w^2)^{1/2}$ and the direction of motion specified by two angles, for example, the compass heading of the motion and its angle with respect to the horizontal plane.

One of the global quantities associated with motion is the kinetic energy, given with the total speed $|\mathbf{v}|$ as

$$K = \int \rho \frac{|\mathbf{v}|^2}{2} \, dV \tag{19}$$

We have defined three energies so far, and in sum they give the total energy

$$E = K + I + P \tag{20}$$

In studying atmospheric motion, we shall be concerned with the atmosphere's energy budget and the question of how the static forms of energy, I and P, are converted into kinetic energy.

Parcels of air When we speak of the velocity, or pressure, or density of the air, we are referring to a "parcel" of air large enough to contain millions of molecules so that the statistical properties such as pressure and density are well defined. But we are restricting ourselves to parcels of air which are small enough so that the statistical properties are the same in all parts of a parcel. Thus the pressure and density are macroscopic descriptions of the molecular properties and motions of the gas, and the velocity of the parcel is the average of all the molecular velocities.

It is, of course, being a bit vague to say that a parcel is big enough to be one thing and yet small enough to be another. This very loose definition, however, is in agreement with the measurements we can make, because the operation of barographs, thermometers, and anemometers depends on the interaction of the sensing surface of the instrument with the millions of air molecules that impinge on that surface.

PROBLEM 2.3.1 Express the temperature as a function of the kinetic energy of molecular motion by showing that $K_{\text{mol}} = \mu \overline{c^2}/2 = 3R^*T/2$.

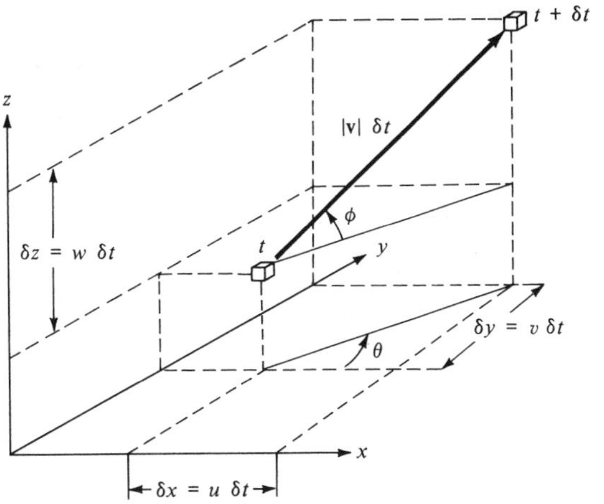

FIGURE 2.2
The velocity components u, v, w, as obtained from the trajectory of a parcel in space. The location of the parcel is shown at times t and $t + \delta t$.

PROBLEM 2.3.2 On the Fahrenheit scale, water boils at 212° and ice melts at 32° at standard pressure. Derive the conversion formula between Fahrenheit and Celsius scales.

PROBLEM 2.3.3 There are approximately 6×10^{23} molecules/mol (Avogadro's number). If the pressure is 10^3 mbar, the temperature 0°C, how many molecules are there in 1 cm^3 of air?

PROBLEM 2.3.4 The constant in Eq. (17) is about 7.2×10^6 in cgs units. What are its dimensions? The mass per unit area of the atmosphere is about 10^3 g/cm^2. With this information and the mean-value theorem, estimate the amount of K, I, and P per unit area. You should find that I is roughly twice P, and that both are very much larger than K.

*PROBLEM 2.3.5 Show that if Eq. (19) is finite, then $\rho|\mathbf{v}|$ is less than $\rho^{1/2}\epsilon(z)$ as z goes to infinity, where $\epsilon(z) \to 0$ as $z \to \infty$.

2.4 TAYLOR'S THEOREM AND NUMERICAL PREDICTION

The best-known practical use of meteorological theory is weather prediction. To predict the weather, we must use the data on what the values of atmospheric variables are now to infer what they will be in the future. Taylor's theorem provides a basis for this effort.

Theorem Let f have $n+1$ continuous derivatives in an interval I containing the point x_0. Let

$$P_n(x,x_0) = f(x_0) + f'(x_0)(x - x_0) + \cdots + f^{(n)}(x_0)\frac{(x-x_0)^n}{n!} \tag{1}$$

Then for x in I,

$$f(x) - P_n(x,x_0) = R_n(x) = \frac{1}{n!}\int_{x_0}^{x} f^{(n+1)}(\xi)(x-\xi)^n \, d\xi \tag{2}$$

PROOF Let $g(\zeta) = P_n(x,\zeta)$. Then

$$R_n(x) = P_n(x,x) - P_n(x,x_0) = g(x) - g(x_0)$$
$$= \int_{x_0}^{x} g'(\xi)\, d\xi \tag{3}$$

But

$$g(\xi) = f(\xi) + f'(\xi)(x - \xi) + \cdots \tag{4}$$

and therefore

$$g'(\xi) = f'(\xi) + [f''(\xi)(x-\xi) - f'(\xi)]$$
$$+ \left[f'''(\xi)\frac{(x-\xi)^2}{2!} - f''(\xi)(x-\xi)\right] +$$
$$\cdots + \left[f^{(n+1)}(\xi)\frac{(x-\xi)^n}{n!} - f^{(n)}(\xi)\frac{(x-\xi)^{n-1}}{(n-1)!}\right] \tag{5}$$

But every term except the first one in the last bracket is paired with an identical term of opposite sign in another bracket, and so

$$g'(\xi) = f^{n+1}(\xi)\frac{(x-\xi)^n}{n!} \tag{6}$$

The theorem thus follows from Eq. (3). ////

Corollary 1 The remainder may be expressed as

$$R_n(x) = f^{n+1}(x^*) \int_{x_0}^{x} \frac{(x-\xi)^n}{n!}\, d\xi$$
$$= f^{n+1}(x^*) \frac{(x-x_0)^{n+1}}{(n+1)!} \tag{7}$$

PROOF In the integral of Eq. (2), $x_0 \leqslant \xi \leqslant x$ and so $(x-\xi)^n \geqslant 0$. According to the mean-value theorem for integrals, there exists a point x^*, where $x_0 \leqslant x^* \leqslant x$ such that the first line of Eq. (7) is true. The second line follows by evaluating the integral. ////

Corollary 2 Under the hypothesis of the theorem, there is a point x^* where $x_0 \leqslant x^* \leqslant x$ such that

$$f(x) = f(x_0) + f'(x_0)(x-x_0) + \cdots + f^{n+1}(x^*) \frac{(x-x_0)^{n+1}}{(n+1)!} \tag{8}$$

PROOF Use of Corollary 1 in Eq. (2) gives the result immediately. ////

For some applications, including one in the next section, we need Taylor's theorem extended to several variables. In this case we have:

Theorem Let $f(x_1,x_2, \ldots, x_N)$ be continuously differentiable in each variable to order $n+1$. Then

$$f(x_1+\xi_1, x_2+\xi_2, \ldots, x_N+\xi_N) = f(x_1,x_2,\ldots,x_N) + \sum_{m=1}^{n} \left. \frac{[\xi_1(\partial/\partial x_1) + \cdots + \xi_N(\partial/\partial x_N)]^m f}{m!} \right|_{\mathbf{x}} + R_{n+1} \tag{9}$$

where

$$R_{n+1} = \left. \frac{[\xi_1(\partial/\partial x_1) + \cdots + \xi_N(\partial/\partial x_N)]^{n+1} f}{(n+1)!} \right|_{\mathbf{x}^*} \tag{10}$$

and where $\mathbf{x}^*$ is on the line segment joining $\mathbf{x}$ and $\mathbf{x}+\boldsymbol{\xi}$ so that $x_k^* = x_k + \lambda\xi_k$ for $0 \leqslant \lambda \leqslant 1$.

Here $\mathbf{x} = (x_1, \ldots, x_N)$ and we used the convention that $(\partial/\partial x)^n f = \partial^n f/\partial x^n$. The proof is omitted. ////

For one type of meteorological application, we use Eq. (8) in the form

$$f(x) = f(x_0) + f'(x^*)(x-x_0) \tag{11}$$

to estimate a value at x given data at x_0. Because we almost always know $f'(x_0)$ rather than $f'(x^*)$, we simply use $f'(x_0)$ in Eq. (11) to obtain an approximate value.

For weather prediction, we must determine a value of f a time increment Δt in the future from data known at time t. We may write the exact expression

$$f(x,y,z,t+\Delta t) = f(x,y,z,t) + \left[\frac{\partial f(x,y,z,t)}{\partial t}\right]\Delta t + \left(\left.\frac{\partial f}{\partial t}\right|_{t*} - \left.\frac{\partial f}{\partial t}\right|_{t}\right)\Delta t \tag{12}$$

Thus if the time steps Δt are kept small enough so that f does not change rapidly in the interval Δt, we have the approximate equation

$$f(x,y,z,t+\Delta t) = f(x,y,z,t) + \left[\frac{\partial f(x,y,z,t)}{\partial t}\right]\Delta t \tag{13}$$

This is essentially the method used in modern numerical weather prediction, but to apply it, we must be able to find the rates of change of the variables. This information is given by the equations of atmospheric motion that we shall be studying in this book. When we have the equations, then we can use Eq. (13) to compute values for the variables at each point at time $t + \Delta t$, and these can then be used in turn to obtain values for $t + 2\Delta t$. We may continue in this way until the errors have accumulated and the prediction is grossly wrong. In present global, numerical prediction models, the time step Δt is 10 minutes, and the predictions appear to correspond to large-scale features adequately over periods as long as 5 days in advance.

In order to see better how this scheme works, we need to consider how the wind fields are described and to study the relations between various types of derivatives.

2.5 THE TOTAL OR MATERIAL DERIVATIVE

Let us suppose that we can identify a particular parcel of air, and that we can measure how some property of the parcel, such as its temperature, changes with time. This operation would produce a new derivative–a derivative following the parcel–which is quite different from the partial derivatives in time and space. In this section, we shall show how these three derivatives are related.

If we consider a parcel of air moving in the (x,y,z) coordinate system, then the trajectory of the parcel is specified by its coordinates as functions of time and we have the velocities

$$u = \frac{dx(t)}{dt} \qquad v = \frac{dy(t)}{dt} \qquad w = \frac{dz(t)}{dt} \tag{1}$$

Now we can calculate a derivative of the property of the parcel as the parcel moves in time δt from the point $(x(t),y(t),z(t))$ to the point $(x(t+\delta t),\ y(t+\delta t),\ z(t+\delta t))$. Applying the basic definition of a derivative, we start from a finite difference

$$\delta f = f(x(t+\delta t), y(t+\delta t), z(t+\delta t), t+\delta t) - f(x,y,z,t) \tag{2}$$

Now with Taylor's theorem, Eq. (9) in Sec. 2.4, applied to each independent variable, we have

$$\delta f = \frac{\partial f}{\partial t}\delta t + \frac{\partial f}{\partial x}[x(t+\delta t) - x(t)] + \frac{\partial f}{\partial y}[y(t+\delta t) - y(t)]$$
$$+ \frac{\partial f}{\partial z}[z(t+\delta t) - z(t)] + \frac{\partial^2 f}{\partial t^2}\frac{(\delta t)^2}{2} + \cdots \qquad (3)$$

and using the theorem again in the form of Eq. (2) in Sec. 2.4 on the differences of trajectory coordinates gives

$$\delta f = \left(\frac{\partial f}{\partial t} + \frac{\partial f}{\partial x}\frac{dx}{dt} + \frac{\partial f}{\partial y}\frac{dy}{dt} + \frac{\partial f}{\partial z}\frac{dz}{dt}\right)\delta t + (\ \ )(\delta t)^2 + \cdots \qquad (4)$$

The derivative following the parcel is

$$\frac{df}{dt} = \lim_{\delta t \to 0} \frac{\delta f}{\delta t} \qquad (5)$$

and so we have established *Euler's relation*

$$\frac{df}{dt} = \frac{\partial f}{\partial t} + u\frac{\partial f}{\partial x} + v\frac{\partial f}{\partial y} + w\frac{\partial f}{\partial z} \qquad (6)$$

A graphical derivation is given in Fig. 2.3.

It is essential to grasp the distinctions between the total rate of change df/dt, the local rate of change $\partial f/\partial t$, and what is called the advection, A_f, of f, given by

$$A_f = -\left(u\frac{\partial f}{\partial x} + v\frac{\partial f}{\partial y} + w\frac{\partial f}{\partial z}\right) \qquad (7)$$

The local rate of change at a point is determined by

$$\frac{\partial f}{\partial t} = \frac{df}{dt} + A_f \qquad (8)$$

and is composed of: (1) the rate of change occurring in the moving parcel which happens to be over the point, and (2) the rate of change A_f, which is due to the fact that parcels with different values of f are being blown over the point by the wind.

The main result can be derived by a less cumbersome, but also less revealing, method. We have a meteorological variable

$$f = f(x,y,z,t) \qquad (9)$$

and this function possesses a *total differential* that takes account of all sources of variation in the form

$$df = \frac{\partial f}{\partial t}dt + \frac{\partial f}{\partial x}dx + \frac{\partial f}{\partial y}dy + \frac{\partial f}{\partial z}dz \qquad (10)$$

If we associate the variable f with a parcel whose trajectory is specified by $(x(t), y(t), z(t))$, then we may use Eq. (1) to interpret ratios such as dx/dt. Hence we may divide Eq. (10) by dt and obtain Eq. (6). The crucial point is that the concept of

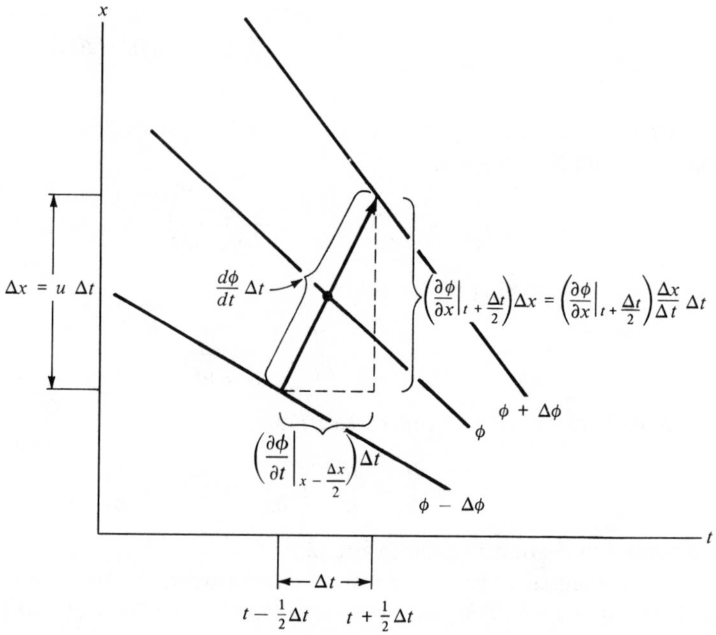

FIGURE 2.3
Illustration of Euler's relation. The isolines indicate the value of ϕ found in the parcels at the points (x,t). The arrow indicates the trajectory of the parcel for which we are computing the material derivative. From the drawing

$$\frac{d\phi}{dt} = \left.\frac{\partial\phi}{\partial t}\right|_{x-\frac{\Delta x}{2}} + \left.\frac{\partial\phi}{\partial x}\right|_{t+\frac{1}{2}\Delta t}\frac{\Delta x}{\Delta t}$$

for any Δt, so that as $\Delta t \to 0$ we obtain

$$\frac{d\phi}{dt} = \frac{\partial\phi}{\partial t} + u\frac{\partial\phi}{\partial x}$$

Taking account of symmetry in x, y, z then gives the complete equation.

a moving parcel permits us to identify the resulting df/dt as a derivative following the parcel because the derivative dx/dt is a parcel velocity. Thus the two points $(x + dx, y + dy, z + dz)$ and (x,y,z) at which we take the difference in Eq. (10), are the locations of the parcel at $t + dt$ and t. Note that we cannot use Eq. (10) to define df/dx because we have not defined dt/dx.

This derivative following a parcel is perhaps best referred to as a *material derivative* because the name suggests the association with the moving matter. It is also called a *total* or *substantive* derivative. The material derivative is sometimes denoted by Df/Dt, but our usage is now common in meteorology.

2.6 FROM PARCELS TO POINTS

One major aim of dynamic meteorology is to use the data on the state of the atmosphere at time t to predict the state at time $t + dt$. These attempts are, at the present time, based on equations such as Newton's second law, which in a form appropriate to the atmosphere is

$$\frac{du}{dt} = F_x \qquad \frac{dv}{dt} = F_y \qquad \frac{dw}{dt} = F_z \tag{1}$$

Thus the acceleration of a parcel of air is determined by the components (F_x, F_y, F_z) of the forces per unit mass acting parallel to the coordinate axes. We shall be concerned later with the exact specification of these forces that act on parcels in the atmosphere, but for now we expand the total derivative with the Euler relation, Eq. (6) in Sec. 2.5, to obtain

$$\begin{aligned}
\frac{\partial u}{\partial t} + u\frac{\partial u}{\partial x} + v\frac{\partial u}{\partial y} + w\frac{\partial u}{\partial z} &= F_x \\
\frac{\partial v}{\partial t} + u\frac{\partial v}{\partial x} + v\frac{\partial v}{\partial y} + w\frac{\partial v}{\partial z} &= F_y \\
\frac{\partial w}{\partial t} + u\frac{\partial w}{\partial x} + v\frac{\partial w}{\partial y} + w\frac{\partial w}{\partial z} &= F_z
\end{aligned} \tag{2}$$

We have just taken what is perhaps the most significant step in fluid mechanics and atmospheric science. Newton's laws apply to parcels of air, and the second of them says that the parcels will be accelerated if forces are applied. To use the law in this form, we should have to keep track of parcels and follow each of them in their wanderings around the globe. But with the Euler relation, we can transform Newton's law so that it applies at a point. Thus we can measure wind velocities and other variables at various points and predict new values of the variables at those points. This is certainly the greatest simplification ever introduced in fluid mechanics and its implications are still being explored, for it does create a serious problem. Still, we are well justified in viewing the Euler relation as one of the fundamental theorems of atmospheric science.

To find the local accelerations $\partial u/\partial t$, $\partial v/\partial t$, and $\partial w/\partial t$, we must compute all the other terms from our analysis of the meteorological variables. We see immediately that we must have wind analyses at more than one level because derivatives with respect to the vertical occur, and we must also have information on the vertical velocity components. These equations are coupled between the components: we have three partial differential equations in three unknowns and they are nonlinear. That is, if u_1, v_1, and w_1 are solutions and u_2, v_2, and w_2 are another set of solutions, the sums

$u_1 + u_2$, $v_1 + v_2$, $w_1 + w_2$ are not solutions because of the products of velocities and derivatives in the advective terms. Most of the major difficulties of atmospheric science and fluid mechanics stem from this nonlinearity.

This is the serious problem introduced by the Euler relation. Either we must keep track of each parcel or, in transforming to the local derivative form, we encounter this nonlinear form of the equations of motion. Thus we have replaced the intuitively difficult and practically impossible parcel representation with a nonlinear system that still defies our analytical capability. Fluid dynamics made the choice in its infancy to confront the nonlinearity as the simpler approach. Some progress has been made, but we still are largely in doubt whether the problems raised by nonlinearity will ever be overcome analytically.

Even though the nonlinearity is a formidable obstacle to analytic study of the equations, it does not prevent numerical prediction over short enough time periods. Thus with Eq. (2) we may find temporal derivatives of the wind fields and, as explained in Sec. 2.4, integrate the equations forward in time.

Thus we must know how the forces arise, and so we turn next to the subject of thermodynamics, for it is the thermal properties and phenomena of the atmosphere that give it life and motion.

PROBLEM 2.6.1 Prove that the equations (2) are nonlinear in the velocities u,v,w. Can you think of any special cases in which they would be linear?

PROBLEM 2.6.2 Given the wind data at point P shown in the figure for time t and the fact that $w = F_x = F_y = 0$, find the values of u and v at P at $t + \Delta t$ where $\Delta t = 10$ min.

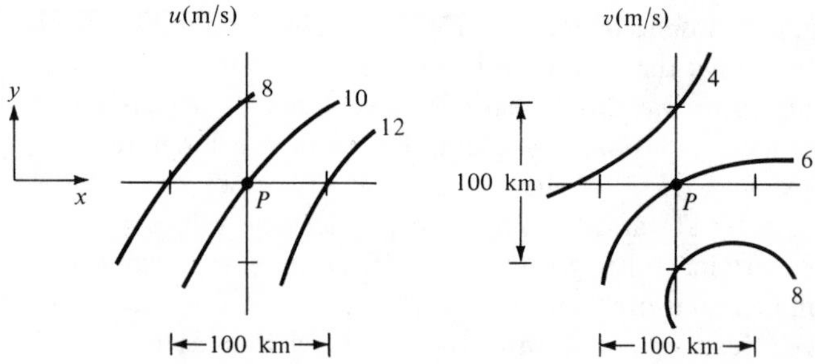

BIBLIOGRAPHIC NOTES

The material of this chapter is fairly standard; additional details or presentations with different emphasis may be found in works on calculus or advanced calculus, physics (especially mechanics), and texts on fluid mechanics. The proof of Taylor's theorem follows that in the book by R. C. Buck cited in the bibliography to Chap. 5.

3

INTRODUCTION TO ATMOSPHERIC THERMODYNAMICS

The forces that drive the ceaseless motion of the atmosphere are created by the heating of the earth and air by the sun. This solar heating is differential–because of the earth's spherical shape–thus creating thermodynamic imbalances. The motions result as the atmosphere attempts to return to thermodynamic equilibrium.

The elementary theory in this chapter will provide sufficient understanding of thermodynamics for our initial investigations of the motion of the atmosphere. As we consider more complex problems in our progress through the theory of dynamic meteorology, we shall return again to thermodynamics, adding to our considerations the modifications necessary to deal with water vapor and the processes of frictional energy dissipation and heat conduction.

3.1 BASIC THERMODYNAMICS†

Thermodynamics is a collection of concepts and equations that describe our accumulated experience with processes that involve fluxes of heat and changes in

†An alternative approach based on less sophisticated mathematics and reasoning is provided in Appendix 1 for use in beginning courses.

energy content. It is often a confusing subject, mainly because written explanations have generally preceded the equations, and such verbal descriptions of thermodynamic experience are subject to misinterpretation.

The same problem has been banished from mechanics by Newton's laws, which abstracted the common experience with motion in a compact form. We shall later use Newton's laws as the postulates for the development of the theory of the mechanical properties of atmospheric motion. In a similar spirit, we now present an axiomatic approach to thermodynamics expounded recently by Prof. Clifford Truesdell. The advantage of this approach lies in the economy with which experience is assimilated in the axioms, and the fact that once the basic axioms are stated, the rest of the theory follows from mathematical arguments without the need for introducing additional physical arguments.

We focus attention on a particular system of total mass M. The system is presumed to be homogeneous, identifiable in some sense, and its mass is constant, even though it may interact with its environment in other ways. The reader who desires a specific example to keep in mind may imagine a balloon filled with air.

We shall denote the *net mechanical power* applied to the system as $N = N_i - N_o$, where N_i (power *in*; $N_i \geqslant 0$) is the mechanical power applied to the system by the environment or by internal mechanical processes such as friction, and N_o (power *out*; $N_o \geqslant 0$) is the power applied by the system to the environment. For example, mechanical power is being applied to a system when it is compressed. Similarly, we shall denote the *net thermal power* by $Q = Q_i - Q_o$. For example, thermal power is applied to the balloon full of air when a bright light shines on it.

The system may have kinetic energy K deriving from the translational motion of the entire system.

The three basic axioms codifying our experience are:

1 Not all the power applied to the system produces kinetic energy. Thus any power applied which does not result in an increase of K is stored in a form called the internal energy and denoted by I. From this definition we have

$$\dot{I} = N + Q - \dot{K} \tag{1}$$

where the overdot denotes a time rate of change. This equation is usually called the *first law of thermodynamics.* As an example, if a tethered balloon is heated by sunlight, it does not increase its kinetic energy, but the temperature and pressure of the air inside increase.

2 It is possible to choose a scale for a thermometer so that $T \geqslant 0$. In discussing fundamental physical quantities, we defined temperature as the property measured with a thermometer. Now we assert that we may graduate our thermometer so that T is never negative; the Kelvin scale provides an example.

The final axiom is an ingenious one and distinguishes the present approach from classical presentations:

3 The thermal power is bounded. That is,

$$Q \leqslant B \tag{2}$$

where the bound B depends on the properties of the system under consideration.

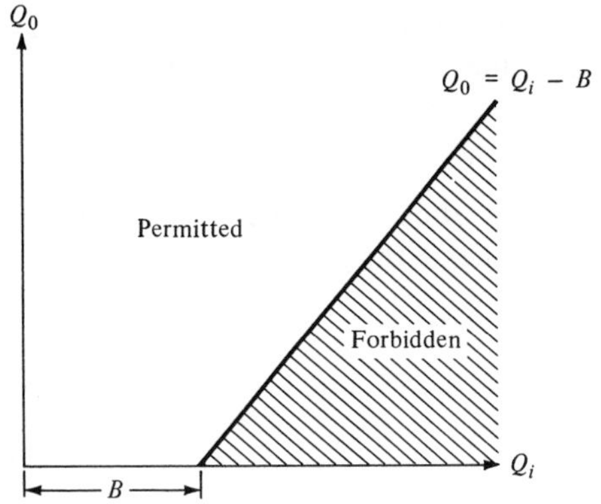

FIGURE 3.1
Implication of the assumption that the thermal power Q is bounded by B.

The physical content of this axiom is revealed by rewriting it as

$$Q_i - Q_o \leqslant B \qquad \text{or} \qquad Q_o \geqslant Q_i - B \tag{3}$$

Thus it says that when the thermal power applied to the system exceeds the bound B, then the system necessarily begins to exert thermal power back on the environment. In more commonplace terms, Eq. (3) says that we cannot apply heat to a system at a rate exceeding B without the system simultaneously giving off heat. This axiom is illustrated in Fig. 3.1.

With the aid of the bound B, we define the *entropy* or the *heat content* as

$$S = \int \frac{B}{T} dt \tag{4}$$

and thus with Eq. (2) we have

$$\dot{S} = \frac{B}{T} \geqslant \frac{Q}{T} \tag{5}$$

or

$$Q \leqslant T\dot{S} \tag{6}$$

This last result is usually called the *second law of thermodynamics.*

In most cases, we need more variables than the temperature to describe the state of our thermodynamic system. We denote the other variables now by $Y_1, \ldots, Y_n$; such variables relevant to the balloon are the pressure inside and its volume. Thus a *thermodynamic process* is described by specifying the history of the variables T, Y_1, $\ldots$, Y_n, I, S, Q, N, and K. Very often N and K appear only as the difference $W = N - \dot{K}$, which we shall call the *work storage rate.* The history of the set of variables listed above is subject to the three axioms, and very often from measurements we shall know all of them except Q, which we may then find from Eq. (1) in the form

$$\dot{I} = W + Q \tag{7}$$

Since Q is often unavailable from direct measurements, it is useful to eliminate it from the equations. If we combine Eqs. (7) and (6), we shall have

$$\dot{I} \leqslant W + T\dot{S} \tag{8}$$

This equation would be easier to use if it involved $\dot{T}$ rather than $\dot{S}$. Toward this end, let us define the *free energy*

$$\Psi = I - TS \tag{9}$$

which in combination with Eq. (8) gives

$$\dot{\Psi} - W + \dot{T}S \leqslant 0 \tag{10}$$

This is known as the *reduced dissipation inequality* (reduced because Q has been eliminated; dissipation because frictional dissipation is one of the causes leading to inequality).

Special restrictions on the various quantities we have been considering are often imposed. The most frequent, their definitions, and some of their implications are listed in Table 3.1.

In order to proceed further, we must specify relations between W, Ψ, and S and the state variables $T, Y_1, \ldots, Y_n$. To do so, we assume that the past and present values of these state variables determine the present values of W, Ψ, and S; this assumption is known as the *principle of thermodynamic determinism.*

The general approach is to deduce (by whatever means we can) a relation $\Psi = \Psi(T, Y_1, \ldots, Y_n)$ and an expression for the work storage in the form

$$W = W(T, Y_1, \ldots, Y_n) \tag{11}$$

The relation (11) is known as a *constitutive equation* and may be quite arbitrary except that it must satisfy the reduced dissipation inequality.

Table 3.1 DESCRIPTION OF THERMODYNAMIC PROCESSES

Description	Definition	Implication
Isothermal	$\dot{T} = 0$	$\dot{\Psi} - W \leqslant 0$
Adiabatic	$Q = 0$	$\dot{I} = W$
Isentropic	$\dot{S} = 0$	$\dot{\Psi} = \dot{I} - \dot{T}S$
Reversible	$Q = T\dot{S}$	Equality in Eqs. (6), (8), (10)
Reversible and adiabatic	$0 = T\dot{S}$	Isentropic: $\dot{S} = 0$
Irreversible	$Q < T\dot{S}$	Inequality in Eqs. (6), (8), (10)
Irreversible and adiabatic	$0 < T\dot{S}$	$\dot{S} > 0$
Irreversible and isentropic	$Q < 0$	$\dot{I} < W$

Let us consider the case of a system composed of a compressible substance. We shall hold it fixed in place so that $\dot{K} = 0$ and let compression by the environment ($N_i > 0$) and expansion of the system ($N_o > 0$) be the only components of the mechanical power. The thermodynamic state is described by T, the pressure p, and volume V. For simplicity we consider a spherical system with constant internal pressure supplying the force per unit area that resists compression or produces expansion. The mechanical power per unit area is

$$\frac{N}{A} = -p\dot{r} \tag{12}$$

where $\dot{r}$ is the rate of change of radius. If $\dot{r} < 0$, then the system is being compressed and $N_i = -p\dot{r}A$ while $N_o = 0$; if $\dot{r} > 0$, then the system is expanding and $N_i = 0$ while $N_o = p\dot{r}A$. But for a sphere, $A = 4\pi r^2$, so we have for either case

$$N = -p(4\pi r^2 \dot{r}) = -p\dot{V} \tag{13}$$

For $\dot{K} = 0$, we may now write Eq. (10) as

$$\dot{\Psi} + p\dot{V} + \dot{T}S \leqslant 0 \tag{14}$$

With the assumption that there is some equation of state $F(p,T,V) = 0$ relating T, V, and p, we may reduce $\Psi = \Psi(T,p,V)$ to $\Psi = \Psi(T,V)$ and so by the chain rule Eq. (14) becomes

$$\frac{\partial \Psi}{\partial T}\dot{T} + \frac{\partial \Psi}{\partial V}\dot{V} + p\dot{V} + \dot{T}S \leqslant 0 \tag{15}$$

which we write as

$$\left(p + \frac{\partial \Psi}{\partial V}\right)\dot{V} + \left(S + \frac{\partial \Psi}{\partial T}\right)\dot{T} \leqslant 0 \tag{16}$$

Now the rates $\dot{V}$ and $\dot{T}$ are completely arbitrary, so that the only way the inequality may be satisfied for all choices of $\dot{V}$ and $\dot{T}$ is for the equality to hold, which requires that

$$p = -\left(\frac{\partial \Psi}{\partial V}\right)_T \qquad S = -\left(\frac{\partial \Psi}{\partial T}\right)_V \tag{17}$$

Because of equality, we have only reversible processes in this case. From the obvious fact that

$$\frac{\partial^2 \Psi}{\partial V\, \partial T} = \frac{\partial^2 \Psi}{\partial T\, \partial V} \tag{18}$$

it also follows that

$$\left(\frac{\partial S}{\partial V}\right)_T = \left(\frac{\partial p}{\partial T}\right)_V \tag{19}$$

This is one of Maxwell's relations.

With these results, a host of other thermodynamic relations may be obtained. They are displayed in Table 3.2.

The thermodynamic relations summarized in Table 3.2 are of more general validity than the derivation given would indicate. To see this, let us augment the mechanical power by a frictional resistance to compression so that

$$N = -p\dot{V} + \lambda\left(\frac{\dot{V}}{V}\right)^2 \tag{20}$$

This choice is made to illustrate the mathematical technique of dealing with dissipative processes. Now Eq. (10), written in the form of Eq. (16), becomes

$$\left(\frac{\partial\Psi}{\partial T} + S\right)\dot{T} + \left(\frac{\partial\Psi}{\partial V} + p\right)\dot{V} - \lambda\left(\frac{\dot{V}}{V}\right)^2 \leqslant 0 \tag{21}$$

This relation must again remain true for arbitrary choices of the rates, and to discover the implications of this fact, we multiply each rate by a factor γ so that we have

$$\gamma\left[\left(\frac{\partial\psi}{\partial T} + S\right)\dot{T} + \left(\frac{\partial\Psi}{\partial V} + p\right)\dot{V}\right] - \gamma^2\left[\lambda\left(\frac{\dot{V}}{V}\right)^2\right] \leqslant 0 \tag{22}$$

The only way this can be true for all γ is for Eq. (17) to be valid, so that the first

Table 3.2 THERMODYNAMIC RELATIONS FOR A COMPRESSIBLE SYSTEM

Definitions	The first law	Derivative relations
Free energy $\Psi = \Psi(T,V)$ $\Psi = I - TS$	$\dot{\Psi} + p\dot{V} + \dot{T}S = 0$	$p = -\left(\frac{\partial\Psi}{\partial V}\right)_T \quad S = -\left(\frac{\partial\Psi}{\partial T}\right)_V$ $\left(\frac{\partial p}{\partial T}\right)_V = \left(\frac{\partial S}{\partial V}\right)_T$
Free enthalpy $G = G(p,T)$ $G = \Psi + pV$ $= I - TS + pV$	$\dot{G} - \dot{p}V + \dot{T}S = 0$	$S = -\left(\frac{\partial G}{\partial T}\right)_p \quad V = \left(\frac{\partial G}{\partial p}\right)_T$ $\left(\frac{\partial S}{\partial p}\right)_T = -\left(\frac{\partial V}{\partial T}\right)_p$
Enthalpy $H = H(p,S)$ $H = I + pV$ $= G + TS$	$\dot{H} - T\dot{S} - \dot{p}V = 0$	$T = \left(\frac{\partial H}{\partial S}\right)_p \quad V = \left(\frac{\partial H}{\partial p}\right)_S$ $\left(\frac{\partial T}{\partial p}\right)_S = \left(\frac{\partial V}{\partial S}\right)_p$
Internal energy $I = I(V,S)$ $I = H - pV$	$\dot{I} + p\dot{V} - T\dot{S} = 0$	$T = \left(\frac{\partial I}{\partial S}\right)_V \quad p = -\left(\frac{\partial I}{\partial V}\right)_S$ $\left(\frac{\partial T}{\partial V}\right)_S = -\left(\frac{\partial p}{\partial S}\right)_V$

term vanishes, and for λ to be positive, so that the second term satisfies the inequality. Hence all the relations in the table are still valid. Moreover, by comparing the first law of thermodynamics, which now is

$$\dot{I} + p\dot{V} = Q + \lambda\left(\frac{\dot{V}}{V}\right)^2 \tag{23}$$

with the general statement (see Table 3.2) that

$$\dot{I} + p\dot{V} - T\dot{S} = 0 \tag{24}$$

we see that

$$T\dot{S} = Q + \lambda\left(\frac{\dot{V}}{V}\right)^2 \tag{25}$$

which shows that dissipative processes, such as the one modeled here, are responsible for the validity of the second law.

*PROBLEM 3.1.1 Prove each of the relations in Table 3.2.

*PROBLEM 3.1.2 Let a system of mass M and its environment be affected by the force of gravity. Both the total energy E of the system and the work rate W must be changed. Show that, even so, the equation for $\dot{I}$ remains the same.

PROBLEM 3.1.3 (Truesdell) Consider a compressible system described by the variables T, p, V, and X_i ($i = 1, \ldots, n$). Let the work rate be

$$W = N - \dot{K} = -p\dot{V} + \sum_{i=1}^{n}\sum_{j=1}^{n} \lambda_{ij}\dot{X}_i\dot{X}_j$$

Use the theory of positive definite quadratic forms to determine explicit conditions for the coefficients λ_{ij}. Show that $\partial\Psi/\partial X_i = 0$ and that the thermodynamic pressure is still specified by Eq. (17). Do the relations of Table 3.2 remain valid? What is the expression for the rate of change of the entropy?

3.2 THERMODYNAMICS OF AIR AS AN IDEAL GAS

The relations and consequences of the thermodynamic theory just developed become particularly simple for an ideal gas, whose equation of state may be written from Eq. (16) in Sec. 2.3 as

$$pV = MRT = M\frac{R^*}{\mu}T \tag{1}$$

First we have:

Theorem The internal energy of an ideal gas depends only on temperature.

PROOF From Table 3.2 or Eq. (9) in Sec. 3.1 we have

$$I = \Psi(T,V) + TS \tag{2}$$

where Eq. (17) in Sec. 3.1 shows that $S = S(T,V)$. Hence we may use $I = I(T,V)$. From Eq. (2) in this section and Eq. (17) in Sec. 3.1

$$\left(\frac{\partial I}{\partial T}\right)_V = \left(\frac{\partial \Psi}{\partial T}\right)_V + S + T\left(\frac{\partial S}{\partial T}\right)_V = T\left(\frac{\partial S}{\partial T}\right)_V \tag{3}$$

and

$$\left(\frac{\partial I}{\partial V}\right)_T = \left(\frac{\partial \Psi}{\partial V}\right)_T + T\left(\frac{\partial S}{\partial V}\right)_T = -p + T\left(\frac{\partial p}{\partial T}\right)_V = 0 \tag{4}$$

in which we have used Eq. (1) to deduce that $(\partial p/\partial T)_V = MR/V$, concluding the proof. ////

Now we have $I = I(T)$, and we use the result in Eq. (3) to define the coefficient, known as *specific heat*,

$$\left(\frac{\partial I}{\partial T}\right)_V = C_V(T) = T\left(\frac{\partial S}{\partial T}\right)_V \tag{5}$$

and therefore

$$I = I_0 + \int_{T_0}^{T} C_V(T')\,dT' \tag{6}$$

A companion relation follows from

$$dS = \left(\frac{\partial S}{\partial T}\right)_V dT + \left(\frac{\partial S}{\partial V}\right)_T dV \tag{7}$$

so that integrating from T_0 to T with V constant at V_0 and then from V_0 to V with T constant gives

$$S = S_0 + \int_{T_0}^{T} \frac{C_V(T')}{T'}\,dT' + MR \ln \frac{V}{V_0} \tag{8}$$

If we now write the first law as

$$C_V(T)\dot{T} + p\dot{V} = Q \tag{9}$$

and substitute the derivative of Eq. (1), we have

$$[C_V(T) + MR]\dot{T} - V\dot{p} = Q \tag{10}$$

so that, because $H = I + pV = G + TS$, we may define another specific heat by

$$\left(\frac{\partial H}{\partial T}\right)_p = C_p(T) = C_V(T) + MR = T\left(\frac{\partial S}{\partial T}\right)_p \tag{11}$$

These are the basic relations for an ideal gas; to proceed further, we must use some assumptions about the molecular properties of the gas to find explicit values for the specific heats and to show that they are constant. The molecules of the gases of which air is composed are primarily diatomic, and they may be conceived of as arranged like a dumbbell with an atom at each end of an imaginary axis. They have three degrees of freedom in their translational motion, and they can rotate simultaneously around their orthogonal axes. Thus there are six degrees of freedom in all, but it is assumed that there is negligible energy in the rotation around the axis connecting the two ends of the dumbbell, leaving effectively five degrees of freedom. We assume that the internal energy is equally distributed among these five degrees of freedom. Thus we have the internal energy as

$$I = K_t + K_r \tag{12}$$

with

$$K_t = \tfrac{3}{5}I \qquad K_r = \tfrac{2}{5}I \tag{13}$$

From Problem 2.3.1 we have

$$K_t = \tfrac{3}{2}R^*T\frac{M}{\mu} \tag{14}$$

and hence

$$I = \tfrac{5}{2}R^*T\frac{M}{\mu} \tag{15}$$

which implies that

$$C_V = \frac{\partial I}{\partial T} = \tfrac{5}{2}R^*\frac{M}{\mu} \qquad C_p = C_V + RM = \tfrac{7}{2}R^*\frac{M}{\mu} \tag{16}$$

With this simplification, Eqs. (6) and (8) become

$$I = I_0 + C_V(T - T_0) \tag{17}$$

and

$$S = S_0 + C_V \ln\frac{T}{T_0} + MR \ln\frac{V}{V_0} \tag{18}$$

In our future work with these thermodynamic relations, we shall make two changes. First we shall express our quantities so that they have dimensions of energy/unit mass. We have then

$$R = \frac{R^*}{\mu} \qquad c_V = \frac{C_V}{M} \qquad c_p = \frac{C_p}{M} \qquad \frac{c_V}{c_p} = \frac{5}{7} \qquad \frac{R}{c_p} = \frac{2}{7}$$

$$c_V(T - T_0) = \frac{I - I_0}{M} \qquad s = \frac{S}{M} \qquad q = \frac{Q}{M} \qquad w = \frac{W}{M} \tag{19}$$

For air, $c_p = 1$ J/g • deg for nearly all practical purposes.

The second change is more significant. We need to have relations that apply locally in the atmosphere or to parcels of air. Thus we *assume* that the results derived so far apply to *individual parcels of air*, provided that we differentiate following the motion. For such parcels, we insist that the mass be constant and so the basic thermodynamic relations become the equation of state (1), the first law,

$$c_V \frac{dT}{dt} + p \frac{d\alpha}{dt} = T \frac{ds}{dt} = q + f \tag{20}$$

where $f = w + p\, d\alpha/dt$ and thus represents the mechanical dissipative processes, and the second law

$$T \frac{ds}{dt} \geqslant q \tag{21}$$

which requires that

$$f \geqslant 0 \tag{22}$$

We shall find explicit representations for q and f later; for now we note only that q is composed of effects due to radiation, heat conduction, and change of phase of water, and that f is the frictional dissipation of kinetic energy.

PROBLEM 3.2.1 Show that $p \geqslant 0$.

PROBLEM 3.2.2 Suppose that a gram of air is alternately held at constant volume and constant pressure and is heated in each case so that its temperature increases 1 K. How much energy was required in each case? If this same energy were applied to a gram of water, how much would the water temperature increase? Use your result to discuss quantitatively what happens when cold air blows over a warm lake.

PROBLEM 3.2.3 Obtain expressions for $s = s(T,\alpha)$, $s = s(T,p)$, and $s = s(\alpha,p)$.

PROBLEM 3.2.4 Consider a gas confined in a small vessel. The gas is allowed to escape into an infinitely large reservoir that is at pressure $p = 0$. Use this experiment to argue that $I = I(T)$. What approximation must be made for a finite reservoir?

PROBLEM 3.2.5 The coefficients of volume expansion at constant pressure and temperature are, respectively,

$$\epsilon = \frac{1}{V}\left(\frac{\partial V}{\partial T}\right)_p \qquad \eta = -\frac{1}{V}\left(\frac{\partial V}{\partial p}\right)_T$$

What are their values for an ideal gas?

PROBLEM 3.2.6 What is the most general form of an equation of state that will yield $(\partial I/\partial V)_T = 0$?

*PROBLEM 3.2.7 Nearly incompressible substances such as water are often described by an equation of state, with ϵ and η constant, of the form

$$\alpha = \alpha_0[1 + \epsilon(T - T_0) - \eta(p - p_0)]$$

where the subscript zero denotes some reference state. How does the internal energy depend on volume in this case?

3.3 INTEGRALS AND ENTROPY

Often we are interested in the net changes of thermodynamic quantities that develop over a time interval rather than their rates of change. To illustrate, we consider a thermodynamic system composed of an ideal gas (it may be a parcel of air). Its state variables and energy functions are specified as a function of time. We take the first law, Eq. (20) in Sec. 3.2 (with $f = 0$), and integrate from time t_1 to time t_2 to obtain

$$c_V(T_2 - T_1) + \int_{t_1}^{t_2} p\, d\alpha(t) = \int_{t_1}^{t_2} q(t)\, dt \tag{1}$$

Two quite different kinds of quantities are involved here. We had dT/dt, which is the derivative of the function $T(t)$. Thus the integral of dT/dt depends only on the temperatures of the system at t_1 and t_2; it does not depend on temperatures at intermediate times. In contrast, $p\, d\alpha/dt$ is the product of a function and a derivative, and it cannot in general be written as a derivative of a single function. Thus its integral in Eq. (1) will generally depend on the values both p and α take at all times between t_1 and t_2. To see this, note that in Fig. 3.2 the values of p along the lower curve from A to B are different from those along the upper curve. The α values are the same on both curves, so the integrals over the two paths from A to B will be different.

When the integral of a quantity along a curve depends only on its values at the end points of a curve, that quantity is called an *exact differential*. Hence dT/dt is an exact differential, $p\, d\alpha/dt$ is not, and so $q(t)$ is not either. Thus we have terms in the first law that depend on the history a process has had in reaching any particular state.

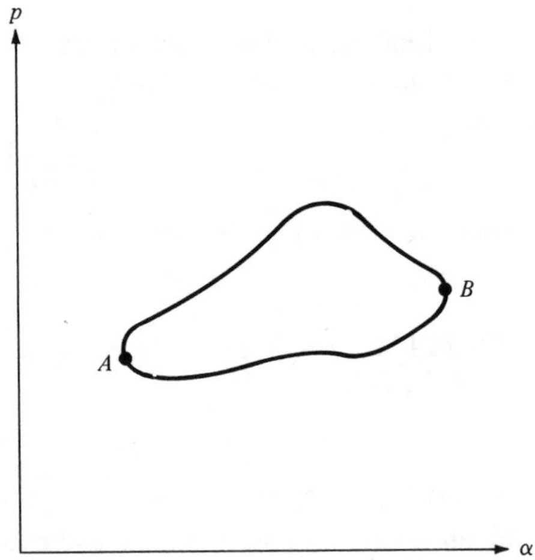

FIGURE 3.2
Path of a thermodynamic process.

For example, let us consider the cyclic process shown in Fig. 3.2. Suppose we start at A, proceed by the upper curve to B and back to A via the lower curve. For the integral over the closed curve in the direction described we have from Eq. (1)

$$\oint q(t)\,dt = \oint p\,d\alpha > 0 \tag{2}$$

so that it was necessary to apply thermal power to the system to force it through this cycle. This thermal power provides the energy necessary for the system to expand at higher pressure than it contracts; in other words, the environment provides the thermal power that makes it possible for the system to exert mechanical power.

*PROBLEM 3.3.1 Show that $\oint p\,d\alpha$ is equal to the area enclosed on a (p,α) diagram by the curve specifying a cyclic process.

PROBLEM 3.3.2 Write out in detail the steps sketched below in the proof of:

Theorem The necessary and sufficient condition that the differential expression defined by

$$df = M(x,y)\,dx + N(x,y)\,dy \tag{3}$$

be an exact differential is that

$$\frac{\partial M}{\partial y} = \frac{\partial N}{\partial x} \tag{4}$$

PROOF 1. Necessity: Consider the square described by $((x_0,y_0),(x,y_0),(x,y),(x_0,y))$. Let df be exact so that its integral around the square vanishes.

Compare the integrals from (x_0,y_0) to (x,y) via both (x,y_0) and (x_0,y) to show that

$$\int_{x_0}^{x}\int_{y_0}^{y}\left(\frac{\partial M}{\partial y}-\frac{\partial N}{\partial x}\right)dx\,dy = 0 \tag{5}$$

2. Sufficiency: Assume Eq. (4) and put

$$g(x,y) = \int M(x,y)\,dx \tag{6}$$

Show next that $\partial g/\partial x = M$ and, with a suitable choice of a function of integration, that $\partial g/\partial y = N$. Hence there exists a function $g(x,y)$ such that

$$df = \frac{\partial g}{\partial x}\,dx + \frac{\partial g}{\partial y}\,dy \tag{7}$$

and thus $\oint df = 0$. ////

PROBLEM 3.3.3 Use the test of Problem 3.3.2 to show that dT for an ideal gas is an exact differential in (p,α) space. Show that $p\,d\alpha$ is not.

The only exact differential that occurs in the first law for an ideal gas is the internal energy, and being a function of temperature only, it cannot describe the state of a thermodynamic system. A natural question, then, is whether we can modify the first law to yield a quantity which is both an exact differential and a function of two variables.

We combine the first law, Eq. (20) in Sec. 3.2, and the equation of state to obtain

$$\frac{c_V}{T}\frac{dT}{dt} + \frac{R}{\alpha}\frac{d\alpha}{dt} = \frac{q+f}{T} \tag{8}$$

which is an exact differential because we may write

$$c_V\,d\ln T + R\,d\ln\alpha = \frac{q+f}{T}\,dt \tag{9}$$

Now both M and N are constants, so we may define the exact differential

$$ds = c_V\,d\ln T + R\,d\ln\alpha \tag{10}$$

Integration yields

$$s - s_0 = c_V\ln\frac{T}{T_0} + R\ln\frac{\alpha}{\alpha_0} \tag{11}$$

The function s is a thermodynamic function of only the initial and final states of a system, and depends on two of the variables. It is obviously the entropy of Eq. (18) in Sec. 3.2 (expressed per unit mass) and thus is subject to the second law.

Despite the general practice in the literature, it is essential to introduce the constants of integration, s_0, T_0, and α_0 in Eq. (11) because logarithms must have dimensionless arguments. The constant need not appear in expressions such as $d \ln T$ because what is meant is dT/T and the constant in $d \ln (T/T_0)$ cancels.

Entropy is the most important thermodynamic function, not only because it is an exact differential which completely describes the thermodynamic state, but because it has a predictive function as well.

To illustrate this predictive function, let α_0 and p_0 be the present values in a thermodynamic system. With the equation of state (or the results of Problem 3.2.3) we may write

$$s - s_0 = c_p \ln \left[\frac{T}{T_0}\left(\frac{p_0}{p}\right)^{R/c_p}\right] \tag{12}$$

so that for any process beginning at (T_0, p_0) we have

$$T = T_0\left(\frac{p}{p_0}\right)^{R/c_p} e^{(s-s_0)/c_p} \tag{13}$$

The lines of constant entropy described by this equation are *isentropes* and are shown on a (p,α) diagram in Fig. 3.3. The regions in which $s - s_0 > 0$ and $s - s_0 < 0$ are easily found from Eq. (12). For isentropic processes, the path of the process must remain on the isentrope passing through the initial point.

If we write the second law as

$$T\frac{ds}{dt} = q + f \geqslant q \tag{14}$$

we see that adiabatic processes ($q = 0$) will, because of friction ($f \geqslant 0$), move into the $s - s_0 \geqslant 0$ region. A process can move into the $s - s_0 < 0$ region only if heat is extracted from the system.

In other words, because of the second law, the $s - s_0 < 0$ region is forbidden to adiabatic processes and to diabatic processes for which $q + f \geqslant 0$. In these two cases, the processes can only proceed along, or to the right of, the isentrope through the initial point. Therefore the second law gives thermodynamic processes a definite direction that can be altered only by exchange of heat with the environment.

Thus it is important to distinguish between adiabatic—which means no exchange of heat with the environment—and isentropic—which means constant entropy. For reversible processes, the two terms have the same meaning. But all natural processes are irreversible, and then they do not.

In meteorology we are almost always more concerned with whether processes are isentropic than whether they are adiabatic, and in most cases when meteorologists say adiabatic, they are assuming reversibility and really mean isentropic. The correct terminology is obviously more precise.

The concept of entropy appears in both meteorological theory and synoptic practice in modified form. If we define a function $\theta = \theta(p,T)$ by

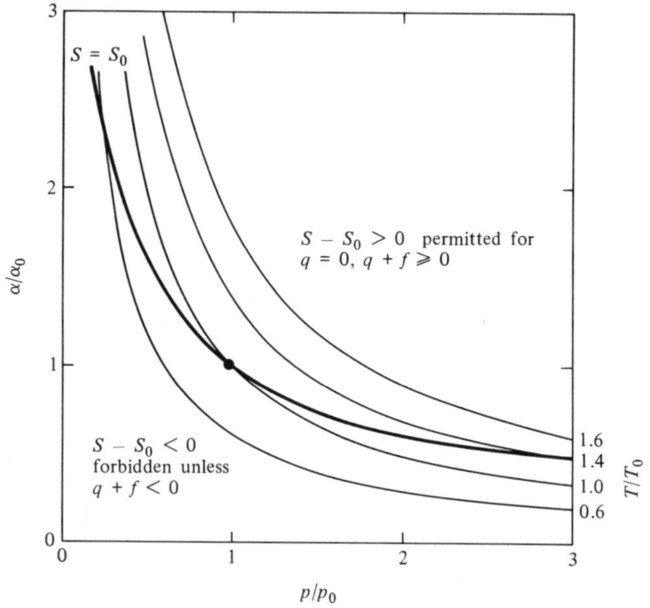

FIGURE 3.3
Diagram illustrating the implications of the second law of thermodynamics. The parcel initially has pressure p_0 and specific volume α_0. Isotherms for an ideal gas are shown as light curves. In isentropic processes, the values of pressure, temperature, and specific volume will follow the isentrope $S = S_0$ denoted by the heavy curve.

$$c_p \ln \frac{\theta}{\theta_0} = s - s_0 \tag{15}$$

then in comparison with Eq. (12) we have

$$\theta = \frac{\theta_0}{T_0} T \left(\frac{p_0}{p}\right)^{R/c_p} \tag{16}$$

We are free to choose the constants; the usual convention is to let p_0 be 1000 mbar and to let $\theta_0 = T_0$. This gives *Poisson's equation*

$$\theta = T \left(\frac{1000}{p}\right)^{R/c_p} \qquad p \text{ in millibars} \tag{17}$$

It is clear from Eq. (15) that θ does not change in an isentropic process. Thus Eq. (17) shows that θ is the temperature a parcel would have if it were moved isentropically to 1000 mbar, and so θ is called the *potential temperature.*

The relation (17) is used for the construction of a thermodynamic diagram used frequently in the analysis of atmospheric processes. The abscissa of the diagram is

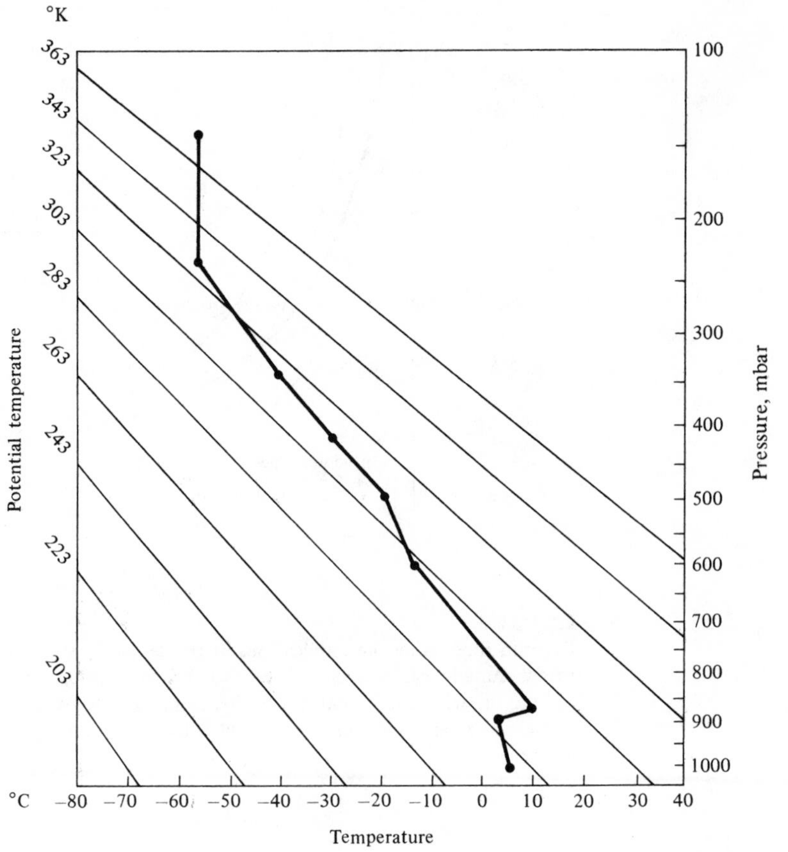

FIGURE 3.4
Illustration of a typical meteorological thermodynamic diagram (Stüve chart). The ordinate is calibrated in $(p/1000)^{R/c_p}$, the abscissa in temperature T. The isentropes are the slanting straight lines, and a typical atmospheric sounding is depicted by the curve as shown.

temperature, and the vertical coordinate is $(p/1000)^{R/c_p}$, but is usually marked off as a function of p giving rise to a semilogarithmic scale. On such a chart, the isentropes are straight lines, as illustrated in Fig. 3.4. When a vertical sounding of temperature is plotted on the chart, the values of potential temperature at each point are easily read.

The potential temperature shares with entropy the property of being an exact differential that depends on two of the thermodynamic variables, and the second law of thermodynamics may be stated for adiabatic processes as

$$\frac{d\theta}{dt} = \frac{\theta_0}{c_p} \exp\left(\frac{s - s_0}{c_p}\right) \frac{ds}{dt} \geqslant 0 \tag{18}$$

For the special case of isentropic motion we have $ds/dt = 0$ and consequently $d\theta/dt = 0$, so surfaces of constant values of potential temperature move along with, or are frozen into, isentropic motion.

Thus we have the alternative forms of either entropy or potential temperature to incorporate the implications of the second law in meteorological theory. It is well to remember that the first law prescribes what may happen; the second law predicts what must happen. This idea is illustrated by a quotation from the English physicist Robert Emden: "... energy and entropy. In the course of advancing knowledge the two seem to me to have exchanged places. In the huge manufactory of natural processes, the principle of entropy occupies the position of manager, for it dictates the manner and method of the whole business, whilst the principle of energy merely does the bookkeeping, balancing credits and debits."

PROBLEM 3.3.4 Show that

$$\theta\rho = \frac{p^{c_V/c_p}(1000)^{R/c_p}}{R}$$

PROBLEM 3.3.5 Beginning students often say that the potential temperature is defined only for isentropic processes. Explain why they are wrong. Why do we use θ instead of $\phi = T(p_0/p)^{R/c_V}$?

PROBLEM 3.3.6 Show with the first law that for isentropic processes

$$T = T_0\left(\frac{p}{p_0}\right)^{R/c_p}$$

$$T = T_0\left(\frac{\alpha_0}{\alpha}\right)^{R/c_V}$$

$$\alpha = \alpha_0\left(\frac{p_0}{p}\right)^{c_V/c_p}$$

PROBLEM 3.3.7 Use the first of the relations in Problem 3.3.6 to derive Poisson's equation.

PROBLEM 3.3.8 (Emden) Compute the total thermodynamic energy in a heated room. If the internal pressure equals a constant outdoor pressure, show that the energy *decreases* as the room is heated. [Hint: Begin with Eq. (17) in Sec. 3.2, where for even slightly moist air $I_0 \gg C_V T_0$ because of the large latent energy contained by water vapor.] Thus with heating, the temperature of the room increases but the total energy in the room decreases. What is happening to the thermal power supplied by the furnace?

The following three problems concern some classical experiments that illustrate thermodynamic processes.

PROBLEM 3.3.9 *An Adiabatic and Reversible Process.* An ideal gas (with $f = 0$) is confined in an adiabatic cylinder by a weightless piston free to move up and down. A collection of little weights is available so that the total weight resting on the piston may be adjusted. The weights are moved one at a time so that equilibrium is maintained. Let $m(t)$ be the total mass on the piston at time t. Show that $dp/p = dm/m$ and that $\theta(t) = \text{constant}$.

PROBLEM 3.3.10 *An Adiabatic and Irreversible Process.* An ideal gas (with $f = 0$) is confined in an adiabatic cylinder by a piston. The piston is moved so rapidly that the volume instantaneously increases from V_0 to $V_1 = \lambda V_0$. Then the piston is slowly pushed back down to its original position. Justify the approximation that the temperature after the expansion is $T_1 = T_0$. Express the temperature and pressure as functions of time, and show that the final entropy S_F is

$$S_F - S_0 = MR \ln \lambda > 0$$

*PROBLEM 3.3.11 *Mixing of Gases.* Consider two identical gases, each of mass M and equal pressure p, but with temperatures T_1 and T_2. If the two gases are mixed adiabatically without any mechanical power applied, then show that the mixture temperature is $T = (T_1 + T_2)/2$ and that the difference between the final entropy and the sum of the initial entropies is

$$\Delta s = \frac{S - S_1 - S_2}{2M} > 0$$

[Hint: $(a - b)^2 \geqslant 0$ so that $a^2 + b^2 \geqslant 2ab$, Cauchy's inequality.]

PROBLEM 3.3.12 Generalize the results of Problem 3.3.11 to apply to the case of initial masses M_1 and M_2.

3.4 OPEN SYSTEMS

Elementary thermodynamic theory applies to *closed systems* in which the mass is constant. But in atmospheric science we often have occasion to consider *open systems*

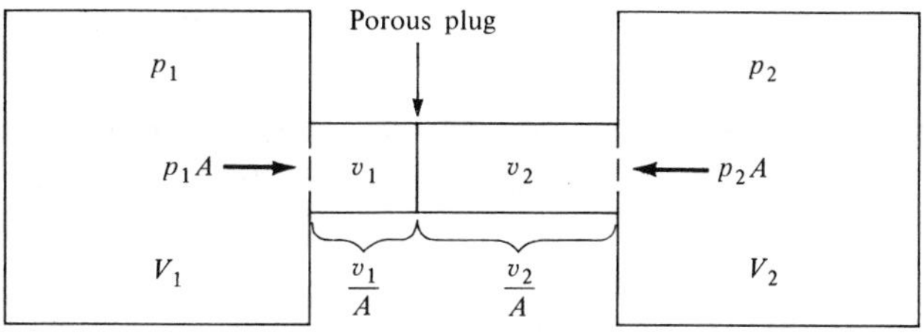

FIGURE 3.5
Arrangement of the Joule-Kelvin porous-plug experiment.

in which mass exchange between the system and surroundings does occur. A cloud is a good example of an open system.

The development of a suitable theory will be based upon the *Joule-Kelvin porous-plug experiment*. The arrangement is depicted in Fig. 3.5. On the left side a gas of mass M_1, pressure p_1, and temperature T_1 occupies a volume $V_1 + v_1$. Conditions on the right are denoted by subscript 2. The connecting pipe is of cross-sectional area A and has a porous plug that allows the gas to seep slowly from the reservoir with higher pressure p_1 to that with lower pressure p_2. We assume that the pressure changes from p_1 to p_2 in an infinitesimal distance at the infinitesimal plug.

We analyze what happens to a mass m of gas passing from left to right. In the pipe, but on the left of the plug, the mass occupies a volume v_1, so that it is pushed a distance v_1/A by a force p_1A; thus mechanical work in the amount p_1v_1 is done on the gas. On the other side of the plug, the mass m is retarded by a force p_2A as it passes from the plug to the reservoir, so that the moving gas does work p_2v_2. Thus the net work done on the gas (for $f = 0$) is

$$\int_{t_1}^{t_2} N\,dt = p_1v_1 - p_2v_2 \tag{1}$$

On the assumption that no thermal power is applied and that the process is so slow that $\dot{K} \cong 0$, the first law

$$\dot{I} = W \tag{2}$$

shows that

$$I_2 - I_1 = p_1v_1 - p_2v_2 \tag{3}$$

in which I_1 is the internal energy of the gas leaving the reservoir on the left, I_2 the internal energy as it passes into the reservoir on the right. The gas entering the reservoir is assumed to have adapted to its conditions of p_2 and v_2, and hence

$$I_2 + p_2v_2 = I_1 + p_1v_1 \tag{4}$$

implies that the specific enthalpy

$$h = \frac{I + pv}{m} \tag{5}$$

is constant in the gas moving from one reservoir to another. For ideal gases, this implies that the temperature is constant; Joule and Kelvin verified experimentally that air is nearly an ideal gas in the sense that it exhibits negligible temperature change in this experiment.

In this arrangement, the separate systems on either side of the plug are examples of open systems. More generally now, we let an open system have mass m, pressure p, temperature T, and specific volume $\alpha = V/m$. We consider the specific energy e, enthalpy h, and rate of heat addition q, and we take $f = 0$. Mass-weighted quantities are defined by

$$I = me \qquad H = mh \qquad Q = mq \tag{6}$$

Then the first law applied to the mass of gas in the system at time t gives

$$\dot{h} - \alpha\dot{p} = \dot{e} + p\dot{\alpha} = q \tag{7}$$

At time $t + \Delta t$ we imagine that the mass of the system is $m = m_0 + m_e$, where m_e is the mass gained or lost by flux through the boundary and m_0 is the mass at time t. Clearly then

$$\Delta m = m(t + \Delta t) - m_0 = m_e \tag{8}$$

On the basis of the Joule-Kelvin experiment, we may suppose that the mass of gas carried into or out of the system arrives or leaves without a change in its enthalpy h_e. Thus the enthalpy difference from t to $t + \Delta t$ for the open system is

$$\Delta H = m_0 h(t + \Delta t) + m_e h_e - m_0 h(t) = m_0 \Delta h + h_e\, \Delta m \tag{9}$$

Thus in the limit we have

$$\dot{H} = m\dot{h} + h_e \dot{m} \tag{10}$$

and with the aid of Eq. (7) this becomes

$$\dot{H} = mq + m\alpha\dot{p} + h_e\dot{m} = Q + V\dot{p} + h_e\dot{m} \tag{11}$$

We may calculate with Eq. (5) that $\dot{H} = \dot{I} + d(pV)/dt$ so that Eq. (11) may be expressed as

$$\dot{H} - V\dot{p} = \dot{I} + p\dot{V} = Q + h_e\dot{m} \tag{12}$$

This is the first law of thermodynamics for open systems. It is worth noting that $p\dot{V}$ no longer represents mechanical power because the volume can expand owing to mass flux without work being done.

Next we turn to the question of how the entropy of an open system behaves. As before, we define $S = ms$ and from Eq. (7) we obtain

$$T\dot{s} = \dot{h} - \alpha\dot{p} = \dot{e} + p\dot{\alpha} \tag{13}$$

Now we must recognize that although enthalpy h is constant as it fluxes into or out of the system, the entropy of the fluxing gas may change. We denote by s_f the value of the entropy of the fluxing gas as it crosses the boundary of the open system. Then from Eq. (13) we have

$$T\dot{S} = m\dot{h} - V\dot{p} + Ts_f\dot{m} = \dot{H} - V\dot{p} - \dot{m}h_e + T\dot{m}s_f \tag{14}$$

In many cases the fluxing gas will have the same temperature and pressure as the gas in the system (especially if the flux is out of the system). Then we introduce the factor

$$\mu = h - Ts \tag{15}$$

which is known as the *chemical potential* and is important in physical chemistry, and Eq. (14) becomes

$$T\dot{S} = \dot{H} - V\dot{p} - \mu\dot{m} \tag{16}$$

which is Gibbs' law applied to a single constituent.

In the more general case, s_f is not equal to s or s_e because the fluxing gas must change pressure. For an ideal gas, however, simplification is possible. If the mass is fluxing into the system, we have by the Joule-Kelvin experiment that $\dot{h}_e = 0$ during the process so that

$$\dot{s}_e = -\frac{\alpha_e \dot{p}_e}{T_e} = -\frac{R\dot{p}_e}{p_e} \tag{17}$$

and by integration

$$s_f = s_e - R \ln \frac{p}{p_e} \tag{18}$$

on the assumption that the fluxing gas adopts the internal pressure p at the system boundary.

Thus for an ideal gas, we have from Eqs. (12) and (14) the law

$$\dot{S} = \frac{Q}{T} + \dot{m}\left(s_e - R \ln \frac{p}{p_e}\right) \tag{19}$$

Now we return to the porous-plug experiment to analyze entropy change in open systems. We may apply Eq. (12) to the reservoirs of fixed volume ($\dot{V} = 0$) on either side of the porous plug. Because $Q = 0$,

$$\dot{I}_1 = h_1\dot{m}_1 < 0 \tag{20}$$

and

$$\dot{I}_2 = h_1\dot{m}_2 > 0 \tag{21}$$

No mass is lost in the flux from one reservoir to the other so

$$\dot{m}_1 = -\dot{m}_2 \tag{22}$$

and therefore

$$\dot{I}_1 = -\dot{I}_2 \tag{23}$$

so that the total internal energy does not change. But for the entropy, Eq. (19) shows that

$$\dot{S}_1 = s_1 \dot{m}_1 \tag{24}$$

and

$$\dot{S}_2 = -\dot{m}_1\left(s_1 - R \ln \frac{p_2}{p_1}\right) \tag{25}$$

For the entropy of the total system, then, we find that

$$\dot{S} = \dot{S}_1 + \dot{S}_2 = -\dot{m}_1 R \ln \frac{p_1}{p_2} = \dot{m}_2 R \ln \frac{p_1}{p_2} \tag{26}$$

But this is positive because of the initial assumption controlling the direction of mass flow which gives $p_1 > p_2$. The conclusion is:

Theorem Let S be the total entropy of an adiabatic system composed of two identical ideal gases. Then $\dot{S} > 0$, if mass flux is allowed from a subsystem with high pressure to a subsystem with low pressure, regardless of the arrangement of the two initial temperatures and regardless of the initial mass distribution.

////

We have crossed the frontier from thermostatics to thermodynamics in considering how the properties of gases behave when dynamical forces such as pressure gradients are imposed to initiate mass fluxes. The porous-plug experiment suggests a model (see Problem 3.4.4) of what happens in the general circulation of the atmosphere. Pressure and temperature differences are created by differential heating, the atmosphere is set in motion, and the mixing that ensues is accompanied by increasing entropy. But in contrast to the isobaric equilibrium reached in the Joule-Kelvin experiment, this entropy is then again destroyed by differential heating, the process never stops, and equilibrium is never reached.

PROBLEM 3.4.1 Let a closed system be at temperature T and pressure p and have constituents with masses m_j. Derive Gibbs' law

$$T\dot{S} = \dot{H} - V\dot{p} - \sum_j \mu_j \dot{m}_j$$

for this case. Under what conditions will this law hold for open systems?

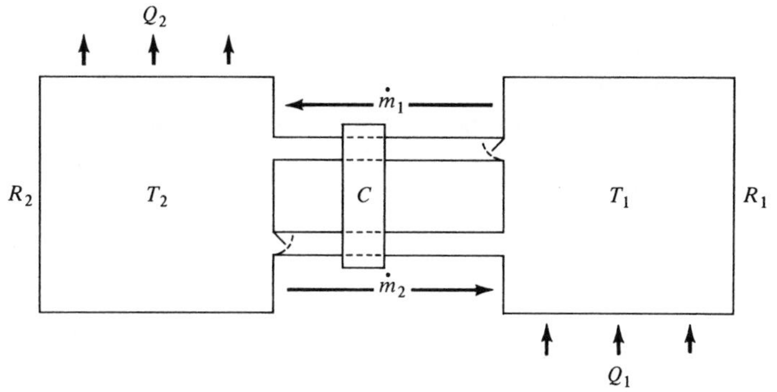

FIGURE 3.6
A thermodynamic system that provides a crude model of the general circulation. See Problem 3.4.4.

PROBLEM 3.4.2 An open system of constant volume composed of an ideal gas with energy $I = M[c_V(T - T_0) + e_0]$ is heated at rate Q. If the pressure remains equal to a constant environmental pressure, then show that the system loses mass at rate $-Q/c_pT$.

PROBLEM 3.4.3 Reanalyze the situation given in Problem 3.3.8, treating the heated room explicitly as an open system. Find an expression for the rate of change of the total energy I that depends on the heating rate. Was the conclusion reached in Problem 3.3.8 correct?

PROBLEM 3.4.4 A thermodynamic system is composed of an ideal gas confined to reservoirs R_1 and R_2 and connecting pipes with valves to permit flow only in the direction shown in Fig. 3.6. Let the mass M_1 in R_1 be equal to the mass M_2 in R_2, and let the gases be well mixed so that T_1 and T_2 are constant. The gas in R_1 is heated at the rate Q_1, R_2 is cooled at the rate Q_2, and we have $T_1 > T_2$. Let $\dot{m}_1$ be the rate at which mass flows into (out of) the upper pipe from R_1 (into R_2); $\dot{m}_2$ is the rate at which mass flows into (out of) the lower pipe from R_2 (into R_1). The controller C ensures that $\dot{M}_1 = \dot{M}_2 = 0$. We assume that the pressure distribution in each reservoir is controlled by $\partial \ln p/\partial z = -g/RT$ [as in Eq. (13) in Sec. 4.1 of the next chapter]. Finally, assume that friction, heat conduction, phase changes, and the rate of kinetic energy change may all be neglected.

Show that:

(*a*) $\dot{M}_1 = 0$ implies that $\dot{m}_1 = \dot{m}_2$.

(*b*) If the total internal energy I is constant, then $Q_1 + Q_2 = 0$.

(*c*) The total entropy change $\dot{S} = \dot{S}_1 + \dot{S}_2$ is composed of two components, one due to the heating, which decreases the entropy, and one due to the mass flux, which increases the entropy.

(*d*) The mass flux rate which would prevail when $\dot{S} = 0$ depends on the heating rate Q_1 and the difference Δz in elevation between the two pipes.

In this problem, we have constructed a model that contains the essential thermodynamics of the general circulation.

BIBLIOGRAPHIC NOTES

3.1 This section is based entirely on:

Truesdell, C., 1968: "The Nonlinear Field Theories in Mechanics," in N. J. Zabusky (ed.), *Topics in Nonlinear Physics*, Springer-Verlag OHG, Berlin.

With some trepidation, we have changed some of Prof. Truesdell's original terminology to accord more fully with conventional meteorological usage.

3.2 The results given here are classical. For further discussion of the statistical theory of gases see the standard reference:

Chapman, S., and T. G. Cowling, 1939: *The Mathematical Theory of Non-Uniform Gases*, Cambridge University Press, Cambridge, 431 pp. Reprinted 1952.

A very readable account is given by:

Sommerfeld, Arnold, 1950: *Thermodynamics and Statistical Mechanics*, vol. V of *Lectures on Theoretical Physics*, Academic Press, Inc., New York, 401 pp. Reprinted in paperback by Academic Press, 1964.

Further discussion of thermodynamic theory presented in these sections can be found in Sommerfeld's book or in the excellent text:

Callen, H. B., 1960: *Thermodynamics*, John Wiley & Sons, Inc., New York, 376 pp.

3.3 The concept of entropy is discussed by virtually all works on thermodynamics; potential temperature seems to be a concept peculiar to meteorology. The discussion surrounding Fig. 3.3 is based on results of Constantin Caratheodory's axiomatic approach to thermodynamics; an account in English and references to the original work are given by:

Chandrasekhar, S., 1939: *An Introduction to the Study of Stellar Structure*, University of Chicago Press, Chicago, 509 pp. Reprinted by Dover Publications.

The quotation from Emden is part of a delightful little paper which appeared as:

Emden, Robert, 1938: "Why Do We Have Winter Heating?" *Nature,* **141**:908.

and is quoted by Sommerfeld as well.

The isentropic relations of this section are standard in meteorology. Extensive discussion of meteorological charts and diagrams is given in the book:

Saucier, Walter J., 1955: *Principles of Meteorological Analysis*, University of Chicago Press, Chicago, 438 pp.

3.4 The Joule-Kelvin experiment is discussed in Sommerfeld's book; open systems are discussed by:

Van Mieghem, J., 1951: "Application of the Thermodynamics of Open Systems to Meteorology" in T. F. Malone (ed.), *Compendium of Meteorology*, pp. 531–538, American Meteorological Society, Boston.

4

THE VERTICAL STRUCTURE OF THE ATMOSPHERE

The most pronounced changes in atmospheric variables occur in the vertical. The pressure falls by half every 6 km, the temperature decreases to −20°C in about 6 km and to −50°C in about 10 km, and the density decreases an order of magnitude in 18 km. From the surface to the core of the jet stream at about 10 km, the wind may increase from zero speed to 100 m/s. Horizontal derivatives this large are never observed in large-scale patterns, although they may occur in severe storms or in very small-scale flows.

These strong vertical variations, reflecting the fact that the atmosphere is a shallow layer of air on the planet, result in the atmosphere being stratified in layers that have small horizontal variability compared to the variations in the vertical. To illustrate, the part of the atmosphere usually of interest in weather prediction is about 20 km deep whereas a mature cyclone may have a diameter of about 3000 km.

Despite the fact that horizontal variations are small, it is precisely these variations that lead to motion and weather-producing systems. The vertical variations, although not a prime cause of atmospheric events, profoundly affect the motions, whose patterns and currents are molded into the shapes we see in part by the vertical stratification of the mass field.

4.1 HYDROSTATIC PRESSURE VARIATIONS

With density defined as mass per unit volume, it is obvious that the total mass M of the atmosphere is given by

$$M = \int_V \rho \, dV \tag{1}$$

a three-dimensional integral over the entire volume V occupied by the atmosphere. If we assume that this total mass is finite, then the density must become vanishingly small at large heights. We can similarly define the mass per unit area above altitude z in a column of the atmosphere as

$$m(z) = \int_z^\infty \rho \, dz' \tag{2}$$

According to Newton's laws, the force p_h per unit area obtained from the product of this mass and the acceleration of gravity will be

$$p_h(z) = gm(z) = g \int_z^\infty \rho \, dz' \tag{3}$$

In Eq. (3) we have used a constant but representative value for the acceleration of gravity and have used p_h to denote force per unit area. We name this force per unit area the *static pressure* (or hydrostatic pressure); it is the weight of the air in an atmosphere at rest.

By differentiating Eq. (3), we find that

$$\frac{\partial p_h}{\partial z} = -g\rho \tag{4}$$

This is the *hydrostatic equation*.

PROBLEM 4.1.1 Show that in equilibrium conditions the hydrostatic pressure is exerted equally in all directions. (Hint: Consider the balance of forces on a prism embedded in the fluid. Note that the result must be true for any such prism.)

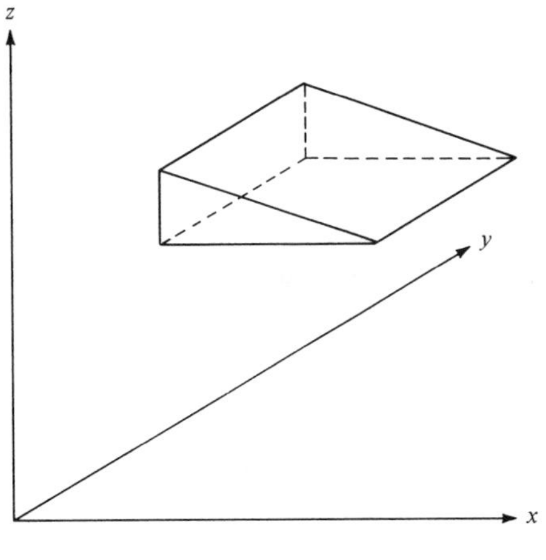

PROBLEM 4.1.2 Estimate the mass per unit area of the atmosphere. What is the total mass?

PROBLEM 4.1.3 Show that if $M < \infty$, then $\rho(z) \to 0$ as $z \to \infty$ faster than $z^{-1} \to 0$.

The next step is to determine the relation between the hydrostatic pressure p_h and the thermodynamic pressure, which we shall denote for now by p_T, that we discussed in the previous chapter. Consider a column of area A isolated by rigid sides but filled with air. We put weightless horizontal dividers in the column at height $z - \Delta z$ and height $z + \Delta z$ to isolate the air in the section of thickness $2\,\Delta z$. The hydrostatic pressure on the upper divider is

$$p_h = \int_{z+\Delta z}^{\infty} g\rho\, dz' \tag{5}$$

The air between the dividers obeys the equation of state and has a thermodynamic pressure, which according to Taylor's theorem varies across the thickness $2\Delta z$ by

$$p_T(z + \Delta z) = p_T(z - \Delta z) + \left(\frac{\partial p_T}{\partial z}\right)_\zeta 2\,\Delta z \qquad z - \Delta z \leqslant \zeta \leqslant z + \Delta z \tag{6}$$

so that the pressure at the bottom divider is

$$\begin{aligned} p(z - \Delta z) &= p_h(z + \Delta z) + [p_T(z - \Delta z) - p_T(z + \Delta z)] \\ &= p_h(z + \Delta z) - \left(\frac{\partial p_T}{\partial z}\right)_\zeta 2\,\Delta z \end{aligned} \tag{7}$$

According to Newton's laws, if the air between the dividers is of mass M, then the motion of its midpoint z is governed by

$$\ddot{z} = -g + [p(z - \Delta z) - p_h(z + \Delta z)]\frac{A}{M} = -g - \left(\frac{\partial p_T}{\partial z}\right)_\zeta \frac{2\,\Delta z A}{M} \tag{8}$$

which in the limit $\Delta z \to 0$ gives

$$\ddot{z} = -g - \frac{1}{\rho}\frac{\partial p_T}{\partial z} \tag{9}$$

Thus we have

$$p_T = p_h \qquad \text{for } \ddot{z} = 0 \tag{10}$$

where the usual constant of integration is eliminated by the condition that both p_T and p_h must vanish at the top of the atmosphere where the density vanishes. We may rewrite Eq. (9) as

$$\ddot{z} = \frac{1}{\rho}\frac{\partial}{\partial z}(p_h - p_T) \tag{11}$$

which shows that vertical accelerations will occur when the hydrostatic and thermal pressures have different derivatives. Moreover, our example shows that the derivative of thermal pressure occurs explicitly in the equation of motion (9), while that of the hydrostatic pressure is represented by $g\rho$. Thus the thermal pressure in the first law and in the equation of motion (9) are denoted by p henceforth. When the thermodynamic pressure is used in Eq. (4), we have the *hydrostatic approximation*†

$$\frac{\partial p}{\partial z} = -g\rho \tag{12}$$

As shown by Eq. (9), the hydrostatic relation is exact in an atmosphere at rest. We shall see later that the hydrostatic approximation makes a fundamental change in the equations of motion. Without it, the full set of equations contains a prognostic equation that determines the rate of change of the vertical velocity w; with the approximation, the prognostic feature is lost and w must be determined from the other variables by a diagnostic equation deduced from the remaining equations. In this case, w changes with time, but only as specified by the diagnostic equation. The important point is that making the hydrostatic approximation does *not* imply that either w or dw/dt vanishes; it means that they must be found by diagnostic techniques and that the resulting values are precisely those needed to maintain hydrostatic equilibrium.

The obvious question, then, is to what extent it is valid for the actual atmosphere. As shown by Problem 4.1.4 (page 64), it almost always gives a good representation of the relation between pressure and density in large-scale systems, but may be suspect in small-scale flows or severe weather. It is exactly the failure of the approximation that creates vertical accelerations as the usually small difference between two large terms of nearly equal size. Thus $g\rho$ will generally give a good estimate of $-\partial p/\partial z$, but if we want to know the acceleration, we must know both terms on the right of Eq. (11) very accurately.

†The hydrostatic approximation is often written as

$$dp = -g\rho\, dz$$

This is not a good practice, because it is misleading. To see why, note that the expression (3) can be written for the situation at time t in the column centered on the point (x,y) as

$$p(x,y,z,t) = g\int_z^\infty \rho(x,y,z',t)\, dz'$$

so that we have

$$\frac{\partial p(x,y,z,t)}{\partial z} = -g\rho(x,y,z,t)$$

We know from Eq. (10) in Sec. 2.5 that

$$dp = \frac{\partial p}{\partial t}dt + \frac{\partial p}{\partial x}dx + \frac{\partial p}{\partial y}dy + \frac{\partial p}{\partial z}dz$$

and thus the first equation is correct only when the pressure varies solely in the vertical. The content of the hydrostatic approximation is the relation between p and ρ in the vertical, and this is best indicated by the partial derivative notation.

PROBLEM 4.1.4 Ascertain how well the hydrostatic approximation relates pressure and density in the two situations:

(*a*) In the 12 hours it takes a parcel to travel through a cyclone, its vertical velocity may reverse from 10 cm/s down to 10 cm/s up.

(*b*) In the growth phase of a thunderstorm, a parcel may accelerate from rest to a speed of 10 m/s in a minute or so.

PROBLEM 4.1.5 To what accuracy must the density and pressure gradient be measured to compute the accelerations given in Problem 4.1.4 within 10 percent accuracy?

The fact that the hydrostatic approximation almost always gives an adequate representation between pressure and density makes it a potent concept, especially for analysis of large-scale flow. It provides one relation between p and ρ and the equation of state provides another between p, ρ, and T, and hence one may be eliminated. For example,

$$\frac{1}{p}\frac{\partial p}{\partial z} = -\frac{g}{RT} \tag{13}$$

shows how pressure varies as a function of temperature under hydrostatic conditions.

PROBLEM 4.1.6 Show that

$$\ln\frac{p(z_2)}{p(z_1)} = -\int_{z_1}^{z_2}\frac{g}{RT}\,dz \tag{14}$$

Let $\bar{T}$ be the average temperature in the layer (z_1,z_2). Define $T' = T - \bar{T}$. Use the binomial expansion to determine whether the often-used statement

$$\ln\frac{p_2}{p_1} = -\frac{g(z_2 - z_1)}{R\bar{T}} \tag{15}$$

is a good approximation.

*PROBLEM 4.1.7 Let $h(p)$ be the height at which pressure p occurs. Derive the *hypsometric equation*

$$h(p_2) - h(p_1) = -\int_{p_1}^{p_2}\frac{RT}{g}\,d\ln p \tag{16}$$

[Hint: We show in Chap. 5 that the obvious statement $(\partial h/\partial p) = (\partial p/\partial z)^{-1}$ is justified.] Estimate the height of the 500-mbar surface. Justify the claim that about half the mass of the atmosphere lies above and half lies below this surface.

4.2 GEOPOTENTIAL AND GEOPOTENTIAL ALTITUDE

In the previous section we employed a constant value of the acceleration of gravity in order to derive the hydrostatic approximation, ignoring the fact that this parameter actually varies in the vertical.

The acceleration of gravity decreases with the square of the distance from the earth's center, so that letting g_ϕ be the latitudinally variable value at sea level, we may express the acceleration g at any altitude and latitude by

$$g = g(\phi, z) = g_\phi\left(\frac{a}{a+z}\right)^2 \tag{1}$$

in which a is the radius of the earth and z is altitude above sea level. If we raise a unit mass from sea level, we must work against the force of gravity, and in raising the mass to height z we give it a potential energy per unit mass of

$$\Phi = \int_0^z g_\phi\left(\frac{a}{a+z_1}\right)^2 dz_1 + \Phi(0) \tag{2}$$

In meteorology, we usually take Φ to be zero at sea level. Hence

$$\Phi(z) = g_\phi a^2\left(\frac{1}{a} - \frac{1}{a+z}\right) = g_\phi \frac{az}{a+z} \tag{3}$$

Because Φ is the potential energy of a unit mass in the earth's gravity field, it is called the *geopotential*. There are actually two components to the observed or apparent force of gravity on the earth. The first is the attraction toward the center of the earth due to the interaction of masses; the second is the centrifugal force due to the earth's rotation, a force perpendicular to the earth's axis and away from the earth. This will be discussed further in Chap. 7, but for now we observe that the surface value g_ϕ combining both these forces is given on the average as a function of latitude by the empirical formula

$$g_\phi = 980.6160[1 - 2.64 \times 10^{-3} \cos 2\phi + 5.9 \times 10^{-6} \cos^2 2\phi] \text{ cm/s}^2 \tag{4}$$

Thus g_ϕ is less at the equator than at the poles.

In meteorology, especially in the reduction of upper-air data, it is convenient to use the surfaces on which the geopotential is constant as the level or horizontal surfaces instead of using surfaces parallel to mean sea level. This is sometimes accomplished by using the geopotential itself as a measure of altitude. We define the *geopotential height* Z from Eq. (3) as

$$Z = \frac{\Phi}{g_{38}} = \frac{g_\phi a}{g_{38}}\left(\frac{z}{a+z}\right) \qquad g_{38} = 980 \text{ cm/s}^2 \tag{5}$$

The variable Z has dimensions of length; it is usually expressed in geopotential meters by expressing a and z in units of geometric meters.

The main advantage of geopotential height appears when we examine the hydrostatic approximation with the variable value of g employed. Then by the chain rule we have

$$\frac{\partial p}{\partial Z}\frac{\partial Z}{\partial z} = -g\rho = -g_\phi\left(\frac{a}{a+z}\right)^2\rho \tag{6}$$

But

$$\frac{\partial Z}{\partial z} = \frac{g_\phi a}{g_{38}}\left[\frac{1}{a+z} - \frac{z}{(a+z)^2}\right] = \frac{g_\phi}{g_{38}}\left(\frac{a}{a+z}\right)^2 \tag{7}$$

and hence

$$\frac{\partial p}{\partial Z} = -980\rho \qquad [980] = \text{cm/s}^2 \tag{8}$$

Thus if we consider our height coordinate to be in geopotential units, we may express the hydrostatic approximation with the constant acceleration 980 cm/s^2. In meteorology, it is conventional to use z for Z, g for 980 cm/s^2 (if only to remind ourselves of the proper dimensions), and to write the hydrostatic approximation in the usual form

$$\frac{\partial p}{\partial z} = -g\rho \tag{9}$$

Except in special cases, it is thus simply assumed that the transformation to geopotential height has been made even if it is not mentioned.

PROBLEM 4.2.1 Show that the surfaces of constant geopotential are higher at the equator than at the poles.

PROBLEM 4.2.2 Determine whether there is a value of the constant c and an altitude range such that $Z = cz$ is accurate within 1 percent for all latitudes.

4.3 TEMPERATURE, LAPSE RATE, AND MODEL ATMOSPHERES

Near the surface of the earth, the temperature generally decreases with increasing altitude, and certain aspects of the atmospheric structure and stability depend quite critically on the rate of change of temperature in the vertical.

This rate of change is usually characterized by the *lapse rate* of temperature, defined by

$$\gamma = -\frac{\partial T}{\partial z} \tag{1}$$

The sign is chosen so that, as implied by the name, the lapse rate is positive when the temperature decreases.

Soundings taken in the atmosphere and plotted on thermodynamic charts usually give the impression that the temperature is linear over quite broad ranges of altitude. But this impression should not be converted into a general assumption that γ is constant. First, many of these charts do not use z as a vertical coordinate (p^{R/c_p} and $\ln p$ are common alternatives). Second, as is easily seen from Eq. (1), γ can be a function of all the independent variables so that

$$\gamma = \gamma(x, y, z, t) \tag{2}$$

The lapse rate often appears in equations for the vertical rates of change of other atmospheric variables. For example, by logarithmically differentiating Poisson's equation

$$\theta = T\left(\frac{p_{00}}{p}\right)^{R/c_p} \tag{3}$$

we find

$$\frac{1}{\theta}\frac{\partial \theta}{\partial z} = \frac{1}{T}\frac{\partial T}{\partial z} - \frac{R}{c_p p}\frac{\partial p}{\partial z} \tag{4}$$

Combination of this result with the hydrostatic approximation, the equation of state, and Eq. (1) gives

$$\frac{1}{\theta}\frac{\partial \theta}{\partial z} = \frac{1}{T}\left(\frac{g}{c_p} - \gamma\right) \tag{5}$$

Thus the potential temperature will be constant in the vertical when

$$\gamma = \frac{g}{c_p} \qquad \cong 10 \text{ K/km} \tag{6}$$

This rate is known as the *dry adiabatic lapse rate* and is generally denoted by

$$\gamma_d = \frac{g}{c_p} \tag{7}$$

Thus when $\gamma < \gamma_d$, the potential temperature increases with height; when $\gamma > \gamma_d$, the potential temperature decreases with height. As will be shown later, the stability of the atmosphere to vertical displacement of dry parcels depends upon the lapse rate γ being less than the dry adiabatic rate g/c_p.

When the temperature is increasing upward, then γ is negative, and as seen from Eq. (5) the potential temperature will increase rather rapidly with height. This condition of increasing temperature occurs frequently near the ground at night. It also occurs where warmer air lies over colder air; then the intermediate layer with rapidly increasing temperature is referred to as an *inversion.* Finally, the *stratosphere* is by definition the region in which the temperature no longer decreases as it does in the troposphere, but instead becomes constant or increases slightly with height, thus leading to rapidly increasing potential temperature.

PROBLEM 4.3.1 Show that the density will increase with height in a hydrostatic atmosphere if $\gamma > g/R$. The quantity $\gamma_a = g/R$ is known as the *autoconvective lapse rate*, because we may assume that immediate overturning would result if heavier air were on top of light air. What is the numerical value of γ_a?

PROBLEM 4.3.2 Let the temperature be given everywhere by a constant lapse rate as $T = T_0 - \gamma z$, with T_0 constant. Show that if $\gamma = g/c_V$, then the sum of the internal and potential energies is $I + P = c_V M T_0$, where M is the total mass of the atmosphere.

Important features of the possible vertical variation of atmospheric variables are revealed by model atmospheres constructed as simplifications of the actual situation. The general idea is that the hydrostatic equation provides a relation between any two of the thermodynamic variables, and then the specification of any one of them in simple form allows us to determine all of them analytically.

Properties of the model atmospheres most often encountered are given in Table 4.1.

Vertical structure is often characterized by scale heights of various types. The three most common are

$$H^{-1} = -\frac{1}{\rho}\frac{\partial \rho}{\partial z} \tag{8}$$

$$H_i = \frac{p}{g\rho} = \frac{RT}{g} \tag{9}$$

and

$$H_\theta^{-1} = \frac{1}{\theta}\frac{\partial \theta}{\partial z} \tag{10}$$

Of these, H_θ is the most important because it characterizes the effect of stratification on perturbations and disturbances of otherwise-balanced situations, and thus indicates how the stratification will affect changes in the motion field.

PROBLEM 4.3.3 Verify the relations shown in Table 4.1.

PROBLEM 4.3.4 Show that

$$H^{-1} = \frac{1}{T}(\gamma_a - \gamma)$$

and that H represents the distance in which density falls by a factor e^{-1} in an isothermal atmosphere. Show that

$$H_\theta^{-1} = H^{-1} - \frac{c_V}{c_p} H_i^{-1}$$

and that for $\gamma \cong 2\gamma_d/3$, $H_\theta \cong 75$ km. Show that H_i represents the depth of a constant density fluid with density and surface pressure equal to the values used in Eq. (9).

4.4 STATIC STABILITY AND PARCEL MOTION

Many of the most interesting meteorological phenomena are associated with strong or even violent vertical motion. Grouped under the general term of convection, these include cumulus development, thunderstorms, dust devils, and possibly tornadoes. These phenomena suggest that we attempt to determine the conditions that will yield strong vertical motions. The classical approach to this problem is to investigate what will occur when a small parcel is vertically displaced from its original position.

Two assumptions are made in order to permit the problem to be attacked by elementary means:

1 The pressure p_p of the parcel will always be identical to the ambient pressure p of the air surrounding the parcel at any given instant or position.
2 The parcel moves isentropically so that its potential temperature θ_p is constant at its initial value.

Thus in effect we consider the parcel as insulated from the surrounding air: its mass

Table 4.1 MODEL ATMOSPHERES

	Constant variable and main use			
Variable	**Temperature: stratosphere; top for models**	**Density*: fluid dynamics oceanography**	**Entropy: convection studies**	**Lapse rate: modeling layers**
$T =$	T_0		$T_0 - \gamma_d z$	$T_0 - \gamma z$
$p =$	$p_0 e^{-gz/RT_0}$	$p_0 - g\rho_0 z$	$p_0\left(1 - \frac{\gamma_d z}{T_0}\right)^{c_p/R}$	$p_0\left(1 - \frac{\gamma z}{T_0}\right)^{\gamma_a/\gamma}$
$\rho =$	$\rho_0 e^{-gz/RT_0}$	ρ_0	$\rho_0\left(1 - \frac{\gamma_d z}{T_0}\right)^{c_V/R}$	$\rho_0\left(1 - \frac{\gamma z}{T_0}\right)^{\gamma_a/\gamma - 1}$
$\theta =$	$\theta_0 e^{gz/c_p T_0}$		θ_0	$\theta_0\left(1 - \frac{\gamma z}{T_0}\right)^{1 - \gamma_d/\gamma}$
Top $p = 0$ at	$z = \infty$	$z = \frac{p_0}{g\rho_0}$	$z = \frac{T_0}{\gamma_d}$	$z = \frac{T_0}{\gamma}$
$H = -\left(\frac{1}{\rho}\frac{\partial\rho}{\partial z}\right)^{-1}$	$\frac{RT_0}{g} \sim 8$ km	∞	$\frac{R}{c_V}\frac{T_0}{\gamma_d}\left(1 - \frac{\gamma_d z}{T_0}\right)$	$\frac{T_0}{\gamma}\left(\frac{\gamma_a}{\gamma} - 1\right)^{-1}\left(1 - \frac{\gamma z}{T_0}\right)$

Note: $p_0 = p(0)$, $T_0 = T(0)$, $\rho_0 = \rho(0)$, $\theta_0 = \theta(0)$
*Not an ideal gas.

does not mix with the mass of the ambient air and no heat flows between the parcel and its surroundings. Such a situation is obviously a simplified abstraction of what actually occurs with parcels of air moving in the atmosphere.

At any point, the net vertical force on the parcel is given by the Archimedean principle as the difference between the force of gravity on the parcel mass M_p and the mass M of the ambient air displaced by the parcel. Thus this force, taken to be positive upward, is

$$F = g(M - M_p) \tag{1}$$

According to Newton's second law, the acceleration of the parcel is therefore given by

$$M_p \frac{d^2z}{dt^2} = F = g(M - M_p) \tag{2}$$

Division of both sides by the volume of the parcel produces the result

$$\frac{d^2z}{dt^2} = g\left(\frac{\rho - \rho_p}{\rho_p}\right) \tag{3}$$

Clearly then, the parcel will accelerate upward if it is less dense than the ambient air, accelerate downward if it is more dense.

PROBLEM 4.4.1 Show that under the two assumptions above

$$\frac{\rho}{\rho_p} = \frac{T_p}{T} = \frac{\theta_p}{\theta} \tag{4}$$

and hence that Eq. (3) may be written as

$$\frac{d^2z}{dt^2} = g\left(\frac{T_p - T}{T}\right) = g\left(\frac{\theta_p - \theta}{\theta}\right) \tag{5}$$

Equation (5) reveals the crucial importance of whether the ambient potential temperature increases with height. Let us start with a parcel in complete equilibrium with its surroundings so that its pressure and temperature, and hence its density and potential temperature, all equal the corresponding ambient variables. Now if the parcel is displaced upward, its potential temperature is constant according to the second assumption. If θ increases with height, then after displacement we have $\theta_p - \theta < 0$ so that the parcel will be accelerated from its displaced position back toward its original position. This is referred to as *stability*.

But if θ decreases with height, then $\theta_p - \theta > 0$ after displacement and the parcel will be accelerated in the upward direction, even farther away from its initial position. This is an *unstable* situation. Finally, if θ is constant with height, no force at all will be imposed on the parcel in its displaced location, a case which is called *neutral stability*.

The results apply to an atmosphere which is at rest except for the movements of the parcel. Thus we have the criteria for unsaturated air:

	Static stability	Hydrostatic stability
Stable:	$\partial\theta/\partial z > 0$	$\gamma < \gamma_d$
Neutral:	$\partial\theta/\partial z = 0$	$\gamma = \gamma_d$
Unstable:	$\partial\theta/\partial z < 0$	$\gamma > \gamma_d$

The criteria for hydrostatic stability are obtained with Eq. (5) in Sec. 4.3.

4.4.1 Parcel Motion (Simple Theory)

Let us investigate the vertical movements of a parcel under these conditions more completely. We take z_0 to be the initial height of the parcel, so that the potential temperature at this height θ_0 is the potential temperature θ_p of the parcel. With Taylor's theorem we may expand the potential temperature θ of the surrounding air in the form

$$\theta - \theta_0 = \left(\frac{\partial\theta}{\partial z}\right)_0 (z - z_0) + \left(\frac{\partial^2\theta}{\partial z^2}\right)_0 \frac{(z - z_0)^2}{2} + \cdots \tag{6}$$

Thus for small enough intervals, $z - z_0$, Eq. (5) for the acceleration of the parcel may be written as

$$\frac{d^2(z - z_0)}{dt^2} = -g\left(\frac{\theta - \theta_0}{\theta}\right) \cong -\frac{g}{\theta}\left(\frac{\partial\theta}{\partial z}\right)_0 (z - z_0) \cong -g\left(\frac{1}{\theta}\frac{\partial\theta}{\partial z}\right)_0 (z - z_0) \tag{7}$$

Now we use $\zeta = z - z_0$ and take as our (approximate) equation

$$\frac{d^2\zeta}{dt^2} + \left(\frac{g}{\theta}\frac{\partial\theta}{\partial z}\right)_0 \zeta = 0 \tag{8}$$

If we try solutions of the form $\zeta = Ae^{rt}$, we obtain the characteristic equation

$$r^2 + \left(\frac{g}{\theta}\frac{\partial\theta}{\partial z}\right)_0 = 0 \tag{9}$$

so that the definition (for $\partial\theta/\partial z > 0$) of the *Brunt-Väisälä frequency*

$$\omega_g = \left[\left(\frac{g}{\theta}\frac{\partial\theta}{\partial z}\right)_0\right]^{1/2} \tag{10}$$

gives

$$r = \pm i\omega_g \qquad i = \sqrt{-1} \tag{11}$$

Thus the solution, which must contain two arbitrary constants, is

$$\zeta = Ae^{i\omega_g t} + Be^{-i\omega_g t} \tag{12}$$

The initial conditions that $\zeta = 0$ and $d\zeta/dt = w_0$ at $t = 0$ now imply that

$$\zeta = \frac{w_0}{2i\omega_g}\left(e^{i\omega_g t} - e^{-i\omega_g t}\right) \tag{13}$$

If $\partial\theta/\partial z$ is positive, then ω_g is real and we have

$$\zeta = \frac{w_0}{\omega_g} \sin \omega_g t \tag{14}$$

so that the parcel oscillates in the vertical with a frequency ω_g. Note that as ω_g increases, the frequency of oscillation increases and the amplitude of the oscillation decreases for the same initial displacement velocity.

But if $\partial\theta/\partial z < 0$, then Eq. (10) may be written

$$\omega_g = i\left|\left(\frac{g}{\theta}\frac{\partial\theta}{\partial z}\right)_0\right|^{1/2} \tag{15}$$

and the solution (13) gives

$$\zeta = \frac{w_0}{2}\left[\exp\left(\left|\frac{g}{\theta}\frac{\partial\theta}{\partial z}\right|^{1/2} t\right) - \exp\left(-\left|\frac{g}{\theta}\frac{\partial\theta}{\partial z}\right|^{1/2} t\right)\right]\left|\frac{g}{\theta}\frac{\partial\theta}{\partial z}\right|^{-1/2} \tag{16}$$

The first term thus implies that the displacement will increase exponentially with time, clearly a case of instability.

Equation (14) has important implications if viewed as a relation predicting maximum displacement for a given initial velocity. The more stable the atmosphere, the less the maximum displacement. Thus smokestack effluent may be trapped near the ground in stable conditions, or under an inversion at night, whereas when less stable conditions prevail, the effluent is dispersed at higher altitudes in the atmosphere.

PROBLEM 4.4.2 The circular frequency ω_g (radians per second) is related to a frequency f_g (cycles per second) and a period of oscillation T_g (seconds) by

$$\omega_g = 2\pi f_g = \frac{2\pi}{T_g}$$

Show that for typical tropospheric values of $\gamma \cong 2\gamma_d/3$, T_g is about 500 s, and that for the stratosphere it is about 300 s.

PROBLEM 4.4.3 To provide an initial displacement acceleration for a parcel, assume that owing to heating, its initial potential temperature θ_p is greater than the ambient θ_0. Show that for $\partial\theta/\partial z > 0$ the solution to Eq. (5) is

$$\zeta = \frac{g}{{\omega_g}^2}\,\frac{\theta_p - \theta_0}{\theta_0}(1 - \cos \omega_g t)$$

(Hint: The differential equation is no longer homogeneous, but it has an obvious particular solution.)

4.4.2 Parcel Motion (Nonlinear Theory)

The approximate equation (8) contains a number of assumptions that are conventionally made. Let us now try to treat the vertical motion of a parcel more rigorously to determine how its temperature changes and whether the stability criteria already outlined are good approximations.

The first law of thermodynamics applied to the isentropic motion of the parcel predicts that

$$c_p \frac{dT_p}{dt} - \alpha_p \frac{dp_p}{dt} = 0 \tag{17}$$

We assume that the local rate of change and horizontal advection of pressure are negligible, so that the requirement that the parcel adjust to the ambient pressure gives

$$\frac{dp_p}{dt} = \frac{dp}{dt} = w_p \frac{\partial p}{\partial z} = -g\rho w_p \tag{18}$$

in which the last equality is obtained by making the assumption that the ambient atmosphere is hydrostatic.

PROBLEM 4.4.4 Show that the terms ignored in dp/dt could be of the same order as the term retained.

With Eq. (18), Eq. (17) becomes

$$c_p \frac{dT_p}{dt} + g w_p \alpha_p \rho = 0 \tag{19}$$

With the aid of Eq. (4) we obtain

$$c_p \frac{dT_p}{dt} + g w_p \frac{T_p}{T} = 0 \tag{20}$$

and the companion equation from Eq. (5) is

$$\frac{dw_p}{dt} - g\left(\frac{T_p - T}{T}\right) = 0 \tag{21}$$

This system of equations could be solved for the velocities (and hence the displacements) of a parcel moving in an atmosphere with arbitrary temperature profile $T(z)$.

*PROBLEM 4.4.5 Eliminate T_p/T from the system to show that

$$\frac{dT_p}{dt} + w_p\gamma_d = -\frac{1}{2c_p}\frac{dw_p{}^2}{dt} \tag{22}$$

Equation (22) demonstrates that the temperature of a vertically moving parcel will change with height at the dry adiabatic rate, if the change in kinetic energy is small. This assumption is usually made implicitly in meteorological reasoning.

Now let us return to the stability problem. If we assume that the ambient temperature is a function of height only, then we can eliminate T_p from the system (20) and (21) to obtain the equation, where now $w = w_p$,

$$\frac{d^2w}{dt^2} + \frac{w}{T}\frac{dw}{dt}(\gamma_d - \gamma) + \frac{wg}{T}(\gamma_d - \gamma) = 0 \tag{23}$$

which is more conveniently written as

$$\frac{d}{dt}\left\{\exp\left[\int_0^t \frac{w}{T}(\gamma_d - \gamma)\,d\tau\right]\frac{dw}{dt}\right\} + \frac{wg}{T}(\gamma_d - \gamma)\exp\left[\int_0^t \frac{w}{T}(\gamma_d - \gamma)\,d\tau\right] = 0 \tag{24}$$

If this equation is multiplied by w and the terms are rearranged, we have

$$\frac{d}{dt}\left\{\exp\left[\int_0^t \frac{w}{T}(\gamma_d - \gamma)\,d\tau\right]\frac{1}{2}\frac{dw^2}{dt}\right\}$$
$$+ \exp\left[\int_0^t \frac{w}{T}(\gamma_d - \gamma)\,d\tau\right]\left[\frac{w^2g}{T}(\gamma_d - \gamma) - \left(\frac{dw}{dt}\right)^2\right] = 0 \tag{25}$$

Now let us apply the initial conditions that at $t = 0$ the parcel velocity is zero and that the acceleration is finite so that $dw^2/dt = w\,dw/dt = 0$ and integrate from $t = 0$ to time t. The result is

$$\frac{dw^2}{dt} = 2\exp\left[-\int_0^t \frac{w}{T}(\gamma_d - \gamma)\,d\tau\right]\int_0^t\left\{\exp\left[\int_0^\tau \frac{w}{T}(\gamma_d - \gamma)\,d\tau'\right]\left[\left(\frac{dw}{d\tau}\right)^2\right.\right.$$
$$\left.\left. - \frac{w^2g}{T}(\gamma_d - \gamma)\right]\right\}d\tau \tag{26}$$

If $\gamma_d < \gamma$ over the entire trajectory from the initial point, then both terms on the right are positive and the kinetic energy of the parcel is increasing. Thus $\gamma_d < \gamma$ ensures instability if the parcel is displaced. In contrast, if we assume the ambient conditions are stable to parcel displacements, then it must be true that

$$\left(\frac{dw}{dt}\right)^2 - \frac{w^2 g}{T}(\gamma_d - \gamma) < 0 \tag{27}$$

along at least part of the trajectory, and hence $\gamma < \gamma_d$ somewhere.

These conditions are typical of the results of stability analyses in that guarantees of both stability and instability of the type found earlier in the section with the approximate equation are rarely found from more careful analyses.

An important additional factor needed to explain actual atmospheric phenomena is the effect of water vapor, most particularly the latent heating which results from condensation. Although this will be considered in Chap. 8, we note that the criterion for instability of a saturated parcel becomes a lapse rate of about 6.5 K/km rather than 10 K/km, because a rising parcel that reaches saturation is warmed by the condensation occurring in it. Thus a rising saturated parcel has an additional impetus to vertical acceleration compared with a dry or unsaturated parcel. It is such instability of moist air that leads to the violent phenomena mentioned at the beginning of this section.

Another important modification must be made when parcels move in a wind field with vertical variation. We consider this case next.

4.5 STABILITY AND WIND SHEAR

When both buoyancy forces and those created by horizontal air motion affect a vertically moving parcel, the problem becomes more complicated than the development of Sec. 4.4.

The situation in which we are now interested occurs when there are perturbations that produce an oscillatory, wavelike motion of the air, as illustrated in Fig. 4.1. Wave motion of this kind, which requires an analysis that utilizes the full set of equations of motion because all the variables change as the motion evolves, is treated in Chaps. 12 and 13.

At this point, we shall avoid the complication of using all the equations and instead try to abstract the essential features of the problem with a simplified case that permits an analysis based on energy considerations. The use of the concept of energy is a substitute for the equations of motion, because as we shall see later, the conservation of energy is a prediction of the equations.

The simple model we shall use is also illustrated in Fig. 4.1. Instead of a wavelike oscillation, we consider two parcels that interchange places in the vertical. Thus the two parcels may be considered to represent those whose motion is depicted by arrows in the illustration of wave motion. The plan for this stability analysis is to show that if two parcels are interchanged as shown, then the sum of their potential energy increases, and thus the interchange requires a source of energy. We hypothesize that the variable wind field is such a source.

If the two moving parcels cannot extract energy from the wind field as fast as their potential energy increases, then the interchange is prohibited by conservation of

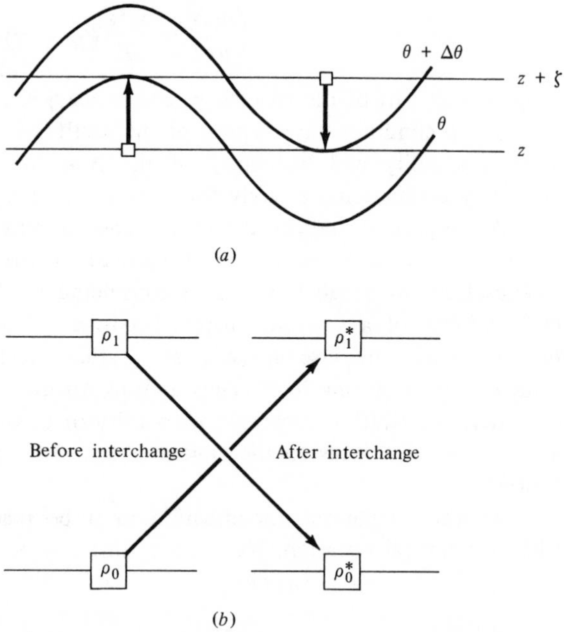

FIGURE 4.1
The top sketch shows how two parcels may effectively interchange heights when the potential temperature surfaces are perturbed in a wavelike motion. The lower sketch illustrates the virtual exchange model of such a process that is analyzed in the text.

energy and we would assume that the situation is stable. In contrast, if the moving parcels can extract more than enough energy to compensate for the increase in potential energy, then we may assume that the initial perturbations leading to the interchange would be unstable and would grow.

Thus we shall examine what is perhaps the most basic atmospheric stability criterion–whether small perturbations will be damped or whether they might grow to provide mechanisms for mixing and smoothing of wind and temperature profiles.

In attacking the problem, we shall assume that displacements are small enough so that we need to consider only linear vertical variation of the ambient variables. Thus we shall represent the potential temperature, density, and wind speed U in the neighborhood of an initial height z in the form

$$\theta(z+\zeta) = \theta(z) + \zeta\left(\frac{\partial \theta}{\partial z}\right)_0 = \theta_0 + \zeta\theta_0' \tag{1}$$

$$\rho(z+\zeta) = \rho(z) + \zeta\left(\frac{\partial \rho}{\partial z}\right)_0 = \rho_0 - \zeta|\rho_0'| \tag{2}$$

and

$$U(z+\zeta) = U(z) + \zeta\left(\frac{\partial U}{\partial z}\right)_0 = U_0 + \zeta U_0' \tag{3}$$

We consider a parcel initially at height z with density ρ_0 and a parcel initially at height $z+\zeta$ with density $\rho_1 = \rho_0 - \zeta|\rho_0'|$. The *potential energy* (per unit volume) of these two parcels is

$$\begin{aligned} P &= g[\rho_0 z + \rho_1(z+\zeta)] = g[\rho_0 z + (\rho_0 - \zeta|\rho_0'|)(z+\zeta)] \\ &= g[2\rho_0 z + \rho_0\zeta - \zeta|\rho_0'|(z+\zeta)] \end{aligned} \tag{4}$$

Now let us interchange these two parcels isentropically, and, as before, we insist that their pressures adjust instantly to the ambient pressure. Then because of Poisson's equation (Problem 3.3.4), we know that the parcel variables and those of the environment, to be denoted with a subscript e, are always related by

$$\theta_p\rho_p = \theta_e\rho_e \tag{5}$$

Therefore the parcel at $z+\zeta$ after interchange has the new density ρ_1^*, given by

$$\rho_1^* = \frac{\theta_1\rho_1}{\theta_0} = \frac{(\theta_0 + \zeta\theta_0')(\rho_0 - \zeta|\rho_0'|)}{\theta_0} = \rho_0 - \zeta|\rho_0'| + \zeta\frac{\rho_0\theta_0'}{\theta_0} \tag{6}$$

in which we have dropped the term in ζ^2 in accordance with the linear hypothesis based on small values of ζ. Similarly the parcel at z after the interchange came from $z+\zeta$ and has the new density

$$\rho_0^* = \frac{\rho_0\theta_0}{\theta_1} = \frac{\rho_0\theta_0}{\theta_0[1+\zeta(\theta_0'/\theta_0)]} = \rho_0\left(1 - \zeta\frac{\theta_0'}{\theta_0}\right) \tag{7}$$

in which we have used the binomial theorem and then discarded terms of higher order than the first.

Thus the potential energy of the two parcels after the interchange is

$$\begin{aligned} P^* = g[\rho_1^*(z+\zeta) + \rho_0^* z] &= g\left[\left(\rho_0 - \zeta|\rho_0'| + \zeta\frac{\rho_0\theta_0'}{\theta_0}\right)(z+\zeta) + \rho_0\left(1 - \zeta\frac{\theta_0'}{\theta_0}\right)z\right] \\ &= g\left[2\rho_0 z + \rho_0\zeta - \zeta|\rho_0'|(z+\zeta) + \zeta^2\frac{\rho_0\theta_0'}{\theta_0}\right] \end{aligned} \tag{8}$$

The increase in the potential energy of the two parcels brought about by the interchange is therefore

$$\Delta P = P^* - P = g\zeta^2\rho_0\left(\frac{1}{\theta}\frac{\partial\theta}{\partial z}\right)_0 = \zeta^2\rho_0\omega_g{}^2 \tag{9}$$

The internal energy of the parcels is given by $c_V \rho T = (c_V/R)p$ and thus depends only on the ambient pressure. Therefore the combined internal energy will be the same after the interchange as it was initially.

PROBLEM 4.5.1 Show that we avoid the complication of internal energy changes occurring during the interchange process by assuming that at any time the two parcels are always displaced an equal distance from their initial position, so that with the assumed linear variation of the ambient pressure field in the neighborhood of z, the total internal energy is always $(c_V/R)(2p_0 + \zeta p'_0)$.

Energy cannot be created, so we are faced with the question of determining the source of the potential energy [Eq. (9)] gained in the interchange. Some of it could be derived from the kinetic energy imparted to the parcels when the interchange was set in motion. If this were the only source, then the parcels would oscillate so that the sum k of the kinetic energy of their vertical motion and the gain in potential energy was governed by

$$k + \Delta P = k_0 \tag{10}$$

where k_0 is the initial kinetic energy. This would limit the parcels to an oscillation of the type considered in Sec. 4.4, and the maximum displacement would occur when k was zero at the end points of the oscillation; we would have

$$(\zeta_{\max})^2 = \frac{k_0}{\omega_g{}^2 \rho_0} \tag{11}$$

Now we suppose that the parcels are embedded in a wind field governed by Eq. (3). On the assumption that variations in wind speed over the interval ζ are more significant to the kinetic energy than those of density, we may express the initial horizontal kinetic energy (per unit volume) of the two parcels as

$$K = \frac{\rho_0}{2}[(U_0 + \zeta U'_0)^2 + U_0{}^2] = \rho_0\left[U_0{}^2 + \zeta U_0 U'_0 + \tfrac{1}{2}\zeta^2(U'_0)^2\right] \tag{12}$$

The parcel moving upward will be accelerated horizontally by the faster-moving air it encounters, but we may expect that it cannot instantaneously adjust to that air and will lag behind. We hypothesize that parcels initially in equilibrium in the wind field will have at each subsequent point of their vertical motion the horizontal velocity of the average wind field through which they passed. Thus the parcel starting at z has an initial speed of U_0. The wind varies linearly from z to $z + \zeta$ and the average over this interval is $U_0 + \frac{1}{2}\zeta U'_0$. Hence at $z + \zeta$ this parcel will have the speed $U_0 + \frac{1}{2}\zeta U'_0$ compared with the ambient air speed of $U_0 + \zeta U'_0$. The parcel moving down from $z + \zeta$ to z has moved through a wind field with the same average speed and thus arrives at z with exactly the same velocity. Disregarding density effects again, the horizontal kinetic energy after interchange is

$$K^* = \frac{\rho_0}{2}\left[2\left(U_0 + \tfrac{1}{2}\zeta U_0'\right)^2\right] = \rho_0\left[U_0^2 + \zeta U_0 U_0' + \frac{\zeta^2}{4}(U_0')^2\right] \tag{13}$$

and thus the change in the horizontal kinetic energy is

$$\Delta K = K^* - K = -\frac{\zeta^2}{4}(U_0')^2 \rho_0 \tag{14}$$

Equating the initial total energy and that after interchange allows us to write the equation

$$k_0 = k + \Delta P + \Delta K = k + \rho_0 \zeta^2\left[\omega_g^2 - \frac{1}{4}\left(\frac{\partial U}{\partial z}\right)_0^2\right] \tag{15}$$

If the motion is stable, its kinetic energy must be decreasing, and so we require that the vertical kinetic energy k cannot exceed k_0 as a *sufficient condition for stability*. This sufficient condition becomes

$$k - k_0 = \zeta^2 \rho_0\left[\frac{1}{4}\left(\frac{\partial U}{\partial z}\right)_0^2 - \omega_g^2\right] \leqslant 0 \tag{16}$$

which implies that

$$\tfrac{1}{4} \leqslant \frac{\omega_g^2}{(\partial U/\partial z)_0^2} \tag{17}$$

This ratio is known as the Richardson number Ri,

$$Ri = \frac{\omega_g^2}{(\partial U/\partial z)^2} = \frac{(g/\theta)(\partial\theta/\partial z)}{(\partial U/\partial z)^2} \tag{18}$$

and on the basis of this analysis, instability is prohibited if Ri exceeds $\frac{1}{4}$. Notice that Ri is proportional to the ratio of the energy extracted by buoyancy forces to the energy gained from the shear of the large-scale wind field.

Conversely, if instability is to occur, then it is *necessary* that

$$Ri < \tfrac{1}{4} \tag{19}$$

But we cannot be certain that Richardson numbers less than $\frac{1}{4}$ will actually cause instability, because the parcels, or more accurately, the air being perturbed, may not actually utilize the energy available in the wind shear and so the vertical oscillation would not amplify.

The Richardson number thus provides important quantitative information on the relation between the stabilizing effect of buoyancy and the destabilizing effect of wind shear. Sharp gradients and small Richardson numbers are continually created by large-scale processes. The atmosphere generally takes advantage of these small Richardson numbers to create amplifying motions that break down into turbulence and thus stir and mix the air, a process that reduces the gradients and restores the stability associated with large Richardson numbers. The atmosphere, then, finds the thermodynamic profits quite handsome indeed in unstable regions, and so it carries on much of its business where Ri is small.

PROBLEM 4.5.2 The theory presented here does not apply at the nose of a jet where $U_0' = 0$. Analyze this situation to determine whether such a region is stable if $\omega_g{}^2 > 0$. (Hint: Let $U_0' = 0$ at z. Interchange parcels between $z + \zeta/2$ and $z - \zeta/2$.)

PROBLEM 4.5.3 Consider a layer of initial thickness $(\Delta z)_0$ with a difference in wind speed of ΔU and a difference of potential temperature of $\Delta\theta$ across the layer at time t_0. Show that as the layer is squeezed to smaller thickness, Ri will decrease provided that ΔU and $\Delta\theta$ do not change.

PROBLEM 4.5.4 Suppose that the squeezing process in Problem 4.5.3 proceeds until $Ri < \frac{1}{4}$ and that turbulence breaks out at time t_1. Then suppose that the layer thickness remains constant, but that both ΔU and $\Delta\theta$ are reduced by the mixing by a fraction $\epsilon(t) < 1$ of their value at t_1. Show that if $\partial\epsilon/\partial t < 0$, then eventually Ri will become greater than $\frac{1}{4}$ again.

4.6 VERTICAL AND LATITUDINAL VARIATION OF ATMOSPHERIC STRUCTURE

Most meteorologists are fascinated by the never-ending variations they see in the atmosphere, by the evolving panorama of the sky, by the storms and blizzards that come their way, by the magic of the wind. But these phenomena that compose the fleeting features of the weather are small variations on a large-scale structure that is remarkably constant.

Pressure and density always decrease with height, and at very nearly the same rate in both summer and winter. Temperature generally decreases with latitude in the lower atmosphere, but reverses this pattern in the stratosphere. Pressure, too, tends to decrease with latitude in a broad belt between the tropics and the polar regions, a gradient that is associated with the westerly winds of mid-latitude. This belt of westerlies reaches its maximum intensity at the boundary between the troposphere and the stratosphere in the jet stream, a river of air that, while not always present, tends to circle the globe in a sinuous pattern—curving equatorward over cold continents in the winter and poleward over warm ones in the summer. At the surface, there are the subtropical anticyclones on the east side of the major oceans; like the jet stream, they move poleward in the summer, but they are nearly always present and are the engine of the trade winds.

It is important to keep the basic structure of the atmosphere in mind, however deeply we delve into the study of the details of those changes and perturbations that make meteorology interesting. The weather forecaster has an obvious need to know and consider the normal or average state of the atmosphere whose future he would foretell. This knowledge is important, too, for the theoretician and dynamic meteorologist for two significant reasons.

The first is that the observed structure is a manifestation of the mode of response that the atmosphere has chosen to satisfy the thermal forcing that sets it in motion. We cannot explain, otherwise than by reference to the equations of motion, why the atmosphere has made this particular choice. And we do not know whether the present one is the only one possible given the present external conditions such as the rate of solar energy output, the distribution of continents and oceans, and the composition of the air. There is a strong suggestion in the known variations of climate in the past that it may not be. Thus there is an ever-present challenge to the theoretician to explain why the atmosphere has the structure it does.

The second reason is, as we shall see, that in order to accomplish anything at all, the dynamicist is forced to make approximations when he works with the equations governing the evolution of the atmosphere. These approximations are useful only if they reflect reality. Thus they must be judged by the extent to which they represent the observed state of the atmosphere and the extent to which the predictions made with their assistance are also consistent with observations. So the dynamicist must find both inspiration and justification for his approximations by studying the structure of the atmosphere.

In this section, then, we shall present some figures that are intended to portray the main elements of the vertical and latitudinal structure of the atmosphere. They have been chosen and constructed with the aims of allowing the student to obtain an intuitive feeling and understanding of that structure and of giving him the information with which to make the numerical estimates of wind speeds and shears, temperature gradients and lapse rates, and all the other quantities that are required to convert mathematical statements into quantitative ones that apply to the earth's atmosphere.

4.6.1 Vertical Variations

As pointed out at the beginning of the chapter, the most pronounced variations in atmospheric variables occur in the vertical direction. Figures 4.2 to 4.5 show the variations with height of pressure, temperature, density, and potential temperature that are characteristic of the tropics, the middle latitudes, and the wintertime polar regions.

The curves given in Fig. 4.2 for pressure as a function of height are useful in several contexts. Meteorologists have a recurring need to convert altitudes expressed in geometric units to altitudes in pressure units, and vice versa, and this figure allows that to be done easily. It is also of interest to see that the equatorial pressure at a given level is a little higher than that of middle latitudes or the winter polar regions. As can be seen from the figure, the 500-mbar surface that divides the atmosphere into two portions of nearly equal mass has a latitudinal altitude variation of about $\frac{1}{2}$ kilometer.

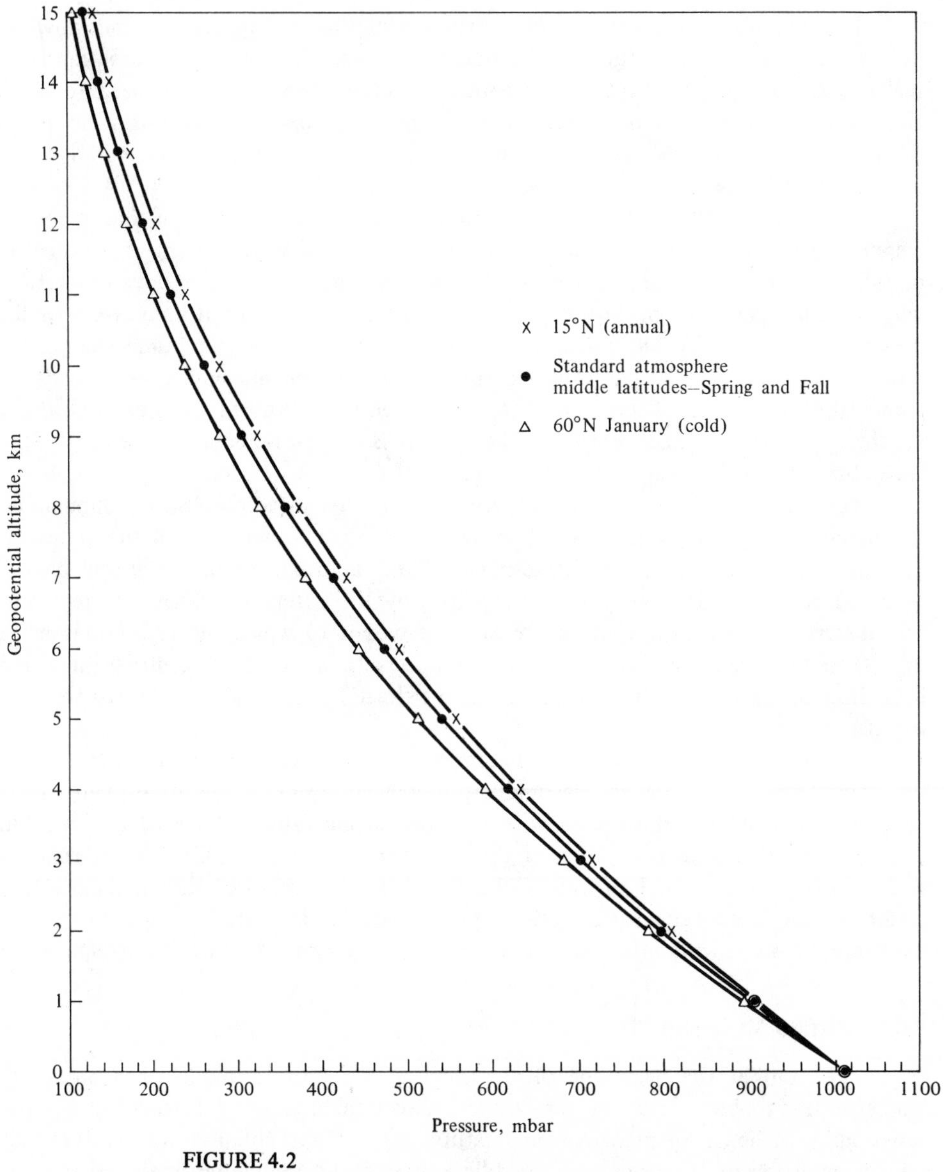

FIGURE 4.2
Pressure variation with height and latitude in the *U.S. Standard Atmosphere*.

Finally, in comparison with Fig. 4.4 representing the density variation with height, the curves show that the pressure has a stronger vertical gradient in the regions where the density is larger, as required by the hydrostatic relation.

The curves for temperature in Fig. 4.3 demonstrate both the vertical decrease of temperature in the troposphere and the variation with latitude of the atmosphere's thermal structure. The middle-latitude curve shows a linear decrease of temperature at a rate of 6.5 K/km from the surface to 11 km. The isothermal region above 11 km is in the stratosphere, and so the tropopause occurs on this average profile at 11 km. In comparison, the equatorial tropopause on the standard-atmosphere curve occurs at 16.5 km, and that for the cold, winter polar regions at 8.5 km. Furthermore, the curves show that the equatorial stratosphere is colder than the stratosphere in middle latitudes and even colder than the wintertime polar stratosphere. Finally, the curve for the polar regions reflects quite clearly the loss of heat that occurs at the surface, a loss that is strong enough to create the low-altitude inversion in which the temperature increases with height.

Despite the appearance of strong variation in temperature when we use Celsius units, it is important to recognize that the percentage variations of temperature shown are not large in terms of the absolute units that appear in the equations of motion. Thus the mean temperature on the graph is roughly 250 K so that the maximum excursions from that mean of about 50 K represent only a 20 percent variation about the average temperature.

The curves for density, given in Fig. 4.4, show quite clearly the contraction of mass that occurs in the colder regions. Still, the surface pressure varies over the globe by only a few percent, so that to the extent that hydrostatic equilibrium prevails, these curves must have nearly the same integral in the vertical. Thus they cross at about 7 km or 400 mbar, producing a level in the atmosphere on which the density is nearly constant. It is thus a feature of atmospheric structure in middle altitudes that isobaric and isothermal surfaces slope downward toward the poles while the isopycnic surfaces representing constant density are quasi-horizontal.

The variations of potential temperature are shown in Fig. 4.5. As was shown in Sec. 4.4, the static stability of the atmosphere is proportional to the vertical gradient of potential temperature. Thus two stability regimes are evident: the troposphere and the more stable stratosphere.

The annual average of the vertical distribution of water vapor in middle latitudes is shown in Fig. 4.6. Here we use mixing ratio, the ratio of the mass of water vapor to the mass of dry air in a volume. Clearly, the vapor in the troposphere is concentrated near the earth's surface. The increase in mixing ratio shown in the upper stratosphere appears to be an actual feature of atmospheric structure, but humidity measurements at these altitudes are both difficult and rare.

It is appropriate to end this discussion by emphasizing that these curves represent average conditions. The variations of individual soundings from these averages may be pronounced; the most striking ones are often the result of air from these different geographical regions being superimposed in a layered structure and separated by the fronts associated with mid-latitude cyclone systems.

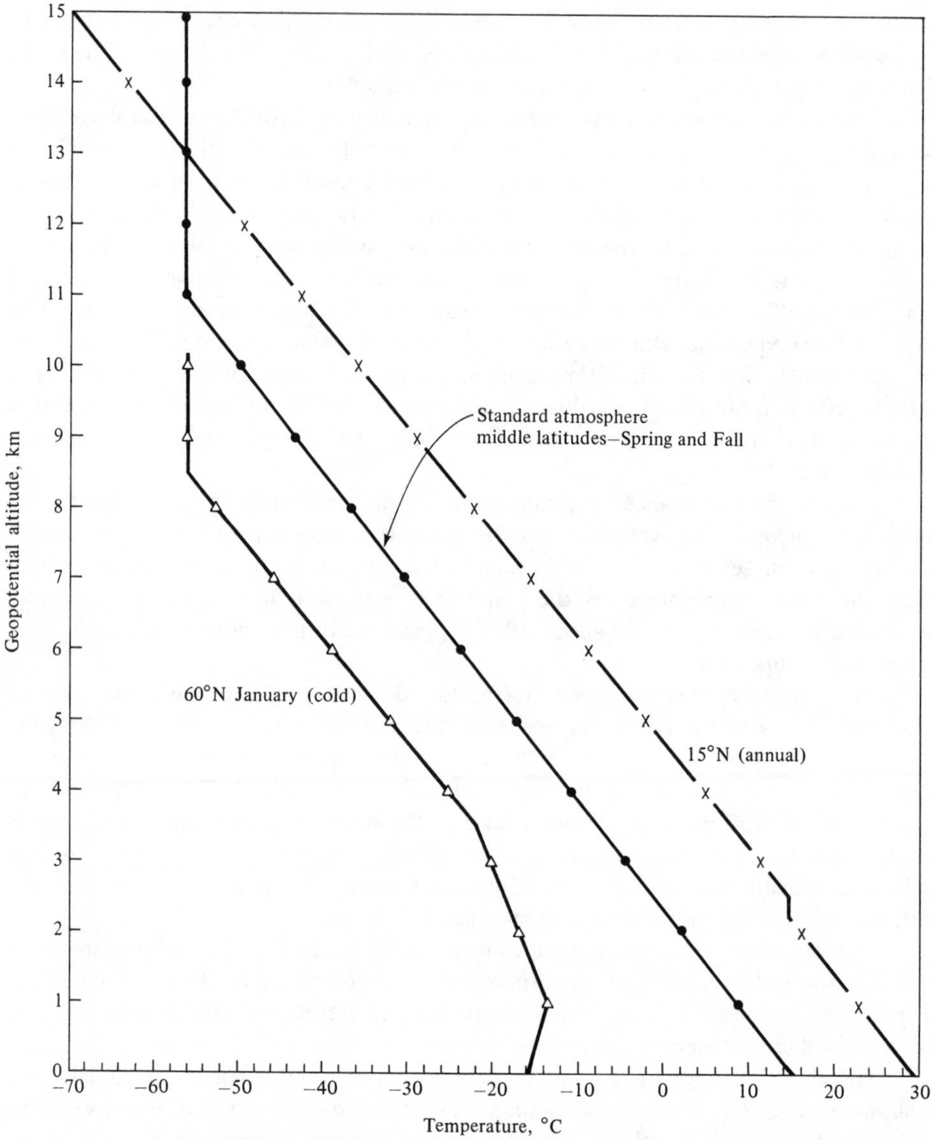

FIGURE 4.3
Temperature variations with height and latitude in the *U.S. Standard Atmosphere.*

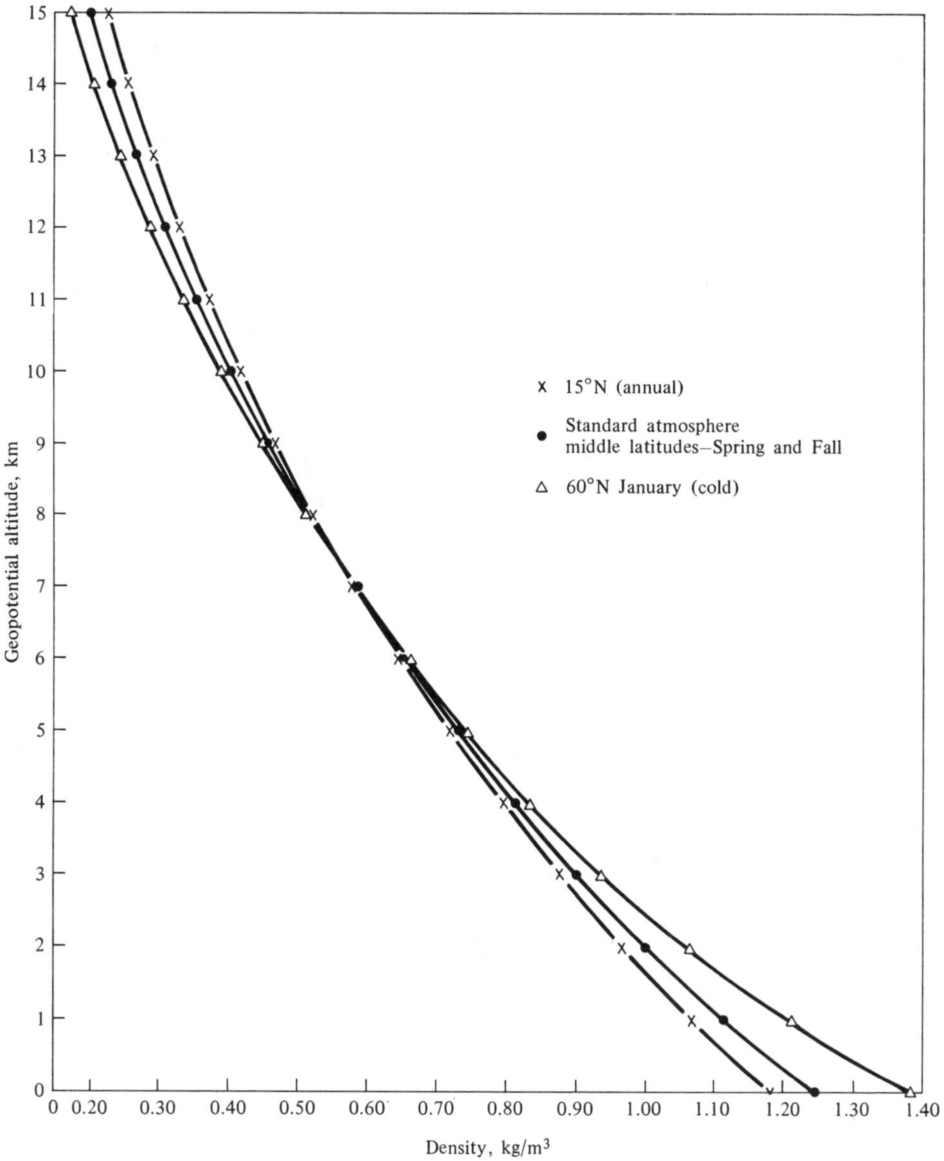

FIGURE 4.4
Density variation with height and latitude in the *U.S. Standard Atmosphere.*

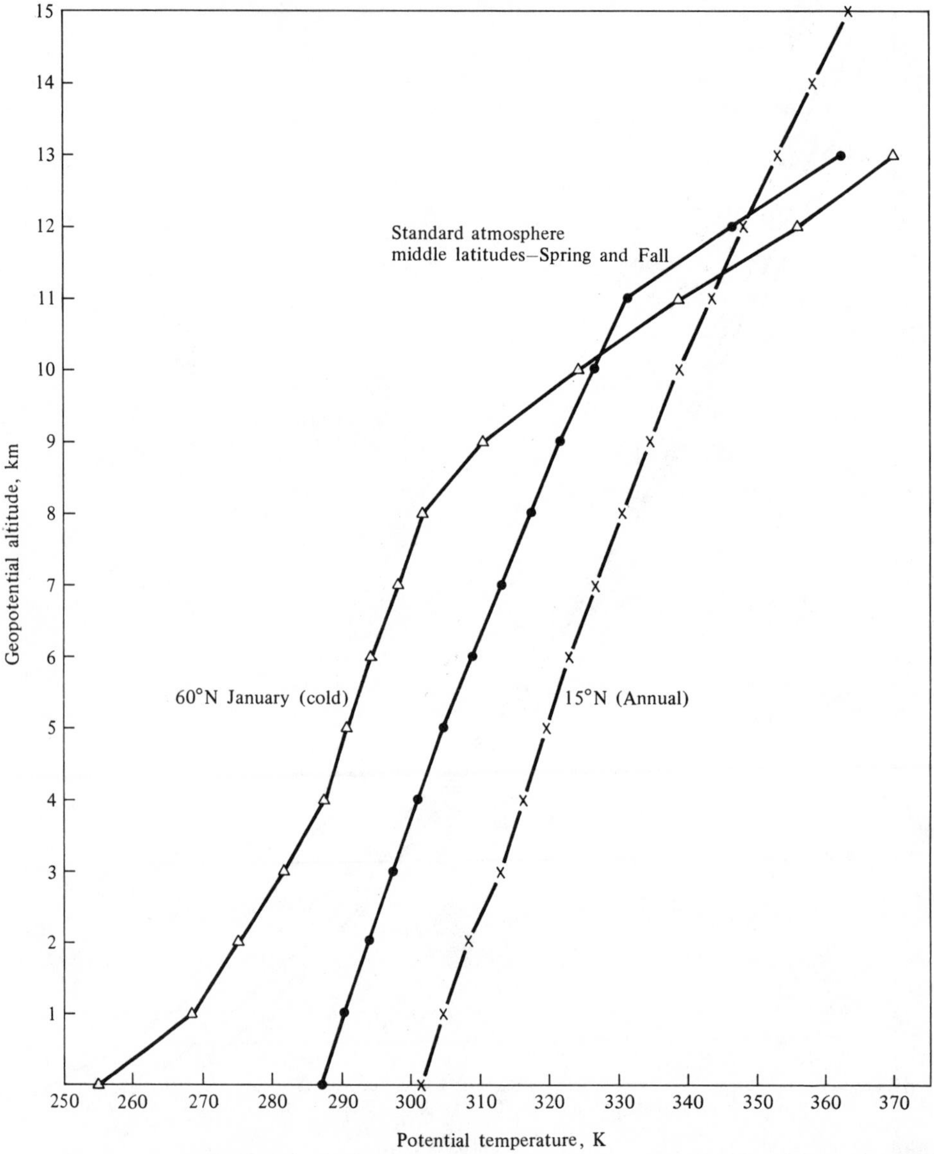

FIGURE 4.5
Potential temperature variations with height and latitude in the *U.S. Standard Atmosphere.*

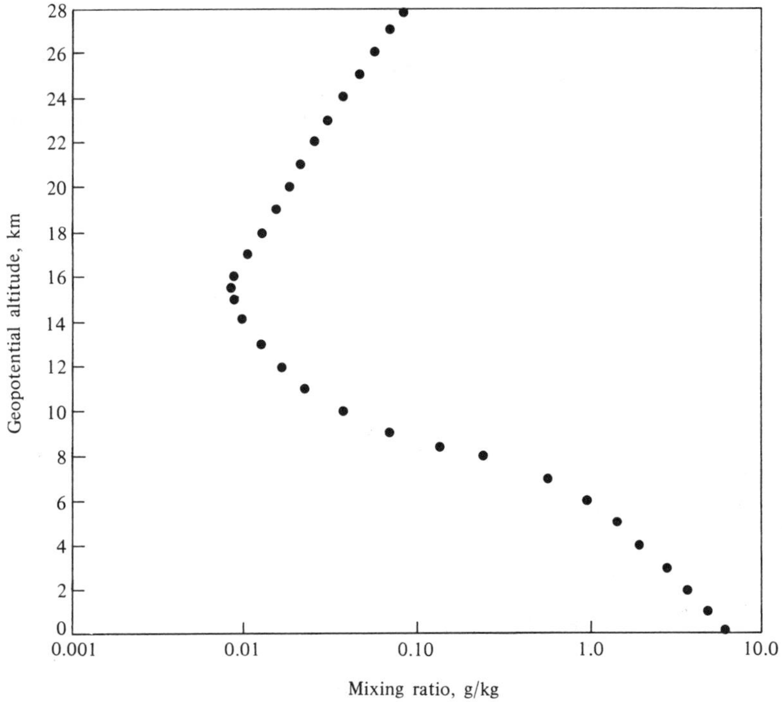

FIGURE 4.6
Annual mean mixing ratio profile in middle latitudes. (*From Handbook of Geophysics and Space Environments.*)

4.6.2 The Motion Field

We turn now from the vertical structure of the thermodynamic variables to the variations of the wind with height and latitude. But before looking at the structure of the motion field, some preliminary comments are in order.

First, we shall use pressure as a linear vertical coordinate rather than geometric altitude, because then equal vertical intervals represent equal amounts of mass, as shown by the hydrostatic equation. Thus the isopleths of a plotted quantity and the area they cover combine to depict the contribution from that region to the mass-weighted integral of the quantity over the whole atmosphere. This is particularly useful with the wind, because we measure and usually analyze velocity but are really more interested in the momentum. By the same reasoning, we should use the sine of the latitude as the horizontal coordinate, but for the present purposes this procedure would lead to an excessive loss of detail in the polar regions.

Second, we shall present average conditions along 80°W rather than longitudinal averages of the structure. There are two reasons for this choice. The first is that there

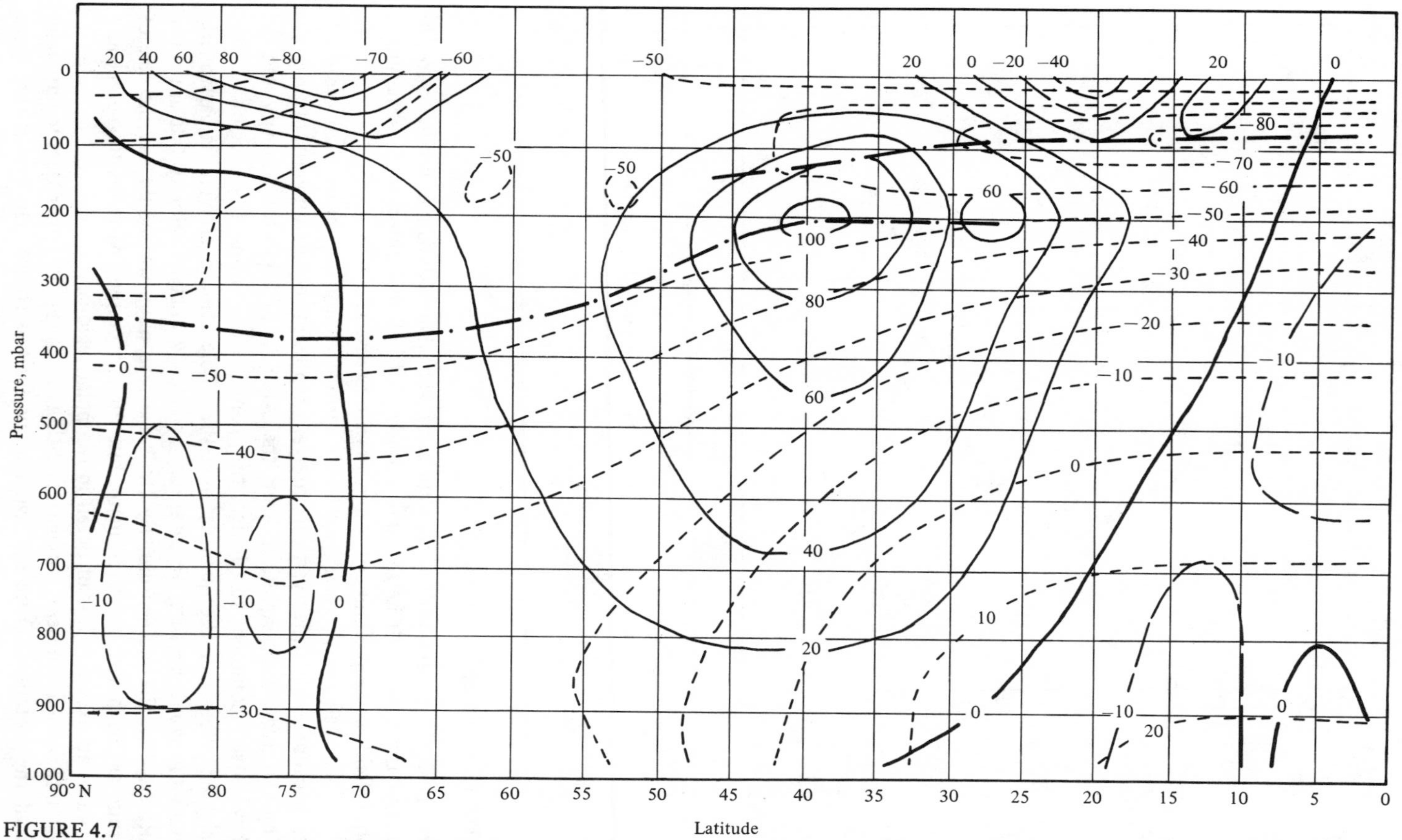

FIGURE 4.7
January average of the zonal component of the geostrophic wind at 80°W in the Northern Hemisphere along with isotherms in degrees Celsius. Solid lines are isotachs in knots with positive values for westerly winds (broken lines for easterlies); dashed lines are isotherms; the heavy broken line denotes the tropopause. (*After Kochanski, 1955.*)

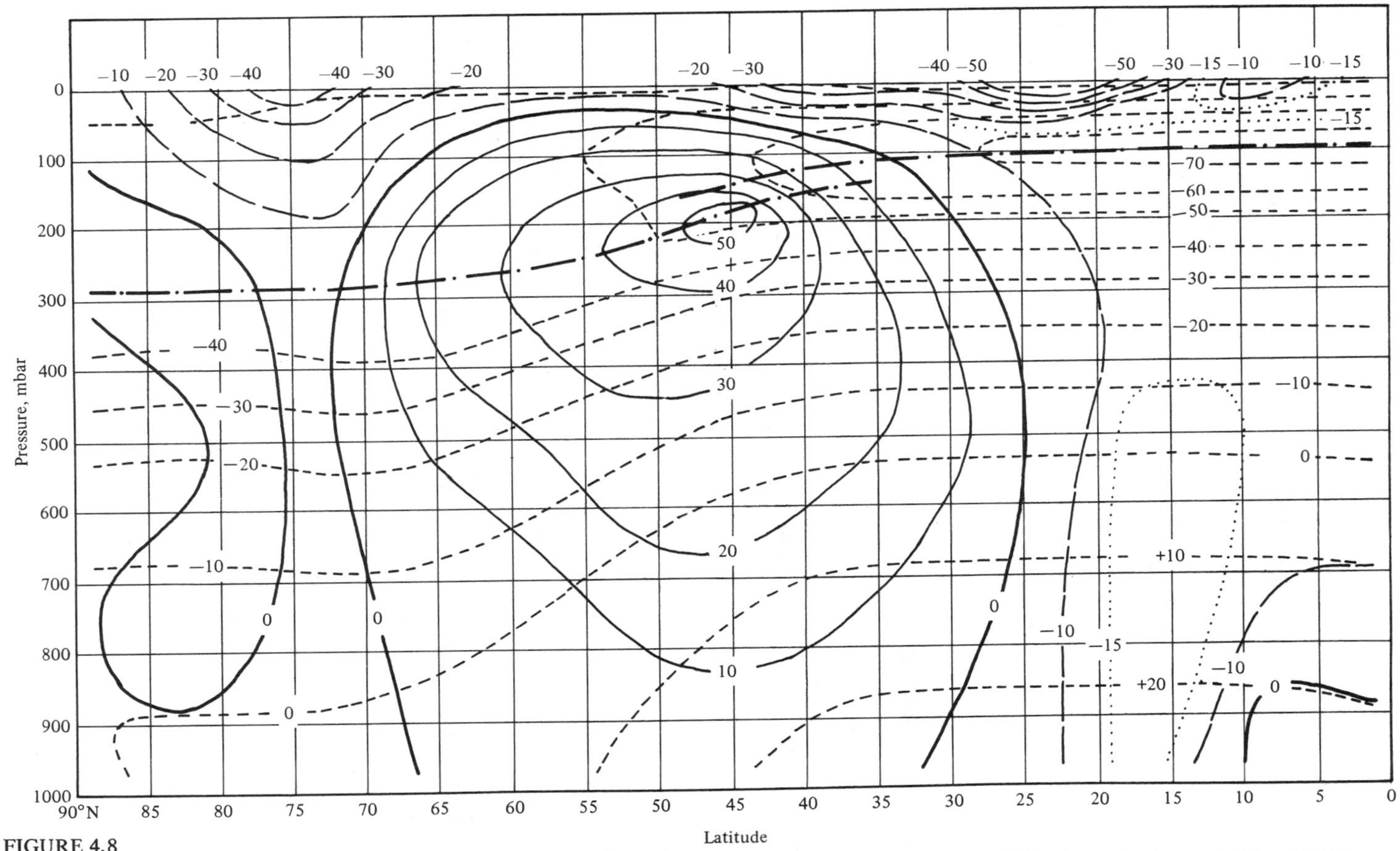

FIGURE 4.8
July average of the zonal component of the geostrophic wind at 80°W in the Northern Hemisphere along with isotherms in degrees Celsius. Solid lines are isotachs in knots with positive values for westerly winds (broken lines for easterlies); dashed lines are isotherms; the heavy broken line denotes the tropopause. (*After Kochanski, 1955.*)

is an unusually high observational density along this meridian so that latitudinal cross sections have both more detail and reliability than is usually the case. But perhaps of more significance is the fact that atmospheric structure is, in some respects, quite dependent upon longitude because of the relation of patterns in the flow to geographical features of the earth's surface. Thus the same physical features occur at different latitudes, and when they are averaged by longitude, details are lost and the gross features may be misrepresented.

The cross sections of the zonal component of the (geostrophic) wind are shown for January in Fig. 4.7 and for July in Fig. 4.8. The dominant feature of the atmospheric circulation in terms of momentum is the large belt of westerly winds in mid-latitude, winds that increase in speed with height and toward the center of the belt to form the mid-latitude westerly jet stream at the tropopause. The jet doubles in intensity from summer to winter and moves from a summer position of about 50°N to a winter position of about 35 to 40°N. It is also obvious from the two figures that the entire westerly regime expands considerably in the winter season. In the winter, the mid-latitude jet shown here has two maxima, reflecting a separation of the flow that occurs often, but not always. The stronger one at 40°N is referred to as the polar-front jet, the weaker one at about 28°N as the subtropical jet.

There are two other jet stream phenomena shown on the cross sections. The stratospheric jet at 20°N is shown as being composed of easterly winds of 40 kt (about 20 m/s). However, this current oscillates in both intensity and direction in a 26-month cycle, changing from an easterly jet with core velocities of 30 m/s to a westerly jet with velocity of 15 m/s. The high-altitude polar easterlies of summer reverse in winter to become a strong, westerly current known as the night polar jet.

Near the surface, the easterly components associated with the trade winds of the subtropics provide the only strong contrast to the generally westerly winds of mid-latitudes. Comparison of the two figures shows that this easterly regime expands poleward in the summer, reflecting the increase in intensity of the subtropical anticyclones. From the chart, it is possible to see that in summer the equatorial easterlies below 400 mbar have momentum about equal to half that of the core ($V > 30$ kt) of the mid-latitude jet.

The structure of the meridional component of the motion is more difficult to determine because geostrophic relations cannot be used to construct cross sections. Thus direct upper-air wind observations must be used, and we are forced to consider averaging series of observations of nearly equal magnitude but varying sign. Two cross sections representing the average meridional components for the Northern Hemisphere summer and winter are shown in Figs. 4.9 and 4.10. The dark arrows on the figures indicate the directions of the mass circulations derived from the data. In the winter, the dominant feature is the circulation cell between 30°N and the equator, a feature known as a Hadley cell. The weaker countercirculation of the Ferrel cell in mid-latitudes is also apparent. Both of these features are much weaker in the summer, with the Southern Hemisphere Hadley cell extending markedly into the Northern Hemisphere.

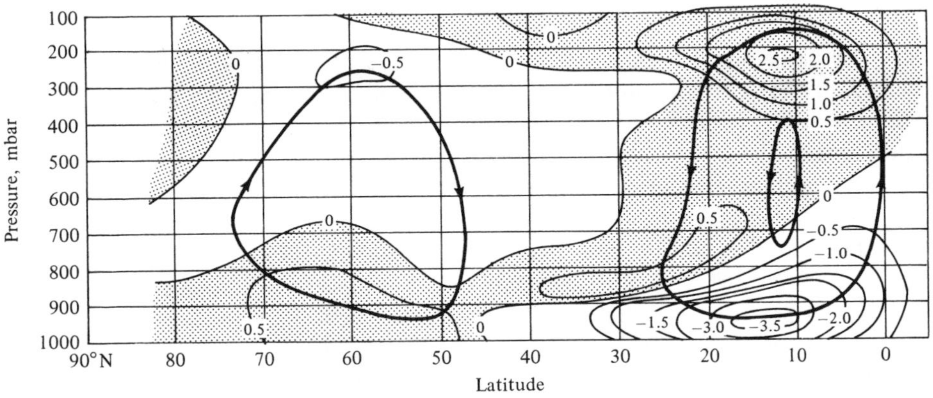

FIGURE 4.9
Winter average meridional circulation in the Northern Hemisphere (December–February). Isotachs in meters per second with positive values for south winds. The dark arrows indicate the direction of the mass transport as computed from the data. (*After Palmén and Vuorela, 1963.*)

However, the meridional circulations determined by these methods may be misleading. There are strong meridional circulations in the systems of synoptic scale in mid-latitudes, but they are quasi-horizontal and do not appear in these averages. Thus the necessary transports of heat and momentum in the atmosphere are not accomplished solely by the weak, mean meridional circulations shown here.

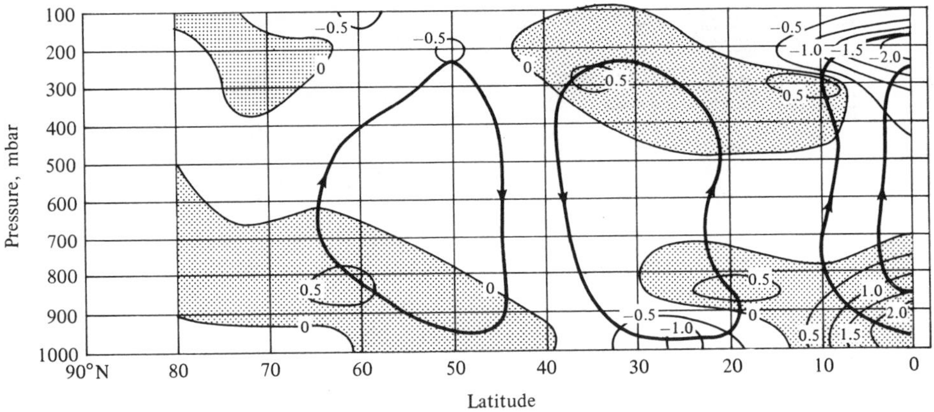

FIGURE 4.10
Summer average meridional circulation in the Northern Hemisphere (June–August). Isotachs in meters per second with positive values for south winds. The dark arrows indicate the direction of the mass transport as computed from the data. (*After Vuorela and Tuominen, 1964.*)

4.6.3 The Thermal Structure

The thermal structures along 80°W existing in conjunction with the motion field studied in the previous subsection are also shown in Figs. 4.7 and 4.8. The essential features depicted on both cross sections are the general increase of temperature southward in the troposphere and poleward in the stratosphere, with the exception of the polar regions in winter. But perhaps of more interest is the seasonal variation in the intensity and location of the temperature gradient.

In Chap. 9 it will be shown that the vertical gradient of the geostrophic wind is proportional to the horizontal gradient of the temperature. From a comparison of the thermal patterns with those of the zonal wind field, it can be seen easily that the mid-latitude jet stream lies above the region in which the tropospheric temperature gradient is concentrated, and then loses intensity in the stratosphere as the gradient is reversed.

When there are temperature variations on an isobaric surface, the atmosphere is said to be *baroclinic*; if there are no temperature variations, it is referred to as *barotropic*. In this sense, baroclinity is revealed by the slope of the isotherms with respect to the isobaric surfaces. Thus the cross sections show that the equatorial and polar regions are nearly barotropic, but are separated by a strongly baroclinic region in mid-latitudes, a baroclinic region whose intensity increases markedly in the winter.

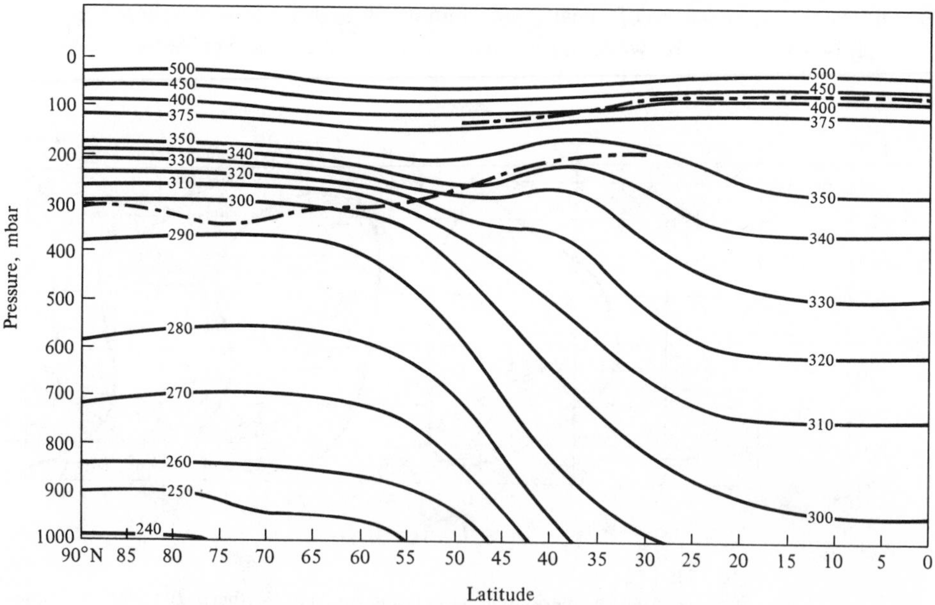

FIGURE 4.11
January average of the potential temperature field along 80°W in the Northern Hemisphere as constructed from isotherms in Fig. 4.7.

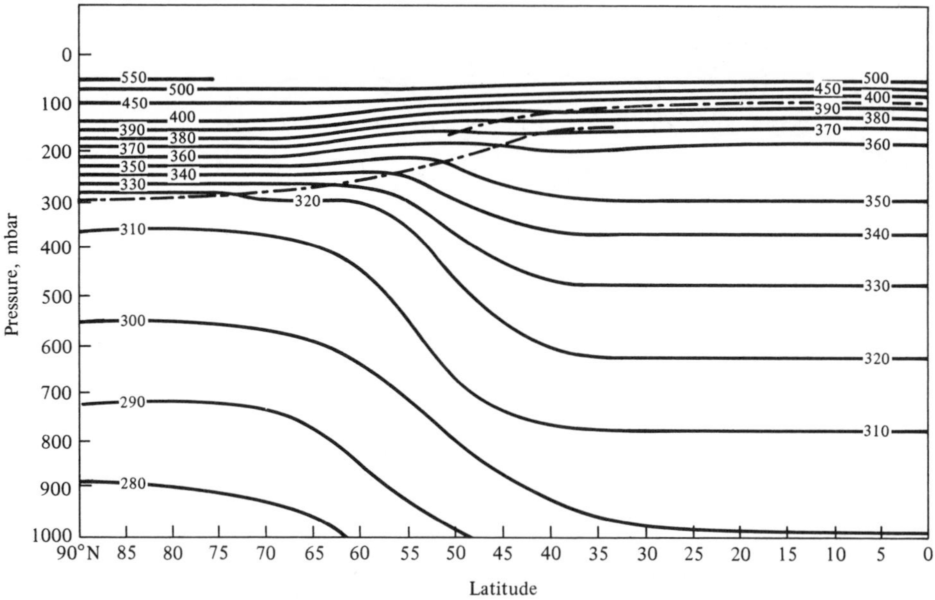

FIGURE 4.12
July average of the potential temperature field along 80°W in the Northern Hemisphere as constructed from the isotherms in Fig. 4.8.

This strongly baroclinic region thus separates subtropical structure from that of the subpolar regions and is often called the polar front; the polar-front jet stream is formed above this strongly baroclinic region.

The tendency for the baroclinity to be concentrated in a relatively narrow zone can perhaps be seen more clearly on Figs. 4.11 and 4.12, which depict cross sections of potential temperature derived from those for temperature. On these figures, the tight packing of the isentropes makes the location of the stratosphere evident and the concentrations of baroclinity at about 45°N in the winter and 50 to 55°N in the summer are quite clearly revealed.

The variations between the summer and winter thermal structure are depicted in Fig. 4.13, which gives the difference between the July and January values of potential temperature. It is evident that there is little change in the equatorial and subtropical regions, and that the greatest variations occur in the polar regions, both near the surface and near the top of the atmosphere. The variation in the strength and location of the main baroclinic zone is emphasized by the vertical band of isopleths in middle latitudes.

We point out again that the variations of potential temperature in the mid-latitude and tropical troposphere are not greater than about 10 percent. But still, even such relatively small variations in thermal structure are associated with strong variations in the motion field, as shown by Figs. 4.7 and 4.8.

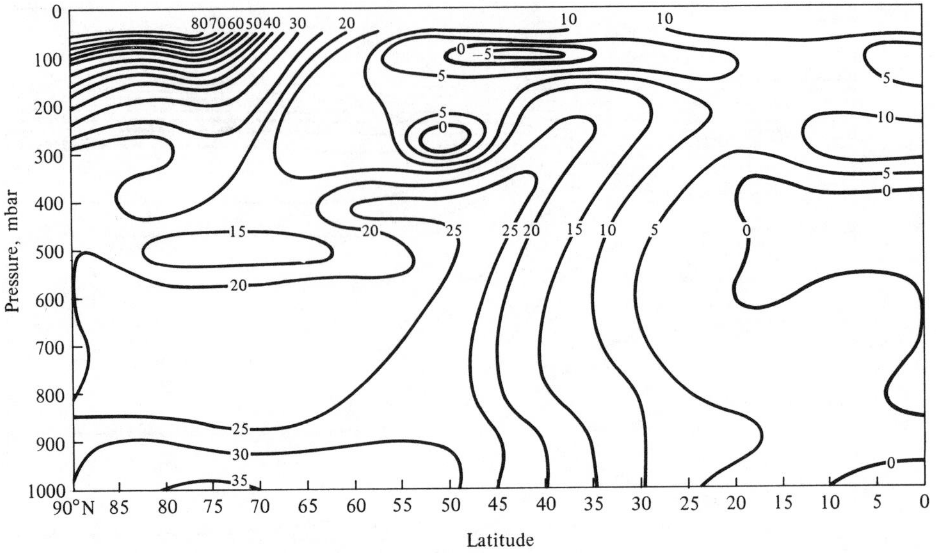

FIGURE 4.13
The difference between July and January values of potential temperature along 80°W in the Northern Hemisphere.

BIBLIOGRAPHIC NOTES

4.1 The first part of this section presents standard material that is discussed in all meteorological texts, although the question of the relation between thermodynamic and hydrostatic pressure is usually ignored.

4.2 Additional details and a variety of numerical information about geopotential quantities may be found in:

Smithsonian Meteorological Tables, R. J. List (ed.), Smithsonian Miscellaneous Collections, vol. 114, 1963, Washington, 527 pp.

This book presents a wealth of meteorological information and is an invaluable reference.

4.3-4.4 Again, standard materials, except for the analysis of nonlinear stability, which is believed to be new.

4.5 The discussion of this section will be amplified considerably when wave motion is discussed in Chaps. 12 and 13.

4.6 The material presented in this section was compiled from a variety of sources. They include:

U.S. Standard Atmosphere, 1962, U.S. Government Printing Office, Washington, 278 pp.

U.S. Standard Atmosphere Supplements, 1966, U.S. Government Printing Office, Washington, 289 pp.

Kochanski, A., 1955: "Cross Sections of the Mean Zonal Flow and Temperature along 80°W," *J. Meteorol.*, **12**:95–106.

Palmén, E., and L. A. Vuorela, 1963: "On the Mean Meridional Circulations in the Northern Hemisphere during the Winter Season," *Quart. J. Roy. Meteorol. Soc.*, **89**:131–138.

Valley, S. L. (ed.), 1965: *Handbook of Geophysics and Space Environments*, McGraw-Hill Book Company, New York, 621 pp.

Vuorela, L. A , and I. Tuominen, 1964: "On the Mean Zonal and Meridional Circulations and the Flux of Moisture in the Northern Hemisphere during the Summer Season," *Pure and Appl. Geophys.*, **57**:167–180.

Additional information on the basic structure of the atmosphere may be found in the books:

Lorenz, E. N., 1967: *The Nature and Theory of the General Circulation of the Atmosphere*, World Meteorological Organization, No. 218 TD 115, Geneva, 161 pp.

Palmén, E., and C. W. Newton, 1969: *Atmospheric Circulation Systems*, Academic Press, Inc., New York, 603 pp.

5

VECTOR AND TENSOR ANALYSIS AND THE FUNDAMENTAL KINEMATICS OF FLUID FLOW

The patterns of the wind are complex and ever-changing, but they are patterns, and thus there is hope of comprehending them. Our present understanding of fluid motion, including that of the atmosphere, is based upon the equations of motion, which assemble the information contained in the spatial relations of the wind's patterns at one instant to predict how they will evolve in the next.

In this chapter we assemble the mathematical concepts that will allow us to proceed in a reasonably straightforward manner, concentrating on the physics, once we begin to derive and discuss the equations of motion. It may be reassuring to the beginner to emphasize that certainly not all the mathematics in this chapter is needed for an initial acquaintance with atmospheric motion or the equations that describe it. But everything here is used later in the book as more advanced topics are discussed and as we consider the implications of spatial and geometric relations that can be described only mathematically.

We would like the mathematical concepts and operations we use to be founded upon the geometric intuition with which we perceive the physical situation. This is the main advantage of the vector and tensor concepts we shall summarize here and use later; the geometric aspects of the problem receive the emphasis.

Vectors provide a representation of quantities, like the wind velocity, that have both a magnitude and a direction. In order to analyze such quantities we must develop methods for differentiating and integrating them, and we present the classical theorems that facilitate the application of these two operations to physical problems. Fluid-mechanical examples are used to illustrate the concepts under discussion.

Next we consider how certain integrals that appear often in fluid dynamics can be differentiated with respect to time, and this necessitates consideration of some of the properties of transformations and the geometry of curves, surfaces, and volumes.

Another major topic is curvilinear coordinate systems, which will appear in Chap. 7; they provide an introduction to the basic concepts of tensor analysis. Tensors cannot be avoided in fluid dynamics because they represent physical quantities, more complicated than vectors, that often appear. The presentation here is restricted to development of results needed later, and is designed with the hope of encouraging the student to learn more about the details that have been stripped away in an attempt to make the main concepts apparent.

Returning to fluid motion, we show how the divergence and vorticity patterns of an arbitrary flow can be used to recover the velocity field, and we show how a given flow can be resolved into a sum of pure types of motion. This decomposition of flows is an important part of the process of obtaining dynamic inferences from weather charts.

Finally, we consider the fundamental question of which mathematical structure and concepts provide a model of fluid flow. The answer is that temporally varying, differentiable coordinate transformations are the proper kinematic abstraction of both fluid motion and the motion of the atmosphere. This point of view yields important information about the hypotheses that are necessary to study fluid motion mathematically.

5.1 VECTOR ANALYSIS

In atmospheric science we encounter a variety of physical quantities, some of which are quite different in character. There are the variables such as temperature or pressure that are quantified by a number; they are called *scalars*. Some of the variables associated with motion, such as velocities, accelerations, or forces, are different because they involve both a magnitude and a direction; they are called *vectors*. A vector quantity can also be constructed from a scalar field, by specifying, for example, at what rate it changes in a certain direction.

Vectors can be represented either geometrically or analytically. Their geometric form is that of an arrow whose direction corresponds to that of the motion or force, for example, and whose length is proportional to the magnitude. The analytical representation of vectors takes its simplest form in the usual rectangular cartesian coordinate system. But the vector operations and relations we shall introduce are defined so that they remain valid when they are evaluated in any right-handed coordinate system.

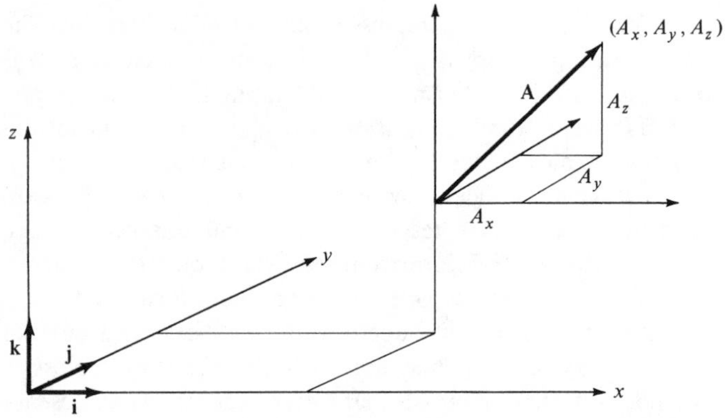

FIGURE 5.1
Illustration of the two coordinate systems used in vector analysis: the location of a vector is described in the **x** system, and its direction and magnitude in the **A** system.

To begin, we assume that we have a cartesian coordinate system with which we determine position in a three-dimensional space. The (x,y,z) axes of this system determine three orthogonal directions, and we shall use these directions in an analytical representation of vectors.

Let there be a vector quantity at point (x,y,z) at time t. To represent it, we draw an arrow of proper length in the proper direction, putting its tail at the point (x,y,z). In another coordinate system whose origin is at (x,y,z), we may determine the coordinates (A_x,A_y,A_z) of the tip of the arrow **A** (Fig. 5.1).

The basic directions established by the original coordinate system are given by the *unit vectors* **i**, **j**, and **k**, which are parallel to the coordinate axes and of unit length. The operations of multiplication of a vector by a scalar and of vector addition are defined in Table 5.1, and with them we have the relation

$$\mathbf{A} = A_x\mathbf{i} + A_y\mathbf{j} + A_z\mathbf{k} \tag{1}$$

which is the basic analytical representation of vectors in rectangular cartesian coordinates.

The definition of scalar product in Table 5.1 implies that

$$\mathbf{i}\cdot\mathbf{i} = \mathbf{j}\cdot\mathbf{j} = \mathbf{k}\cdot\mathbf{k} = 1 \qquad \mathbf{i}\cdot\mathbf{j} = \mathbf{i}\cdot\mathbf{k} = \mathbf{j}\cdot\mathbf{k} = 0 \tag{2}$$

so we can find the components A_x, A_y, A_z of a vector **A** by the operations

$$A_x = \mathbf{i}\cdot\mathbf{A} \qquad A_y = \mathbf{j}\cdot\mathbf{A} \qquad A_z = \mathbf{k}\cdot\mathbf{A} \tag{3}$$

With the Pythagorean theorem we may calculate that the length $|\mathbf{A}| = A$ of **A** is given by

Table 5.1 **BASIC VECTOR OPERATIONS**

Operation		Definition and geometric significance		Analytic form	Properties
$\mathbf{C} = \alpha\mathbf{A}$	Multiplication by a scalar α	**C** is $\lvert\alpha\rvert$ times as long as **A** and in the same direction if $\alpha > 0$, the opposite direction if $\alpha < 0$.	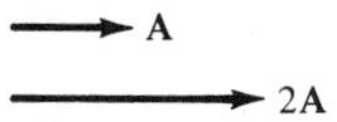	$\mathbf{C} = \alpha\mathbf{A} = \alpha A_x\mathbf{i} + \alpha A_y\mathbf{j} + \alpha A_z\mathbf{k}$	$\alpha\mathbf{A} = \mathbf{A}\alpha$ $(\alpha + \beta)\mathbf{A} = \alpha\mathbf{A} + \beta\mathbf{A}$
$\mathbf{C} = \mathbf{A} + \mathbf{B}$	Addition	Move tail of **B** to tip of **A**; the sum is the vector from the tail of **A** to the tip of **B**.	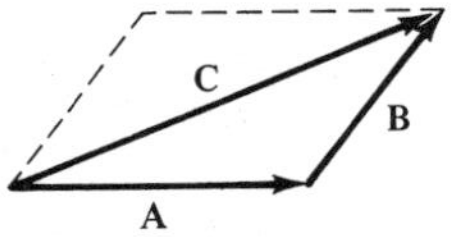	$\mathbf{C} = (A_x + B_x)\mathbf{i} + (A_y + B_y)\mathbf{j} + (A_z + B_z)\mathbf{k}$	$\alpha(\mathbf{A} + \mathbf{B}) = \alpha\mathbf{A} + \alpha\mathbf{B}$ $\mathbf{A} + \mathbf{B} = \mathbf{B} + \mathbf{A}$ $\mathbf{A} + (\mathbf{B} + \mathbf{C}) = (\mathbf{A} + \mathbf{B}) + \mathbf{C}$
$\mathbf{C} = \mathbf{A} - \mathbf{B}$	Subtraction	Add **A** and **–B**.	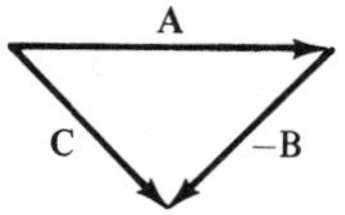	$\mathbf{C} = (A_x - B_x)\mathbf{i} + (A_y - B_y)\mathbf{j} + (A_z - B_z)\mathbf{k}$	$\mathbf{A} - \mathbf{B} = -\mathbf{B} + \mathbf{A}$ $(\mathbf{A} - \mathbf{B}) + \mathbf{C} = (\mathbf{A} + \mathbf{C}) - \mathbf{B}$
$C = \mathbf{A} \cdot \mathbf{B}$	Scalar product (dot product)	$C = AB\cos\theta$, so the length of **B** projected onto **A** is $(\mathbf{A} \cdot \mathbf{B})/A$. When **A** is a unit vector, $\mathbf{A} \cdot \mathbf{B}$ is the component B_A of **B** in the direction of **A**.	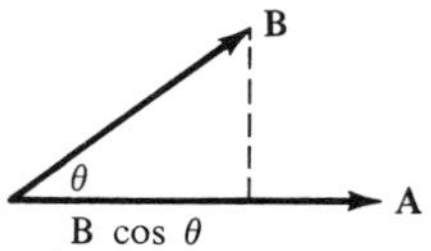	$C = A_xB_x + A_yB_y + A_zB_z$	$\mathbf{A} \cdot \mathbf{B} = \mathbf{B} \cdot \mathbf{A}$ $\mathbf{A} \cdot (\mathbf{B} + \mathbf{C}) = \mathbf{A} \cdot \mathbf{B} + \mathbf{A} \cdot \mathbf{C}$ $\mathbf{A} \cdot \mathbf{A} = A^2$
$\mathbf{C} = \mathbf{A} \times \mathbf{B}$	Vector product (cross product)	**C** is orthogonal to the plane containing **A** and **B**, is of length $AB\sin\theta$, and points from the plane in the direction of the right thumb when the right hand is parallel to **A** and the fingers curl from **A** to **B**.	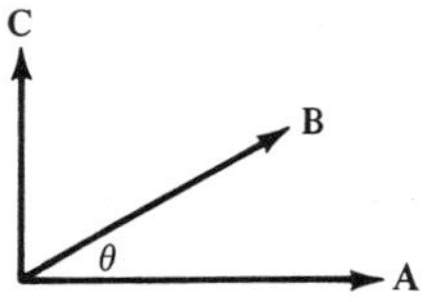	$\mathbf{C} = \begin{vmatrix} \mathbf{i} & \mathbf{j} & \mathbf{k} \\ A_x & A_y & A_z \\ B_x & B_y & B_z \end{vmatrix}$	$\mathbf{A} \times \mathbf{B} = -\mathbf{B} \times \mathbf{A}$ $\mathbf{A} \times (\mathbf{B} + \mathbf{C}) = \mathbf{A} \times \mathbf{B} + \mathbf{A} \times \mathbf{C}$ $\mathbf{A} \times \mathbf{A} = 0$

Note: These operations are defined for arbitrary vectors, but the vectors often must be moved into the required correspondence with each other. In doing so, we move them without altering either their length or their direction.

$$A^2 = (A_x^2 + A_y^2 + A_z^2) = \mathbf{A} \cdot \mathbf{A} \tag{4}$$

in which the second equality follows from Eqs. (1) and (2).

PROBLEM 5.1.1 Show that the angles ϕ, θ, and ψ between $\mathbf{A}$ and the (x,y,z) axes are given by

$$\phi = \cos^{-1}\frac{A_x}{A} \qquad \theta = \cos^{-1}\frac{A_y}{A} \qquad \psi = \cos^{-1}\frac{A_z}{A}$$

PROBLEM 5.1.2 Two vectors $\mathbf{A}$ and $\mathbf{B}$ are equal if their directions and magnitudes are equal. Show that $\mathbf{A} = \mathbf{B}$ if and only if $A_x = B_x$, $A_y = B_y$, and $A_z = B_z$.

PROBLEM 5.1.3 Let m be the mass of an object resting on an inclined plane. The unit vector $\mathbf{n}$ is normal to the plane and the unit vector $\mathbf{t}$ is orthogonal to $\mathbf{n}$ and directed down the plane. Find the forces F_t and F_n imposed on the object by the acceleration g of gravity.

PROBLEM 5.1.4 Let a vector $\mathbf{A}$ make an angle ϕ with the (x,y) plane, and let its projection A_H on that plane make an angle θ with the x axis. Find A_x, A_y, and A_z and prove trigonometrically that Eq. (4) is true.

The vector product defined in Table 5.1 arises often, especially in physical applications that involve rotation, and its essential feature is that the vector product of two vectors is orthogonal to both of them. From the definition we have

$$\begin{array}{lll} \mathbf{i} \times \mathbf{i} = 0 & \mathbf{i} \times \mathbf{j} = \mathbf{k} & \mathbf{i} \times \mathbf{k} = -\mathbf{j} \\ \mathbf{j} \times \mathbf{i} = -\mathbf{k} & \mathbf{j} \times \mathbf{j} = 0 & \mathbf{j} \times \mathbf{k} = \mathbf{i} \\ \mathbf{k} \times \mathbf{i} = \mathbf{j} & \mathbf{k} \times \mathbf{j} = -\mathbf{i} & \mathbf{k} \times \mathbf{k} = 0 \end{array} \tag{5}$$

and thus

$$\begin{aligned} \mathbf{A} \times \mathbf{B} &= (A_x\mathbf{i} + A_y\mathbf{j} + A_z\mathbf{k}) \times (B_x\mathbf{i} + B_y\mathbf{j} + B_z\mathbf{k}) \\ &= \mathbf{i}(A_yB_z - A_zB_y) + \mathbf{j}(A_zB_x - A_xB_z) + \mathbf{k}(A_xB_y - A_yB_x) \end{aligned} \tag{6}$$

*PROBLEM 5.1.5 Show by expanding the determinant definition of the vector product given in Table 5.1 that it is equivalent to Eq. (6).

PROBLEM 5.1.6 Let $\mathbf{A}$ be an arbitrary vector in the horizontal plane and $\mathbf{k}$ the vertical unit vector. Show that $\mathbf{k} \times (\mathbf{k} \times \mathbf{A}) = -\mathbf{A}$.

PROBLEM 5.1.7 Let $\mathbf{C} = \mathbf{A} \times \mathbf{B}$. Show that $C/2$ is the area of the triangle formed by connecting the tips of **A** and **B**. Show that $\mathbf{A} \cdot (\mathbf{B} \times \mathbf{C})$ is the volume of the parallelepiped with edges **A**, **B**, **C** for any vectors **A**, **B**, **C**.

PROBLEM 5.1.8 Consider a disk rotating at angular velocity Ω. Let **r** be a radial vector on the disk and let $\mathbf{\Omega}$ be a vector parallel to the axis of rotation, and directed toward the observer when the disk appears to be rotating counterclockwise. Show that the tangential velocity of a point on the disk is $\mathbf{V}_T = \mathbf{\Omega} \times \mathbf{r}$.

Combinations of the scalar and vector product appear often. For the mixed product we have the identity

$$\mathbf{A} \cdot (\mathbf{B} \times \mathbf{C}) = (\mathbf{A} \times \mathbf{B}) \cdot \mathbf{C} \tag{7}$$

which is described by saying that the dot and cross may be interchanged provided that the cross product is evaluated first. For the triple vector product, we find

$$\mathbf{A} \times (\mathbf{B} \times \mathbf{C}) = \mathbf{B}(\mathbf{A} \cdot \mathbf{C}) - \mathbf{C}(\mathbf{A} \cdot \mathbf{B}) \tag{8}$$

which is easily remembered as the *BAC–CAB* formula.

PROBLEM 5.1.9 Prove Eqs. (7) and (8). (Hint: Use the determinant definition of a cross product.)

PROBLEM 5.1.10 Let an object of mass m be lying on the disk of Problem 5.1.8. Show that the centrifugal force on the object is $\mathbf{F} = -m\mathbf{\Omega} \times (\mathbf{\Omega} \times \mathbf{r})$.

PROBLEM 5.1.11 The angular momentum **L** of an object of mass m with respect to a point is $\mathbf{L} = m\mathbf{r} \times \mathbf{v}$, where **r** is the radial vector from the point to the object with velocity **v**. What are the components of angular momentum parallel and orthogonal to $\mathbf{\Omega}$ for an object lying on the rotating disk of Problem 5.1.8?

PROBLEM 5.1.12 Let $\mathbf{a} \cdot \mathbf{b} = n$ and $\mathbf{a} \times \mathbf{c} = \mathbf{m}$. Show that $\mathbf{a} = (\mathbf{b} \times \mathbf{m} + n\mathbf{c})/\mathbf{c} \cdot \mathbf{b}$.

It is possible to write a great deal of inadvertent nonsense in vector notation, and the beginner must guard against doing so. In the first place, the order of operations is important. For example,

$$\mathbf{A} \times (\mathbf{B} \times \mathbf{C}) \neq (\mathbf{A} \times \mathbf{B}) \times \mathbf{C}$$

In the second place, care must be taken to ensure that undefined expressions are not written. Some examples are

$$\mathbf{A} \times \mathbf{B} \times \mathbf{C} \qquad (\mathbf{A} \cdot \mathbf{B}) \times \mathbf{C} \qquad (\mathbf{A} \cdot \mathbf{B}) \cdot \mathbf{C}$$

Another possibility for nonsense occurs because there is no operation of division in vector analysis. Hence something like $\mathbf{A}/\mathbf{B}$ is just not defined. Moreover, $\mathbf{A} \cdot \mathbf{B} = \mathbf{A} \cdot \mathbf{C}$ does not imply $\mathbf{B} = \mathbf{C}$ because for any vector $\mathbf{D}$ which is perpendicular to $\mathbf{A}$ we have $\mathbf{A} \cdot (\mathbf{C} + \mathbf{D}) = \mathbf{A} \cdot \mathbf{B}$ and both $\mathbf{C}$ and $\mathbf{C} + \mathbf{D}$ cannot equal $\mathbf{B}$.

The vector quantities that are of interest in meteorology, such as wind velocities or forces, are, like the scalar quantities such as pressure or temperature, generally functions of both time and the spatial coordinates. Thus we would write

$$\mathbf{A} = \mathbf{A}(x,y,z,t) = \mathbf{A}(\mathbf{x},t) \tag{9}$$

where

$$\mathbf{x} = \mathbf{i}x + \mathbf{j}y + \mathbf{k}z \tag{10}$$

Partial derivatives of a vector are defined with a difference and a limit exactly as they are for a scalar. When the unit vectors are constant, then the partial derivative becomes a sum of partial derivatives of the components of the vector. Using partial differentiation with respect to time as an example, we have

$$\frac{\partial \mathbf{A}}{\partial t} = \lim_{\Delta t \to 0} \frac{\mathbf{A}(\mathbf{x}, t + \Delta t) - \mathbf{A}(\mathbf{x},t)}{\Delta t} = \mathbf{i}\frac{\partial A_x}{\partial t} + \mathbf{j}\frac{\partial A_y}{\partial t} + \mathbf{k}\frac{\partial A_z}{\partial t} \tag{11}$$

when $\mathbf{i}$, $\mathbf{j}$, $\mathbf{k}$ are constant. Analogous formulas hold for spatial derivatives. With these results we may apply the expansion, Eq. (6) in Sec. 2.5, of the total derivative to obtain

$$\frac{d\mathbf{A}}{dt} = \frac{\partial \mathbf{A}}{\partial t} + u\frac{\partial \mathbf{A}}{\partial x} + v\frac{\partial \mathbf{A}}{\partial y} + w\frac{\partial \mathbf{A}}{\partial z} \tag{12}$$

and in particular, the vector velocity is expressed in accordance with Eq. (1) in Sec. 2.5 as

$$\mathbf{v} = \frac{d\mathbf{x}}{dt} = u\mathbf{i} + v\mathbf{j} + w\mathbf{k} \tag{13}$$

Thus $d\mathbf{v}/dt$ is the material acceleration of a parcel, and it represents a rate of change of velocity along the trajectory of the motion.

We mention now for completeness a case that will occur later. If the unit vectors, $\mathbf{i}_1$, $\mathbf{i}_2$, and $\mathbf{i}_3$ are functions of the independent variables

$$\mathbf{i}_j = \mathbf{i}_j(x,y,z,t) \qquad j = 1, 2, 3 \tag{14}$$

then we must differentiate them too, so that, for example,

$$\begin{aligned} \frac{d\mathbf{A}}{dt} &= \mathbf{i}_1\frac{dA_1}{dt} + A_1\frac{d\mathbf{i}_1}{dt} + \mathbf{i}_2\frac{dA_2}{dt} + A_2\frac{d\mathbf{i}_2}{dt} + \mathbf{i}_3\frac{dA_3}{dt} + A_3\frac{d\mathbf{i}_3}{dt} \\ &= \sum_{j=1}^{3}\left(\mathbf{i}_j\frac{dA_j}{dt} + A_j\frac{d\mathbf{i}_j}{dt}\right) \end{aligned} \tag{15}$$

PROBLEM 5.1.13 Show that

$$\frac{d}{dt}(\mathbf{A}\cdot\mathbf{B}) = \frac{d\mathbf{A}}{dt}\cdot\mathbf{B} + \mathbf{A}\cdot\frac{d\mathbf{B}}{dt}$$

$$\frac{d}{dt}(\mathbf{A}\times\mathbf{B}) = \frac{d\mathbf{A}}{dt}\times\mathbf{B} + \mathbf{A}\times\frac{d\mathbf{B}}{dt}$$

PROBLEM 5.1.14 Show that the rate of change of the specific kinetic energy $v^2/2$ may be written for $v = |\mathbf{v}|$ as

$$\frac{1}{2}\frac{dv^2}{dt} = \mathbf{v}\cdot\frac{d\mathbf{v}}{dt}$$

Describe a situation in which this vanishes when neither the velocity nor acceleration vanishes.

With the vector operations we have been considering, it is possible to simplify expressions of the type in Eq. (12) for the total derivative. We know now that for both scalars and vectors, the operator $d(\ )/dt$ may be expressed as

$$\frac{d}{dt} = \frac{\partial}{\partial t} + u\frac{\partial}{\partial x} + v\frac{\partial}{\partial y} + w\frac{\partial}{\partial z} \tag{16}$$

If we define, for cartesian coordinates, the vector differentiation operator

$$\nabla = \mathbf{i}\frac{\partial}{\partial x} + \mathbf{j}\frac{\partial}{\partial y} + \mathbf{k}\frac{\partial}{\partial z} \tag{17}$$

then Eq. (16) may be written more economically as

$$\frac{d}{dt} = \frac{\partial}{\partial t} + \mathbf{v}\cdot\nabla \tag{18}$$

The operator ∇ is called *del* or, more formally, the *gradient operator.* It behaves like a vector, but it is a differentiation operator, and it operates on quantities to its right. The gradient operator may be applied directly to a scalar function or may operate with a scalar or vector product on a vector.

For example, if $f = f(x,y,z,t)$ is a scalar function, we have

$$\nabla f = \mathbf{i}\frac{\partial f}{\partial x} + \mathbf{j}\frac{\partial f}{\partial y} + \mathbf{k}\frac{\partial f}{\partial z} \tag{19}$$

This quantity is called the *gradient of f.* For a vector function **A**, we can also define

$$\nabla\cdot\mathbf{A} = \frac{\partial A_x}{\partial x} + \frac{\partial A_y}{\partial y} + \frac{\partial A_z}{\partial z} \tag{20}$$

which is known as the *divergence* of **A**, and finally we can perform the operation

$$\nabla \times \mathbf{A} = \begin{vmatrix} \mathbf{i} & \mathbf{j} & \mathbf{k} \\ \dfrac{\partial}{\partial x} & \dfrac{\partial}{\partial y} & \dfrac{\partial}{\partial z} \\ A_x & A_y & A_z \end{vmatrix} \tag{21}$$

which is called the *curl* of **A**. When applied to the velocity vectors of air motion, the divergence is positive when the parcels are expanding, and the curl is nonzero if they are spinning.

Now let us turn our attention to ∇f, and for convenience we consider a function $f = f(x,y,z)$, which is independent of time. Then by the chain rule of differentiation we have

$$df = \frac{\partial f}{\partial x} dx + \frac{\partial f}{\partial y} dy + \frac{\partial f}{\partial z} dz = d\mathbf{x} \cdot \nabla f \tag{22}$$

From Eq. (19) we see that ∇f is a vector and its magnitude is

$$|\nabla f| = (\nabla f \cdot \nabla f)^{1/2} = \left[\left(\frac{\partial f}{\partial x}\right)^2 + \left(\frac{\partial f}{\partial y}\right)^2 + \left(\frac{\partial f}{\partial z}\right)^2 \right]^{1/2} \tag{23}$$

But in which direction does ∇f point?

Theorem The vector ∇f points in the direction of most rapid increase of f.

PROOF Consider a unit vector **h** that points in an arbitrary direction, and let h represent distance in this direction. Then df/dh will be the rate of change of f in the direction **h**. From Eq. (22) we have

$$\frac{df}{dh} = \frac{d\mathbf{x}}{dh} \cdot \nabla f \tag{24}$$

But since

$$\frac{d\mathbf{x}}{dh} = \lim_{\delta h \to 0} \frac{[\mathbf{x} + (\delta h)\mathbf{h}] - \mathbf{x}}{\delta h} = \mathbf{h} \tag{25}$$

we may use the definition of a scalar product to write

$$\frac{df}{dh} = \mathbf{h} \cdot \nabla f = \cos\theta \, |\nabla f| \tag{26}$$

in which θ is the angle between ∇f and the unit vector **h**.

Clearly, then, df/dh will be a maximum when θ is zero, that is, if **h** is a unit vector in the direction of ∇f. Therefore we may conclude that ∇f points in the direction of maximum rate of increase of f and that the derivative df/dh in this direction is in fact $|\nabla f|$, thus completing the proof. ////

Corollary The vector ∇f is perpendicular to the surface $f =$ const.

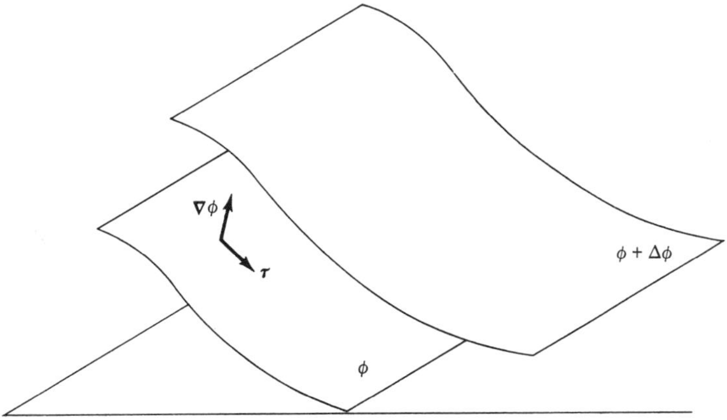

FIGURE 5.2
The normal, $\nabla\phi$, points in the direction of most rapidly increasing ϕ; the tangent $\boldsymbol{\tau}$ is a vector in the tangent plane to ϕ.

PROOF In Eq. (26), let $\mathbf{h} = \boldsymbol{\tau}$ be a vector tangent to the surface $f =$ const (see Fig. 5.2). Then $df/d\tau = 0$ and so

$$\boldsymbol{\tau} \cdot \nabla f = \cos\theta\,|\nabla f| = 0 \tag{27}$$

implies that $\theta = \pm\pi/2$. Therefore since $\boldsymbol{\tau}$ was tangent to $f =$ const and ∇f is perpendicular to every such $\boldsymbol{\tau}$, it must be perpendicular to $f =$ const. ////

The gradient may be used to demonstrate the geometric meaning of *advection*, which was defined in Eq. (7) in Sec. 2.5 for a field ψ as $-\mathbf{v} \cdot \nabla\psi$ and appears in the local rate of change as

$$\frac{\partial\psi}{\partial t} = -\mathbf{v} \cdot \nabla\psi + \frac{d\psi}{dt} \tag{28}$$

Consider the case shown in Fig. 5.3, in which we have a pattern of ψ on a chart and a wind field denoted by arrows. At point A, $\mathbf{v} \cdot \nabla\psi$ is negative and has magnitude $|\mathbf{v}|\,|\nabla\psi|$ so that according to Eq. (28) the local rate of change of ψ due to $\mathbf{v} \cdot \nabla\psi$ will be large. At point B, there will be no advective part in the local rate of change because $\mathbf{v}$ and $\nabla\psi$ are perpendicular.

For conservative quantities (like potential temperature, in some cases) the only local change is due to advection since then $d\psi/dt$ is zero. In other cases there may be a contribution to the local rate of change due to changes that occur in the moving parcels. For example, in the figure let ψ be temperature. Then at A we have warm advection. If the air flowing past A is absorbing radiation so that $dT/dt > 0$, then $\partial T/\partial t$ will be greater than the advection. If the parcels are emitting radiation so that $dT/dt < 0$, then $\partial T/\partial t$ will be less than the advection.

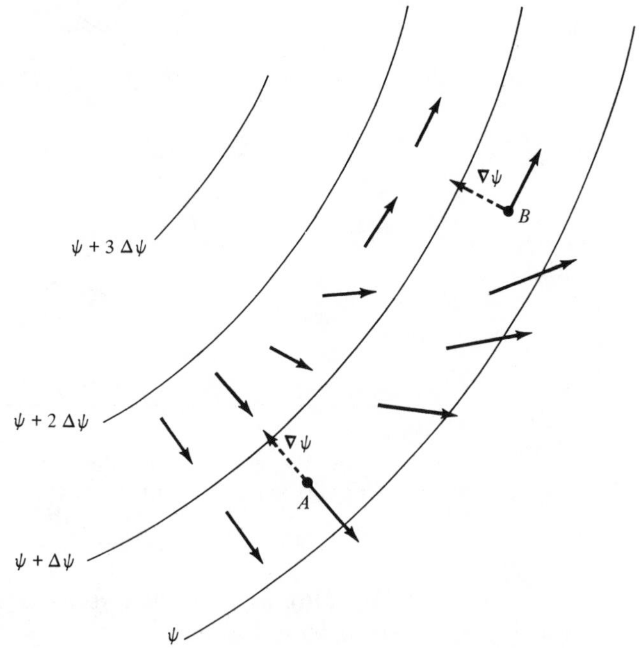

FIGURE 5.3
Advection of larger values of ψ at A; no advection of ψ at B.

Many of the changes in the weather at a point are due to advection. For example, the warm fall days of the mid-latitudes are due in part to warm advection from the south; the bitter cold days of the winter are due to advection of cold air from arctic to temperate regions. Thus the forecaster must be concerned about the advection of various quantities, and the changes that will occur in the air as it moves over his station in order to make an accurate forecast of the local change.

PROBLEM 5.1.15 Let the isotherms near station A be aligned east-west, with the temperature decreasing 1°C in every 100 km to the north. What is the local change of temperature due to advection at A for a south wind of 10 m/s, a north wind of 15 m/s, and an east wind of 20 m/s?

PROBLEM 5.1.16 Let $\boldsymbol{\tau}$ be a unit vector tangent to a parcel trajectory, let s be distance along the trajectory, and let the parcel speed be $v(s)$. Show that the advection of ψ due to the parcel motion is

$$A = -v\frac{\partial \psi}{\partial s}$$

and that the parcel acceleration is

$$\begin{aligned}\frac{d\mathbf{v}}{dt} &= \frac{\partial \mathbf{v}}{\partial t} + v\frac{\partial \mathbf{v}}{\partial s} \\ &= v\left(\frac{\partial \boldsymbol{\tau}}{\partial t} + v\frac{\partial \boldsymbol{\tau}}{\partial s}\right) + \boldsymbol{\tau}\left(\frac{\partial v}{\partial t} + v\frac{\partial v}{\partial s}\right)\end{aligned}$$

Show that the first term in the second line represents the effects of trajectory curvature.

Numerous identities are useful in manipulations with the del operator. We collect here the most important of them, leaving the proofs as exercises, for now and for Sec. 5.5.

For arbitrary vector functions **A** and **B**, a scalar function ϕ, and the position vector **x**, we have

$$\nabla \times \nabla\phi = 0 \tag{29}$$

$$\nabla \cdot (\nabla \times \mathbf{A}) = 0 \tag{30}$$

$$\nabla \cdot \mathbf{x} = 3 \tag{31}$$

$$\nabla \times \mathbf{x} = 0 \tag{32}$$

$$\nabla \cdot (\phi\mathbf{A}) = \phi\nabla \cdot \mathbf{A} + (\mathbf{A} \cdot \nabla)\phi \tag{33}$$

$$\begin{aligned}\nabla \times (\phi\mathbf{A}) &= \phi\nabla \times \mathbf{A} + (\nabla\phi) \times \mathbf{A} \\ &= \phi\nabla \times \mathbf{A} - \mathbf{A} \times (\nabla\phi)\end{aligned} \tag{34}$$

$$\nabla \cdot (\mathbf{A} \times \mathbf{B}) = \mathbf{B} \cdot (\nabla \times \mathbf{A}) - \mathbf{A} \cdot (\nabla \times \mathbf{B}) \tag{35}$$

$$\nabla(\mathbf{A} \cdot \mathbf{B}) = (\mathbf{A} \cdot \nabla)\mathbf{B} + (\mathbf{B} \cdot \nabla)\mathbf{A} + \mathbf{A} \times (\nabla \times \mathbf{B}) + \mathbf{B} \times (\nabla \times \mathbf{A}) \tag{36}$$

$$\nabla \times (\mathbf{A} \times \mathbf{B}) = \mathbf{A}(\nabla \cdot \mathbf{B}) + (\mathbf{B} \cdot \nabla)\mathbf{A} - \mathbf{B}(\nabla \cdot \mathbf{A}) - (\mathbf{A} \cdot \nabla)\mathbf{B} \tag{37}$$

$$\nabla \times (\nabla \times \mathbf{A}) = \nabla(\nabla \cdot \mathbf{A}) - \nabla^2\mathbf{A} \tag{38}$$

$$\mathbf{A} \times (\nabla \times \mathbf{B}) - (\mathbf{A} \times \nabla) \times \mathbf{B} = \mathbf{A}\nabla \cdot \mathbf{B} - (\mathbf{A} \cdot \nabla)\mathbf{B} \tag{39}$$

The rules that are helpful when operating with del are:

1 Each term of an expression must have the same vector character as all the others, just as each term in an equation must have the same physical dimensions.
2 Each factor or term that appears immediately on the right of del must be differentiated because the partial derivatives in del operate on each of these factors and terms, just as an ordinary partial derivative would.
3 If the unit vectors vary in space, they must be differentiated too.

PROBLEM 5.1.17 Prove the identities (29) to (34) by direct evaluation.

5.2 INTEGRALS OF VECTOR QUANTITIES

All the significance and subtleties of the integration of scalar quantities are also associated with integration of vector quantities. We define the integral of a vector quantity by applying the integral to its scalar components, and the analogs of integration by parts we shall discuss will reveal the physical significance of the divergence and curl of velocity fields.

For a vector field $\mathbf{A} = \mathbf{A}(\mathbf{x},t)$, the integral in any number of dimensions, definite or indefinite, is defined by

$$\int \mathbf{A} = \mathbf{i}\int A_x + \mathbf{j}\int A_y + \mathbf{k}\int A_z \tag{1}$$

Two particular forms appear. The first is the *line integral*

$$I = \int_C \mathbf{A} \cdot \boldsymbol{\tau}\, ds \tag{2}$$

in which $\boldsymbol{\tau}$ is tangent to a curve C and s is arc length along C. The positive direction is generally taken to be counterclockwise.

The second type of integral involves a surface S. Either the surface completely encloses a volume, or it does not. In the first case we shall take a unit vector $\boldsymbol{\eta}$, a normal to the surface, to be directed from inside the volume to the outside. Thus $\boldsymbol{\eta}$ is an *exterior unit normal vector* (Fig. 5.4). When S does not enclose a volume, we draw a curve C on S, let the tangent to C be $\boldsymbol{\tau}$, and choose $\boldsymbol{\eta}$ so that the vector

$$\mathbf{n} = \boldsymbol{\tau} \times \boldsymbol{\eta} \tag{3}$$

is directed from the central part of S toward its outside boundary. For the quasi-horizontal surfaces of meteorology, we may take $\boldsymbol{\eta}$ to be upward, and then $\boldsymbol{\tau}$ will point in the counterclockwise direction. Thus the *surface integral* is

$$J = \int_S \mathbf{A} \cdot \boldsymbol{\eta}\, d\sigma \tag{4}$$

in which $d\sigma$ is the differential element of area on S and we use the notation $\int_S (\ )\, d\sigma = \iint_S (\ )\, d\sigma$.

Before proceeding, we consider one of those subtleties of integration theory that is often overlooked in practical applications. Specifically, we have been writing integrals over volumes and surfaces as distinct entities without saying how they are to be evaluated. The usual trick, of course, is to convert such an integral into an iterated integral. Thus if the rectangle R is bounded by $a_1 \leqslant x \leqslant a_2$, $b_1 \leqslant y \leqslant b_2$, we would write

$$\iint_R f(x,y)\, dR = \int_{a_1}^{a_2}\int_{b_1}^{b_2} f(x,y)\, dx\, dy = \int_{a_1}^{a_2}\left[\int_{b_1}^{b_2} f(x,y)\, dy\right] dx$$
$$= \int_{b_1}^{b_2}\left[\int_{a_1}^{a_2} f(x,y)\, dx\right] dy \tag{5}$$

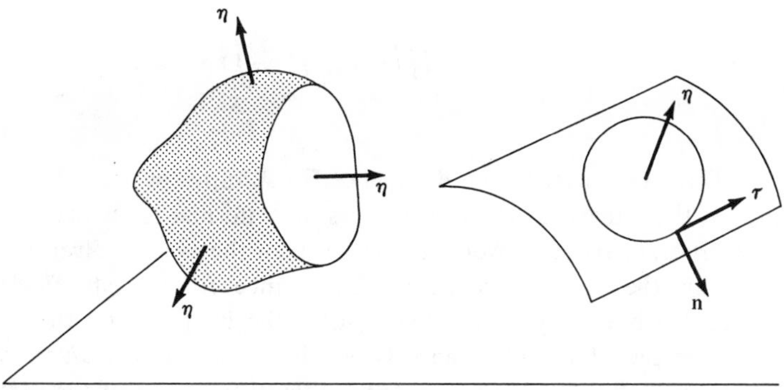

FIGURE 5.4
Illustration of the exterior normal vector $\boldsymbol{\eta}$ on a volume and the relation between the normal $\boldsymbol{\eta}$ on a surface, the tangent $\boldsymbol{\tau}$, and the exterior vector **n**.

A sufficient condition for Eq. (5) to be valid when we have bounded and closed domains of integration is that $f(x,y)$ be continuous. Another sufficient condition for Eq. (5) to be true is that either

$$\int_{a_1}^{a_2}\left[\int_{b_1}^{b_2}|f(x,y)|\,dy\right]dx<\infty \qquad \text{or} \qquad \int_{b_1}^{b_2}\left[\int_{a_1}^{a_2}|f(x,y)|\,dx\right]dy<\infty \tag{6}$$

This test works for infinite domains as well and is of frequent value in certain specialties of atmospheric science.

An additional use of these theorems derives from the fact that when Eq. (5) is valid, for either finite or infinite domains, the order of iterated integration is obviously immaterial, and thus the conditions given also justify interchanges in the order of integration.

In the remainder of the section we shall need to interpret certain volume and surface integrals as iterated ones, and the restrictions on continuity given in the theorems are intended to permit this interpretation. Whenever these theorems are used, we must be sure that the conditions are satisfied.

5.2.1 The Divergence Theorem

Recognition that an integrand is an exact derivative permits immediate evaluation of integrals of scalar quantities. The divergence theorem, and Stokes' theorem to follow, are vector analogs of this process.

Theorem The divergence theorem If **A** is a vector field with continuous first spatial derivatives, and if V is a suitably behaved volume bounded by a surface S with exterior unit normal vector $\boldsymbol{\eta}$, then

$$\iiint_V \nabla \cdot \mathbf{A}\, dV = \iint_S \mathbf{A} \cdot \boldsymbol{\eta}\, d\sigma \tag{7}$$

////

Thus no matter how a divergence $\nabla \cdot \mathbf{A}$ varies over a volume, its integral depends only on the integral of the components of $\mathbf{A}$ normal to the surface of the boundary. If these integrate to zero over the surface, then the average divergence must also be zero.

This theorem has particular importance for velocity fields $\mathbf{v}$. If the net flow through the boundary S is outward, then the integral over the surface is positive and so the average of the divergence is positive–a case called *divergent* motion. If the net flow through the boundary is inward, then the average of the divergence is negative–a case called *convergent* motion.

To illustrate the significance of these results, suppose that the volume V is always composed of the same fluid particles so that the boundary S is frozen into the motion field and moves with the fluid. We choose a small spherical volume of radius r with vector $\mathbf{r}$ from the center of the volume to the bounding surface and apply the mean-value theorem to Eq. (7). Then when V is small enough,

$$V\nabla \cdot \mathbf{v} = \iiint_V \nabla \cdot \mathbf{v}\, dV = \iint_S \mathbf{v} \cdot \boldsymbol{\eta}\, d\sigma = \iint_S \frac{d\mathbf{r}}{dt} \cdot \boldsymbol{\eta}\, d\sigma \tag{8}$$

where $\mathbf{r}$ is the position vector. Applying the mean-value theorem to the last integral over the surface of area A, we find

$$\iint_S \frac{d\mathbf{r}}{dt} \cdot \boldsymbol{\eta}\, d\sigma = \frac{dr}{dt}A = 4\pi r^2 \frac{dr}{dt} = \frac{4\pi}{3}\frac{dr^3}{dt} = \frac{dV}{dt} \tag{9}$$

and so

$$\nabla \cdot \mathbf{v} = \frac{1}{V}\frac{dV}{dt} \tag{10}$$

Hence we have shown that the divergence of a velocity field is directly equal to the fractional rate of expansion of the volume occupied by the fluid particles. In other words, with divergent motion the volume occupied by the fluid is expanding. With convergent motion, it is contracting. Recalling from Eq. (13) in Sec. 3.1 that the mechanical power term in the first law of thermodynamics, Eq. (1) in Sec. 3.1, was $N = -p\dot{V}$, we can perceive the significance of Eq. (10) to atmospheric science, because we are obviously on the verge of linking a property of the motion fields to the thermodynamics of the fluid. Furthermore we are within one step of deriving one of the basic equations of fluid motion.

To do so, let us recall that if M is the mass of the fluid in the small volume, then the density is $\rho = M/V$. The basic principle of conservation of mass requires that the fluid parcel enclosed in V cannot lose or gain mass because the boundaries of V are frozen into the fluid motion. Thus $dM/dt = 0$ and so

$$\frac{1}{\rho}\frac{d\rho}{dt} = \frac{d}{dt}\ln\frac{M}{V} = -\frac{1}{V}\frac{dV}{dt} \tag{11}$$

With this result, Eq. (10) becomes

$$\frac{d\rho}{dt} + \rho\nabla \cdot \mathbf{v} = 0 \tag{12}$$

and is known as the *equation of mass continuity*. It states that because the mass of a parcel is constant, the density must decrease if the flow diverges and conversely.

Now that some of the significance of the divergence theorem has been explained, let us prove it. First we consider a rectangular box bounded by $a_1 \leqslant x \leqslant a_2$, $b_1 \leqslant y \leqslant b_2$, $c_1 \leqslant z \leqslant c_2$, and by direct computation we find that

$$\begin{aligned}\iiint_R \nabla \cdot \mathbf{A}\, dR &= \int_{a_1}^{a_2}\int_{b_1}^{b_2}\int_{c_1}^{c_2}\left(\frac{\partial A_x}{\partial x} + \frac{\partial A_y}{\partial y} + \frac{\partial A_z}{\partial z}\right) dx\, dy\, dz \\ &= \int_{b_1}^{b_2}\int_{c_1}^{c_2} [A_x(a_2,y,z) - A_x(a_1,y,z)]\, dy\, dz \\ &\quad + \int_{a_1}^{a_2}\int_{c_1}^{c_2} [A_y(x,b_2,z) - A_y(x,b_1,z)]\, dx\, dz \\ &\quad + \int_{a_1}^{a_2}\int_{b_1}^{b_2} [A_z(x,y,c_2) - A_z(x,y,c_1)]\, dx\, dy \end{aligned} \tag{13}$$

Each of the last three integrals is the difference of two surface integrals over the sides of the box perpendicular to the direction in which the integration was performed, and the signs are the same as would be obtained by taking the scalar product of **A** and the exterior unit normals on those two surfaces. Thus for any box R,

$$\iiint_R \nabla \cdot \mathbf{A}\, dR = \iint_{S(R)} \mathbf{A} \cdot \boldsymbol{\eta}\, d\sigma(R) \tag{14}$$

If we have any volume that can be represented or approximated as the sum of many small boxes, we can apply the theorem to each little box separately. But where the sides of two little boxes touch, the two exterior unit normal vectors will be in opposite directions so that the surface integral over the side of one box will cancel the integral over the adjacent side of the next box. Thus only the integral over the outside of the outside boxes will remain and we have established Eq. (7).

If the volume of interest had holes in the middle, we could apply the theorem to the holes and subtract the result from the result we obtain for the entire region including the holes and the volume. In this case we would note that the negative direction for the exterior unit normal vector of the holes is in fact the exterior unit normal vector for the volume, and so Eq. (7) holds for this case too when the surface integral is extended over the exterior boundary of the volume and the interior boundary delineating the holes.

Finally, we must consider the situation in which the volume is infinite, as would be true for integrations over the entire atmosphere. Then we shall usually have the volume integral of either $\nabla \cdot \rho \mathbf{v}$ or $\nabla \cdot p\mathbf{v}$. Applying the divergence theorem in this case to the volume between the earth's surface and a spherical surface at altitude Z, we obtain the difference of the integrals over these two surfaces. But the integrand of the integral over the upper surface may be assumed to vanish as Z goes to infinity (see Problem 2.3.5). The lower integral also vanishes, since $\boldsymbol{\eta} \cdot \mathbf{v}$ must be zero at the earth's surface.

PROBLEM 5.2.1 Assuming for now that

$$\frac{\partial}{\partial t}\int_{\text{atm}} \rho\, dV = \int_{\text{atm}} \frac{\partial \rho}{\partial t}\, dV$$

show that the mass of the entire atmosphere is constant.

PROBLEM 5.2.2 Prove *Green's theorem*

$$\iiint_V (f\,\nabla^2 g - g\,\nabla^2 f)\, dV = \iint_S (f\,\nabla g - g\,\nabla f) \cdot \boldsymbol{\eta}\, d\sigma$$

(Hint: Consider the identity $\nabla \cdot f\nabla g = f\nabla^2 g + \nabla f \cdot \nabla g$.)

PROBLEM 5.2.3 Define the horizontal divergence by $\nabla_H \cdot \mathbf{v} = \partial u/\partial x + \partial v/\partial y$. Show that if A is a material area, then

$$\nabla_H \cdot \mathbf{v} = \frac{1}{A}\frac{dA}{dt}$$

PROBLEM 5.2.4 Prove that the integral of $\nabla\phi \cdot (\nabla \times \mathbf{A})$ over a volume can always be reduced to an integral over the surface of the volume.

5.2.2 Stokes' Theorem: Integration of the Curl

We come now to the integration formula that applies to integrals of the curl (or $\nabla\times$) of vector fields.

Theorem Stokes' theorem Let **B** be a vector field, with continuous first spatial derivatives, defined on a surface S, not enclosing a volume. Let $\boldsymbol{\eta}$ be the normal to S, and $\boldsymbol{\tau}$ the tangent vector to the curve C bounding S. With the orientation convention that $\boldsymbol{\tau} \times \boldsymbol{\eta}$ is directed outside S, we have

$$\iint_S \boldsymbol{\eta} \cdot (\nabla \times \mathbf{B})\, d\sigma = \int_C \boldsymbol{\tau} \cdot \mathbf{B}\, ds \tag{15}$$

////

Once again this theorem has important physical implications when **B** is the velocity field **v**. The line integral on the right sums the component of **v** parallel to the curve C at each point, and thus tells us whether, on the average, there is a tendency for the flow to be circulating around C; hence we call this integral the *circulation around* C. If the circulation is positive, then there is a net flow in the positive direction around C. The theorem says that this circulation is the product of the area of the surface times the average, over the surface, of the normal component of $\nabla \times \mathbf{v}$. We call $\nabla \times \mathbf{v}$ the *vorticity vector* of the motion, and $\boldsymbol{\eta} \cdot (\nabla \times \mathbf{v})$ is the component of the vector parallel to the normal $\boldsymbol{\eta}$. In meteorology we usually consider horizontal surfaces, so that $\boldsymbol{\eta} \cdot (\nabla \times \mathbf{v})$ is the vertical component of the vorticity. When the component of vorticity normal to a surface has a nonzero average, then there will be a circulation around the bounding curve, and conversely.

Thus evaluation of the quantity $\nabla \cdot \mathbf{v}$ tells us whether the motion is contracting or expanding whereas $\nabla \times \mathbf{v}$ reveals whether the fluid is spinning or circulating. These two quantities are thus central to the study of fluid motion and the structure of flow patterns; we shall become intimately acquainted with them as we proceed.

Although Stokes' theorem can be proved directly in a manner analogous to the divergence theorem, we will leave such a proof for an exercise and obtain it here as a consequence of the previous theorem. This has the advantage of providing experience with vector operations, even though it is perhaps more complicated than the direct approach.

Clearly we can apply the divergence theorem to a small plane surface S_i, in the form

$$\iint_{S_i} \nabla_{S_i} \cdot \mathbf{B}\, d\sigma = \int_C \mathbf{B} \cdot \mathbf{n}\, ds \tag{16}$$

where $\mathbf{n} = \boldsymbol{\tau} \times \boldsymbol{\eta}$ and ∇_{S_i} is restricted to operate in the surface and thus has no component normal to S_i. From the vector identity (35) in Sec. 5.1 we find

$$\nabla_{S_i} \cdot (\boldsymbol{\eta} \times \mathbf{B}) = \nabla \cdot (\boldsymbol{\eta} \times \mathbf{B}) = -\boldsymbol{\eta} \cdot (\nabla \times \mathbf{B}) + \mathbf{B} \cdot (\nabla \times \boldsymbol{\eta}) \tag{17}$$

and so

$$\begin{aligned}\iint_{S_i} \boldsymbol{\eta} \cdot (\nabla \times \mathbf{B})\, d\sigma &= -\iint_{S_i} \nabla_{S_i} \cdot (\boldsymbol{\eta} \times \mathbf{B})\, d\sigma + \iint_{S_i} \mathbf{B} \cdot (\nabla \times \boldsymbol{\eta})\, d\sigma \\ &= -\int_{C_i} \mathbf{n} \cdot (\boldsymbol{\eta} \times \mathbf{B})\, dS + \iint_{S_i} \mathbf{B} \cdot (\nabla \times \boldsymbol{\eta})\, d\sigma \end{aligned} \tag{18}$$

Now an interchange of the cross and dot products in the first integral on the right gives

$$\mathbf{n} \cdot (\boldsymbol{\eta} \times \mathbf{B}) = (\mathbf{n} \times \boldsymbol{\eta}) \cdot \mathbf{B} = -\boldsymbol{\tau} \cdot \mathbf{B} \tag{19}$$

so we must show that the second integral vanishes.

As shown in Sec. 5.1, $\nabla f/|\nabla f|$ is the unit normal vector to the surface with equation $f(x,y,z) = \text{const}$. Let the surface S_i have an equation of the form $z + \alpha x + \beta y = \text{const}$ so that $\boldsymbol{\eta} = \mathbf{k} + \alpha\mathbf{i} + \beta\mathbf{j}/(1 + \alpha^2 + \beta^2)^{1/2}$. Thus in the last integral on the right we have $\nabla \times \boldsymbol{\eta} = 0$, and so we have shown that Stokes' theorem is valid on a plane surface S_i. If we approximate the arbitrary surface S by a collection of little plane surfaces each tangent at some point to S, then we may sum Eq. (18) applied to each plane to obtain Eq. (15), because the line integrals cancel each other at every interior intersection of the approximating plane surfaces and have only the line integral around the exterior of S. Thus we have proved Stokes' theorem.

The basic forms of the divergence theorem and Stokes' theorem can be generalized to apply to more cases than those already considered; the proofs depend upon use of an arbitrary, but constant, vector $\mathbf{c}$.

For the divergence theorem we have

$$\int_V \nabla \circ \mathbf{A}\, dV = \int_S \boldsymbol{\eta} \circ \mathbf{A}\, d\sigma \tag{20}$$

in which $\mathbf{A}$ may be a scalar or a vector. When $\mathbf{A}$ is a scalar, the operation $\circ$ denotes ordinary multiplication [proof: let $\mathbf{A} = f\mathbf{c}$ in Eq. (7)]; when $\mathbf{A}$ is a vector, the operation $\circ$ may be a dot [proof: Eq. (7)] or a cross [proof: let $\mathbf{A} = \mathbf{B} \times \mathbf{c}$ in Eq. (7)], provided the same operation is used in both places $\circ$ occurs.

For Stokes' theorem, we similarly let $\mathbf{B} = f\mathbf{c}$ and $\mathbf{B} = \mathbf{A} \times \mathbf{c}$ to obtain

$$\int_S (\boldsymbol{\eta} \times \nabla) \circ \mathbf{A}\, d\sigma = \int_C \boldsymbol{\tau} \circ \mathbf{A}\, ds \tag{21}$$

We can now easily prove a result of some importance. Consider S to be a connected closed surface enclosing a volume. We may draw a curve C around the volume separating S into two parts, S_1 and S_2. But then

$$\begin{aligned}\int_S (\boldsymbol{\eta} \times \nabla) \circ \mathbf{A}\, d\sigma &= \left(\int_{S_1} + \int_{S_2}\right)(\boldsymbol{\eta} \times \nabla) \circ \mathbf{A}\, d\sigma \\ &= \int_C (\boldsymbol{\tau}_1 \circ \mathbf{A} + \boldsymbol{\tau}_2 \circ \mathbf{A})\, ds = 0 \end{aligned} \tag{22}$$

The last integral vanishes because at C the normals $\boldsymbol{\eta}_1$ and $\boldsymbol{\eta}_2$ will be in the same direction for both S_1 and S_2, but the exterior vectors $\mathbf{n}_1$ and $\mathbf{n}_2$ are in opposite directions and so $\boldsymbol{\tau}_1 = -\boldsymbol{\tau}_2$. We shall have an application of this theorem when we study vorticity in Chap. 10, but it obviously says the average vorticity normal to a closed surface vanishes.

PROBLEM 5.2.5 Give a proof of Stokes' theorem constructed in a direct manner analogous to that used for the divergence theorem.

PROBLEM 5.2.6 Prove analytically that $(\mathbf{A} \times \nabla) \cdot \mathbf{B} = \mathbf{A} \cdot (\nabla \times \mathbf{B})$ and hence that Stokes' theorem may be written

$$\int_S (\eta \times \nabla) \cdot \mathbf{B}\, d\sigma = \int_C \tau \cdot \mathbf{B}\, ds$$

PROBLEM 5.2.7 Suppose you put a little cream in a cup of coffee and stir it to produce whirls of cream. Despite the whirls you see, prove that the average vorticity component normal to the coffee surface is zero. (Hint: Friction prevents any motion at the sides and bottom of the cup.)

PROBLEM 5.2.8 Prove the generalized formulas (20) and (21).

5.3 DIFFERENTIATION OF INTEGRALS; THE TRANSPORT THEOREM

The study of atmospheric motion often involves integrals or averages computed over part or all of the atmosphere. Such integrals are functions of time, and we will want to investigate their temporal rates of change. The attempt to establish the necessary results on a reasonably secure foundation will lead us briefly into the study of the parametric representation of curves, surfaces, and volumes. Although this may seem to be a digression in a book on atmospheric motion, the powerful concepts of this theory yield effective methods for deriving the necessary results. Furthermore, much of meteorology is permeated with the study of surfaces composed of constant values of one variable or another, and a presentation of the mathematical properties of such surfaces will aid us in understanding their physics. Finally, these concepts are essential for comprehension of the significance of the last section of the chapter.

We begin with Leibniz' rule, which provides the prototype for the results of this section. If the function $f(x,t)$ is continuously differentiable with respect to time and if the derivatives $\alpha'(t)$ and $\beta'(t)$ exist, then the integral

$$F(t) = \int_{\alpha(t)}^{\beta(t)} f(x,t)\, dx \tag{1}$$

has the derivative

$$\frac{\partial F}{\partial t} = \int_{\alpha(t)}^{\beta(t)} \frac{\partial f}{\partial t}\, dx + f(\beta(t),t)\beta'(t) - f(\alpha(t),t)\alpha'(t) \tag{2}$$

PROBLEM 5.3.1 Justify Leibniz' rule. [Hint: Apply the chain rule to $F(t) = G(t,\alpha(t),\beta(t))$.]

5.3.1 The Transport Theorem and Its Consequences

The case of integration over all three spatial variables is more complicated than the one-dimensional case, but it occurs frequently in meteorology, most particularly in the study of the general circulation and the global motions of the atmosphere.

We define the integral

$$F(t) = \int_{V(t)} f(x,y,z,t)\, dV \tag{3}$$

in which the volume $V(t)$ may be variable in time in the sense that its boundary may change shape, expand, or contract, or the volume may move from one place to another. This integral can have two sources of change. The first is that f may change within the volume, the second is that the boundary may move so that new values of f come inside the volume and so that values which were inside are left behind. If each point on the boundary S has velocity $\mathbf{w}$, then we have

$$\frac{\partial F}{\partial t} = \int_{V(t)} \frac{\partial f}{\partial t}\, dV + \int_{S} f(S)\boldsymbol{\eta} \cdot \mathbf{w}\, d\sigma \tag{4}$$

as illustrated in Fig. 5.5. The divergence theorem thus implies that

$$\frac{\partial F}{\partial t} = \int_{V(t)} \frac{\partial f}{\partial t}\, dV + \int_{\underline{V(t)}} \nabla \cdot f\mathbf{w}\, dV \tag{5}$$

where $\mathbf{w}$ is now the velocity of the points inside the volume. A rigorous proof of this result is given later in the section.

In many situations we shall deal with *material curves*, *surfaces*, or *volumes*, geometric figures that are composed of groups of molecules that are part of the material of the fluid itself. In other words, material curves, surfaces, or volumes are always composed of the same parcels of air and they move with the motion field. An obvious consequence of this definition is that a material volume always contains the same air and thus the mass inside the volume cannot change (unless we encounter some mathematical disaster like condensation so that rain may fall out of the volume).

If the volume $V(t)$ is a material one, then its boundary moves with the velocity $\mathbf{v}$ of the fluid and we may expand the second integral to obtain

$$\frac{\partial F}{\partial t} = \int_{V(t)} \left(\frac{\partial f}{\partial t} + \mathbf{v} \cdot \nabla f\right) dV + \int_{V(t)} f\nabla \cdot \mathbf{v}\, dV = \int_{V(t)} \left(\frac{df}{dt} + f\nabla \cdot \mathbf{v}\right) dV \tag{6}$$

This result is frequently called the *transport theorem.*

In meteorological applications, it is common to have density appear as a factor in integrals. To determine the consequences of this, we put $f = \rho g$, where g is now an

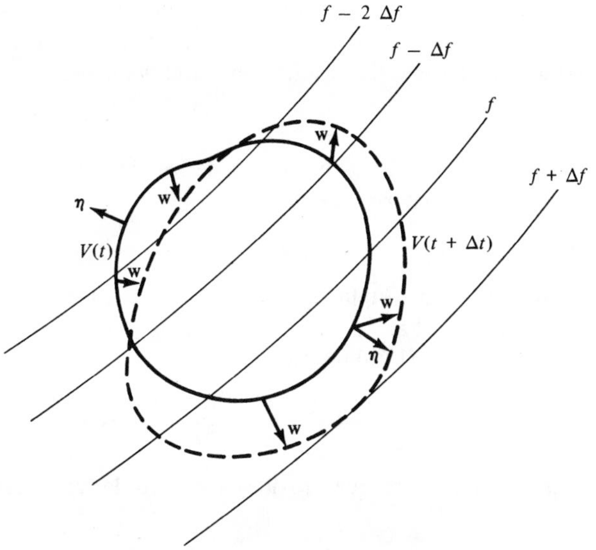

FIGURE 5.5
Illustration of the contribution of boundary motion to Eq. (4) in Sec. 5.3. The values of f where $\mathbf{w} \cdot \eta > 0$ are added, those where $\mathbf{w} \cdot \eta < 0$ are subtracted.

arbitrary function, and substitute in Eq. (5). Then use of the equation of continuity (12) in Sec. 5.2 gives

$$\frac{\partial F}{\partial t} = \int_{V(t)} \rho \frac{dg}{dt}\, dV + \int_{V(t)} \nabla \cdot [\rho g(\mathbf{w} - \mathbf{v})]\, dV \tag{7}$$

For a material volume only the first term remains. Only the first term is applicable when we integrate over the entire atmosphere as well, because we may assume that the entire atmosphere is a material volume.

In general, then, for material volumes V_M and for the entire atmosphere we have

$$\frac{\partial}{\partial t}\int_{V_M} \rho g\, dV = \int_{V_M} \rho \frac{dg}{dt}\, dV \tag{8}$$

where it is obvious that the function g may be either a scalar or a vector. This is a particularly important result for atmospheric science because it greatly simplifies the computation of rates of change of important integrals.

PROBLEM 5.3.2 Prove Eq. (7).

*PROBLEM 5.3.3 Let $\int_V f\,dV$ vanish for every volume. Show that $f = 0$ except on a set of zero volume and that $f = 0$ everywhere if f is continuous. (Hint: Develop a contradiction.) Extend the result to material volumes.

PROBLEM 5.3.4 Use Eq. (6) to derive the equation of continuity by applying it to the mass in a material volume.

PROBLEM 5.3.5 Suppose we are able to reason from basic principles that a quantity $F(t) = \int_V f\,dV$ has a rate of change given for every material volume by $\partial F(t)/\partial t = \int_V Q\,dV$. Show that

$$\frac{\partial f}{\partial t} + \nabla \cdot f\mathbf{v} = Q$$

Note that this technique provides a way to derive dynamic equations.

5.3.2 Parametric Representation of Curves and Surfaces

It is frequently useful to employ a general representation that specifies the coordinates of a line or surface as a function of some other independent coordinates. These representations are called *parametric equations*.

In general, the parametric representation of a curve in a three-dimensional space is

$$x = \phi(\xi) \qquad y = \psi(\xi) \qquad z = \theta(\xi) \tag{9}$$

where we now take ξ to be in some domain D. We shall have a lot of differentiating to do, and we adopt the convention that

$$x_{,\xi} = \frac{\partial x}{\partial \xi} \tag{10}$$

Of particular interest is that the line integral I along a curve given by Eq. (9) is

$$I = \int_C f(x,y,z)\,ds = \int_D f(x(\xi),y(\xi),z(\xi))\left[(\phi_{,\xi})^2 + (\psi_{,\xi})^2 + (\theta_{,\xi})^2\right]^{1/2} d\xi \tag{11}$$

It is obvious that we are obtaining the same values of f in each case; the distance differential ds is represented by

$$ds = \left[(\phi_{,\xi})^2 + (\psi_{,\xi})^2 + (\theta_{,\xi})^2\right]^{1/2} d\xi = \left|\frac{\partial \mathbf{x}}{\partial \xi}\right| d\xi \tag{12}$$

where $\mathbf{x}$ is the vector

$$\mathbf{x} = \phi(\xi)\mathbf{i} + \psi(\xi)\mathbf{j} + \theta(\xi)\mathbf{k} = (\phi, \psi, \theta) \tag{13}$$

If the parametric representation is a function of time so that the curve may move in the three-dimensional space, then we would have

$$\mathbf{x}(t) = \phi(\xi,t)\mathbf{i} + \psi(\xi,t)\mathbf{j} + \theta(\xi,t)\mathbf{k} \tag{14}$$

and so the velocity of a point on the curve would be

$$\mathbf{u} = \left(\frac{\partial \mathbf{x}}{\partial t}\right)_\xi \tag{15}$$

PROBLEM 5.3.6 Show that a circle on an (x,y) plane with radius r may be represented as a function of ξ where $0 \leqslant \xi \leqslant 1$ by the equations $x = r \cos 2\pi\xi$, $y = r \sin 2\pi\xi$.

PROBLEM 5.3.7 Show that Eq. (12) is correct. [Hint: The points $\mathbf{x}_0 = \mathbf{x}(\xi_0)$ and $\mathbf{x}_1 = \mathbf{x}(\xi_1)$ are on the curve and the distance between them is $(\Delta s)^2 = |\mathbf{x}_1 - \mathbf{x}_0|^2$. Use a Taylor series.]

PROBLEM 5.3.8 Let the curve C be specified by Eq. (9) for $0 \leqslant \xi \leqslant 1$. What is its length?

*PROBLEM 5.3.9 The vector $\mathbf{x}$ given by Eq. (13) traces out the curve. Show that the unit tangent vector to the curve is $\boldsymbol{\tau} = (\partial \mathbf{x}/\partial \xi)/|\partial \mathbf{x}/\partial \xi|$.

The parametric equations for a surface are given by

$$x = \phi(\xi,\zeta) \qquad y = \psi(\xi,\zeta) \qquad z = \theta(\xi,\zeta) \tag{16}$$

We have shown in Sec. 5.1 that when ξ and ζ are eliminated to obtain an equation for the surface of the form $F(x,y,z) = \text{const}$, then the normal to the surface is given by $\eta = \nabla F/|\nabla F|$.

Now we consider surfaces which do not become perpendicular to the (x,y) plane at any point. The more steeply the surface is inclined to the vertical, the greater the ratio of its actual area to the projection on the (x,y) plane. It is easily seen that this ratio is in fact $\sec\theta$, where θ is the angle between an element of the surface and the (x,y) plane (Fig. 5.6). For this case we may write the equation of the surface in the form

$$z - h(x,y,c) = 0 \tag{17}$$

where c is a constant. Thus the normal vector has components

$$\eta = \left(-\mathbf{i}\frac{\partial h}{\partial x} - \mathbf{j}\frac{\partial h}{\partial y} + \mathbf{k}\right)\left[1 + \left(\frac{\partial h}{\partial x}\right)^2 + \left(\frac{\partial h}{\partial y}\right)^2\right]^{-1/2} \tag{18}$$

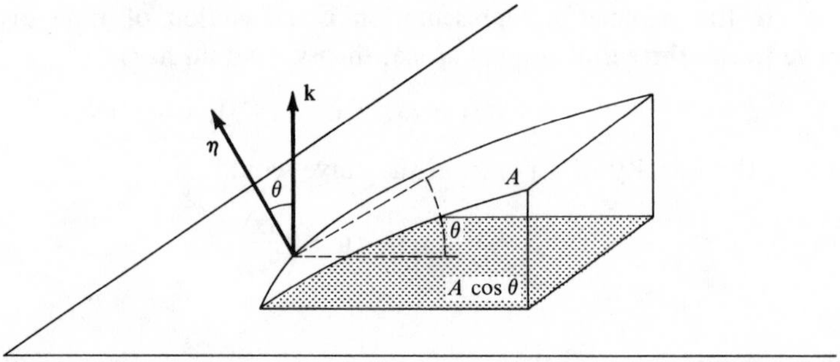

FIGURE 5.6
Illustration of the geometry involved in projecting area on a surface onto the horizontal plane.

Now we have

$$\boldsymbol{\eta} \cdot \mathbf{k} = \cos\theta = \left[1 + \left(\frac{\partial h}{\partial x}\right)^2 + \left(\frac{\partial h}{\partial y}\right)^2\right]^{-1/2} \tag{19}$$

and thus the area differential may be written as

$$d\sigma = \sec\theta \, dx \, dy = \left[1 + \left(\frac{\partial h}{\partial x}\right)^2 + \left(\frac{\partial h}{\partial y}\right)^2\right]^{1/2} dx \, dy \tag{20}$$

and this clearly reduces to the correct differential when the surface is confined to the (x,y) plane.

More generally, we can show that the vector

$$\mathbf{N} = \mathbf{i}\left|\frac{\partial(y,z)}{\partial(\xi,\zeta)}\right| + \mathbf{j}\left|\frac{\partial(z,x)}{\partial(\xi,\zeta)}\right| + \mathbf{k}\left|\frac{\partial(x,y)}{\partial(\xi,\zeta)}\right| \tag{21}$$

is normal to the surface. Here the determinants are those of the Jacobian

$$\left|\frac{\partial(f,g)}{\partial(\xi,\zeta)}\right| = \begin{vmatrix} f_{,\xi} & g_{,\xi} \\ f_{,\zeta} & g_{,\zeta} \end{vmatrix} \tag{22}$$

PROBLEM 5.3.10 Prove that **N** in Eq. (21) is normal to the surface. [Hint: With a relation $\zeta = \gamma(\xi)$ between ζ and ξ, Eqs. (16) become functions of ξ alone and thus describe a curve C embedded in the surface. The tangent vector to this surface has direction

$$\mathbf{T} = \left(\frac{\partial \mathbf{x}}{\partial \xi}\right) + \left(\frac{\partial \mathbf{x}}{\partial \zeta}\right)_{\xi} \frac{\partial \gamma}{\partial \xi}$$

The scalar product of the normal **N** and tangent **T** may be written in a determinant form that vanishes because of linearly dependent rows. Thus the vector defined by Eq. (21) is perpendicular to the tangent vector to every curve lying in the surface and may be justifiably called a normal to the surface.]

To obtain a more general definition of surface area than that permitted by Eq. (20)–which required that the surface not become perpendicular to the horizontal plane–we may use the expression

$$A(S) = \iint |\mathbf{N}|\, d\xi\, d\zeta \tag{23}$$

It cannot be proved by elementary means that this is actually the area of the surface because there is no elementary definition of surface area. However, we can easily see that if the surface is in the (x,y) plane, the result is proportional to its area, with the constant given by

$$c = \left|\frac{\partial(x,y)}{\partial(\xi,\zeta)}\right| \tag{24}$$

For the majority of meteorological applications the surfaces we consider will be quasi-horizontal so that the parametric representations

$$x = x \qquad y = y \qquad z = h(x,y) \tag{25}$$

may be used. Thus definition (23) reduces to the same area as would be obtained from Eq. (20).

PROBLEM 5.3.11 The parametric equations for a spherical surface are $x = r \cos\phi \cos\lambda$, $y = r\cos\phi \sin\lambda$, $z = r\sin\phi$. Determine appropriate limits for ϕ and λ, and use Eq. (23) to find the surface area of the sphere.

5.3.3 Temporal Differentiation of Line Integrals

We now have the mathematical concepts at hand to determine the derivative of the line integral

$$I_C(t) = \int_C f \circ \boldsymbol{\tau}\, ds \tag{26}$$

where C is a curve with the parametric representation

$$\mathbf{x} = \mathbf{x}(\xi, t) \tag{27}$$

and f may be a vector or scalar with the $\circ$ denoting a suitable operation on the tangent vector $\boldsymbol{\tau}$. The integral is a function of time; variations in its value may arise from changes in f along the curve or because the curve may move in space and thus encounter different values of f.

Upon changing Eq. (26) to parametric form, we have

$$I_C(t) = \int_D f(\mathbf{x}(\xi, t), t) \cdot \boldsymbol{\tau} \left|\frac{\partial \mathbf{x}}{\partial \xi}\right| d\xi = \int_D f(\mathbf{x}(\xi, t), t) \circ \frac{\partial \mathbf{x}}{\partial \xi} d\xi \tag{28}$$

and thus the rate of change of the integral becomes

$$\frac{\partial I_C(t)}{\partial t} = \int_D \left[\left(\frac{\partial f}{\partial t_x} + \frac{\partial \mathbf{x}}{\partial t} \cdot \nabla f\right) \circ \frac{\partial \mathbf{x}}{\partial \xi} + f \circ \frac{\partial^2 \mathbf{x}}{\partial \xi\, \partial t}\right] d\xi \tag{29}$$

According to Eq. (15) we may interpret $\mathbf{x}_{,t}$ as the velocity $\mathbf{u}$ of a point on the curve and so we have with the aid of Problem 5.3.9,

$$\frac{\partial^2 \mathbf{x}}{\partial t\, \partial \xi} = \mathbf{x}_{,t\xi} = \mathbf{u}_{,\xi} = \mathbf{x}_{,\xi} \cdot \nabla \mathbf{u} = (\boldsymbol{\tau} \cdot \nabla \mathbf{u}) \left|\frac{\partial \mathbf{x}}{\partial \xi}\right| \tag{30}$$

Upon making this substitution and changing the integral back to its original form, we obtain

$$\frac{\partial I_C(t)}{\partial t} = \int_C \left(\frac{\partial f}{\partial t} + \mathbf{u} \cdot \nabla f\right) \circ \boldsymbol{\tau}\, ds + \int_C f \circ (\boldsymbol{\tau} \cdot \nabla \mathbf{u})\, ds \tag{31}$$

It is often convenient to replace the local derivative of f with the total derivative and the advection, which gives

$$\frac{\partial I_C(t)}{\partial t} = \int_C \left[\frac{df}{dt} + (\mathbf{u} - \mathbf{v}) \cdot \nabla f\right] \circ \boldsymbol{\tau}\, ds + \int_C f \circ (\boldsymbol{\tau} \cdot \nabla \mathbf{u})\, ds \tag{32}$$

This expression is vital in the study of the dynamics of vorticity in the atmosphere and will play a central role when we consider that topic in Chap. 10.

5.3.4 The Transport Theorem Revisited

With the development of the concepts of parametric representation, we are now able to give a general and elegant proof of the transport theorem. We suppose that the volume of interest is obtained by a transformation of an elementary volume V_ξ into a volume V according to the equations

$$x = \phi(\xi, \eta, \zeta, t) \qquad y = \psi(\xi, \eta, \zeta, t) \qquad z = \theta(\xi, \eta, \zeta, t) \tag{33}$$

where the vector $\boldsymbol{\xi} = (\xi, \eta, \zeta)$ is in V_ξ; for points in the volume V and on its boundary we have

$$\mathbf{x} = \mathbf{x}(\boldsymbol{\xi}, t) \tag{34}$$

For each point $\boldsymbol{\xi}$ (which is not a function of time) there is a velocity of the point $\mathbf{x}$ associated with $\boldsymbol{\xi}$, given by

$$\mathbf{u} = \left(\frac{\partial \mathbf{x}}{\partial t}\right)_{\boldsymbol{\xi}} \tag{35}$$

In parametric form the integral of Eq. (3) becomes

$$F(t) = \int_{V(t)} f(\mathbf{x}, t)\, dV_x = \int_{V_\xi} f(\mathbf{x}(\boldsymbol{\xi}, t), t) J\, dV_\xi \qquad dV_\xi = d\xi\, d\eta\, d\zeta \tag{36}$$

where J is the determinant of the Jacobian

$$J = \left|\frac{\partial(x, y, z)}{\partial(\xi, \eta, \zeta)}\right| \tag{37}$$

[The proof of Eq. (36) is set as a problem at the end of Sec. 5.5 after the necessary concepts are available.]

The derivative may now be written as

$$\frac{\partial F}{\partial t} = \int_{V_\xi} \left[\left(\frac{\partial f}{\partial t}\right)_x + \mathbf{u} \cdot \nabla f\right] J\, dV_\xi + \int_{V_\xi} f \frac{\partial J}{\partial t_\xi}\, dV_\xi \tag{38}$$

We may proceed to calculate

$$\frac{\partial J}{\partial t_\xi} = \begin{vmatrix} u_{,\xi} & y_{,\xi} & z_{,\xi} \\ u_{,\eta} & y_{,\eta} & z_{,\eta} \\ u_{,\zeta} & y_{,\zeta} & z_{,\zeta} \end{vmatrix} + \begin{vmatrix} x_{,\xi} & v_{,\xi} & z_{,\xi} \\ x_{,\eta} & v_{,\eta} & z_{,\eta} \\ x_{,\zeta} & v_{,\zeta} & z_{,\zeta} \end{vmatrix} + \begin{vmatrix} x_{,\xi} & y_{,\xi} & w_{,\xi} \\ x_{,\eta} & y_{,\eta} & w_{,\eta} \\ x_{,\zeta} & y_{,\zeta} & w_{,\zeta} \end{vmatrix} \tag{39}$$

With the chain rule, we have $u_{,\xi} = (\nabla u) \cdot \mathbf{x}_{,\xi}$ so that the first determinant, D_u, on the right in Eq. (39) becomes

$$D_u = \nabla u \cdot \left|\frac{\partial(\mathbf{x}, y, z)}{\partial(\xi, \eta, \zeta)}\right| = \frac{\partial u}{\partial x} J \tag{40}$$

because the determinants multiplying $\partial u/\partial y$ and $\partial u/\partial z$ have two identical columns. Similarly, the second determinant in Eq. (39) gives $(\partial v/\partial y)J$ and the third gives $(\partial w/\partial z)J$. Hence

$$\frac{\partial J}{\partial t_\xi} = (\nabla \cdot \mathbf{u})J \tag{41}$$

Restoration of the original form of the integral produces

$$\frac{\partial F}{\partial t} = \int_V \left(\frac{\partial f}{\partial t} + \mathbf{u} \cdot \nabla f\right) dV + \int f \nabla \cdot \mathbf{u}\, dV = \int_V \left(\frac{\partial f}{\partial t} + \nabla \cdot f\mathbf{u}\right) dV \tag{42}$$

and upon application of the divergence theorem and the use of $\mathbf{w}$ for the values of $\mathbf{u}$ at the boundary we have

$$\frac{\partial F}{\partial t} = \int \frac{\partial f}{\partial t}\, dV + \int_S f\mathbf{w} \cdot \boldsymbol{\eta}\, d\sigma \tag{43}$$

which was the result (4).

PROBLEM 5.3.12 Show that the temporal derivative of the surface integral

$$I_S(t) = \int_S f\,d\sigma$$

is

$$\frac{\partial I_S(t)}{\partial t} = \int_S \left[\frac{df}{dt} + (\mathbf{u} - \mathbf{v}) \cdot \nabla f\right] d\sigma + \int_S f \nabla_S \cdot \mathbf{u}\,d\sigma$$

[Hint: Calculate the derivative

$$\frac{\partial |\mathbf{N}|}{\partial t} = \frac{\partial}{\partial t}(\mathbf{N} \cdot \mathbf{N})^{1/2} = |\mathbf{N}|^{-1}\mathbf{N} \cdot \frac{\partial \mathbf{N}}{\partial t} = [(\nabla \cdot \mathbf{u})|\mathbf{N}|^2 - \mathbf{N} \cdot (\mathbf{N} \cdot \nabla \mathbf{u})]|\mathbf{N}|^{-1}$$

where $\mathbf{N}$ may be written from Eq. (21) as a determinant. Use $\nabla(\ ) = \nabla_S(\ ) + \mathbf{n}\partial(\ )/\partial h$ where $\mathbf{n}$ is the unit normal $\mathbf{N}/|\mathbf{N}|$.]

5.4 CURVILINEAR COORDINATES AND THE ∇ OPERATOR

It was stressed earlier that vector operations, and those involving del in particular, are independent of the choice of coordinate system. But to apply the results of a vector computation in a coordinate system other than a rectangular cartesian one, we must specifically evaluate the vector quantities we have obtained. This generally involves determination of the unit vectors in the new system, determination of the components of a vector with respect to these unit vectors, and determination of the divergence, curl, and Laplacian of a vector. All these quantities depend explicitly on the form of the new coordinate system, and to approach the problem efficiently, we first need to assemble some facts about coordinate transformations.

It is convenient to assume that we always have available for comparison, or for stating basic definitions, a rectangular cartesian coordinate system with right-hand orientation of the unit vectors. The implication of rectangular cartesian is that the system is the usual (x,y,z) or (x_1,x_2,x_3) system we have been employing. The right-hand orientation requires that $\mathbf{i}_1 \times \mathbf{i}_2 = \mathbf{i}_3$ or, in other words, that x increases toward the east, y toward the north, and z upward.

We consider all transformations to a new coordinate system $(\hat{x}_1,\hat{x}_2,\hat{x}_3)$ that can be specified by the equations

$$\hat{x}_i = f_i(x_1,x_2,x_3) \qquad i = 1, 2, 3 \tag{1}$$

or in vector form

$$\hat{\mathbf{x}} = \mathbf{f}(\mathbf{x}) = \hat{\mathbf{x}}(\mathbf{x}) \tag{2}$$

where the function $\mathbf{f}$ prescribes one and only one value of $\hat{\mathbf{x}}$ for each value of $\mathbf{x}$ and is such that the three coordinates are independent of each other. Naturally, we assume continuity of the function **f**, and that the relation (1) can be inverted by a function **g** to yield

$$\mathbf{x} = \mathbf{g}(\hat{\mathbf{x}}) \tag{3}$$

The new coordinates $\hat{\mathbf{x}}$ are called *curvilinear*, and we have restricted ourselves here to *holonomic transformations*; that is, the transformation is specified in functional form. Later we shall encounter *anholonomic transformations* that are given by differential relations.

Although the terminology in the literature is not uniform, it will be convenient to adopt some classification scheme for distinguishing coordinate systems according to their characteristics.

Curvilinear coordinates are a family of variables $(\hat{x}_1,\hat{x}_2,\hat{x}_3)$ such that each point of the space is associated with one and only one set of values of $(\hat{x}_1,\hat{x}_2,\hat{x}_3)$. Curvilinear coordinates are thus defined here to be any set that can be obtained by equations of the form of Eq. (1) with the restrictions given. If the surfaces $\hat{x}_i = \text{const}$ are orthogonal, then we refer to orthogonal curvilinear coordinates.

Cartesian coordinates are those curvilinear systems in which the positions of points are determined by their distance from intersecting planes, and thus they need not be orthogonal; those that are orthogonal will be called *rectangular*.

Thus the usual (x,y,z) coordinates are rectangular cartesian; cylindrical polar and spherical coordinates are orthogonal curvilinear coordinates but are not cartesian. Most systems encountered in physics and meteorology are orthogonal, and this has distinct advantages.

The condition for the transformation (1) to be uniquely invertible is that the Jacobian determinant

$$|J_x^{\hat{x}}| = \left|\frac{\partial(\hat{x}_1,\hat{x}_2,\hat{x}_3)}{\partial(x_1,x_2,x_3)}\right| \tag{4}$$

does not vanish for any value of **x**. This condition also ensures that the new coordinates are independent. If $|J_x^{\hat{x}}|$ does not vanish, then we can also define the reciprocal Jacobian, $\{J_{\hat{x}}^{x}\}$, which has the determinant

$$|J_{\hat{x}}^{x}| = \left|\frac{\partial(x_1,x_2,x_3)}{\partial(\hat{x}_1,\hat{x}_2,\hat{x}_3)}\right| \tag{5}$$

and we shall show that

$$|J_x^{\hat{x}}| \cdot |J_{\hat{x}}^{x}| = 1 \tag{6}$$

To see this, note that the element P_{ij} in the ith row and jth column of the matrix product $\{J_x^{\hat{x}}\} \ \{J_{\hat{x}}^{x}\}$ is given by

$$P_{ij} = \frac{\partial \hat{x}_1}{\partial x_i}\frac{\partial x_j}{\partial \hat{x}_1} + \frac{\partial \hat{x}_2}{\partial x_i}\frac{\partial x_j}{\partial \hat{x}_2} + \frac{\partial \hat{x}_3}{\partial x_i}\frac{\partial x_j}{\partial \hat{x}_3} = \frac{\partial x_j}{\partial x_i} = \begin{cases} 1 & i = j \\ 0 & i \neq j \end{cases} \tag{7}$$

in which we have used the chain rule. Thus the product matrix has ones on the diagonals and zeros elsewhere, and therefore its determinant has the value unity. This proves Eq. (6).

It is worth noting that we did not cancel the partial derivatives that appeared in numerator and denominator of the factors in the sum of Eq. (7). This temptation should usually be resisted, but a case in which derivatives can be canceled or inverted occurs frequently in meteorology. Suppose we have a transformation of only the vertical coordinate in the form

$$\hat{x}_1 = x_1 \qquad \hat{x}_2 = x_2 \qquad \hat{x}_3 = f(x_1, x_2, x_3) \tag{8}$$

Then

$$|J_x^{\hat{x}}| = \begin{vmatrix} 1 & 0 & 0 \\ 0 & 1 & 0 \\ f_{,x_1} & f_{,x_2} & f_{,x_3} \end{vmatrix} = \frac{\partial \hat{x}_3}{\partial x_3} \qquad |J_{\hat{x}}^{x}| = \frac{\partial x_3}{\partial \hat{x}_3} \tag{9}$$

so we have proved that

$$\frac{\partial x_3}{\partial \hat{x}_3} = \left(\frac{\partial \hat{x}_3}{\partial x_3}\right)^{-1} \tag{10}$$

in this special case.

Finding a set of unit vectors in the new system is quite easy. The normal to the surface $\hat{x}_i = \text{const}$ will point in the direction of increasing $\hat{x}_i$, and hence the unit vectors are

$$\hat{\mathbf{i}}_i = h_i \, \nabla \hat{x}_i \qquad h_i^{-1} = |\nabla \hat{x}_i| \qquad i = 1, 2, 3 \tag{11}$$

PROBLEM 5.4.1 Write out the matrices $\{J_{\hat{x}}^{x}\}$ and $\{J_x^{\hat{x}}\}$ and verify Eq. (7).

PROBLEM 5.4.2 Show that the unit vectors of Eq. (11) are *basis vectors* in the sense that they are *linearly independent*. [Hint: Linear independence means that there are no constants α and β such that $\hat{\mathbf{i}}_1 = \alpha \hat{\mathbf{i}}_2 + \beta \hat{\mathbf{i}}_3$. Show that linear dependence implies that

$$\hat{\mathbf{i}}_3 \cdot (\hat{\mathbf{i}}_2 \times \hat{\mathbf{i}}_1) = -|J_x^{\hat{x}}| h_1 h_2 h_3$$

vanishes, which is a contradiction when the Jacobian determinant does not vanish.]

5.4.1 Orthogonal Coordinates

Next we must test this set of unit vectors to determine whether they are *orthogonal*, that is, whether

$$\hat{\mathbf{i}}_i \cdot \hat{\mathbf{i}}_j = \delta_{ij} \qquad i, j = 1, 2, 3 \tag{12}$$

in which δ_{ij} is the *Kronecker delta* and is defined by

$$\delta_{ij} = \begin{cases} 1 & i = j \\ 0 & i \neq j \end{cases} \tag{13}$$

PROBLEM 5.4.3 If the unit vectors are orthogonal, then the components are given by $\hat{A}_i = \hat{\mathbf{i}}_i \cdot \mathbf{A}$ and $\mathbf{A} = \sum_{i=1}^{3} \hat{\mathbf{i}}_i \hat{A}_i$. Show that these two relations are incompatible if the unit vectors are not orthogonal.

The concept of distance is defined in the original $\mathbf{x}$ system in the usual way, and we must translate results from there into an orthogonal $\hat{\mathbf{x}}$ system to provide a calibration of distance in the $\hat{\mathbf{x}}$ system. We have the three unit vectors $\hat{\mathbf{i}}_1, \hat{\mathbf{i}}_2, \hat{\mathbf{i}}_3$, and so the differential position vector can be projected on them with the scalar product to determine its components with respect to the new unit vectors. Thus with $(\nabla\hat{x}_1) \cdot d\mathbf{x} = d\hat{x}_1$,

$$d\mathbf{x} = (\hat{\mathbf{i}}_1 \cdot d\mathbf{x})\hat{\mathbf{i}}_1 + (\hat{\mathbf{i}}_2 \cdot d\mathbf{x})\hat{\mathbf{i}}_2 + (\hat{\mathbf{i}}_3 \cdot d\mathbf{x})\hat{\mathbf{i}}_3 = h_1\, d\hat{x}_1\hat{\mathbf{i}}_1 + h_2\, d\hat{x}_2\hat{\mathbf{i}}_2 + h_3\, d\hat{x}_3\hat{\mathbf{i}}_3 \tag{14}$$

PROBLEM 5.4.4 Show that for orthogonal systems

$$|J_{\hat{x}}^{x}| = h_1 h_2 h_3$$

(Hint: See Problem 5.4.2.)

The differential volume element to be used in integration after the change of variables is the same element of volume as in the original coordinates. But it has a different representation in the $\hat{x}$ coordinates. Denoting the invariant element of volume by $dV_x = d\hat{V}_{\hat{x}}$, we have (note carefully the definitions contained in these equations, including the convention that $|J|$ is the absolute value of the determinant)

$$dV_x = d\hat{V}_{\hat{x}} = |J_{\hat{x}}^{x}|\, d\hat{\mathbf{x}} = |J_{\hat{x}}^{x}|\, d\hat{x}_1\, d\hat{x}_2\, d\hat{x}_3 = |J_{\hat{x}}^{x}|\, dV_{\hat{x}} \tag{15}$$

or

$$dV_x = d\mathbf{x} = dx_1\, dx_2\, dx_3 = h_1 h_2 h_3\, d\hat{x}_1\, d\hat{x}_2\, d\hat{x}_3 = h_1 h_2 h_3\, d\hat{\mathbf{x}} = h_1 h_2 h_3\, dV_{\hat{x}} \tag{16}$$

which is in agreement with the interpretation of differential length in the $\hat{\mathbf{i}}_1$ direction being $h_1\, d\hat{x}_1$. The relations (15) are also valid for nonorthogonal coordinates, as shown in Problem 5.5.20.

The definitions given previously for the gradient, divergence, and curl depended explicitly on the fact that the unit vectors did not vary in space. For orthogonal coordinates, we have the relations given below. They are obtained by either rewriting the definitions for rectangular cartesian coordinates in the orthogonal coordinates and then taking account of variations in unit vectors, or by applying the divergence theorem and Stokes' theorem to rectangles in the new coordinates. Thus:

$$\nabla f = \frac{\hat{\mathbf{i}}_1}{h_1}\frac{\partial f}{\partial \hat{x}_1} + \frac{\hat{\mathbf{i}}_2}{h_2}\frac{\partial f}{\partial \hat{x}_2} + \frac{\hat{\mathbf{i}}_3}{h_3}\frac{\partial f}{\partial \hat{x}_3} \tag{17}$$

$$\nabla \cdot \mathbf{A} = \frac{1}{h_1 h_2 h_3}\left[\frac{\partial}{\partial \hat{x}_1}(h_2 h_3 \hat{A}_1) + \frac{\partial}{\partial \hat{x}_2}(h_1 h_3 \hat{A}_2) + \frac{\partial}{\partial \hat{x}_3}(h_1 h_2 \hat{A}_3)\right] \tag{18}$$

$$\nabla \times \mathbf{A} = \frac{1}{h_1 h_2 h_3}\begin{vmatrix} h_1\hat{\mathbf{i}}_1 & h_2\hat{\mathbf{i}}_2 & h_3\hat{\mathbf{i}}_3 \\ \dfrac{\partial}{\partial \hat{x}_1} & \dfrac{\partial}{\partial \hat{x}_2} & \dfrac{\partial}{\partial \hat{x}_3} \\ h_1\hat{A}_1 & h_2\hat{A}_2 & h_3\hat{A}_3 \end{vmatrix} \tag{19}$$

$$\nabla^2 f = \frac{1}{h_1 h_2 h_3}\left[\frac{\partial}{\partial \hat{x}_1}\left(\frac{h_2 h_3}{h_1}\frac{\partial f}{\partial \hat{x}_1}\right) + \frac{\partial}{\partial \hat{x}_2}\left(\frac{h_1 h_3}{h_2}\frac{\partial f}{\partial \hat{x}_2}\right) + \frac{\partial}{\partial \hat{x}_3}\left(\frac{h_1 h_2}{h_3}\frac{\partial f}{\partial \hat{x}_3}\right)\right] \tag{20}$$

PROBLEM 5.4.5 Spherical coordinates are often used to define positions on the earth. We start with a rectangular cartesian coordinate system (x,y,z) whose origin is at the center of the earth. Then for radius r, longitude λ, and latitude ϕ we have

$$x = r\cos\phi\cos\lambda \qquad y = r\cos\phi\sin\lambda \qquad z = r\sin\phi$$

Show that

$$\hat{x}_1 = r = (x^2+y^2+z^2)^{1/2}$$

$$\hat{x}_2 = \lambda = \arctan\frac{y}{x}$$

$$\hat{x}_3 = \phi = \arctan\frac{z}{(x^2+y^2)^{1/2}}$$

$$h_1 = |\nabla r|^{-1} = 1$$

$$h_2 = |\nabla\lambda|^{-1} = (x^2+y^2)^{1/2} = r\cos\phi$$

$$h_3 = |\nabla\phi|^{-1} = (x^2+y^2+z^2)^{1/2} = r$$

$$|J| = h_1 h_2 h_3 = r^2\cos\phi$$

$$\nabla f = \mathbf{i}_r\frac{\partial f}{\partial r} + \frac{\mathbf{i}_\lambda}{r\cos\phi}\frac{\partial f}{\partial\lambda} + \frac{\mathbf{i}_\phi}{r}\frac{\partial f}{\partial\phi}$$

$$\nabla \cdot \mathbf{A} = \frac{1}{r^2}\frac{\partial}{\partial r}(r^2 A_r) + \frac{1}{r\cos\phi}\frac{\partial A_\lambda}{\partial\lambda} + \frac{1}{r\cos\phi}\frac{\partial}{\partial\phi}(\cos\phi)A_\phi$$

$$\nabla \times \mathbf{A} = \frac{1}{r^2\cos\phi}\begin{vmatrix} \mathbf{i}_r & (r\cos\phi)\mathbf{i}_\lambda & r\mathbf{i}_\phi \\ \dfrac{\partial}{\partial r} & \dfrac{\partial}{\partial\lambda} & \dfrac{\partial}{\partial\phi} \\ A_r & r\cos\phi\, A_\lambda & rA_\phi \end{vmatrix}$$

and

$$\nabla^2 f = \frac{1}{r^2}\frac{\partial}{\partial r}\left(r^2 \frac{\partial f}{\partial r}\right) + \frac{1}{r^2 \cos^2\phi}\frac{\partial^2 f}{\partial \lambda^2} + \frac{1}{r^2 \cos\phi}\frac{\partial}{\partial \phi}(\cos\phi)\frac{\partial f}{\partial \phi}$$

In these equations $\mathbf{i}_r$, $\mathbf{i}_\lambda$, and $\mathbf{i}_\phi$ are unit vectors pointing toward the zenith, toward the east, and toward the north, respectively. Show that they are orthogonal.

PROBLEM 5.4.6 Work out the relations of the preceding problem for cylindrical coordinates.

5.4.2 Nonorthogonal Curvilinear Coordinates

When the coordinates fail to be orthogonal, we must use two sets of basis vectors and two sets of components in order to be able to determine components with scalar products. Moreover, there is no longer any advantage to having basis vectors of unit length, and instead the magnitudes of the basis vectors will carry the necessary information on distance scaling.

Before beginning, let us establish one of the great notational conveniences of all time—the *summation convention*, which requires that we sum on repeated indices when they appear on two quantities that are multiplied by each other. For example,

$$\mathbf{A} \cdot \mathbf{B} = \sum_{k=1}^{3} A_k B_k = A_k B_k$$

$$(\mathbf{A} \cdot \mathbf{B})(\mathbf{C} \cdot \mathbf{D}) = A_i B_i C_j D_j \tag{21}$$

The key to success with the summation convention is to use lots of different letters for indices, because if more than two of the same letter ever appear in the same term, we shall undoubtedly become confused. As an example, if for cartesian coordinates we write $\mathbf{A} = \mathbf{i}_j A_j$ and $\mathbf{B} = \mathbf{i}_k B_k$, then it is obvious that $\mathbf{A} \cdot \mathbf{B} = (\mathbf{i}_j \cdot \mathbf{i}_k) A_j B_k = \delta_{jk} A_j B_k = A_j B_j$.

We shall also anticipate results to come, and change the position of the index on a coordinate from a subscript to a superscript, and agree that a superscript appearing in a denominator is equivalent to a subscript.

Now assume that coordinates $\hat{\mathbf{x}}(\mathbf{x})$ are defined by Eq. (2). With the chain rule we have the two expansions

$$d\mathbf{x} = \frac{\partial \mathbf{x}}{\partial \hat{x}^j} d\hat{x}^j \qquad d\hat{x}^i = \frac{\partial \hat{x}^i}{\partial x^j} dx^j = (\nabla \hat{x}^i) \cdot d\mathbf{x} \tag{22}$$

and thus the two vectors

$$\hat{\boldsymbol{\tau}}_j = \frac{\partial \mathbf{x}}{\partial \hat{x}^j} \qquad \hat{\boldsymbol{\eta}}^i = \nabla \hat{x}^i \tag{23}$$

have appeared naturally. We know already that $\hat{\boldsymbol{\eta}}^i$ is normal to the surface $\hat{x}^i = \text{const.}$

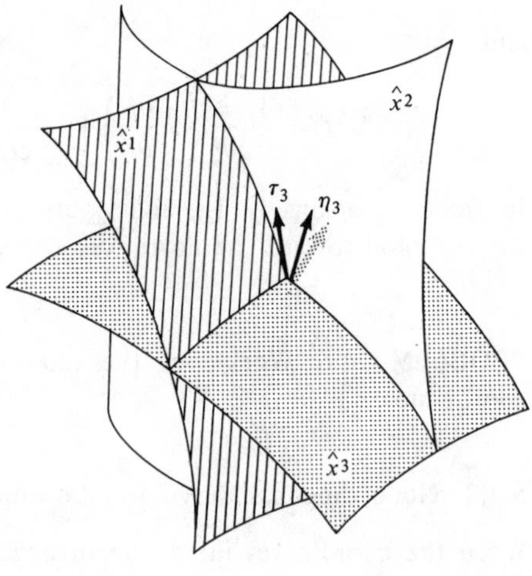

FIGURE 5.7
Illustration of nonorthogonal coordinates showing that τ_3 and η^3 need not be identical, but that τ_3 will be orthogonal to η^1 and η^3. The shadow behind η^3 indicates that it is not in the surface $\hat{x}^2 = \text{const.}$

According to the definition of a partial derivative, the vector $\hat{\tau}_j$ reveals the variation of the position vector as it traces out a curve in which $\hat{x}^j$ varies and the other two coordinates are constant. Hence $\hat{\tau}_j$ is tangent to the curve along which only $\hat{x}^j$ varies.

When the coordinates are orthogonal, these two sets of vectors must become identical in direction. But for nonorthogonal coordinates (see Fig. 5.7), the tangent vector τ_3 to the curve on which $\hat{x}^1$ and $\hat{x}^2$ are constant does not have to coincide with the normal to the surface $\hat{x}^3 = \text{const}$. But it is quite clear that the vector $\hat{\tau}_3$ must be orthogonal to the vectors $\hat{\eta}^1$ and $\hat{\eta}^2$ that are normal to the surfaces on which $\hat{x}^1$ and $\hat{x}^2$ are constant.

In fact, with the aid of the chain rule we have the orthogonality relation

$$\hat{\tau}_j \cdot \hat{\eta}^k = \frac{\partial \mathbf{x}}{\partial \hat{x}^j} \cdot \nabla \hat{x}^k = \frac{\partial x^l}{\partial \hat{x}^j} \frac{\partial \hat{x}^k}{\partial x^l} = \frac{\partial \hat{x}^k}{\partial \hat{x}^j} = \delta_j{}^k = \begin{cases} 1 & j = k \\ 0 & j \neq k \end{cases} \tag{24}$$

and this permits us to define components of vectors. If we let $\mathbf{A}$ be represented by

$$\mathbf{A} = \hat{A}_k \hat{\eta}^k = \hat{A}^k \hat{\tau}_k \tag{25}$$

then we may recover the components with the aid of Eq. (24) as

$$\hat{A}^i = \hat{\eta}^i \cdot \mathbf{A} = \hat{A}^k \hat{\eta}^i \cdot \hat{\tau}_k \qquad \hat{A}_i = \hat{\tau}_i \cdot \mathbf{A} = \hat{A}_k \hat{\tau}_i \cdot \hat{\eta}^k \tag{26}$$

Thus the general definition of scalar product becomes

$$\mathbf{A} \cdot \mathbf{B} = A_k B^k \tag{27}$$

The two types of components, then, must always occur together in scalar products, but they must not be mixed in vector addition because the unit vectors are different.

PROBLEM 5.4.7 How do the relative lengths of the vectors $\hat{\tau}_i$ and $\hat{\eta}^i$ adjust as the angle between them changes?

PROBLEM 5.4.8 Show that $\hat{\boldsymbol{\tau}}_1 = (\hat{\boldsymbol{\eta}}^2 \times \hat{\boldsymbol{\eta}}^3)M, \hat{\boldsymbol{\tau}}_2 = (\hat{\boldsymbol{\eta}}^3 \times \hat{\boldsymbol{\eta}}^1)M, \hat{\boldsymbol{\tau}}_3 = (\hat{\boldsymbol{\eta}}^1 \times \hat{\boldsymbol{\eta}}^2)M$ where $M^{-1} = \hat{\boldsymbol{\eta}}^1 \cdot (\hat{\boldsymbol{\eta}}^2 \times \hat{\boldsymbol{\eta}}^3)$. [Hint: The components of $\hat{\boldsymbol{\tau}}_i$ with respect to the cartesian unit vectors appear in the matrix $\{J^x_{\hat{x}}\}$, and thus may be extracted from the inverse matrix $\{J^{\hat{x}}_x\}$.]

PROBLEM 5.4.9 Let Greek letters be constants and define

$$\hat{x}^1 = \alpha x + \beta y \qquad \hat{x}^2 = \gamma y + \delta z \qquad \hat{x}^3 = z$$

What is the inverse transformation, and what are the normal and tangent vectors in x coordinates? This system is not orthogonal, but is it cartesian?

The next step is to determine how the basis vectors behave under further transformation. We define another set of curvilinear coordinates with the relations

$$\bar{z}^i = \bar{z}^i(x^1, x^2, x^3) \qquad i = 1, 2, 3 \tag{28}$$

and, as before, the position vector differential becomes

$$d\mathbf{x} = \frac{\partial \mathbf{x}}{\partial \bar{z}^i} d\bar{z}^i = \bar{\boldsymbol{\tau}}_i d\bar{z}^i \tag{29}$$

while the coordinate differential is

$$d\bar{z}^i = (\nabla \bar{z}^i) \cdot d\mathbf{x} = \bar{\boldsymbol{\eta}}^i \cdot d\mathbf{x} \tag{30}$$

But the coordinates $\hat{x}^i$ are also functions of $\mathbf{x}$ and the relation can be inverted to give $\mathbf{x} = \mathbf{x}(\hat{\mathbf{x}})$. Thus we may find the appropriate functions so that the $\bar{z}^i$ may be expressed as a transformation of the $\hat{\mathbf{x}}$ coordinates in the form

$$\bar{z}^i = \bar{z}^i(\hat{x}_1, \hat{x}_2, \hat{x}_3) \tag{31}$$

Now we may apply the chain rule to calculate that

$$\bar{\boldsymbol{\tau}}_i = \frac{\partial \mathbf{x}}{\partial \bar{z}^i} = \frac{\partial \mathbf{x}}{\partial \hat{x}^k} \frac{\partial \hat{x}^k}{\partial \bar{z}^i} = \frac{\partial \hat{x}^k}{\partial \bar{z}^i} \hat{\boldsymbol{\tau}}_k \tag{32}$$

This relation shows that the tangent vectors have a specific law of transformation whose characteristics are revealed by the placement of variables and indices in the derivative $\partial \hat{x}^k / \partial \bar{z}^i$—the position of variables controlling the differentiation and the placement of indices controlling the summation. In a similar manner we find

$$\bar{\boldsymbol{\eta}}^i = \nabla \bar{z}^i = \mathbf{i}_l \frac{\partial \bar{z}^i}{\partial x^l} = \mathbf{i}_l \frac{\partial \bar{z}^i}{\partial \hat{x}^k} \frac{\partial \hat{x}^k}{\partial x^l} = \frac{\partial \bar{z}^i}{\partial \hat{x}^k} \hat{\boldsymbol{\eta}}^k \tag{33}$$

Therefore the normal vectors also have their own law of transformation, which is evidently different from Eq. (32) in both the differentiation and the summation.

It is customary to choose the law of transformation of the tangent vectors as the one to compare with other types, and hence quantities that transform like the tangent vectors are called *covariant*. For a scalar ϕ, the quantity $\partial \phi / \partial \bar{z}^i$ becomes

$$\frac{\partial \phi}{\partial \bar{z}^i} = \frac{\partial \phi}{\partial \hat{x}^k} \frac{\partial \hat{x}^k}{\partial \bar{z}^i} \tag{34}$$

and so is covariant. In contrast we have

$$d\bar{z}^i = \frac{\partial \bar{z}^i}{\partial x^k} dx^k = \frac{\partial \bar{z}^i}{\partial \hat{x}^j} \frac{\partial \hat{x}^j}{\partial x^k} dx^k = \frac{\partial \bar{z}^i}{\partial \hat{x}^j} d\hat{x}^j \tag{35}$$

so that the coordinate differentials transform like the normal vectors. Such quantities are called *contravariant* to indicate that they behave like the normal vectors, not like the tangent vectors.

PROBLEM 5.4.10 Show that $\bar{A}_i$ is covariant and that $\bar{A}^i$ is contravariant.

PROBLEM 5.4.11 Show that

$$\hat{A}_k = (\hat{\boldsymbol{\tau}}_i \cdot \hat{\boldsymbol{\tau}}_k)\hat{A}^i \qquad \hat{A}^i = (\hat{\boldsymbol{\eta}}^i \cdot \hat{\boldsymbol{\eta}}^k)\hat{A}_k$$

PROBLEM 5.4.12 Show that

$$(\hat{\boldsymbol{\tau}}_i \cdot \hat{\boldsymbol{\tau}}_j)(\hat{\boldsymbol{\eta}}^i \cdot \hat{\boldsymbol{\eta}}^k) = \delta_j^{\,k}$$

Finally, we must face the question of what a vector looks like in one of these nonorthogonal coordinate systems. The answer, of course, is that it looks exactly as it did in the original system; the coordinate system cannot alter physical reality. We have had to define the tangential and normal vectors, the covariant and contravariant components, and the transformation laws in a way that ensures that reality is preserved.

Unit vectors are not used because the unnormalized basis vectors have simple transformation laws when the coordinates are changed. If we tried to use unit vectors, we would need transformation laws for both direction and magnitude, and it is obviously easier to work with the vectors $\hat{\boldsymbol{\tau}}_i$ and $\hat{\boldsymbol{\eta}}^i$ that carry the full information about the coordinate system.

This does introduce a complication in seeing how a vector $\mathbf{A}$ is distributed by components in the various directions, because the projection of $\mathbf{A}$ on one of the basis vectors, for example $\hat{\boldsymbol{\eta}}^i$, depends on the size of $\hat{\boldsymbol{\eta}}^i$. The actual component of $\mathbf{A}$ in a given direction, as would be determined by measurements in the cartesian system, is called the *physical component* in that direction. To find the physical component, we simply project on a unit vector. Thus if we want the physical component in a given direction, we take a unit vector $\boldsymbol{\lambda}$, in that direction, and we have either

$$A_{(\lambda)} = \boldsymbol{\lambda} \cdot \mathbf{A} = (\boldsymbol{\lambda} \cdot \hat{\boldsymbol{\tau}}_i)\hat{A}^i \tag{36}$$

or

$$A_{(\lambda)} = \boldsymbol{\lambda} \cdot \mathbf{A} = (\boldsymbol{\lambda} \cdot \hat{\boldsymbol{\eta}}^i)\hat{A}_i = (\boldsymbol{\lambda} \cdot \hat{\boldsymbol{\eta}}^i)(\hat{\boldsymbol{\tau}}_i \cdot \hat{\boldsymbol{\tau}}_k)\hat{A}^k \tag{37}$$

The point here is that the components and basis vectors do in fact contain the information necessary to obtain a correct physical perception of the situation.

PROBLEM 5.4.13 Show that in orthogonal coordinates with the vector $\boldsymbol{\lambda}$ in the direction of one of the unit vectors $\hat{\boldsymbol{\eta}}^I$ we have $A_{(\lambda)} = h_I\hat{A}^I = \hat{A}_I/h_I$ (no sum).

We have introduced many of the concepts of the tensor calculus in this subsection in an attempt to explain the properties of nonorthogonal coordinate systems and to see how vectors can be represented in them. Now we shall extend some of these concepts and consider how to handle the objects, more complicated than vectors, that do appear in studies of atmospheric motion. The concepts and methods which permit us to accomplish this task efficiently are known as tensor analysis.

5.5 INTRODUCTION TO TENSOR ANALYSIS

Once learned and understood, the tensor calculus is one of the most effective methods ever devised for rapid penetration of the complicated mysteries of physical systems. Its genius is that if the notational rules are observed, mistakes are almost impossible.

With vector calculus, the emphasis is placed on the geometric aspects of the problems. But when the geometrical situation becomes too complicated, the techniques of vector analysis are overwhelmed and we must turn to the tensor calculus with its advantage of notational simplicity. The physical and geometric significance is not lost, even though the intuitively appealing arrows of vector analysis are replaced by the at first less communicative indices of the tensor calculus.

There are three reasons why tensor analysis is necessary in the study of atmospheric motion. First, many calculations with vector quantities can be performed more easily with the tensor calculus, even in rectangular cartesian systems; this is especially true when we have several iterations of the del operator. Second, true tensors of more complicated character than vectors actually appear in the study of fluid motion; for example, the family of velocity derivatives $\partial u_i/\partial x_k$ has nine members because it depends on two directions simultaneously. Third, we shall encounter nonorthogonal coordinate systems, and then tensor methods are mandatory—as shown in the last section, where we presented an introduction to the basic concepts of tensor analysis.

5.5.1 Cartesian Tensor Analysis

First we shall develop a tensor calculus for rectangular cartesian coordinates and, in particular, shall stay in the familiar (x,y,z) coordinate system. The directions signified by the unit vectors $(\mathbf{i},\mathbf{j},\mathbf{k})$ will be denoted by the numbers 1, 2, and 3, and the component A_i may be A_1, A_2, A_3. Because these are orthogonal coordinates, we have only one type of component and the position of indices becomes unimportant *as long as we stay in this coordinate system.*

The summation convention is employed [see Eq. (21) in Sec. 5.4] and the Kronecker delta ($\delta_{ij} = 1$ if $i = j$, $\delta_{ij} = 0$ if $i \neq j$) participates in the summation convention so that

$$A_i \, \delta_{ij} = A_j \tag{1}$$

PROBLEM 5.5.1 Show that $\delta_{ii} = 3$.

Quantities that transform in certain ways when coordinates are changed and that depend on directions as revealed by their indices are called *tensors.* A scalar ϕ is a zero-order tensor. A vector **A** or, as now represented, A_i, has one index on its components and is a first-order tensor. A second-order tensor has two indices, and a fluid-mechanical example is the derivative already mentioned, $\partial u_i/\partial x_k$. We can obviously go on forming higher-order tensors as long as our velocity fields are differentiable. Thus $\partial^2 u_i/\partial x_k \, \partial x_j$ has 27 components, $\partial^3 u_i/\partial x_j \, \partial x_k \, \partial x_l$ has 81, and so on. There are, of course, other ways to form higher-order tensors.

A tensor is a mathematical or numerical analog of a geometric object. For example, there is a one-to-one correspondence between the set of all arrows beginning at the origin and the arrays of numbers (A_1, A_2, A_3). It is hard to conceive of more complicated geometric objects than the elementary ones in three-space, but tensors represent such objects even though we may not be able to draw pictures of them as simply as we can draw an arrow.

A tensor may always be represented as a matrix of the same order. Thus **A** becomes a column or row matrix of its components; a second-order tensor is a 3 x 3 matrix, a third-order tensor a 3 x 3 x 3 array. Therefore only tensors of the same order may be added or subtracted, and the sum is the tensor having as elements in its matrix the sum of the elements of the two tensors in the same position.

There are several tensor products in rectangular cartesian coordinates:

Outer product	$P_{ij \cdots nrs \cdots w} = T_{ij \cdots n} S_{rs \cdots w}$
Inner product	Set two indices equal in the outer product, for example, $P_{ii \cdots nrs \cdots w}$, and sum.
Scalar product	An inner product with the equal indices occurring at the same position in the two factors, for example, $T_{ij \cdots n} S_{rj \cdots w}$

Now clearly the outer product of tensors of order n and m produces a tensor of order $n + m$; both the inner and the scalar products of tensors of order n and m are of order $n + m - 2$. The operation of setting two indices equal to obtain a sum–as when the outer product is reduced to the inner product–is called a *contraction.* Thus we can define the magnitude of a tensor by contracting the outer product of the tensor with itself to a scalar and thus

$$|T_{i,j,\ldots,n}| = (T_{i,j,\ldots,n} \cdot T_{i,j,\ldots,n})^{1/2} \tag{2}$$

To reduce a vector expression to one in cartesian tensor form, we use the summation convention for scalar products ($\mathbf{A} \cdot \mathbf{B} = A_i B_i$) and in the case of vectors we simply take the scalar product with $\mathbf{i}_i$ and give the result subscript i. Thus $\mathbf{A} = \alpha\mathbf{B} + \beta\mathbf{C}$ becomes $A_i = \alpha B_i + \beta C_i$, an equivalence we shall denote by $\mathbf{A} \overset{\mathrm{T}}{=} A_i$.

PROBLEM 5.5.2 Establish the following tensor equivalents:

$$\nabla\phi \overset{\mathrm{T}}{=} \frac{\partial \phi}{\partial x_i} \qquad \nabla \cdot \mathbf{A} = \frac{\partial A_i}{\partial x_i}$$

$$(\mathbf{B} \cdot \nabla)\mathbf{A} \overset{\mathrm{T}}{=} B_k \frac{\partial A_i}{\partial x_k} \qquad \nabla^2\phi = \frac{\partial^2 \phi}{\partial x_i^{\,2}}$$

$$\nabla^2\mathbf{A} \overset{\mathrm{T}}{=} \frac{\partial^2 A_i}{\partial x_k^{\,2}} \qquad |A|^2 = A_i^{\,2}$$

To express cross products we use the *alternating unit tensor* or *permutation symbol* ϵ_{ijk}:

$$\epsilon_{ijk} = \begin{cases} 0 & \text{if } i = j, \text{ or } j = k, \text{ or } i = k \\ 1 & \text{if } i, j, k \text{ are an even permutation of } 1, 2, 3 \\ -1 & \text{if } i, j, k \text{ are an odd permutation of } 1, 2, 3 \end{cases} \tag{3}$$

Thus $\epsilon_{1,2,3} = 1$, $\epsilon_{1,3,2} = -1$, $\epsilon_{3,1,2} = 1$, $\epsilon_{3,2,1} = -1$, $\epsilon_{3,3,1} = \epsilon_{2,1,2} = 0$, and so on. The ith component of a cross product may be written

$$\mathbf{i}_i \cdot (\mathbf{A} \times \mathbf{B}) = \epsilon_{ijk} A_j B_k \tag{4}$$

and for the curl

$$\mathbf{i}_i \cdot (\nabla \times \mathbf{A}) = \epsilon_{ijk} \frac{\partial A_k}{\partial x_j} \tag{5}$$

The perhaps strange properties of ϵ_{ijk} suggest that we study the behavior of indices a little more carefully. For a second-order tensor T_{ik}, we have three possibilities. If $T_{ik} = T_{ki}$, it is called *symmetric*; if $T_{ik} = -T_{ki}$, it is *antisymmetric* (or skew-symmetric). The third possibility is that there is no relation between T_{ik} and T_{ki}.

Furthermore, any second-order tensor can be written as the sum of symmetric and antisymmetric parts. To demonstrate this, it is necessary only to write out the identity

$$T_{ik} = \tfrac{1}{2}(T_{ik} + T_{ki}) + \tfrac{1}{2}(T_{ik} - T_{ki}) \tag{6}$$

The first term is obviously symmetric; the second is antisymmetric. Writing tensors in this form facilitates computations, because products of the symmetric and antisymmetric components may be immediately set to zero.

*PROBLEM 5.5.3 Let N_{ik} be antisymmetric. Show that $N_{11} = N_{22} = N_{33} = 0$. Let S_{ik} be symmetric. Show that $S_{ik}N_{ik} = 0$.

PROBLEM 5.5.4 Show that

$$\nabla \cdot (\nabla \times \mathbf{A}) = \epsilon_{ijk} \frac{\partial^2 A_k}{\partial x_i \, \partial x_j} = 0$$

*PROBLEM 5.5.5 Show that $\epsilon_{ijk}\epsilon_{imn} = \delta_{jm}\,\delta_{kn} - \delta_{km}\,\delta_{jn}$.

PROBLEM 5.5.6 Use cartesian tensor methods to prove the vector identities (35) to (39) in Sec. 5.1.

PROBLEM 5.5.7 Let the equation of motion be $d\mathbf{v}/dt = \mathbf{F}$. Find an expression for $d(\nabla \cdot \mathbf{v})/dt$ and explain why the result cannot be conveniently expressed in vector notation.

5.5.2 Tensor Analysis

The efficient notation of cartesian tensor analysis and the concepts of Sec. 5.4 relating to nonorthogonal coordinates can now be combined to produce a tensor calculus that is valid for any holonomic coordinate system. One of the main advantages of tensor analysis lies in this point: if the mathematical treatment of physical reality is done correctly in tensor format, the results are true in any holonomic coordinate system. Conversely, something that can be proved to be true in one coordinate system is true in any other when the result is correctly translated by tensor methods.

We shall assume that we are operating in *Euclidean space*, which means that there is some rectangular cartesian coordinate system $\mathbf{x}$ in which distance is measured in accordance with the Pythagorean theorem, so that if ds is the differential element of distance, we have

$$(ds)^2 = (dx^1)^2 + (dx^2)^2 + (dx^3)^2 \tag{7}$$

The distance between two points in a Euclidean space must always agree with Eq. (7), no matter what coordinate system is used to represent the position of points; it is by making distance come out correctly in various coordinate systems that the transformation properties are determined. Not all spaces are Euclidean, but these are most often used for nonrelativistic fluid mechanics.

Let us suppose then that we have a basic system of rectangular cartesian coordinates x^i, for which Eq. (7) is true. A new set of curvilinear coordinates, $\hat{x}^i$, is now defined as before by

$$\hat{x}^i = \hat{x}^i(x^1, x^2, x^3) \qquad i = 1, 2, 3 \tag{8}$$

Because (and only because) of the original cartesian system we may define the two sets of basis vectors as before by

$$\hat{\boldsymbol{\eta}}^i = \nabla \hat{x}^i \qquad \hat{\boldsymbol{\tau}}_i = \frac{\partial \mathbf{x}}{\partial \hat{x}^i} \tag{9}$$

The properties of these vectors are considered in detail in Sec. 5.4.2. In the spirit of tensor methods we shall generally ignore these vectors now, and instead consider the differential

$$dx^i = \frac{\partial x^i}{\partial \hat{x}^j} d\hat{x}^j \tag{10}$$

where the summation convention is in force, as it will be throughout. We are also now adopting the convention of using superscripts on coordinates and on all the contravariant quantities that transform like η^i and subscripts on all covariant quantities that transform like τ_i, but the position of the index can be varied in orthogonal coordinates because there is only one type of component. Combination of Eqs. (7) and (10) produces the expression for distance

$$(ds)^2 = dx_i\, dx^i = \frac{\partial x^i}{\partial \hat{x}^j} \frac{\partial x^i}{\partial \hat{x}^k} d\hat{x}^j\, d\hat{x}^k = \hat{g}_{jk}\, d\hat{x}^j\, d\hat{x}^k \tag{11}$$

where we define the second-order tensor

$$\hat{g}_{jk} = \frac{\partial x^i}{\partial \hat{x}^j} \frac{\partial x^i}{\partial \hat{x}^k} \tag{12}$$

Because of its obvious role in the measurement of distance, this quantity $\hat{g}_{ij}$ is called the *metric tensor*. Observe that $\hat{g}_{ij}$ is a symmetric tensor, and that it may be interpreted as a 3 x 3 matrix. As such it has an inverse matrix that we will denote by $\hat{g}^{ik}$ and, because the product of a matrix and its inverse must be the identity matrix with ones on the diagonal and zeros elsewhere, we may write

$$\hat{g}_{ij}\hat{g}^{ik} = \delta_j{}^k \tag{13}$$

PROBLEM 5.5.8 Show that $\hat{g}^{ij} = \hat{\eta}^i \cdot \hat{\eta}^j$. (Hint: Refer to Problem 5.4.12.)

PROBLEM 5.5.9 Show that $\hat{g}_{ij} = \{J_{\hat{x}}^{x}\}^T \{J_{\hat{x}}^{x}\}$ in which the transpose A^T is obtained from the matrix A by interchanging rows and columns. Show further that

$\hat{g}^{ij} = \{J^{\hat{x}}_{x}\}\{J^{\hat{x}}_{x}\}^T$ and that the determinant of the metric tensor is

$$g = |\hat{g}_{ij}| = |J^{x}_{\hat{x}}|^2$$

Here we use $\{J^{x}_{\hat{x}}\} = \{\partial x^i/\partial \hat{x}_j\}$, where i denotes row, j denotes column.

PROBLEM 5.5.10 Show that for an orthogonal coordinate system

$$\hat{g}^{ij} = \begin{bmatrix} h_1^{-2} & 0 & 0 \\ 0 & h_2^{-2} & 0 \\ 0 & 0 & h_3^{-2} \end{bmatrix}$$

What is $\hat{g}_{ij}$?

*PROBLEM 5.5.11 Determine the metric tensor for a cartesian coordinate system (orthogonal or not) and show that it cannot be a function of position in the coordinate system. (Hint: Any cartesian system is specified by $\hat{x}^i = A^i{}_k x^k + \hat{x}_0{}^i$. What properties does the definition of cartesian coordinates impart to $A^i{}_k$ and $\hat{x}_0{}^i$?)

Upon the introduction of another set of coordinates $\bar{z}^i = \bar{z}^i(\mathbf{x})$, which may be written as $\bar{z}^i = \bar{z}^i(\hat{\mathbf{x}})$, we are ready to proceed to the transformation laws of tensor calculus. Repeating the calculations of Eqs. (32) and (33) in Sec. 5.4, we find

$$\bar{\boldsymbol{\tau}}_i = \frac{\partial \hat{x}^k}{\partial \bar{z}^i}\hat{\boldsymbol{\tau}}_k \qquad \bar{\boldsymbol{\eta}}^i = \frac{\partial \bar{z}^i}{\partial \hat{x}^k}\hat{\boldsymbol{\eta}}^k \tag{14}$$

It is worth repeating that any quantity that transforms upon coordinate change by the same rule as the one given by Eq. (14) for the tangent vectors $\bar{\boldsymbol{\tau}}_i$ is called *covariant*; any quantity that transforms by a rule like Eq. (14) for the normal vectors $\bar{\boldsymbol{\eta}}^i$ is called *contravariant*. The transformation characteristic is indicated by the position of the index: a subscript for covariant and a superscript for contravariant.

The same rules apply to tensors of higher order. Thus we have

$$\bar{T}^{ij} = \frac{\partial \bar{z}^i}{\partial \hat{x}^k}\frac{\partial \bar{z}^j}{\partial \hat{x}^l}\hat{T}^{kl} \qquad \bar{T}^i{}_j = \frac{\partial \bar{z}^i}{\partial \hat{x}^k}\frac{\partial \hat{x}^l}{\partial \bar{z}^j}\hat{T}^k_l \qquad \bar{T}_{ij} = \frac{\partial \hat{x}^k}{\partial \bar{z}^i}\frac{\partial \hat{x}^l}{\partial \bar{z}^j}\hat{T}_{kl} \tag{15}$$

Higher-order tensors obviously transform by iterations of these rules.

Because the covariant and contravariant components are different quantities, they cannot be added together. Hence the rule is that only tensors of the same order and the same transformation characteristic may be added. But in a contraction, both types must occur, and thus a permitted inner product is $P_{ijk}{}^{lmn} = T_{ijkr}S^{rlmn}$ (the common index r could occur at any position in either of the tensors T and S) and the magnitude of a tensor is given by $|T_{ijk\cdots n}| = (T_{ijk\cdots n}T^{ijk\cdots n})^{1/2}$.

The metric tensor and its inverse are endowed with the power to change the transformation characteristics of a tensor. To show how this occurs, let us write the expression

$$\hat{A}_i = \frac{\partial x^k}{\partial \hat{x}^i} A_k \tag{16}$$

Multiplication by $\hat{g}^{ij}$ gives

$$\hat{g}^{ij} \hat{A}_i = \left(\frac{\partial \hat{x}^i}{\partial x^l} \frac{\partial \hat{x}^j}{\partial x^l} \right) \frac{\partial x^k}{\partial \hat{x}^i} A_k = \frac{\partial \hat{x}^j}{\partial x^k} A_k = \hat{A}^j \tag{17}$$

By a direct calculation it can also be shown that the inverse operation works as well. As a more efficient demonstration, we may use Eq. (13) along with Eq. (17) to obtain

$$\hat{g}_{jl} \hat{A}^j = \hat{g}_{jl} \hat{g}^{ij} \hat{A}_i = \delta^i_l \hat{A}_i = \hat{A}_l \tag{18}$$

The same property holds, naturally, for higher-order tensors, with one metric tensor or its inverse required for each index to be changed.

An important and crucial consequence of the fact that indices can be raised and lowered with the metric tensor and its inverse is that knowledge of the metric tensor and either the covariant or contravariant component will yield the other component–perhaps after calculating the inverse of the metric tensor. For example, the magnitude of a tensor may now be written as

$$|T_{ijk \cdots n}| = (g_{ir} g_{js} g_{kt} \cdots g_{nw} T^{rst \cdots w} T^{ijk \cdots n})^{1/2}$$

PROBLEM 5.5.12 Show that the metric tensor has a covariant transformation law and that its inverse is contravariant. [Hint: Be careful; you cannot just apply Eq. (15) but must show that it is true.]

PROBLEM 5.5.13 Propose a definition of the Kronecker delta $\delta_i{}^j$ that has the suitable transformation properties and gives the same numerical components in all coordinate systems.

Not every quantity with indices affixed to it is a tensor; a tensor must obey either the covariant or contravariant transformation law for each of its indices. This fact becomes the crucial one in the attempt to define a derivative. In Eq. (34) in Sec. 5.4 we saw that the derivative $\partial\phi/\partial\hat{x}^k$ was covariant, and the quantity $\partial\hat{A}_i/\partial\hat{x}^k$ at first glance appears as if it might be a second-order covariant tensor. But on changing to $\bar{z}$ coordinates we find

$$\frac{\partial \bar{A}_i}{\partial \bar{z}^k} = \frac{\partial}{\partial \bar{z}^k} \left(\frac{\partial \hat{x}^l}{\partial \bar{z}^i} \hat{A}_l \right) = \frac{\partial \hat{x}^l}{\partial \bar{z}^i} \frac{\partial \hat{A}_l}{\partial \bar{z}^k} + \hat{A}_l \frac{\partial^2 \hat{x}^l}{\partial \bar{z}^k \, \partial \bar{z}^i} \tag{19}$$

This shows that $\partial\hat{A}_i/\partial\hat{x}_k$ is not a tensor because it does not transform like a tensor.

To solve this apparent problem let us start from orthogonal cartesian coordinates with a covariant vector of the form $\hat{A}_i = (\partial x^l/\partial \hat{x}^i)A_l$ and differentiate by $\hat{x}^k$ to obtain

$$\frac{\partial \hat{A}_i}{\partial \hat{x}^k} = \frac{\partial^2 x^l}{\partial \hat{x}^k\, \partial \hat{x}^i} A_l + \frac{\partial x^l}{\partial \hat{x}^i} \frac{\partial A_l}{\partial \hat{x}^k} \tag{20}$$

But we want A_l differentiated in its own coordinate system so that

$$\frac{\partial A_l}{\partial \hat{x}^k} = \frac{\partial x^m}{\partial \hat{x}^k} \frac{\partial A_l}{\partial x^m} \tag{21}$$

and we compute that

$$\frac{\partial \hat{x}^t}{\partial x^l} \hat{A}_t = \frac{\partial \hat{x}^t}{\partial x^l} \frac{\partial x^j}{\partial \hat{x}^t} A_j = A_l \tag{22}$$

The substitution of Eqs. (21) and (22) in Eq. (20) produces the result

$$\frac{\partial \hat{A}_i}{\partial \hat{x}^k} - \frac{\partial^2 x^l}{\partial \hat{x}^k\, \partial \hat{x}^i} \frac{\partial \hat{x}^t}{\partial x^l} \hat{A}_t = \frac{\partial x^l}{\partial \hat{x}^i} \frac{\partial x^m}{\partial \hat{x}^k} \frac{\partial A_l}{\partial x^m} \tag{23}$$

The significance of this equation is that the quantity on the left is derived from the cartesian derivative $\partial A_l/\partial x^m$ by a covariant transformation law. Therefore it must be a covariant tensor of second order, even though it is the sum of two terms, neither of which is a tensor. Because this quantity is obtained by a covariant transformation of a cartesian derivative, we are justified in interpreting it as a *covariant derivative* and we shall denote it with the symbol $\hat{A}_{i;k}$.

The quantity that appears in the covariant derivative is known as the *Christoffel symbol* and is defined by

$$\hat{\Gamma}_{ki}{}^t = \frac{\partial \hat{x}^t}{\partial x^l} \frac{\partial^2 x^l}{\partial \hat{x}^k\, \partial \hat{x}^i} \tag{24}$$

where this definition depends upon the fact that the x^i coordinates are the original cartesian set. Note that the symbol is symmetric in k and i.

It is obvious from Eq. (24) that the Christoffel symbol may be evaluated directly in $\bar{z}^i$ coordinates and so

$$\bar{A}_{i;k} = \frac{\partial \bar{A}_i}{\partial \bar{z}^k} - \bar{\Gamma}_{ki}{}^t \bar{A}_t \tag{25}$$

Computations similar to that leading to Eq. (23) show that

$$\hat{A}^i{}_{;k} = \frac{\partial \hat{A}^i}{\partial \hat{x}^k} + \hat{\Gamma}_{kj}{}^i \hat{A}^j \tag{26}$$

and

$$\hat{T}^i{}_{j;k} = \frac{\partial \hat{T}^i{}_j}{\partial \hat{x}^k} - \hat{\Gamma}_{jk}{}^t \hat{T}^i{}_t + \hat{\Gamma}_{ks}{}^i \hat{T}^s{}_j \tag{27}$$

and for higher-order tensors we shall obtain one term with a Christoffel symbol for each index.

PROBLEM 5.5.14 Because of the central role of the metric tensor, we may expect that the Christoffel symbol is dependent on it. Show that

$$\hat{\Gamma}_{ki}{}^{j} = \tfrac{1}{2}\hat{g}^{jl}\left(\frac{\partial \hat{g}_{il}}{\partial \hat{x}^{k}} + \frac{\partial \hat{g}_{kl}}{\partial \hat{x}^{i}} - \frac{\partial \hat{g}_{ik}}{\partial \hat{x}^{l}}\right)$$

*PROBLEM 5.5.15 Show that the Christoffel symbol vanishes in cartesian coordinates.

PROBLEM 5.5.16 The Christoffel symbol is not a tensor, but it does have a transformation law. Show that

$$\bar{\Gamma}_{ki}{}^{t} = \frac{\partial \bar{z}^{t}}{\partial \hat{x}^{r}} \frac{\partial \hat{x}^{s}}{\partial \bar{z}^{k}} \frac{\partial \hat{x}^{u}}{\partial \bar{z}^{i}} \hat{\Gamma}_{su}{}^{r} + \frac{\partial \bar{z}^{t}}{\partial \hat{x}^{r}} \frac{\partial^{2} \hat{x}^{r}}{\partial \bar{z}^{k}\, \partial \bar{z}^{i}}$$

Now it is true that in cartesian coordinates

$$g^{ij}{}_{;k} = g_{ij;k} = 0 \tag{28}$$

so this covariant derivative must have the same value in any coordinate system.

The application of Eq. (27) to the product A^iB_j shows that the product rule applies to covariant differentiation:

$$(\hat{A}^{i}\hat{B}_{j})_{;k} = \hat{B}_{j}\hat{A}^{i}{}_{;k} + \hat{A}^{i}\hat{B}_{j;k} \tag{29}$$

Of equal significance is the fact that Eqs. (28) and (29) combine to show that

$$\hat{A}^{i}{}_{;k} = (\hat{g}^{il}\hat{A}_{l})_{;k} = \hat{g}^{il}\hat{A}_{l;k} \tag{30}$$

and hence the covariant derivative of a contravariant component may be obtained by computing the inner product of the metric tensor and a covariant derivative of a covariant component. The inverse result is, of course, also valid.

PROBLEM 5.5.17 The derivation of Eq. (28) used a logical step often useful in tensor analysis. Verify by direct computation of $\hat{g}_{ij;k}$ that the logic was indeed correct.

As a meteorological example of the application of these ideas, let us consider the total derivative, which in cartesian coordinates is written

$$\frac{d\mathbf{A}}{dt} = \frac{\partial \mathbf{A}}{\partial t} + \mathbf{v} \cdot \nabla \mathbf{A} \tag{31}$$

Because of the properties of the metric tensor and covariant differentiation in cartesian coordinates, this may be written in tensor format for these coordinates as

$$\frac{dA_i}{dt} = \frac{\partial A_i}{\partial t} + v^k \frac{\partial A_i}{\partial x^k} = \frac{\partial A_i}{\partial t} + v^k A_{i;k} \tag{32}$$

These expressions reduce to the correct form for the original coordinates so they are the correct form for the total or material derivative in any holonomic coordinate system where the component A_i, metric tensor, velocity v^r, and x^k all refer to the same system.

Similarly, the divergence in the original cartesian coordinates is

$$\nabla \cdot \mathbf{A} = \frac{\partial A^i}{\partial x^i} = A^i{}_{;i} = g^{ik} A_{k;i} \tag{33}$$

The two expressions on the right are the correct forms for the divergence in any coordinate system because, once again, they are correct in the original system. To simplify them, let us observe from Eq. (26) that the covariant derivative $\hat{A}^i{}_{;i}$ will involve the Christoffel symbol in the form $\hat{\Gamma}_{ik}{}^i$, and because there is now a product of symmetric and antisymmetric terms we have

$$\hat{\Gamma}_{ik}{}^i = \tfrac{1}{2} \hat{g}^{il} \frac{\partial \hat{g}_{il}}{\partial \hat{x}^k} \tag{34}$$

The matrix $\hat{g}^{il}$, as the inverse of $\hat{g}_{il}$, may be expressed as the cofactor $\hat{C}_{li}$ of the element in the ith row and lth column of $\hat{g}_{il}$ divided by the value of the determinant $\hat{g} = |\hat{g}_{il}|$. Thus

$$\hat{\Gamma}_{ik}{}^i = \frac{1}{2} \frac{\hat{C}_{li}}{\hat{g}} \frac{\partial \hat{g}_{il}}{\partial \hat{x}^k} \tag{35}$$

But the derivative of the determinant $\hat{g}$ is given by the derivative of each element times its cofactor, so that with Eq. (35) we see that

$$\frac{\partial \hat{g}}{\partial \hat{x}^k} = C_{li} \frac{\partial \hat{g}_{il}}{\partial \hat{x}^k} = 2\hat{g}\hat{\Gamma}_{ik}{}^i \tag{36}$$

and finally

$$\hat{\Gamma}_{ik}{}^i = \frac{\partial}{\partial \hat{x}^k} \ln \hat{g}^{1/2} = \frac{\partial}{\partial \hat{x}^k} \ln |J^x_{\hat{x}}| \tag{37}$$

in which we have used Problem 5.5.9. Now the covariant version of the divergence becomes

$$\hat{A}^i{}_{;i} = \frac{\partial \hat{A}^i}{\partial \hat{x}^i} + \frac{\hat{A}^s}{\sqrt{\hat{g}}} \frac{\partial}{\partial \hat{x}^s} (\sqrt{\hat{g}}) = \frac{1}{\sqrt{\hat{g}}} \frac{\partial}{\partial \hat{x}^i} (\sqrt{\hat{g}} \hat{A}^i) \tag{38}$$

PROBLEM 5.5.18 Show that the mass continuity equation may be written in invariant form as

$$\frac{\partial \rho}{\partial t} + \frac{1}{\sqrt{\hat{g}}} \frac{\partial}{\partial \hat{x}^i} (\rho \sqrt{\hat{g}}\, \hat{v}^i) = \frac{d\rho}{dt} + \frac{\rho}{\sqrt{\hat{g}}} \frac{\partial}{\partial \hat{x}^i} (\sqrt{\hat{g}}\, \hat{v}^i) = 0$$

Finally, the curl in cartesian coordinates is

$$\mathbf{i}_i \cdot (\nabla \times \mathbf{A}) = \epsilon_{ijk} \frac{\partial A_k}{\partial x^j} = \epsilon^{ijk} \frac{\partial A_k}{\partial x^j} \tag{39}$$

The last expression is a suitable form, provided that we determine how ϵ^{ijk} behaves under transformation of coordinates.

The ϵ_{ijk} symbol arises in cartesian coordinates in the expansion of a determinant $|a_{ij}|$, which may be written as

$$|a_{ij}| = a_{i1} a_{j2} a_{k3} \epsilon_{ijk} \tag{40}$$

and thus we may use the expansion

$$|J_{\hat{x}}^{x}| = \frac{\partial x^i}{\partial \hat{x}^1} \frac{\partial x^j}{\partial \hat{x}^2} \frac{\partial x^k}{\partial \hat{x}^3} \epsilon_{ijk} \tag{41}$$

With this we see that

$$\frac{\partial x^i}{\partial \hat{x}^r} \frac{\partial x^j}{\partial \hat{x}^s} \frac{\partial x^k}{\partial \hat{x}^t} \epsilon_{ijk} = |J_{\hat{x}}^{x}| \epsilon_{rst} \tag{42}$$

because if r, s, t are all different, we obtain Eq. (41) with appropriate sign changes if rows or columns have been interchanged; otherwise the left side vanishes because we have a determinant with two identical rows or columns.

Let us define for the moment

$$\hat{E}_{ijk} = |J_{\hat{x}}^{x}| \epsilon_{ijk} \tag{43}$$

where for the original coordinates $|J| = 1$ so that $E_{ijk} = \epsilon_{ijk}$. Now Eq. (42) may be written as

$$\hat{E}_{rst} = \frac{\partial x^i}{\partial \hat{x}^r} \frac{\partial x^j}{\partial \hat{x}^s} \frac{\partial x^k}{\partial \hat{x}^t} E_{ijk} \tag{44}$$

and thus $\hat{E}_{rst}$ transforms as a third-order covariant tensor and therefore is the correct version of E_{ijk} in any coordinates. Hence we use

$$\hat{E}_{ijk} = \sqrt{\hat{g}}\, \epsilon_{ijk} \tag{45}$$

in transformation of the permutation symbol ϵ_{ijk}. A quantity which transforms according to $\hat{R}_{ijk\cdots n} = |J_{\hat{x}}^{x}|^W R_{ijk\cdots n}$ is called a *relative tensor* of weight W.

Raising the indices of Eq. (45) gives

$$\hat{E}^{ijk} = \hat{g}^{ir} \hat{g}^{js} \hat{g}^{kt} \hat{E}_{rst} = \sqrt{\hat{g}}\, \hat{g}^{ir} \hat{g}^{js} \hat{g}^{kt} \epsilon_{rst} = \frac{\epsilon_{ijk}}{\sqrt{\hat{g}}} \tag{46}$$

where we have again used the argument that

$$\hat{g}^{ir}\hat{g}^{js}\hat{g}^{kt}\epsilon_{rst} = |\hat{g}^{mn}|\epsilon_{ijk} = \frac{1}{\hat{g}}\epsilon_{ijk} \tag{47}$$

which is obviously true according to Eq. (40) if i, j, k are all different. If two are the same, say i and j, then we have the product of the form $g^{ir}g^{is}$ symmetric in r and s and ϵ_{rst}, which is antisymmetric in r and s.

Thus referring back to Eq. (39) we write the ith component of the curl of **A** as $\omega^i(\mathbf{A})$ and we have

$$\omega^i(\mathbf{A}) = \epsilon^{ijk}A_{k;j} = E^{ijk}A_{k;j} \tag{48}$$

so that the correct tensor form is

$$\hat{\omega}^i(\mathbf{A}) = \frac{\epsilon^{ijk}}{\sqrt{\hat{g}}}\hat{A}_{k;j} = \frac{\epsilon^{ijk}}{\sqrt{\hat{g}}}\frac{\partial \hat{A}_k}{\partial \hat{x}_j} \tag{49}$$

and the covariant component is thus obtained from

$$\hat{\omega}_i(\mathbf{A}) = \hat{g}_{ij}\hat{\omega}^j(\mathbf{A}) = \frac{\hat{g}_{ij}}{\sqrt{\hat{g}}}\epsilon^{jrs}\frac{\partial \hat{A}_s}{\partial \hat{x}^r} \tag{50}$$

PROBLEM 5.5.19 Show that the invariant form of $\nabla^2\phi$ is

$$\nabla^2\phi = \frac{1}{\sqrt{\hat{g}}}\frac{\partial}{\partial \hat{x}^k}\left(\sqrt{\hat{g}}\,\hat{g}^{ks}\frac{\partial \phi}{\partial \hat{x}^s}\right)$$

PROBLEM 5.5.20 Consider a holonomic transformation $\hat{\mathbf{x}}(\mathbf{x})$ as given by Eq. (8), where the $\mathbf{x}$ are rectangular cartesian coordinates. Show that:

(*a*) Distance ds_L in the direction $\hat{\boldsymbol{\tau}}_L$ is measured by

$$ds_L = |\hat{\boldsymbol{\tau}}_L|d\hat{x}^L \qquad \text{no sum on } L$$

(*b*) Area $d\sigma$ on the surface $\hat{x}^k = \text{const}$ is measured by

$$d\sigma = |\hat{\boldsymbol{\eta}}^k|\sqrt{\hat{g}}\,d\hat{x}^i\,d\hat{x}^j \qquad i \neq j \neq k$$

(*c*) Volume dV is measured by

$$dV = \sqrt{\hat{g}}\,d\hat{x}^1\,d\hat{x}^2\,d\hat{x}^3 = |J^{\mathbf{x}}_{\hat{\mathbf{x}}}|\,d\hat{x}^1\,d\hat{x}^2\,d\hat{x}^3$$

Here ds_L, $d\sigma$, and dV are the invariant measures with units identical to those of the original system. [Hints: For (*b*): project the area outlined by $\hat{\boldsymbol{\tau}}_I\,d\hat{x}^I$ and $\hat{\boldsymbol{\tau}}_J\,d\hat{x}^J$ onto a tangent plane with Problem 5.1.7 ($I=i$, $J=j$, but no sum on I or J). For (*c*): first method, extend (*b*) to another dimension; second method, calculate the components dx^i for the three vectors obtained in the original space by taking each of $d\hat{x}^1$, $d\hat{x}^2$, $d\hat{x}^3$ separately nonzero and use Problem 5.1.7.]

PROBLEM 5.5.21 Verify that the divergence theorem and Stokes' theorem are true in arbitrary holonomic coordinates.

PROBLEM 5.5.22 Let x be cartesian coordinates and consider anholonomic coordinates defined by

$$d\mathbf{x} = \bar{\boldsymbol{\tau}}_i \, d\bar{x}^i \qquad d\bar{x}^i = \bar{\boldsymbol{\eta}}^i \cdot d\mathbf{x}$$

in which the basis vectors $\bar{\boldsymbol{\tau}}_i$ are a given linearly independent set (so that the vectors $\bar{\boldsymbol{\eta}}^i$ exist) and are *not* defined as $\bar{\boldsymbol{\tau}}_i = \partial\mathbf{x}/\partial\bar{x}^i$. The vectors $\bar{\boldsymbol{\tau}}_i$ have components $\bar{\tau}_i{}^l$, and $\bar{\boldsymbol{\eta}}^i$ has components $\bar{\eta}^i_l$. In holonomic coordinates it is always true that

$$\frac{\partial \bar{\tau}_i{}^k}{\partial \bar{x}^j} = \frac{\partial \bar{\tau}_j{}^k}{\partial \bar{x}^i}$$

(why?), but this relation is not generally valid in anholonomic coordinates. Show that $\partial\phi/\partial\bar{x}^k = \bar{\tau}_k{}^i(\partial\phi/\partial x^i)$ and that

$$\Gamma^t_{ik} = \frac{\partial \bar{\tau}_i{}^l}{\partial \bar{x}^k}\bar{\eta}^t_l = -\bar{\tau}_i{}^l \frac{\partial \bar{\eta}_l{}^t}{\partial \bar{x}^k}$$

5.6 REPRESENTATIONS OF VELOCITY FIELDS

Every velocity field has associated with it the fields of divergence, $D = \nabla \cdot \mathbf{v}$, and vorticity, $\boldsymbol{\zeta} = \nabla \times \mathbf{v}$, derived from it by differentiation. In this section, we first show that the velocity field can be determined by integrations of the vorticity and divergence over the entire domain. The second topic concerns representations of velocity fields near a point, and we show that flows may be considered as superpositions of certain pure types.

The ideas here are kinematical ones; they are mathematical relations that are true for all velocity fields but take no account of the actual forces that produce the motion. Because of their value in representing or portraying the complicated patterns we encounter in the study of fluid motion, they are widely used in both fluid dynamics and atmospheric science.

As shown in Table 5.2, when the divergence vanishes throughout a region, we refer to the flow as *solenoidal*; when the vorticity vanishes, it is *irrotational*. Now for any scalar field, we know that $\nabla \times (\nabla\phi) = 0$, so when $\mathbf{v}$ is irrotational, it is natural to try to find a function ϕ that will give $\mathbf{v} = \nabla\phi$. The assumption that such a ϕ exists leads to the differential equation

$$\nabla^2\phi = \nabla \cdot \mathbf{v} \tag{1}$$

that we can attempt to solve for ϕ, which is called a *velocity potential*. Similarly, for any vector field $\nabla \cdot (\nabla \times \boldsymbol{\psi}) = 0$, so we might try to represent solenoidal flow as $\mathbf{v} = \nabla \times \boldsymbol{\psi}$, where now

Table 5.2 SOLENOIDAL AND IRROTATIONAL COMPONENTS

Type of flow	Defining conditions	Representation	Equations	Solutions
Irrotational and solenoidal	$\nabla \times \mathbf{v} = 0$ $\nabla \cdot \mathbf{v} = 0$	$\mathbf{v} = \nabla\phi$	$\nabla^2 \phi = 0$	$\phi(\mathbf{x}) = \phi(\mathbf{x}_0) + \int_{\mathbf{x}_0}^{\mathbf{x}} \mathbf{v}(\mathbf{x}') \cdot \boldsymbol{\tau}\, ds'$
Solenoidal (incompressible)	$\nabla \cdot \mathbf{v} = 0$	$\mathbf{v} = \nabla \times \boldsymbol{\psi}$	$\nabla^2 \boldsymbol{\psi} = -\nabla \times \mathbf{v}$ $(\nabla \cdot \boldsymbol{\psi} = 0)$	$\boldsymbol{\psi}(\mathbf{x}) = \frac{1}{4\pi} \int_V \mathbf{v} \times \nabla\left(\frac{1}{\lvert\mathbf{x}' - \mathbf{x}\rvert}\right) dV_{x'}$
Irrotational	$\nabla \times \mathbf{v} = 0$	$\mathbf{v} = \nabla\phi$	$\nabla^2 \phi = \nabla \cdot \mathbf{v}$	$\phi(\mathbf{x}) = \frac{1}{4\pi} \int_V \mathbf{v} \cdot \nabla\left(\frac{1}{\lvert\mathbf{x}' - \mathbf{x}\rvert}\right) dV_{x'}$
General flows	$\nabla \times \mathbf{v} \neq 0$ $\nabla \cdot \mathbf{v} \neq 0$	$\mathbf{v} = \nabla\phi + \nabla \times \boldsymbol{\psi}$	$\nabla^2 \phi = \nabla \cdot \mathbf{v}$ $\nabla^2 \boldsymbol{\psi} = -\nabla \times \mathbf{v}$ $(\nabla \cdot \boldsymbol{\psi} = 0)$	$\phi = -\nabla \cdot \mathbf{P}$ $\boldsymbol{\psi} = \nabla \times \mathbf{P}$ $\mathbf{P} = \frac{1}{4\pi} \int_V \frac{\mathbf{v}}{\lvert\mathbf{x}' - \mathbf{x}\rvert}\, dV_{x'}$

$$\nabla \times (\nabla \times \boldsymbol{\psi}) = -\nabla^2 \boldsymbol{\psi} + \nabla(\nabla \cdot \boldsymbol{\psi}) = \nabla \times \mathbf{v} \tag{2}$$

The field $\boldsymbol{\psi}$ is called a *vector potential* for **v**. For more general flows, we combine these hypotheses and try to represent **v** as the sum of solenoidal and irrotational components formed from a potential and a vector potential. We shall take up this topic in detail after looking first at some elementary applications of the concepts already discussed.

5.6.1 Elementary Applications

Solenoidal flow appears in an important special case. If the density of each parcel is constant during the motion, then $d\rho/dt = 0$ and the equation of continuity (12) in Sec. 5.2 implies that the flow is solenoidal. If the density does not change, then the fluid parcels cannot expand or contract, and so this phenomenon is referred to as *incompressible* flow. Although no real substance is actually incompressible, the equation of continuity is often replaced with the nondivergence condition $\nabla \cdot \mathbf{v} = 0$, and in effect only the solenoidal or rotational component of the motion is considered. This is frequently a good approximation for liquids such as water, and we shall see in Chap. 13 that it is even a reasonable approximation for certain atmospheric conditions.

For some scales of atmospheric motion it is frequently justifiable to make even more stringent assumptions—although their validity should always be justified for each case considered. If we assume that the flow is incompressible and that the vertical component either vanishes or is constant in the vertical through a layer, then the continuity equation becomes

$$\nabla \cdot \mathbf{v} = \nabla_z \cdot \mathbf{v} + \frac{\partial w}{\partial z} = \nabla_z \cdot \mathbf{v} = 0 \tag{3}$$

where the subscript implies that z is held constant and thus restricts the del operator to horizontal surfaces. (For cartesian coordinates we write $\nabla_z = \mathbf{i}\ \partial/\partial x + \mathbf{j}\ \partial/\partial y$.) In this case, the two-dimensional velocity $\mathbf{v} = u\mathbf{i} + v\mathbf{j}$ is solenoidal and we see that taking $\boldsymbol{\psi} = (0,0,-\psi)$ will give

$$u = -\frac{\partial \psi}{\partial y} \qquad v = \frac{\partial \psi}{\partial x} \tag{4}$$

From these equations we find that

$$\frac{\partial^2 \psi}{\partial x^2} + \frac{\partial^2 \psi}{\partial y^2} = \nabla_z{}^2 \psi = \frac{\partial v}{\partial x} - \frac{\partial u}{\partial y} = \zeta \tag{5}$$

so that ψ is determined, provided we can solve the partial differential equation, by the vertical component ζ of the vorticity. The scalar ψ is called a *stream function.*

PROBLEM 5.6.1 Verify that $\mathbf{v} = \mathbf{k} \times \nabla\psi$ for two-dimensional, solenoidal **v**. Lines of constant ψ are called *streamlines*. Show that the vector **v** is tangent to the streamlines with large values of ψ on its right, and that the magnitude of $\mathbf{v}$ increases as the distance between lines of constant ψ decreases.

A second application concerns flows for which both the divergence and the vorticity vanish; it is particularly useful for demonstrating the important role of boundary conditions in the solution of partial differential equations.

We denote the irrotational and solenoidal velocity as $\mathbf{v}_0$, attempt the representation $\mathbf{v}_0 = \nabla\phi$, and because of the solenoidal property we have

$$\nabla^2\phi = 0 \tag{6}$$

which is *Laplace's equation*. Its solutions are called *harmonic functions* and depend on the conditions at the boundary of the region in which Eq. (6) holds. To see this, we write $\mathbf{v}_0 \cdot \mathbf{v}_0 = \nabla\phi \cdot \nabla\phi = \nabla \cdot (\phi\,\nabla\phi) - \phi\,\nabla^2\phi$. The last term vanishes by Eq. (6), and so upon applying the divergence theorem, we have

$$\int_V \mathbf{v}_0 \cdot \mathbf{v}_0\, dV = \int_V \nabla \cdot (\phi\,\nabla\phi)\, dV = \int_S \phi\mathbf{v}_0 \cdot \boldsymbol{\eta}\, d\sigma \tag{7}$$

This simple computation thus yields a dramatic result: If the normal component of the velocity vanishes on the boundary, then the integral of $\mathbf{v}_0 \cdot \mathbf{v}_0$ must vanish, and this can happen only if $\mathbf{v}_0$ vanishes everywhere. The computation also proves the unique dependence of the solutions on the boundary conditions.

PROBLEM 5.6.2 Let $\mathbf{v}_1$ and $\mathbf{v}_2$ be derived from two solutions to Eq. (6) obtained with the same boundary conditions. Show that $\mathbf{v}_1$ and $\mathbf{v}_2$ are identical. (Hint: The equation is linear.)

PROBLEM 5.6.3 Show that

$$\int_C \mathbf{v}_0 \cdot \boldsymbol{\tau}\, ds = 0$$

for any closed curve C. Thus show that an integral of $\mathbf{v}_0 \cdot \boldsymbol{\tau}$ along a curve between two end points depends only on the end points.

PROBLEM 5.6.4 We have been guilty of the common failing of deriving consequences of Eq. (6) without knowing whether or not a solution actually exists. This is easily rectified by exhibiting a solution. Show that

$$\phi(\mathbf{x}) = \phi(\mathbf{x}_0) + \int_{\mathbf{x}_0}^{\mathbf{x}} \mathbf{v}_0(\mathbf{x}') \cdot \boldsymbol{\tau}\, ds'$$

in which the integration is along any curve between the points $\mathbf{x}_0$ and $\mathbf{x}$, is a solution to Eq. (6) when the velocity field is differentiable.

PROBLEM 5.6.5 Let $\mathbf{m}_0 = (\rho \mathbf{v})_0$ be a momentum field which is irrotational and solenoidal throughout the entire atmosphere. Show that $\mathbf{m}_0$ vanishes identically when the total kinetic energy is finite.

5.6.2 Decompositions of Flows into Solenoidal and Irrotational Components

In this section we show how the divergence and vorticity fields of a given velocity field will yield potentials that allow us to recapture the velocity field. The fact that this can be done is known as *Helmholtz' theorem*, and in proving it we shall establish the representations shown in Table 5.2 for solenoidal flows, for irrotational flows, and for the general flows in which neither the divergence nor the vorticity vanishes identically.

We shall let r denote the magnitude $|\mathbf{x}' - \mathbf{x}|$, and ∇ without subscript will be in the $\mathbf{x}'$ coordinates while ∇_x will operate in the $\mathbf{x}$ coordinates. Our plan is to show that the potential

$$\mathbf{P} = \frac{1}{4\pi}\int \frac{\mathbf{v}}{r}\, dV(x') \tag{8}$$

yields the functions

$$\begin{aligned} \phi &= \frac{1}{4\pi}\int_V \mathbf{v} \cdot \nabla \frac{1}{r}\, dV(x') = -\nabla_x \cdot \left[\frac{1}{4\pi}\int_V \frac{\mathbf{v}}{r}\, dV(x')\right] \\ \boldsymbol{\psi} &= \frac{1}{4\pi}\int_V \mathbf{v} \times \nabla \frac{1}{r}\, dV(x') = \nabla_x \times \left[\frac{1}{4\pi}\int_V \frac{\mathbf{v}}{r}\, dV(x')\right] \end{aligned} \tag{9}$$

which give the representation

$$\mathbf{v} = \nabla\phi + \nabla \times \boldsymbol{\psi} \tag{10}$$

The following problems establish results used in the proof.

PROBLEM 5.6.6 Show that when $r > 0$, $\nabla r^{-1} = -\mathbf{r}/r^3$, $\nabla^2 r^{-1} = 0$, and $\nabla_x r^{-1} = -\nabla r^{-1}$. (Hint: Use cartesian coordinates.)

PROBLEM 5.6.7 (Gauss' theorem) Let V be a volume enclosed between an outer closed surface S_1 and an inner closed surface S_0. The origin is inside S_0 and $\mathbf{r}$ is a position vector. Show that

$$\int_{S_1} \eta_1 \cdot \nabla r^{-1}\, d\sigma + \int_{S_0} \eta_0 \cdot \nabla r^{-1}\, d\sigma = 0$$

Thus if S_0 is a sphere of unit radius

$$\int_{S_1} \frac{\boldsymbol{\eta} \cdot \mathbf{r}}{r^3}\, d\sigma = 4\pi$$

so $\boldsymbol{\eta} \cdot \mathbf{r}/r^3$ projects the area of S_1 onto the surface of the unit sphere. (Hint: The vectors $\boldsymbol{\eta}_0$ and $\mathbf{r}$ are in opposite directions because $\boldsymbol{\eta}_0$ is an exterior unit normal to V.)

First we must show that $\mathbf{P}$ actually exists in the sense that the integral converges. This follows from an assumed differentiability of $\mathbf{v}$, which implies continuity and thus boundedness over any finite volume. For the remaining part of the volume in case the entire region is infinite, we must assume that the positive numbers, ϵ, c, and R exist such that

$$|\mathbf{v}| < \frac{c}{r^{2+\epsilon}} \tag{11}$$

when $r > R$. Thus the integral I, which remains in Eq. (8) after subtracting the integral over a finite spherical volume of radius R, may be estimated as

$$|I| < c \int_R^\infty \frac{dr}{r^{1+\epsilon}} = \frac{c}{\epsilon} R^{-\epsilon} \tag{12}$$

Hence the entire integral is finite.

The next step is to show that $\mathbf{P}$ is continuously differentiable so that we can apply the divergence theorem to it, and to show the derivatives ∇_x may be taken under the integral sign. This is the crucial step in the proof, but it is too lengthy a task for this book; we refer the reader to Phillips (1933, p. 126). With these results established by reference, we then have, for example,

$$\nabla_x \cdot \mathbf{P} = \frac{1}{4\pi} \int_V \mathbf{v} \cdot \nabla_x \frac{1}{r}\, dV(x') \tag{13}$$

Now let us take a unit normal vector on the boundary of a region W enclosed within V and use the component P_i of $\mathbf{P}$, as obtained from Eq. (8), to write

$$\int_{S(W)} \boldsymbol{\eta} \cdot \nabla_x P_i\, d\sigma(x) = \frac{1}{4\pi} \int_V v_i \left[\int_{S(W)} \boldsymbol{\eta} \cdot \nabla_x \frac{1}{r}\, d\sigma(x) \right] dV(x') \tag{14}$$

If a point $\mathbf{x}'$ in V is outside of W, then $r \neq 0$ inside W and we may apply the divergence theorem to obtain

$$\int_{S(W)} \boldsymbol{\eta} \cdot \nabla_x \frac{1}{r}\, d\sigma(x) = \int_W \nabla_x{}^2 \frac{1}{r}\, dW(x) = 0 \tag{15}$$

which vanishes because $\nabla^2(1/r)$ vanishes everywhere except at the origin. If a point $\mathbf{x}'$ is inside W, then according to Gauss' theorem the inner integral of Eq. (14) has value -4π. Thus Eq. (14) becomes

$$\int_{S(W)} \boldsymbol{\eta} \cdot \nabla P_i \, d\sigma(x) = -\int_W v_i \, dW(x) \tag{16}$$

Because **P** is continuously differentiable, we may apply the divergence theorem to obtain

$$\int_W \nabla^2 P_i \, dW(x) = -\int_W v_i \, dW(x) \tag{17}$$

This equation is valid for every volume W inside the region, so it must be true that $\nabla^2 P_i = -v_i$, and thus

$$\nabla^2 \mathbf{P} = -\mathbf{v} \tag{18}$$

To recapitulate, what has been accomplished so far is to show that **P** is differentiable under the integral sign and that the expression for **P** given by Eq. (8) is a solution to the Poisson's equation (18).

Because **P** is differentiable, we may define the functions

$$\phi = -\nabla_x \cdot \mathbf{P} = \frac{1}{4\pi} \int_V \mathbf{v} \cdot \nabla \frac{1}{r} \, dV(x') \tag{19}$$

and

$$\boldsymbol{\psi} = \nabla_x \times \mathbf{P} = \frac{1}{4\pi} \int \mathbf{v} \times \nabla \frac{1}{r} \, dV(x') \tag{20}$$

These integrals converge if $|\mathbf{v}| \to 0$ as r^{-2} at infinity even though the integral for **P** would not in this case. Now the application of the vector identity $\nabla^2 \mathbf{P} = \nabla(\nabla \cdot \mathbf{P}) - \nabla \times (\nabla \times \mathbf{P})$ shows that

$$\mathbf{v} = \nabla\phi + \nabla \times \boldsymbol{\psi} \tag{21}$$

Thus we have shown that we can recover **v** from the potential **P**, but we have not yet shown that the functions ϕ and $\boldsymbol{\psi}$ are determined by the divergence and vorticity patterns. Toward this end, consider a volume $V - V_0$, where V_0 is a small volume enclosing the origin. If we let V_0 be a sphere of radius ϵ, then

$$\left| \int_{S_0} \frac{\mathbf{v} \cdot \boldsymbol{\eta}}{r} \, d\sigma \right| \leqslant \frac{\max\,[\mathbf{v} \cdot \boldsymbol{\eta}]}{\epsilon} \int_{S_0} d\sigma = 4\pi\epsilon \max\,[\mathbf{v} \cdot \boldsymbol{\eta}] \tag{22}$$

In $V - V_0$, the function $\mathbf{v}/r$ is continuously differentiable and so

$$\int_{V-V_0} \nabla \cdot \frac{\mathbf{v}}{r} \, dV(x') = \int_{S_1} \frac{\boldsymbol{\eta} \cdot \mathbf{v}}{r} \, d\sigma + \int_{S_0} \frac{\boldsymbol{\eta} \cdot \mathbf{v}}{r} \, d\sigma \tag{23}$$

in which S_1 is the outer boundary of V. But by Eq. (22), the second integral vanishes in the limit, and so the expression for ϕ in Eq. (9) may be written

$$\phi(\mathbf{x}) = -\frac{1}{4\pi}\int_V \frac{\nabla \cdot \mathbf{v}}{r}\, dV(x') + \frac{1}{4\pi}\int_{S_1} \frac{\mathbf{v} \cdot \boldsymbol{\eta}}{r}\, d\sigma(x') \tag{24}$$

Similarly,

$$\boldsymbol{\psi}(\mathbf{x}) = \frac{1}{4\pi}\int_V \frac{\boldsymbol{\zeta}(x')}{r}\, dV(x') - \frac{1}{4\pi}\int_{S_1} \frac{\boldsymbol{\eta} \times \mathbf{v}}{r}\, d\sigma(x') \tag{25}$$

We have shown then that the divergence and vorticity patterns generate velocity potentials that can be used to separate the flow into its solenoidal and irrotational components. The main point of all this for meteorology is not that velocities are found from the potentials in practice, but that we have demonstrated that the flow fields can be separated into their solenoidal and irrotational components when that will prove to be of advantage in the study of atmospheric motions.

PROBLEM 5.6.8 Show that the potentials **P**, ϕ, and $\boldsymbol{\psi}$ are not unique, in the sense that once determined, they may be modified by addition of suitably chosen arbitrary functions and still yield the result (10).

PROBLEM 5.6.9 Show that $\boldsymbol{\psi}$ is solenoidal.

PROBLEM 5.6.10 We cannot be sure that the velocity fields in the entire atmosphere meet the conditions at infinity necessary to apply Helmholtz' theorem, but we may assume that the momentum fields $\mathbf{m} = \rho\mathbf{v}$ do. Determine suitable forms of potentials that will give the momentum field from the mass divergence $M = \nabla \cdot (\rho\mathbf{v})$ and the mass vorticity $\boldsymbol{\mu} = \nabla \times (\rho\mathbf{v})$.

5.6.3 Pure Types of Motion: Two-dimensional Flow

The decomposition of a two-dimensional flow into components of pure motion types is particularly important to meteorology. The analyst of weather patterns and the forecaster are always faced with understanding and interpreting precisely what is occurring in the flows represented on weather charts; having the pure types of motion in mind is a considerable aid to the intuitive understanding of the dynamics of the flow.

We consider a flow with only horizontal velocity components. Then by Taylor's theorem, the velocities surrounding a point $\mathbf{x}_0$ are given by the expansion

$$\begin{aligned}\mathbf{u}(\mathbf{x}_0 + \mathbf{r}) &= \mathbf{u}(\mathbf{x}_0) + r_1\left(\frac{\partial \mathbf{u}}{\partial x_1}\right)_{\mathbf{x}_0} + r_2\left(\frac{\partial \mathbf{u}}{\partial x_2}\right)_{\mathbf{x}_0} + O(r^2) \\ &= \mathbf{u}(\mathbf{x}_0) + \mathbf{r} \cdot (\nabla\mathbf{u})_{x_0} + O(r^2)\end{aligned} \tag{26}$$

where the term $O(r^2)$ vanishes at the rate r^2 as r approaches zero. Writing out the components of Eq. (26), we put $\mathbf{x} = \mathbf{x}_0 + \mathbf{r}$ and leave implicit the fact that derivatives are to be evaluated at $\mathbf{x}_0$, and by adding and subtracting the same quantities in a series of steps we arrive at

$$u_1(\mathbf{x}) = u_1(\mathbf{x}_0) + \frac{r_1}{2}D + \frac{r_1}{2}T + \frac{r_2}{2}H - \frac{r_2}{2}\zeta$$
$$u_2(\mathbf{x}) = u_2(\mathbf{x}_0) + \frac{r_2}{2}D - \frac{r_2}{2}T + \frac{r_1}{2}H + \frac{r_1}{2}\zeta \qquad (27)$$

where

$$D = \frac{\partial u_1}{\partial x_1} + \frac{\partial u_2}{\partial x_2} \qquad \zeta = \frac{\partial u_2}{\partial x_1} - \frac{\partial u_1}{\partial x_2}$$

HORIZONTAL DIVERGENCE   VERTICAL COMPONENT OF THE VORTICITY   (28)

$$H = \frac{\partial u_1}{\partial x_2} + \frac{\partial u_2}{\partial x_1} \qquad T = \frac{\partial u_1}{\partial x_1} - \frac{\partial u_2}{\partial x_2}$$

SHEARING DEFORMATION   STRETCHING DEFORMATION

Equations (27) give the *linear variation in velocity* near a point. The velocity pattern arising from them by letting all terms except one vanish is considered a pure type of motion.

PROBLEM 5.6.11 Verify that Eqs. (27) follow from Eq. (26) and the definitions in Eq. (28).

PROBLEM 5.6.12 Show that the value of the divergence, vorticity, and the deformations of $\mathbf{u}$ near a point $\mathbf{x}_0$ are given by the constant values in Eqs. (27). (Hint: $\partial/\partial x_i = \partial/\partial r_i$.)

The simplest pure type is *translation*, obtained by retaining only the first terms in each of the equations (27) so that

$$\mathbf{u} = u_1(\mathbf{x}_0)\mathbf{i} + u_2(\mathbf{x}_0)\mathbf{j} \qquad (29)$$

The character of the other types is most easily revealed by using polar coordinates (r,θ) with θ increasing counterclockwise from the x_1 axis. Then $r_1 = r\cos\theta$, $r_2 = r\sin\theta$, and we have the orthogonal unit vectors $\mathbf{l}$ in the radial direction and $\mathbf{t}$ in the direction of increasing θ. Then $\mathbf{l}\cdot\mathbf{i} = \mathbf{t}\cdot\mathbf{j} = \cos\theta$ and $\mathbf{l}\cdot\mathbf{j} = -\mathbf{t}\cdot\mathbf{i} = \sin\theta$; the velocity component u_r is in the radial direction and u_θ is in the direction of $\mathbf{t}$. Denoting the vector velocity field associated with each pure type with the appropriate subscript, we have now

$$\begin{aligned}
\mathbf{u}_D(\mathbf{x}) &= u_r\mathbf{l} = \frac{rD}{2}\,\mathbf{l} \\
\mathbf{u}_\zeta(\mathbf{x}) &= u_\theta\mathbf{t} = \frac{r\zeta}{2}\,\mathbf{t} \\
\mathbf{u}_T(\mathbf{x}) &= (\mathbf{l}\cos 2\theta - \mathbf{t}\sin 2\theta)\frac{rT}{2} \\
\mathbf{u}_H(\mathbf{x}) &= (\mathbf{l}\sin 2\theta + \mathbf{t}\cos 2\theta)\frac{rH}{2}
\end{aligned} \tag{30}$$

Thus divergent motion has radial components only (expansion for $D > 0$, convergence for $D < 0$), rotational motion has tangential components only (*cyclonic* for $\zeta > 0$, *anticyclonic* for $\zeta < 0$); in stretching deformation the velocity is outward along the axes at 0 and π, and inward along those at $\pi/2$ and $3\pi/2$ so that parcels are compressed in the y direction and stretched in the x direction, and shearing deformation is an identical pattern rotated by $\pi/4$ in the counterclockwise direction. These patterns are illustrated in Fig. 5.8.

PROBLEM 5.6.13 Verify the results in Eqs. (30).

PROBLEM 5.6.14 Show that a motion field with only vorticity ζ is equivalent to that of a solid body rotation with angular velocity $\omega = \zeta/2$.

Principal axes of deformation Combination of the two components of deformation from Eqs. (30) gives

$$\mathbf{u}(\mathbf{x}) = [\mathbf{l}(T\cos 2\theta + H\sin 2\theta) + \mathbf{t}(H\cos 2\theta - T\sin 2\theta)]\frac{r}{2} \tag{31}$$

We can let

$$\begin{aligned}
T &= F\sin\psi \qquad F = (T^2 + H^2)^{1/2} \\
H &= F\cos\psi \qquad \psi = \tan^{-1}\frac{T}{H}
\end{aligned} \tag{32}$$

so Eq. (31) becomes

$$\mathbf{u}(\mathbf{x}) = [\mathbf{l}\sin(\psi + 2\theta) + \mathbf{t}\cos(2\theta + \psi)]\frac{Fr}{2} \tag{33}$$

Now the axes of outward radial motion are at $\theta = (\pi/2 - \psi)/2$ and $\theta = (3\pi/2 - \psi)/2$. Rotation of the $\mathbf{i}$ and $\mathbf{j}$ axis system counterclockwise through the angle $(\pi/2 - \psi)/2$ will result in $\mathbf{i}$ coinciding with the axis of outward motion and $\mathbf{j}$ coinciding with the axis of inward motion. These new axes are the *principal axes of deformation*. This choice of axes clearly does not affect the values of the vorticity or the divergence.

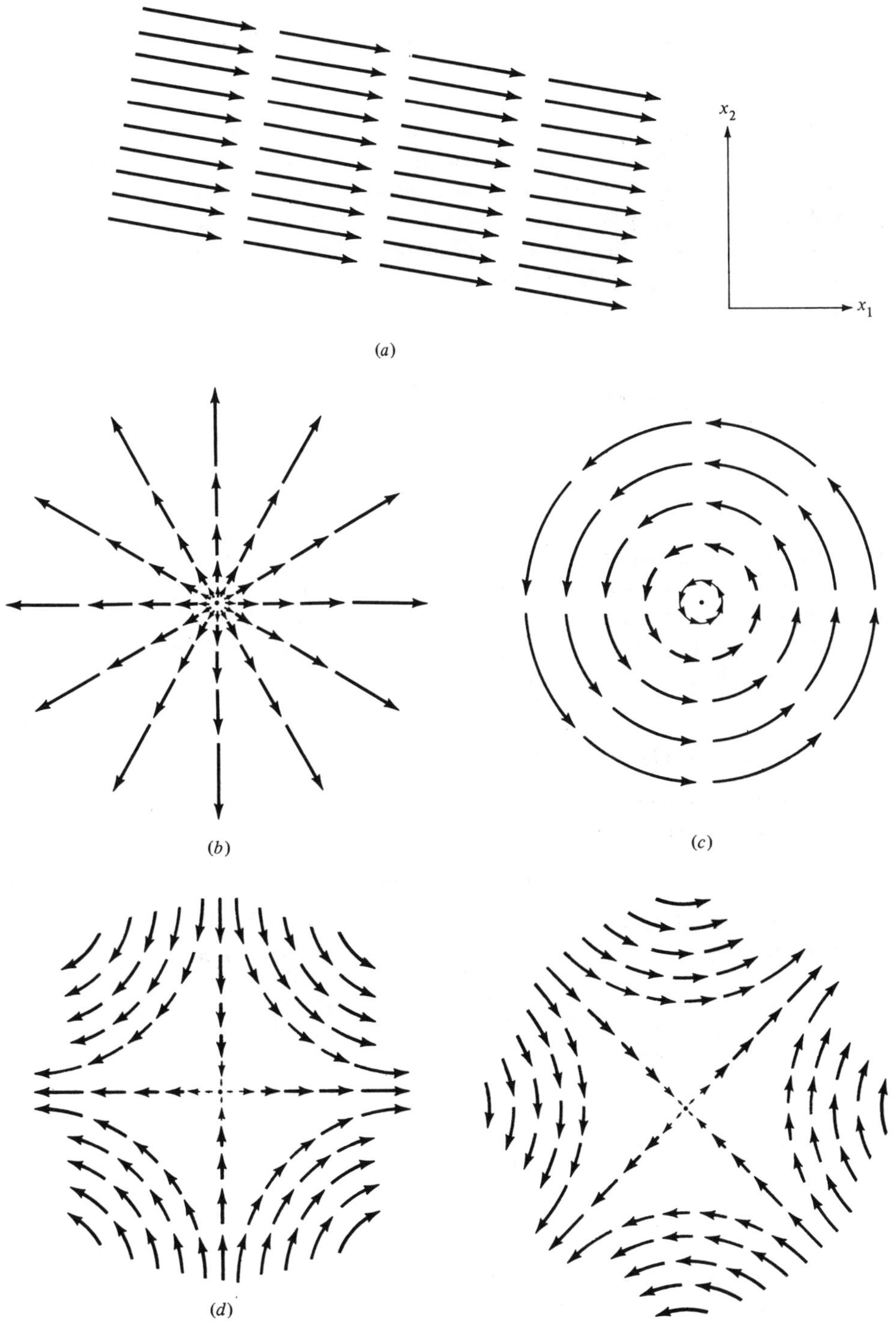

FIGURE 5.8
Pure types of motion at the point x_0. Translatory (*a*); divergent motion (direction reverses for convergence) (*b*); rotational motion with positive or cyclonic vorticity (direction reverses for anticyclonic vorticity) (*c*); stretching deformation (*d*); shearing deformation (*e*). All arrow lengths are proportional to speed.

The decomposition given above shows that an arbitrary two-dimensional velocity field may be considered to be composed of the four types of motion–translation, divergence, deformation, and rotation–and illustrates that the study of patterns of atmospheric motion is indeed a complicated one. It is not at all easy to look at a weather map and detect the pure types of motion; deformation is often confused with divergence, and shearing motion involving both components may appear rotational when in fact the average vorticity near a point is zero.

PROBLEM 5.6.15 (Truesdell) The relative significance of rotation is measured by the quantity

$$R = \left(\frac{\zeta^2}{D^2 + T^2 + H^2} \right)^{1/2}$$

Show that $R = \infty$ for rigid body rotation and that $R = 1$ for a pure shear flow of the type $u = kr_2$, $v = 0$. Show that when $u = hr_2$ and $v = kr_1$, then $R = [(k-h)^2/(k+h)^2]^{1/2}$ so that $h = k$ is equivalent to irrotational flow while $h = -k$ is equivalent to rigid body rotation. Show that the vorticity does indeed vanish in the case $h = k$. The parameters h and k are constants.

5.6.4 Pure Types of Motion: Three-dimensional Flow

Analysis of the components of three-dimensional flow proceeds in an analogous, but more complicated, manner. Starting again at a point $\mathbf{x}_0$, we have

$$\mathbf{u}(\mathbf{x}) = \mathbf{u}(\mathbf{x}_0) + \mathbf{r} \cdot (\nabla \mathbf{u})_{\mathbf{x}_0} + O(r^2) \tag{34}$$

and dropping the subscript on the derivative, we may write the ith component of the first-order term as

$$\mathbf{i}_i \cdot (\mathbf{r} \cdot \nabla \mathbf{u}) = r^k \frac{\partial u_i}{\partial x^k} = r^k \left[\frac{1}{2}\left(\frac{\partial u_i}{\partial x^k} + \frac{\partial u_k}{\partial x^i} \right) + \frac{1}{2}\left(\frac{\partial u_i}{\partial x^k} - \frac{\partial u_k}{\partial x^i} \right) \right] \tag{35}$$

PROBLEM 5.6.16 Show that

$$\mathbf{i}_i \cdot (\mathbf{r} \times \boldsymbol{\zeta}) = -r^j \left(\frac{\partial u_i}{\partial x^j} - \frac{\partial u_j}{\partial x^i} \right)$$

and hence that the components of the second term of Eq. (35) may be interpreted as the tangential velocities arising from a solid body rotation with vector angular velocity $\boldsymbol{\zeta}/2$.

By analogy with the decomposition of two-dimensional motion, we may conclude that the symmetric tensor

$$e_{ij} = \frac{1}{2}\left(\frac{\partial u_i}{\partial x_j} + \frac{\partial u_j}{\partial x_i}\right) \tag{36}$$

known as the *rate of strain tensor*, represents the sum of three-dimensional divergence and deformation. The three derivatives that form the divergence obviously appear on the main diagonal of e_{ij}. With the identity $\partial r^l/\partial r^k = \delta^l_k$, we may write

$$r^k e_{ik} = \frac{1}{2}\frac{\partial}{\partial r^i}(r^m r^n e_{mn}) \tag{37}$$

in which we use the symmetry of e_{ij} and have dropped the subscript $\mathbf{x}_0$.

Thus the definition of the scalar quantity $E_0 = \frac{1}{2}r^m r^n (e_{mn})_{\mathbf{x}_0}$ allows us to put Eq. (34) in the form

$$\mathbf{u}(\mathbf{x}) = \mathbf{u}(\mathbf{x}_0) + \nabla E_0 + \tfrac{1}{2}\boldsymbol{\zeta}_0 \times \mathbf{r} \tag{38}$$

The surfaces such that $E_0 = \frac{1}{2}r^m r^n (e_{mn})_{\mathbf{x}_0} = \text{const}$ involve second powers of the coordinates r^j and are therefore called *quadric surfaces*. According to Eq. (38), the vector $\mathbf{u}(\mathbf{x})$ clearly has a component normal to the quadric surface passing through the point $\mathbf{x} = \mathbf{x}_0 + \mathbf{r}$.

For two-dimensional motion we were able to find the principal axes of deformation. The three-dimensional equivalent is to find *principal axes of strain* so that the strain or deformation may be considered as expansion or contraction parallel to these axes. We change to $\hat{r}^k$ coordinates, and according to the covariant law,

$$\hat{e}_{ij} = \frac{\partial r^l}{\partial \hat{r}^i}\frac{\partial r^m}{\partial \hat{r}^j} e_{lm} \tag{39}$$

Let $\{r_{ij}\}$ denote the matrix $\partial r^i/\partial \hat{r}^j$; then Eq. (39) may be written as a matrix product in the form

$$\hat{e}_{ij} = \frac{\partial r^l}{\partial \hat{r}^i} e_{lm} \frac{\partial r^m}{\partial \hat{r}^j} = \{r_{li}\}^T \{e_{lm}\}\{r_{mj}\} \tag{40}$$

where T denotes a transpose. It is proved in matrix theory that any square symmetric matrix admits a diagonalization in the form of Eq. (40) by an *orthogonal matrix*, that is, one whose transpose equals its inverse. An orthogonal matrix $\{r_{li}\}$ has the property that $r_{li}r_{lj} = \delta_{ij}$, so that

$$r_{li}r_{lj} = \frac{\partial r^l}{\partial \hat{r}^j}\frac{\partial r^l}{\partial \hat{r}^i} = \hat{g}_{ij} = \delta_{ij} \tag{41}$$

Thus the coordinates that yield a diagonal form for $\hat{e}_{ij}$ are orthogonal cartesian coordinates with the same distance scaling as the original ones, and they are equivalent to a rotation of the original coordinate axes. Furthermore, we now may conclude from Eq. (39) that

$$\hat{e}_{ii} = \frac{\partial r^l}{\partial \hat{r}^i}\frac{\partial r^m}{\partial \hat{r}^i} e_{lm} = \hat{g}^{lm} e_{lm} = e_{ll} \tag{42}$$

because $\hat{g}^{lm}$ is the inverse of $\hat{g}_{lm}$ and therefore is the matrix δ^{lm}. Thus the sum $\hat{e}_{ii}$ of the elements on the diagonal of $\{\hat{e}_{ij}\}$ is the divergence $\partial u_i/\partial x^i$. In these new coordinates the scalar function E_0 becomes

$$2E_0 = \hat{r}_m\hat{r}_n(\hat{e}_{mn})_{\mathbf{x}_0} = \hat{r}_1^{\,2}(\hat{e}_{11})_0 + \hat{r}_2^{\,2}(\hat{e}_{22})_0 + \hat{r}_3^{\,2}(\hat{e}_{33})_0 \tag{43}$$

where we have shown in Eq. (42) that the factors, $\hat{e}_{11}$, $\hat{e}_{22}$, and $\hat{e}_{33}$, sum to the divergence.

The conclusion is that the contribution from ∇E_0 to the vector $\mathbf{u}(\mathbf{x})$ is in the direction of the normal to the surface $E_0 = \text{const}$, and is the sum of expansions or contractions in the direction of the unit vectors at the rates $\hat{r}_1(\hat{e}_{11})_0$, $\hat{r}_2(\hat{e}_{22})_0$, and $\hat{r}_3(\hat{e}_{33})_0$. This leads to the important conclusion that a thin filament of fluid originally aligned along the $\hat{r}_1$ axis will be stretched further along that axis at the rate $\hat{e}_{11}$; the same is true for axes $\hat{r}_2$ and $\hat{r}_3$.

In the $\hat{r}$ coordinates we have

$$\nabla E_0 = \mathbf{i}_1\hat{r}_1(\hat{e}_{11})_0 + \mathbf{i}_2\hat{r}_2(\hat{e}_{22})_0 + \mathbf{i}_3\hat{r}_3(\hat{e}_{33})_0 \tag{44}$$

and there is no contribution to the vorticity because $\nabla \times (\nabla E_0) = 0$; the divergence is

$$\nabla \cdot \nabla E_0 = (\hat{e}_{11})_0 + (\hat{e}_{22})_0 + (\hat{e}_{33})_0 \tag{45}$$

in accordance with Eq. (42).

A final point is that the relation (42) *does not* imply that $\hat{e}_{11} = e_{11}$. The factors $\hat{e}_{11}$ are the characteristic roots or eigenvalues of the matrix $\{e_{ij}\}$ and are not in general equal to the diagonal elements.

PROBLEM 5.6.17 To apply these ideas to large-scale atmospheric flow, assume that horizontal shears are small compared to vertical shears and that the vertical component of the velocity vanishes. Then the only remaining terms of e_{ij} are $u_{1;3}$ and $u_{2;3}$. Show that the eigenvalues of the matrix $\{e_{ij}\}$ are $\lambda_1 = -\lambda_3 = [(u_{1;3})^2 + (u_{2;3})^2]^{1/2}$, $\lambda_2 = 0$, and find the associated eigenvectors ξ_n as solutions of the equations $\{e_{ij}\}\,\xi_n = \lambda_n\xi_n$ where ξ_n is a column vector. Find the unit vectors τ_i of the $\hat{r}$ system and show that there is stretching along $\hat{\tau}_1$, contraction along $\hat{\tau}_3$, and no strain along $\hat{\tau}_2$. Illustrate your results with a suitable figure. Use the case $u_{2;3} = 0$ and $u_{1;3} > 0$ to show how a line of parcels initially vertical is stretched out by the wind shear. What does the vorticity of the flow do to the line of parcels?

5.7 A MATHEMATICAL MODEL OF FLUID FLOW; EULERIAN AND MATERIAL DESCRIPTIONS

The motions of actual fluids are usually too complex for theoretical study, so we must develop a mathematical model that represents the essential features of the physical phenomenon. Thus we show now which mathematical concepts provide an abstract foundation on which we can base the theoretical study of fluids in motion.

At some initial time t_0, each parcel of the fluid is considered to be at some definite point whose coordinates may be specified by the vector $\boldsymbol{\xi}$. At some later time, the parcel that was at $\boldsymbol{\xi}$ at time t_0 is now at a new position denoted by $\mathbf{x}(t)$. On the assumption that each parcel that could be observed at time t_0 has a definite position at time t, we clearly have a transformation of the form

$$\mathbf{x}(t) = \mathbf{x}(\boldsymbol{\xi}, t) \tag{1}$$

The study of fluid motion thus has its foundation in the study of such transformations of three-dimensional space into itself.

Naturally, we assume that the Jacobian $\{J_{\xi}^{x}\}$ has a nonvanishing determinant so that, theoretically at least, we could find the inverse transformation

$$\boldsymbol{\xi} = \boldsymbol{\xi}(\mathbf{x}, t) \tag{2}$$

that would give the initial coordinates of the parcel of fluid, which at time t is at $\mathbf{x}$. We have therefore assumed that the spatial coordinates $\boldsymbol{\xi}$ of the parcel at t_0 provide a means of identifying each fluid parcel for all time. These initial coordinates are called *material* or *Lagrangian coordinates*.

If we hold t fixed in Eq. (1), we obtain the coordinates $\mathbf{x}$ at time t of the parcels which were at $\boldsymbol{\xi}$ at time t_0. We assume that this is a continuous transformation, and even more stringently, we shall require that $\mathbf{x}(\boldsymbol{\xi},t)$ be differentiable at least three times in order to write the equations of motion. The continuity of the transformation requires that

$$\lim_{\Delta\boldsymbol{\xi}\to 0} |\mathbf{x}(\boldsymbol{\xi} + \Delta\boldsymbol{\xi}, t) - \mathbf{x}(\boldsymbol{\xi}, t)| = 0 \tag{3}$$

or that parcels which are initially neighbors remain neighbors. Thus a line of fluid particles at time t_0 must remain an unbroken line at time t, no matter how it is distorted by the motion. A further consequence is that parcels inside a closed surface at t_0 are forever separated from those outside. Thus we are not permitting obstacles such as knife-edges that split the material lines or surfaces of the flow. An illustration of the evolution of a flow is given in Fig. 5.9, as originally presented by Welander (1955). These examples show how convoluted an initial pattern can become at subsequent times.

If $\boldsymbol{\xi}$, rather than t, is held constant in Eq. (1), the equation will give the trajectory of the fluid parcel that was at $\boldsymbol{\xi}$ at t_0 as a function of t. Thus the velocity of the particle is defined by

$$\mathbf{v} = \frac{\partial}{\partial t_{\xi}} \mathbf{x}(\boldsymbol{\xi}, t) \tag{4}$$

where the subscript denotes a derivative with $\boldsymbol{\xi}$ held constant.

Any property of the fluid or its motion may be referred to either set of coordinates. So far in this book, we have been using the spatial or Eulerian $\mathbf{x}$ coordinates to identify $f(\mathbf{x},t)$ as the value of f for the parcel at point $\mathbf{x}$ at time $\mathbf{t}$. We can equally well use the transformation (1) to write

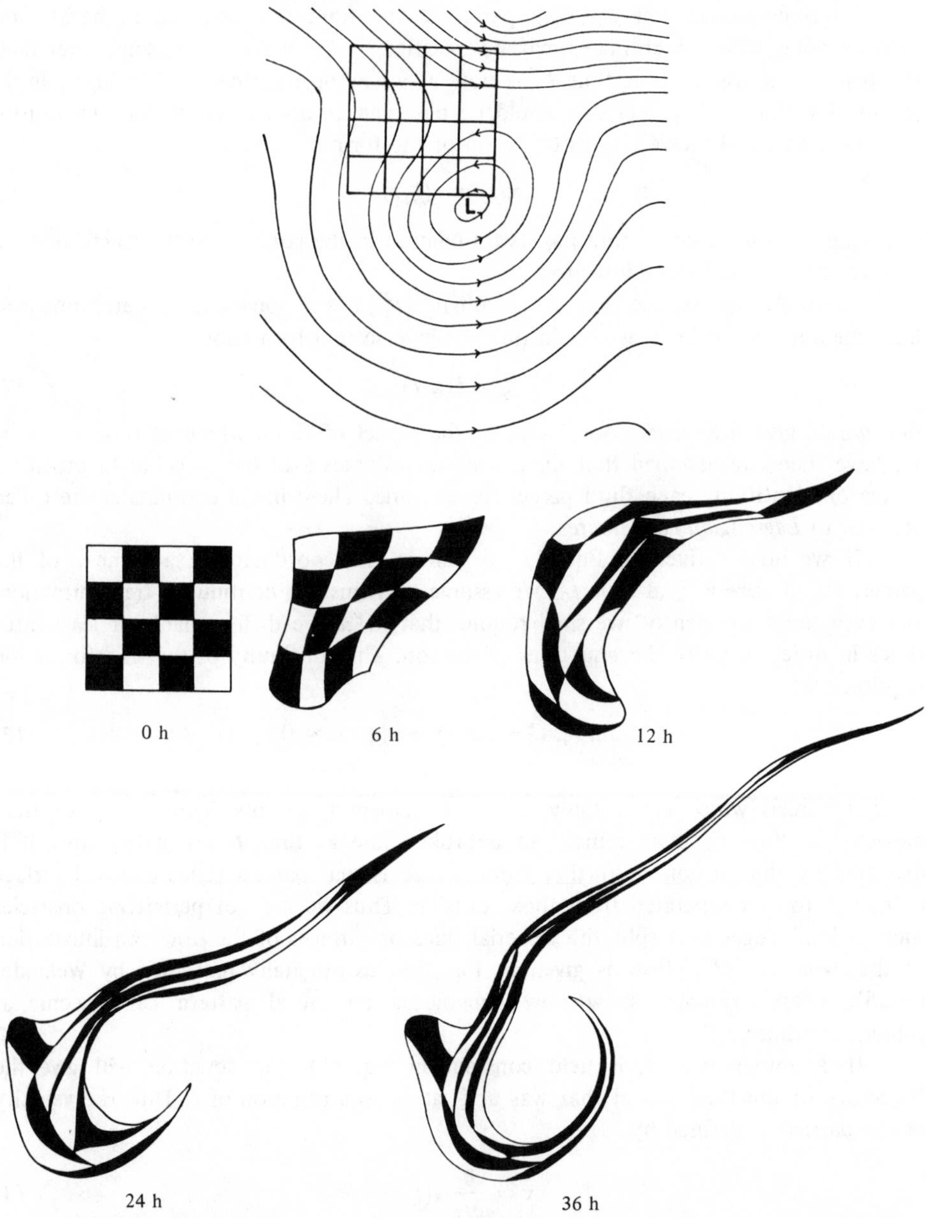

FIGURE 5.9
Evolution of a pattern in (*above*) a numerically simulated and (*next page*) an experimental flow (Welander, 1955). The numerical example was developed with a two-dimensional barotropic model that simulated the evolution of the initial 500-mbar pattern as shown by the streamlines. The experimental case was obtained by observing the history of patches of colored water initially confined by a grid at the surface of a rotating tank.

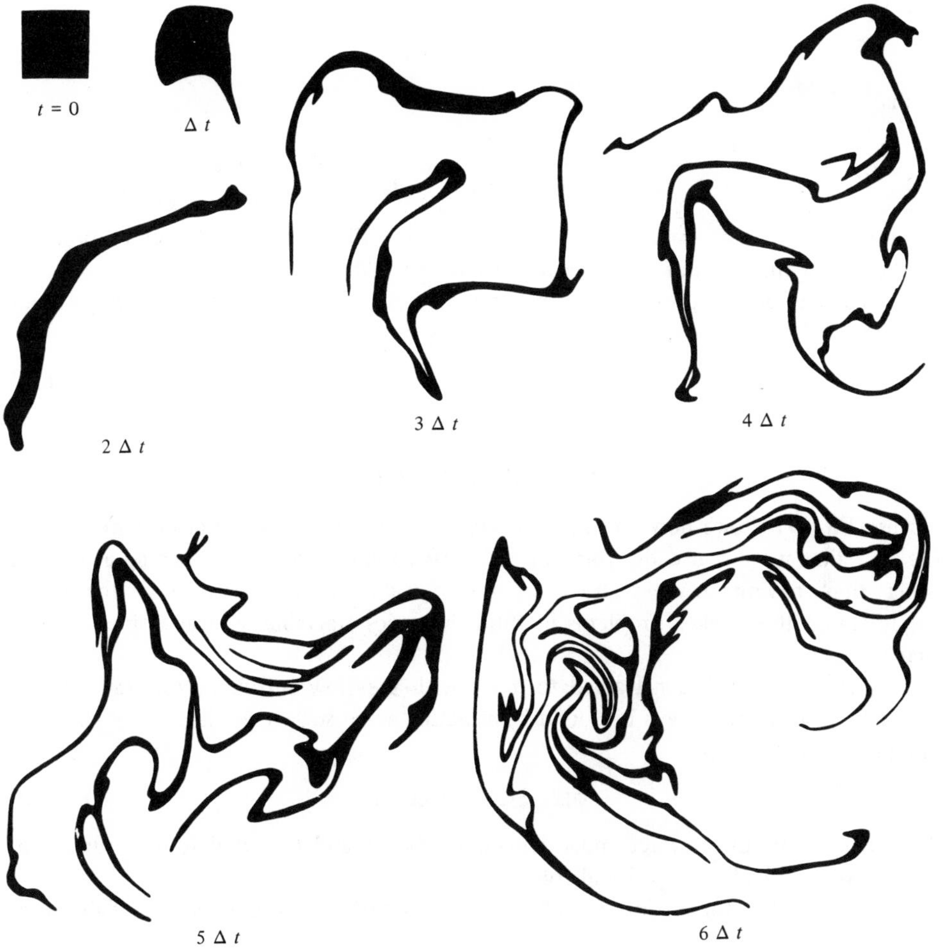

FIGURE 5.9 (*Continued*)

$$f(\mathbf{x}, t) = f(\mathbf{x}(\boldsymbol{\xi}, t), t) = f(\boldsymbol{\xi}, t) \tag{5}$$

An important consequence of the model is that the initial coordinate $\boldsymbol{\xi}$ is an invariant of the motion. Thus the crucial expression for fluid mechanics is that

$$\frac{d\boldsymbol{\xi}}{dt} = 0 \tag{6}$$

With this we may now consider the total derivative

$$\frac{df}{dt} = \frac{df(\boldsymbol{\xi}, t)}{dt} = \frac{\partial f}{\partial t_{\xi}} + \frac{\partial f}{\partial \xi^i}\frac{d\xi^i}{dt} = \frac{\partial f}{\partial t_{\xi}} \tag{7}$$

This last expression may be further expanded by considering

$$f = f(\mathbf{x}(\boldsymbol{\xi}, t), t) \tag{8}$$

so that

$$\frac{\partial f}{\partial t_\xi} = \frac{\partial f}{\partial t_x} + \frac{\partial f}{\partial x^i}\frac{\partial x^i}{\partial t_\xi} = \frac{\partial f}{\partial t_x} + v^i \frac{\partial f}{\partial x^i} = \frac{\partial f}{\partial t_x} + \mathbf{v} \cdot \nabla f \tag{9}$$

Because of the fundamental constraint (6), total derivatives with respect to time are identical with the *material* or *convective derivative*, $\partial/\partial t_\xi$.

The relation (9) provides the means for passing from the material description of the flow to the spatial description. Thus we may consider the functions

$$f = f(\mathbf{x}, t) \tag{10}$$

instead of

$$f = f(\mathbf{x}(\boldsymbol{\xi}, t), t) = f(\boldsymbol{\xi}, t) \tag{11}$$

because Eq. (9) shows how the rates of change of the two representations are related. The fact that we can use the form of Eq. (10) to study what is happening at a point rather than having to keep track of the motion of each particle is the major simplification that makes problems in fluid dynamics tractable, as was pointed out in Chap. 2.

A surface that is constrained to move with the flow is called a *material surface*. Material surfaces can always be found by isolating some surface at time t_0 by choosing an equation of the form

$$\phi(\xi_1, \xi_2, \xi_3) = \text{const} \tag{12}$$

The particles on this surface must remain neighbors and thus will form a surface at time t, however badly warped and folded.

Suppose that for an interval of time such a surface has an equation of the form

$$F(\mathbf{x}, t) = \text{const} \tag{13}$$

Then the total derivative of the surface is given by

$$\frac{dF}{dt} = \frac{\partial F}{\partial t_x} + \frac{\partial F}{\partial x^i}\left(\frac{\partial x^i}{\partial t_\xi} + \frac{\partial x^i}{\partial \xi^j}\frac{d\xi^j}{dt}\right) = 0 \tag{14}$$

But $d\xi^j/dt$ is zero for the parcels on a material surface. Thus the necessary and sufficient condition for Eq. (13) to represent a material surface is that

$$\frac{\partial F}{\partial t_\xi} = \frac{\partial F}{\partial t_x} + v^i \frac{\partial F}{\partial x^i} = 0 \tag{15}$$

If instead we let

$$G(\mathbf{x}, t) = \text{const} \tag{16}$$

be the equation of a nonmaterial surface, we will have

$$\left(\frac{dG}{dt}\right)_G = \frac{\partial G}{\partial t_x} + \frac{\partial G}{\partial x^i}\frac{dx^i}{dt} = 0 \tag{17}$$

where now dx^i/dt represents the velocity w_i of the nonmaterial surface and $(\ )_G$ denotes a derivative following the surface G. Thus with Eq. (17) we conclude that

$$\frac{\partial G}{\partial t} = -\mathbf{w} \cdot \nabla G \tag{18}$$

But Eq. (9) must hold for $G(\mathbf{x},t)$, and so we have

$$\frac{dG}{dt} = (\mathbf{v} - \mathbf{w}) \cdot \nabla G \tag{19}$$

as the total derivative of a nonmaterial surface.

Finally, we derive again a result we used in a slightly more general setting in the derivation of the transport theorem. We have

$$\{J_\xi^x\} = \left\{\frac{\partial x^i}{\partial \xi^j}\right\} \tag{20}$$

where i denotes row and j denotes column. With an obvious modification of Eq. (41) in Sec. 5.5 we may write out an expansion

$$|J_\xi^x| = \frac{\partial x^1}{\partial \xi^i}\frac{\partial x^2}{\partial \xi^j}\frac{\partial x^3}{\partial \xi^k}\epsilon_{ijk} \tag{21}$$

Thus

$$\frac{\partial}{\partial t_\xi}|J_\xi^x| = \left(\frac{\partial v^1}{\partial \xi^i}\frac{\partial x^2}{\partial \xi^j}\frac{\partial x^3}{\partial \xi^k} + \frac{\partial v^2}{\partial \xi^j}\frac{\partial x^1}{\partial \xi^i}\frac{\partial x^3}{\partial \xi^k} + \frac{\partial v^3}{\partial \xi^k}\frac{\partial x^1}{\partial \xi^i}\frac{\partial x^2}{\partial \xi^j}\right)\epsilon_{ijk} \tag{22}$$

The first term becomes

$$\frac{\partial v^1}{\partial \xi^i}\frac{\partial x^2}{\partial \xi^j}\frac{\partial x^3}{\partial \xi^k}\epsilon_{ijk} = \frac{\partial v^1}{\partial x^l}\left(\frac{\partial x^l}{\partial \xi^i}\frac{\partial x^2}{\partial \xi^j}\frac{\partial x^3}{\partial \xi^k}\epsilon_{ijk}\right) = \frac{\partial v^1}{\partial x^1}|J_\xi^x| \tag{23}$$

because the only term in l which does not vanish is $l = 1$. Combining identical results from the other two terms of Eq. (22), we have

$$\frac{d}{dt}|J_\xi^x| = \frac{\partial v^i}{\partial x^i}|J_\xi^x| \tag{24}$$

or

$$\frac{d}{dt}\ln|J_\xi^x| = \nabla \cdot \mathbf{v} \tag{25}$$

As shown earlier, this is the basic result that leads to the equation of mass continuity. It is worth noting that it depends explicitly on the assumption that $|J_\xi^x|$ neither vanishes nor becomes infinite; in other words, that the transformation (1) is invertible and differentiable, which is equivalent to the statement that fluid particles are distinct throughout the motion and that no singular points are permitted at which the positions $\mathbf{x}$ are not differentiable functions of $\boldsymbol{\xi}$.

*PROBLEM 5.7.1 Show that the density of a parcel obeys $\rho|J_\xi^x| = \rho_0$, where ρ_0 is the density at t_0. This result is known as the equation of continuity in Lagrangian coordinates.

PROBLEM 5.7.2 Let C be a segment of a material curve. Can C intersect itself as the motion proceeds?

PROBLEM 5.7.3 Let L be the length of a material curve, A the area of a material surface, and V the volume of a material volume. Calculate the rates of change ot L, A, and V. Explain the structure of motion fields required for these quantities to be constant.

BIBLIOGRAPHIC NOTES

5.1 The material of this section is standard and is discussed to some extent in almost all works on dynamic meteorology, fluid mechanics, and mechanics, as well as in mathematical texts on vector and tensor analysis. Among well-known references are the books:

Hay, G. E., 1953: *Vector and Tensor Analysis*, Dover Publications, Inc., New York, 193 pp.

Phillips, H. B., 1933: *Vector Analysis*, John Wiley & Sons, Inc., New York, 236 pp.

5.2 Much of this material also is standard. Interchange in the order of integration is a topic that has been studied extensively in mathematics; the main results are associated with the theorems of Fubini and Tonelli. A discussion of the elements of the theory is given by:

Titchmarsh, E. C., 1939: *The Theory of Functions*, 2d ed., Oxford University Press, Oxford, 454 pp.

The approach to the divergence theorem follows Phillips; the fluid-mechanical consequences are well known. The proof of Stokes' theorem is original (although not necessarily new).

5.3 The transport theorem is of major importance in modern approaches to fluid mechanics. The first proof is given for pedagogical purposes; the final proof follows that of:

Serrin, James, 1959: "Mathematical Principles of Classical Fluid Mechanics," in S. Flügge (ed.), *Handbuch der Physik*, vol. 8, no. 1, pp. 125–263, Springer-Verlag OHG, Berlin.

The material on parametric representations of curves, surfaces, and volumes, and on the properties of coordinate transformations was developed with the considerable aid of a book that has proved to be a faithful friend:

Buck, R. C., 1956: *Advanced Calculus*, McGraw-Hill Book Company, New York, 423 pp.; 2d ed., 1965, 527 pp.

The time rates of change of line and surface integrals have not generally received the amount of attention devoted to them here, but we shall use the results later. The derivations were conceived in obvious analogy with the proof of the transport theorem given by Serrin.

5.4–5.5 The results presented about transformation of the del operator are standard and are discussed to some extent in almost all works on vector analysis.

The discussion of the properties of nonorthogonal coordinates was developed from the outline presented in the invaluable book:

Richards, Paul I., 1959: *Manual of Mathematical Physics*, Pergamon Press, New York, 486 pp.

This book cannot be commended too enthusiastically; it is one of those rare books which, once used, becomes an essential reference.

Richards' book also served as a guide for organizing the basic concepts of tensor analysis. Other presentations of varying levels of difficulty are given in the book by Hay and the works:

Aris, R., 1962: *Vectors, Tensors, and the Basic Equations of Fluid Mechanics*, Prentice-Hall, Inc., Englewood Cliffs, N.J., 286 pp.

Defrise, Pierre, 1964: "Tensor Calculus in Atmospheric Science," in H. Landsberg and J. van Miegham (eds.), *Advances in Geophysics*, vol. 10, pp. 261–314, Academic Press, Inc., New York.

Jeffreys, Harold, 1931: *Cartesian Tensors*, Cambridge University Press, Cambridge, 92 pp. Reprinted 1952.

5.6 The subject of solenoidal and irrotational vector fields has been studied for many years. In developing the presentation given here we referred mainly to the book by Phillips and to:

Batchelor, G. K., 1967: *An Introduction to Fluid Dynamics*, Cambridge University Press, Cambridge, 615 pp.

The decomposition of flows into pure motion types in two dimensions is standard meteorological material. A fairly detailed account emphasizing the difficulties of analysis is given by:

Saucier, Walter J., 1955: *Principles of Meteorological Analysis*, University of Chicago Press, Chicago, 438 pp.

The three-dimensional problem has been studied in fluid dynamics; our discussion derives from that given in Batchelor's book. The problem at the end of the section was conceived as an illustration of the ideas and serves the useful purpose of showing how an eigenvalue problem arises.

5.7 The concepts of this section were emphasized by Serrin and discussed further by Aris. Consideration of the mathematical foundations of the study of fluids in motion serves, in my opinion, to clarify many points—the most important example being the relation between material and Eulerian derivatives.

The examples in Fig. 5.9 are from the article:

Welander, Pierre, 1955: "Studies on the General Development of Motion in a Two-dimensional, Ideal Fluid," *Tellus*, 7:141–156.

PART TWO

The Equations of Atmospheric Motion

Nearly all analytical and numerical study of atmospheric motion is based upon the system of partial differential equations that we shall derive in this part of the book. The equations of atmospheric motion are obtained by applying the general equations of fluid motion to an ideal gas, and then taking account of the rotation of the earth and the presence of water vapor and its changes of phase. In this part, we derive the equations, set boundary conditions for them, examine some simplifications that are possible when certain constraints are applied, develop the coordinate transformations upon which much of meteorological thought is based, and analyze in some detail the consequences of the fact that there is water in the air.

6
THE EQUATIONS OF MOTION IN INERTIAL COORDINATES

The study of how fluids are excited into motion by thermodynamic or mechanical disturbances, of the reasons why certain flow patterns develop, and of how balances are reached so that the motions are in harmony with continuing disturbances constitutes the science of fluid dynamics. The application of the concepts and discoveries of this study to the motions of the atmosphere is the special task of dynamic meteorology, and we turn now to the determination of the laws that govern the motion of the atmosphere.

The essential step in the derivations given here is the conversion of Newton's second law into a form that will reveal how fluid accelerations are caused by the distributions of mass, momentum, and thermodynamic energy. Effects of the earth's rotation—an obvious factor in the motion of the atmosphere—are touched upon here and are considered in further detail in the next chapter.

It will become apparent that there are two major categories of endeavor in the study of dynamics. The first involves the discovery of the dynamic laws themselves, and the consideration of how they may be formulated in more fundamental or more tractable forms. The second is the application of the laws to specific problems, or to a class of problems, in an attempt to understand or to predict the evolution of the actual phenomena observed in nature.

In this chapter, we shall concentrate on the first task, reserving most of the discussion of the implications of the equations for atmospheric problems for later chapters. Thus the main effort will be to derive the equations of motion and to discover what modifications of them may be possible in an attempt to use them to reveal the physics of fluid motion.

The equations of motion are, in a sense, full of surprises; one of the rewards of advanced studies in meteorology and fluid dynamics is the continuing discovery of the amazing properties of these equations, and how despite nonlinearities, they are arranged so that the coupling between the equations always turns out to be infallible. They would not represent fluid flow correctly if this were not the case; still, they are a work of both science and art.

6.1 COORDINATE SYSTEMS: A PRELUDE TO THE EQUATIONS OF MOTION

The concept of motion is based upon the observation that the position of objects changes in time. On a blustery fall day, we see the leaves tumbling on their travels in the wind and conclude that the air itself must be in motion. In this and similar cases, we rely on our visual perception to detect changes in the distance or azimuth of the objects from our eyes.

In order to discuss the motion of any body or parcel of air quantitatively, we need to have formal systems for detecting changes of position. By equating the geometric reality of position with the arithmetic specification provided by coordinate systems, we have reduced the problem to one that can be handled numerically. The fly walking on a corner of a ceiling that excited Descartes's realization that it was "doing analytic geometry" was also in motion, and the coordinate system provided a way to record its trajectory.

Before choosing a coordinate system, however, we must define a *frame of reference* or a *reference body* that we can use as the basis for our measurements. Our usual reference body in daily experience is naturally the earth, but here we shall start from a more basic and more abstract frame of reference.

The choice of the frame of reference is crucial. But once we have a reference body and a basic coordinate system, we are at liberty to represent geometrical relations with any other coordinate system that can be derived from the basic one. The point is that distance and orthogonality, for example, will be established by the original frame of reference.

It is presumably possible to express the laws of physics in any frame of reference. But although the truth of a law is not altered by the choice of the frame of reference or the coordinate system, its mathematical form may be changed considerably.

A natural attempt for a dynamicist, then, is to search for frames of reference in which physical laws take their simplest form. This is the essential content of Newton's first law: *There exist frames of reference in which a body not interacting with any other body will stay at rest or will stay in uniform motion if initially in such motion.* Such a frame of reference is called a *Galilean* or *inertial* coordinate system. The term *inertial* derives from the fact that the expression of the physical law of inertia takes its simplest form in such a coordinate system: if no forces act on a body, it will be unaccelerated.

Let us assume then that we have an inertial system composed of a reference body or point and an orthogonal cartesian coordinate system (x,y,z). If we let the position of a body not interacting with any other body be denoted by $\mathbf{x}$, then it must be true in this inertial system that

$$\frac{d^2\mathbf{x}}{dt^2} = \ddot{\mathbf{x}} = 0 \tag{1}$$

Now let us transform to another cartesian coordinate system with the relations

$$x_i' = A_{ij}x_j \qquad i,j = 1,2,3; \text{sum on } j \tag{2}$$

in which the A_{ij} are all constant. Obviously in this system it is true that

$$\ddot{\mathbf{x}}' = \ddot{x}' = \ddot{y}' = \ddot{z}' = 0 \tag{3}$$

because of Eq. (1). This new system is also inertial, because here too accelerations do not occur in the absence of forces.

Let us try another transformation, this one given by

$$\begin{aligned} x' &= x \cos \omega t + y \sin \omega t \\ y' &= -x \sin \omega t + y \cos \omega t \\ z' &= z \end{aligned} \tag{4}$$

In this case, the new coordinate system rotates around the z axis of the original system with an angular frequency $\omega = 2\pi/T$, where T is the time required for one complete rotation. Now we have

$$\begin{aligned} \ddot{x}' &= -2\omega(\dot{x} \sin \omega t - \dot{y} \cos \omega t) - \omega^2(x \cos \omega t + y \sin \omega t) \\ \ddot{y}' &= -2\omega(\dot{x} \cos \omega t + \dot{y} \sin \omega t) + \omega^2(x \sin \omega t - y \cos \omega t) \\ \ddot{z}' &= 0 \end{aligned} \tag{5}$$

So, perhaps to our surprise, in this system the body appears to be accelerated even though there are no forces acting upon it (Fig. 6.1). This rotating system, then, is not inertial. There are two kinds of *apparent forces* appearing in Eqs. (5): the first are those that involve the product of ω and the velocities $(\dot{x},\dot{y},\dot{z})$, and are called *Coriolis forces*; the second are the products of position coordinates and ω^2, and are called *centrifugal forces*. These forces are perfectly real and observable to an observer rotating with the $\mathbf{x}'$ system; they are called apparent because they are not due to interactions with another body.

In order to apply Newtonian mechanics to the atmosphere, we must find an inertial system so that we can determine the difference between actual accelerations

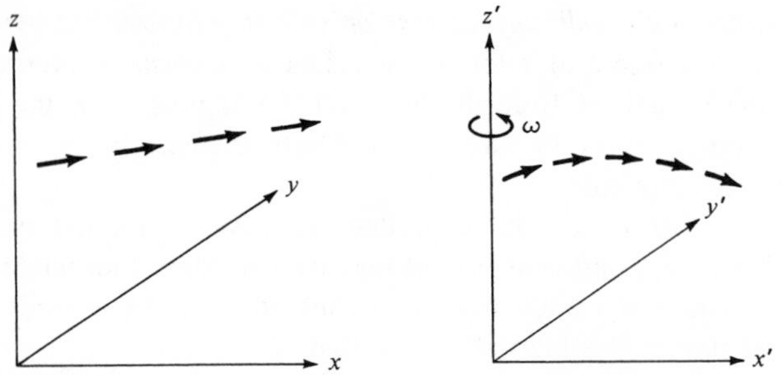

FIGURE 6.1
An unaccelerated motion in the inertial system **x** appears accelerated in the rotating system **x**′.

and those that result from coordinate transformations. Thus it is generally presumed that there exists an inertial system somewhere in space, and that choosing a frame of reference based on the "fixed" stars will yield an inertial system. Even with this assumption, we will have to account for the rotation of the earth with respect to this inertial system if we want to study the motions of the atmosphere in a coordinate system that uses the earth as a frame of reference.

The detailed consideration of the effects of the earth's rotation on atmospheric motion and the derivation of suitable modifications of the equations of motion will be reserved for Chap. 7. For now, we turn to the other forces that affect the motion of fluids, and concentrate on developing an understanding of motion in inertial coordinates.

PROBLEM 6.1.1 Let the **x** coordinates be inertial. What are the necessary and sufficient conditions that the coordinates $x_i' = A_{ij}x_j + v_j(\mathbf{x},t)t$ be inertial for constant A_{ij}? What are they for $x_i' = f_i(\mathbf{x},t)$?

PROBLEM 6.1.2 Let the **x** coordinates be inertial. What are the necessary and sufficient conditions that the anholonomic transformation. $dx_i' = f_{ij}(\mathbf{x},t)\, dx_j$ produces an inertial set of coordinates?

6.2 FORCES THAT ACCELERATE FLUIDS

If bodies are observed from an inertial system to be in accelerated motion, we may assume, in accordance with Newton's first law, that they are interacting with some other body. The second law says that these accelerations are produced by *forces* and thus constitutes a definition of forces as natural phenomena that produce accelerations.

Assuming then that we have discovered an inertial system, we will use the second law to write

$$m\mathbf{a} = \mathbf{F} \tag{1}$$

where $\mathbf{F}$ is the sum of the forces. In meteorology, it is convenient to use *specific forces*, that is, the force per unit mass, and so we have $\mathbf{a} = \mathbf{f} = \mathbf{F}/m$.

It is important to recognize that we usually determine the forces in order to calculate the accelerations. If we are consistently in error in our predictions of accelerations, we can assume that other forces must be present.

The forces important in fluid dynamics are all part of our daily experience. Newton studied gravity, and this is a familiar force to us. We know from our experience that things in motion appear to slow down and stop, and we attribute these decelerations to the force of friction. (Note that we have not really accomplished anything here except to give a name to a force that we observe slows things in motion.) We know that water comes out of pipes, toothpaste out of the tube, and air out of balloons, and these accelerations must be due to a force. The common factor in these cases is a spatial difference in pressure and so we attribute the force involved to pressure gradients.

Thus we have the forces due to pressure gradients, gravitation, and friction. There are others (electromagnetic forces in ionized fluids, for example), but these three are the most important in the macroscopic description of fluid motions in an inertial coordinate system. When we turn in Chap. 7 to coordinate systems rotating with the earth, then we shall encounter the Coriolis and centrifugal forces of Sec. 6.1 again.

Pressure gradient force Consider a volume of air as shown in Fig. 6.2. Pressure is the force per unit area due to molecular movement, as discussed in Chap. 3, and at each point it is the same in all directions. Thus at the left side of the box we have the force in the x direction of magnitude $p_0\,\Delta y\,\Delta z$. If $p_0 = p_1$, the air will not move in the x direction, so it must be, as with toothpaste, that accelerations will result from differences in pressure along the x axis. Let us assume that Δx is small enough that we can use Taylor's theorem to write

$$p_1 = p_0 + \left(\frac{\partial p}{\partial x}\right)_0 \Delta x \tag{2}$$

Then the net force due to the difference in pressure will be

$$F_x = (p_0 - p_1)\,\Delta y\,\Delta z = -\left(\frac{\partial p}{\partial x}\right)_0 \Delta x\,\Delta y\,\Delta z \tag{3}$$

so that putting $\Delta x\,\Delta y\,\Delta z = \Delta V$, where ΔV is the increment of volume, and letting ΔM be the increment of mass in the box, we have for the specific force

$$\frac{F_x}{\Delta M} = -\left(\frac{\partial p}{\partial x}\right)_0 \frac{\Delta V}{\Delta M} \tag{4}$$

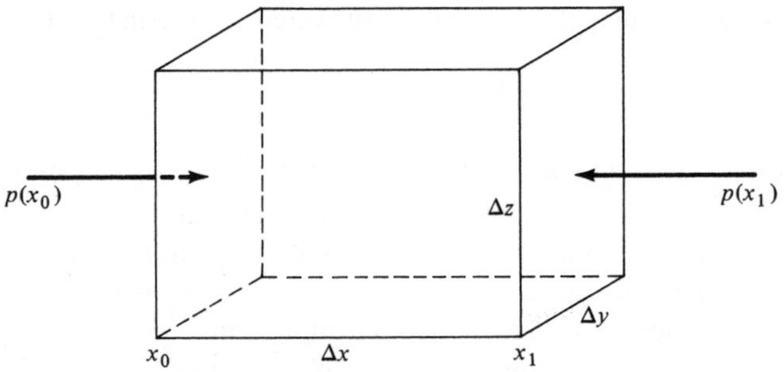

FIGURE 6.2
Diagram for derivation of pressure gradient force.

so that in general, because $\Delta M/\Delta V$ becomes ρ in the limit,

$$f_x = -\frac{1}{\rho}\frac{\partial p}{\partial x} \tag{5}$$

Similarly, for the other directions $f_y = -\alpha\, \partial p/\partial y$ and $f_z = -\alpha\, \partial p/\partial z$. In vector notation, we may combine these results to express the pressure gradient force as

$$\mathbf{f} = -\frac{1}{\rho}\nabla p \tag{6}$$

According to the theorem of Sec. 5.1, ∇p is directed in the direction of most rapid increase of pressure. Thus when isobars connecting the values of equal pressure are drawn on a weather chart, the horizontal pressure gradient force will be normal to them and directed toward lower pressure.

The force of gravitation According to Newton's law of gravitation, the force of attraction between two bodies is

$$\mathbf{F} = \frac{GMm}{r^3}\mathbf{r} \tag{7}$$

where $\mathbf{r}$ is the vector expressing the position of one body with respect to the other, M and m are the masses, and G is the universal gravitation constant.

For *gravity*, or *terrestrial gravitation*, we take M to be the mass of the earth, $\mathbf{r}$ to be the vector from the earth's center to the other body, and so the specific force on a body becomes

$$\mathbf{f} = -\frac{GM}{r^3}\mathbf{r} \tag{8}$$

If we put $g_0 = GM/a^2$, where a is the earth's radius and $r = a + z$, we have

$$\mathbf{f} = -g_0 \left[\frac{a^2}{(a+z)^2}\right] \frac{\mathbf{r}}{r} = -g_e \frac{\mathbf{r}}{r} \qquad g_e = g_0 \frac{a^2}{(a+z)^2} \tag{9}$$

Thus the force of gravity always accelerates objects and parcels of air directly toward the center of the earth.

The force of friction The force of friction is due to collisions between molecules, but is manifested in the atmosphere by spatial variations in wind speed, with the strongest frictional effects being due to wind shears on the scale of less than centimeters. We shall show later that the most common approximation to the force of friction for fluids with properties similar to air is

$$\mathbf{f}_r = \frac{1}{\rho} [\nabla \cdot (\mu \nabla \mathbf{v}) + \nabla(\lambda \, \nabla \cdot \mathbf{v})] \tag{10}$$

where μ and λ are coefficients of viscosity. For most of our purposes, it will be sufficient to write the force of friction in the equation of motion as $\mathbf{f}_r$ and not give it an explicit form.

The three forces important in an inertial system have thus been described and written quantitatively in vector notation, and thus these expressions are valid in any coordinate system.

PROBLEM 6.2.1 Show that for inertial coordinates, the sum of forces in a frictionless, hydrostatic atmosphere is horizontal.

PROBLEM 6.2.2 Exner's function is defined as $\pi = c_p(p/p_{00})^{R/c_p}$. Show that $\alpha \nabla p = \theta \nabla \pi$ for ideal gases. Convert this into an expression involving only temperature and entropy, thus manifesting the thermodynamic origin of the pressure gradient force.

6.3 THE FUNDAMENTAL EQUATIONS OF MOTION

In Chap. 2 we showed how the rate of change of the position of a parcel of air is defined as its velocity. In vector notation we may write

$$\mathbf{v} = \frac{d\mathbf{x}}{dt} \tag{1}$$

and the *acceleration* **a** is obtained by another differentiation,

$$\mathbf{a} = \frac{d\mathbf{v}}{dt} \tag{2}$$

Newton's law referred to interactions between bodies; we are now interpreting the bodies to be parcels of air and are thus concerned with the situation as it would be experienced by a moving parcel. The total or material derivative of Eq. (2) is therefore the correct way to define the acceleration we shall use in applying Newton's law.

Adding up the forces discussed in Sec. 6.2, we have the *equation of motion* for a parcel in an inertial system (Fig. 6.3):

$$\frac{d\mathbf{v}}{dt} = -\frac{1}{\rho}\nabla p - g\frac{\mathbf{r}}{r} + \mathbf{f}_r \qquad g = g_e \tag{3}$$

Thus if we know the pressure distribution in the neighborhood of a parcel, and decide to neglect friction, we know immediately that the resulting acceleration will be given by the first two terms of Eq. (3). Consider the tube of toothpaste again. When we pick it up, the contents are at equal pressure with the environment. We squeeze the tube, raising the pressure inside until the pressure gradient is strong enough to overcome friction. The toothpaste comes out of the tube, is accelerated downward immediately, and lands on the toothbrush.

6.3.1 The System of Equations

The equation of motion (3) is not generally useful by itself. As soon as the air started to move toward lower pressure, the new mass collecting in that region would result in rising pressure. Thus we must know how the pressure changes in time too. The first law of thermodynamics, Eq. (20) in Sec. 3.2, and the equation of mass continuity (12) in Sec. 5.2 are therefore used along with Eq. (3), and the connection between p, ρ, and T is established by the equation of state. Thus we have the *system of equations*

$$\begin{aligned} &\frac{d\mathbf{v}}{dt} = -\frac{1}{\rho}\nabla p - g\frac{\mathbf{r}}{r} + \frac{1}{\rho}[\nabla \cdot (\mu \nabla \mathbf{v}) + \nabla(\lambda \nabla \cdot \mathbf{v})] \\ &c_V\frac{dT}{dt} + p\frac{d\alpha}{dt} = q + f \\ &\frac{d\rho}{dt} + \rho\nabla \cdot \mathbf{v} = 0 \\ &p = \rho R T \end{aligned} \tag{4}$$

This system is generally called the equations of motion in this book. Notice that if the heating q is specified and if the dissipation f depends on the velocities, then there are six variables, u, v, w, p, ρ, and T, and six equations, when we count each of the three component equations contained in the vector equation. Thus the system is complete, in the sense that there are the same number of equations as variables. In most cases we will have $g\mathbf{r}/r = g\mathbf{k}$ because we consider motion over a flat plane or because the vector $\mathbf{k}$ will coincide with $\mathbf{r}/r$.

It is often useful to make a change in the system. From the continuity equation we find easily that

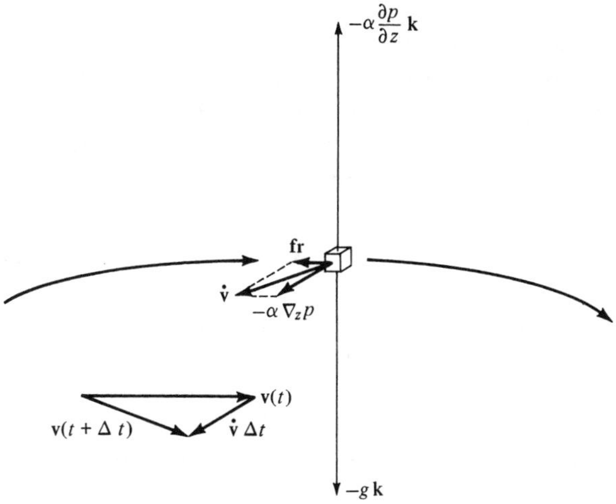

FIGURE 6.3
The forces acting on a moving parcel create accelerations.

$$\frac{d\alpha}{dt} = \alpha \nabla \cdot \mathbf{v} \tag{5}$$

and so the first law may be written as

$$c_V \rho \frac{dT}{dt} + p \nabla \cdot \mathbf{v} = \rho(q + f) \tag{6}$$

This has the distinct advantage of coupling the thermodynamic effects directly with the mechanical ones expressed by the velocity divergence.

These equations are deceptively simple in the form we have written them. But in order to apply them, we would have to keep track of all the different parcels, computing the acceleration for each parcel as it moves along. For almost all applications of the equations, we must expand the total derivative and write the system in the form

$$\begin{aligned}
&\frac{\partial \mathbf{v}}{\partial t} + (\mathbf{v} \cdot \nabla)\mathbf{v} = -\frac{1}{\rho}\nabla p - g\frac{\mathbf{r}}{r} + \frac{1}{\rho}[\nabla \cdot (\mu \nabla \mathbf{v}) + \nabla(\lambda \nabla \cdot \mathbf{v})] \\
&c_V \rho\left(\frac{\partial T}{\partial t} + \mathbf{v} \cdot \nabla T\right) + p \nabla \cdot \mathbf{v} = \rho(q + f) \\
&\frac{\partial \rho}{\partial t} + \mathbf{v} \cdot \nabla \rho + \rho \nabla \cdot \mathbf{v} = 0 \\
&p = \rho R T
\end{aligned} \tag{7}$$

Now we can see how such a set of partial differential equations can be used. Suppose at some time we had values of u, v, w, p, ρ, and T for all points throughout the fluid. Then we could compute all the spatial derivatives in Eqs. (7), so that if we knew $(q + f)$, we could use the equations to find at each point the value of the local derivatives of $\mathbf{v}$, ρ, and T. With these local derivatives we could use Taylor's theorem to calculate at each point (see Sec. 2.4)

$$\mathbf{v}(\mathbf{x}, t + \Delta t) = \mathbf{v}(\mathbf{x}, t) + \left(\frac{\partial \mathbf{v}}{\partial t}\right)_{\mathbf{x},t} \Delta t \tag{8}$$

for some small value of Δt. Then with a similar equation for ρ and T we would know the value of each variable at $t + \Delta t$, and could iterate the procedure to the next time, $t + 2\Delta t$. In other words, we could make forecasts of velocity, pressure, temperature, and density for the whole fluid for the future. This is precisely what is done in numerical weather prediction, in which computers are used to make the millions of computations that are necessary.

These equations are formidable ones; they have been solved analytically in only a few special cases, some of which will be discussed at the end of this chapter. The basic difficulty with the equations of motion, as pointed out in Chap. 2, is that they are *nonlinear*, and two solutions do not sum to a solution.

Nonlinearities appear in any term where products of the variables, such as $\alpha\, \nabla p$, occur. The essential nonlinear feature of fluid or atmospheric motion, however, is due to the advective terms and the presence of the operation $(\mathbf{v} \cdot \nabla)$. These nonlinear terms do not cause prohibitive problems in the numerical solution of the equations. But they are so inhibitive to analytical investigations that in fact fluid dynamicists and meteorologists often undertake computer integration of the equations because it is the only way to find the answers to their questions.

6.3.2 Modifications of the Equations

For various purposes, it is often useful to change the system (4) or (7) by eliminating one of the variables now present or by introducing new ones. When this is done, we must be sure to preserve the completeness property that there are the same number of equations as variables—neither more nor less.

The rules are:

1 If two equations are combined, we may keep the new equation and one of the first two. Thus the other equation is discarded from the set, although it can always be recovered from the two remaining ones.

2 If a new variable is introduced by adding a seventh equation to the set [for example, $\theta = T(p_{00}/p)^{R/c_p}$], then we eliminate one of the present variables in favor of the new variable every place that it occurs, thus obtaining six equations in six unknowns again.

3 If an approximate form of an equation is introduced in place of the original one, then the original one is discarded and cannot be used.

The student is urged to examine each modification we make in the equations in this book to see how the rules are followed. When they are not, then nonsense results.

As an example of a modified set, it is useful to introduce the potential temperature when we have inviscid, isentropic motion so that $\mathbf{f}_r = 0$ and $q + f = 0$. Thus the first law in Eqs. (4) becomes simply $d\theta/dt = 0$ and the equation of state is replaced by Poisson's equation. If this is used to eliminate the density from the equation of continuity, then with the isentropic assumption we have

$$\begin{aligned}
&\frac{d\mathbf{v}}{dt} = -\theta\,\nabla\pi - g\frac{\mathbf{r}}{r} \\
&\frac{d\theta}{dt} = 0 \\
&\frac{d\pi}{dt} + \frac{R}{c_V}\pi\,\nabla\cdot\mathbf{v} = 0 \\
&\pi = c_p\left(\frac{p}{p_{00}}\right)^{R/c_p} = c_p\,\frac{T}{\theta}
\end{aligned} \tag{9}$$

PROBLEM 6.3.1 Demonstrate that Eq. (9) is correct for the assumptions stated. What is the role of the last equation?

*PROBLEM 6.3.2 Rewrite the system (7) so that the local derivative of momentum $\rho\mathbf{v}$ appears in the first equation and so that the local derivative of ρT appears in the first law. Can the advective terms be written in a simpler form now?

PROBLEM 6.3.3 Derive the continuity equation by considering the mass budget of a fixed volume V. (Hint: Take V to be a cube. The flow of mass through one side is given by the product of the density and the velocity component normal to the side. Use Taylor's theorem in analogy with the derivation of the pressure gradient force. Note that because the volume is constant, the density changes in proportion to mass changes.)

*PROBLEM 6.3.4 The system (4) is written for an ideal gas. Assume an equation of state of the form $F(p,\rho,T) = 0$, and follow through the derivation of the equations to obtain the system in a more generally applicable form.

6.4 THE FIRST LAW OF THERMODYNAMICS

In the discussion of the first law of thermodynamics given in Chap. 3, we did not consider the physical processes contributing to the heating rate q or to the dissipation f. Those important in atmospheric science are:

Radiation The most important source or sink of thermodynamic energy for the atmosphere is the gain or loss of heat by radiation, and there are the essentially distinct streams of solar radiation and infrared or long-wave radiation. The solar, or short-wave, radiation is obviously that received from the sun, although it may be traveling in the atmosphere after reflection from the ground or clouds, or after being scattered in the clear atmosphere. The long-wave radiation is emitted by the earth and by the air in proportion to the fourth power of the temperature.

Radiation is the name given to the propagation of energy by electromagnetic waves, and may be represented as a packet of energy moving in a certain direction at the speed of light. Without going into all the details of radiation geometry, let us imagine a small circular disk of unit area (Fig. 6.4). At each point in space, we can find the total amount of energy per unit time passing across one side of the disk in the direction toward the disk due to the motion of the packets of energy. As part of this measurement, we might also determine the total flow of energy toward the other side of the disk. If the flow on one side is larger than that on the other, there is a net flow of energy across the disk. Thus we have a direction perpendicular to the disk and an associated net flow of energy in this direction; this quantity is called the net flux across the disk. It is a vector with magnitude given in energy per unit area per unit time and is normal to the disk in the direction of the net flow of energy across the disk.

The net flux across the disk may change as we change the orientation of the disk. If we now average the net flux vectors obtained for every orientation of the disk, we have the net radiant flux at a point in space. This is a vector in the direction of the net flow of energy at that point and with the magnitude of the net flux through a unit area normal to the direction of energy flow.

In the atmosphere, it is generally true that the radiant transfers of energy in the vertical are most important. The average fluxes of solar and infrared (long-wave) radiation in the upward and downward directions are often denoted by $R_s\uparrow$, $R_l\uparrow$, $R_s\downarrow$, and $R_l\downarrow$. The net fluxes of solar and infrared radiation are $\mathbf{R}_s = \mathbf{R}_s\downarrow - \mathbf{R}_s\uparrow$ and $\mathbf{R}_l = \mathbf{R}_l\uparrow - \mathbf{R}_l\downarrow$, where we have arranged the signs to take account of the fact that the dominant part of the solar radiation is incoming and that the long-wave radiation is dominantly outgoing. The net radiative flux at a point, then, is $\mathbf{R} = \mathbf{R}_s - \mathbf{R}_l$, so that $\mathbf{R}$ is downward if incoming solar and long-wave radiation exceeds the outgoing solar and infrared radiation. Obviously, $\mathbf{R}$ is generally downward during the day and upward at night. We shall let $\mathbf{R}$ denote the net radiant flux vector in general, and on occasion revert to its specialized meteorological usage of indicating only upward or downward net flux.

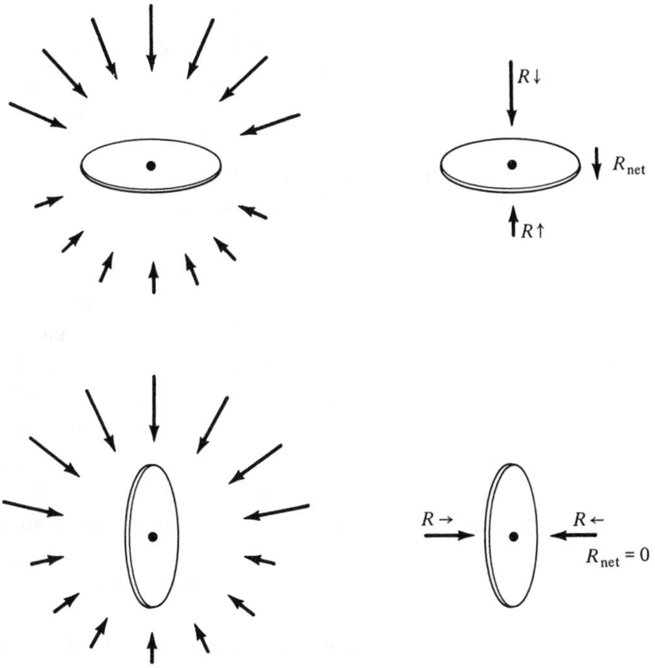

FIGURE 6.4
Measurement of net radiation as a function of direction in the radiation field.

Energy, like mass, must be conserved. If more of one kind of energy enters a volume than leaves, it must be turning into some other form of energy. Because the radiant energy must interact with molecules first of all, we may conclude that it contributes, or derives from, molecular energy.

We can measure the net gain or loss of radiant energy at a point in a manner similar to the way we studied gains and losses of mass. At each point on the surface S of a small volume V, let the net flux vector be $\mathbf{R}$. Then the gain or loss H_R of radiant energy in the volume must be

$$H_R = -\int_S \eta \cdot \mathbf{R}\, d\sigma = -\int_V \nabla \cdot \mathbf{R}\, dV \tag{1}$$

where we have $H_R > 0$ if there is a net gain because the unit vector η is outward. Because the volume is arbitrary, we may conclude that for small enough V we shall have

$$H_R = -V \nabla \cdot \mathbf{R} \tag{2}$$

But if there is a net flux of radiant energy into the volume, it must represent a gain of energy of some other form—and by the molecular association we attribute it to heat energy. Thus one of the components of heat gain or loss in the first law is

$$(\rho q)_R = \frac{H_R}{V} = -\nabla \cdot \mathbf{R} \tag{3}$$

Thus there is heating at a point where the radiant stream in the atmosphere is convergent and cooling where the stream is divergent.

Molecular conduction If there is a temperature gradient in the atmosphere, then there is, according to the definitions of Chap. 3, a gradient in the kinetic energy of molecular motion. In this case, the molecules in the region of high temperature will be colliding with those in the region of lower temperature. In doing so, they will cause the slower molecules to accelerate, and they themselves will decelerate on the average. Thus there will be a transfer of molecular kinetic energy from the region of higher temperature to that of lower temperature, and this represents a transfer of internal energy. The gain or loss in energy at a point is thus assumed to be some function of the net temperature gradient at a point. Let us represent this net gradient by

$$G = \int_S \boldsymbol{\eta} \cdot \nabla T \, d\sigma \tag{4}$$

Fourier's hypothesis is that there is a factor of proportionality k such that the net gain of heat $H(S)$ inside S is

$$H(S) = \int_S \boldsymbol{\eta} \cdot k \, \nabla T \, d\sigma = \int_V \nabla \cdot k \, \nabla T \, dV \tag{5}$$

so that we conclude as before that

$$(\rho q)_{\nabla T} = \nabla \cdot k \, \nabla T \tag{6}$$

The factor k is called the *coefficient of thermal conductivity* and is frequently taken to be a constant in applications of Eq. (6).

Release of latent heat When the water vapor in the air reaches saturation and forms the droplets of fog or clouds, the latent heat acquired by the vapor when it evaporated is given up and appears as heat energy again. Thus we have the third component

$$(\rho q)_{H_2O} = C \tag{7}$$

where C denotes the rate of heating due to latent heat release by condensation. If water droplets are evaporating in the air, then the term C will be negative. Further consideration of this term will be delayed until Chap. 8, where we discuss the processes associated with water vapor in the atmosphere.

Frictional heating The kinetic energy lost to frictional forces must reappear in another form, and it is a common experience that motion against strong frictional forces produces heat. We will be precise about the mathematical form for frictional addition of heat later on; suffice it to say now that we write it as f_h [$=f$ in Eq. (7) in Sec. 6.3] and that f_h is related in a way to be specified to the term $\mathbf{f}_r$ of the equation of motion.

The first law Upon combination of these results, the first law of thermodynamics takes the complete (for meteorology) form

$$\rho c_V \frac{dT}{dt} + p\,\nabla \cdot \mathbf{v} = -\nabla \cdot \mathbf{R} + \nabla \cdot k\,\nabla T + C + f_h \tag{8}$$

The entropy was shown in Chap. 3 to be that perfect differential of the first law given by

$$T\frac{ds}{dt} = c_V \frac{dT}{dt} + p\frac{d\alpha}{dt} \tag{9}$$

With this definition and Eq. (5) in Sec. 6.3 we may express the rate of change of entropy as

$$\rho T\frac{ds}{dt} = -\nabla \cdot \mathbf{R} + \nabla \cdot k\,\nabla T + C + f_h \tag{10}$$

PROBLEM 6.4.1 The local rate of change of temperature due to radiation is often expressed as $-\nabla \cdot \mathbf{R}/\rho c_p$. From what form of the first law does this expression derive, and why was it chosen instead of Eq. (8)?

PROBLEM 6.4.2 Show that Eq. (10) is valid for any equation of state.

PROBLEM 6.4.3 Let the energy flux (erg/cm^2 · s) on a disk normal to the sun's rays be I_0 at the top of the atmosphere. How does the energy flux I, received at the top of the atmosphere from the sun, vary as a function of latitude? Explain why the long-range energy leaving the atmosphere is nearly constant with latitude. Determine what the constant must be, and calculate the difference between the incoming and outgoing radiation as a function of latitude.

6.5 SPECIAL FORMS OF THE EQUATIONS OF MOTION

Two kinds of simplifications of the equations are often made. The first is to reduce one of the equations in the set to a simpler form; the second is to introduce a new variable with special properties. We consider several examples here.

Some of these changes present particular problems in numerical integrations. When there is a prognostic equation (one giving the local rate of change) for each variable, numerical integration is relatively straightforward by the methods discussed in Sec. 6.3.1. But some of the simplifications lead to replacement of a prognostic equation with a diagnostic one that does not contain a local rate of change. Then the new value of one of the variables must be inferred from a suitable combination of the equations that produces a diagnostic equation giving the remaining variable as a function of the others–whose new values are obtained from the prognostic equations in the set. These systems, although usually easier to handle analytically, often present numerical problems. The student is urged to study carefully the problems that arise in both the incompressible and hydrostatic approximations because of the presence of a diagnostic equation.

6.5.1 Incompressible Flow

A frequently used idealization of real fluids assumes that they are incompressible. The validity of this approximation will not be discussed now; rather we concentrate on how to formulate a consistent set of equations to represent this model of fluid behavior.

The incompressibility assumption is introduced by replacing the equation of continuity with the condition

$$\nabla \cdot \mathbf{v} = 0 \tag{1}$$

Once this is done, this is the equation of continuity. We cannot also have the equation $d\rho/dt = 0$ for then we would have one more equation than variables.

Because gases are not generally incompressible, we do not use the ideal-gas equation of state but rather one of the form

$$\rho = \rho_0[1 - \epsilon(T - T_0)] \tag{2}$$

where the subscript denotes some reference density and temperature and ϵ is the coefficient of thermal expansion given by

$$\left(\frac{1}{\rho}\frac{\partial \rho}{\partial T}\right)_0 = -\epsilon \tag{3}$$

Therefore, if density changes do occur in such a fluid, they are due to changes in temperature, not to expansion or contraction of the fluid parcels caused by the fluid motion.

Because we have eliminated the expansion of fluid parcels, the motion is unable to do work against its environment $(p\dot{V} = 0)$ and so the thermodynamic energy equation becomes

$$c_V \rho \frac{dT}{dt} = -\nabla \cdot \mathbf{R} + \nabla \cdot k\, \nabla T + f_h \tag{4}$$

in which we have dropped the term owing to phase changes. The equation of motion is still

$$\frac{\partial \mathbf{v}}{\partial t} + \mathbf{v} \cdot \nabla \mathbf{v} = -\frac{1}{\rho} \nabla p - g\mathbf{k} + \mathbf{f}_r \tag{5}$$

Equations (1), (2), (4), and (5) form a complete set in the variables u, v, w, ρ, T, and p. An apparent problem is how to determine the pressure because it occurs only in the equation of motion. By applying Eq. (1) to Eq. (5) we find

$$\nabla \cdot (\mathbf{v} \cdot \nabla \mathbf{v}) = -\frac{1}{\rho} \nabla^2 p + \frac{1}{\rho^2} \nabla \rho \cdot \nabla p + \nabla \cdot \mathbf{f}_r \tag{6}$$

so that the pressure is determined, upon solving this partial differential equation, solely by the velocity and density fields. Therefore in incompressible flow, pressure is a mechanical, and not a thermodynamic, variable.

In many cases of interest, the possible external sources of heat are negligible, and temperature and density variations are ignored. Thus the system becomes

$$\frac{\partial \mathbf{v}}{\partial t} + \mathbf{v} \cdot \nabla \mathbf{v} = -\frac{1}{\rho_0} \nabla (p + g\rho_0 z) + \mathbf{f}_r \tag{7}$$

and

$$\nabla \cdot \mathbf{v} = 0 \tag{8}$$

These incompressible equations contain the essential nonlinear feature represented by the inertial term $\mathbf{v} \cdot \nabla \mathbf{v}$, and are the simplest model of fluid flow that is not unduly restrictive. For this reason, most studies of the mathematical properties of the equations of motion are devoted to, or at least begin with, systems (7) and (8), even though the kinetic energy lost to friction is not again available to the system through transformations of thermal energy. In the literature, the term $p + g\rho_0 z$ is often denoted by just p and it is determined by Eq. (6) with $\nabla \rho = \nabla \rho_0 = 0$.

PROBLEM 6.5.1 Assume that suitable boundary conditions are specified for Eqs. (7) and (8). Explain in general terms how to integrate the equations on a computer.

PROBLEM 6.5.2 Let an incompressible fluid with constant ϵ be in isothermal motion. Show that the equations of motion are (1), (5), and $d\rho/dt = 0$. Note that we discarded $d\rho/dt = 0$ once. Explain why it appears again in this case.

PROBLEM 6.5.3 Assume the equation of state given in Problem 3.2.7 is valid. What then are the incompressible equations of motion? How can they be integrated numerically?

6.5.2 Autobarotropic Flow

It is possible to imagine a flow that, at some instant, passes through a state in which the density is a function of the pressure in the form

$$\rho = f(p) \tag{9}$$

When such a stratification exists, the fluid is said to be *barotropic*; if it is not barotropic, then it is *baroclinic*. When Eq. (9) is true, then according to the equation of state

$$T = \frac{p}{R\rho} = \frac{p}{Rf(p)} \tag{10}$$

so that the temperature is a function only of the pressure, and the same is true of the potential temperature or the entropy. Clearly, then, there will be no variation of density, temperature, or potential temperature on isobaric surfaces in barotropic stratification.

Just because a fluid is barotropic at some instant does not mean that it will stay that way or ever become barotropic again. In general the motion will convert it back into a baroclinic stratification. Should the flow be such that the stratification remains barotropic, however, we call it *autobarotropic* motion. The condition for autobarotropic flow is that there exists some function so that

$$\rho = f(p, t) = f(p(x,y,z,t), t) \tag{11}$$

Thus density is dependent only on the pressure and time, and at each instant a barotropic stratification prevails, although the barotropic variation of density with respect to pressure may vary.

It is of interest to determine whether a condition like Eq. (11) can be realized in a natural flow governed by the complete system of equations. Obviously the relation may be inserted in the equations every place that density occurs and we have a set of equations in $\mathbf{v}$, p, T and $f(p,t)$. Whether Eq. (11) is possible depends on whether solutions can be found that have this special variation of density.

First of all, it is obvious that $\rho(x,y,z,t) = \text{const}$ is a solution to Eq. (11). Hence the incompressible flow of a homogeneous (constant density) fluid is autobarotropic. Another case of interest is derived from the system (9) in Sec. 6.3 by assuming that θ is initially constant everywhere. Then variations in θ cannot develop because of the condition of isentropic flow and the system of equations reduces to

$$\begin{aligned} \dot{\mathbf{v}} &= -\theta \, \nabla \pi - g\mathbf{k} \\ \dot{\pi} &+ \frac{R}{c_V} \pi \, \nabla \cdot \mathbf{v} = 0 \end{aligned} \tag{12}$$

With θ constant, Poisson's equation shows that knowledge of π will give a relation of the form (11) between pressure and density.

This generalization of homogeneous, incompressible flows to isentropic flows without entropy variation is of some theoretical importance, because both rotational and divergent motion are possible. In the atmosphere, however, a central feature is the vertical stratification and this results in vertical variation of entropy.

PROBLEM 6.5.4 Consider an isentropic ($\dot{\theta} = 0$) flow with $\pi = \pi(\theta,t)$. Use the continuity equation in (12) to show that the divergence must be invariant on isentropic surfaces.

6.5.3 Hydrostatic Motion

That the atmosphere is nearly in hydrostatic balance is the fundamental fact about its vertical stratification. It is therefore of some interest to study the dynamics of purely hydrostatic flow, and this option has attracted considerable attention in meteorology.

To construct a model of purely hydrostatic motion, we simply replace the vertical component of the vector equation of motion with the hydrostatic relation so that the new equation of motion is

$$\frac{\partial \mathbf{v}_H}{\partial t} + \mathbf{v} \cdot \nabla \mathbf{v}_H = -\frac{1}{\rho} \nabla_H p + (\mathbf{f}_r)_H \tag{13}$$

and

$$\frac{\partial p}{\partial z} + g\rho = 0 \tag{14}$$

Note that we no longer have an equation for dw/dt, and so we must find some other means of determining the vertical velocity.

The first law may be combined with the equation of state to give

$$\frac{\partial p}{\partial t} + \mathbf{v}_H \cdot \nabla p - wg\rho - \frac{c_p}{c_V}\frac{p}{\rho}\frac{d\rho}{dt} = \frac{p}{c_V}\frac{ds}{dt} \tag{15}$$

and with the continuity equation this becomes

$$\frac{\partial p}{\partial t} + \mathbf{v}_H \cdot \nabla p - wg\rho + \frac{c_p}{c_V} p\left(\nabla_H \cdot \mathbf{v} + \frac{\partial w}{\partial z}\right) = \frac{p}{c_V}\frac{ds}{dt} \tag{16}$$

The hydrostatic condition may be integrated, using the condition that pressure vanishes at the top of the atmosphere, to obtain

$$p = \int_z^\infty g\rho \, dz \tag{17}$$

and so by differentiation

$$\begin{aligned} \frac{\partial p}{\partial t} &= \int_z^\infty g \frac{\partial \rho}{\partial t} dz = -\int_z^\infty g(\nabla \cdot \rho \mathbf{v}) \, dz \\ &= -\int_z^\infty g \nabla_H \cdot \rho \mathbf{v} \, dz + g\rho w \end{aligned} \tag{18}$$

in which we have used the boundary condition that ρw also vanishes at the top of the atmosphere. With this result Eq. (16) may be rewritten as

$$\frac{\partial w}{\partial z} = -\nabla_H \cdot \mathbf{v} + \frac{c_V}{c_p}\frac{1}{p}\left(\int_z^\infty g\,\nabla_H \cdot \rho\mathbf{v}\,dz' - \mathbf{v}_H \cdot \nabla p + \frac{p}{c_V}\frac{ds}{dt}\right) \tag{19}$$

Now obviously the vertical velocity must vanish at a solid, level lower boundary, so with this condition, inserted here pending a discussion of boundary conditions to come later in the chapter, we have

$$w = \int_0^z \left[-\nabla_H \cdot \mathbf{v} + \frac{c_V}{c_p}\frac{1}{p}\left(\int_{z'}^\infty g\,\nabla_H \cdot \rho\mathbf{v}\,dz'' - \mathbf{v}_H \cdot \nabla p + \frac{p}{c_V}\frac{ds}{dt}\right)\right] dz' \tag{20}$$

This vertical velocity is dependent on the horizontal motion and the pressure field, and is precisely the amount of vertical motion required to maintain hydrostatic equilibrium.

Note that horizontal convergence near the surface will induce upward vertical motion, and that heating as represented by increases in entropy also contributes to upward motion.

The specification of the vertical velocity in hydrostatic flow seems to be unique to meteorology, and applies most accurately to large-scale motion. Equation (20) was discovered by L. F. Richardson in his attempt to design a numerical prediction method long before the computers that make it possible were a reality.

PROBLEM 6.5.5 Finding w is much simpler for the hydrostatic, incompressible flow of a constant density fluid. Derive the equivalent of Eq. (20) for this case. How do we find the pressure now? (Hint: Derive a tendency equation for the height Z of the top surface from the incompressibility condition.)

PROBLEM 6.5.6 Equation (18) is the *pressure tendency equation* and can be used to predict changes in pressure at a given level. Discuss the processes that will lead to pressure rises or falls, and explain why the equation is not generally successful in applications with observations.

PROBLEM 6.5.7 Show that the hydrostatic approximation *does not* imply $dw/dt = 0$ by exhibiting an expression for the vertical acceleration.

6.5.4 Entropy, the Pressure Gradient Force, and Bernoulli's Theorem

The pressure gradient force is a manifestation of spatial variations in molecular kinetic energy, and as such it is related to fundamental thermodynamic quantities like entropy and the internal energy. The specification of this relation in direct form and the investigation of a few of its consequences are the tasks of this subsection.

From the definition of entropy, we have the logarithmic derivative

$$\nabla s = \frac{c_p}{\theta}\nabla\theta = c_p\frac{\nabla T}{T} - \frac{R}{p}\nabla p \tag{21}$$

and so the pressure gradient force may be written with the aid of the equation of state and the definition of enthalpy as $h = c_p T$ as

$$-\frac{1}{\rho}\nabla p = T\,\nabla s - \nabla h \tag{22}$$

By virtue of the identity (36) in Sec. 5.1, we may express the inertial terms as ($v = |\mathbf{v}|$)

$$\mathbf{v}\cdot\nabla\mathbf{v} = \nabla\frac{v^2}{2} - \mathbf{v}\times(\nabla\times\mathbf{v}) = \nabla\frac{v^2}{2} - \mathbf{v}\times\boldsymbol{\zeta} \tag{23}$$

Thus the equation of motion may be written in the new form

$$\frac{\partial\mathbf{v}}{\partial t} - \mathbf{v}\times\boldsymbol{\zeta} = T\,\nabla s - \nabla\left(\frac{v^2}{2} + h + gz\right) + \mathbf{f}_r \tag{24}$$

where we have taken g to be constant. This result, known as the *Crocco-Vazsonyi theorem*, provides considerable insight into the properties of fluid motion.

For example, the steady flow ($\partial\mathbf{v}/\partial t = 0$) of an inviscid fluid must obey the relation

$$T\,\nabla s + \mathbf{v}\times\boldsymbol{\zeta} = \nabla\left(\frac{v^2}{2} + ch + gz\right) \tag{25}$$

If all the variables exhibit the steady behavior, then $d(\ )/dt = \mathbf{v}\cdot\nabla(\ )$ and so Eq. (25) in the form

$$T\mathbf{v}\cdot\nabla s - \mathbf{v}\cdot\nabla\left(\frac{v^2}{2} + h + gz\right) = 0 \tag{26}$$

implies for this case that

$$T\frac{ds}{dt} = \frac{d}{dt}\left(\frac{v^2}{2} + c_p T + gz\right) \tag{27}$$

Thus the steady flow of a gas in isentropic motion obeys the relation

$$\frac{d}{dt}\left(\frac{v^2}{2} + c_p T + gz\right) = 0 \tag{28}$$

and this implies that

$$\frac{v^2}{2} + c_p T + gz = \text{const} \tag{29}$$

along the trajectories of the parcels, and for a steady flow, trajectories are identical to streamlines. This result is known as *Bernoulli's theorem*; we shall encounter more general versions of it later.

The important general conclusion that can be drawn from Eq. (24) is that local accelerations in the fluid result from gradients in the entropy, from gradients in the specific energy form $(v^2/2 + c_p T + gz)$, and from motion patterns in which the velocity and the vorticity vector are not parallel. When these four vectors satisfy Eq. (25), then the acceleration vanishes.

PROBLEM 6.5.8 Find a vector that is always normal to the vorticity in inviscid, steady flow. Use your result to find whether the ratio of the horizontal and vertical components of the vorticity is uniquely determined in inviscid, steady flow.

*PROBLEM 6.5.9 Show that the Bernoulli theorem for the steady, inviscid, incompressible flow of a homogeneous fluid is

$$\frac{v^2}{2} + \frac{p}{\rho_0} + gz = \text{const}$$

Use this result to explain the lift on an airfoil. (Hint: The air passing over the upper curved surface must go farther.)

6.6 ENERGY CONSERVATION IN THE ATMOSPHERE

Now that we have a complete set of equations, we can derive an energy theorem for atmospheric motion. We consider the material volume represented by the entire atmosphere, so that the transport theorem of Sec. 5.3 implies that for a scalar f we have

$$\frac{\partial}{\partial t}\int \rho f \, dV = \int \rho \frac{df}{dt} \, dV \tag{1}$$

For the kinetic energy K, we find from Eq. (4) in Sec. 6.3 that $(v^2 = \mathbf{v} \cdot \mathbf{v})$

$$\frac{\partial K}{\partial t} = \frac{\partial}{\partial t}\int \rho \frac{v^2}{2} \, dV = \int \rho \mathbf{v} \cdot \frac{d\mathbf{v}}{dt} \, dV = -\int \rho \mathbf{v} \cdot \left(\frac{1}{\rho}\nabla p + g\mathbf{k} - \mathbf{f}_r\right) dV \tag{2}$$

and for the internal energy I, Eq. (8) in Sec. 6.4 gives

$$\frac{\partial I}{\partial t} = \frac{\partial}{\partial t}\int \rho c_V T\, dV = \int \rho c_V \frac{dT}{dt}\, dV$$

$$= \int (-p\, \nabla \cdot \mathbf{v} - \nabla \cdot \mathbf{R} + \nabla \cdot k\, \nabla T + C + f_h)\, dV \tag{3}$$

The kinetic energy equation contains the term

$$\int \rho g \mathbf{v} \cdot \mathbf{k}\, dV = \int \rho g w\, dV = \int \rho g \frac{dz}{dt} = \frac{\partial}{\partial t}\int \rho g z\, dV \tag{4}$$

If we therefore define the potential energy as

$$P = \int g \rho z\, dV \tag{5}$$

then Eqs. (2), (3), and (5) sum to

$$\frac{\partial}{\partial t}(K + I + P) = \int [-\nabla \cdot (p\mathbf{v} + \mathbf{R} - k\, \nabla T) + C + f_h + \rho \mathbf{v} \cdot \mathbf{f}_r]\, dV \tag{6}$$

in which we have used the identity $\nabla \cdot p\mathbf{v} = \mathbf{v} \cdot \nabla p + p\nabla \cdot \mathbf{v}$. We may apply the divergence theorem to the first term so that

$$\frac{\partial}{\partial t}(K + I + P) = -\int_S \boldsymbol{\eta} \cdot (p\mathbf{v} + \mathbf{R} - k\, \nabla T)\, d\sigma + \int_V (C + f_h + \rho \mathbf{v} \cdot \mathbf{f}_r)\, dV \tag{7}$$

where S is the exterior boundary of the atmosphere. This boundary is composed of the top of the atmosphere where p vanishes and the earth's surface (assumed flat here) where $\boldsymbol{\eta} \cdot \mathbf{v} = -\mathbf{k} \cdot \mathbf{v} = w = 0$. Thus the total energy of the atmosphere can change only through the thermodynamic or frictional processes represented on the right of Eq. (7). Atmospheric energetics will be discussed more fully in Chap. 11.

PROBLEM 6.6.1 Consider two adjoining, arbitrary material volumes in an inviscid, dry atmosphere. Show how energy can be transported from one volume to another, and illustrate your result with a suitable diagram. Under what conditions will the energy increase in a material volume?

PROBLEM 6.6.2 Show that the Coriolis forces discussed in Sec. 6.1 do not affect the kinetic energy. (Hint: You must find an equation of motion involving only primed variables.)

PROBLEM 6.6.3 Derive an energy conservation theorem for the incompressible flow of constant density fluid. (Hint: First derive a transport theorem.)

PROBLEM 6.6.4 Relax the condition that density be constant and compute energy integrals for the system of Eqs. (1), (2), (4), and (5) in Sec. 6.5. Compare your result to the energy equations for a compressible system with first law $\rho\dot{e} + p\,\nabla \cdot \mathbf{v} = q + f_h$, where e is specific internal energy. What is the source of difficulty in the energetics of incompressible fluids without constant density?

6.7 THE ORIGIN OF FLUID MOTION

All natural fluids seem to be in motion most of the time, however slight the velocities may be. Interest so often centers on what a fluid in motion will do next, or on what dynamical laws and constraints are controlling the evolution of the flow patterns, that attention is seldom given to the question of how the motion began or whether it will ever cease. These questions are important ones in attempting to comprehend the motion of the atmosphere, and in searching for understanding of these fundamental matters we shall enhance our comprehension of the physics of atmospheric motion.

Frictional forces retard motion, and the molecular conduction of heat reduces temperature gradients. It seems intuitively obvious that if there is no external forcing and if these internal forces are allowed to act to their conclusion, then a fluid in motion will reach an equilibrium state in which all motion has vanished. We shall investigate this question more rigorously later; now we simply assume that such an equilibrium is possible. When the velocity and acceleration vanish identically, the equations of motion become

$$\nabla p + g\rho\mathbf{k} = 0 \tag{1}$$

$$c_V\rho\frac{\partial T}{\partial t} = -\nabla \cdot \mathbf{R} + \nabla \cdot k\,\nabla T \tag{2}$$

$$\frac{\partial \rho}{\partial t} = 0 \tag{3}$$

and

$$p = \rho RT \tag{4}$$

in which we have used the yet to be demonstrated fact that both the frictional forces and frictional heating vanish if the velocity does. We have also decided, for simplicity, to study a fluid without phase changes so that $C = 0$. The equation of motion is reduced to Eq. (1), which can be expanded to give

$$\nabla_z p = 0 \tag{5}$$

and

$$\frac{\partial p}{\partial z} = -g\rho \tag{6}$$

This motionless state has no horizontal pressure gradients, and the vertical distribution of pressure is controlled by the hydrostatic relation. We may apply Eq. (5)

to Eq. (6) to show that

$$\frac{\partial}{\partial z}\nabla_z p = -g\,\nabla_z \rho = 0 \tag{7}$$

and so the horizontal gradients of density vanish, and therefore by Eq. (4) those of temperature do as well.

Thus we have shown that when both the velocities and accelerations vanish, so must all horizontal temperature gradients. Upon turning this result around (that is, stating the contrapositive result that if A implies B, then "not B" implies "not A") we have an important theorem that gives a sufficient condition for motion to exist in the atmosphere:

> **Theorem** If there are temperature gradients on any level surface, then there must be motion in the sense that both the velocities and the accelerations cannot vanish identically; the motion or acceleration will persist at least until the gradients of temperature vanish. ////

Now let us consider the implications of the first law of thermodynamics when both velocities and accelerations vanish. Because horizontal temperature gradients must also vanish, Eq. (2) becomes

$$c_V \rho \frac{\partial T}{\partial t} + \nabla \cdot \mathbf{R} = \nabla \cdot k\,\nabla T = \frac{\partial}{\partial z}\left(k\frac{\partial T}{\partial z}\right) \tag{8}$$

But we can also show that the local change of temperature must vanish. With Eqs. (3), (4), and (6) and an interchange in the order of differentiation we find

$$0 = \frac{\partial}{\partial t}\left(\frac{\partial p}{\partial z}\right) = \frac{\partial}{\partial z}\left(R\rho\frac{\partial T}{\partial t}\right) \tag{9}$$

Thus, $R\rho\,\partial T/\partial t$ is a constant function of height, but a function that vanishes at infinity because the density does. The only way this can occur is for $\partial T/\partial t$ to vanish, and so we have the result that

$$T = T(z) \tag{10}$$

Therefore the radiation flux divergence can depend only on altitude and must have a constant value on each level surface.

Thus when we eliminate the motions and accelerations, the only temperature gradients that are permitted are vertical ones that provide the conductional heat transfer necessary to balance nonzero values of the radiation divergence. Let us again turn this around to obtain the contrapositive:

> **Theorem** (Jeffreys, 1925) If temperature gradients are maintained on any level surface in the atmosphere, motion will persist. If the radiational and

conductional heat sources are out of balance and lead to net heating or cooling at any point, motion will persist at least as long as they are not in balance.

////

This theorem may be viewed as a basic result of atmospheric science, for it reveals that the fundamental cause of motion in the atmosphere is the existence of temperature gradients. The temperature gradients that drive the large-scale motion of the atmosphere are caused by the spherical shape of the earth, which results in less solar radiation being received at the poles than at the equator. Smaller-scale motions, such as the turbulence near the ground on a sunny day, are similarly driven by horizontal variations in surface heating and the fact that the radiation and heat conduction fields are not in balance.

Thus in this compact statement, we have a mathematical version of what all meteorologists know intuitively to be the case: The motions of the atmosphere result from differential heating.

Let us see how the physics of the situation are arranged to agree with this theorem.

We shall ignore, in analogy with the atmospheric situation, the possibility of mechanical forces, such as stirring with a paddle, being applied to a fluid at rest. This leaves only two ways to set the fluid in motion. Either the external sources or sinks of radiation change so that the radiation patterns in the fluid are altered, or the conduction of heat across the boundaries is changed by an externally caused change in the temperature of the bounding surface. If either one of these two effects occurs, then local changes in temperature will also occur according to Eq. (2). At the instant this happens, the equation of state says that pressure changes will also occur, and thus there are accelerating forces and the fluid will begin to move. If the disturbance of the radiation or conduction patterns is severe enough, the resulting motion may be sufficiently intense that friction is not able to retard it immediately, and a flow of considerable complexity will develop.

In order to pursue this further, let us see precisely what will occur when heating is imposed on the fluid. We shall treat the problem as though we were integrating the equations numerically in order to obtain the correct sequencing of events. At time step zero, both the velocities and accelerations vanish and the system (1) to (4) describes the fluid. If we impose changes in the radiational heating at time step 1, then we shall have temperature changes develop at step 2. These will, through the equation of state, yield pressure changes at step 2, and hence we shall have accelerations. Thus velocities finally will appear at step 3 when we have integrated forward from step 2.

To determine the character of this initial motion, let us differentiate the equation of motion to obtain a result that is useful at step 1 when $\mathbf{v} = \partial\mathbf{v}/\partial t = 0$. We find

$$\frac{\partial^2 \mathbf{v}}{\partial t^2} = -\frac{\partial}{\partial t}\left(\frac{1}{\rho}\nabla p\right) = -\frac{1}{\rho}\frac{\partial}{\partial t}\nabla p = -\frac{R\,\nabla\rho}{\rho}\frac{\partial T}{\partial t} - R\,\nabla\frac{\partial T}{\partial t} \tag{11}$$

in which we have used Eqs. (3) and (4) and in which $\nabla\rho = \mathbf{k}\,\partial\rho/\partial z$. Thus upon integration of this equation twice we would find the velocity field at step 3.

The result (11) shows that there are two components to the initial motion. The second term on the right makes it clear that the fluid will accelerate away from regions in which heating is occurring and toward those in which cooling occurs. The first term demonstrates the effect of stratification. Assuming that density decreases upward, we see that warmed fluid will rise, at least initially, and that cooled fluid will descend. This vertical preference for the convection is a consequence of gravity.

There are, then, two important aspects of the process of initiating fluid motion that deserve emphasis. The first is that cooling may be just as effective as heating in initiating motion; the crucial factor is that the equilibrium must be disturbed by either heating or cooling. The second is that at the instant heating occurs at one point in the fluid and cooling at another, there will be an acceleration developing from the source of heat toward the sink; the motion obviously tries to restore the equilibrium that was disturbed by external influences.

The attempt to restore equilibrium is most fundamentally viewed as a response to an alteration in the equilibrium entropy field. Accelerations are produced that, at least initially, are in the correct direction to carry air with increasing entropy toward those regions in which the entropy is decreasing. The initial accelerations are obviously attempting to restore the entropy to an equilibrium configuration. But if the external heating or cooling that has set the fluid in motion then changes again, the motion must attempt to reach a balance at a new equilibrium. And if the heating is always differential or changing, the fluid is forever in motion toward an equilibrium it can never reach. This is the cause of the atmosphere's motion; it is in a ceaseless quest for an equilibrium that shall, as surely as the earth is round and seasons pass, prove forever elusive.

PROBLEM 6.7.1 Suppose an atmospheric state with neither motion nor accelerations is heated or cooled uniformly so that $\nabla(\partial T/\partial t) = 0$. Show that only horizontally uniform vertical expansion or contraction will develop initially.

PROBLEM 6.7.2 Let the total kinetic energy K vanish identically in an interval $[t_1, t_2]$ with $t_2 > t_1$. Show that the horizontal temperature gradients vanish in (t_1, t_2). (Hint: Calculate $\partial^2 K/\partial t^2$.)

PROBLEM 6.7.3 Prove the following corollary to Jeffreys' theorem: If the radiation divergence does not vanish identically in (t_1, t_2), then the atmosphere is not in equilibrium in $[t_1, t_2]$. Equilibrium is defined here to mean $\mathbf{v} \equiv \dot{\mathbf{v}} \equiv \nabla T \equiv 0$.

PROBLEM 6.7.4 Show that if any of the entropy S, internal energy I, or potential energy P vary in an interval (t_1, t_2), then K cannot vanish identically in $[t_1, t_2]$.

PROBLEM 6.7.5 Show that if $\dot{I} \neq 0$, then there must be either motions or accelerations.

6.8 THE STRESS TENSOR AND FRICTIONAL FORCES

In the derivation of the equations of motion given earlier in the chapter, we used an elementary approach to determine expressions for the forces that experience dictates must be important in fluid motion. Now we turn to a postulational approach that will yield the previous results and also produce an explicit expression for the frictional forces. The development and use of postulates in the study of fluid motion has a distinct advantage: it makes very clear precisely what we are assuming. The discovery of correct postulates depends on experience and experiment, and results derived from any set of postulates are naturally subject to experimental verification.

6.8.1 The Stress Tensor

To begin, we assume that only two kinds of forces can be exerted on a fluid parcel in an inertial system. The first are external or body forces that act throughout the fluid and are independent of its motion. The only body force with which we are concerned is the force of gravitation. The second kind of forces are those due to internal interactions of the fluid with itself, and they may depend upon the state of motion.

The internal forces in a fluid will be referred to as *stresses*, with stress defined as force per unit area. At a point on the surface of a parcel, the stress may have components both normal to the surface and tangent to the surface. Thus the stress in a fluid is in general a function of position, of time, and of the direction or the orientation of the surface on which it is acting. When the action of the stresses produces a change in the relative position of elements in a fluid parcel, we say that strain is occurring; the rate of strain tensor e_{ij} has already been introduced in Sec. 5.6.

The fundamental postulate or principle of fluid mechanics is an *analogy* with Newton's second law: *The rate of change of (linear) momentum in a material volume equals the resultant of the forces acting on the volume; thus*

$$\frac{d}{dt}\int_V \rho \mathbf{v}\, dV = \int_V \rho \mathbf{f}\, dV + \int_V \rho \mathbf{F}\, dV \tag{1}$$

in which V is the material volume, $\mathbf{f}$ *is the resultant specific body or external force, and* $\mathbf{F}$ *is the resultant specific internal force.*

A further basic postulate, formulated some 150 years ago, is due to A.-L. Cauchy: *All forces exerted on a parcel by the rest of the fluid can be represented as a distribution of stress vectors applied to the bounding surface of the parcel.*

Applying Cauchy's postulate, we may write

$$\int_V \rho \mathbf{F}\, dV = \int_S \mathbf{t}\, d\sigma \tag{2}$$

in which $\mathbf{t}$ is the stress vector and S is the surface of the volume. As noted earlier, we have

$$\mathbf{t} = \mathbf{t}(\mathbf{x}, t; \boldsymbol{\eta}) \tag{3}$$

where $\boldsymbol{\eta}$ is the exterior unit normal to the portion of the surface S at point $\mathbf{x}$; the orientation of the surface on which the stress is acting is thus represented by the direction of the normal to the surface. Application of the transport theorem now gives the equation of motion in the form

$$\int_V \rho \frac{d\mathbf{v}}{dt}\, dV = \int_V \rho \mathbf{f}\, dV + \int_S \mathbf{t}\, d\sigma \tag{4}$$

With the necessary assumption that the density, total acceleration, and external forces are bounded, we may apply the mean-value theorem to Eq. (4) to obtain, for a small volume V, the statement that the average quantities $\overline{(\ )}$ are related by

$$\left(\overline{\rho \frac{d\mathbf{v}}{dt}} - \overline{\rho \mathbf{f}}\right) V = \int_S \mathbf{t}\, d\sigma \tag{5}$$

If we divide both sides by the surface area S and let V shrink down on itself until it vanishes, we have $\lim (V/S) = 0$ and hence

$$\lim_{V \to 0} \frac{1}{S} \int \mathbf{t}\, d\sigma = 0 \tag{6}$$

Thus the stresses acting on any point of the fluid are in equilibrium.

To see the significance of this result, let us consider a small spherical volume. If we divide the sphere into two hemispheres, the normals to the bases of the hemispheres are in opposite directions. But according to Eq. (6), the stresses on the bases are equal to the averages over the spherical part of the hemispheres; furthermore, these two averages are of equal magnitude and opposite sign because the average over the sphere vanishes. Thus the stresses on the bases of the hemispheres are likewise equal in magnitude but opposite in sign, and so we may write

$$\mathbf{t}(\mathbf{x}, t; \boldsymbol{\eta}) = -\mathbf{t}(\mathbf{x}, t; -\boldsymbol{\eta}) \tag{7}$$

But at any point on the surface of a parcel, the exterior normal to the parcel is the negative of the exterior normal to the rest of the fluid. Thus we have proved:

Theorem The stress placed on the parcel by the fluid is equal in magnitude but opposite in sign to the stress that the parcel imposes on the rest of the fluid.

////

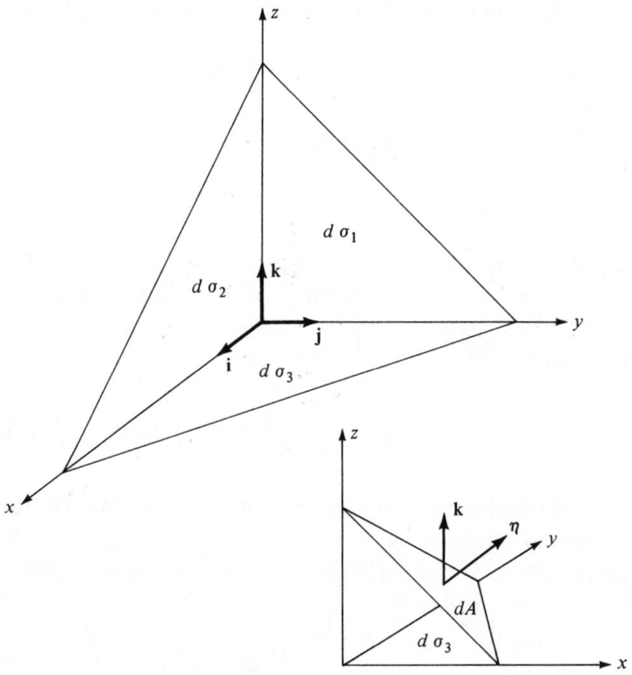

FIGURE 6.5
Geometry of the tetrahedron used in development of the stress tensor. Note that $d\sigma_3 = \mathbf{k} \cdot \boldsymbol{\eta}\, dA = \eta_3\, dA$.

We may use these results to obtain a more specific characterization of the nature of the stresses. Suppose we place a triangular pyramid (a tetrahedron) with its vertex at the point **x**, and with each of the sides parallel to one of the unit vectors **i**, **j**, or **k**. Its base, then, will be a plane surface in the first octant of the coordinate system with vertices of the base on the positive x, y, and z axes (see Fig. 6.5). The normal to the base has the form

$$\boldsymbol{\eta} = \eta_1 \mathbf{i} + \eta_2 \mathbf{j} + \eta_3 \mathbf{k} \tag{8}$$

and according to the results of Problem 5.5.20, the faces of the pyramid have area given by

$$d\sigma_i = \eta_i\, dA \tag{9}$$

where dA is the area of the base and $d\sigma_i$ is equivalent to the projection of that area on the coordinate plane normal to the ith axis.

According to the equilibrium of stresses we must have

$$\mathbf{t}(\boldsymbol{\eta})\, dA + \mathbf{t}(-\mathbf{i})\, d\sigma_1 + \mathbf{t}(-\mathbf{j})\, d\sigma_2 + \mathbf{t}(-\mathbf{k})\, d\sigma_3 = 0 \tag{10}$$

in which $-\mathbf{i}$, $-\mathbf{j}$, and $-\mathbf{k}$, are the exterior normals to the faces of the pyramid. With Eqs. (9) and (7), we may convert Eq. (10) into the expression

$$\mathbf{t}(\boldsymbol{\eta}) = \eta_1 \mathbf{t}(\mathbf{i}) + \eta_2 \mathbf{t}(\mathbf{j}) + \eta_3 \mathbf{t}(\mathbf{k}) \tag{11}$$

Therefore the component of the stress in the $\mathbf{i}_i$ direction is given by (summation convention applied)

$$t_i(\boldsymbol{\eta}) = \mathbf{i}_i \cdot \mathbf{t}(\boldsymbol{\eta}) = \eta_j [\mathbf{i}_i \cdot \mathbf{t}(\mathbf{i}_j)] \tag{12}$$

The scalar product acts as the coefficient of the normal components η_j in a linear expression, and we may express the result as

$$t_i(\boldsymbol{\eta}) = \eta_j T_{ij} \tag{13}$$

where

$$T_{ij} = \mathbf{i}_i \cdot \mathbf{t}(\mathbf{i}_j) \tag{14}$$

is a tensor giving the stress acting in the ith direction on a surface normal to the jth coordinate axis.

The ith component of the equation of motion (4) may be expressed with Eq. (13) as

$$\int_V \rho \frac{dv_i}{dt}\, dV = \int_V \rho f_i \, dV + \int_S \eta_j T_{ij} \, d\sigma \tag{15}$$

The last integral is in the form that appears in the divergence theorem, and because the material volume V is arbitrary, we may write

$$\rho \frac{dv_i}{dt} = \rho f_i + \frac{\partial}{\partial x_j} T_{ij} \tag{16}$$

We know from comparison with Eq. (9) in Sec. 6.2 that

$$f_i = -g\, \delta_{i3} \tag{17}$$

We hypothesize that the only internal force in a fluid at rest and in equilibrium is the normal pressure being exerted equally in all directions (Problem 4.1.1). Hence we must have

$$\mathbf{t}(\boldsymbol{\eta})|_{\mathbf{v}=0} = -p\boldsymbol{\eta} \tag{18}$$

where $\boldsymbol{\eta}$ is the exterior normal to an element of fluid volume. Upon choosing $\boldsymbol{\eta}$ to be $\mathbf{i}_j$, we find that

$$T_{ij}|_{\mathbf{v}=0} = -p\, \delta_{ij} \tag{19}$$

The form of T_{ij} when motion is present is the crucial question to be investigated in this section.

The part of the stress tensor revealed by Eq. (19) is symmetric; the next question is obviously whether the general form of T_{ij} should also be symmetric. But the symmetry of T_{ij}, although plausible as will be shown, must actually be postulated in the general case because it cannot be rigorously deduced except by appeal to another postulate. If T_{ij} is symmetric, then $T_{ij} = T_{ji}$. To see what this means, let us consider the contribution of the stress to the rate of change of the angular momentum,

$$\begin{aligned} \rho \mathbf{r} \times \frac{d\mathbf{v}}{dt} &= \rho \frac{d}{dt}(\mathbf{r} \times \mathbf{v}) - \rho \frac{d\mathbf{r}}{dt} \times \mathbf{v} \\ &= \rho \frac{d}{dt}(\mathbf{r} \times \mathbf{v}) = \rho \frac{d}{dt}(\mathbf{i}_i \epsilon_{ijk} r_j v_k) \end{aligned} \tag{20}$$

because the second term vanishes. Thus the contribution of the stresses is

$$\begin{aligned} \left[\rho \frac{d}{dt}(\epsilon_{ijk} r_j v_k)\right]_{T_{jk}} &= \epsilon_{ijk} r_j \frac{\partial}{\partial x_l} T_{kl} \\ &= \epsilon_{ijk} \frac{\partial}{\partial x_l}(r_j T_{kl}) - \epsilon_{ijk} T_{kj} \end{aligned} \tag{21}$$

Therefore if the stress tensor is symmetric, the last term vanishes. If it is not symmetric, then there is an internal contribution to the total angular momentum that is independent of the origin with respect to which the angular momentum is determined. Thus the assumption of symmetry of T_{ij} and the statement that the angular momentum must be controlled by a conservation law of the form

$$\int_V \rho \frac{d}{dt}(\epsilon_{ijk} r_j v_k)\, dV = \int_V \rho \epsilon_{ijk} r_j f_k \, dV + \int_S \epsilon_{ijk} n_l r_j T_{kl}\, d\sigma \tag{22}$$

are equivalent postulates. For the types of fluid we shall deal with here, the symmetry of the stress tensor will result from still another postulate, and thus Eq. (22) will be true. In advanced studies of fluid mechanics, account is taken of asymmetric stress tensors resulting from internal rotational moments.

PROBLEM 6.8.1 Prove that T_{ij} is a tensor.

PROBLEM 6.8.2 *Mechanical equilibrium* is defined by the requirements that the total force and the moment of the forces with respect to an arbitrary vector **r** both vanish. Show that the necessary and sufficient condition for mechanical equilibrium is that $\dot{\mathbf{v}} = 0$ and that the stress tensor be symmetric. [Hint: According to Cauchy's postulate, the moment of the internal forces is the surface integral of $\mathbf{r} \times \mathbf{t}$; use the first condition and the right side of Eq. (21).]

PROBLEM 6.8.3 What is the fluid-mechanical equivalent of Newton's third law?

6.8.2 The Constitutive Equations

The long years of effort that succeeded in reducing the equations of motion to the elegant form (16) will have produced little of practical value unless a relation between the stress tensor and the other variables can be found so that the system of equations is complete. An expression giving T_{ij} as a function of the other variables of the flow is called a *constitutive equation*. The most common form of constitutive equation in fluid mechanics was originally proposed by Newton; the derivation in complete form is due to Sir George Stokes. The considerations on which Stokes' derivation was based are now expressed in postulational form and permit the discovery of a variety of constitutive relations. Here we take a restricted view, staying with what is essentially the classical approach. These postulates may be expressed as:

1 T_{ij} depends on the velocity field only through the deformation or rate-of-strain tensor e_{ij} and may be written as a polynomial in this tensor.

Thus it is *assumed* that the translational and rotational components of the motion do not induce stress on a parcel; only the expansion, the shearing and stretching of deformation, and the normal forces of pressure contribute to stress. This hypothesis will be specialized to a linear dependence on e_{ij} subsequently.

2 T_{ij} does not depend explicitly on the position $\mathbf{x}$.

3 The properties of the fluid do not have a preferred direction in space; thus T_{ij} can depend on arbitrary orthogonal rotations of the coordinate axes only as e_{ij} does.

This postulate expresses the concept of isotropy; an isotropic quantity is invariant under orthogonal rotations, so that, for example, all scalars are isotropic.

4 When $e_{ij} = 0$, then $T_{ij} = -p\,\delta_{ij}$.

The plausibility of this hypothesis follows from the discussion leading to Eq. (19). Fluids for which $T_{ij} = -p\,\delta_{ij}$ in any case are called *perfect*.

Postulates 1 and 2 imply that

$$T_{ij} = \alpha_{ij} + \beta_{ijkl}e_{kl} + \gamma_{ijklmn}e_{kl}e_{mn} + \cdots \tag{23}$$

in which the coefficients α_{ij}, β_{ijkl}, and γ_{ijklmn} are functions of the properties of the fluid, are independent of the velocity, and do not contain explicit dependence on position $\mathbf{x}$.

According to postulate 4

$$\alpha_{ij} = -p\,\delta_{ij} \tag{24}$$

PROBLEM 6.8.4 Show that α_{ij} is invariant under orthogonal rotations of the coordinates and thus is isotropic.

The application of postulate 3 to Eq. (23) requires that we find isotropic forms for the coefficient tensors β_{ijkl} and γ_{ijklmn}. It can be shown (Jeffreys, 1931) that the most general isotropic fourth-order tensor with the required symmetry in (i,j) and (k,l) is

$$\beta_{ijkl} = \lambda\, \delta_{ij}\, \delta_{kl} + \mu(\delta_{ik}\, \delta_{jl} + \delta_{il}\, \delta_{jk}) \tag{25}$$

in which λ and μ are scalars. Thus we have

$$\beta_{ijkl}e_{kl} = \lambda e_{kk}\, \delta_{ij} + 2\mu e_{ij} = \lambda\, \nabla \cdot \mathbf{v}\, \delta_{ij} + 2\mu e_{ij} \tag{26}$$

We are not going to attempt to find the most general isotropic form of the tensor γ; we retain it here only in order to make the point that the usual linear relation involves an approximation to the implication of the first postulate.

Thus having applied the postulates, we have the expression

$$T_{ij} = -p\, \delta_{ij} + \lambda\, \delta_{ij}\, \nabla \cdot \mathbf{v} + 2\mu e_{ij} + \gamma_{ijklmn}e_{kl}e_{mn} + \cdots \tag{27}$$

The sum of the normal stresses at a point is given by $T_{ii} = T_{11} + T_{22} + T_{33}$. The average of the negative of this quantity is defined to be the *mean pressure* $\bar{p}$; it is thus the average force per unit area acting normal to the surface of a parcel, and is given by

$$\bar{p} = -\tfrac{1}{3}T_{ii} = p - \left(\lambda + \frac{2\mu}{3}\right)\nabla \cdot \mathbf{v} - \tfrac{1}{3}\gamma_{iiklmn}e_{kl}e_{mn} + \cdots \tag{28}$$

It is important to remember that the pressure p is a thermodynamic quantity whose definition depends on thermodynamic equilibrium. The mean normal stress $\bar{p}$ includes this equilibrium pressure as well as forces that arise from divergence and deformation. The constitutive equations provide the specification of the relation between these two pressures and the motion fields that follows from the constitutive hypotheses.

Now we define a scalar μ' by

$$\mu' = \tfrac{3}{2}\left(\lambda + \tfrac{2}{3}\mu\right) \tag{29}$$

Then Eq. (27) becomes, on the linear hypothesis,

$$T_{ij} = -p\, \delta_{ij} - \tfrac{2}{3}(\mu - \mu')\delta_{ij}\, \nabla \cdot \mathbf{v} + 2\mu e_{ij} \tag{30}$$

and Eq. (28) takes the new form

$$\bar{p} = p - \tfrac{2}{3}\mu'\, \nabla \cdot \mathbf{v} \tag{31}$$

The magnitude of μ and μ' and their dependence on thermodynamic quantities must be determined by experiment; so must the question of whether the higher-order terms in e_{ij} can be ignored.

6.8.3 The Viscosity Coefficients

To illustrate the significance of the coefficients in Eq. (30), let us assume the linear relation and consider some possible experiments. Suppose that we have fluid initially at rest and that we set it in motion by applying a tangential stress on its top surface, perhaps with a flat plate that is moved by external means. Letting x be the direction of motion of the plate, we have then an external stress S_{13} being applied. Presumably the first response of the fluid will be shear $\partial u/\partial z$, so that $e_{ii}=0$ and $2e_{13}=\partial u/\partial z$. The external stress will be transmitted into the interior of the fluid, and so we will find from the linear version of Eq. (30) that

$$T_{13} = \mu \frac{\partial u}{\partial z} \tag{32}$$

Thus the coefficient μ is the factor of proportionality between stress and shear in this simple case. It is called the *dynamic coefficient of viscosity* and has units of g/cm • s. Note that when the stress is applied in the positive x direction, we may expect also that u will increase with z and so we deduce that $\mu \geqslant 0$.

For air at standard pressure and a temperature of 20°C, $\mu = 1.81 \times 10^{-4}$ g/cm • s. It is frequently useful to express the equations of motion in a form where μ/ρ appears. This coefficient is denoted by $\nu = \mu/\rho$, has units cm^2/s, and is called the *kinematic viscosity*. At the same pressure and temperature, $\nu = 0.15$ cm^2/s.

The problem of determining the coefficient $\frac{2}{3}\mu'$, called the *bulk viscosity* or, as suggested by Batchelor (1967), the *expansion viscosity*, is more difficult. We might, however, consider a hypothetical device that would compress air radially and yet measure the mean normal resisting force $\bar{p}$ at the same time. If the thermodynamic pressure could be found from

$$p = \frac{MRT}{V} \tag{33}$$

then we could use Eq. (31) in the form

$$p - \bar{p} = \frac{MRT}{V} - \bar{p} = \frac{2}{3}\mu' \frac{1}{V}\frac{dV}{dt} \tag{34}$$

Suppose we stop the compression at a given volume and let the gas come to equilibrium; then we have $p = \bar{p}$. Upon starting compression again, we might expect the gas would resist at a rate $\bar{p} > p$ because there has not been time for the new equilibrium to be achieved and the energy imparted by compression will not have been distributed to the rotational degrees of freedom. For the compression, dV/dt is negative, and so if our intuition is correct, μ' is also positive.

The values of viscosity are measured by more realistic techniques. One method is to study the absorption of sound waves by the gas. According to kinetic theory, μ' vanishes for monatomic gases, but the relaxation phenomena associated with polyatomic molecules induce significant expansion viscosities. Thus for air, the indications from supersonic absorption studies are that μ and μ' are of the same order of magnitude. On this basis it would appear that omitting the middle term of Eq. (30) is preferable to the assumption that μ' vanishes.

6.8.4 The Rate of Dissipation

The stress tensor, on the assumption that stress is linear in rate of strain, may be written as

$$T_{ij} = -p\,\delta_{ij} + \Delta_{ij} \tag{35}$$

in which the *viscous tensor* is defined by

$$\Delta_{ij} = 2\mu e_{ij} - \tfrac{2}{3}(\mu - \mu')e_{kk}\,\delta_{ij} \tag{36}$$

From Eq. (2) in Sec. 6.6, the effect of friction on the kinetic energy is given by

$$\left(\frac{dK}{dt}\right)_{\mathrm{fr}} = \int_V v_i \frac{\partial}{\partial x_j}(\Delta_{ij})\,dV = \int_S \eta_j v_i\,\Delta_{ij}\,d\sigma - \int_V \frac{\partial v_i}{\partial x_j}\Delta_{ij}\,dV \tag{37}$$

in which we have employed the divergence theorem. In our development of the first law of thermodynamics we left the frictional heating rate unspecified in the form f_h, so we now have

$$\left[\frac{d}{dt}(K+I)\right]_{\mathrm{fr}} = \int_S \eta_j v_i\,\Delta_{ij}\,d\sigma + \int_V \left(f_h - \frac{\partial v_i}{\partial x_j}\Delta_{ij}\right) dV \tag{38}$$

The frictional effects in the two equations are now linked by adopting the postulate: *Except for effects at the boundary, any changes in kinetic energy due to friction must be exactly balanced by changes in internal energy.* This energy conservation hypothesis thus yields the result that, because the volume is arbitrary, the *rate of frictional dissipation* is

$$f_h = \frac{\partial v_i}{\partial x_j}\Delta_{ij} \tag{39}$$

The viscous tensor Δ_{ij} is symmetric because e_{ij} is, and so we need to retain only the symmetric part of $\partial v_i/\partial x_j$, which is e_{ij} again. Thus

$$f_h = e_{ij}\left[2\mu e_{ij} - \tfrac{2}{3}(\mu - \mu')e_{kk}\,\delta_{ij}\right] \tag{40}$$

We should expect that f_h is nonnegative–so that there is always a gain in internal energy and a loss of kinetic energy (when boundary effects are vanishingly small). Let us write Eq. (40) as

$$f_h = 2\mu(e_{ij})^2 - \tfrac{2}{3}(\mu - \mu')(e_{kk})^2 \tag{41}$$

and determine constants α and β such that

$$f_h = \alpha(e_{ij} - \beta e_{kk}\,\delta_{ij})^2 = \alpha[(e_{ij})^2 - (2\beta - 3\beta^2)(e_{kk})^2] \tag{42}$$

in which we have used $(\delta_{ij})^2 = \delta_{ii} = 3$.

Obviously then, $\alpha = 2\mu$, and

$$3\beta^2 - 2\beta + \tfrac{1}{3}\left(1 - \frac{\mu'}{\mu}\right) = 0 \tag{43}$$

and so

$$\beta = \tfrac{1}{3}\left(1 \pm \sqrt{\frac{\mu'}{\mu}}\right) \tag{44}$$

We have produced an expression for f_h which is a squared quantity and therefore f_h is nonnegative provided that, as indicated by experiment, the coefficients μ and μ' are always positive. Thus whenever deformation by stretching, shear, or divergence occurs, there is a transfer of energy from the kinetic to the internal form.

This frictional destruction of kinetic energy and production of internal energy is the reason why fluids without external forcing tend to flow toward a state of motionless equilibrium.

PROBLEM 6.8.5 Show that $\mathbf{f}_r = \mu\,\nabla^2\mathbf{v} + \tfrac{1}{3}(\mu + 2\mu')\,\nabla(\nabla \cdot \mathbf{v})$ for μ constant.

6.8.5 The Navier-Stokes Equations

The dynamical equations of motion for a fluid embodying the linear form of the relation between stress and strain are called the *Navier-Stokes equations.* For completeness, we collect here a summary of our results by stating them explicitly. We have

$$\rho\frac{dv_i}{dt} = -\frac{\partial p}{\partial x_i} - \rho g\,\delta_{i3} + \frac{\partial}{\partial x_j}\left[2\mu e_{ij} - \tfrac{2}{3}(\mu - \mu')e_{kk}\,\delta_{ij}\right] \tag{45}$$

$$c_V\rho\frac{dT}{dt} = -p\frac{\partial v_i}{\partial x_i} - \frac{\partial R_i}{\partial x} + \frac{\partial}{\partial x_i}\left(k\frac{\partial T}{\partial x_i}\right) + C + \frac{\partial v_i}{\partial x_k}\left[2\mu e_{ik} - \tfrac{2}{3}(\mu - \mu')e_{jj}\,\delta_{ik}\right] \tag{46}$$

$$\frac{d\rho}{dt} + \rho\frac{\partial v_i}{\partial x_i} = 0 \tag{47}$$

$$p = \rho RT \tag{48}$$

In nearly all applications of these equations to atmospheric problems it has been assumed that μ is constant (or even zero) and that $\mu' = 0$, and that therefore no distinction between p and $\bar{p}$ exists. When the incompressible assumption is made and an equation of state of the form (2) in Sec. 6.5 is used, the pressure that appears is solely a mechanical variable and is determined, as shown in Sec. 6.5.1, from the velocity and density fields.

*PROBLEM 6.8.6 Revise Eqs. (45) to (48) so that they apply to a fluid with equation of state $F(p,\rho,T) = 0$. (See Problem 6.3.4.)

6.9 BOUNDARY CONDITIONS

The equations of motion presented in this chapter must have both boundary and initial conditions associated with them in order to provide a description of the flow. The specification of boundary conditions is essential in both analytical and numerical work with the equations.

These equations are partial differential equations based on three spatial variables (x,y,z) and time t. Thus we must have *boundary conditions* for each variable on each bounding surface. These boundary values may, in some cases, be specified as functions of time. We also need *initial values* for each of the variables for every point at the initial time t_0 in order to compute the temporal rate of change of the variables at t_0 and thus be able to integrate the equations forward in time.

The boundary and initial conditions exert a considerable control on the character of the solutions. Thus, it is always advisable to think of a physical system or the differential equations used to model it as effectively incomplete, even though there is one equation for each variable, until the boundary and initial conditions have been specified and joined to the system.

It is customary to distinguish two kinds of bounding surfaces (Fig. 6.6). *Rigid surfaces* are obviously surfaces that confine the fluid and do not move, regardless of what the fluid does. Examples of rigid boundary surfaces are the walls of a metal container holding a liquid or, in the case of the atmosphere, the continents on the earth's surface. The other kind of boundaries are *free surfaces*, which are considered to be material surfaces of the motion and therefore surfaces that move with the flow itself. Examples of surfaces considered to be free are the upper surface of a liquid in a container with an open top, the surface of a lake or ocean, or the imaginary top of the atmosphere.

6.9.1 Boundary Conditions on a Rigid Surface

The essential characteristic of a rigid surface is that it confines the fluid and prevents it from flowing past the bounding surface. Thus at any rigid boundary the component of the velocity normal to the surface must vanish, that is,

$$\mathbf{v} \cdot \boldsymbol{\eta} = 0 \tag{1}$$

where $\boldsymbol{\eta}$ is the unit normal vector to the boundary with the positive direction now taken into the fluid. The remaining conditions depend on whether molecular effects are important at the boundary.

Free-slip boundaries It is often useful to consider that the molecular effects of friction and heat conduction are unimportant at the boundary; such a situation is

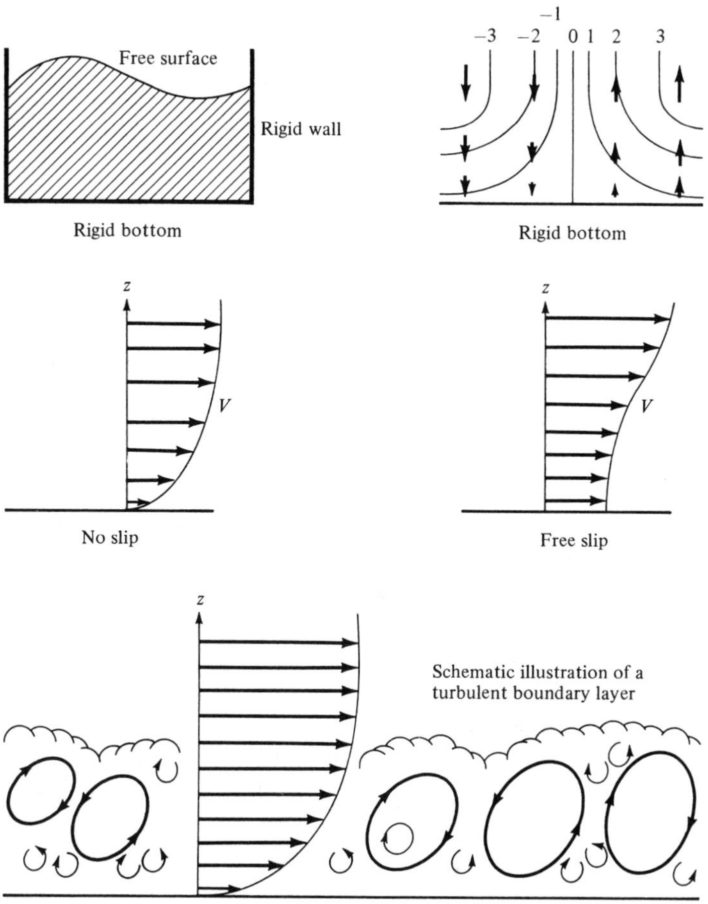

FIGURE 6.6
Illustration of boundary conditions and a boundary layer.

referred to as *free-slip boundary flow*. Because the coefficients of viscosity and thermal conductivity are of the same order of magnitude for air, it is generally consistent in meteorological problems either to include both of these molecular processes or to ignore both. Therefore, it is usually assumed that there is no heat conduction across a free-slip boundary.

If there are no frictional effects at the boundary, then there should be no influence on the flow, and so the suitable condition is that there is no shear at the boundary itself. Thus

$$\frac{\partial \mathbf{v}}{\partial \eta} \cdot \boldsymbol{\tau} = 0 \tag{2}$$

where $\boldsymbol{\tau}$ is any tangent vector to the boundary and η is distance normal to the boundary. By the same reasoning, without conduction of heat across the boundary there should be no effect on the temperature gradient normal to the boundary and hence

$$\boldsymbol{\eta} \cdot \nabla T = 0 \tag{3}$$

Thus we have boundary conditions for the normal and tangential components of the velocity and a condition for the temperature. If we assume that the equation of state holds at the boundary, we need one more condition on either pressure or density, or a condition relating them.

An obvious possibility, if the rigid boundary is horizontal, is to require that the hydrostatic condition holds, so that at the boundary

$$\frac{\partial p}{\partial z} = -g\rho \tag{4}$$

Another possibility on a horizontal rigid surface is to use the equation of continuity, which when differentiated with respect to z with Eq. (2) applied gives

$$\frac{\partial^2 \ln \rho}{\partial t\, \partial z} + \mathbf{v}_H \cdot \nabla \frac{\partial \ln \rho}{\partial z} + \frac{\partial w}{\partial z}\frac{\partial \ln \rho}{\partial z} + \frac{\partial^2 w}{\partial z^2} = 0 \tag{5}$$

Because $\partial T/\partial z$ vanishes, this same equation with $\ln p$ replacing $\ln \rho$ provides a boundary equation for pressure; or Eq. (5) can be solved for density and the equation of state used to find the pressure.

For incompressible flow, boundary conditions are not needed for density because the equation of state (2) in Sec. 6.5 will give boundary values of density once the temperature is determined. In this case it is of interest to note that the nondivergent condition requires, again for a horizontal boundary, that

$$\frac{\partial}{\partial z}(\nabla \cdot \mathbf{v}) = \frac{\partial^2 w}{\partial z^2} = 0 \tag{6}$$

Nonslip surface When frictional forces are important at the boundary, the classical requirement is the *adherence condition* that, in addition to $\mathbf{v} \cdot \eta = 0$, we shall have, for nonmoving surfaces,

$$\mathbf{v} \cdot \boldsymbol{\tau} = 0 \tag{7}$$

To see the intuitive justification for this condition, imagine an enlargement of a picture of the boundary interface. As the degree of enlargement increases, the boundary will seem more and more irregular with sharp peaks and valleys appearing. Clearly the fluid in the valley bottoms cannot move as rapidly as that above the peaks. At immense degrees of enlargement, the fluid molecules will appear trapped between the molecules of the boundary.

For the condition on temperature, it is generally assumed that the boundary and the material behind it represent an infinite reservoir of heat so that the temperature of the boundary surface is either constant or controlled externally and thus

$$T = T(\mathbf{x}_B, t) \tag{8}$$

where the subscript indicates that $\mathbf{x}_B$ ranges over the coordinates of the boundary. With the equations now containing the molecular heat conduction term, the temperatures in the vicinity of the boundary will tend to be quite near the values imposed by Eq. (8).

For an incompressible fluid, these conditions are sufficient, although we note that the condition of no divergence now implies that on horizontal boundaries $\partial w/\partial z = 0$ because u and v are constant at the value zero. On a boundary parallel to the yz plane we have similarly $\partial u/\partial x = 0$.

For compressible flow, the continuity equation implies that

$$\frac{\partial}{\partial t} \ln \rho + \frac{\partial}{\partial \eta} \mathbf{v} \cdot \boldsymbol{\eta} = 0 \tag{9}$$

because the components of the velocity tangent and normal to the boundary vanish identically. Finally, the pressure is determined by

$$p = \rho R T(\mathbf{x}_B, t) \tag{10}$$

PROBLEM 6.9.1 Generalize the conditions for nonslip flow to take account of a boundary moving with velocity $\mathbf{w}$.

PROBLEM 6.9.2 Show that the adherence condition for a stationary boundary implies that the normal component of the vorticity vanishes. (Hint: Apply Stokes' theorem.)

PROBLEM 6.9.3 (Berker) Show that the stress on a stationary, nonslip boundary is given by

$$\mathbf{t} = \left[-p + \left(\tfrac{4}{3}\mu + \tfrac{2}{3}\mu'\right)\nabla \cdot \mathbf{v}\right]\boldsymbol{\eta} + \mu(\boldsymbol{\zeta} \times \boldsymbol{\eta})$$

Show that the tangential stress depends on the tangential component of the vorticity, and that the normal component of the stress exerted on the boundary is

$$-\boldsymbol{\eta} \cdot \mathbf{t} = \bar{p} - \tfrac{4}{3}\mu \nabla \cdot \mathbf{v}$$

[Hint: Apply Stokes' theorem to the normal component of the velocity and use Eq. (39) in Sec. 5.1 and Eq. (35) in Sec. 5.6.]

Boundary layers It is obvious that the frictional forces at the boundary are important in atmospheric flow. But a significant phenomenon that frequently simplifies the situation is the development of a *boundary layer* near a rigid, nonslip boundary (Fig. 6.6). Within this boundary layer the molecular effects are important and the adherence condition holds at the boundary itself. Thus there is strong shear and, by a similar argument, possibly strong heat conduction within the boundary layer. The flow is often very turbulent.

Above the boundary layer, the flow may be relatively insensitive to the adherence condition, and the shears and temperature gradients are not strong enough to make the molecular effects important in the motion. Therefore a frequently used model of atmospheric and other fluid flow includes a rigid nonslip boundary, a boundary layer in which molecular effects are important, and the remaining part of the fluid in which molecular effects are not directly important in large-scale features of the flow.

For many problems, then, the boundary layer is ignored and free-slip conditions are imposed. This is often done in simple numerical prediction schemes in which the thickness of the boundary layer is less than the vertical spacing of the coordinate grid.

Still, boundary layer processes are important for actual flows, and one of the central problems of atmospheric science is the determination of adequate models of these processes. The atmospheric boundary layer will be discussed further in Chap. 11.

6.9.2 Free-Surface Conditions

The conditions at a free surface derive from the fact that it may be considered a material surface of the flow. Thus if the surface is described by an equation of the form

$$F(\mathbf{x}, t) = 0 \tag{11}$$

we must have

$$\frac{dF}{dt} = \frac{\partial F}{\partial t} + \mathbf{v}_H \cdot \nabla F + w\frac{\partial F}{\partial z} = 0 \tag{12}$$

as a boundary condition. Motions of this surface are frequently eliminated, when only internal phenomena are of interest, by the conditions $w_S = 0$, $\nabla F = 0$, and $\partial F/\partial t = 0$. In many problems with liquids, the upper free surface is assumed to be an isobaric surface so that $dp/dt = 0$.

Furthermore, it is reasonable to assume that a material surface at the boundary shares the internal property of the motion that the stress must be differentiable. Therefore it may be assumed that the stress is continuous across the free surface. In many applications this means that the stress at the free surface must equal an externally imposed stress, as, for example, in the case of wind stress driving motions in a lake. In other cases, there is no external stress, and so the condition is that the stress must vanish on the free surface. This then produces a requirement on the derivatives of the velocity field.

PROBLEM 6.9.4 Assume that the tangential stress vanishes on a free surface. What conditions on the velocity are implied?

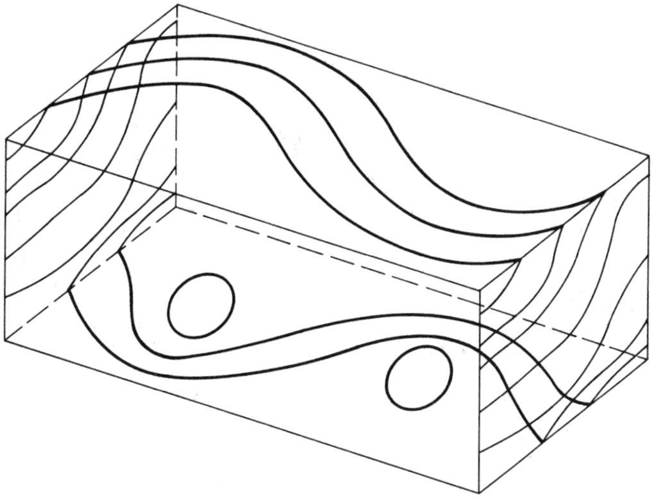

FIGURE 6.7
Schematic illustration of a cyclically continuous channel model for simulating atmospheric flow in the westerlies. Note that the patterns match exactly at the longitudinal boundaries.

6.9.3 Atmospheric Boundary Conditions

It is generally convenient to assume that the entire atmosphere has a rigid lower boundary and an upper free surface at which the pressure and density vanish. Either the free-slip or the adherence condition may be used on the lower boundary depending on the problem under study.

There are, of course, no lateral boundaries in the actual atmosphere, and studies of the atmosphere's motion in spherical coordinate systems encounter no difficulty in this regard. But the attempt to study motion in only one hemisphere or to model the atmospheric flow in cartesian coordinates necessitates the establishment of suitable boundary conditions.

The usual choice is to use a free-slip, rigid boundary simulating walls at constant values of y representing a bounding latitude (Fig. 6.7). Then the condition of *cyclic continuity* is imposed. This means that the solutions are required to be periodic over some basic length L. When numerical integrations are performed over the length L, the boundary condition is that for every variable

$$f(x,y,z,t) = f(x+L,y,z,t) \tag{13}$$

With this condition, the patterns flowing out of the region of integration at one end must flow back in the other. Thus the circular nature of flows around a circle of latitude is correctly modeled by this convention.

The top of the atmosphere For most theoretical analysis of the flow of the entire atmosphere it is sufficient to assume that pressure and density vanish rapidly enough as z becomes infinite so that the energy integrals

$$K = \int_V \rho \frac{v^2}{2}\, dV \qquad I = \int_V \frac{c_V}{R} p\, dV \qquad P = \int_V g\rho z\, dV \tag{14}$$

are all finite. Thus the condition on the velocity becomes that $\rho^{1/2} v$ vanish in the limit and so for the momentum

$$\lim_{z \to \infty} \rho \mathbf{v} = 0 \tag{15}$$

For numerical prediction models these assumptions are not sufficient because the model cannot actually extend to infinity. A rigid, free-slip surface can be employed at some small value of the pressure, or a free interfacial surface can be used to separate the lower part of the atmosphere from an upper portion that contains only a small fraction of the mass and is given special flow properties.

PROBLEM 6.9.5 Show that the condition that P be finite implies that the total mass is finite.

6.10 SOME EXACT SOLUTIONS TO THE EQUATIONS OF MOTION

Although a great deal has been discovered about properties that the solutions to the Navier-Stokes equations must exhibit, in only a few cases have the solutions been found. These exact solutions share two characteristics: First, some assumption is used to restrict conditions so that the nonlinear inertial terms either vanish or are made linear. Second, the number of relevant variables is reduced. A few such examples are presented here; others will be given later.

Couette flow The equation of motion for an incompressible fluid of constant density becomes [see Eq. (7) in Sec. 6.5]

$$\frac{\partial \mathbf{v}}{\partial t} + \mathbf{v} \cdot \nabla \mathbf{v} = -\frac{1}{\rho} \nabla P + \nu \nabla^2 \mathbf{v} \qquad P = p + g\rho z \tag{1}$$

A velocity field with only horizontal components that vary only in the vertical will cause the nonlinear term to vanish and satisfy the condition of no divergence. If the motion is steady, we have $\mathbf{v} = \mathbf{v}_H(z)$, and thus

$$-\frac{1}{\rho} \nabla_H P + \nu \frac{\partial^2 \mathbf{v}_H}{\partial z^2} = 0 \tag{2}$$

The additional conditions

$$\nabla_H P = 0 \tag{3}$$

and

$$\frac{\partial^2 \mathbf{v}_H}{\partial z^2} = 0 \tag{4}$$

clearly provide a field of velocity and pressure that satisfy Eq. (2). Thus for constant $\boldsymbol{\gamma}$ we have

$$\mathbf{v}_H = \mathbf{v}_0 + \boldsymbol{\gamma} z \tag{5}$$

and

$$P = P_0 + f(z) \tag{6}$$

as solutions. Motions with a linear velocity profile like Eq. (5) are generally called *Couette flows.*

If we impose the boundary conditions that the lower surface at $z = 0$ is a rigid, nonmoving adherence surface and that the upper one at $z = z_T$ is also a rigid nonslip surface but one moving with velocity $\mathbf{v}_T$, the solution must be

$$\mathbf{v} = \frac{\mathbf{v}_T z}{z_T} \tag{7}$$

These linear velocity profiles are an exact solution under the conditions given; there is, of course, no assurance that all flows forced by a moving upper surface will select this solution. Experimental evidence agrees with the intuitively reasonable conclusion that Eq. (7) will be the case if the viscosity is large enough and the velocity $\mathbf{v}_T$ small enough.

Poisseuille flow With the same conditions imposed in the previous example to obtain Eq. (2), we may suppose that the pressure gradient is constant at the value

$$\mathbf{G}_0 = -\frac{1}{\rho}\nabla_H P \tag{8}$$

Then Eq. (2) permits second derivatives of $\mathbf{v}_H(z)$ to exist and we have, for constant $\boldsymbol{\alpha}$ and $\boldsymbol{\beta}$,

$$\mathbf{v} = \mathbf{v}_H(z) = \mathbf{v}_0 + \boldsymbol{\alpha} z + \boldsymbol{\beta} z^2 \tag{9}$$

For simplicity, we shall assume here that the gradient (8) is imposed on fluid confined between two rigid surfaces of infinite extent in the x and y direction with pressure decreasing in the x direction. If there are nonslip surfaces at $z = 0$ and $z = z_T$, then

$$\mathbf{v} = \boldsymbol{\beta} z^2 - \boldsymbol{\beta} z z_T \tag{10}$$

to satisfy the boundary conditions and from Eqs. (2) and (8) we find

$$\frac{\partial^2 \mathbf{v}}{\partial z^2} = -\frac{\mathbf{G}_0}{\nu} = 2\boldsymbol{\beta} \tag{11}$$

so that for $\mathbf{G}_0 = \mathbf{i}G_0$ we have

$$\beta = \frac{1}{2\rho\nu}\frac{\partial p}{\partial x} = -\frac{1}{2\rho\nu}\left|\frac{\partial p}{\partial x}\right| \tag{12}$$

and therefore

$$\mathbf{v} = \mathbf{i}u = \mathbf{i}\frac{G_0}{2\nu}(zz_T - z^2) \tag{13}$$

PROBLEM 6.10.1 What are the vorticity patterns for the models of Couette flow and Poisseuille flow considered in the text? Show that the average vorticity vanishes for Poisseuille flow.

PROBLEM 6.10.2 Replace the upper boundary condition used in the Poisseuille example with a free-slip surface. What then is the solution and its vorticity pattern? Note that they are drastically different from the nonslip solution.

PROBLEM 6.10.3 Consider a constant density, incompressible flow confined between vertical cylinders (of infinite extent) rotating with different angular velocities. What would the Couette solution be in this case?

PROBLEM 6.10.4 Consider a homogeneous, incompressible fluid in a cylindrical pipe with a pressure drop of $|\Delta p|$ from entrance to exit. Obtain the Poisseuille solution for this case. (Hint: Eliminate the singularity that appears by a suitable choice of a constant.)

PROBLEM 6.10.5 Find an infinite class of solutions for the autobarotropic equations (12) in Sec. 6.5. What is the essential difference between the autobarotropic equations and the set considered in this section that permits such a set to exist?

Burgers' equation The effects of the nonlinearity of the inertial terms may be studied in simplified form with the equation

$$\frac{\partial u}{\partial t} + u\frac{\partial u}{\partial x} - \nu\frac{\partial^2 u}{\partial x^2} = 0 \tag{14}$$

proposed and investigated by Burgers (1948). The equation has often been considered in developing methods for study of turbulent motions, and its solutions are sometimes called "Burgulence." A more general form is

$$\frac{\partial \mathbf{v}}{\partial t} + \mathbf{v} \cdot \nabla \mathbf{v} - \nu \nabla^2 \mathbf{v} = 0 \tag{15}$$

The absence of the pressure gradient term allows us to consider Eq. (15) as a complete system without imposing the solenoidal condition as an equation of continuity. Several exact solutions to Burgers' equation are known; we give one here as an illustration of how the nonlinearity is circumvented.

An irrotational solution to Eq. (15) can be constructed from the hypothesis that a scalar field Θ exists such that

$$\mathbf{v} = -2\nu \nabla \ln \Theta \tag{16}$$

With cartesian tensor notation, the substitution of Eq. (16) in Eq. (15) gives, after some rearrangement and multiplication by Θ,

$$-2\nu \frac{\partial^2 \Theta}{\partial t\, \partial x_i} + \frac{2\nu}{\Theta} \frac{\partial \Theta}{\partial t} \frac{\partial \Theta}{\partial x_i} + 4\nu^2 \frac{\partial}{\partial x_k}\left(\frac{\partial \Theta}{\partial x_k} \frac{1}{\Theta} \frac{\partial \Theta}{\partial x_i}\right) - 4\nu^2 \frac{1}{\Theta} \frac{\partial \Theta}{\partial x_i} \frac{\partial \Theta}{\partial {x_k}^2} + 2\nu^2 \Theta \frac{\partial^2}{\partial {x_j}^2}\left(\frac{1}{\Theta} \frac{\partial \Theta}{\partial x_i}\right) = 0 \tag{17}$$

Rearrangement of the last term produces

$$\frac{\partial}{\partial x_i}\left(-2\nu \frac{\partial \Theta}{\partial t} + 2\nu^2 \frac{\partial^2 \Theta}{\partial {x_j}^2}\right) + \frac{1}{\Theta} \frac{\partial \Theta}{\partial x_i}\left(2\nu \frac{\partial \Theta}{\partial t} - 2\nu^2 \frac{\partial^2 \Theta}{\partial {x_k}^2}\right) = 0 \tag{18}$$

The hypothesis (16) will therefore be a solution if

$$\frac{\partial \Theta}{\partial t} - \nu \nabla^2 \Theta = 0 \tag{19}$$

The particular solution (16) obviously reduced the problem to that of the linear equation (19), which has the solution

$$\Theta(\mathbf{x}, t) = (4\pi\nu t)^{-3/2} \iiint_{-\infty}^{\infty} \Theta(\mathbf{x}', 0) \exp\left[-|\mathbf{x} - \mathbf{x}'|^2 (4\nu t)^{-1}\right] dx'\, dy'\, dz' \tag{20}$$

Now,

$$\mathbf{v}(\mathbf{x},0) = -2\nu \nabla \ln \Theta(\mathbf{x},0) \tag{21}$$

and thus by Eq. (24) in Sec. 5.6 for $|\mathbf{v}(\mathbf{x},0)| < |\mathbf{x}|^{-2}$ as $|\mathbf{x}| \to \infty$, we have

$$\Theta(\mathbf{x},0) = \exp\left[\frac{1}{8\nu\pi} \iiint_{-\infty}^{\infty} \frac{\nabla \cdot \mathbf{v}(\mathbf{x}',0)}{|\mathbf{x} - \mathbf{x}'|}\, dx'\, dy'\, dz'\right] \tag{22}$$

Therefore

$$\mathbf{v}(\mathbf{x},t) = \frac{t^{-1} \iiint_{-\infty}^{\infty} \Theta(\mathbf{x}',0)(\mathbf{x}-\mathbf{x}') \exp\left[-|\mathbf{x}-\mathbf{x}'|^2(4\nu t)^{-1}\right] dx'\,dy'\,dz'}{\iiint_{-\infty}^{\infty} \Theta(\mathbf{x}',0) \exp\left[-|\mathbf{x}-\mathbf{x}'|^2(4\nu t)^{-1}\right] dx'\,dy'\,dz'} \tag{23}$$

A few points deserve emphasis. First, this is only a particular solution to the equation. Second, even so, it is quite complicated and would still require considerable computer analysis to produce a flow field. Third, it is an irrotational solution, as mentioned.

This last remark is significant because Eq. (15) will yield the same equations for the vorticity $\boldsymbol{\zeta} = \nabla \times \mathbf{v}$ as the Navier-Stokes equations do for incompressible fluids with constant density. Therefore, the point is that to obtain a solution to an equation with the same nonlinear effects of the Navier-Stokes equations, the problem was so much restricted that it is not very relevant to the physics of actual fluids or the atmosphere. Still the solution (23) represents progress and provides some valuable insight into the structure of the Navier-Stokes equations. It also shows how remote general solutions to these equations appear to be at the present time.

PROBLEM 6.10.6 Suppose that in observing the initial data $\mathbf{v}(x,0)$ for the solution of Burgers' equation we unavoidably have errors and uncertainty and obtain $\mathbf{v}(\mathbf{x},0) + \mathbf{u}(\mathbf{x},0)$ instead. How will these errors $\mathbf{u}(\mathbf{x},0)$ affect the solution?

6.11 THE EXISTENCE OF SOLUTIONS

The methods now current in fluid dynamics and meteorology are statements of faith that the equations of motion possess solutions relevant to the physics of fluid flow. Whether they do is a question of crucial significance to atmospheric science. Before investing our intellectual and computational resources in an attempt to understand the atmosphere with the equations of motion, we should be sure that the mathematical model actually has solutions and that the physical realities we hope to grasp are correctly represented by them.

Following the introduction to Ladyzhenskaya's book (1963), we might pose the following questions about the equations of motion:

1 Does the complete system of the equations of motion, boundary conditions, and initial conditions have solutions and are they unique?
2 If the solutions exist, how well do they represent the actual flows that are encountered in the laboratory and in the atmosphere?
3 If the solutions appear to be faithful portrayals of natural flows in some cases and not in others, what are the distinguishing characteristics of the two different types of situations?

4 Are there any realistic bounds on the usefulness of the equations in studying a fluid in which the initial data are so poorly prescribed as they will always be in the atmosphere?

As pointed out earlier, most of the study of the mathematical theory of the equations of motion has been confined to the Navier-Stokes equations for an incompressible fluid of constant density. In this case, at least, the answer to the first question is known for some situations. Ladyzhenskaya (1963) has shown that Eqs. (7) and (8) in Sec. 6.5

$$\frac{\partial \mathbf{v}}{\partial t} + \mathbf{v} \cdot \nabla \mathbf{v} = -\frac{1}{\rho} \nabla(p + g\rho z) + \nu \nabla^2 \mathbf{v} \qquad \nabla \cdot \mathbf{v} = 0 \tag{1}$$

have a unique solution for all time if the initial data are, in effect, smooth enough. If the initial data are, in a sense, turbulent, then it is only known that the unique solution exists for a certain interval of time whose length is dependent on the initial data.

The study of the equations for compressible flow, and especially those forced by differential heating, has not advanced to this stage of mathematical confidence. It is likely that methods of increased sophistication, and surely increased complexity, will have to be developed. But the fact that the incompressible equations have solutions is certainly an indication that the central question for the equations of atmospheric science will not be whether solutions exist, but rather, to specify the conditions under which they exist uniquely and to determine what happens if cases occur when they do not. In fact, Serrin (1959) has shown that solutions to the compressible Navier-Stokes equations are unique when they exist.

The answer to the second question posed above seems to be that the equations portray smooth flows adequately for considerable periods of time; strongly turbulent flows are still an imposing challenge.

The present successes of meteorology in comprehension and prediction of atmospheric motion give intuitive faith, if not mathematical proof, that answers to the remaining questions will eventually be specified exactly. Until then, the incompressible equations provide a testimony that our pursuit of a mathematical vision of the atmosphere's essence is not a hopeless search.

BIBLIOGRAPHIC NOTES

6.1 Further discussion of these topics is given by the two main sources of reference for this section:

Bergman, P. G., 1942: *Introduction to the Theory of Relativity*, Prentice-Hall, Inc., Englewood Cliffs, N.J., 287 pp.

Defrise, Pierre, 1964: "Tensor Calculus in Atmospheric Science," in H. Landsberg and J. van Mieghem (eds.), *Advances in Geophysics*, vol. 10, pp. 261–314, Academic Press, Inc., New York.

6.2–6.3 These sections contain standard material.

6.4 Further information about the concept of radiation and its application to meteorological processes is given in the books:

Chandrasekhar, S., 1950: *Radiative Transfer*, Oxford University Press, London, 393 pp. Reprinted by Dover Publications, Inc., 1961.

Fleagle, R. G., and J. A. Businger, 1963: *An Introduction to Atmospheric Physics*, Academic Press, Inc., New York, 346 pp.

Johnson, J. C., 1954: *Physical Meteorology*, Technology Press of the Massachusetts Institute of Technology, Cambridge, Mass., and John Wiley & Sons, Inc., New York, 393 pp.

6.5 Incompressible flow is treated in a variety of books on fluid dynamics. Among the most notable is the recent book:

Batchelor, G. K., 1967: *An Introduction to Fluid Dynamics*, Cambridge University Press, Cambridge, 615 pp.

An account of L. F. Richardson's thinking about numerical weather prediction and some details of how he came to realize the necessity of the approach taken in the discussion of hydrostatic motion are given by:

Platzman, G. W., 1967: "A Retrospective View of Richardson's Book on Weather Prediction," *Bull. Am. Meteorol. Soc.*, **48**:514–550.

6.7 This section derives from the article:

Jeffreys, H., 1925: "On Fluid Motions Produced by Differences of Temperature and Humidity," *Quart. J. Roy. Meteorol. Soc.*, **51**:347–356.

6.8 Although the material of the first part of this section is largely classical in nature, our account is based upon the lucid exposition given in:

Serrin, James, 1959: "Mathematical Principles of Classical Fluid Mechanics," in S. Flügge (ed.), *Handbuch der Physik*, vol. 8, no. 1, pp. 125–263, Springer-Verlag OHG, Berlin.

The only reference we could find which mentions the relative size of the dynamic and the expansion viscosity for air is:

Tisza, L., 1942: "Supersonic Absorption and Stokes' Viscosity Relation," *Phys. Rev.*, **61**:531.

although the subject is discussed at length for other fluids by various authors in a symposium under the leadership of:

Rosenhead, L., 1954: "A Discussion of the First and Second Viscosities of Fluids," *Proc. Roy. Soc. London*, **A226**:1–69.

The reference to Jeffreys' book is given in the notes for Chap. 5.

6.9 Boundary conditions on certain variables are easily formulated; for others they must be developed as part of the specification of a particular problem. Further discussion may be found in Serrin's monograph.

6.10 The first two cases are considered classical and discussed in nearly all texts on fluid dynamics. The third case is from:

Burgers, J. M., 1948: "A Mathematical Model Illustrating the Theory of Turbulence," *Advances in Appl. Mech.*, **1**:171–199.

Cole, J. D., 1951: "On a Quasi-Linear Parabolic Equation Occurring in Aerodynamics," *Quart. J. Appl. Math.*, **9**:225–236.

Hopf, E., 1950: "The Partial Differential Equation $u_t + uu_x = \mu u_{xx}$," *Comm. Pure Appl. Math.*, **3**:201–230.

6.11 The question of existence and uniqueness of solutions to the Navier-Stokes equations is a difficult one. In this section we refer to:

Ladyzhenskaya, O. A., 1963: *The Mathematical Theory of Viscous Incompressible Flow* (trans. by R. A. Silverman), Gordon and Breach, Science Publishers, Inc., New York, 184 pp.

Serrin, J., 1959: "On the Uniqueness of Compressible Fluid Motions," *Arch. Ratl. Mech. Anal.*, **3**:271–288.

7

METEOROLOGICAL EQUATIONS OF MOTION

The equations of fluid motion derived in the previous chapter are valid for an inertial coordinate system. But in meteorology, we use the rotating earth as a reference body, and thus two alterations in the equations are required. The first change is necessary to take account of the rotation; the second is due to the spherical shape of the earth.

The vertical structure that commonly occurs in the atmosphere permits making further changes in the equations when they are used to study large-scale flow. In this chapter, then, we shall explore the basic modifications of the equations that make them suitable for application to meteorological problems.

7.1 EFFECTS OF THE EARTH'S ROTATION

As illustrated by the example in Sec. 6.1, a coordinate system rotating with the earth is not inertial and apparent forces arise that must be considered by an observer who participates in the rotational motion.

To begin, we take an orthogonal cartesian inertial system and move it so that its origin is at the center of the earth; thus we assume at first that only the earth's rotation, not its revolution or translational motion, is important. We also consider an

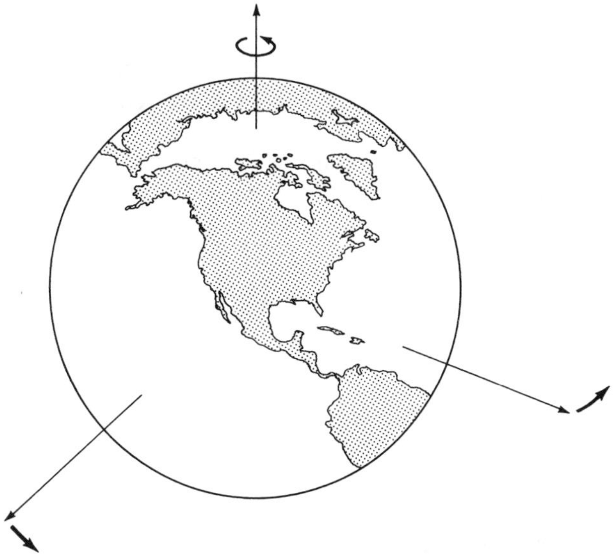

FIGURE 7.1
A cartesian system of axes fixed to the earth and rotating with it.

orthogonal cartesian system whose origin is at the center of the earth, but which rotates with the earth. This rotation is brought into the calculations with an angular velocity vector $\mathbf{\Omega}$, which is parallel to the earth's axis and has magnitude given by the angular speed of the earth. This vector is taken to point from the center of the earth toward the North Pole, and so the counterclockwise rotation seen looking down on the North Pole is considered positive (Fig. 7.1).

7.1.1 Differential Relations

We shall denote the fixed inertial coordinates by $\hat{\mathbf{x}}$ and those rotating with the earth by $\mathbf{x}$. Our calculations are based upon the observation that the velocity $\mathbf{v}_a$ of the air as seen in the fixed or absolute system will be the sum of the velocity $\mathbf{v}_r$, due to the rotation of the earth, and the velocity $\mathbf{v}$, relative to the earth. Thus

$$\mathbf{v}_a = \mathbf{v} + \mathbf{v}_r \tag{1}$$

In Chap. 2 we defined the total or material derivative of a scalar quantity as the derivative obtained following the fluid motion, and thus it should not depend on motions of the coordinate system.

Theorem For a scalar ϕ, the material derivative $(d\phi/dt)_a$ in the absolute system is identical to the material derivative $d\phi/dt$ in the rotating system.

PROOF By definition, we have

$$\left(\frac{d\phi}{dt}\right)_a = \left(\frac{\partial\phi}{\partial t}\right)_{\hat{\mathbf{x}}} + \mathbf{v}_a \cdot \hat{\nabla}\phi \tag{2}$$

Because the $\mathbf{x}$ coordinates are obtained from the $\hat{\mathbf{x}}$ coordinates by a relation of the form $\mathbf{x} = \mathbf{x}(\hat{\mathbf{x}},t)$, we may write $\phi(\mathbf{x},t) = \phi(\mathbf{x}(\hat{\mathbf{x}},t),t)$ and thus the chain rule gives

$$\left(\frac{\partial\phi}{\partial t}\right)_{\hat{\mathbf{x}}} = \left(\frac{\partial\phi}{\partial t}\right)_{\mathbf{x}} + \nabla\phi \cdot \left(\frac{\partial\mathbf{x}}{\partial t}\right)_{\hat{\mathbf{x}}} \tag{3}$$

The velocity $\mathbf{v}_r$ is obtained by measuring the velocity, with respect to the inertial system, of a point $\mathbf{x}$ fixed in the rotating system, and so

$$\mathbf{v}_r = \left(\frac{\partial\hat{\mathbf{x}}}{\partial t}\right)_{\mathbf{x}} \tag{4}$$

(see Fig. 7.2) and conversely

$$\left(\frac{\partial\mathbf{x}}{\partial t}\right)_{\hat{\mathbf{x}}} = -\mathbf{v}_r \tag{5}$$

When Eqs. (3) and (5) are substituted into Eq. (2), we find

$$\left(\frac{d\phi}{dt}\right)_a = \left(\frac{\partial\phi}{\partial t}\right)_{\mathbf{x}} - \mathbf{v}_r \cdot \nabla\phi + \mathbf{v}_a \cdot \hat{\nabla}\phi \tag{6}$$

But the gradient vector is identical in both coordinate systems so that $\nabla\phi = \hat{\nabla}\phi$, and hence Eqs. (6) and (1) imply that

$$\left(\frac{d\phi}{dt}\right)_a = \left(\frac{\partial\phi}{\partial t}\right)_{\mathbf{x}} + \mathbf{v} \cdot \nabla\phi = \frac{d\phi}{dt} \tag{7}$$

completing the proof. ////

Now let us turn to the material derivative of vectors. A unit vector $\mathbf{i}$ in the rotating system will obviously appear to be changing when viewed from the inertial system. As can be seen from Fig. 7.2, a point fixed in the rotating coordinate system with position vector $\mathbf{r}$ from the center of the earth has velocity $\boldsymbol{\Omega} \times \mathbf{r}$. Note here that the direction of the cross product is obviously correct; the tip of the vector $\mathbf{r}$ travels a distance $2\pi r \sin\theta$ in time $2\pi/\Omega$ so that its speed is $\Omega r \sin\theta$, which is the same as the magnitude of $\boldsymbol{\Omega} \times \mathbf{r}$. Hence the velocity of the tip of an arbitrary unit vector $\mathbf{i}$ with tail at $\mathbf{r}$ is $\boldsymbol{\Omega} \times (\mathbf{r} + \mathbf{i})$ and the velocity of the tail is $\boldsymbol{\Omega} \times \mathbf{r}$. But

$$\mathbf{i} = \mathbf{r}_{\text{tip}} - \mathbf{r}_{\text{tail}} \tag{8}$$

so that

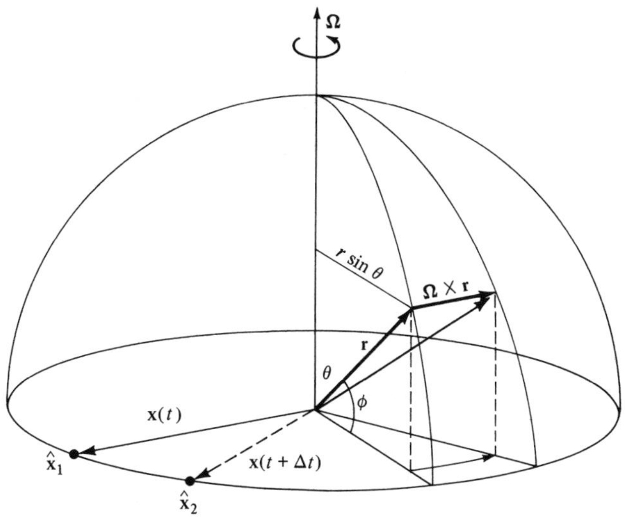

FIGURE 7.2
The basic geometry of rotating coordinate systems. On the left, the figure illustrates that a point x fixed in the rotating system will move from the position $\hat{\mathbf{x}}_1$ to $\hat{\mathbf{x}}_2$ in the nonrotating system in a time interval Δt. Thus $\mathbf{v}_r\,\Delta t = \hat{\mathbf{x}}_2 - \hat{\mathbf{x}}_1 = \hat{\mathbf{x}}(t+\Delta t) - \hat{\mathbf{x}}(t)$, which gives Eq. (4) in the text. On the right, the figure illustrates that the tip of the position vector r will have a velocity $\mathbf{\Omega} \times \mathbf{r}$.

$$\begin{aligned}\left(\frac{d\mathbf{i}}{dt}\right)_a &= \left[\frac{d}{dt}(\mathbf{r}_{\text{tip}} - \mathbf{r}_{\text{tail}})\right]_a = \mathbf{\Omega} \times [(\mathbf{r}+\mathbf{i}) - \mathbf{r}] \\ &= \mathbf{\Omega} \times \mathbf{i}\end{aligned} \tag{9}$$

By definition, the material derivative in the rotating system of a vector **A** is

$$\frac{d\mathbf{A}}{dt} = \mathbf{i}\frac{dA_x}{dt} + \mathbf{j}\frac{dA_y}{dt} + \mathbf{k}\frac{dA_z}{dt} \tag{10}$$

because the vectors are constant in that system. The relation between derivatives is established by:

Theorem For vectors, the material derivatives in an absolute system and in a system rotating with angular velocity $\mathbf{\Omega}$ are related by

$$\left(\frac{d\mathbf{A}}{dt}\right)_a = \frac{d\mathbf{A}}{dt} + \mathbf{\Omega} \times \mathbf{A} \tag{11}$$

PROOF The result follows immediately by writing out the six terms of

$$\left(\frac{d\mathbf{A}}{dt}\right)_a = \left[\frac{d}{dt}(\mathbf{i}A_x + \mathbf{j}A_y + \mathbf{k}A_z)\right]_a \tag{12}$$

and applying Eqs. (9) and (10). ////

Application of this result to a position vector **r** gives

$$\left(\frac{d\mathbf{r}}{dt}\right)_a = \frac{d\mathbf{r}}{dt} + \mathbf{\Omega} \times \mathbf{r} \tag{13}$$

but the crucial task is to derive a formula for the acceleration. Thus from Eq. (13) we find that for $\mathbf{\Omega} = \text{const}$

$$\left(\frac{d^2\mathbf{r}}{dt^2}\right)_a = \left(\frac{d}{dt} + \mathbf{\Omega} \times\right)\left(\mathbf{v} + \mathbf{\Omega} \times \mathbf{r}\right) = \frac{d\mathbf{v}}{dt} + 2\mathbf{\Omega} \times \mathbf{v} + \mathbf{\Omega} \times (\mathbf{\Omega} \times \mathbf{r}) \tag{14}$$

and hence we have the fundamental result:

Theorem The equation of fluid motion

$$\left[\frac{d^2\mathbf{r}}{dt^2}\right]_a = \mathbf{f} \tag{15}$$

in which **f** is the sum of specific forces in an inertial system, becomes

$$\frac{d\mathbf{v}}{dt} = \mathbf{f} - 2\mathbf{\Omega} \times \mathbf{v} - \mathbf{\Omega} \times (\mathbf{\Omega} \times \mathbf{r}) \tag{16}$$

in a system rotating with angular velocity $\mathbf{\Omega}$. ////

Exactly as in the simple case considered in Sec. 6.1, two apparent forces have appeared. The term $-2\mathbf{\Omega} \times \mathbf{v}$ is referred to as the Coriolis force. The vector **v** and the angular velocity vector $\mathbf{\Omega}$ always form a plane surface passing through the earth's axis when the tail of **v** is moved to the axis. According to Eq. (16), the acceleration at the point at which **v** is measured will always be normal and to the right of this plane (Fig. 7.3).†

The remaining term $-\mathbf{\Omega} \times (\mathbf{\Omega} \times \mathbf{r})$ is the centrifugal force. Note that by the right-hand rule $-\mathbf{\Omega} \times (\mathbf{\Omega} \times \mathbf{r})$ is always perpendicular to the earth's axis and directed outward. This force is apparently the cause of the equatorial bulge of the earth itself. However, an observer on the earth cannot distinguish between the gravitational attraction of the earth and the centrifugal force. All he or his instruments experience is

†This statement involves the assumption that we are looking down on the rotation vector, or equivalently, looking down on the Northern Hemisphere. An observer with the Southern Hemisphere and the South Pole in view would see motion to the left of the plane—the earth's rotation is in the opposite sense for such an observer. (Try it with a globe!)

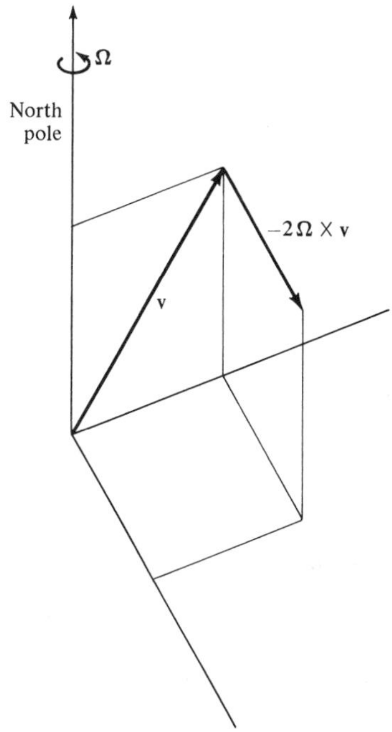

FIGURE 7.3
Diagram illustrating the Coriolis acceleration $-2\mathbf{\Omega} \times \mathbf{v}$ normal to the plane formed by $\mathbf{\Omega}$ and $\mathbf{v}$ and to the right as viewed looking out along the plane from the earth's axis.

the resultant force $-\mathbf{g}$ where

$$\mathbf{g} = g_e \frac{\mathbf{r}}{r} + \mathbf{\Omega} \times (\mathbf{\Omega} \times \mathbf{r}) \tag{17}$$

(See Fig. 7.4.)

We shall term this resultant force the *apparent acceleration of gravity*. Thus the equation of motion may be written as

$$\frac{d\mathbf{v}}{dt} = -\frac{1}{\rho}\nabla p - 2\mathbf{\Omega} \times \mathbf{v} - \mathbf{g} + \mathbf{f}_r \tag{18}$$

We shall generally take $\mathbf{g}$ to be constant, although an improved procedure is to consider the surface that is normal to all the vectors, $\mathbf{g}$, with a given magnitude as a surface of constant height. The difference between these two approximations is not significant in theoretical work, although it may become important in analysis of observational data.

Equation (18), then, is a vector equation of motion valid for any coordinate system rotating with the earth.

PROBLEM 7.1.1 Let the vertical unit vectors $\mathbf{i}_3$ and $\hat{\mathbf{i}}_3$ of the $\mathbf{x}$ and $\hat{\mathbf{x}}$ systems coincide with $\mathbf{\Omega}$. Then show that a position vector $\mathbf{r}$ from the earth's center to a point

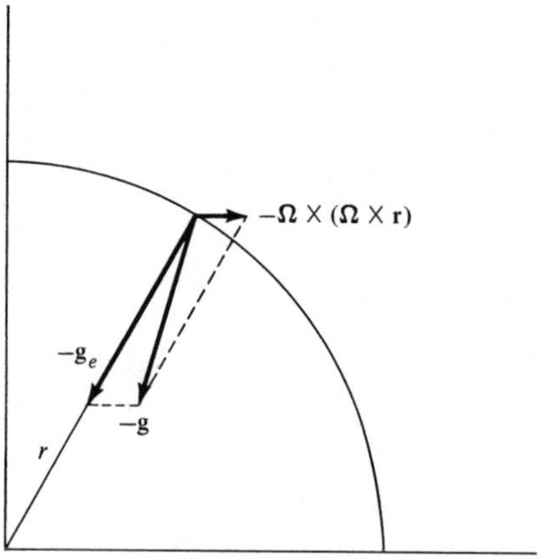

FIGURE 7.4
Resolution of the effective gravity as the sum of the earth's attraction and the centrifugal acceleration.

fixed in the rotating system may be expressed as

$$\mathbf{r} = r \cos\theta \hat{\mathbf{i}}_3 + r \sin\theta (\hat{\mathbf{i}}_1 \cos\Omega t + \hat{\mathbf{i}}_2 \sin\Omega t)$$

Use this relation to verify Eq. (13) for $r = r(t), \theta = \theta(t)$.

PROBLEM 7.1.2 Derive the equations of motion applicable to the atmosphere of a planet whose angular velocity vector varies in time.

PROBLEM 7.1.3 Derive equations of motion that apply to the atmosphere when both the earth's rotation and its revolution are taken into account. By using the ratio between the period of revolution and the period of rotation, show that the revolution could be ignored in meteorological calculations.

7.1.2 Tangential Cartesian Coordinates

The most common meteorological coordinate system and equations are obtained from the previous results by placing an orthogonal cartesian system with its origin at a point on the earth's surface, say $\mathbf{x}_0$. We take the z axis parallel to the radial vector with z increasing away from the earth's surface. The x and y axes form a plane tangent to the earth at $\mathbf{x}_0$, and we let x increase toward the east and let y increase toward the north. The advantage of this system is that it is an orthogonal cartesian system that includes the earth's rotation. Furthermore, it is a system that corresponds to our own natural references, and it is the one that is necessarily appropriate to observations made with radio and radar techniques.

In order to apply the equation of motion (18) in this system we shall need the components of the Coriolis acceleration. From Fig. 7.5 it is easily seen that

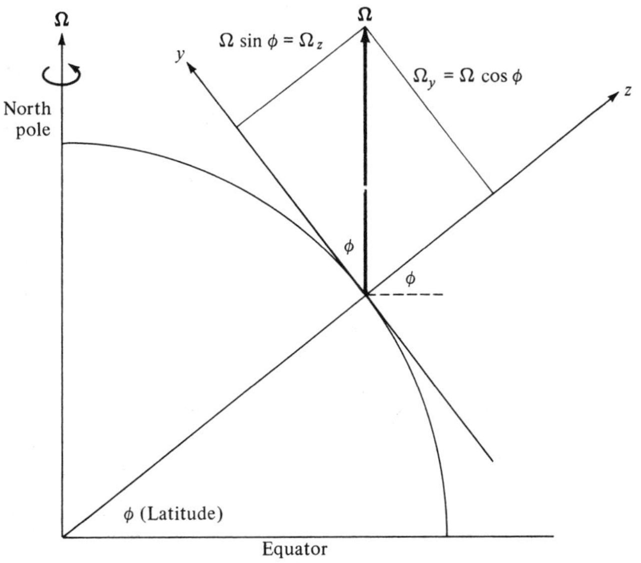

FIGURE 7.5
Projection of the vector **Ω** on the y and z axes to determine its components.

$$\Omega_x = 0 \qquad \Omega_y = \Omega \cos \phi \qquad \Omega_z = \Omega \sin \phi \tag{19}$$

in which ϕ is the latitude.† Thus we have

$$-2\mathbf{\Omega} \times \mathbf{v} = -2\Omega \begin{vmatrix} \mathbf{i} & \mathbf{j} & \mathbf{k} \\ 0 & \cos\phi & \sin\phi \\ u & v & w \end{vmatrix} \tag{20}$$

The scalar equation for any component can be obtained by taking the scalar product of Eqs. (18) and (20) with the appropriate unit vector.

To apply these results we must know the magnitude of Ω as measured with respect to the fixed stars. In Chap. 2 we pointed out that there is one more sidereal day than solar day in each year owing to the earth's revolution around the sun (see Fig. 7.6). Thus we have

$$\begin{aligned} \Omega &= \frac{2\pi}{1 \text{ sidereal day}} \cdot \frac{366.25}{365.25} \frac{\text{sidereal days}}{\text{solar days}} = 2\pi \frac{366.25}{365.25} (\text{solar days})^{-1} \\ &= \frac{2\pi}{86{,}400 \text{ s}} \cdot \frac{366.25}{365.25} = 7.292 \times 10^{-5}/\text{s} \end{aligned} \tag{21}$$

†Note that if we set $\phi = \pi/2$ at the North Pole, $\phi = 0$ at the equator, and $\phi = -\pi/2$ at the South Pole, then the Coriolis force will have the correct sign in both Northern and Southern Hemispheres.

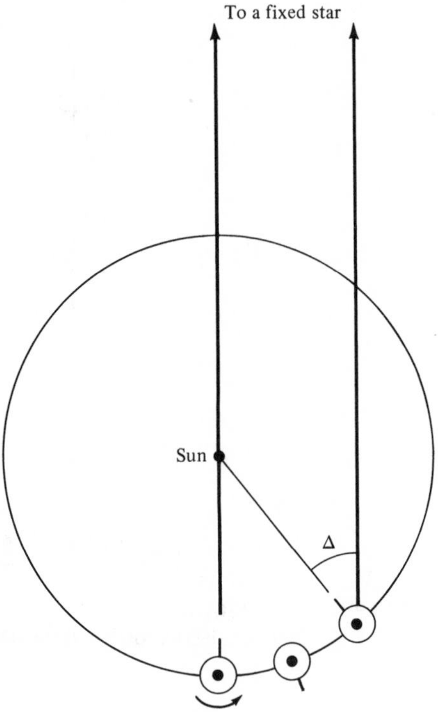

FIGURE 7.6
Schematic comparison of solar and sidereal days. In the interval shown, the earth has made one complete rotation with respect to the sun; hence one solar day has elapsed. But in the process it has rotated once plus the angle Δ with respect to the fixed star. Thus more than one sidereal day has elapsed. As can be seen, Δ will be 2π after one revolution around the sun, giving one more sidereal day in a year than there are solar days.

Therefore we obtain

$$\begin{aligned} 2\Omega_y &= 2\Omega \cos \phi = 1.458 \times 10^{-4} \cos \phi/\text{s} \\ 2\Omega_z &= 2\Omega \sin \phi = 1.458 \times 10^{-4} \sin \phi/\text{s} \end{aligned} \tag{22}$$

and so at $\phi = 45°$, we have $2\Omega_y = 2\Omega_z \cong 10^{-4}$/s.

This coordinate system with ϕ set constant at ϕ_0 is useful for small-scale local studies and provides a simple system in which to perform theoretical analyses of the equations of motion with the effects of rotation included; it is not well suited for practical applications to large-scale problems for two reasons associated with the fact that the earth is spherical. The first is that a purely zonal flow will have only an eastward component at $\mathbf{x}_0$, but as the air proceeds east, it will fall below the tangent plane and acquire a z component of motion in this system. The second is that for large-scale problems we shall have to include the latitudinal variation of the Coriolis forces.

Thus we turn now to the most appropriate coordinate system for studying atmospheric flow—one that takes account of the spherical shape of the earth.

*PROBLEM 7.1.4 Write out the scalar equations of motion on the tangent plane.

PROBLEM 7.1.5 In the development of this section, we included part of the effect of the earth's revolution about the sun. With reference to Problem 7.1.3, explain precisely what assumption was made.

PROBLEM 7.1.6 Let a ball be thrown upward at a velocity w_0 from the surface of a planet without an atmosphere. Where will the ball land with respect to a starting point at latitude 45° if you use numerical values appropriate to earth? (Hint: For reasonable w_0, $\Omega t_1 \ll 1$, where t_1 is the time when the ball returns to the surface. This can be used to simplify the exact solution for t_1. The symbol $\ll$ is read as "is much less than.")

7.2 THE EQUATIONS OF MOTION FOR A SPHERICAL EARTH

The spherical coordinate system is obviously the most natural one to describe atmospheric motions of global scale. In order to obtain the appropriate equations of motion, we shall have to make two coordinate transformations. We denote the fixed inertial system now by (X,Y,Z), and we can translate and rotate this system so that the Z axis coincides with the earth's axis without affecting the inertial character of the system. Using the example of Eq. (4) in Sec. 6.1, we transform to a rectangular cartesian system $(\hat{x},\hat{y},\hat{z})$ rotating with the earth and with $\hat{z}$ coinciding with Z.

Spherical coordinates (λ,ϕ,r), in which λ is longitude, ϕ is latitude, and r is distance from the earth's center, are now defined by the transformation

$$\begin{aligned} \hat{x} &= r\cos\phi\cos\lambda & \qquad r &= [(\hat{x})^2 + (\hat{y})^2 + (\hat{z})^2]^{1/2} \\ \hat{y} &= r\cos\phi\sin\lambda & \tan\lambda &= \frac{\hat{y}}{\hat{x}} \\ \hat{z} &= r\sin\phi & \tan\phi &= \frac{\hat{z}}{[(\hat{x})^2 + (\hat{y})^2]^{1/2}} \end{aligned} \tag{1}$$

Note that we have assumed that $\hat{x} = 0$ occurs at $\lambda = \pm(\pi/2)$ and that $\hat{y} = 0$ corresponds to $\lambda = 0$.

We have an orthogonal set of unit vectors $(\mathbf{i},\mathbf{j},\mathbf{k})$ at each point on the sphere (Fig. 7.7) so that $\mathbf{k}$ is perpendicular to the tangent plane and points away from the center of the earth, $\mathbf{i}$ is parallel to the tangent plane and points eastward along latitude circles, and $\mathbf{j}$ is parallel to the tangent plane and points along the meridians toward increasing latitude.

We could now convert our vector equation of motion (18) in Sec. 7.1 into spherical coordinates, and this is sometimes done. The primary disadvantage of this procedure is that the total derivatives of the independent variables $d\lambda/dt$, $d\phi/dt$, and dr/dt have differing dimensions and thus are not physical components of the velocity vector. We would have to use the scaling factors h_i of Sec. 5.4 to obtain physical components.

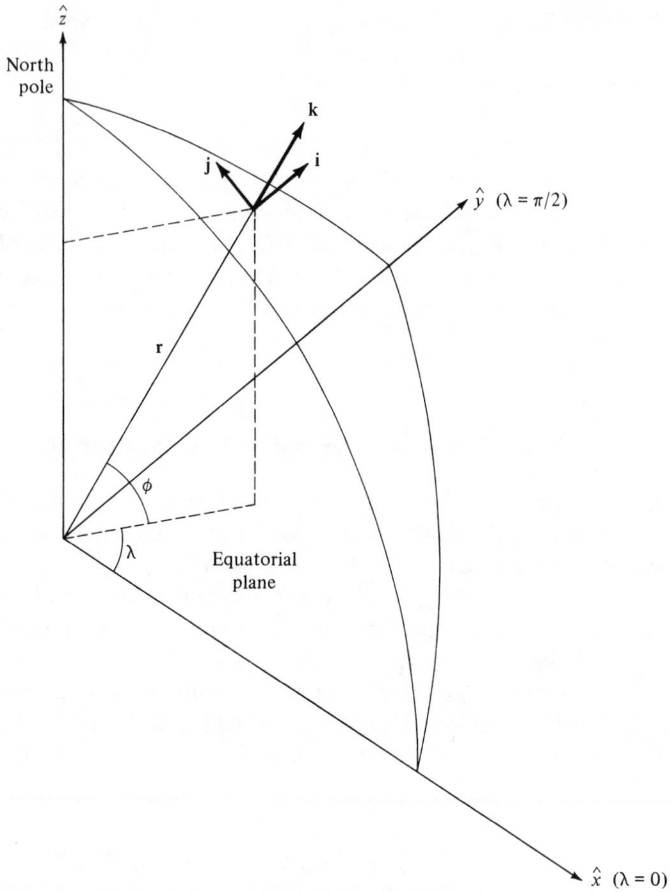

FIGURE 7.7
Illustration of the geometry of the spherical coordinate system.

For this reason, meteorologists develop an anholonomic system in which the transformation is defined by differential relations that lead to suitable velocity vectors.

We define this new system so that x represents distance along the latitude circles and y represents distance along the meridians. The problem is that there are no functions f_1 and f_2 such that we can write

$$x = f_1(\lambda, \phi, r) \qquad y = f_2(\lambda, \phi, r) \tag{2}$$

for such a system. However, it is possible and clear from the definition of the new system that we can write

$$dx = r \cos \phi \, d\lambda \qquad dy = r \, d\phi \qquad dz = dr \tag{3}$$

These definitions are obtained by substituting in the complete formulas of the type

$$dx = \frac{\partial x}{\partial r}dr + \frac{\partial x}{\partial \phi}d\phi + \frac{\partial x}{\partial \lambda}d\lambda \tag{4}$$

the derivatives that express the basic definitions of this system,

$$\begin{aligned} &\frac{\partial x}{\partial \lambda} = r\cos\phi \qquad \frac{\partial y}{\partial \phi} = r \qquad \frac{\partial z}{\partial r} = 1 \\ &\frac{\partial x}{\partial r} = \frac{\partial x}{\partial \phi} = \frac{\partial y}{\partial r} = \frac{\partial y}{\partial \lambda} = \frac{\partial z}{\partial \phi} = \frac{\partial z}{\partial \lambda} = 0 \end{aligned} \tag{5}$$

Thus the new coordinate system is defined by the differential relations (3) or (5) rather than a functional form as in Eq. (2). Note that the Jacobian $\partial(x,y,z)/\partial(\lambda,\phi,r)$ has nonzero elements only along its diagonal. The same must be true of its inverse, and the derivatives $\partial\lambda/\partial x$, $\partial\phi/\partial y$, and $\partial r/\partial z$ will be just the reciprocals of the derivatives in Eq. (5). All other derivatives of λ, ϕ, or r vanish. Velocity components in the new system are now defined in the obvious manner

$$u = \frac{dx}{dt} = r\cos\phi\frac{d\lambda}{dt} \qquad v = \frac{dy}{dt} = r\frac{d\phi}{dt} \qquad w = \frac{dz}{dt} = \frac{dr}{dt} \tag{6}$$

and in vector notation this is

$$\mathbf{v} = \mathbf{i}u + \mathbf{j}v + \mathbf{k}w \tag{7}$$

These velocities thus have satisfactory dimensions.

The difficulty begins with the acceleration, since the unit vectors (**i**,**j**,**k**) will change as a parcel of air flows from one place to another. We may calculate the derivatives as we did in Sec. 7.1, and we find

$$\begin{aligned} \frac{d\mathbf{v}}{dt} &= \mathbf{i}\frac{du}{dt} + \mathbf{j}\frac{dv}{dt} + \mathbf{k}\frac{dw}{dt} + u\frac{d\mathbf{i}}{dt} + v\frac{d\mathbf{j}}{dt} + w\frac{d\mathbf{k}}{dt} \\ &= \left(\frac{d\mathbf{v}}{dt}\right)_C + u\frac{d\mathbf{i}}{dt} + v\frac{d\mathbf{j}}{dt} + w\frac{d\mathbf{k}}{dt} \end{aligned} \tag{8}$$

in which we have singled out in the first term the three terms that are of the same form [Eq. (10) in Sec. 7.1] as the accelerations in the tangential cartesian coordinates.

To find the additional terms, we shall first obtain explicit expressions for the unit vectors in the new curvilinear system; then we can differentiate to find the necessary derivatives in the expressions of the form

$$\frac{d\mathbf{i}}{dt} = u\frac{\partial \mathbf{i}}{\partial x} + v\frac{\partial \mathbf{i}}{\partial y} + w\frac{\partial \mathbf{i}}{\partial z} \tag{9}$$

In order to write the vectors (**i**,**j**,**k**) as functions of the unit vectors $(\hat{\mathbf{i}},\hat{\mathbf{j}},\hat{\mathbf{k}})$ of the inertial system, we note that we will have

$$\mathbf{i} = A(\lambda,\phi,r)\hat{\mathbf{i}} + B(\lambda,\phi,r)\hat{\mathbf{j}} + C(\lambda,\phi,r)\hat{\mathbf{k}} \tag{10}$$

Now we observe the following special cases for **i**:

1 **i** will never have a component in the $\hat{\mathbf{k}}$ direction so that $C = 0$.
2 **i** does not change with changes in either ϕ or r, but only with changes in λ so that $A = A(\lambda)$ and $B = B(\lambda)$.
3 At $\lambda = 0$, $\mathbf{i} = \hat{\mathbf{j}}$; at $\lambda = -\pi/2$, $\mathbf{i} = \hat{\mathbf{i}}$; at $\lambda = \pi/2$, $\mathbf{i} = -\hat{\mathbf{i}}$.

Thus

$$\mathbf{i} = -\sin\lambda\,\hat{\mathbf{i}} + \cos\lambda\,\hat{\mathbf{j}} \tag{11}$$

This relation satisfies the requirement that $|\mathbf{i}| = 1$.

PROBLEM 7.2.1 Show that

$$\mathbf{j} = -\sin\phi\cos\lambda\,\hat{\mathbf{i}} - \sin\phi\sin\lambda\,\hat{\mathbf{j}} + \cos\phi\,\hat{\mathbf{k}}$$

and hence that

$$\mathbf{k} = \mathbf{i} \times \mathbf{j} = \hat{\mathbf{i}}\cos\phi\cos\lambda + \hat{\mathbf{j}}\cos\phi\sin\lambda + \hat{\mathbf{k}}\sin\phi$$

PROBLEM 7.2.2 Show that

$$\frac{\partial\mathbf{k}}{\partial\lambda} = \cos\phi\,\mathbf{i} \qquad \frac{\partial\mathbf{k}}{\partial\phi} = \mathbf{j}$$

$$\frac{\partial\mathbf{j}}{\partial\lambda} = -\sin\phi\,\mathbf{i} \qquad \frac{\partial\mathbf{j}}{\partial\phi} = -\mathbf{k}$$

$$\frac{\partial\mathbf{i}}{\partial\lambda} = \sin\phi\,\mathbf{j} - \cos\phi\,\mathbf{k} \qquad \frac{\partial\mathbf{i}}{\partial\phi} = 0$$

Utilizing the derivatives in Problem 7.2.2 and the remark following Eq. (5) about the Jacobians applied to the chain rule, we have the results that

$$\begin{aligned}
&\frac{\partial\mathbf{k}}{\partial x} = \frac{\partial\mathbf{k}}{\partial\lambda}\frac{\partial\lambda}{\partial x} = \frac{\mathbf{i}}{r} \qquad \frac{\partial\mathbf{k}}{\partial y} = \frac{\partial\mathbf{k}}{\partial\phi}\frac{\partial\phi}{\partial y} = \frac{\mathbf{j}}{r}\\
&\frac{\partial\mathbf{j}}{\partial x} = -\frac{\tan\phi}{r}\mathbf{i} \qquad \frac{\partial\mathbf{j}}{\partial y} = -\frac{\mathbf{k}}{r}\\
&\frac{\partial\mathbf{i}}{\partial x} = \frac{\tan\phi}{r}\mathbf{j} - \frac{\mathbf{k}}{r}
\end{aligned} \tag{12}$$

This gives the expressions

$$\begin{aligned}
&\frac{d\mathbf{i}}{dt} = u\left(\frac{\tan\phi}{r}\mathbf{j} - \frac{\mathbf{k}}{r}\right)\\
&\frac{d\mathbf{j}}{dt} = -u\frac{\tan\phi}{r}\mathbf{i} - \frac{v\mathbf{k}}{r}
\end{aligned} \tag{13}$$

$$\frac{d\mathbf{k}}{dt} = \frac{u\mathbf{i}}{r} + \frac{v\mathbf{j}}{r} \tag{13 (cont'd)}$$

The components of the Coriolis acceleration in this coordinate system are the same as those determined in Sec. 7.1.2, and so upon combining all the results we have

$$\mathbf{i}\left[\left(\frac{du}{dt}\right)_C - \frac{uv\tan\phi}{r} + \frac{uw}{r} + \frac{1}{\rho}\frac{\partial p}{\partial x} + 2\Omega(w\cos\phi - v\sin\phi) + \mathbf{i}\cdot\mathbf{g} - \mathbf{i}\cdot\mathbf{f}_r\right]$$
$$+\mathbf{j}\left[\left(\frac{dv}{dt}\right)_C + \frac{u^2\tan\phi}{r} + \frac{vw}{r} + \frac{1}{\rho}\frac{\partial p}{\partial y} + 2\Omega u\sin\phi + \mathbf{j}\cdot\mathbf{g} - \mathbf{j}\cdot\mathbf{f}_r\right]$$
$$+\mathbf{k}\left[\left(\frac{dw}{dt}\right)_C - \frac{u^2+v^2}{r} + \frac{1}{\rho}\frac{\partial p}{\partial z} + \mathbf{g}\cdot\mathbf{k} - 2\Omega u\cos\phi - \mathbf{k}\cdot\mathbf{f}_r\right] = 0 \tag{14}$$

Now we adjust our coordinate system so that the "horizontal" surfaces become surfaces upon which the acceleration of the apparent gravity is constant so that $\mathbf{i}\cdot\mathbf{g} = \mathbf{j}\cdot\mathbf{g} = 0$ and $\mathbf{k}\cdot\mathbf{g} = g$. We also observe that $r = a + z \cong a$ to a close approximation, where a is the mean radius of the earth. Thus we have the scalar equations [for $d/dt = (d/dt)_C$]

$$\begin{aligned}
\frac{du}{dt} - \frac{uv\tan\phi}{a} + \frac{uw}{a} &= -\frac{1}{\rho}\frac{\partial p}{\partial x} + 2\Omega(v\sin\phi - w\cos\phi) + \mathbf{i}\cdot\mathbf{f}_r \\
\frac{dv}{dt} + \frac{u^2\tan\phi}{a} + \frac{vw}{a} &= -\frac{1}{\rho}\frac{\partial p}{\partial y} - 2\Omega u\sin\phi + \mathbf{j}\cdot\mathbf{f}_r \\
\frac{dw}{dt} - \frac{u^2+v^2}{a} &= -\frac{1}{\rho}\frac{\partial p}{\partial z} - g + 2\Omega u\cos\phi + \mathbf{k}\cdot\mathbf{f}_r
\end{aligned} \tag{15}$$

These are the *basic momentum conservation equations* of dynamic meteorology. Together with the first law of thermodynamics, Eq. (8) in Sec. 6.4, the mass continuity equation (12) in Sec. 5.2, and the equation of state, they form the *system of equations of motion* for the model atmosphere we are considering. The entire system is also referred to as the *primitive equations of motion*. In this book and in the literature, the term *equations of motion* is often used in both senses, first denoting the equations (15), or second, referring to the complete system. The specific meaning is usually clear from the context.

All applications of the system of equations of motion to atmospheric problems must begin with the equations (15), the first law, continuity equation, and equation of state (along with a suitable water substance equation if moist processes are to be considered). The equations in the system are often modified for both theoretical and numerical work because simplifying approximations are available for the problem being considered, because the equations are to be applied in some coordinate system other than the basic spherical one, or because casting them in another form may be appropriate for particular cases.

We illustrate this process now by reducing the equations to the form most often used in elementary studies of large-scale flow.

Orders of magnitude To estimate the orders of magnitude of the terms in the equations, we shall use a typical cyclone at about 45°N as an example. We assume horizontal wind speeds of about 10 m/s, vertical velocities of 5 cm/s, a radius R of the storm of 1000 km, and a difference in pressure of about 10 mbar from the center of the cyclone to the undisturbed surroundings. We may estimate that du/dt and dv/dt are equal to the centripetal acceleration V^2/R. Thus for the first equation we have

$$\begin{array}{cccccccc}
\dfrac{du}{dt} & -\dfrac{uv\tan\phi}{a} & +\dfrac{uw}{a} & = -\dfrac{1}{\rho}\dfrac{\partial p}{\partial x} & +2\Omega(v\sin\phi & -w\cos\phi) & +\mathbf{i}\cdot\mathbf{f}_r & \\
\dfrac{10^6}{10^8} & \dfrac{10^6}{6\times 10^8} & \dfrac{5\times 10^3}{6\times 10^8} & \dfrac{10^3\cdot 10^4}{10^8} & 10^{-4}\,(10^3 & 5) & ? & \dfrac{\text{cm}}{\text{s}^2} \\
10^{-2} & 1.6\times 10^{-3} & 10^{-5} & 10^{-1} & 10^{-1} & 5\times 10^{-4} & ? & \dfrac{\text{cm}}{\text{s}^2}
\end{array} \qquad (16)$$

Therefore, to an accuracy of about 10 percent we need to retain only the first term on the left, and the first, second, and last terms on the right. (We are retaining the friction term for the moment, because we cannot estimate its size from the data given.) For the vertical equation, we estimate that the vertical velocity might change sign in 12 hours and that the 500-mbar surface is at about 6 km. Thus

$$\begin{array}{ccccccc}
\dfrac{dw}{dt} & -\dfrac{u^2+v^2}{a} & = -\dfrac{1}{\rho}\dfrac{\partial p}{\partial z} & -g & +2\Omega u\cos\phi & +\mathbf{k}\cdot\mathbf{f}_r & \\
\dfrac{10}{4.3\times 10^4} & \dfrac{2\times 10^6}{6\times 10^8} & \dfrac{10^3\times 5\times 10^5}{6\times 10^5} & 10^3 & 10^{-4}\cdot 10^3 & ? & \dfrac{\text{cm}}{\text{s}^2} \\
2.5\times 10^{-4} & 3\times 10^{-3} & 10^3 & 10^3 & 10^{-1} & ? & \dfrac{\text{cm}}{\text{s}^2}
\end{array} \qquad (17)$$

We need to retain only the first two terms on the right–the hydrostatic approximation. We can also observe that the magnitude of the horizontal friction term must be considerably less than the Coriolis acceleration since the cross-isobar component of the wind is generally only about 10 percent of the total magnitude. We neglect the vertical component of friction since it is certainly much smaller than the acceleration of gravity.

Thus the approximate equations for large-scale flows are

$$\begin{aligned}
\frac{du}{dt} &= -\frac{1}{\rho}\frac{\partial p}{\partial x} + fv \\
\frac{dv}{dt} &= -\frac{1}{\rho}\frac{\partial p}{\partial y} - fu \\
\frac{\partial p}{\partial z} &= -g\rho
\end{aligned} \qquad (18)$$

where $f = 2\Omega \sin\phi$.

Since we have assumed frictional effects are negligible, we may also ignore the heat conduction term and we have

$$
\begin{aligned}
&c_V \rho \frac{dT}{dt} + p \nabla \cdot \mathbf{v} = -\nabla \cdot \mathbf{R} \\
&\frac{d\rho}{dt} + \rho \nabla \cdot \mathbf{v} = 0 \\
&p = \rho R T
\end{aligned}
\tag{19}
$$

This is the hydrostatic set of equations considered in Sec. 6.5.3, and the vertical velocity is found from Richardson's equation.

This system of equations contains the essential physics of large-scale atmospheric motion in mid-latitudes, but it must be verified as being appropriate for special phenomena (we have not shown that it applies to tornadoes or hurricanes, for example). The effect of the earth's rotation appears as the linear term in u and v, and thus these equations are not mathematically very different from the Navier-Stokes equations studied in Chap. 6. Friction is, of course, actually present in the atmosphere, but the assumption implicit in Eqs. (18) and (19) is that it acts on a much smaller scale than that of the motions we would study with these equations. Also present in the actual atmosphere and not contained in these equations is the effect of water vapor and its changes of phase.

Much of the theory of the dynamics of atmospheric motion has been based on the approximate equations (18) and (19). In the usual application, the sphericity of the earth is included by two further assumptions. First, we require cyclic continuity (periodicity) in x, thus modeling the closed nature of atmospheric flow along a latitude circle (see Sec. 6.9.3). Second, we replace the variable Coriolis force with a linear approximation. We choose a central latitude ϕ_0 (at which y now vanishes) and use Taylor's theorem to write

$$
f = f_0 + \left.\frac{\partial f}{\partial y}\right|_{y=0} y \tag{20}
$$

But

$$
\left.\frac{\partial f}{\partial y}\right|_{y=0} = \left.\frac{\partial}{a\,\partial\phi}(2\Omega \sin\phi)\right|_{y=0} = \frac{2\Omega}{a}\cos\phi_0 = \beta
$$

Hence the constant β gives the rate of change of f with latitude.

The equations of motion are now those of Eq. (18) with $f = f_0 + \beta y$. These two devices eliminate the difficulties with the tangential cartesian coordinates for large-scale flow, but the equations are formally those for a plane with the addition of the β term. We refer to this model as flow occurring on the β *plane*.

*PROBLEM 7.2.3 Determine the effect on the kinetic energy of the sphericity terms on the left of Eq. (15) and the Coriolis force terms on the right.

PROBLEM 7.2.4 Show that an object moving northward over the face of the earth has an absolute angular momentum component parallel to the earth's axis of $(u + \Omega r \cos \phi) r \cos \phi$. Show that if $r = \text{const}$, then the hypothesis that this angular momentum is conserved will give the Coriolis force and sphericity terms (with $w = 0$) of the $\dot{u}$ equation in (15).

PROBLEM 7.2.5 Under the hypothesis of Problem 7.2.4, show that the assumption that the object conserves its kinetic energy will then give the appropriate terms of the $\dot{v}$ equation in (15).

PROBLEM 7.2.6 Generalize the results of the two preceding problems to all three dimensions. In particular, show that the appropriate terms of the $\dot{u}$ and $\dot{v}$ equations follow by assuming that the components of the absolute angular momentum $\mathbf{r} \times \mathbf{v}_a$ with respect to **K**, **I**, and **J** are conserved where **K**, **I**, and **J** are unit vectors of the inertial system fixed at the earth's origin with **K** parallel to the axis. Then assuming conservation of kinetic energy, derive the appropriate terms of the $\dot{w}$ equation. (Hint: Express **I**, **J**, and **K** as functions of the unit vectors in the rotating, anholonomic system. The **K** component will give $\dot{u}$, and the **I** and **J** components will give a relation of the form $\dot{P} = \Omega Q$, which will imply the $\dot{v}$ equation, with the appropriate centrifugal force included.)

*PROBLEM 7.2.7 Show that if $\nabla \cdot \mathbf{R} = 0$, then for the entire atmosphere the system (18) and (19) conserves the sum of the kinetic energy of the horizontal velocity components, the internal energy, and the potential energy.

7.3 ISOBARIC AND ISENTROPIC COORDINATES

The typical vertical structure of the atmosphere suggests that it might be useful to replace the usual independent variable z with one of the thermodynamic variables and then to use the height of surfaces of constant value of that variable as a dependent variable. As we will show, certain simplifications result from this procedure.

It is important to realize precisely what we are doing now. The new coordinates to be introduced have the same variables x and y as the original space, and they have the same unit vector **k** in the vertical. But now we shall denote vertical position with a new variable.

Assume, for example, that we are going to use isobaric coordinates. Location in such coordinates is specified by the triple (x,y,p) rather than (x,y,z). To obtain the new representation, we find the appropriate values of x and y by projecting onto the x,y plane, and we use the pressure at a point as a measure of its vertical position. The coordinates (x,y,p) are thus an orthogonal cartesian system, which we have constructed

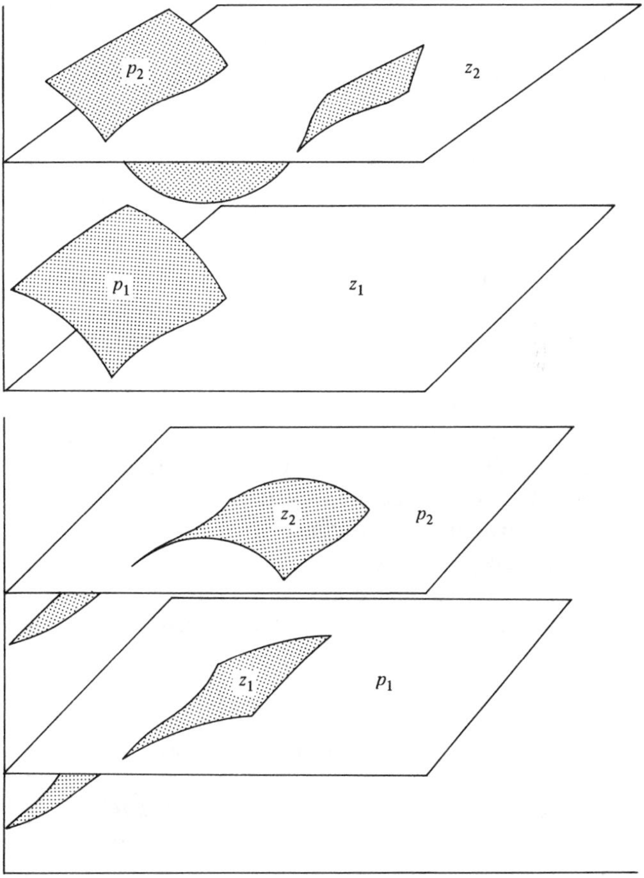

FIGURE 7.8
The top sketch shows two isobaric surfaces as they might appear in (x,y,z) coordinates. The bottom sketch shows the equivalent height fields, as they would appear in isobaric coordinates. Note that because p decreases upward and z increases upward, the region of low pressure on z_2 corresponds to the region of low heights on p_2.

according to the above definition rather than a system obtained by formal transformation from the original (x,y,z) coordinates.

The height $h(x,y,p,t)$ of the surface with pressure p becomes a dependent variable in the new system. If the isobaric surfaces are viewed in the original (x,y,z) system, they have slopes and they move with time. But in the (x,y,p) system, the isobaric surfaces are everywhere flat and parallel and they do not move; it is now the height fields that move and that have slopes (Fig. 7.8). The horizontal components of the velocities in the new system are exactly the same as they were, because differences in position are measured by projection on the xy plane and time is the same in the two systems, as are all three unit vectors. The appropriate choice for a vertical velocity will appear as we proceed.

7.3.1 Isobaric Coordinates

The fact that pressure is always observed to decrease monotonically with height makes it a suitable candidate as a replacement for z. In fact, the radiosonde observations that reveal the structure of the atmosphere are actually taken in the (x,y,p) system because the basic height measurement is the pressure as revealed by the baroswitch in the radiosonde instrument. The height is derived from the temperature and is a dependent variable.

Transformation of the equations of motion to the new system necessitates finding a suitable method of determining derivatives. Vertical derivatives will be measured parallel to **k**, with position specified by pressure, and horizontal derivatives of a variable will be measured by taking a difference in that variable *along an isobaric surface* and dividing by a distance given by projection of the actual difference interval on the x,y plane (Fig. 7.9).

Any scalar function ϕ can be represented in either system and the value of ϕ at point (x,y,p) is the value it has in the original system at the point (x,y,z) with pressure given by $p(x,y,z,t)$. Thus

$$\phi(x,y,z,t) = \phi(x,y,p(x,y,z,t),t) \tag{1}$$

According to the chain rule we therefore have

$$\left(\frac{\partial\phi}{\partial z}\right)_{x,y,t} = \frac{\partial\phi}{\partial p}\left(\frac{\partial p}{\partial z}\right)_{x,y,t} \tag{2}$$

This equation makes it clear that an essential requirement is that $\partial p/\partial z \neq 0$, for otherwise the derivative $\partial\phi/\partial p$ becomes indeterminate. For horizontal derivatives, we find

$$\left(\frac{\partial\phi}{\partial x}\right)_{y,z,t} = \left(\frac{\partial\phi}{\partial x}\right)_p + \frac{\partial\phi}{\partial p}\left(\frac{\partial p}{\partial x}\right)_{y,z,t} \tag{3}$$

Thus with a similar result for y, we have

$$\nabla_z\phi = \nabla_p\phi + \frac{\partial\phi}{\partial p}\nabla_z p \tag{4}$$

where we define the operator ∇_p so that

$$\nabla_p\phi = \mathbf{i}\left(\frac{\partial\phi}{\partial x}\right)_p + \mathbf{j}\left(\frac{\partial\phi}{\partial y}\right)_p \tag{5}$$

In a like manner we find

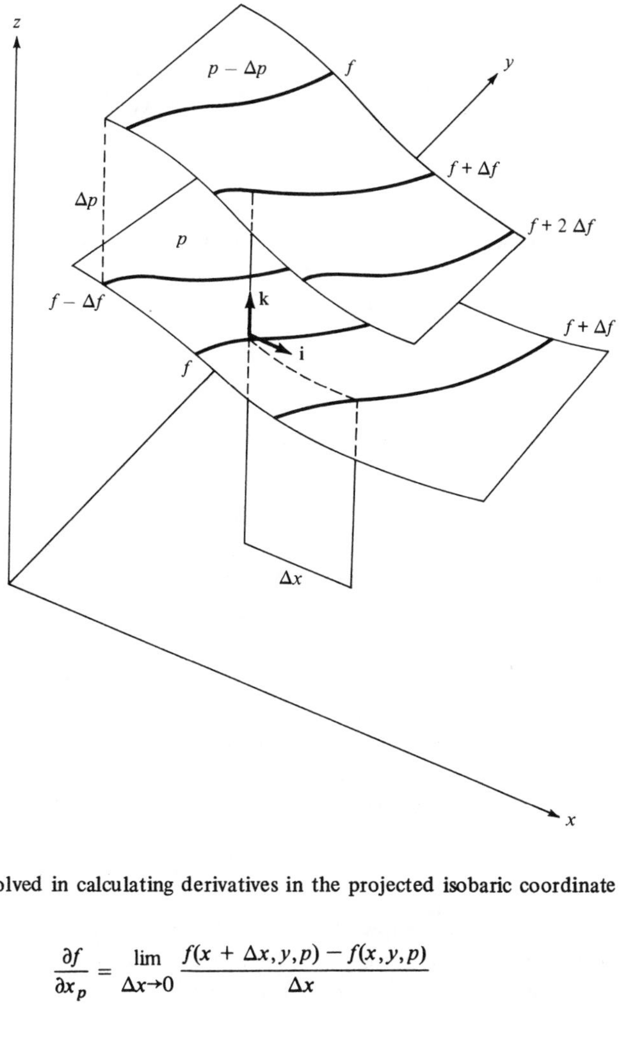

FIGURE 7.9
Geometry involved in calculating derivatives in the projected isobaric coordinate system. Here,

$$\frac{\partial f}{\partial x_p} = \lim_{\Delta x \to 0} \frac{f(x + \Delta x, y, p) - f(x, y, p)}{\Delta x}$$

and

$$\frac{\partial f}{\partial p} = \lim_{\Delta p \to 0} \frac{f(x, y, p + \Delta p) - f(x, y, p)}{\Delta p}$$

$$\left(\frac{\partial \phi}{\partial t}\right)_{x,y,z} = \left(\frac{\partial \phi}{\partial t}\right)_{x,y,p} + \frac{\partial \phi}{\partial p}\left(\frac{\partial p}{\partial t}\right)_{x,y,z} \tag{6}$$

The total derivative

$$\frac{d\phi}{dt} = \left(\frac{\partial \phi}{\partial t}\right)_{x,y,z} + \mathbf{v} \cdot \nabla_z \phi + w\left(\frac{\partial \phi}{\partial z}\right)_{x,y,t} \tag{7}$$

may now be written with the aid of these identities as

$$\begin{aligned}\frac{d\phi}{dt} &= \left(\frac{\partial\phi}{\partial t}\right)_{x,y,p} + \mathbf{v}\cdot\nabla_p\phi + \frac{\partial\phi}{\partial p}\left[\left(\frac{\partial p}{\partial t}\right)_{x,y,z} + \mathbf{v}\cdot\nabla_z p + w\left(\frac{\partial p}{\partial z}\right)_{x,y,t}\right] \\ &= \frac{\partial\phi}{\partial t_p} + \mathbf{v}\cdot\nabla_p\phi + \frac{dp}{dt}\frac{\partial\phi}{\partial p}\end{aligned} \tag{8}$$

in which we have used $\partial\phi/\partial t_p$ as an abbreviation for the first term on the right of the first line. This shows that dp/dt is to be used like a vertical velocity in calculating total derivatives; it occurs so often we shall denote it by ω. We can now write

$$\frac{d\phi}{dt} = \frac{\partial\phi}{\partial t_p} + (\mathbf{v}\cdot\nabla\phi)_p \tag{9}$$

where we use the subscript to indicate that

$$(\mathbf{v}\cdot\nabla\phi)_p = (\mathbf{i}u + \mathbf{j}v + \mathbf{k}\omega)\cdot\left(\nabla_p\phi + \mathbf{k}\frac{\partial\phi}{\partial p}\right) \tag{10}$$

Note here that we obtain a dimensionally correct expression after taking the scalar product even though neither factor can stand alone because both are dimensionally inhomogeneous.

Reciprocal relations, which allow us to find derivatives in the (x,y,z) systems from those in the (x,y,p) system may be obtained in the same manner. We first observe that the dependent variable $h(x,y,p,t)$ gives the value of the coordinate z through which the isobaric surface with pressure p passes at the point (x,y). For the height z of a given surface we have $z = h(x,y,p,t)$ and so

$$\frac{\partial h}{\partial z} = 1 \qquad \nabla_z h = 0 \tag{11}$$

These relations follow from the basic definitions of the partial derivatives. For example, $\nabla_z(\ )$ requires that derivatives be taken in the horizontal with z held fixed.

Use of h for ϕ in Eq. (2) now gives

$$\frac{\partial h}{\partial p}\frac{\partial p}{\partial z} = 1 \tag{12}$$

a result that is used often in isobaric coordinates. The same procedure with Eq. (4) yields

$$\nabla_p h = -\frac{\partial h}{\partial p}\nabla_z p \tag{13}$$

The forcing term in the vector equation of motion

$$\frac{1}{\rho}\nabla p + g\mathbf{k} = \frac{1}{\rho}\nabla_z p + \mathbf{k}\left(\frac{1}{\rho}\frac{\partial p}{\partial z} + g\right) \tag{14}$$

may now be transformed to become

$$\frac{1}{\rho}\nabla p + g\mathbf{k} = -\frac{1}{\rho}\left(\frac{\partial h}{\partial p}\right)^{-1}\nabla_p h + \mathbf{k}\left[g + \frac{1}{\rho}\left(\frac{\partial h}{\partial p}\right)^{-1}\right] \tag{15}$$

But if the atmosphere is hydrostatic, then Eq. (12) implies

$$\left(\frac{\partial h}{\partial p}\right)^{-1} = -g\rho \tag{16}$$

and thus Eq. (15) becomes

$$\frac{1}{\rho}\nabla p + g\mathbf{k} = g\,\nabla_p h \tag{17}$$

The hydrostatic relation for isobaric coordinates is generally written in the equivalent form

$$g\frac{\partial h}{\partial p} = -\alpha \tag{18}$$

Because g and h occur together, we define the potential $\phi_p = gh(x,y,p,t)$. The equations of motion for the horizontal velocity $\mathbf{v}_H$ are thus

$$\frac{\partial \mathbf{v}_H}{\partial t} + (\mathbf{v}\cdot\nabla\mathbf{v}_H)_p = -\nabla_p\phi_p - f\mathbf{k}\times\mathbf{v}_H + (\mathbf{f}_{r,H})_p \tag{19}$$

along with Eq. (18). We have simplified the Coriolis force term in accordance with Eq. (18) in Sec. 7.2.

There are a number of ways to obtain a continuity equation in isobaric coordinates. We shall use a basic approach that can be generalized to other coordinates. If we can find a suitable Jacobian determinant, then we may write the mass in a volume V as

$$M = \int_V \rho\,dV = \int_V \rho\,dx\,dy\,dz = \int_{V_p} \rho J\,dx\,dy\,dp \tag{20}$$

For this purpose we may consider that the relation between the coordinates $(\hat{x},\hat{y},\hat{p})$ and (x,y,z) is

$$x = \hat{x} \qquad y = \hat{y} \qquad z = h(\hat{x},\hat{y},p,t) \tag{21}$$

and so $J = \partial h/\partial p$.

Thus the mass in isobaric coordinates is

$$M = \int_{V_p} \rho\frac{\partial h}{\partial p}\,dx\,dy\,dp \tag{22}$$

where we assume that

$$\int_{V_p} = \int_{x_0}^{x_0+\Delta x} \int_{y_0}^{y_0+\Delta y} \int_{p_0}^{p_0-\Delta p} \tag{23}$$

so that M is positive. For hydrostatic conditions Eq. (20) thus takes the simple form

$$M = -\frac{1}{g} \int_{V_p} dx\, dy\, dp = \frac{V_p}{g} \tag{24}$$

in which V_p denotes the volume as specified in isobaric coordinates. Because the mass in a material volume must be constant, we have

$$\frac{dM}{dt} = \frac{dV_p}{dt} = 0 \tag{25}$$

The general result that

$$\frac{1}{V} \frac{dV}{dt} = \nabla \cdot \mathbf{v} \tag{26}$$

is true in any coordinate system, so that from Eq. (24) we have

$$(\nabla \cdot \mathbf{v})_p = 0 \tag{27}$$

Thus with our conventions

$$(\nabla \cdot \mathbf{v})_p = \nabla_p \cdot \mathbf{v} + \frac{\partial \omega}{\partial p} = 0 \tag{28}$$

Finally, we must find an appropriate form of the first law. We may use Eq. (8) to write

$$c_p \left[\left(\frac{\partial T}{\partial t} \right)_p + \mathbf{v} \cdot \nabla_p T + \omega \frac{\partial T}{\partial p} \right] - \alpha \omega = q + f_h \tag{29}$$

and this form is sometimes used. Because we often make the isentropic assumption, another commonly used form is

$$\frac{\partial \theta}{\partial t_p} + \mathbf{v} \cdot \nabla_p \theta + \omega \frac{\partial \theta}{\partial p} = \frac{\theta}{c_p T}(q + f_h) \tag{30}$$

For completeness, we write out the full set of equations as

$$\begin{aligned}
&\frac{\partial \mathbf{v}_H}{\partial t_p} + \mathbf{v}_H \cdot \nabla_p \mathbf{v}_H + \omega \frac{\partial \mathbf{v}_H}{\partial p} = -\nabla_p \phi_p - f \mathbf{k} \times \mathbf{v}_H + (\mathbf{f}_{r,H})_p \\
&\frac{\partial \phi_p}{\partial p} = -\alpha \\
&(\nabla \cdot \mathbf{v})_p = 0 \\
&\left(\frac{d\theta}{dt} \right)_p = \frac{\theta}{c_p T}(q + f_h)
\end{aligned} \tag{31}$$

$$\theta = T\left(\frac{1000}{p}\right)^{R/c_p} \qquad (31)$$

$$p\alpha = RT \qquad (cont'd)$$

Two basic simplifications in comparison with the equations in cartesian coordinates have been achieved; both derive from the use of the hydrostatic relation. The first is that the pressure gradient term has become the *linear* expression $\nabla_p \phi_p$, and thus ϕ_p is a potential for the pressure gradient force. The second is that the continuity equation has the form of that obtained for an incompressible flow.

To see the significance of this, let us write the equations for a hydrostatic incompressible flow with constant density in the form

$$\frac{d\mathbf{v}_H}{dt} = -\nabla_z \frac{p}{\rho} - f\mathbf{k} \times \mathbf{v}_H + (\mathbf{f}_r)_H$$

$$\frac{\partial p/\rho}{\partial z} = -g \qquad (32)$$

$$\nabla \cdot \mathbf{v} = 0$$

By comparison of Eqs. (31) and (32) we see that the isobaric equations are formally equivalent to those for an incompressible fluid of constant density but with a variable α replacing gravity. The advantages of isobaric coordinates, to be revealed as we progress through the later chapters, depend on this analogy, which derives from the hydrostatic approximation. When hydrostatic conditions do not prevail, the isobaric equations are just as complicated as the original set. It is important to keep in mind the central role of the hydrostatic approximation in these equations; they may give misleading answers when significant departures from hydrostatic equilibrium are present.

PROBLEM 7.3.1 Verify Eq. (26) for isobaric coordinates directly. [Hint: Consider $V_p = \Delta p \, \Delta x \, \Delta y = (p_1 - p_0)(x_1 - x_0)(y_1 - y_0)$.]

PROBLEM 7.3.2 How can the vertical velocity ω be found when the equations (31) are integrated numerically?

PROBLEM 7.3.3 You may see the expression $\omega = -g\rho w$ in the literature. Estimate the accuracy of this approximation.

PROBLEM 7.3.4 What is the continuity equation in isobaric coordinates when the hydrostatic approximation is *not* valid?

PROBLEM 7.3.5 The system (31) involves seven equations in seven variables. Can you reduce it to a five-equation system in the variables $\mathbf{v}_H$, ω, ϕ_p, and θ?

PROBLEM 7.3.6 Derive a system of equations, governing hydrostatic flow, that utilizes Exner's function $\pi = c_p(p/p_{00})^{R/c_p}$ as the vertical coordinate. Here we shall generally take $p_{00} = 1000$ mbar.

7.3.2 Isentropic Coordinates

Another commonly employed meteorological coordinate system utilizes the potential temperature as the measure of vertical position. The surfaces of constant potential temperature are also surfaces on which the entropy is constant and thus they are called isentropic. Hence one of the advantages of isentropic coordinates is that parcels in isentropic motion remain on the coordinate surface. Another is that representation in isentropic coordinates provides maximum resolution in the areas of greatest interest such as baroclinic zones and frontal areas, as illustrated in Fig. 7.10.

Many of the relations we derived for isobaric coordinates are valid with only slight modification. In this case the essential requirement for the validity of the transformation is that $\partial\theta/\partial z > 0$; in other words, that the stratification is everywhere statically stable.

Thus we have $z = h_\theta(x,y,\theta,t)$ along with the differential relations for a scalar function ϕ

$$\left(\frac{\partial\phi}{\partial z}\right)_{x,y,t} = \frac{\partial\phi}{\partial\theta}\left(\frac{\partial\theta}{\partial z}\right)_{x,y,t} = \frac{\partial\phi}{\partial\theta}\left(\frac{\partial h_\theta}{\partial\theta}\right)^{-1} \tag{33}$$

$$\nabla_z\phi = \nabla_\theta\phi + \frac{\partial\phi}{\partial\theta}\nabla_z\theta \tag{34}$$

and

$$\frac{d\phi}{dt} = \frac{\partial\phi}{\partial t_\theta} + \mathbf{v}\cdot\nabla_\theta\phi + \frac{d\theta}{dt}\frac{\partial\phi}{\partial\theta} \tag{35}$$

Furthermore, we have again

$$\frac{\partial h_\theta}{\partial z} = 1 \qquad \nabla_z h_\theta = 0 \tag{36}$$

and thus

$$\nabla_\theta h_\theta = -\frac{\partial h_\theta}{\partial\theta}\nabla_z\theta \tag{37}$$

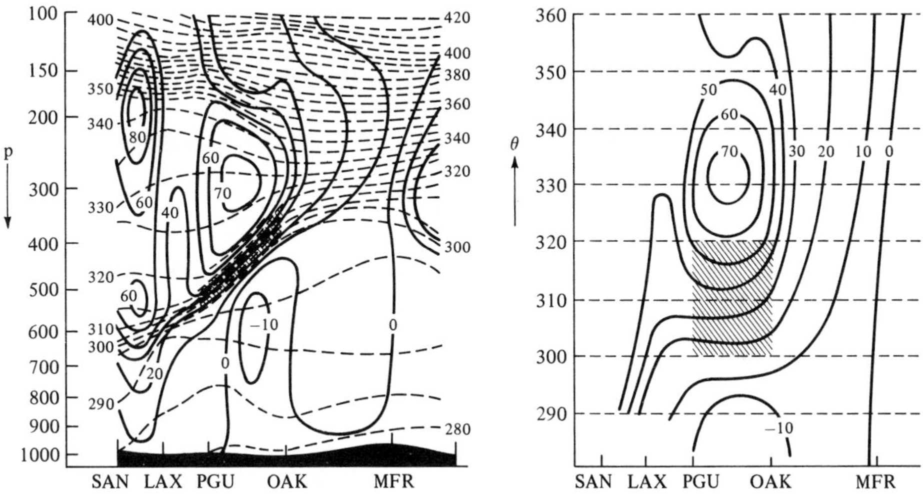

FIGURE 7.10
Comparison of resolution in isobaric and isentropic coordinates. The cross section on the left is the traditional form of presenting isotachs and isentropes with pressure as the vertical coordinate. An upper-level front or baroclinic zone is identified by the shaded area on the cross section. The figure on the right shows the same area as it would be represented in isentropic coordinates, making it clear that the dynamics of this structure could be analyzed or predicted more accurately in isentropic coordinates. *(Illustration provided by Dr. Rainer Bleck, National Center for Atmospheric Research. The figure on the left is from M. A. Shapiro and J. T. Hastings, 1973: "Objective Cross-Section Analysis by Hermite Polynomial Interpolation on Isentropic Surfaces," J. Appl. Meteorol.,* **12**:*753–762.)*

Use of this equation in Eq. (34) provides the result

$$\nabla_z \phi = \nabla_\theta \phi - \left(\frac{\partial h_\theta}{\partial \theta}\right)^{-1} \frac{\partial \phi}{\partial \theta} \nabla_\theta h_\theta = \nabla_\theta \phi - \frac{\partial \phi}{\partial z} \nabla_\theta h_\theta \tag{38}$$

In order to evaluate the forcing term in the equation of motion, we need to compute $\nabla_z p$, and with Eq. (38) we have

$$\nabla_z p = \nabla_\theta p - \frac{\partial p}{\partial z} \nabla_\theta h_\theta \tag{39}$$

and with Poisson's equation

$$\theta = T\left(\frac{p_{00}}{p}\right)^{R/c_p} \tag{40}$$

we find that

$$c_p \nabla_\theta T = \frac{T}{p} R \nabla_\theta p = \frac{1}{\rho} \nabla_\theta p \tag{41}$$

so that with the aid of the hydrostatic equation, Eq. (39) may be written

$$\frac{1}{\rho} \nabla_z p = c_p \nabla_\theta T + g \nabla_\theta h_\theta = \nabla_\theta (c_p T + g h_\theta) \tag{42}$$

This potential, known as the *Montgomery stream function*, is given a symbol of its own, $\Psi = c_p T + g h_\theta$.

When hydrostatic conditions prevail, we will want to express the fact in isentropic coordinates and not have to rely on the (x,y,z) representation. With Eq. (33) we find that in hydrostatic conditions

$$\frac{\partial p}{\partial \theta} = \frac{\partial p}{\partial z} \frac{\partial h_\theta}{\partial \theta} = -g\rho \frac{\partial h_\theta}{\partial \theta} \tag{43}$$

and the logarithmic derivative of Eq. (40) with respect to θ gives

$$\frac{1}{\theta} = \frac{1}{T} \frac{\partial T}{\partial \theta} - \frac{R}{c_p p} \frac{\partial p}{\partial \theta} \tag{44}$$

With Eq. (43) this may be rearranged as

$$c_p \frac{\partial T}{\partial \theta} + g \frac{\partial h_\theta}{\partial \theta} = c_p \frac{T}{\theta} = c_p \left(\frac{p}{p_{00}} \right)^{R/c_p} \tag{45}$$

Thus we have derived a hydrostatic equation for isentropic coordinates in the form

$$\frac{\partial \Psi}{\partial \theta} = c_p \left(\frac{p}{p_{00}} \right)^{R/c_p} \tag{46}$$

The continuity equation in these coordinates follows from the basic principle of mass conservation, just as it did in isobaric coordinates. Thus for a material volume we have

$$M = \int_{V_\theta} \rho \frac{\partial h_\theta}{\partial \theta} \, dx \, dy \, d\theta \tag{47}$$

and so for a small volume

$$M = \rho \frac{\partial h_\theta}{\partial \theta} V_\theta \tag{48}$$

Hence $dM/dt = 0$ implies that [see Eq. (26)]

$$\frac{d}{dt}\left(\rho \frac{\partial h_\theta}{\partial \theta} \right) + \rho \frac{\partial h_\theta}{\partial \theta} \frac{1}{V_\theta} \frac{dV_\theta}{dt} = \frac{d}{dt}\left(\rho \frac{\partial h_\theta}{\partial \theta} \right) + \rho \frac{\partial h_\theta}{\partial \theta} (\nabla \cdot \mathbf{v})_\theta = 0 \tag{49}$$

Upon noting that the correct form of the divergence must be $(\nabla \cdot \mathbf{v})_\theta = \nabla_\theta \cdot \mathbf{v} + \partial\dot{\theta}/\partial\theta$, where $\dot{\theta} = d\theta/dt$, we have

$$\frac{\partial}{\partial t_\theta}\left(\rho\frac{\partial h_\theta}{\partial \theta}\right) + \nabla_\theta \cdot \left(\rho\frac{\partial h_\theta}{\partial \theta}\mathbf{v}\right) + \frac{\partial}{\partial \theta}\left(\rho\frac{\partial h_\theta}{\partial \theta}\frac{d\theta}{dt}\right) = 0 \tag{50}$$

For purely hydrostatic motion, Eq. (43) implies that this may be expressed as

$$\frac{\partial}{\partial t_\theta}\left(\frac{\partial p}{\partial \theta}\right) + \nabla_\theta \cdot \left(\frac{\partial p}{\partial \theta}\mathbf{v}\right) + \frac{\partial}{\partial \theta}\left(\frac{\partial p}{\partial \theta}\frac{d\theta}{dt}\right) = 0 \tag{51}$$

The first law of thermodynamics in isentropic coordinates is simply

$$\frac{d\theta}{dt} = \frac{\theta}{c_p T}(q + f_h) \tag{52}$$

In many cases in the study of large-scale flow the fact that motions tend to be isentropic allows Eq. (52) to be replaced by $\dot{\theta} = 0$.

The full set of equations for hydrostatic flow in isentropic coordinates is

$$\begin{aligned}
&\frac{\partial \mathbf{v}_H}{\partial t_\theta} + \mathbf{v}_H \cdot \nabla_\theta \mathbf{v}_H + \frac{d\theta}{dt}\frac{\partial \mathbf{v}_H}{\partial \theta} = -\nabla_\theta \Psi - f\mathbf{k} \times \mathbf{v}_H + (\mathbf{f}_{r,H})_\theta \\
&\frac{\partial \Psi}{\partial \theta} = c_p\left(\frac{p}{p_{00}}\right)^{R/c_p} \\
&\frac{\partial}{\partial t_\theta}\left(\frac{\partial p}{\partial \theta}\right) + \nabla_\theta \cdot \left(\frac{\partial p}{\partial \theta}\mathbf{v}\right) + \frac{\partial}{\partial \theta}\left(\frac{\partial p}{\partial \theta}\frac{d\theta}{dt}\right) = 0 \\
&\frac{d\theta}{dt} = \frac{\theta}{c_p T}(q + f_h)
\end{aligned} \tag{53}$$

Note that T and h_θ have been combined in Ψ and do not have to appear explicitly in order to complete the system. We can of course add the equations

$$\theta = T\left(\frac{p}{p_{00}}\right)^{-R/c_p} \qquad \Psi = c_p T + g h_\theta \tag{54}$$

to determine T and h_θ explicitly.

The equations in isentropic coordinates share the feature with those in isobaric coordinates that the forcing term is linear—in this case, linear in the Montgomery stream function Ψ. The price for this linearity, however, is the nonlinearity that appears in the hydrostatic equation with the R/c_p power of pressure. A major advantage of these equations is that the vertical velocity term appearing in the cartesian coordinate equations is replaced with a direct function of the heating, $d\theta/dt$.

The advantages of both isobaric and isentropic coordinates will appear as we proceed in the next two chapters. It is worth emphasizing again that these advantages are created primarily by the simplification introduced by the hydrostatic assumption.

PROBLEM 7.3.7 Ascertain the validity of the propositions:

(*a*) A parcel in isentropic motion is confined to its initial θ surface.

(*b*) In isentropic motion, the θ surfaces cannot change position.

(*c*) In isentropic, autobarotropic flow, we must have $\nabla_\theta \cdot \mathbf{v} = 0$.

(*d*) In steady, isentropic flow the quantity $v^2/2 + \Psi$ is conserved.

PROBLEM 7.3.8 Find an expression for the mass between two isentropes. Show that over the entire atmosphere this mass is conserved in isentropic flow.

In the following problems, be sure to find the simplest possible expressions.

PROBLEM 7.3.9 Express the hydrostatic equations of motion with the entropy $s = c_p \ln (\theta/\theta_c)$, where θ_c is a constant, as the vertical coordinate.

PROBLEM 7.3.10 Express the hydrostatic equations of motion with the density as the vertical coordinate.

PROBLEM 7.3.11 Express the hydrostatic equations of motion with the potential density $\sigma = \rho(p_{00}/p)^{c_V/c_p}$ as the vertical coordinate.

7.4 GENERALIZED COORDINATES

In this section we will apply some of the ideas of the preceding sections to develop a generalized set of equations in which the vertical coordinate is any monotonic function $\vartheta(x,y,z,t)$ of the height z. The choice of ϑ as either pressure or potential temperature will allow us to reproduce the results of the previous section. The concepts of the tensor calculus introduced in Chap. 5 will be employed here, and we will specifically investigate the role of the hydrostatic approximation in the usual meteorological coordinates. Thus our aim in this section is to study the mathematical and physical significance of the derivation of the meteorological coordinate systems.

7.4.1 Basic Geometry

On the assumption that a scalar function $\vartheta(x,y,z,t)$, which is always monotonic in the variable z, exists, we may define the relations between a new set of coordinates $(\hat{x}^1,\hat{x}^2,\hat{x}^3,t)$ with the equations

$$\begin{aligned} \hat{x}^1 &= x \qquad & x &= \hat{x}^1 \\ \hat{x}^2 &= y \qquad & y &= \hat{x}^2 \end{aligned} \tag{1}$$

$$\hat{x}^3 = \vartheta(x,y,z,t) \qquad z = h(\hat{x}^1,\hat{x}^2,\hat{x}^3,t)$$
$$\hat{t} = t \qquad t = \hat{t}$$

The basis vectors of this new system with respect to the original orthogonal cartesian system [see Eq. (23) in Sec. 5.4] are

$$\begin{aligned} \boldsymbol{\eta}^1 &= \mathbf{i} & \boldsymbol{\tau}_1 &= \mathbf{i} + \frac{\partial h}{\partial \hat{x}^1}\mathbf{k} \\ \boldsymbol{\eta}^2 &= \mathbf{j} & \boldsymbol{\tau}_2 &= \mathbf{j} + \frac{\partial h}{\partial \hat{x}^2}\mathbf{k} \\ \boldsymbol{\eta}^3 &= \nabla\vartheta & \boldsymbol{\tau}_3 &= \mathbf{k}\frac{\partial h}{\partial \vartheta} \end{aligned} \tag{2}$$

As clearly revealed by these basis vectors, this new system of $(\hat{x}^1,\hat{x}^2,\vartheta,t)$ coordinates is *not* an orthogonal system.

PROBLEM 7.4.1 Compute the Jacobian matrices $\{J_{\hat{x}}^{x}\}$ and $\{J_{x}^{\hat{x}}\}$ for this transformation. Show that $|J_{\hat{x}}^{x}| = \partial h/\partial\vartheta$, that $(\partial h/\partial\vartheta)(\partial\vartheta/\partial z) = 1$, and that $\nabla_z\vartheta = -(\nabla_\vartheta h)(\partial\vartheta/\partial z)$, where $\nabla_\vartheta = \mathbf{i}\partial/\partial\hat{x}^1 + \mathbf{j}\partial/\partial\hat{x}^2$. (Hint: The matrix product of the Jacobian matrices must be the identity matrix.)

PROBLEM 7.4.2 Compute the metric tensors $\hat{g}_{ij}$ and $\hat{g}^{ij}$ for the new coordinates.

PROBLEM 7.4.3 Show that the only nonzero components of the Christoffel symbol are

$$\hat{\Gamma}^3_{ki} = \frac{\partial\vartheta}{\partial x^3}\frac{\partial^2 h}{\partial\hat{x}^k\,\partial\hat{x}^i} = \left(\frac{\partial h}{\partial\vartheta}\right)^{-1}\frac{\partial^2 h}{\partial\hat{x}^k\,\partial\hat{x}^i}$$

[Hint: Use the definition (24) in Sec. 5.5.]

The results of these problems show that the new coordinates are not orthogonal because the metric tensor has elements off the diagonal, and that they are not cartesian because the metric tensor varies as a function of position as well as time. Moreover, the Christoffel symbol term will arise in covariant derivatives.

The components $\hat{v}^i = \boldsymbol{\eta}^i \cdot \mathbf{v}$ and $\hat{v}_i = \boldsymbol{\tau}_i \cdot \mathbf{v}$ of the velocity vector $\mathbf{v}$ in the original system are found directly to be

$$\hat{v}^1 = v_1 \qquad \hat{v}^2 = v_2 \qquad \hat{v}^3 = \mathbf{v}\cdot\nabla\vartheta = \frac{d\vartheta}{dt} - \frac{\partial\vartheta}{\partial t} \tag{3}$$

and

$$\hat{v}_1 = v_1 + \frac{\partial h}{\partial \hat{x}^1} v_3 \qquad \hat{v}_2 = v_2 + \frac{\partial h}{\partial \hat{x}^2} v_3 \qquad \hat{v}_3 = \frac{\partial h}{\partial \vartheta} v_3 \tag{4}$$

PROBLEM 7.4.4 Verify that the specific kinetic energy is an invariant in the sense that $\hat{v}_i \hat{v}^i = v_i^2$.

The equations of motion in the original orthogonal cartesian system may be written as

$$\frac{\partial v_i}{\partial t} + v_k \frac{\partial v_i}{\partial x_k} = -\frac{1}{\rho}\frac{\partial p}{\partial x_i} - \frac{\partial \Phi}{\partial x_i} - 2\epsilon_{ijk}\Omega_j v_k + f_{ri} \tag{5}$$

in which we are explicitly using the fact that $v_i = v^i$. The geopotential Φ is such that $\nabla\Phi = g\mathbf{k}$ when $\mathbf{k}$ is orthogonal to the surfaces on which the acceleration of gravity is constant.

When there is a distinction between contravariant and covariant quantities as in ϑ coordinates, we may write Eq. (5) as

$$\frac{\partial \hat{v}^i}{\partial t} + \hat{v}^k \hat{v}^i{}_{;k} = -\hat{g}^{ik}\left(\frac{1}{\rho}\frac{\partial p}{\partial \hat{x}^k} + \frac{\partial \Phi}{\partial \hat{x}^k}\right) - 2\hat{\epsilon}^{ijk}\hat{\Omega}_j \hat{v}_k + \hat{f}_r{}^i \tag{6}$$

When the coordinates $(\hat{x}^1, \hat{x}^2, \hat{x}^3)$ are orthogonal, this reduces to the original equations, so Eq. (6) is correct for any system of coordinates. For the present we will retain the original definition that

$$\frac{\partial}{\partial t} = \left(\frac{\partial}{\partial t}\right)_{x,y,z} \tag{7}$$

For the $(\hat{x}^1, \hat{x}^2, \hat{x}^3)$ coordinates, we may now apply the results of Problems 7.4.2 and 7.4.3 to determine the component equations

$$\frac{\partial \hat{v}^1}{\partial t} + \hat{v}^1 \frac{\partial \hat{v}^1}{\partial \hat{x}^1} + \hat{v}^2 \frac{\partial \hat{v}^1}{\partial \hat{x}^2} + \hat{v}^3 \frac{\partial \hat{v}^1}{\partial \vartheta} = -\frac{1}{\rho}\frac{\partial p}{\partial \hat{x}^1} - \frac{\partial \Phi}{\partial \hat{x}^1} - \frac{\partial \vartheta}{\partial x^1}\left(\frac{1}{\rho}\frac{\partial p}{\partial \vartheta} + \frac{\partial \Phi}{\partial \vartheta}\right) - 2\hat{\epsilon}^{1jk}\hat{\Omega}_j \hat{v}_k + \hat{f}_r{}^1$$

$$\frac{\partial \hat{v}^2}{\partial t} + \hat{v}^1 \frac{\partial \hat{v}^2}{\partial \hat{x}^1} + \hat{v}^2 \frac{\partial \hat{v}^2}{\partial \hat{x}^2} + \hat{v}^3 \frac{\partial \hat{v}^2}{\partial \vartheta} = -\frac{1}{\rho}\frac{\partial p}{\partial \hat{x}^2} - \frac{\partial \Phi}{\partial \hat{x}^2} - \frac{\partial \vartheta}{\partial x^2}\left(\frac{1}{\rho}\frac{\partial p}{\partial \vartheta} + \frac{\partial \Phi}{\partial \vartheta}\right) - 2\hat{\epsilon}^{2jk}\hat{\Omega}_j \hat{v}_k + \hat{f}_r{}^2$$

$$\frac{\partial \hat{v}^3}{\partial t} + \hat{v}^1 \frac{\partial \hat{v}^3}{\partial \hat{x}^1} + \hat{v}^2 \frac{\partial \hat{v}^3}{\partial \hat{x}^2} + \hat{v}^3 \frac{\partial \hat{v}^3}{\partial \vartheta} + \hat{\Gamma}^3_{jk}\hat{v}^j\hat{v}^k = -\frac{\partial \vartheta}{\partial x^1}\left(\frac{1}{\rho}\frac{\partial p}{\partial \hat{x}^1} + \frac{\partial \Phi}{\partial \hat{x}^1}\right) - \frac{\partial \vartheta}{\partial x^2}\left(\frac{1}{\rho}\frac{\partial p}{\partial \hat{x}^2} + \frac{\partial \Phi}{\partial \hat{x}^2}\right)$$

$$- |\nabla\vartheta|^2\left(\frac{1}{\rho}\frac{\partial p}{\partial \vartheta} + \frac{\partial \Phi}{\partial \vartheta}\right) - 2\hat{\epsilon}^{3jk}\hat{\Omega}_j \hat{v}_k + \hat{f}_r{}^3 \tag{8}$$

7.4.2 Meteorological Modifications

As could be expected from the basis vectors, the major problem with nonorthogonality appears in the equation for the vertical component. These difficulties will obviously vanish if the third equation is replaced by the hydrostatic approximation, but rather than take such a drastic step, let us define a new scalar function χ by the relation

$$\frac{\partial p}{\partial z} = -\left(\frac{\partial \Phi}{\partial z} - \chi\right)\rho \tag{9}$$

Thus χ, which we shall call the *hydrostatic defect*, gives at each point in the atmosphere the actual sum of $\alpha\, \partial p/\partial z$ and $\partial\Phi/\partial z$ and vanishes at any point at which the pressure and density are hydrostatically related. Because the atmosphere tends to be hydrostatic, we may expect χ to be much less in absolute value than $\partial\Phi/\partial z \cong g$. Although χ cannot be determined from the observational data now available, it is useful in theoretical studies.

The definition (9) may be expressed as

$$\frac{\partial p}{\partial \vartheta} = -\left(\frac{\partial \Phi}{\partial z} - \chi\right)\rho\,\frac{\partial h}{\partial \vartheta} = -\rho\,\frac{\partial \Phi}{\partial \vartheta} + \chi\rho\,\frac{\partial h}{\partial \vartheta} \tag{10}$$

Now using Problem 7.4.1 to replace the terms $\partial\vartheta/\partial x^\alpha$ with $-(\partial h/\partial \hat{x}^\alpha)(\partial h/\partial\vartheta)^{-1}$ for $\alpha = 1, 2$, we may write the system (8) as

$$\begin{aligned}
&\frac{\partial \hat{v}^1}{\partial t} + \hat{v}^1\frac{\partial \hat{v}^1}{\partial \hat{x}^1} + \hat{v}^2\frac{\partial \hat{v}^1}{\partial \hat{x}^2} + \hat{v}^3\frac{\partial \hat{v}^1}{\partial \vartheta} = -\frac{1}{\rho}\frac{\partial p}{\partial \hat{x}^1} - \frac{\partial \Phi}{\partial \hat{x}^1} + \chi\frac{\partial h}{\partial \hat{x}^1} - 2\hat{\epsilon}^{1jk}\hat{\Omega}_j\hat{v}_k + \hat{f}_r^{\,1}\\
&\frac{\partial \hat{v}^2}{\partial t} + \hat{v}^1\frac{\partial \hat{v}^2}{\partial \hat{x}^1} + \hat{v}^2\frac{\partial \hat{v}^2}{\partial \hat{x}^2} + \hat{v}^3\frac{\partial \hat{v}^2}{\partial \vartheta} = -\frac{1}{\rho}\frac{\partial p}{\partial \hat{x}^2} - \frac{\partial \Phi}{\partial \hat{x}^2} + \chi\frac{\partial h}{\partial \hat{x}^2} - 2\hat{\epsilon}^{2jk}\hat{\Omega}_j\hat{v}_k + \hat{f}_r^{\,2}\\
&\frac{1}{\rho}\frac{\partial p}{\partial \vartheta} + \frac{\partial \Phi}{\partial \vartheta} = \chi\frac{\partial h}{\partial \vartheta}
\end{aligned} \tag{11}$$

Two problems remain. We may use the chain rule to obtain, for example,

$$\frac{\partial \hat{v}^1}{\partial t} = \frac{\partial \hat{v}^1}{\partial t_\vartheta} + \frac{\partial \vartheta}{\partial t}\frac{\partial \hat{v}^1}{\partial \vartheta} = \frac{\partial \hat{v}^1}{\partial t_\vartheta} + \left(\frac{d\vartheta}{dt} - \hat{v}^3\right)\frac{\partial \hat{v}^1}{\partial \vartheta} \tag{12}$$

and we may use Eq. (4) to replace $\hat{v}^1$ and $\hat{v}^2$ with the original velocity components v_1 and v_2. The second problem is the appearance in the Coriolis force term of the factors involving $\hat{v}_3$ that arise from the covariant components $\hat{v}_k$.

Expansion of these terms with the aid of the facts that $\Omega_1 = 0$ and $\hat{\epsilon}^{ijk} = \epsilon^{ijk}(\partial h/\partial\vartheta)^{-1}$ gives

$$2\hat{\epsilon}^{1jk}\hat{\Omega}_j\hat{v}_k = 2\left[\left(\Omega_2 + \frac{\partial h}{\partial \hat{x}^2}\Omega_3\right)v_3 - \Omega_3\left(v_2 + \frac{\partial h}{\partial \hat{x}^2}v_3\right)\right] \tag{13}$$

$$2\hat{\epsilon}^{2jk}\hat{\Omega}_j\hat{v}_k = 2\left[\Omega_3\left(v_1 + \frac{\partial h}{\partial \hat{x}^1}v_3\right) - \frac{\partial h}{\partial \hat{x}^1}\Omega_3 v_3\right] \tag{13 (cont'd)}$$

The standard meteorological approximations, valid for large-scale motion in mid-latitudes, that Ω_2 is of the same order as Ω_3 and that v_1 and v_2 are both much greater in magnitude than v_3, allow Eq. (13) to be written

$$\begin{aligned} 2\hat{\epsilon}^{1jk}\hat{\Omega}_j\hat{v}_k &= 2\Omega_2 v_3 - 2\Omega_3 v_2 \cong -fv_2 \\ 2\hat{\epsilon}^{2jk}\hat{\Omega}_j\hat{v}_k &= 2\Omega_3 v_1 = fv_1 \end{aligned} \tag{14}$$

With $\mathbf{v}_H = \mathbf{i}\hat{v}^1 + \mathbf{j}\hat{v}^2 = \mathbf{i}v_1 + \mathbf{j}v_2$, we may formally combine the component equations of (11) in the form

$$\begin{aligned} \frac{\partial \mathbf{v}_H}{\partial t_\vartheta} + \mathbf{v}_H \cdot \nabla_\vartheta \mathbf{v}_H + \frac{d\vartheta}{dt}\frac{\partial \mathbf{v}_H}{\partial \vartheta} &= -\frac{1}{\rho}\nabla_\vartheta p - \nabla_\vartheta \Phi + \chi \nabla_\vartheta h - f\mathbf{k} \times \mathbf{v}_H + (\mathbf{f}_{r,H})_\vartheta \\ \frac{1}{\rho}\frac{\partial p}{\partial \vartheta} + \frac{\partial \Phi}{\partial \vartheta} &= \chi \frac{\partial h}{\partial \vartheta} \end{aligned} \tag{15}$$

With these equations there is actually no need to specify the relation of the third unit vector to **i** and **j** because neither this vector nor its direction appears in the equations. (The **k** in the Coriolis force term simply serves to give the correct components.) For this reason, we are free to choose the third basis vector to be **k**. The assumption that χ vanishes will now yield the results of Sec. 7.3 for isobaric and isentropic coordinates.

The equations (8) are equations that are the result of rigorous application of tensor methods, and are valid for any variable ϑ such that $\partial\vartheta/\partial z$ never vanishes. But the meteorological modifications made above have destroyed the tensorial variation properties of these equations. For example, the vertical component of the vorticity computed from these equations is not the same component as could be obtained by a correct transformation of the vorticity in the original coordinates into the nonorthogonal set $(\hat{x}^1, \hat{x}^2, \hat{x}^3)$.

The introduction of two basic observed characteristics of large-scale motion has permitted us to simplify the complicated equations (8) into the less formidable system given by Eqs. (15). But in general, the use of a transformation of the vertical coordinate $\hat{x}^3 = \hat{x}^3(x,y,z,t)$ leads to a nonorthogonal coordinate system. The facts that the atmosphere is usually hydrostatic and that vertical velocities are much smaller than horizontal velocities allowed us to circumvent the nonorthogonality and obtain an approximate set of equations with desirable features. When any of the assumptions are not valid, the system (8) must be used; then only for special purposes will it be more useful than the equations in an orthogonal or a spherical coordinate system.

The continuity equation follows from the transport theorem. The mass in a material volume is given by

$$M(t) = \int_V \rho |J_{\hat{x}}^{x}|\, d\vartheta\, dx\, dy = \int_V \rho \frac{\partial h}{\partial \vartheta}\, d\vartheta\, dx\, dy \tag{16}$$

and so with the requirement that mass is conserved, we have from Eq. (5) in Sec. 5.3 applied to a material volume that

$$\frac{dM}{dt} = \int_V \left\{ \frac{\partial}{\partial t_\vartheta}\left(\rho \frac{\partial h}{\partial \vartheta}\right) + \left[\nabla \cdot \left(\rho \frac{\partial h}{\partial \vartheta}\mathbf{v}\right)\right]_\vartheta \right\} d\vartheta\, dx\, dy = 0 \tag{17}$$

Because the equation must be true for any material volume, we thus have the generalized continuity equation

$$\frac{\partial}{\partial t_\vartheta}\left(\rho \frac{\partial h}{\partial \vartheta}\right) + \left[\nabla \cdot \left(\rho \frac{\partial h}{\partial \vartheta}\mathbf{v}\right)\right]_\vartheta = \frac{\partial}{\partial t_\vartheta}\left(\rho \frac{\partial h}{\partial \vartheta}\right) + \nabla_\vartheta \cdot \left(\rho \frac{\partial h}{\partial \vartheta}\mathbf{v}_H\right) + \frac{\partial}{\partial \vartheta}\left(\rho \frac{\partial h}{\partial \vartheta}\frac{d\vartheta}{dt}\right) = 0 \tag{18}$$

The first law of thermodynamics may be written as

$$c_V \rho \frac{dT}{dt} - \frac{p}{\rho}\frac{d\rho}{dt} = \rho(q + f_h) \tag{19}$$

and from the continuity equation we find that for $J_\vartheta = \partial h/\partial \vartheta$,

$$J_\vartheta \frac{1}{\rho}\frac{d\rho}{dt} = -\left[\frac{\partial}{\partial t_\vartheta} J_\vartheta + \nabla_\vartheta \cdot (J_\vartheta \mathbf{v}_H) + \frac{\partial}{\partial \vartheta}\left(J_\vartheta \frac{d\vartheta}{dt}\right)\right] \tag{20}$$

Thus the first law becomes

$$c_V \rho \frac{dT}{dt} + \frac{p}{J_\vartheta}\left[\frac{\partial}{\partial t_\vartheta}(J_\vartheta) + \nabla_\vartheta \cdot (J_\vartheta \mathbf{v}_H) + \frac{\partial}{\partial \vartheta}\left(J_\vartheta \frac{d\vartheta}{dt}\right)\right] = \rho(q + f_h) \tag{21}$$

The system of generalized meteorological equations is therefore

$$\frac{\partial \mathbf{v}_H}{\partial t_\vartheta} + \mathbf{v}_H \cdot \nabla_\vartheta \mathbf{v}_H + \frac{d\vartheta}{dt}\frac{\partial \mathbf{v}_H}{\partial \vartheta} = -\frac{1}{\rho}\nabla_\vartheta p - \nabla_\vartheta \Phi + \chi \nabla_\vartheta h - f\mathbf{k} \times \mathbf{v}_H + (\mathbf{f}_{r,H})_\vartheta$$

$$\frac{1}{\rho}\frac{\partial p}{\partial \vartheta} + \frac{\partial \Phi}{\partial \vartheta} = \chi \frac{\partial h}{\partial \vartheta}$$

$$\frac{\partial}{\partial t_\vartheta}(\rho J_\vartheta) + \nabla_\vartheta \cdot (\rho J_\vartheta \mathbf{v}_H) + \frac{\partial}{\partial \vartheta}\left(\rho J_\vartheta \frac{d\vartheta}{dt}\right) = 0 \tag{22}$$

$$c_V \rho \frac{dT}{dt} + \frac{p}{J_\vartheta}\left[\frac{\partial J_\vartheta}{\partial t_\vartheta} + \nabla_\vartheta \cdot (J_\vartheta \mathbf{v}_H) + \frac{\partial}{\partial \vartheta}\left(J_\vartheta \frac{d\vartheta}{dt}\right)\right] = \rho(q + f_h)$$

$$p = \rho R T$$

To complete the system, we may add the transformation law relating ϑ and h in the form

$$\nabla_z \vartheta + \frac{\partial \vartheta}{\partial z}\nabla_\vartheta h = 0 \tag{23}$$

and we need, finally, to specify ϑ as a function of the other variables that occur. The choices considered so far in this chapter are $\vartheta = z$, $\vartheta = p$, and $\vartheta = \theta$.

PROBLEM 7.4.5 Deduce equations for isobaric and isentropic coordinates from the system (22).

PROBLEM 7.4.6 Using the results in Sec. 4.2 on the geopotential let $\vartheta = Z = \Phi/g_{38}$. What then are the equations of motion? Obtain the approximate form valid when $z \ll a$. (Recall from Chap. 4 that $g_{38} = 980\ \text{cm/s}^2$.)

PROBLEM 7.4.7 The variable

$$\vartheta = \int_0^z \rho\left(\frac{\mathbf{V}\cdot\mathbf{V}}{2} + c_pT + \Phi\right) dz_1$$

is a monotonic function of height. Can it be used as a vertical coordinate?

PROBLEM 7.4.8 Find a coordinate transformation that produces a system in which the motion is formally incompressible without the necessity of making the hydrostatic approximation. (Hint: Suppose $\rho J_\vartheta = 1$.)

PROBLEM 7.4.9 What is the relation between the hydrostatic defect χ and the vertical acceleration?

*PROBLEM 7.4.10 Show that the transport theorem may be written in ϑ coordinates for a material volume V_ϑ as

$$\frac{dF}{dt} = \int_{V_\vartheta} \rho J_\vartheta \frac{df}{dt}\, dV_\vartheta \tag{24}$$

where

$$F = \int_{V_\vartheta} \rho J_\vartheta f\, dV_\vartheta \qquad dV_\vartheta = d\hat{x}^1\, d\hat{x}^2\, d\vartheta \tag{25}$$

BIBLIOGRAPHIC NOTES

Some of the material of this chapter appears in varying form in all dynamic meteorology texts. An attempt has been made to present the subject with a unified approach and a dependence on formal transformation theory in order to clarify some points that always seem difficult for teachers to explain and for students to grasp.

Additional information may be found in the following publications:

Defrise, Pierre, 1964: "Tensor Calculus in Atmospheric Science," in H. Landsberg and J. van Mieghem (eds.), *Advances in Geophysics*, vol. 10, pp. 261–314, Academic Press, Inc., New York.

Eliassen, Arnt, 1948: "The Quasi-static Equations of Motion with Pressure as Independent Variable," *Geofys. Publikasjoner*, 17:5–44.

McVittie, G. C., 1951: "Coordinate Systems in Dynamic Meteorology," *J. Meteorol.*, 8:161–167.

Starr, Victor P., 1945: "A Quasi-Lagrangian System of Hydrodynamical Equations," *J. Meteorol.*, 2:227–237.

8
WATER IN THE AIR

Water is the only substance that most of us will ever meet as a solid, a liquid, and a gas. The water of the oceans covers more than three-quarters of our planet, and water makes up nearly three-quarters of the structure of living organisms. Its unusual properties make life on earth possible in the form we know it. Water and its changes of form are a most important part of the study of atmospheric motion too; we have delayed so long in discussing it in this book, not because water is unimportant, but because it is so special.

Life on earth came to be in water, and water is essential to both plants and animals. But for the study of the atmosphere, it is the changes of form of water that are of the most interest. Evaporation requires energy, energy that is stored in the vapor and released when it later condenses. This released energy helps to drive the wind on its ceaseless journey, and water provides part of the fuel for cyclones, hurricanes, and thunderstorms.

Thus water is important in the atmosphere, but it is most important when it is changing form, and so most of the time we can actually ignore it when we study the motions of the air. As long as the air is unsaturated, it behaves essentially like dry air even though it is moist. But when the air is saturated, when it can hold no more water vapor, then the changes of phase begin and we must concentrate on the effects they will produce.

In this chapter we shall investigate some of the properties of water as an isolated substance, we shall consider what happens when water is mixed with air, and we shall study some of the methods that are used to deal with moist air.

8.1 WATER VAPOR AND MOIST AIR

All air in the atmosphere contains some water vapor, but as indicated by the fact that clouds occupy a small fraction of the atmosphere, it is only rarely that the air becomes saturated. In this section we begin with some properties of the mixture of dry air and water vapor that we shall refer to as *moist air.*

In a volume V, we shall assume that we have a mass m_d of dry air and a mass m_v of water vapor. Thus we may define the total density of the moist air as

$$\rho_m = \frac{m_d + m_v}{V} = \rho_d + \rho_v \tag{1}$$

where ρ_d is the density of dry air and ρ_v is the density of the vapor. According to Dalton's law, each gas in a mixture obeys its own equation of state. If the water vapor is not near condensation, then we may use the ideal-gas law as an equation of state and we have

$$e = \rho_v R_v T_v \tag{2}$$

where e is the partial pressure of the vapor and R_v is the gas constant for the vapor. The dry air has the equation of state

$$p_d = \rho_d R_d T_d \tag{3}$$

where p_d is the partial pressure of the dry air. If the two gases are thoroughly mixed, then it must be true that T_d and T_v are equal and so

$$\begin{aligned} p = p_d + e &= (\rho_d R_d + \rho_v R_v)T \\ &= \rho_m \left(\frac{m_d R_d + m_v R_v}{m_d + m_v}\right) T \\ &= \rho_m R_d \left[\frac{1 + (m_v/m_d)(R_v/R_d)}{1 + (m_v/m_d)}\right] T \end{aligned} \tag{4}$$

The quantity

$$r = \frac{m_v}{m_d} \tag{5}$$

appearing in the equation of state for moist air is known as the *mixing ratio.* The ratio R_v/R_d is given by the ratio of molecular masses $\mu_d/\mu_v = \frac{28.97}{18.002} = 1.61$. Thus we have

$$p = \rho_m R_d \left(\frac{1 + 1.61r}{1 + r}\right) T = \rho_m R_d \left(1 + \frac{0.61r}{1 + r}\right) T \tag{6}$$

The ratio

$$\sigma = \frac{r}{1 + r} = \frac{m_v}{m_v + m_d} \tag{7}$$

is called the *specific humidity*.

Mixing ratios may become as large as 40 g/kg but generally are smaller. Thus $r \ll 1$ and so

$$p = \rho_m R_d (1 + 0.61\sigma) T \cong \rho_m R_d (1 + 0.61r) T \tag{8}$$

The gas constant for moist air is obviously

$$R_m = (1 + 0.61\sigma) R_d \tag{9}$$

but meteorologists prefer to make the adjustment in the temperature rather than the gas constant. Thus we define a *virtual temperature*

$$T_V = (1 + 0.61\sigma) T \tag{10}$$

and use the equation of state as

$$p = \rho_m R_d T_V \tag{11}$$

If σ has the typical value 10 g/kg $= 10^{-2}$, then the correction is of order 1.6 K. In modern practice, the temperature is usually assumed to be the virtual temperature whether or not it is specifically labeled as such.

The total mass of water vapor in a column of unit area extending from level z to the top of the atmosphere is given by

$$W(z) = \int_z^\infty \rho_v \, dz' \tag{12}$$

The quantity W is called the *precipitable water*, because if all the vapor in the column could be condensed and collected, it would form a pool of liquid in the base of the column of depth $D = W/\rho_{\text{liq}} = W$ (cm) since the density of liquid water is approximately 1 g/cm^3.

The relation of W to the specific humidity is easily obtained. For equilibrium conditions, we obtain a hydrostatic relation exactly as in Chap. 4 because the force exerted by the mass of the vapor may be represented by a pressure p_v, just as the weight of the dry air gives a pressure p_a. Hence

$$p_v = gW(z) \qquad p_a = g\int_z^\infty \rho_d \, dz' \tag{13}$$

and so

$$\frac{\partial p_v}{\partial z} = -g\rho_v \qquad \frac{\partial p_a}{\partial z} = -g\rho_d \tag{14}$$

Thus we have as the hydrostatic approximation for moist air

$$\begin{aligned}\frac{\partial p}{\partial z} = \frac{\partial}{\partial z}(p_d + e) &= -g(\rho_d + \rho_v) = -g\rho_m \\ &= -g\rho_d(1 + r)\end{aligned} \tag{15}$$

Here we have used the fact that $p_a + p_v = p_d + e$; but note that $p_v \neq e$ in general (see Problem 8.1.4).

The definition (12) can be transformed into an integral over pressure as the vertical coordinate. We find

$$W(p) = \int_p^0 \rho_v \frac{\partial z}{\partial p'} dp' = -\frac{1}{g}\int_p^0 \frac{\rho_v}{\rho_d + \rho_v}\, dp' \tag{16}$$

and hence we have

$$W(p) = \frac{1}{g}\int_0^p \sigma(p')\, dp' \tag{17}$$

Just as for dry air, we must be able to account for any changes of the mass of water in a parcel. To derive a mass continuity equation, we write

$$m_v = \rho_v V \tag{18}$$

and so by differentiation

$$\frac{1}{V}\frac{dm_v}{dt} = \frac{d\rho_v}{dt} + \frac{\rho_v}{V}\frac{dV}{dt} \tag{19}$$

But as shown in Sec. 5.2.1, $d \ln V/dt = \nabla \cdot \mathbf{v}$, so we have

$$\frac{d\rho_v}{dt} + \rho_v \nabla \cdot \mathbf{v} = \frac{1}{V}\frac{dm_v}{dt} \tag{20}$$

In the equivalent derivation for dry air, we were able to eliminate the term on the right by the requirement that the mass in a parcel must be conserved. But water may change phase, droplets forming from the vapor or droplets in the air evaporating into vapor, and this would represent a change of the mass of the vapor.

For the dry air we have

$$\frac{dm_d}{dt} = 0 \tag{21}$$

and so with this condition we find

$$\frac{d\sigma}{dt} = \frac{d}{dt}\left(\frac{m_v}{m_d + m_v}\right) = \frac{m_d}{(m_d + m_v)^2}\frac{dm_v}{dt} = \frac{1-\sigma}{m_d + m_v}\frac{dm_v}{dt} \tag{22}$$

Obviously then, σ is a conservative property of a parcel of moist air if changes of phase do not occur. More significantly, we can now write Eq. (20) as

$$\frac{d\rho_v}{dt} + \rho_v \nabla \cdot \mathbf{v} = \frac{\rho_m}{1-\sigma} \frac{d\sigma}{dt} \tag{23}$$

Addition of Eq. (23) and the continuity equation for dry air gives the equation of continuity for moist air as

$$\frac{d\rho_m}{dt} + \rho_m \nabla \cdot \mathbf{v} = \frac{\rho_m}{1-\sigma} \frac{d\sigma}{dt} \tag{24}$$

If we want to study the motions of moist air, then clearly we are going to have to be able to predict when either m_v or σ will change owing to phase changes of water taking place in the moving parcels. To do this, we must investigate the physical processes and thermodynamics of the phase change mechanism. But first we study the thermodynamics of the mixture of dry air and vapor for the case in which no phase changes are occurring.

A mixture of unsaturated moist air may be considered as an ideal gas, and so the results obtained in the study of the first law of Sec. 3.2 may be applied. We know that the internal energy is $I = I(T)$, and because the internal energy of moist air is proportional to the sum of the energies of the molecules, we have

$$I_m = m_d e_d + m_v e_v \tag{25}$$

where e is the internal energy per unit mass. By the definitions of Sec. 3.2

$$\begin{aligned} C_{Vm} &= \left(\frac{\partial I_m}{\partial T}\right)_V = m_d\left(\frac{\partial e_d}{\partial T}\right)_V + m_v\left(\frac{\partial e_v}{\partial T}\right)_V \\ &= m_d c_{Vd} + m_v c_{Vv} \end{aligned} \tag{26}$$

and therefore

$$c_{Vm} = \frac{C_{Vm}}{m_d + m_v} = \frac{1}{1+r} c_{Vd} + \frac{r}{1+r} c_{Vv} \tag{27}$$

By exactly the same argument we have

$$c_{pm} = \frac{1}{1+r} c_{pd} + \frac{r}{1+r} c_{pv} \tag{28}$$

For water vapor we have $c_{Vv} = 1.35 \times 10^7$ erg/g·deg and $c_{pv} = 1.81 \times 10^7$ erg/g·deg. Thus we have

$$c_{Vm} = \frac{1}{1+r} c_{Vd}(1 + 1.89r) = c_{Vd}(1 + 0.89\sigma) \tag{29}$$

and

$$c_{pm} = c_{pd}(1 + 0.81\sigma) \tag{30}$$

The first law may now be written as

$$c_{Vd}(1+0.89\sigma)\frac{dT}{dt}+p\frac{d\alpha_m}{dt}=q^*+f_H \tag{31}$$

where we are assuming that $dm_v/dt=0$ and that $p=p_d+e$ so that no phase change is occurring. Thus q^*+f_H represents the heating rate due to all the diabatic processes except those involving change of phase.

If we now restate the equations for moist air in which phase changes are *not* occurring, we have Eq. (31) for the first law, Eq. (8) for the equation of state, and Eq. (24) for the continuity equation in the form

$$\frac{d\rho_m}{dt}+\rho_m\,\nabla\cdot\mathbf{v}=0 \tag{32}$$

Because the specific humidity, which we are assuming constant, is of order 10^{-2} to 10^{-3}, we shall have a more than adequate approximation if we use p, T, and α_m as variables along with constants c_{Vd} and R_d and in effect take $\sigma=0$.

Thus if phase changes are not occurring, we may treat moist air as though it were dry. This conclusion has an obvious basis: dry air is after all itself a mixture of gases, and when we add 1 to 4 percent water vapor to the mixture, it does not significantly alter the properties of the mixture as long as it stays in its gaseous phase.

*PROBLEM 8.1.1 Find and justify a precise version of the statement "Moist air is lighter than dry air."

PROBLEM 8.1.2 Check Eqs. (27) and (28) with an appropriate thermodynamic identity.

PROBLEM 8.1.3 Estimate the precipitable water below 700, 500, and 300 mbar.

PROBLEM 8.1.4 Explain why $p_a+p_v=p_d+e$, but $p_v\neq e$ in general. (Hint: Consider a case with very moist air aloft, but dry air at the surface.)

8.2 CONDENSATION AND THE VAN DER WAALS EQUATION OF STATE

The equation of state for an ideal gas, $p\alpha=RT$, has been applied throughout the book to describe the properties of a gas. It is a suitable approximation at low enough densities that the molecules are able to move about essentially independently of each other.

As the density increases, however, the molecules are pressed close enough for cohesive forces to begin to be effective. If this compression is continued, the cohesive forces will restrict the molecular motion and the gas will change phase and become a liquid. The ideal-gas law cannot describe this phenomenon.

The van der Waals equation of state takes a qualitative account of the cohesive forces. It can be derived in an elegant manner with the modern quantum approaches of statistical physics. Here we present a derivation that emphasizes the fact that the van der Waals equation represents an approximation to the actual behavior.

First, as the gas is compressed and the cohesive forces come into play, we would expect that the energy would be reduced from that of an ideal gas because the cohesion reduces the freedom of molecular movement. We denote by e_i the energy of an identical ideal gas at the same temperature and make a correction, linear in the density, of the form

$$e = e_i - A\rho \tag{1}$$

where A is a positive constant to be estimated from experimental data.

Second, we observe that as the compression proceeds, the molecules of the gas will occupy a significant fraction of the total volume so that the molecules cannot move about at will, but can utilize only the space not occupied by other molecules. Thus the effective volume V_e in which the molecules can move is the total volume less the volume V_m occupied by the molecules, where $V_m = BM = B\rho V$, in which M is the total mass and B is a constant. Thus

$$V_e = V - V_m = V(1 - B\rho) \tag{2}$$

and so the effective density of the gas becomes

$$\rho_e = \frac{M}{V_e} = \frac{M}{V(1 - B\rho)} = \frac{\rho}{1 - B\rho} \tag{3}$$

For an ideal gas, the entropy is

$$s = c_V \ln T - R \ln \rho + \text{const} \tag{4}$$

If we use the effective density to modify the entropy, we have

$$s = s_i + R \ln\left(1 - \frac{B}{\alpha}\right) + \text{const} \tag{5}$$

This equation contains the restriction that $\alpha > B$ so that the argument of the logarithm remains positive and thus compression past the specific volume limit B is not permitted.

Now that we have expressions for e and s, we can form the free energy f and determine the pressure with the aid of Table 3.2. First, by way of example, we carry out the procedure for an ideal gas. From the table, we have

$$f_i = c_V T - T(c_V \ln T + R \ln \alpha + \text{const}) \tag{6}$$

and so

$$p = -\left(\frac{\partial f_i}{\partial \alpha}\right)_T = \frac{RT}{\alpha} \tag{7}$$

For the more general case, we have

$$f = e_i - \frac{A}{\alpha} - T\left[s_i + R\ln\left(1 - \frac{B}{\alpha}\right) + \text{const}\right] = f_i - TR\ln\left(1 - \frac{B}{\alpha}\right) - \frac{A}{\alpha} \tag{8}$$

and so now

$$p = -\left(\frac{\partial f}{\partial \alpha}\right)_T = \frac{RT}{\alpha} + \frac{RT}{1 - B/\alpha}\frac{B}{\alpha^2} - \frac{A}{\alpha^2} \tag{9}$$

or

$$p = \frac{RT}{\alpha - B} - \frac{A}{\alpha^2} \tag{10}$$

This is the *van der Waals* equation, which clearly reduces to the ideal-gas law for large enough specific volumes.

The equation may be written as

$$\alpha^2(\alpha - B) - \frac{RT\alpha^2}{p} + \frac{A(\alpha - B)}{p} = 0 \tag{11}$$

Thus we have a cubic equation for α as a function of p and T. This is the clue that we may have a new phenomenon occurring in the van der Waals gas that does not occur in an ideal gas. For some values of p and T there will be three different values of specific volume. The largest corresponds to a gas, the smallest to a liquid, and the intermediate value to a mixture. Hence the van der Waals equation allows us to explore the process of condensation or evaporation analytically, even though we shall find that it does not give an exact representation of the process as it occurs in water.

We shall begin by studying the isotherms on a (p,α) diagram (Fig. 8.1), taking account of the three roots of the cubic equation (11). Let us follow along an isotherm and see how the pressure varies. Thus from Eq. (10) we have

$$\left(\frac{\partial p}{\partial \alpha}\right)_T = -\frac{RT}{(\alpha - B)^2} + \frac{2A}{\alpha^3} \tag{12}$$

Now obviously, if T is large enough, then $(\partial p/\partial \alpha)_T < 0$ for all α along the isotherm. This is the behavior exhibited by an ideal gas. If α is large enough, the first term dominates for any T, and so at large α we also have $(\partial p/\partial \alpha)_T < 0$. For α near B, the first term will also dominate for all T. But for small enough T, we may have a midrange of α where $(\partial p/\partial \alpha)_T > 0$. Thus at low temperatures, the pressure along the isotherm may actually increase with increasing α in a certain region. Therefore on such an isotherm we start at small α with $(\partial p/\partial \alpha)_T < 0$, the derivative changes to positive,

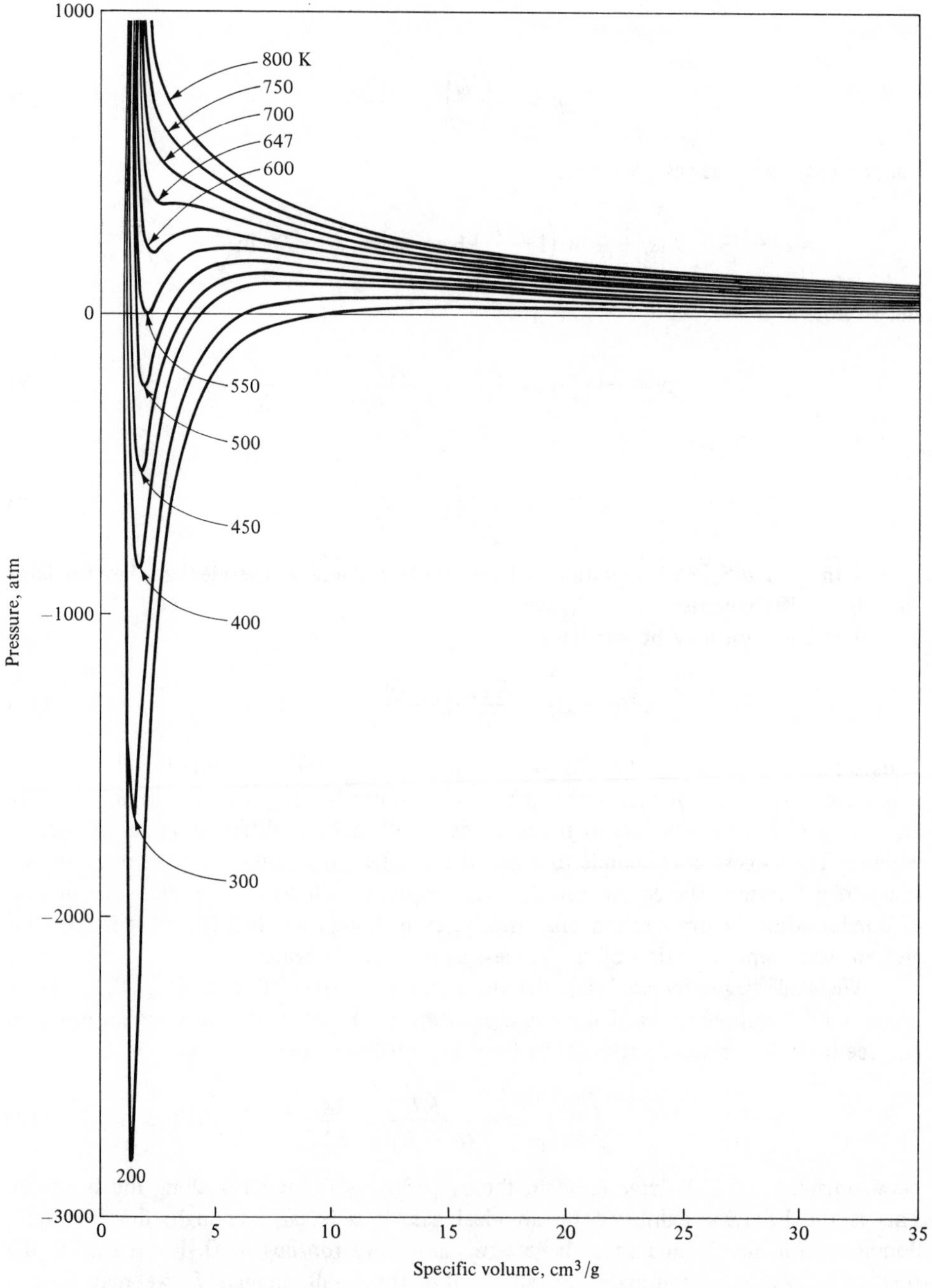

FIGURE 8.1
Isotherms of water vapor as predicted by the van der Waals equation. Values used in the figure are $T_{cr} = 647$ K, $A = 1.05 \times 10^{10}$ erg·cm^3·g^{-2}, $B = 1.033$ cm^3/g. The ordinate is given in atmospheres, where 1 atm = 1013.3 mbar.

and then back to negative. This implies that $(\partial^2 p/\partial\alpha^2)_T$ must also change sign in midrange of α. Thus the *critical point* in the (p,α) diagram separates the region in which $(\partial p/\partial\alpha)_T < 0$ for all α from that in which it has a change of sign; it is specified by

$$\left(\frac{\partial p}{\partial\alpha}\right)_T = 0 \quad \text{and} \quad \left(\frac{\partial^2 p}{\partial\alpha^2}\right)_T = 0 \tag{13}$$

From Eq. (12), we have the critical point conditions

$$\frac{RT}{(\alpha - B)^2} = \frac{2A}{\alpha^3} \qquad \frac{RT}{(\alpha - B)^3} = \frac{3A}{\alpha^4} \tag{14}$$

which imply that

$$\alpha_{cr} = 3B \qquad RT_{cr} = \frac{8A}{27B} \tag{15}$$

The pressure at the critical point obtained from Eq. (10) is

$$p_{cr} = \frac{A}{27B^2} \tag{16}$$

From measured values T_{cr} and p_{cr} at the critical point, we can therefore determine the constants A and B.

For water, the experimentally observed values for the critical point are $T_{cr} = 647$ K, $\alpha_{cr} = 3.1$ cm^3/g, and $p_{cr} = 218.4$ atm $\cong 220$ bar. Hence we have $B = \alpha/3 = 1.033$ cm^3/g and $A = 27BRT/8 = 1.04 \times 10^{10}$ erg·cm^3/g^2. We have used $R = R^*/(18$ g/mol$) = 8.32 \times 10^7/18$ erg/g·deg $= 4.62 \times 10^6$ erg/g·deg. Thus the critical pressure predicted by the van der Waals equation is $p_{cr} = A/27B^2 = 360$ bar. The approximation is clearly not too accurate in the neighborhood of the critical point.

To further compare these results to reality, let us consider the experiment shown in Fig. 8.2. The cylinder with the piston is embedded in a very large reservoir with a temperature T less than T_{cr}. The thermal contact with the reservoir will ensure that the contents of the cylinder stay at temperature T so that we can perform our experiment isothermally. As we compress the gas (suppose it is water vapor) with the piston, eventually some drops of liquid will form. With further compression, the pressure will remain the same, but the liquid will form an increasing fraction of the mixture. Finally there will be only liquid and the pressure will rise rapidly with additional compression.

The oscillations of the pressure along the isotherm in Fig. 8.1 do not appear in the experiment and, obviously, neither do the negative pressures. Thus the oscillations at positive pressure represent unstable states of supersaturation on the vapor side and undercooling on the liquid side that can be reached only with careful experiments using pure gases. Hence we must find a way to make our analysis agree with the experiment; that is, we must determine a method for placing a straight line at the correct place through the oscillating part of the isotherms in Fig. 8.1.

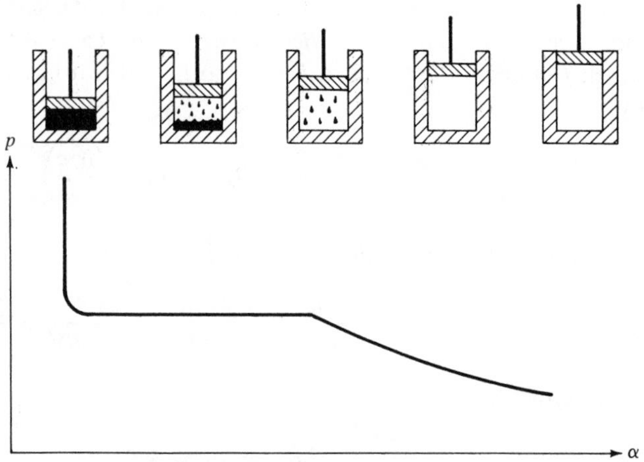

FIGURE 8.2
Schematic diagram showing the formation of liquid by isothermal compression and the associated increase in pressure until the first drops form. After that, the pressure remains constant until there is only liquid in the cylinder, rising rapidly if this liquid is compressed.

The free enthalpy provides the proper procedure. When the mixture of gas and liquid exists in the cylinder, the temperature and pressure of both gas and liquid are the same because the mixture is in thermal and mechanical equilibrium. Thus a small mass of gas passing to the liquid state (or vice versa) will do so without changing pressure or temperature. But then, as shown in Table 3.2, the free enthalpy of the mass will not change. Therefore the free enthalpy of the gas and the free enthalpy of the liquid must be the same.

From the equality of the free enthalpy $g = f + p\alpha$ at the points A and C in Fig. 8.3, we have

$$f_A - f_C = p_A(\alpha_C - \alpha_A) \tag{17}$$

But now $p = -(\partial f/\partial \alpha)_T$ implies that for an integral along the isotherm

$$\int_A^C p_{T=\text{const}}\, d\alpha = -\int_A^C \left(\frac{\partial f}{\partial \alpha}\right)_{T=\text{const}} d\alpha = f_A - f_C = p_A(\alpha_C - \alpha_A) \tag{18}$$

Therefore, the area $f_A - f_C$ under the isotherm equals the area in the rectangle formed by the straight line from the point A to C. The straight line must be placed to equalize the areas formed above and below it by the oscillation in the isotherms. This places it uniquely, and the resulting line AC with $p =$ const is known as *Maxwell's line.*

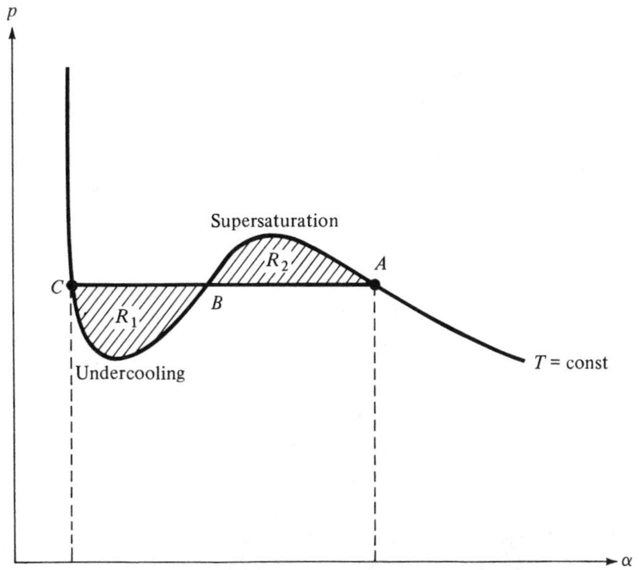

FIGURE 8.3
Illustration of Maxwell's line AC, which equalizes the areas R_1 and R_2.

The isotherms of water vapor according to the van der Waals equation are replotted in Fig. 8.4 with the Maxwell lines in place. Now the negative pressures that appeared in Fig. 8.1 have disappeared along with the oscillations.

More significantly, we have shown that the thermodynamic work done on the gas in the compression from vapor to liquid is given by Eq. (18). Because (for water) $\alpha_C \sim 1\ \text{g/cm}^3 \ll \alpha_A$, the sign is indeed negative, indicating that work has been done on the gas.

For a van der Waals gas, we may calculate for the internal energy e that

$$\int_A^C de_{T=\text{const}} = e_C - e_A = A\left(\frac{1}{\alpha_A} - \frac{1}{\alpha_C}\right) \tag{19}$$

in which we have used Eq. (1) and $de_i = 0$. Thus in the transition from gas to liquid, the internal energy difference is also negative ($\alpha_C \ll \alpha_A$). Furthermore, we have

$$\begin{aligned} \int_{t(A)}^{t(C)} q_{T=\text{const}}\, dt &= \int_A^C de_{T=\text{const}} + \int_A^C p_{T=\text{const}}\, d\alpha \\ &= A\left(\frac{1}{\alpha_A} - \frac{1}{\alpha_C}\right) + p(\alpha_C - \alpha_A) \end{aligned} \tag{20}$$

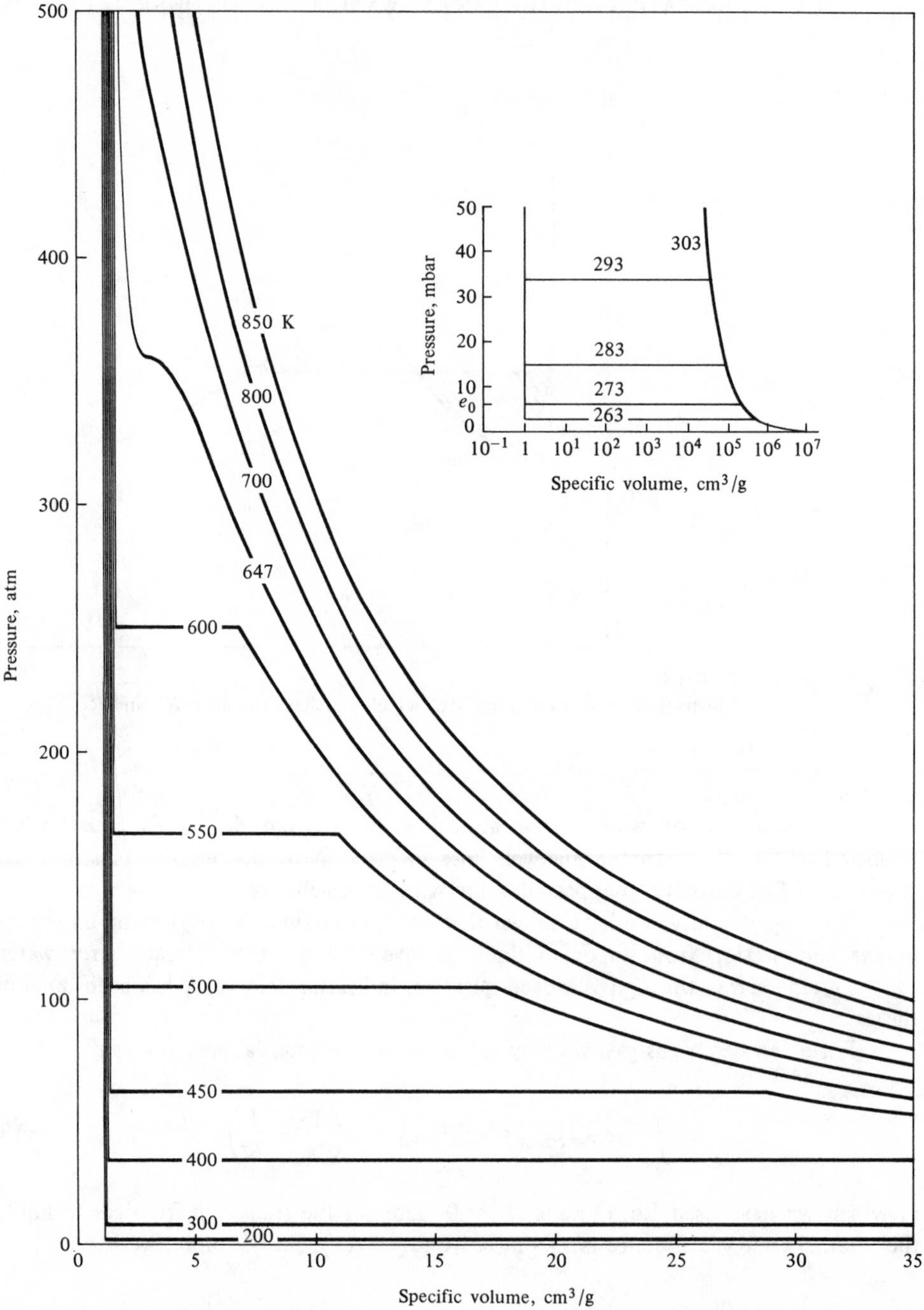

FIGURE 8.4
Isotherms of water vapor as predicted by the van der Waals equation, with the Maxwell line eliminating both the oscillations and negative pressures that appeared in Fig. 8.1.

We conclude the amount of heat specified by Eq. (20) is necessarily released in the condensation; the opposite transition of evaporation from liquid to gas requires that this heat be supplied. This quantity of heat is referred to as *latent* because it is supplied in evaporation and carried by the vapor until it is released by condensation. From Eq. (20), then, we have for the latent heat of a van der Waals gas that

$$L_{lv} = \frac{A}{\alpha_l} + p\alpha_v \tag{21}$$

when we use the approximation $\alpha_l = 1\ \text{cm}^3/\text{g} \ll \alpha_v$. We may use the ideal-gas law for the vapor state and so

$$L_{lv} = \frac{A}{\alpha_l} + R_v T \tag{22}$$

Thus for water at 27°C, we find that $L_{lv} \cong 280$ cal/g, again a wide divergence from the experimental value of $597 - 0.57(T-273)$ cal/g. Even though the numerical values are in error, the van der Waals equation does show that the majority of the latent heat is used to overcome the cohesive forces holding the molecules together in the liquid form.

The latent heat released by condensation of water vapor mixed in the air is released to the air and may be used to increase its temperature or expand its volume. The initial evaporation of liquid water occurs at the earth's surface, where the heat needed is generally supplied by sunlight or warm air near the surface.

PROBLEM 8.2.1 Use the observed value of L_{lv} to determine A. Can the van der Waals equation now be modified to give a more accurate representation of the condensation of water? What must be sacrificed?

PROBLEM 8.2.2 Calculate the entropy difference between the vapor and the liquid state.

8.3 SATURATION, MELTING, AND THE CLAUSIUS-CLAPEYRON EQUATION

In the experiment depicted in Fig. 8.2, we found that condensation began at a particular pressure. If we changed the temperature of both the apparatus and the gas in the cylinder, then the condensation would occur at a different pressure. If we decreased the pressure below this value again, we would pass from a mixture with only a few liquid drops back to a purely gaseous state. This pressure at the boundary between the gaseous state and a mixture of liquid and gas is called the *saturation vapor pressure* and is usually denoted by e_s. The experimental results show that it depends only on the temperature and so $e_s = e_s(T)$. We can find the functional form of this relation with the concepts of the preceding section.

In the mixture, we have for the free enthalpy

$$g_l = g_v \tag{1}$$

where l denotes liquid and v denotes vapor. The condition that the mixture be in mechanical and thermodynamic equilibrium is that $p_l = p_v$ and $T_l = T_v$, and so by Eq. (1), for any change in the mixture that maintains equilibrium, we have

$$dg_l = dg_v \qquad dT_l = dT_v \qquad dp_l = dp_v \tag{2}$$

We know that

$$dg = -s\,dT + \alpha\,dp \tag{3}$$

and thus we find by applying Eq. (3) to both components and subtracting the result that

$$(s_l - s_v)\,dT = (\alpha_l - \alpha_v)\,dp \tag{4}$$

where T and p are the temperature and pressure of the mixture. If we apply the result (4) at saturation, we obtain the relation

$$(s_l - s_v)\,dT = (\alpha_l - \alpha_v)\,de_s \tag{5}$$

But $e_s = e_s(T)$, and so we may write Eq. (5) as

$$\frac{de_s(T)}{dT} = \frac{s_l - s_v}{\alpha_l - \alpha_v} \tag{6}$$

If we define the latent heat as before by

$$L_{lv} = \int_{t(\text{liq})}^{t(\text{vap})} q_{T=\text{const}}\,dt = T\int_{\text{liq}}^{\text{vap}} ds = T(s_v - s_l) \tag{7}$$

then Eq. (6) becomes

$$\frac{de_s(T)}{dT} = \frac{L_{lv}}{T(\alpha_v - \alpha_l)} \tag{8}$$

This is the *Clausius-Clapeyron equation* specifying the temperature dependence of the saturation vapor pressure. But again we know that $\alpha_v \gg \alpha_l$, and we use the equation of state for the vapor form as

$$e_s\alpha_v = R_vT \tag{9}$$

so an approximate form of Eq. (8) is

$$\frac{1}{e_s}\frac{de_s}{dT} = \frac{L_{lv}}{e_sT\alpha_v} = \frac{L_{lv}}{R_vT^2} \tag{10}$$

Integration of this equation on the assumption that $L_{lv} = \text{const}$ gives

$$\begin{aligned} e_s &= e_0 \exp\left[\frac{L_{lv}}{R_v}\left(\frac{1}{T_0} - \frac{1}{T}\right)\right] \\ &= e_0 \exp\left(\frac{L_{lv}}{R_v T_0}\right) \exp\left(-\frac{L_{lv}}{R_v T}\right) \end{aligned} \tag{11}$$

where e_0 and T_0 are reference values. Precisely the same reasoning applies to the ice-liquid transformation, and we have for the pressure e_m of melting

$$\frac{de_m}{dT} = \frac{L_{il}}{T(\alpha_l - \alpha_i)} \tag{12}$$

But in this case

$$\begin{aligned} \alpha_l - \alpha_i &= (1.00 - 1.091)\ \text{cm}^3/\text{g} \\ &= -0.091\ \text{cm}^3/\text{g} \end{aligned} \tag{13}$$

and the latent heat of fusion $L_{il} = 80$ cal/g. Thus we have

$$e_m = e_0 - \ln \frac{T}{T_0}\left(\frac{L_{il}}{0.091\ \text{cm}^3/\text{g}}\right) \tag{14}$$

Finally, for the vapor-ice transformation, we may simply use the latent heat of sublimation L_{iv} in Eq. (11) to obtain the vapor pressure e_S at which sublimation occurs as

$$e_S = e_0 \exp\left(\frac{L_{iv}}{R_v T_0}\right) \exp\left(-\frac{L_{iv}}{R_v T}\right) \tag{15}$$

In writing the three equations (11), (14), and (15), we have assumed that a point (e_0, T_0) common to all three exists. To determine whether it does, we could take a mixture of ice and water and seal it in an evacuated, insulated chamber, leaving the upper part of the chamber empty. We would expect that after a while an equilibrium would be established with water vapor above a mixture of ice and water. Carefully performed experiments of this type show that the temperature is 0.0100°C and that the pressure is 6.11 mbar. A triple-point diagram for water constructed from the three equations using these values is shown in Fig. 8.5.

This figure reveals an interesting situation, and leads to the Bergeron-Findeisen theory of precipitation formation. If a mixture of air and water droplets is cooled below the freezing temperature, it is referred to as supercooled. This often happens in clouds, and the vapor pressure will lie between the curves e_s and e_S where, for $T < 273$ K, $e_s > e_S$. Thus the air is not saturated with respect to liquid water but is saturated with respect to ice. Hence the water drops will evaporate and their vapor will sublimate on any ice particles that are present. Therefore the formation of ice in a cloud with supercooled drops provides a mechanism for creating particles large enough to fall from the cloud as precipitation, perhaps melting in warmer air encountered during their descent.

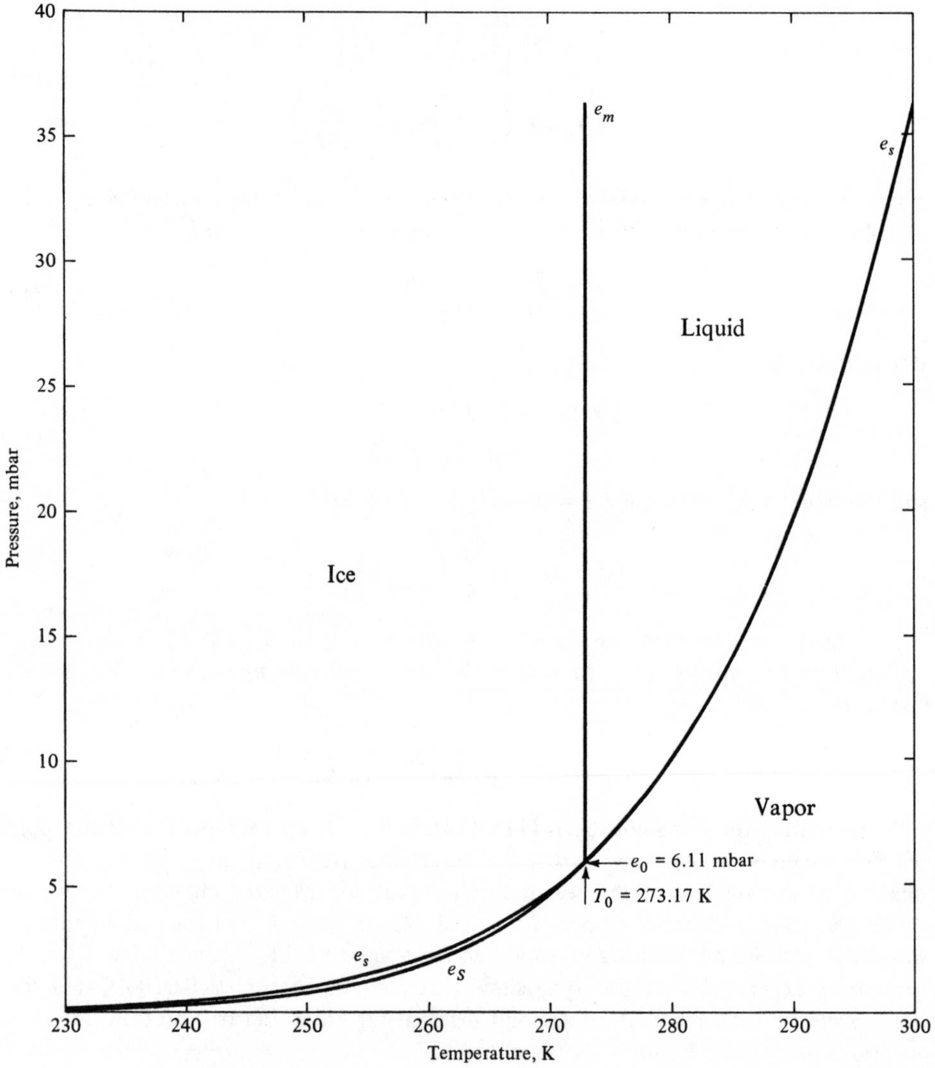

FIGURE 8.5
Phase diagram for water showing the triple point. The implication of the fact that the curves e_S for sublimation and e_s for saturation have different values at temperatures below freezing is discussed in the text.

PROBLEM 8.3.1 In the integration of Eq. (10), we assumed L_{lv} was constant. Can it be integrated if we use the empirical values $L_{lv} = 597 - 0.57(T - 273)$ cal/g instead? How good an approximation is Eq. (11)?

PROBLEM 8.3.2 At what point does the maximum difference between e_s and e_S occur?

8.4 DEW POINT AND HUMIDITY

Several quantities that measure the amount of water vapor present in moist air have been introduced in this chapter, but we have not yet indicated how they can be determined by an explicit measurement procedure.

Nature has given us a clue to a successful technique by providing the example of dew formation. In this case the ground is cooled by radiation and the air near the surface reaches saturation. Dewdrops are then deposited on the grass. If we determined the air temperature at the instant the first dewdrops formed, then we could find the saturation vapor pressure from Eq. (11) in Sec. 8.3. We could assume that this was the vapor pressure in the air. If we also know the total pressure $p = p_d + e$, then from the equation of state for dry air and vapor we could find, assuming that the temperature of the air and vapor were identical, that

$$\frac{e}{p_d} = \frac{R_v}{R_d}\frac{\rho_v}{\rho_d} = \frac{R_v}{R_d}\frac{m_v}{m_d} = \frac{R_v}{R_d}r \tag{1}$$

Then with $R_v/R_d = \mu_d/\mu_v = 1.61$ we would have

$$r = \frac{0.621e}{p_d} = \frac{0.621e}{p - e} \tag{2}$$

and thus would know precisely how much water was contained in the air.

In analogy with this natural process, we shall define the *dew-point temperature* as the temperature at which saturation will occur if moist air is cooled isobarically and at constant vapor pressure (or equivalently, constant mixing ratio). The dew-point temperature may be found experimentally by determining when fog begins to form on a mirror with a controllable temperature.

According to the definition, we have saturation at the dew point, so from Eq. (11) in Sec. 8.3 we may write

$$e_s(T_D) = 6.11 \exp\left[\frac{L_{lv}}{R_v}\left(\frac{1}{273} - \frac{1}{T_D}\right)\right] \tag{3}$$

where we have used the triple-point values for e_0 and T_0 and denoted the dew-point temperature by T_D. Note that the definition requires that $e_s(T_D) = e$, the actual vapor pressure.

The *relative humidity* h is defined by

$$h = \frac{r}{r_s} \tag{4}$$

where r is the actual mixing ratio and r_s is the mixing ratio at saturation at the same temperature. From Eq. (2) we have

$$h = \frac{e}{e_s} = \frac{e_s(T_D)}{e_s(T)} \tag{5}$$

and so from Eq. (3) of this section and Eq. (11) of Sec. 8.3

$$h = \exp\left[\frac{L_{lv}}{R_v}\left(\frac{1}{T} - \frac{1}{T_D}\right)\right] \tag{6}$$

Relative humidity is rarely used for scientific purposes because it is temperature-dependent. Instead, once the vapor pressure is found from Eq. (3), it is used in Eq. (2) to give the mixing ratio, which may be converted into specific humidity if desired.

These computations are generally performed on thermodynamic diagrams adapted for meteorological use. In addition to pressure and temperature curves, the diagrams have saturation mixing ratio lines that indicate the saturation mixing ratio for each (p,T) point. Thus when the curves for temperature versus pressure and dew point versus pressure are plotted, both temperature curves intersect a saturation mixing ratio curve at each pressure. At a given pressure, then, the intersection of a saturation mixing ratio curve and the dew point is the actual mixing ratio of the air; the temperature curve intersection gives the saturation mixing ratio. Hence we know both r and r_s from the diagram.

8.5 ASCENT AND STABILITY OF MOIST PARCELS

The presence of water vapor in moist air introduces a formidable complication in the behavior of parcels rising in the atmosphere. As they rise, they cool and will eventually reach saturation. Then condensation begins and the latent heat of vaporization is released, providing heat to the air and the droplets. Some of the droplets may fall out as rain, carrying away some of the energy.

Initially, the parcel temperature decreases at the dry adiabatic rate during its ascent. Once condensation begins, the rate of decrease of temperature is less because of the latent heat that is released. If the parcel carries along some drops and rises high enough, then eventually the drops will freeze, releasing the latent heat of fusion and providing additional heat to the air.

Thus the analysis of temperature change in rising parcels and the determination of whether they are more or less dense than their environment must take account of the complications introduced by the phase changes of water. Certain simplifications are generally made in order to obtain expressions that can be easily handled.

We shall consider now the case in which all waterdrops fall out of the parcel as soon as they condense. This is a *pseudoadiabatic process*, because the energy contained in the liquid water is carried away from the parcel. It is often called *irreversible* as well, on the grounds that the drops cannot be restored to the parcel, and therefore when it descends a different temperature will result because no heat is needed for evaporation. In a *reversible moist adiabatic process*, according to this nomenclature, the liquid water is retained in the air parcel and will evaporate during descent.

Because the drops fall out in the irreversible case, the heat released by condensation is assumed to be supplied to the moist air. Thus we have

$$Q = L\frac{dm_l}{dt} = -L\frac{dm_v}{dt} \qquad L = L_{lv} \tag{1}$$

and so the first law, Eq. (31) of Sec. 8.1, may be written as

$$\begin{aligned} c_{pm}\frac{dT}{T} - \frac{\alpha_m\,dp}{T} &= -\frac{L}{T}\frac{dm_v}{m_d + m_v} \\ &= -\frac{L}{T}\frac{d\sigma}{1-\sigma} \end{aligned} \tag{2}$$

We adopt the approximations that $c_{pm} = c_p$, that

$$\frac{\alpha_m}{T} = \frac{R_d}{p} \tag{3}$$

and that

$$\frac{dm_v}{m_d + m_v} \cong \frac{dm_v}{m_d} \cong dr \tag{4}$$

All three depend on the fact that $r \ll 1$. Now Eq. (2) may be written as

$$c_p\frac{dT}{T} - \frac{R_d}{p}\,dp = -d\left(\frac{Lr}{T}\right) - \frac{Lr}{T^2}\,dT \tag{5}$$

or

$$\left(c_p + \frac{Lr}{T}\right)\frac{dT}{T} - \frac{R_d}{p}\,dp = -d\,\frac{Lr}{T} \tag{6}$$

But

$$\frac{Lr}{c_pT} \sim \frac{600 \times 4.2 \times 10^7 \times 10^{-2}}{10^7 \times 250} \sim 10^{-1} \tag{7}$$

so as a further approximation

$$c_p \frac{dT}{T} - \frac{R_d}{p}\,dp = -d\frac{Lr}{T} \tag{8}$$

These approximations may seem to be rather gross and to be destroying the accuracy of the calculation. But we are constructing only an approximate model of real processes anyway, because we are ignoring the effect of mixing by entrainment, we are not retaining any of the waterdrops, and we are ignoring the actual physics of nucleation. The result (8), although approximate, still contains the essential physics of the process—that the loss of vapor by condensation provides energy to the moist air.

The differential equation (8) can be integrated immediately, and we find

$$c_p \ln \frac{T}{T_0} - R_d \ln \frac{p}{p_0} = \left(\frac{Lr}{T}\right)_0 - \frac{Lr}{T} \tag{9}$$

We let the reference point denoted by subscript zero be 1000 mbar and denote T_0 by θ_w. Thus

$$\theta_w(r_0) = \theta \exp\left[\frac{Lr}{c_pT} - \left(\frac{Lr}{c_pT}\right)_0\right] \tag{10}$$

where θ is the usual potential temperature.

The new potential temperature $\theta_w(r_0)$ defined by Eq. (10) has the disadvantage that it depends on the amount of moisture in the parcel when it reaches 1000 mbar. To avoid this problem, we shall say instead that we determine it by carrying the parcel pseudoadiabatically upward until all the vapor condenses and falls out, and by then bringing the parcel isentropically to 1000 mbar. Thus the parcel is dry on its descent and $r_0 = 0$. Therefore we have the *equivalent potential temperature*

$$\theta_E = \theta \exp\left(\frac{Lr}{c_pT}\right) \tag{11}$$

As is obvious from the derivation, θ_E is a conservative property of a pseudoadiabatic process. From the estimate (7) we see that

$$\theta_E \sim \theta\left(1 + \frac{Lr}{c_pT}\right) \tag{12}$$

and thus a typical value would be $\theta_E \sim 1.1\theta$.

It is possible to use these results (or their more accurate equivalents) to construct pseudoadiabats that will describe the behavior of moist parcels on a thermodynamic diagram just as the dry adiabats describe dry air. A rising parcel then will cool at the dry adiabatic rate until saturation is encountered. Up to this point, its temperature and pressure will follow a line of constant θ. At saturation, p and T will be governed by Eq. (8), and the parcel will track along a moist adiabat specified by the solution (9) with the reference level taken to be the point at which saturation began. (On most diagrams these adiabats are labeled with the value of θ_w that would be obtained after descent to 1000 mbar.)

We assume that once saturation is reached, the parcel remains saturated so that the mixing ratio r in Eq. (9) is the saturation mixing ratio appropriate to the pressure and temperature the parcel has at any point after saturation is reached. Thus Eq. (9) reveals that the temperature of the parcel will be higher after saturation is reached than a dry parcel at the same pressure. Therefore the crucial question for stability of whether the parcel is warmer than its environment is determined by the slope of the environmental temperature profile compared to that of the adiabat which the parcel follows.

The environmental temperature profile often has a lapse rate less than a dry adiabat, but greater than the moist adiabat. Thus this situation is unstable if the parcel rises far enough and enough condensation occurs for its temperature to become warmer than that of the environment. Therefore stability in this case depends on the amount of the lifting and resulting condensation, and the parcel is referred to as *conditionally unstable*. An illustration is given in Fig. 8.6. If we denote the slope of the moist adiabat by γ_m, then with the results of Sec. 4.4 we have

Absolutely stable: $\gamma < \gamma_m$

Conditionally unstable: $\gamma_m < \gamma < \gamma_d$

Absolutely unstable: $\gamma_d < \gamma$

Another form of instability occurs when entire layers are lifted. If the bottom of the layer is moist and the top is dry, the temperature profile may change markedly as the top air follows a dry adiabat and the lower air follows a moist one. If the lifting of the layer produces an unstable lapse rate by this process, the layer is called *potentially* or *convectively unstable*. An illustration is given in Fig. 8.7.

In order to estimate the moist adiabatic lapse rate, we return to Eq. (2) and write it with the approximations already used as

$$c_p \frac{dT}{T} - R_d \frac{dp}{p} = -\frac{L_{lv}\, dr}{T} \tag{13}$$

Invoking the assumption that the parcel remains saturated, we have $dr = dr_s$. In Eq. (2) in Sec. 8.4 we replace p_d by p for simplicity, and then with Eq. (11) in Sec. 8.3 we use the triple-point values to obtain

$$r_s = \frac{0.621 e_s}{p} = \frac{0.621 \times 6.11}{p} \exp\left[\frac{L_{lv}}{R_v}\left(\frac{1}{273} - \frac{1}{T}\right)\right] \tag{14}$$

Thus

$$\frac{dr_s}{r_s} = -\frac{dp}{p} + \frac{L_{lv}}{R_v}\frac{dT}{T^2} \tag{15}$$

With this result Eq. (13) becomes, for $L = L_{lv}$,

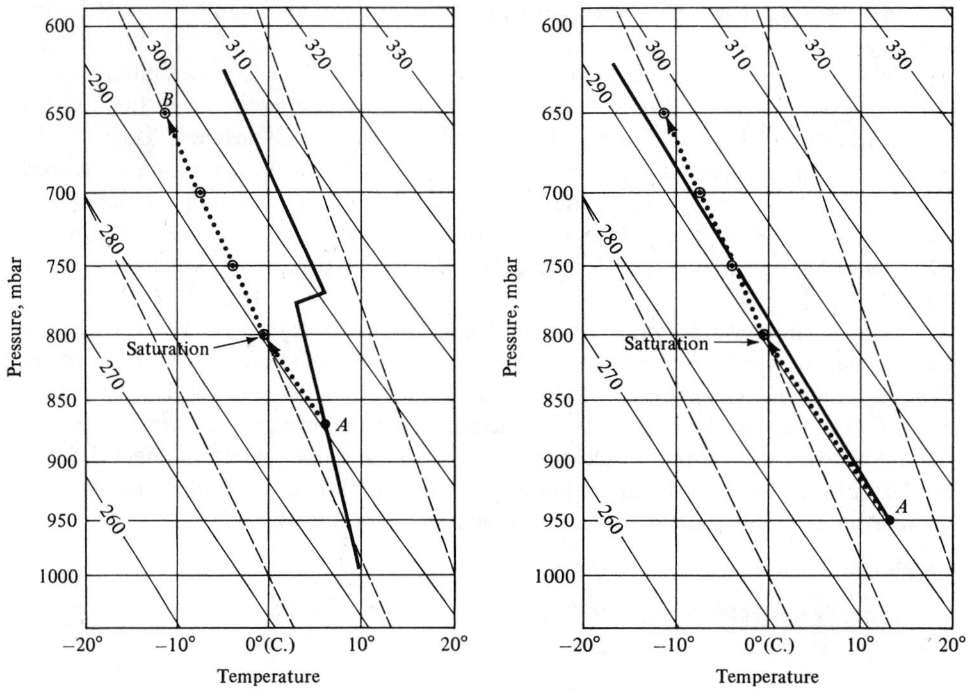

FIGURE 8.6
Illustration of absolute and conditional stability. On the left, the rising parcel remains cooler than the environment even though condensation occurs. On the right, the parcel becomes warmer than the environment, owing to condensational heating, if it rises high enough.

$$\left(c_p + \frac{L^2}{R_v}\frac{r_s}{T^2}\right)\frac{dT}{T} - \left(R_d + \frac{Lr_s}{T}\right)\frac{dp}{p} = 0 \tag{16}$$

and so

$$\left(c_p + \frac{L^2 r_s}{R_v T^2}\right)\frac{dT}{dt} = \alpha\left(1 + \frac{Lr_s}{R_d T}\right)\frac{dp}{dt} \tag{17}$$

Exactly as in Sec. 4.4, we assume that the parcel, to be denoted now by subscript p, adjusts to the ambient pressure p and so

$$\frac{dp_p}{dt} = w_p\,\frac{\partial p}{\partial z} = -w_p g\rho \tag{18}$$

with the hydrostatic approximation. Now Eq. (17) is

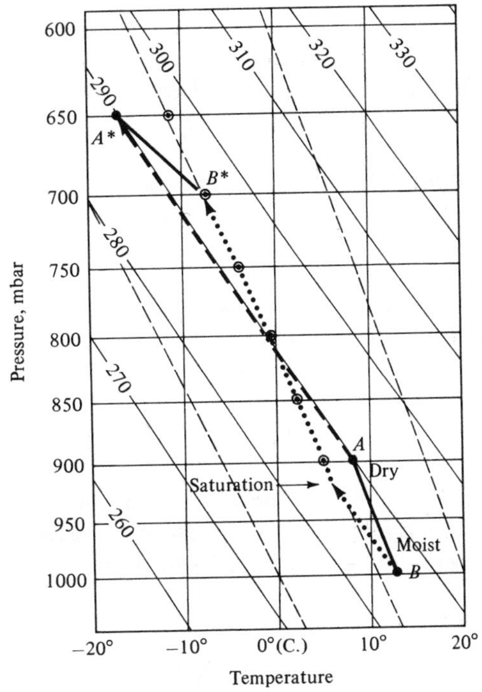

FIGURE 8.7
Illustration of convective instability. The layer AB is carried by lifting to A^*B^*, but the lower part of the layer cools at the moist adiabatic rate for most of the ascent, producing an absolutely unstable profile.

$$\left(c_p + \frac{L^2 r_s}{R_v T^2}\right)\frac{dT}{dt} = -\left(1 + \frac{L r_s}{R_d T}\right)\alpha_p \rho g w_p \tag{19}$$

The complication caused by $\alpha_p \rho$ was discussed in detail in Sec. 4.4 and we ignore it here. Thus we have

$$\gamma_m = \gamma_d\left(1 + \frac{L r_s}{R_d T}\right)\left(1 + \frac{L^2 r_s}{c_p R_v T^2}\right)^{-1} \tag{20}$$

For $T \sim 273$ K we have $L = 600$ cal/g so that $L/c_p T \cong 10$ and thus

$$1 + \frac{L r_s}{R_d T} \cong 1 + \frac{c_p}{R_d} 10 r_s \cong 1 + 35 r_s \tag{21}$$

Similarly we have $c_p/R_v = c_p/1.61 R_d$ and so

$$1 + \frac{L^2 r_s}{c_p R_v T^2} = 1 + 2.2 r_s \times 10^2 \tag{22}$$

and so

$$\gamma_m = \gamma_d \frac{1 + 35r_s}{1 + 2.2r_s \times 10^2} \tag{23}$$

Thus for $r_s = 5 \times 10^{-3}$ we have $\gamma_m \cong 5.5$ K/km.

PROBLEM 8.5.1 Verify the three approximations made in arriving at Eq. (5) from Eq. (2). Can you estimate the total error made as a function of r and other variables?

*PROBLEM 8.5.2 In the atmosphere, it is almost always true that $\theta = \theta(z)$ is a single-valued function. Under what condition would $\theta_E = \theta_E(z)$ have multiple values? Would you expect this condition to be met often? Can θ_E be used as a vertical coordinate?

PROBLEM 8.5.3 Tornadoes often form when a warm, moist lower layer is topped by a dry layer. There is usually an inversion between the layers. Why would tornadoes form in this situation, and what is necessary to start the process?

8.6 THERMODYNAMIC THEORY OF PHASE CHANGES IN MOIST AIR

The major complication introduced by the phase changes of water that may occur in moist air is that some of the waterdrops or ice particles may precipitate from the parcel, carrying away both mass and energy. To investigate the phenomenon of phase changes in moist air rigorously, we must therefore turn to the thermodynamic theory of open systems that was introduced in Sec. 3.4.

We shall assume that we have several constituents, each contributing mass m_j to the total mass

$$m = \sum_j m_j \tag{1}$$

Because of phase changes, the amount of mass m_j may change internally by an amount $d_i m_j$ or there may be an external mass flux, $d_e m_j$, into or out of the parcel. Thus

$$dm_j = d_i m_j + d_e m_j \tag{2}$$

The internal mass changes are subject to the obvious conditions that

$$d_i m = \sum_j d_i m_j = 0 \qquad dm = \sum_j d_e m_j \tag{3}$$

The first law of thermodynamics for open systems as expressed by Eq. (12) in Sec. 3.4

may be written

$$dH = Q + V\,dp + h\,dm \tag{4}$$

The enthalpy H is $H(T,p,m_1, \ldots, m_n)$ for $j = 1, \ldots, n$ and by definition

$$H = \sum_j m_j h_j \tag{5}$$

Thus we may calculate

$$dH = \frac{\partial H}{\partial T} dT + \frac{\partial H}{\partial p} dp + \sum_j h_j dm_j \tag{6}$$

and because the changes in the first two terms are internal changes, we have

$$d_i H = \frac{\partial H}{\partial T} dT + \frac{\partial H}{\partial p} dp + \sum_j h_j d_i m_j \tag{7}$$

Hence

$$d_e H = \sum_j h_j d_e\, m_j \tag{8}$$

as expected.

If we split Q into components Q_i and Q_e, we may use Eq. (4) to define

$$Q_i = d_i H - V\,dp \tag{9}$$

and then from Eqs. (7) and (8) follows

$$Q_e = d_e H - h\,dm = \sum_j (h_j - h)\, d_e m_j \tag{10}$$

The appropriate entropy law is Gibbs' relation from Problem 3.4.1,

$$T\,dS = dH - V\,dp - \sum_j \mu_j dm_j \tag{11}$$

where the *chemical potential* is $\mu = h - Ts$. We define the internal entropy change as

$$T\,d_i S = d_i H - V\,dp - \sum_j \mu_j\, d_i m_j = Q_i - \sum_j \mu_j\, d_i m_j \tag{12}$$

Subtracting Eq. (12) from Eq. (11) produces

$$T\,d_e S = d_e H - \sum_j \mu_j\, d_e m_j = \sum_j T s_j\, d_e m_j \tag{13}$$

a result that again is intuitively obvious.

As the specific definition of a *pseudoadiabatic process* we require that $Q_i = 0$ and hence the only energy gain or loss is due to external mass fluxes $d_e m_j$. Thus we have the two fundamental equations

$$d_i H = dH - d_e H = dH - \sum_j h_j \, d_e m_j = V \, dp \tag{14}$$

from Eqs. (8) and (9) and

$$d_i S = dS - dS_e = dS - \sum_j s_j \, d_e m_j = -\sum_j \frac{\mu_j}{T} d_i m_j \tag{15}$$

from Eqs. (12) and (13).

The first of these shows that the internal enthalpy changes are directly related to changes in the pressure. In this sense the mixture behaves like an ordinary thermodynamic system undergoing adiabatic transformations. The second equation permits explicit calculations of the effects of phase changes. We consider a mixture composed of m_d grams of dry air, m_v of water vapor, m_l of liquid water, and m_i of ice. We require that

$$d_i m_d = d_e m_d = 0 \qquad d_i m_v + d_i m_l + d_i m_i = 0 \qquad d_e m_v = 0 \tag{16}$$

For this case, Eq. (15) becomes

$$\begin{aligned} d_i S &= -\frac{\mu_l}{T} d_i m_l - \frac{\mu_i}{T} d_i m_i - \frac{\mu_v}{T} d_i m_v \\ &= \frac{\mu_l - \mu_v}{T} d_i m_v + \frac{\mu_l - \mu_i}{T} d_i m_i \end{aligned} \tag{17}$$

The latent heat required to pass from liquid to vapor is

$$\begin{aligned} L_{lv} &= \int_l^v de + \int_l^v p \, d\alpha \\ &= \int_l^v dh - \int_l^v \alpha \, dp \end{aligned} \tag{18}$$

We have seen earlier that the phase transformation takes place at constant pressure, and so

$$L_{lv} = h_v - h_l \tag{19}$$

This result allows us to construct the identity

$$\begin{aligned} T(s_v - s_l) &= L_{lv} + h_l - h_v + T(s_v - s_l) \\ &= L_{lv} + \mu_l - \mu_v \\ &= L_{lv} + A_{lv} \end{aligned} \tag{20}$$

where we have used the last two lines to define the *affinity of vaporization*, A_{lv}. In precisely the same way

$$T(s_l - s_i) = L_{il} + A_{il} \tag{21}$$

Tracing back through the definitions, we can show that these affinities are the differences of the free enthalpy per unit mass of the two phases. Hence they vanish when the two phases are in equilibrium.

With these new definitions, Eq. (17) may be written as

$$d_i S = \frac{A_{lv}}{T}\, d_i m_v - \frac{A_{il}}{T}\, d_i m_i \tag{22}$$

We shall combine Eq. (22) with Eq. (15), using the fact that

$$S = m_d s_d + m_v s_v + m_l s_l + m_i s_i \tag{23}$$

and that the only external fluxes permitted are those of ice and liquid. Thus we find

$$d(m_d s_d + m_v s_v) + s_l\, d_i m_l + s_i\, d_i m_i + m_l\, ds_l + m_i\, ds_i = \frac{A_{lv}}{T}\, d_i m_v - \frac{A_{il}}{T}\, d_i m_i \tag{24}$$

We use Eq. (20) to replace s_v in the parentheses and find that the equation can then be expressed as

$$d\left(m_d s_d + \frac{m_v L_{lv}}{T}\right) + m_v d\left(\frac{A_{lv}}{T}\right) + (m_v + m_l)\, ds_l + s_l(dm_v + d_i m_l) + s_i\, d_i m_i + m_i\, ds_i = -\frac{A_{il}}{T}\, d_i m_i \tag{25}$$

But now with Eqs. (21) and (16) we have

$$d\left(m_d s_d + \frac{m_v L_{lv}}{T}\right) + m_v d\left(\frac{A_{lv}}{T}\right) + (m_v + m_l)\, ds_l - \frac{L_{il}}{T}\, d_i m_i + m_i\, ds_i = 0 \tag{26}$$

A final use of Eq. (21) to replace s_i gives

$$d\left(m_d s_d + \frac{m_v L_{lv}}{T}\right) - d_i\left(\frac{m_i L_{il}}{T}\right) + m_v d\left(\frac{A_{lv}}{T}\right) - m_i d\left(\frac{A_{il}}{T}\right) + (m_v + m_l + m_i)\, ds_l = 0 \tag{27}$$

If we assume that the liquid is incompressible, we have

$$ds_l = c_l \frac{dT}{T} \tag{28}$$

By definition

$$s_d = c_{pd} \ln T - R_d \ln p_d + \text{const} \tag{29}$$

Thus with the mixing ratios $r_j = m_j/m_d$ we may write Eq. (27) as

$$d\left\{c_{pd}\ln T - R_d \ln p_d + \frac{r_v L_{lv}}{T}\right\} - d_i\left(\frac{r_i L_{il}}{T}\right) + r_v\, d\left(\frac{A_{lv}}{T}\right) - r_i\, d\left(\frac{A_{il}}{T}\right)$$
$$+ (r_v + r_l + r_i)c_l \frac{dT}{T} = 0 \quad (30)$$

As a first application, let m_i be zero throughout the process and let all waterdrops fall out immediately ($r_l = 0$). Furthermore we assume saturation so that $A_{lv\cdot} = 0$. Then with the mean-value theorem to give

$$\int r_v c_l \frac{dT}{T} = \overline{r_v} c_l \ln T \quad (31)$$

where $\overline{r_v}$ is a suitable mean value, we have

$$(c_{pd} + \overline{r_v} c_l) \ln T - R_d \ln p_d + \frac{r_v L_{lv}}{T} = \text{const} \quad (32)$$

This is thus an exact equation for the pseudoadiabatic process described by the assumptions.

As another example, we could assume that we had a mixture of dry air, vapor, and drops at the triple point. We let all the drops freeze at the triple point so that $dT = 0$ and so that $dr_i = -dr_l$ while $dr_v = 0$. Thus in Eq. (30) we have $A_{lv} = A_{il} = 0$ because we stay at the triple point and so

$$-R_d \ln p_d + \frac{r_v L_{lv}}{T_0} - r_i \frac{L_{il}}{T_0} = \text{const} \quad (33)$$

where T_0 is the temperature of the triple point. We let superscript 1 denote initial conditions, superscript 2 denote final conditions. Hence

$$R_d \ln p_d^{(2)} + r_i^{(2)} \frac{L_{il}}{T_0} = R_d \ln p_d^{(1)} \quad (34)$$

Thus the pressure must drop in such a process, and so if it is to occur, it must happen in an ascending parcel.

In both these examples, we specified in advance how the mixing ratios would behave and we were careful to specify equilibrium conditions so that the affinities vanished. To complete the set of equations, we actually must develop additional equations that will yield the rates of change of mixing ratios from knowledge of the temperature and pressure. They must take account of the effects of surface tensions, and of droplet coalescence and collision, and so they must be stochastic equations.

These comments indicate the formidable difficulties that water in the air poses to dynamic meteorologists. It must be taken into account, and yet the mathematical description of the results of phase changes is clearly beyond analytic capability. The behavior of the drops themselves depends on complicated phenomena and must be treated with statistical models that require extensive computer routines for their solution.

The computer models of cloud processes that take account of these factors are far too complicated to use in large-scale weather prediction. At present, then, we can only set forth the basic principles involved in moist thermodynamics and use them as a guide for incorporating knowledge about cloud processes gained from observation and numerical models into the analytical and numerical techniques of diagnosing and predicting the state of the atmosphere.

PROBLEM 8.6.1 With a suitable expression for $r_v = r_v(T, p_d)$ for the saturated case, produce a version of Eq. (32) in which the integral in Eq. (31) is explicitly present. Can you evaluate the resulting integral analytically?

PROBLEM 8.6.2 Show that Eq. (32) is consistent with Eq. (9) of Sec. 8.5.

8.7 THESE FEW SYMBOLS

We arrive at the end of the second part of the book, having combined the symbols listed in Chap. 1 into the equations that describe the motion of the atmosphere, and find ourselves overwhelmed by the complexity of any attempt to model phase changes exactly.

It is easier to explain how it might be done for large-scale flow than to write out the equations. The water vapor may be advected along with the wind, and then the mixing ratio at each step and grid point compared to the saturation mixing ratio for that pressure and temperature. If the air is unsaturated, the process continues. But if we find that the advected mixing ratio exceeds the saturation value, then we reduce it to that value, letting, in effect, precipitation fall from an imaginary cloud. From the two ratios we can compute how much water was condensed, and so we know how much heat must be added to the moist air.

The cloud must be accounted for, too, in the radiation budget if we are to have a realistic model. We have not considered how to model the radiation divergence that appears in the first law. That may seem strange, for after all, the radiation is the cause of the motion, heating the tropics and cooling the polar regions, leaving behind work for the wind. But the story of radiation is complicated and perhaps even as difficult as that of water. We have tried to emphasize its role, and yet have not attempted to explain the details of radiative processes. These two processes—phase changes of water and radiative heating—set meteorology apart from the study of simpler fluids, and they both provide forcing for the wind.

Although we cannot yet construct a practical model of atmospheric motion that includes reasonably exact versions of all the processes in the atmosphere, we can learn a great deal by studying the relations between variables and the properties of simplified models provided by the equations of motion. We turn now, in the final part of the book, to the task of using our few symbols, arranged in the equations of motion, to try to understand the wind.

BIBLIOGRAPHIC NOTES

8.1 Standard material.

8.2–8.4 Most of this material also is standard. Further discussion may be found in the book by Sommerfeld cited in the Bibliographic Notes to Chap. 3 or in most meteorological or thermodynamics texts.

8.5 This section introduces the subject of moist convection and contains standard material.

8.6 The material of this section follows the presentation of phase changes in open systems given in the article by Van Mieghem cited in the Bibliographic Notes to Chap. 3.

PART THREE

The Theory of Atmospheric Motion

No general solution of the equations of atmospheric motion is known or anticipated. Thus the theory of atmospheric motion has been constructed by considering the implications of the equations in a variety of special cases. Particular solutions that would obtain in limited circumstances and models developed with the aid of simplifying assumptions, often based on observational evidence, are both widely used. The theory is fragmented and dependent in part on restriction to phenomena in certain ranges of temporal and spatial scales. A large fraction of it was created by circumventing the nonlinearity of the equations of motion by one means or another. Even so, we have an elegant collection of mathematical models that seem to produce replicas of the actual patterns and processes of atmospheric motion.

9

AIR IN MOTION: MODELS OF THE WINDS

The motions of the atmosphere arise in its quest for thermal equilibrium. Differential heating causes changes in the distribution of mass and temperature, and this creates the imbalances of atmospheric forces that lead to motion. The resulting wind patterns are complex–ranging from the great sweeps of the subpolar jet across continents and oceans to the little gusts that swirl along near the ground or between the waves.

We shall try to capture some of these winds that give the atmosphere both life and mystery, and attempt to abstract their evanescent form and patterns with models that portray their main features. Such models cannot be an always faithful reproduction of reality, but they do provide conceptual understanding of the atmosphere. They are the starting point for putting our equations to work at the job of helping us comprehend atmospheric motion.

9.1 THE GEOSTROPHIC WIND

Charts depicting patterns in the middle and upper troposphere reveal that the wind is nearly parallel to the isobars, low pressure on the left, and that its speed increases in proportion to the pressure gradient. This observation raises the question of why nature does not fill the partial vacuums of low pressure by having the wind blow directly from regions of high pressure to those of lower pressure.

The answer is that the rotation of the earth prevents this simple sort of flow. For if an air parcel is accelerated into motion toward lower pressure, then the Coriolis force acts upon it, turning it to the right (in the Northern Hemisphere). Thus the parcels of air end up circling around the center of lower pressure. In this section we shall investigate this phenomenon and thus study the most basic model of the wind.

Let us refer to the estimates of the orders of magnitude given in Eq. (16) in Sec. 7.2. If we retain only the dominant terms of order 10^{-1} cm/s^2 in this equation and the equivalent terms in the equation for the v component, then we will have the two equations

$$fv = \frac{1}{\rho}\frac{\partial p}{\partial x} \qquad fu = -\frac{1}{\rho}\frac{\partial p}{\partial y} \tag{1}$$

With these approximate results as a motivation, we may use vector notation and *define* the *geostrophic wind* as

$$\mathbf{v}_g = \frac{1}{\rho f}\mathbf{k} \times \nabla_z p \qquad f \neq 0 \tag{2}$$

The name is derived from the Greek *geo-* implying "earth" and *strophe* implying a "turning."

Now the definition (2) can always be used to replace the pressure gradient with an equivalent velocity; the question is under what conditions the geostrophic wind is a good approximation to the actual horizontal wind. By taking a cross product with the unit vector **k**, we may write the pressure gradient force as

$$-\frac{1}{\rho}\nabla_z p = f\mathbf{k} \times \mathbf{v}_g \tag{3}$$

and so the equation for the horizontal component of the wind, now denoted by **v**, may be written with a friction term, **F**, as

$$\frac{d\mathbf{v}}{dt} = f\mathbf{k} \times (\mathbf{v}_g - \mathbf{v}) + \mathbf{F} \tag{4}$$

Clearly then, the actual wind will be geostrophic whenever $d\mathbf{v}/dt - \mathbf{F} = 0$, and a sufficient condition for this to be true is that the parcels are unaccelerated and that the frictional force vanishes. These two conditions are rarely met exactly in the atmosphere, but they are often approximately verified a kilometer or more above the earth's surface and so the flow there is often nearly geostrophic.

The concept of geostrophic flow is important in meteorology for it provides a direct relation between the mass field and a hypothetical velocity field which is often close to the actual one. Indeed, we have arrived at the simplest nontrivial solution to the equations of large-scale motion, and one that has been central to meteorological thought. If the vertical distribution of pressure and density obeys the hydrostatic relation, then a geostrophic current, constant with longitude but varying in latitude and altitude, is a solution (see Problem 9.1.1 below). This solution represents the basic balance present in the large-scale flow of middle latitudes. The solution does not actually appear, however, for the differential heating induces continual readjustments in the mass field, and hence in the geostrophic wind.

Moreover, slight lapses from geostrophic equilibrium will lead to changes in the entire motion field. This can be seen in Eq. (4), where departures from geostrophic equilibrium will in general produce accelerations. Thus it is the deviations from geostrophic winds that are important in the evolution of the atmosphere's motions.

The recent history of dynamic meteorology involves attempts to use as much of the geostrophic concept as is possible without producing patently incorrect results. As will be shown, care in applying the assumption of geostrophic balance will lead to satisfactory results; indiscriminate use of the assumption yields nonsense.

*PROBLEM 9.1.1 Consider the system of equations (18) and (19) in Sec. 7.2. Let $u = u_g(y,z)$, $v = w = 0$, $p = p(y,z)$, $T = T(y,z)$, $\nabla \cdot \mathbf{R} = 0$. Show that this geostrophic, zonal current is a solution for the system. Show that this current is not a solution if $\nabla \cdot \mathbf{R} \neq 0$.

To develop some of the relations between the geostrophic wind and the actual wind, it is convenient to assume that friction causes acceleration in the direction opposite to the wind and in proportion to its speed so that $\mathbf{F} = -k\mathbf{v}$, a situation we shall refer to as the assumption of *linear friction.*

*PROBLEM 9.1.2 Assume that the actual wind is unaccelerated and that linear friction applies with $k \ll f$. Show that (1) the actual wind is directed slightly to the left of the geostrophic wind; (2) the actual wind speed is less than the geostrophic speed. (Hint: Solve for u and v as functions of u_g and v_g; orient the coordinate axes so that $v_g = 0$.)

So far we have examined the geostrophic wind only in cartesian coordinates and have seen that it depends both on the pressure gradient field and the density. In the isobaric and isentropic coordinate systems studied in Chap. 7, the expressions are somewhat simpler. From the equations of motion in these coordinate systems we see immediately that the geostrophic wind is given by

Isobaric: $$\mathbf{v}_g = \frac{g}{f}\mathbf{k} \times \nabla_p h_p \tag{5}$$

and

Isentropic: $$\mathbf{v}_g = \frac{1}{f}\mathbf{k} \times \nabla_\theta \Psi \tag{6}$$

The horizontal divergence and vorticity of the wind are important in a variety of meteorological contexts, and an important question is whether these properties of the geostrophic wind can be used as approximations for the actual values. In practice it is difficult to compute both vorticity and divergence from the reported wind observations owing to lack of precision in the data and to the large spacing between the stations. But if the geostrophic values could be used, then we would be computing the divergence and vorticity from the height or pressure fields, which are more accurately measured and more easily analyzed. The divergence is the more difficult of the two to measure from the wind, but, as we shall show now, it cannot be replaced by the geostrophic divergence. In contrast, the geostrophic vorticity is a useful approximation to the actual vorticity.

In computing the divergence, we shall have, for Φ representing p, h_p, or Ψ, and ∇_2 representing the appropriate two-dimensional del operator,

$$\nabla_2 \cdot (\mathbf{k} \times \nabla_2 \Phi) = -\mathbf{k} \cdot [\nabla_2 \times (\nabla_2 \Phi)] = 0 \tag{7}$$

Thus the results are

Cartesian:
$$\nabla_z \cdot \mathbf{v}_g = -\left(\frac{1}{\rho^2 f} \nabla_z \rho + \frac{1}{\rho f^2} \nabla_z f\right) \cdot \mathbf{k} \times \nabla_z p$$
$$= -\left(\frac{1}{\rho} \nabla_z \rho + \frac{1}{f} \frac{\partial f}{\partial y} \mathbf{j}\right) \cdot \mathbf{v}_g \tag{8}$$

Isobaric:
$$\nabla_p \cdot \mathbf{v}_g = -\frac{1}{f} \frac{\partial f}{\partial y} \mathbf{j} \cdot \mathbf{v}_g = -\frac{1}{f} \frac{\partial f}{\partial y} v_g \tag{9}$$

and

Isentropic:
$$\nabla_\theta \cdot \mathbf{v}_g = -\frac{1}{f} \frac{\partial f}{\partial y} \mathbf{j} \cdot \mathbf{v}_g = -\frac{1}{f} \frac{\partial f}{\partial y} v_g \tag{10}$$

We observe again that $\partial/\partial y = \partial/a\ \partial\phi$ so that

$$\frac{1}{f} \frac{\partial f}{\partial y} = \frac{\cot \phi}{a} \tag{11}$$

Thus we have shown that the divergence of the geostrophic wind, in the isobaric or isentropic coordinates generally used in meteorological analysis, results only from the convergence of poleward flow on the spherical earth or the divergence of equatorward flow. But the patterns of divergence and convergence associated with long waves and cyclones have a more complicated structure and are an important part of atmospheric dynamics. Thus we must conclude that the divergence of the geostrophic wind is *not* a good approximation for the divergence of the actual wind.

In the calculation of the geostrophic vorticity we shall have [Eq. (37) in Sec. 5.1]

$$\nabla \times (\mathbf{k} \times \nabla_2 \Phi) = \mathbf{k}\, \nabla_2{}^2 \Phi - (\mathbf{k} \cdot \nabla)\, \nabla_2 \Phi \tag{12}$$

Thus when we calculate the vertical component of the geostrophic vorticity

$$\zeta_g = \mathbf{k} \cdot (\nabla \times \mathbf{v}_g) \tag{13}$$

we obtain $\nabla_2{}^2\Phi$ on the right of Eq. (12) and $\mathbf{k}\cdot(\mathbf{k}\cdot\nabla)\,\nabla_2\Phi$ vanishes. Thus

$$\text{Cartesian:}\qquad \mathbf{k}\cdot(\nabla\times\mathbf{v}_g)=\frac{1}{\rho f}\nabla_z{}^2 p+\mathbf{k}\cdot\left[\mathbf{v}_g\times\left(\frac{1}{\rho}\nabla_z\rho+\mathbf{j}\frac{\cot\phi}{a}\right)\right] \tag{14}$$

$$\text{Isobaric:}\qquad \mathbf{k}\cdot(\nabla\times\mathbf{v}_g)=\frac{g}{f}\nabla_p{}^2 h_p+\mathbf{k}\cdot\left(\mathbf{v}_g\times\mathbf{j}\frac{\cot\phi}{a}\right) \tag{15}$$

$$=\frac{g}{f}\nabla_p{}^2 h_p+u_g\frac{\cot\phi}{a}$$

and

$$\text{Isentropic:}\qquad \mathbf{k}\cdot(\nabla\times\mathbf{v}_g)=\frac{1}{f}\nabla_\theta{}^2\Psi+u_g\frac{\cot\phi}{a} \tag{16}$$

These geostrophic vorticity expressions give results that are in good agreement with the actual vorticity for large-scale flow; in fact, the first numerical weather prediction schemes were based upon them. In these expressions, the last terms with the earth's radius are usually small and are neglected.

9.2 SCALE ANALYSIS AND THE GEOSTROPHIC APPROXIMATION

One of the most effective means of determining the conditions under which an approximation may be useful is to analyze the relative size of the terms in the relevant equations. The horizontal equation of motion, upon using the assumption that the kinematic and expansion viscosities are equal (see Sec. 6.8), becomes

$$\frac{\partial\mathbf{v}_H}{\partial t}+\mathbf{v}\cdot\nabla\mathbf{v}_H=f\mathbf{k}\times(\mathbf{v}_g-\mathbf{v}_H)+\nu\,\nabla^2\mathbf{v}_H+\nu\,\nabla_z(\nabla\cdot\mathbf{v}) \tag{1}$$

in which $\mathbf{v}_H$ is the horizontal component of the wind velocity vector

$$\mathbf{v}=\mathbf{v}_H+\mathbf{k}w \tag{2}$$

Let us define the quantities V and W as typical magnitudes or amplitudes of $\mathbf{v}_H$ and w, and let us consider flow patterns with horizontal wavelengths of size L, vertical wavelengths of a typical size D, and period T.† Then we can define nondimensional variables

$$\mathbf{u}^*=\frac{\mathbf{v}_H}{V}\qquad x^*=\frac{x}{L} \tag{3}$$

†Strictly speaking, the quantities L, D, and T should be somewhat less than the wavelength or period since we are estimating derivatives. If ϕ is harmonic, then $|\partial\phi/\partial x|\cong(2\pi/L_x)|\phi|$. Thus we set this derivative equal to $|\phi|/L$ so that $L=L_x/2\pi$, where L_x is the actual wavelength and we refer to L as a scale. The details are discussed in Chaps. 12 and 13, where it is shown that there is good reason to take $L=L_x/4$. In making numerical estimates, we shall treat L, D, and T as scales.

$$w^* = \frac{w}{W} \quad y^* = \frac{y}{L}$$

$$\mathbf{u}_g^* = \frac{\mathbf{v}_g}{V} \quad z^* = \frac{z}{D}$$

$$t^* = \frac{t}{T} \qquad (3)\ (cont'd)$$

Upon making these substitutions in Eq. (1) and multiplying by $(Vf)^{-1}$, we find

$$\frac{1}{fT}\frac{\partial \mathbf{u}^*}{\partial t^*} + \frac{V}{fL}\mathbf{u}^* \cdot (\nabla_z \mathbf{u})^* + \frac{W}{fD}w^*\frac{\partial \mathbf{u}^*}{\partial z^*} = \mathbf{k} \times (\mathbf{u}_g^* - \mathbf{u}^*) + \frac{\nu}{f}\left[\frac{1}{L^2}(\nabla_z{}^2\mathbf{u})^* + \frac{1}{D^2}\frac{\partial^2 \mathbf{u}^*}{\partial z^{*2}}\right]$$
$$+ \frac{\nu}{f}\left\{\frac{1}{L^2}[\nabla_z(\nabla_z \cdot \mathbf{u})]^* + \frac{W}{DLV}\left(\nabla_z \frac{\partial w}{\partial z}\right)^*\right\} \qquad (4)$$

Two of the factors in this equation appear often and have been named after the scientists who studied their implications. The *Rossby number Ro* is defined as

$$Ro = \frac{V}{fL} \qquad (5)$$

and the Reynolds number *Re* as

$$Re = \frac{VL}{\nu} \qquad (6)$$

Now we may put the equation in the form

$$\frac{1}{fT}\frac{\partial \mathbf{u}^*}{\partial t^*} + Ro\mathbf{u}^* \cdot (\nabla_z \mathbf{u})^* + \frac{W}{V}\frac{L}{D}Ro w^*\frac{\partial \mathbf{u}^*}{\partial z^*} = \mathbf{k} \times (\mathbf{u}_g^* - \mathbf{u}^*) + \frac{Ro}{Re}(\nabla_z{}^2\mathbf{u})^*$$
$$+ \frac{L^2}{D^2}\frac{Ro}{Re}\frac{\partial^2 \mathbf{u}^*}{\partial z^{*2}} + \frac{Ro}{Re}(\nabla_z\nabla_z \cdot \mathbf{u})^* + \frac{W}{V}\frac{L}{D}\frac{Ro}{Re}\nabla_z\frac{\partial w}{\partial z}^* = 0 \qquad (7)$$

If we apply the equation to motions whose sizes are approximately those of the scaling parameters, we may assume that all the nondimensional variables are of order unity in magnitude. Thus the importance of the various terms is revealed by the size of the nondimensional numbers that appear. For large-scale flow in the middle atmosphere we have the estimates

$$Ro = \frac{V}{fL} \sim \frac{10\ \text{m/s}}{10^{-4}\ \text{s}^{-1}\ 10^6\ \text{m}} \sim 0.1$$

$$\frac{WL}{VD} \sim \frac{(0.01 - 0.10)\text{m/s}\ 10^6\ \text{m}}{10\ \text{m/s}\ 5 \times 10^3\ \text{m}} \sim 0.2 - 2.0$$

$$Re \sim \frac{VL}{\nu} = \frac{10\ \text{m/s}\ 10^6\ \text{m}}{0.15 \times 10^{-4}\ \text{m}^2/\text{s}} \sim 6.6 \times 10^{11}$$

$$\frac{D^2 Re}{L^2} \sim \left(\frac{5}{1000}\right)^2 6.6 \times 10^{11} \sim 1.6 \times 10^7$$

$$|\mathbf{u}_g^*| \sim |\mathbf{u}^*| \sim 1$$

Thus for large-scale flow in the atmosphere we may apparently drop the friction terms because of the immense size of the Reynolds number. Furthermore if the Rossby number is small, we are left with only

$$\frac{1}{fT}\frac{\partial \mathbf{u}^*}{\partial t^*} = \mathbf{k} \times (\mathbf{u}_g^* - \mathbf{u}^*) \tag{8}$$

and thus for time scales T such that

$$T \gg \frac{1}{f} \sim \frac{24\text{ h}}{4\pi \sin\phi} \sim \frac{6\text{ h}}{\pi \sin\phi} \sim \frac{2\text{ h}}{\sin\phi} \tag{9}$$

the first term is also small, and the flow is approximately geostrophic.

The criteria for the applicability of the geostrophic assumption, then, are that Ro be small, that Re be large, and that the flows be slowly varying. There are two cases where it is obviously inappropriate. The first is the tropics, where the Coriolis parameter becomes very small and vanishes at the equator. The second is small-scale motion such as atmospheric turbulence, which is not affected significantly by the rotation of the earth.

Let us consider the physical meaning of these nondimensional numbers. The Rossby number is essentially the ratio of the inertial term $\mathbf{v}_H \cdot \nabla \mathbf{v}_H$ to the Coriolis term because, using $|\ |_M$ to denote order of magnitude, we have

$$\left|\frac{\mathbf{v}_H \cdot \nabla \mathbf{v}_H}{f\mathbf{k} \times \mathbf{v}_H}\right|_M = \frac{V^2/L}{fV} = \frac{V}{fL} = Ro \tag{10}$$

while the Reynolds number is a ratio of the inertial term to the viscous term

$$\left|\frac{\mathbf{v}_H \cdot \nabla \mathbf{v}_H}{\nu \nabla_H{}^2 \mathbf{v}_H}\right|_M = \frac{V^2/L}{\nu V/L^2} = \frac{VL}{\nu} = Re \tag{11}$$

The largest of the four friction terms in Eq. (7) is $[Ro(D^2 Re/L^2)^{-1}]\ \partial^2 \mathbf{u}^*/\partial z^{*2}$, which suggests that the appropriate Reynolds number is $D^2 Re/L^2$, a form that takes account of the fact that vertical derivatives are much larger than those in the horizontal directions.

We shall return to a more thorough examination of approximate equations of motion and their dependence on the Rossby number in Chap. 14.

PROBLEM 9.2.1 Determine a sufficient condition for the Coriolis force term $-2\Omega \times \mathbf{v}$ to be neglected in the equation of motion (18) in Sec. 7.1.

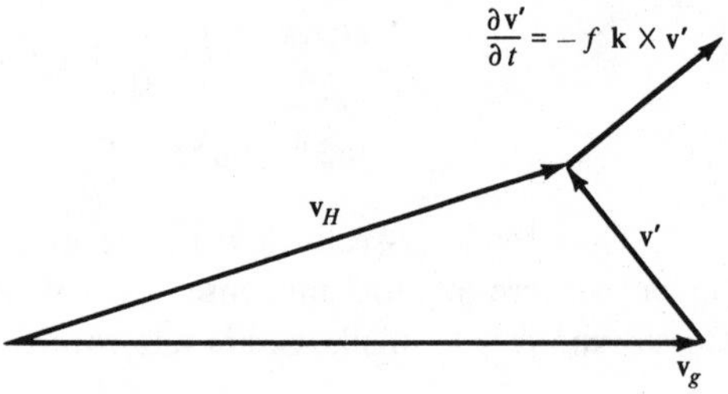

FIGURE 9.1
Illustration of Eq. (3), showing how the ageostrophic component $\mathbf{v}'$ will be accelerated toward its right.

PROBLEM 9.2.2 It is often said that the geostrophic velocity becomes infinite at the equator. Is this strictly correct? [Hint: Compare Eqs. (1) and (2) in Sec. 9.1.]

9.3 INERTIAL OSCILLATIONS AND GEOSTROPHIC ADJUSTMENT WITH LINEAR FRICTION

The analysis of the preceding section showed that when the Rossby number is sufficiently small, the equation of motion may be written as

$$\frac{\partial \mathbf{v}_H}{\partial t} = f\mathbf{k} \times (\mathbf{v}_g - \mathbf{v}_H) \tag{1}$$

where we have restored Eq. (8) in Sec. 9.2 to its original form in the notation of Eq. (1) in Sec. 9.2. We may define the *ageostrophic component* of the wind as

$$\mathbf{v}' = \mathbf{v}_H - \mathbf{v}_g \tag{2}$$

and if $\mathbf{v}_g$ is steady ($\partial \mathbf{v}_g/\partial t = 0$), then Eq. (1) may be written as

$$\frac{\partial \mathbf{v}'}{\partial t} = -f\mathbf{k} \times \mathbf{v}' \tag{3}$$

The equation is illustrated in Fig. 9.1, and it is clear that $\mathbf{v}'$ will be accelerated toward its right. We obviously have

$$\mathbf{v}' \cdot \frac{\partial \mathbf{v}'}{\partial t} = -f\mathbf{v}' \cdot (\mathbf{k} \times \mathbf{v}') = 0 \tag{4}$$

so that $\mathbf{v}'$ does not change in magnitude. Thus the tip of the vector $\mathbf{v}'$ will go around in a circle, and hence $\mathbf{v}_H$ will oscillate to either side of the geostrophic vector.

PROBLEM 9.3.1 Solve Eq. (3) subject to initial conditions $u'(0) = u_0'$ and $v'(0) = v_0'$. Show that

$$u' = u_0' \cos ft - v_0' \sin ft$$
$$v' = v_0' \cos ft + u_0' \sin ft$$

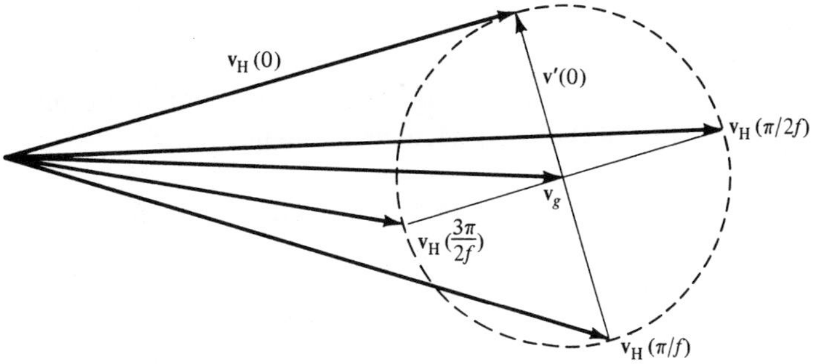

FIGURE 9.2
Schematic illustration of the changes in the true wind vector $\mathbf{v}_H$ as the ageostrophic component $\mathbf{v}'$ executes the rotation of the inertial oscillation in a period of time equal to $2\pi/f$.

[Hint: Obtain the component equations of (3) and then differentiate to find a second-order equation in one of the velocities.]

As shown by this problem, the ageostrophic vector completes one revolution as t advances an amount

$$\Delta t = \frac{2\pi}{f} = \frac{2\pi}{2\Omega \sin\phi} = \frac{2\pi}{4\pi \sin\phi/24\text{ h}} = \frac{12\text{ h}}{\sin\phi} \tag{5}$$

The time required for a complete rotation of a Foucault pendulum is

$$T_F = \frac{2\pi}{\Omega \sin\phi} = \frac{24\text{ h}}{\sin\phi} \tag{6}$$

Thus the period of the rotation of the ageostrophic component is exactly half the period of the Foucault pendulum and is therefore referred to as a *half pendulum day*. This oscillation of the ageostrophic wind in response to the Coriolis force is referred to as an *inertial oscillation* and is depicted in Fig. 9.2. Pure inertial oscillations are rarely found in the atmosphere, presumably because they are damped by friction or because other forces also control the evolution of the motion. In addition, inertial oscillations may interact with phenomena with different time scales and thus be masked.

To determine what effect friction might have, we again assume that the frictional force is linear in the velocity, that $\partial \mathbf{v}_g/\partial t = 0$, and we write the equation as

$$\frac{\partial \mathbf{v}'}{\partial t} + f\mathbf{k} \times \mathbf{v}' + k(\mathbf{v}' + \mathbf{v}_g) = 0 \tag{7}$$

The component equations are

$$\frac{\partial u'}{\partial t} + ku' - fv' + ku_g = 0 \qquad \frac{\partial v'}{\partial t} + kv' + fu' + kv_g = 0 \tag{8}$$

and we could obtain a second-order equation exactly as in Problem 9.3.1, but a simpler procedure is available. Let us multiply the second equation by $i\,(=\sqrt{-1})$ and notice that $i(u' + iv') = iu' - v'$. The definition of the complex quantity $W = u' + iv'$ allows us to add the two equations to obtain

$$\frac{\partial W}{\partial t} + (k + if)W + kW_g = 0 \qquad W_g = u_g + iv_g \tag{9}$$

This first-order equation has the solution

$$W(t) = e^{-(k+if)t}W(0) + kW_g\left(\frac{e^{-(k+if)t} - 1}{k + if}\right) \tag{10}$$

Observe that as t goes to infinity, the factor exp $(-kt)$ will vanish so that

$$W(\infty) = -\frac{kW_g}{k + if} = -\frac{kW_g(k - if)}{k^2 + f^2}$$
$$= -\frac{k^2u_g + kfv_g + i(k^2v_g - kfu_g)}{k^2 + f^2} \tag{11}$$

Thus we find for the *total velocity* $\mathbf{u}$ that

$$u(\infty) = \frac{f^2u_g - kfv_g}{k^2 + f^2} \qquad v(\infty) = \frac{f^2v_g + kfu_g}{k^2 + f^2} \qquad \mathbf{u} = \mathbf{u}_g + \mathbf{u}' \tag{12}$$

which agrees with the solution to Problem 9.1.2.

Thus whatever the initial conditions, with linear friction the ageostrophic component will have an oscillation, of frequency f, which tends with exponential damping from the initial conditions toward the steady-state solution (12).

PROBLEM 9.3.2 Determine the component equations of (10).

9.4 INERTIAL INSTABILITY

Now we turn to the question of how a parcel, originally in geostrophic balance, will behave when its motion is perturbed and it is thrust into a region where the forces constituting the geostrophic balance are other than they were at the original position of the parcel. Thus we shall study horizontal displacement of parcels, in analogy with the vertical displacements we studied at the end of Chap. 4. We shall find that the horizontal wind shear can provide destabilization, just as vertical wind shear can for vertical displacements.

The problem is most appropriately treated in isentropic coordinates, because it is convenient to assume that the displaced parcel moves isentropically, conserving its initial potential temperature. We start with a zonal current in geostrophic balance, so that $\mathbf{v}_g = \mathbf{i}u_g(y,\theta)$. Thus the flow varies in the north-south direction and (in the vertical) with potential temperature. In these coordinates, $f\mathbf{v}_g = \mathbf{k} \times \nabla_\theta \Psi$, and the equations of motion for a parcel in the flow are

$$\frac{du}{dt} - fv = 0 \qquad \frac{dv}{dt} + f(u - u_g) = 0 \qquad v_g = 0 \tag{1}$$

Now let us assume that a parcel can be displaced and move through the geostrophic flow without disturbing it. Suppose we start with a northward displacement from y_0 (the same results are obtained for a southward displacement) so that initially $dv/dt > 0$ and $v > 0$. Then the speed u of the parcel will increase from its initial value of $u_g(y_0,\theta)$ at y_0. As an approximation, we let the eastward speed of the parcel at $y_0 + \Delta y$ be $u = u_g(y_0,\theta) + f_0 v_0 \,\Delta t$. The geostrophic current at the same point has speed $u_g(y + \Delta y,\theta) = u_g(y_0,\theta) + (\partial u_g/\partial y)_0 v_0 \,\Delta t$. Thus from Eq. (1) we see that $dv/dt \gtreqless 0$ at $y_0 + \Delta y$ if

$$u_g(y_0 + \Delta y, \theta) - u(y_0 + \Delta y, \theta) = \left[\left(\frac{\partial u_g}{\partial y}\right)_{y_0,\theta} - f_0\right] v_0 \,\Delta t \gtreqless 0 \tag{2}$$

We started with $v_0 > 0$, so that if $dv/dt > 0$, we would say the parcel is unstable because it is accelerating away from its initial position; if $dv/dt < 0$, it is stable because it is decelerating after the initial perturbation. Thus Eq. (2) gives the criteria for inertially unstable motion, inertially neutral motion, and inertially stable motion respectively.

These stability criteria are of most interest in the vicinity of the jet axis where shears are strong. In a region without a jet, a typically large shear might be 10 m/s in 100 km or $\partial u_g/\partial y \sim 10^{-4}$ s^{-1} $\sim f$. Thus regions of the atmosphere without jets are probably inertially stable or inertially neutral. On the north side of a jet axis, the winds decrease with increasing latitude and so $\partial u_g/\partial y < 0$, and this region is also stable. But on the south side, $\partial u_g/\partial y > 0$, so that if the shears are strong, the region could be inertially unstable. We might thus expect that the south sides of strong jets could be regions of meteorological activity.

The simple approach given above is valid only as a *linear* approximation near y_0. If the parcel is displaced a significant distance, the Coriolis parameter will change, and we must take this into account. To do so, we will convert Eq. (1) into one differential equation in the velocity v. Let us differentiate the v equation again to obtain

$$\frac{d^2v}{dt^2} + \frac{df}{dt}(u - u_g) + f\left(\frac{du}{dt} - \frac{du_g}{dt}\right) = 0 \tag{3}$$

Now $f = 2\Omega \sin \phi$, and from Chap. 7 we know that $\partial/\partial y = \partial/a\,\partial\phi$, so that

$$\frac{df}{dt} = \frac{2\Omega\, v \cos\phi}{a} = v\beta \qquad \beta = a^{-1}\frac{\partial f}{\partial \phi} \tag{4}$$

because f is not a function of t, x, or θ. We have agreed to consider only isentropic displacements and zonal currents that are not changing in time or with longitude, so that following the displaced parcel we have

$$\frac{du_g}{dt} = v\left(\frac{\partial u_g}{\partial y}\right)_\theta \tag{5}$$

Upon substituting for du/dt and $u - u_g$ from Eq. (1) and using Eqs. (4) and (5), Eq. (3) becomes

$$\frac{d^2v}{dt^2} - \frac{\beta}{f}v\frac{dv}{dt} + vf\left[f - \left(\frac{\partial u_g}{\partial y}\right)_\theta\right] = 0 \tag{6}$$

Now we observe that $\beta/f = \cot\phi/a$ and thus

$$\frac{d^2v}{dt^2} - \frac{\cot\phi}{a}v\frac{dv}{dt} + vf\left[f - \left(\frac{\partial u_g}{\partial y}\right)_\theta\right] = 0 \tag{7}$$

This is a badly nonlinear equation, because the velocity $v = a\,d\phi/dt$ is a rate of change of latitude and $\cot\phi$, f, and $(\partial u_g/\partial y)_\theta$ are all functions of latitude.

We convert the equation into one involving the latitude of the displaced parcel and obtain

$$\frac{d^3\phi}{dt^3} - \cot\phi\,\frac{d\phi}{dt}\frac{d^2\phi}{dt^2} + \frac{d\phi}{dt}f\left[f - \left(\frac{\partial u_g}{\partial y}\right)_\theta\right] = 0 \tag{8}$$

This equation is closer to a standard Sturm-Liouville form when it is rewritten as

$$\frac{d}{dt}\left(\csc\phi\,\frac{d^2\phi}{dt^2}\right) + 2\Omega\frac{d\phi}{dt}\left[f - \left(\frac{\partial u_g}{\partial y}\right)_\theta\right] = 0 \tag{9}$$

As in Sec. 4.4.2, we shall not attempt to obtain a solution to this equation; rather we use the classical technique associated with Sturm-Liouville equations for analyzing stability indirectly. Upon multiplying by $d\phi/dt$, we may put the equation in the form

$$\frac{d}{dt}\left[\frac{\csc\phi}{2}\frac{d}{dt}\left(\frac{d\phi}{dt}\right)^2\right] - \csc\phi\left(\frac{d^2\phi}{dt^2}\right)^2 + 2\Omega\left(\frac{d\phi}{dt}\right)^2\left[f - \left(\frac{\partial u_g}{\partial y}\right)_\theta\right] = 0 \tag{10}$$

in which we used

$$\frac{d\phi}{dt}\frac{d^2\phi}{dt^2} = \frac{1}{2}\frac{d}{dt}\left(\frac{d\phi}{dt}\right)^2 \tag{11}$$

Now we can integrate from an initial time t_0 to a later time t to obtain

$$\frac{\csc\phi}{2}\frac{d}{dt}\left(\frac{d\phi}{dt}\right)^2 = \left[\frac{\csc\phi}{2}\frac{d}{dt}\left(\frac{d\phi}{dt}\right)^2\right]_{t_0} + \int_{t_0}^{t}\left\{\csc\phi\left(\frac{d^2\phi}{d\tau^2}\right)^2 - 2\Omega\left(\frac{d\phi}{d\tau}\right)^2\left[f - \left(\frac{\partial u_g}{\partial y}\right)_\theta\right]\right\} d\tau \tag{12}$$

To aid in the interpretation of our result, let us restore the velocity. This produces

$$\frac{\csc\phi}{2}\frac{dv^2}{dt} = \left(\frac{\csc\phi}{2}\frac{dv^2}{dt}\right)_{t_0} + \int_{t_0}^{t}\left[\csc\phi\left(\frac{dv}{d\tau}\right)^2 - 2\Omega v^2\left(f - \frac{\partial u_g}{\partial y}\right)_\theta\right] d\tau \tag{13}$$

We have obtained an equation relating the rate of change of the specific kinetic energy of the parcel to a series of terms that involve the initial increase in kinetic energy and positive factors except for $f-(\partial u_g/\partial y)_\theta$. Since the parcel is assumed initially in equilibrium, it has $v(t_0)=0$, and so if perturbed either northward or southward, $(dv^2/dt)_{t_0} > 0$. Thus the motion will be stable if the initial kinetic energy is lost, or in other words, if $dv^2/dt < 0$. If $dv^2/dt > 0$, then the parcel continues to gain kinetic energy, and its motion will be considered unstable.

Notice in Eq. (13) that if

$$\left[\left(\frac{\partial u_g}{\partial y}\right)_\theta - f\right] > 0 \tag{14}$$

then every term is positive and the motion is unstable, in agreement with the linear result (2). However, it does not appear possible to obtain a simple criterion that guarantees stability in the nonlinear case, because this depends in part on how the displacement is initiated. It is of course possible to use Eq. (13) to conclude that if the displacement is stable, then it must be true that

$$\left[f - \left(\frac{\partial u_g}{\partial y}\right)_\theta\right] > 0 \tag{15}$$

somewhere along the trajectory. This is thus a necessary, but not sufficient, condition for stability. Finally, we note that if $f-(\partial u_g/\partial y)_\theta = 0$ all along the trajectory, then the motion will be unstable.

PROBLEM 9.4.1 The transformation from Eq. (8) to Eq. (9) is found by multiplying Eq. (8) by an arbitrary function h, converting the first term to $d(hd^2\phi/dt^2)/dt$, and then determining a function h so that the remaining terms in $d^2\phi/dt^2$ vanish. Work out the details, and show that Eq. (9) follows.

PROBLEM 9.4.2 Suppose the motion of the parcel is not isentropic. Can nonlinear stability criteria then be obtained? [Hint: The result (5) must be modified.]

9.5 VERTICAL SHEAR OF V_g: THE THERMAL WIND

The dominant influence in the atmosphere is the differential heating that produces warm equatorial regions and cold polar regions, and a consequent latitudinal variation in the mass field. These variations would be expected to have a strong influence on the winds of the atmosphere, and they do: among other effects, they control the *vertical variation* of the geostrophic wind field. Thus an explanation of the main features of the vertical structure of the atmosphere may be found in the horizontal variations of the temperature field.

As an elementary approach, let us consider a surface pressure field decreasing at a uniform rate from a southerly warm region to a cold region in the north (Fig. 9.3). From the hydrostatic equation we know that

$$\frac{1}{p}\frac{\partial p}{\partial z} = -\frac{g}{RT} \tag{1}$$

and so the pressure will decrease more rapidly with height in the cold region than it will in the warm. Thus at an upper level, the horizontal pressure gradient will be stronger than at the surface and we would expect that the westerly geostrophic winds will increase in speed with increasing height. Notice that if the temperature contrast is strong enough, then even easterly surface geostrophic winds would be turned around to become strong westerlies at a sufficient distance from the surface. The main concept of this section and its implication for the atmosphere is summarized in the phrase: "The westerlies increase with height because it's colder toward the poles."†

9.5.1 The Taylor-Proudman Theorem

Rather than proceed directly to analyze the effect of temperature gradients on the geostrophic wind in the atmosphere, let us study first a simpler case–one of interest because in contrast to the atmosphere, there are no vertical variations of the geostrophic wind.

We consider steady motion of a rotating, incompressible fluid of constant density. A container of water placed at the center of a turntable would be an example. The equation of motion is

$$\mathbf{v} \cdot \nabla \mathbf{v} = -\frac{1}{\rho_0}\nabla p - g\mathbf{k} - 2\mathbf{\Omega} \times \mathbf{v} - \nu \nabla^2 \mathbf{v} \tag{2}$$

in which we assume the axis of rotation is parallel to the vertical direction.

†An apt summary used in lectures by Prof. Reid A. Bryson.

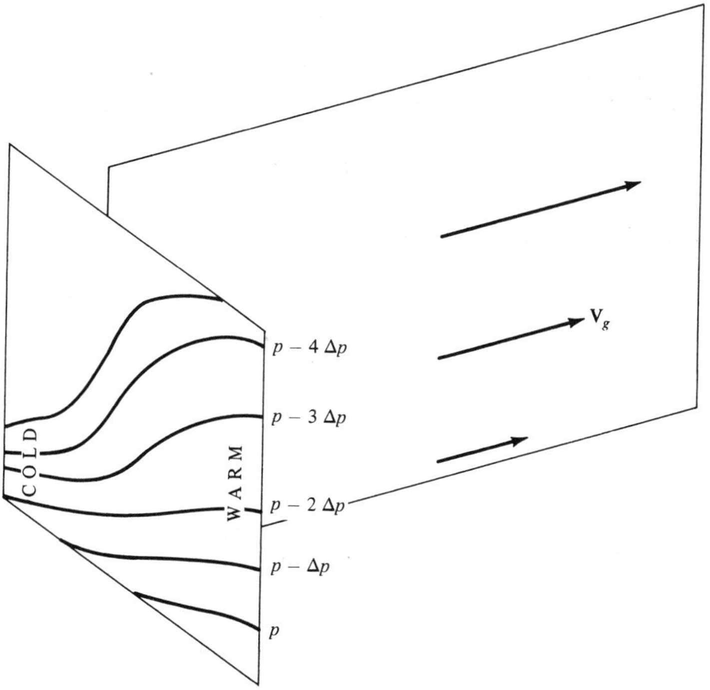

FIGURE 9.3
The latitudinal temperature gradient causes an increase in the latitudinal pressure gradient with increasing height, and thus the geostrophic wind speed also increases with height.

Applying the scale analysis of Sec. 9.2, we find that the nonlinear term $\mathbf{v} \cdot \nabla \mathbf{v}$ may be ignored in comparison with $2\boldsymbol{\Omega} \times \mathbf{v}$ if the Rossby number

$$Ro \sim \left| \frac{\mathbf{v} \cdot \nabla \mathbf{v}}{2\boldsymbol{\Omega} \times \mathbf{v}} \right|_M \sim \frac{v}{2\Omega L} = \frac{v}{2v_{\text{tan}}} \tag{3}$$

is small. Because $\Omega L = v_{\text{tan}}$ is of the order of the tangential velocity of the rotating fluid, the criterion is that the relative motion be small in comparison with the linear speeds due to rotation. Similarly we have a Reynolds number

$$\frac{1}{Re} \sim \left| \frac{\nu \nabla^2 \mathbf{v}}{2\boldsymbol{\Omega} \times \mathbf{v}} \right|_M \sim \frac{\nu}{2\Omega L^2} \sim \frac{\nu}{2v_{\text{tan}} L} \tag{4}$$

which, if large, allows us to neglect the friction term. Thus we consider flows with large Reynolds number and small Rossby number.

With the definition (recall $\rho_0 = \text{const}$) $P = p/\rho_0 + gz$, the equation of motion becomes

$$\nabla P = -2\boldsymbol{\Omega} \times \mathbf{v} \tag{5}$$

Because $\boldsymbol{\Omega} = \Omega\mathbf{k}$, it must always be true that $\mathbf{k} \cdot (2\boldsymbol{\Omega} \times \mathbf{v}) = 0$ so that $\mathbf{k} \cdot \nabla P = 0$ and this implies that

$$\frac{\partial P}{\partial z} = \frac{1}{\rho_0}\frac{\partial p}{\partial z} + g = 0 \tag{6}$$

which is the hydrostatic equation. Furthermore

$$\frac{\partial}{\partial z}(\nabla_z P) = \frac{\partial}{\partial z}\frac{1}{\rho_0}\nabla_z p = \frac{1}{\rho_0}\nabla_z\frac{\partial p}{\partial z} = -\nabla_z g = 0 \tag{7}$$

so that there is no *vertical* variation of the horizontal pressure gradient force.

From $\mathbf{k} \cdot \nabla P = 0$ we can conclude that

$$\nabla_z P = -2\boldsymbol{\Omega} \times \mathbf{v} = -2\boldsymbol{\Omega} \times \mathbf{v}_H \tag{8}$$

and thus Eq. (7) yields

$$2\boldsymbol{\Omega} \times \frac{\partial \mathbf{v}_H}{\partial z} = 0 \tag{9}$$

Thus the motion is invariant with height.

If relative motion is created in a rotating container by heating or by stirring and if an obstacle is placed on the bottom of the tank so that the moving fluid must flow around it, then the streamlines of the flow will form a column, going around the obstacle as if it extended to the top of the water. This phenomenon is called a *Taylor column.*

We have shown that the geostrophic motion of a rotating, constant density fluid does not vary with height. This result is no longer true in a stratified fluid in which $\rho = \rho(z)$, and thus we can anticipate that the arrangement of the stratification of the atmosphere will have a profound effect on its motion.

PROBLEM 9.5.1 What vertical velocity field accompanies the motion governed by the Taylor-Proudman theorem?

9.5.2 The Thermal Wind in the Atmosphere

Under the same assumptions that the Rossby number is small, that the Reynolds number is large, and that the flows are slowly varying, the equations for the large-scale horizontal flow in the atmosphere reduce to

$$\mathbf{v}_H = \mathbf{v}_g = \frac{1}{\rho f}\mathbf{k} \times \nabla_z p \tag{10}$$

Thus for the vertical rate of change of the *geostrophic momentum* we have

$$\frac{\partial}{\partial z}\rho \mathbf{v}_g = \frac{1}{f}\mathbf{k} \times \nabla_z \frac{\partial p}{\partial z} = -\frac{g}{f}\mathbf{k} \times \nabla_z \rho \tag{11}$$

in which we have used the hydrostatic approximation to introduce the variable density that distinguishes the atmospheric case from that of the Taylor-Proudman theorem. Expansion of the term on the left gives

$$\frac{\partial \mathbf{v}_g}{\partial z} = -\frac{g}{\rho f}\mathbf{k} \times \nabla_z \rho - \frac{1}{\rho}\frac{\partial \rho}{\partial z}\mathbf{v}_g \tag{12}$$

With the logarithmic derivative of the equation of state and another use of Eq. (10) and the hydrostatic approximation, we have

$$\frac{\partial \mathbf{v}_g}{\partial z} = \frac{g}{f}\frac{\mathbf{k} \times \nabla_z T}{T} + \frac{1}{T}\frac{\partial T}{\partial z}\mathbf{v}_g \tag{13}$$

The ratio of the two terms is about

$$\frac{|f(\partial T/\partial z)\mathbf{v}_g|}{g\,|\nabla_z T|} \sim \frac{10^{-4}\ \mathrm{s}^{-1}\ (5\ \mathrm{K/km})\ 20\ \mathrm{m/s}}{10\ \mathrm{m/s^2}\ (5\ \mathrm{K/100\ km})} \sim 2 \times 10^{-2} \tag{14}$$

so we have shown, the first term in Eq. (13) being dominant, that the shear of the geostrophic wind is proportional to the horizontal temperature gradient. The geostrophic wind will increase with height at a given level in a direction parallel to the isotherms, with cold air on the left of the shear vector.

As a linear approximation, we may write that

$$\mathbf{v}_g(z + \Delta z) = \mathbf{v}_g(z) + \frac{\partial \mathbf{v}_g}{\partial z}\Delta z = \mathbf{v}_g(z) + \mathbf{v}_T(z, \Delta z) \tag{15}$$

Thus the thermal wind $\mathbf{v}_T(z,\Delta z)$ is a shear vector that gives the change of the geostrophic wind in a height increment Δz.

Meteorologists talk about the thermal wind as though it were an actual wind composed of air in motion. They say, for example, that the thermal wind is strong, that it is westerly, or that it is increasing. They say it blows with cold air on its left, meaning $\mathbf{v}_T$ is parallel to the mean isotherms and directed with respect to them as $\mathbf{v}_g$ is to the isobars. But the thermal wind is composed of a shear vector and a height increment, and just as $\mathbf{v}_g$ provides an economical representation of the pressure gradient forces, so $\mathbf{v}_T$ is a useful representation of the horizontal temperature gradients and their effect upon the wind structure in the vertical. It is not a real wind, but a quantity, with dimensions of a velocity, that represents the horizontal temperature gradient. Nevertheless, it is a quantity useful in estimating actual wind shear.

In Sec. 9.1 we observed that the geostrophic and actual winds would coincide whenever $d\mathbf{v}/dt - \mathbf{F} = 0$. Here we see that the shear of the actual wind will be exactly equal to the shear of the geostrophic wind when $\partial(d\mathbf{v}/dt - \mathbf{F})/\partial z$ vanishes. Thus we can expect the thermal wind to give a good representation of the shear even in some cases where the winds are not approximately geostrophic.

In both isobaric and isentropic coordinates the geostrophic wind is linear in the gradient of the stream function, and consequently the expression for the thermal wind is somewhat simpler than in cartesian coordinates.

PROBLEM 9.5.2 Show that in isobaric coordinates

$$\frac{\partial \mathbf{v}_g}{\partial z} = \frac{g}{fT}\mathbf{k} \times \nabla_p T$$

PROBLEM 9.5.3 Show that in isentropic coordinates

$$\frac{\partial \mathbf{v}_g}{\partial \theta} = \frac{c_p}{f\theta}\mathbf{k} \times \nabla_\theta T$$

PROBLEM 9.5.4 Demonstrate that the thermal wind must vanish in a hydrostatic, barotropic atmosphere.

9.5.3 Applications of the Thermal Wind Concept

The relation between vertical structure and horizontal temperature gradients is one of the most useful concepts available for analyzing and understanding atmospheric structure. In this subsection, we present a number of examples.

The jet stream Because it is generally colder toward the poles in the troposphere, the thermal wind equation leads us to expect that the wind will have an increasing westerly component with height. If the tropospheric temperature gradient were maintained to the top of the atmosphere, the westerlies would continue to increase in speed. But as seen in Sec. 4.6, the gradient is reversed in the stratosphere, with warmer air toward the poles. Thus the geostrophic wind shear reverses, and so the westerlies reach a maximum at the level of the mid-latitude tropopause.

Furthermore, the analysis of the tropospheric temperature field in mid-latitudes reveals that the gradient is concentrated in a baroclinic zone—the polar front. Thus the thermal wind is most pronounced in this baroclinic zone, leading to a latitudinal maximum in the westerlies. The result is that the westerlies reach a maximum speed at the tropopause above the polar front—thus creating the polar-front jet stream.

*PROBLEM 9.5.5 Show that over a continent, an upper-air trough may be expected in winter, an upper-air ridge in summer.

PROBLEM 9.5.6 Determine whether the jet core speeds shown in Figs. 4.7 and 4.8 are consistent with the latitudinal temperature gradient.

Smoothness of patterns aloft Comparison of surface and upper-air meteorological charts shows that the complicated and often contorted patterns of the surface chart become much smoother with height. This is in obvious contrast to the Taylor column phenomenon, and it is another consequence of the fact that it is colder toward the poles.

Whatever the surface wind pattern, the sum of the surface wind vector and the thermal wind will generally produce a vector with a strong westerly component, thus smoothing the surface pattern into one dominated by westerly flow.

To demonstrate this analytically, let us consider a streamline drawn parallel to the wind vectors in horizontal flow. Then this streamline will have a slope in (x,y) coordinates of

$$S = \frac{dy}{dx} = \frac{v}{u} \tag{16}$$

On the assumption that the thermal wind represents the actual wind shear, the slope of the streamline aloft will be

$$S = \frac{v + v_T}{u + u_T} \tag{17}$$

so that if $u_T > 0$ and if $|u_T| \gg |v_T|$, as is usually the case, then there is a reduction in the slope of the streamline with increasing altitude.

PROBLEM 9.5.7 Draw a circular cyclone with a circular anticyclone to the west. Let it be colder to the north, and draw the resulting upper-air pattern.

Vertical alignment of disturbances The basic relation involved in the thermal wind concept allows us to infer, from the horizontal temperature field, the vertical alignment of surface cyclones and anticyclones with troughs and ridges aloft.

From the hypsometric equation (16) in Sec. 4.1, we find immediately that

$$\nabla_p(h_2 - h_1) = -\int_{p_1}^{p_2} \frac{R\,\nabla_p T}{g}\,d\ln p \tag{18}$$

Thus the thickness between two isobaric surfaces will increase most rapidly in the direction toward warm air and decrease most rapidly toward cold air. Hence a surface cyclone may be expected to lean with increasing altitude toward cold air; an anticyclone will lean toward warm air.

In mid-latitudes, the coldest air is usually found to the northwest of the surface cyclone, so that the associated upper-air trough is found aloft to the northwest. As an idealized situation, we consider surface cyclones and anticyclones forming an alternating pattern from west to east with cold air over the anticyclones and warm air over the cyclones. The cyclones would lean back over the anticyclones with increasing height, so that aloft the ridges would lie over the cyclones, the troughs over the anticyclones.

PROBLEM 9.5.8 Do Problem 9.5.7 again, letting the coldest air be northwest of the surface cyclone and northeast of the surface anticyclone.

PROBLEM 9.5.9 Show that a warm-core high intensifies with height; a cold-core high disappears. What are the equivalent rules for low-pressure centers?

Temperature and stability advection The thermal wind concept is useful in weather prediction because knowledge of the wind shear allows us to infer certain characteristics of the advection at a station. Such an approach to weather prediction was essential before the advent of global or nationwide data collection and dissemination networks; today the art and science of single-station forecasting is no longer emphasized, despite its obvious value as a diagnostic and a pedagogical device.

We shall consider the advection of temperature and stability as revealed by the thermal wind. In practice, these computations are usually made graphically rather than mathematically.

The horizontal advection A of temperature is defined in isobaric coordinates (see Sec. 5.1) by

$$A = -\mathbf{v} \cdot \nabla_p T \tag{19}$$

with $A > 0$ called "warm" advection (see Fig. 9.4). Because $\mathbf{v} \cong \mathbf{v}_g$ in the upper-air flow, we shall consider the geostrophic advection

$$A_g = -\mathbf{v}_g \cdot \nabla_p T \tag{20}$$

From Problem 9.5.2 we see that

$$\nabla_p T = \frac{fT}{g} \frac{\partial \mathbf{v}_g}{\partial z} \times \mathbf{k} \tag{21}$$

so we have

$$A_g = -\frac{fT}{g}\left[\mathbf{v}_g \cdot \left(\frac{\partial \mathbf{v}_g}{\partial z} \times \mathbf{k}\right)\right] = -\frac{fT}{g}\left(\mathbf{v}_g \times \frac{\partial \mathbf{v}_g}{\partial z}\right) \cdot \mathbf{k} \tag{22}$$

Thus the advection is negative when the wind is backing (turning counterclockwise with increasing height) because the cross product $\mathbf{v}_g \times \partial \mathbf{v}_g/\partial z$ will point upward. Similarly warm air is being advected over the station when the wind veers (turns clockwise with increasing height). These conclusions apply to the derivative $\partial \mathbf{v}_g/\partial z$;

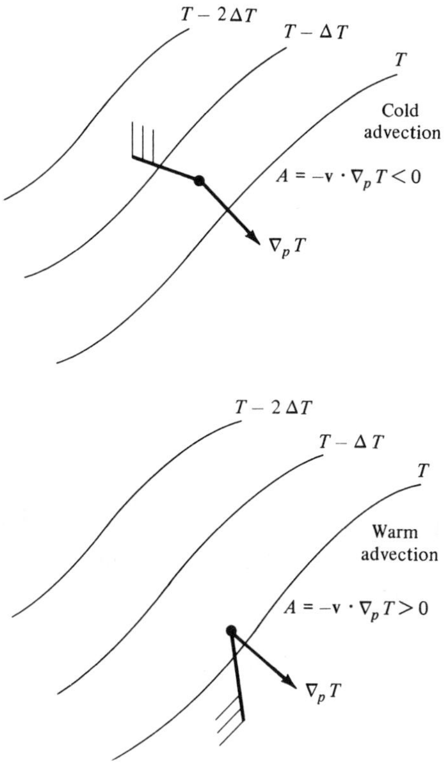

FIG 9.4
Illustration of warm and cold advection.

care must be taken when layer differences $\Delta \mathbf{v}_g/\Delta z$ are used so that a correct estimate of the derivative is obtained.

PROBLEM 9.5.10 As a measure of static stability we may use $\partial T/\partial z = -\gamma$, with increasing stability represented by decreasing lapse rate so that the advection of stability is $S = \mathbf{v} \cdot \nabla_p \gamma$. Show that the advection of stability by the geostrophic wind is

$$S_g = -\frac{fT}{g}\left(\mathbf{v}_g \times \frac{\partial^2 \mathbf{v}_g}{\partial z^2}\right) \cdot \mathbf{k} - \frac{\gamma}{T} A_g$$

and express the first term in a form involving only the gradient of temperature and height.

Baroclinity and stability Baroclinity is associated with geostrophic wind shear and might be expected to be important in stability of the flow to perturbations as measured by the Richardson number, which was discussed in Sec. 4.5 and will appear again in Chap. 13.

For the development we shall utilize isentropic coordinates, and from Problem 9.5.3 we find

$$\frac{\partial \mathbf{v}_g}{\partial z} = \frac{\partial \mathbf{v}_g}{\partial \theta}\frac{\partial \theta}{\partial z} = \frac{1}{f\theta}\frac{\partial \theta}{\partial z} c_p \mathbf{k} \times \nabla_\theta T \tag{23}$$

But from the definition of potential temperature we have

$$\nabla_\theta T = \frac{R}{c_p}\frac{T}{p}\nabla_\theta p = \frac{\alpha}{c_p}\nabla_\theta p \tag{24}$$

and upon replacement of the specific volume by the hydrostatic approximation, Eq. (23) becomes

$$\frac{\partial \mathbf{v}_g}{\partial z} = -\frac{g}{\theta}\frac{\partial \theta}{\partial z}\frac{\mathbf{k} \times \nabla_\theta p}{f(\partial p/\partial z)} \tag{25}$$

In Chap. 4 we defined the Brunt-Väisälä frequency as $\omega_g{}^2 = (g/\theta)\,\partial\theta/\partial z$ and the Richardson number as $Ri = \omega_g{}^2\,|\partial \mathbf{v}/\partial z|^{-2}$. Now we shall use a geostrophic Richardson number, Ri_g, obtained by using the geostrophic wind shear. With Eq. (25) we have then

$$Ri_g = \frac{\omega_g{}^2}{|\partial \mathbf{v}_g/\partial z|^2} = \frac{f^2}{\omega_g{}^2|\nabla_\theta p(\partial p/\partial z)^{-1}|^2} \tag{26}$$

Now clearly $\nabla_\theta p$ is a measure of baroclinity, but the factor containing it can be put in a more revealing form. With the chain rule $\nabla_p(\;) = \nabla_\theta(\;) + \nabla_p\theta\,\partial(\;)/\partial\theta$ we find $\nabla_\theta p = -(\partial p/\partial\theta)\nabla_p\theta$ and then with $\nabla_z\theta = \nabla_p\theta + (\partial\theta/\partial p)\nabla_z p$, we have

$$\boldsymbol{\mu} = \frac{\nabla_\theta p}{\partial p/\partial z} = -\frac{\nabla_z\theta}{\partial\theta/\partial z} + \frac{\nabla_z p}{\partial p/\partial z} = \nabla_\theta h_\theta - \nabla_p h_p \tag{27}$$

in which we have used Eqs. (13) and (37) in Sec. 7.3 to obtain the term on the right. Now by definition, $|\nabla_\theta h_\theta| = [(\partial h/\partial x)_\theta{}^2 + (\partial h/\partial y)_\theta{}^2]^{1/2}$ is the slope of the isentrope, so that

$$\mu^2 = \left(\frac{\partial h_\theta}{\partial x_\theta} - \frac{\partial h_p}{\partial x_p}\right)^2 + \left(\frac{\partial h_\theta}{\partial y_\theta} - \frac{\partial h_p}{\partial y_p}\right)^2 \tag{28}$$

is a form that can be used in cross-section analysis. We assume that the slope of the isobaric surface is small and that the x axis is aligned along the section. Then we use

$$Ri_g = \frac{f^2}{\omega_g{}^2\mu^2} \tag{29}$$

with the slope of the isentrope taken relative to the isobaric surface. This technique produces values of Ri_g greater than the actual value because of the missing contribution from the slope normal to the cross section.

Now we can see that Ri_g will decrease as the vertical stability increases and as the difference in slopes of the isentropes and isobars increases. Thus strongly sloping, stable baroclinic layers have small values of Ri_g, and presumably the actual gradient Richardson number is also small.

If we hold the isobaric surface at the same slope, we can most easily analyze the cause of this situation. As the isentropes are tilted up at a steeper slope, the vertical stability decreases, but the thermal wind increases and is more important because it appears as a square in Eq. (26). If the sloping isentropes are now packed more tightly together, the stability will increase, but the square of the thermal wind will increase more rapidly.

In summary, we have shown that if Ri_g is a guide to the value of Ri, then the steeply sloping baroclinic layers that appear in the upper atmosphere may be quite unstable to perturbations. It appears that these regions represent a situation in which large-scale processes are controlling the slope and packing of the isentropes, and small-scale motion, such as turbulence, feeds on the energy thus made available. This process appears to be important in the creation of situations in which clear air turbulence forms.

PROBLEM 9.5.11 Develop a method for using wind profile data to infer the slopes of isentropic surfaces being drawn in a cross-section analysis.

9.6 THE GRADIENT WIND

The geostrophic wind is defined as the hypothetical motion, parallel to the isobars, that would provide a balance between pressure gradient and Coriolis forces. Thus the effects of friction are ignored and, more significantly, any possible curvature of the actual flow, which would lead to centripetal accelerations of the parcels, is also neglected.

Therefore the first step in defining a model wind that is a better approximation to the actual wind than the geostrophic velocity is to include the effects of curvature in the motion of the parcel. The resulting quantity is called the *gradient wind.*

For frictionless, horizontal flow the equation of motion is

$$\frac{d\mathbf{v}}{dt} = -f\mathbf{k} \times (\mathbf{v} - \mathbf{v}_g) \tag{1}$$

In defining the gradient wind we assume that the differences of the pressure gradient and Coriolis forces are in balance with any centrifugal force arising because of the curvature of the trajectory of the parcel. The centrifugal force will have magnitude v^2/R, where $R = R_T$ is the radius of curvature of the trajectory, and act in a direction normal to the motion at a point on the curved trajectory. Alternatively, the imbalance of the forces on the right of Eq. (1) produces a centripetal acceleration $d\mathbf{v}/dt$ on the left.

Thus for a parcel in motion around a low-pressure area, the geostrophic vector will have the low pressure on its left, and the Coriolis force will be directed to the right. If the parcel is moving in a circle around the center of low pressure, it must be accelerating toward the center and we have (Fig. 9.5)

$$\frac{d\mathbf{v}}{dt} = \frac{v\mathbf{k} \times \mathbf{v}}{R} \tag{2}$$

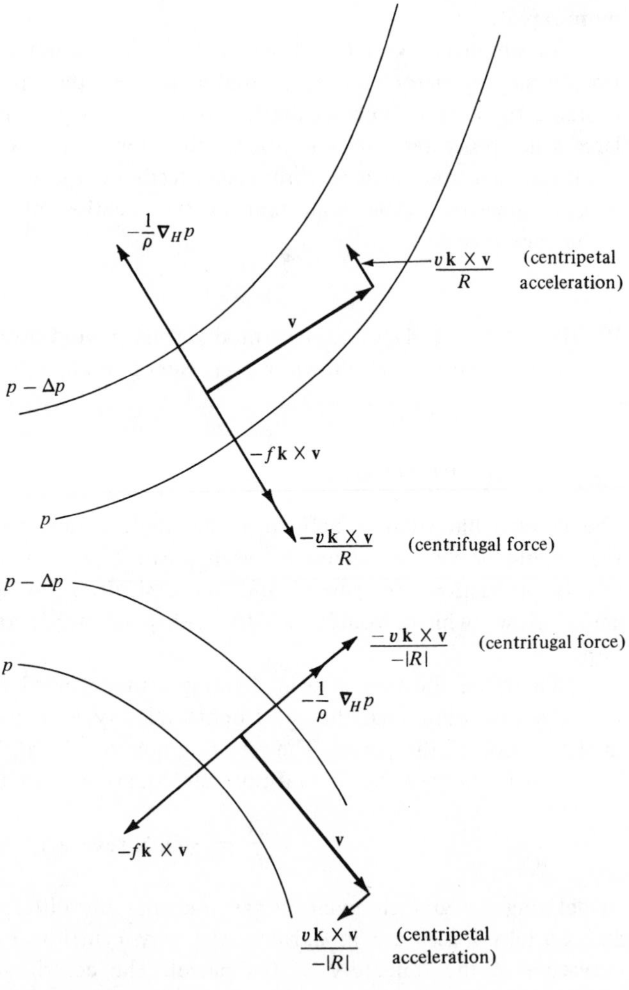

FIGURE 9.5
In gradient flow, the pressure gradient, Coriolis, and centrifugal forces balance. Thus there must be a centripetal acceleration due to the imbalance of the pressure gradient and Coriolis forces.

Hence the equation of motion becomes

$$\frac{v\mathbf{k} \times \mathbf{v}}{R} + f\mathbf{k} \times \mathbf{v} - f\mathbf{k} \times \mathbf{v}_g = 0 \tag{3}$$

If the parcel were circling around a high, then the centripetal acceleration would be in the same direction as the Coriolis force and we would have to write Eq. (2) as $d\mathbf{v}/dt = -(v\mathbf{k} \times \mathbf{v})R^{-1}$.

As a simplification, we shall instead attach the negative sign to the radius of curvature when the Coriolis force and centripetal acceleration are in the same direction, so that Eq. (3) is valid as the equation of motion in both cases. Thus we define *cyclonic flow* by $R > 0$ when the Coriolis force $(-f\mathbf{k} \times \mathbf{v})$ and the centripetal acceleration are in opposite directions; *anticyclonic flow* by $R < 0$ when they are in the same direction. These definitions are valid in both the Northern and Southern Hemispheres.

Denoting the horizontal wind vector by $\mathbf{v}$, we may deduce from Eq. (3) by taking a cross product with $\mathbf{k}$ that

$$\frac{v\mathbf{v}}{R} + f\mathbf{v} - f\mathbf{v}_g = 0 \tag{4}$$

The vector $\mathbf{v}$ in Eq. (4) is called the gradient wind $\mathbf{v}_{gr}$, and we have

$$\mathbf{v}_{gr} = \frac{\mathbf{v}_g}{1 + v_{gr}/fR} \tag{5}$$

where the magnitude obeys the equation

$$\frac{v_{gr}^2}{R} + fv_{gr} - fv_g = 0 \qquad 1 + \frac{v_{gr}}{fR} > 0 \tag{6}$$

Note that as the curvature disappears $(R \to \infty)$, $\mathbf{v}_{gr}$ becomes identical to $\mathbf{v}_g$.

For large-scale flow in the atmosphere, v_{gr} does not differ greatly from v_g. To exploit this fact, we restrict ourselves now to cases where $|v_{gr}/fR| < 0.1$. Then with the binomial theorem, we have from Eq. (5) the approximation

$$v_{gr} \cong v_g\left(1 - \frac{v_{gr}}{fR}\right) \tag{7}$$

and thus

$$v_{gr} \cong \frac{v_g}{1 + (v_g/fR)} \tag{8}$$

which may be written in the vector form

$$\mathbf{v}_{gr} = \frac{\mathbf{v}_g}{1 + (v_g/fR)} \tag{9}$$

Now while it would be possible to differentiate Eq. (5) with height to obtain a complicated expression linking $\partial \mathbf{v}_{gr}/\partial z$ and $\partial \mathbf{v}_g/\partial z$, little is to be gained by such a procedure. Instead, Eq. (9) shows that the gradient wind shear will be essentially equal to the geostrophic shear in large-scale flow, with some contribution to its change made by variations of the term in the denominator.

The restriction we used to obtain Eq. (9) is that what we shall call the *gradient Rossby number*, $Ro_{gr} = v_{gr}/fR$, is small in magnitude. Thus we might expect that the appearance of this term in the denominator of Eq. (9) will not have a significant effect on the shear of the gradient wind. In particular, we find

$$\frac{\partial \mathbf{v}_{gr}}{\partial z} = \frac{\partial \mathbf{v}_g/\partial z}{1 + Ro_{gr}} - \frac{\mathbf{v}_g}{(1 + Ro_{gr})^2} \frac{\partial Ro_{gr}}{\partial z} \tag{10}$$

so that if Ro_{gr} is approximately constant with height, then the gradient wind shear, and thus in many cases the actual wind shear, will be well represented by the geostrophic shear. Thus the efficacy of the thermal wind in studying wind shear and advection of temperature or stability arises not from the fact that the winds are geostrophic, but in part from the fact that the shear of the gradient wind is given approximately by the geostrophic shear. This argument depends on the assumption that Ro_{gr} varies slowly with height—an assumption verified at least intuitively by the observation that both wind speed and radius of curvature tend to increase with height.

In any case, this discussion points out that the appearance of the trajectory radius of curvature in the definition of the gradient wind adds a complication not present in the definition of the geostrophic wind. The radius of curvature cannot be discussed without further consideration of the properties of the streamlines and trajectories of atmospheric flow.

*PROBLEM 9.6.1 Show that $v_{gr} < v_g$ for cyclonic flow and that $v_{gr} > v_g$ for anticyclonic flow.

PROBLEM 9.6.2 Solve Eq. (6) for the gradient wind speed and determine the correct sign to appear before the radical in the cases $R > 0$ and $R < 0$. (Hint: Consider $v_g = 0$.)

PROBLEM 9.6.3 Show that the intensity of gradient anticyclones is limited by determining an inequality between R and v_g that bounds the pressure gradient. Find the limiting gradient wind speed. Explain what parcel motion may be expected if the limiting values are surpassed.

PROBLEM 9.6.4 Show that anticyclonic flow around a low is possible. [Hint: This case violates the condition associated with Eq. (6); use cylindrical coordinates to express Eq. (4) in component form.]

PROBLEM 9.6.5 A parcel moving through a region without a pressure gradient is said to be in *inertial motion*. Describe the character of such motion. How do the trajectories in mid-latitudes differ from those at the equator?

9.7 STREAMLINES AND TRAJECTORIES

The notions of streamlines and trajectories are important conceptual aids in both the intuitive comprehension and the mathematical analysis of atmospheric flow.

A *streamline*, as its name implies, is a line representing the direction of flow at a fixed instant of time. Thus a streamline is everywhere tangent to the velocity vectors. A particularly useful set of streamlines is one constructed so that in addition to being everywhere parallel to the flow, the spacing between the lines is inversely proportional to the wind speed. Such a set of streamlines simultaneously represents the wind speed and direction.

As an elementary example of streamlines let us consider horizontal geostrophic flow. In any local region we may take the Coriolis parameter as constant, so that with this restriction, in isobaric coordinates we have

$$\mathbf{v}_g = \mathbf{k} \times \nabla_p \left(\frac{gh_p}{f} \right) \tag{1}$$

Now let us consider lines on the isobaric surface along which $\Phi = gh_p/f$ has constant values. Such lines are perpendicular to the gradient $\nabla_p \Phi$ or are parallel to the direction of the geostrophic flow. Moreover, $|\mathbf{v}_g|$ increases as the spacing between these lines decreases. Thus the lines along which Φ is constant are streamlines, and Φ is therefore called a *stream function*. If the flow is not geostrophic, determination of a suitable stream function is obviously going to be more complicated.

A *trajectory* is a line representing the path followed by an air parcel in its motion. Thus a trajectory is a curve, located in (x,y,z,t) space, that reveals the coordinates the parcel had (or will have) for every time t. At a given instant of time, each parcel in the flow is moving on its trajectory, or parallel to the tangent to its trajectory, so that the streamlines at that instant are all tangent to parcel trajectories. But as time passes, the parcels all move along their trajectories, and the streamlines must change to adjust to the flow. Thus the parcels do not move parallel to the streamlines; rather, the streamlines are parallel to the flow at a particular time. While a streamline field can be drawn on a synoptic chart representing observed wind data at a given level, it is a far more complicated task to determine or deduce the trajectory of an air parcel, especially with observational data available only at 12-hour intervals.

Many of the interesting aspects of atmospheric motion involve processes associated with curvature of both the streamlines and the trajectories. To investigate these phenomena, and to determine more precisely the role of the radius of curvature in the gradient wind equation, we need to establish some mathematical properties of the curvature of lines.

For simplicity we shall use the term *flow line* to mean either a streamline or a trajectory, and we shall consider now only horizontal flow so that flow lines are confined to the (xy) plane. Suppose then we have a flow line, and in the vicinity of point P on the flow line we let it have the equation $y = y(x)$. Now with a compass, we find an origin and a radius for a circle that best fits the curve $y(x)$ in the neighborhood of P. We let the circle be of radius r and have its origin at (x_0, y_0). (See Fig. 9.6.) In the neighborhood of P then, the line has the equation

$$(y - y_0)^2 + (x - x_0)^2 = r^2 \tag{2}$$

so that we may deduce the relations (for $y > y_0$)

$$y = y_0 + [r^2 - (x - x_0)^2]^{1/2} \tag{3}$$

$$\frac{dy}{dx} = y' = -\frac{x - x_0}{[r^2 - (x - x_0)^2]^{1/2}} \tag{4}$$

and

$$y'' = -\frac{r^2}{[r^2 - (x - x_0)^2]^{3/2}} \tag{5}$$

These three equations suffice to determine r, x_0, and y_0 as functions of y, y', and y'' at P. Thus even if our original circle found with the compass does not give an accurate portrayal of the curvature, we have an analytical way of determining an appropriate circle tangent to $y(x)$ at P. The combination of Eqs. (4) and (5) gives

$$y'' = -r^2\left\{r^2 - \left[\frac{r^2(y')^2}{1 + (y')^2}\right]\right\}^{-3/2} = -\frac{[1 + (y')^2]^{3/2} r^2}{r^3} \tag{6}$$

and so in general

$$\pm\frac{1}{r} = \frac{y''}{[1 + (y')^2]^{3/2}} \tag{7}$$

In the study of curves, the quantity r given by Eq. (7) with $|y''|$ in the numerator is defined as the radius of curvature. If Eq. (7) with the positive sign used on the left is used to define $1/r$, then the sign of the radius of curvature will be positive for cyclonic curvature and negative for anticyclonic curvature—in agreement with the convention used in the gradient wind equation.

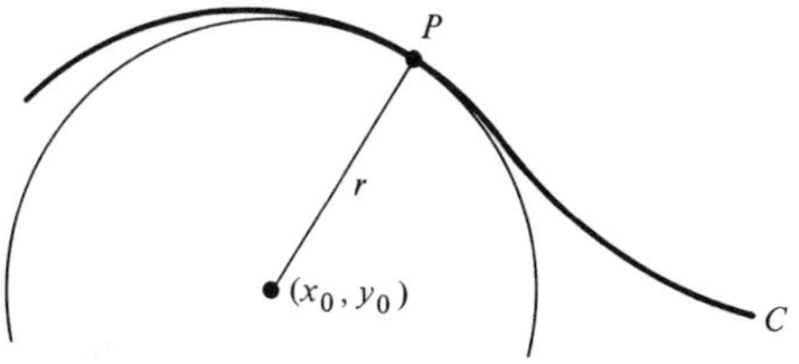

FIGURE 9.6
A circle constructed tangent to the curve C at point P.

These results would be awkward to use on a synoptic chart, where we have wind speeds and directions rather than $y(x)$. So let α be the angle the flow line makes with the x axis. Then in the neighborhood of P we have $y' = \tan \alpha$ and $y'' = (\sec^2 \alpha)\alpha'$, and so from Eq. (7) we may compute

$$\frac{1}{r} = \frac{\sec^2 \alpha(\partial\alpha/\partial x)}{(1 + \tan^2 \alpha)^{3/2}} = \frac{1}{\sec \alpha}\frac{\partial\alpha}{\partial x} \tag{8}$$

But by elementary trigonometry $\delta s = \sec \alpha\, \delta x$ where δs is distance along the flow line. Finally then

$$\frac{1}{r} = \frac{\partial\alpha}{\partial s} \tag{9}$$

If the flow line is a streamline, then when θ is the wind direction (compass direction from which the wind is blowing), we have $\partial\theta/\partial s = -\partial\alpha/\partial s$ and so the radius of curvature of the streamline is

$$\frac{1}{R_S} = -\frac{\partial\theta}{\partial s} \tag{10}$$

In the case of a trajectory, the angle θ is a function only of distance along the trajectory, and so for radius of curvature we have

$$\frac{1}{R_T} = -\frac{d\theta}{ds} \tag{11}$$

Now let us see how the rate of change of the wind direction is related to flow-line curvature. For horizontal flow, the chain rule applied at point P and time t gives

$$\frac{d\theta}{dt} = \left(\frac{\partial\theta}{\partial t}\right)_s + \frac{\partial\theta}{\partial s}\frac{ds}{dt} \tag{12}$$

where $(\partial\theta/\partial t)_s$ is the local rate of change, $\partial\theta/\partial t$, of wind direction, $\partial\theta/\partial s$ is the negative of the curvature of the streamline tangent to the trajectory at point P at time t, and ds/dt is, by definition, the parcel speed v. Along the trajectory we have

$$\frac{d\theta}{dt} = \frac{d\theta}{ds}\frac{ds}{dt} \tag{13}$$

and so

$$-\frac{1}{R_T}v = \frac{\partial\theta}{\partial t} - \frac{1}{R_S}v \tag{14}$$

or

$$\frac{\partial\theta}{\partial t} = v\left(\frac{1}{R_S} - \frac{1}{R_T}\right) \tag{15}$$

This last result is known as *Blaton's equation*. It shows that if the flow is steady, so that wind directions are locally constant, then the radii of curvature of streamlines and trajectories are identical, and hence trajectories and streamlines coincide. We shall return to more discussion of this point later in the section. Perhaps of more significance to meteorological analysis is the fact that when wind directions are changing, the streamlines and trajectories may have unequal radii of curvature, or even curvature of opposite sign in some cases.

*PROBLEM 9.7.1 Show that the gradient wind equation may be written as

$$\frac{v_{gr}^{\ 2}}{R_S} + \left(f - \frac{\partial\theta}{\partial t}\right)v_{gr} - fv_g = 0$$

Determine when the gradient wind equation (6) in Sec. 9.6 may be applied with $R = R_S$, a radius that for most purposes may be estimated from the curvature of the geostrophic streamlines.

PROBLEM 9.7.2 Let s be arc length and $\mathbf{r}(s)$ the position vector tracing out a curve. The unit vector tangent to the curve is $\boldsymbol{\tau} = d\mathbf{r}(s)/ds$. Show that $d\boldsymbol{\tau}/ds = \boldsymbol{\eta}/|R|$ where $\boldsymbol{\eta}$ is a unit vector known as the *principal normal* and R is the radius of curvature. The third unit vector $\boldsymbol{\beta} = \boldsymbol{\tau} \times \boldsymbol{\eta}$ is known as the *binormal*. Show that $\boldsymbol{\tau}$, $\boldsymbol{\eta}$, and $\boldsymbol{\beta}$ are orthogonal. (See illustration.)

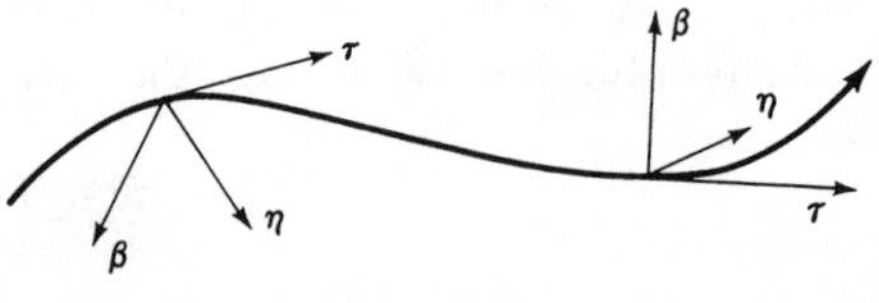

The principal triad, composed of tangent vector $\boldsymbol{\tau}$, principal normal $\boldsymbol{\eta}$, and the binormal $\boldsymbol{\beta}$.

PROBLEM 9.7.3 Let $\mathbf{r}(s)$ describe the trajectory of a parcel in motion. Define $d\boldsymbol{\tau}/ds = \boldsymbol{\eta}/R_T$, so that $\boldsymbol{\eta}$ is always to the left of $\boldsymbol{\tau}$. Then the unit vectors $(\boldsymbol{\tau},\boldsymbol{\eta},\boldsymbol{\beta})$ and the associated distances (s,n,b) are known as *natural coordinates* and the vector velocity is given by $\mathbf{v} = v\boldsymbol{\tau}$. Show that $d\boldsymbol{\tau}/dt = v\boldsymbol{\eta}/R_T$ and find an expression for the acceleration in natural coordinates. Explain the significance of the terms that appear.

PROBLEM 9.7.4 For horizontal motion, show that the divergence is $\partial v/\partial s - v\,\partial\theta/\partial n$, where θ is the wind direction.

In the remainder of this section we turn to the mathematical questions associated with whether flow lines can be found for an arbitrary wind field.

Because a streamline is tangent to the flow at each point, the components of its differential tangent vector, $d\mathbf{r} = \mathbf{i}\,dx + \mathbf{j}\,dy + \mathbf{k}\,dz$, must have the same ratios as the wind components (u,v,w). Thus the streamline is represented by the equations

$$\frac{dx}{u} = \frac{dy}{v} = \frac{dz}{w} \tag{16}$$

If we solve the equation on the left, we will obtain $y = f(x,z,c_1)$ while that on the right will produce $y = g\,(x,z,c_2)$. When these two are arranged in the form

$$F(x,y,z) = c_1 \qquad G(x,y,z) = c_2 \tag{17}$$

they are called first integrals of the system (16). The equations in (17) each represent a surface, and the intersection of the surfaces is the streamline.

The simplest case is that of two-dimensional flow (we use the illustration $w = 0$) so that Eq. (16) becomes

$$v\,dx - u\,dy = 0 \tag{18}$$

Now if $\partial v/\partial y = -\partial u/\partial x$, then the equation is exact, and so the condition for exactness is that the two-dimensional motion be nondivergent. We can then find the function

$$\psi(x,y) = \int_{x_0}^{x} v(\xi,y)\,d\xi - \int_{y_0}^{y} u(x_0,\eta)\,d\eta \tag{19}$$

which satisfies Eq. (18) because

$$v = \frac{\partial\psi}{\partial x} \qquad u = -\frac{\partial\psi}{\partial y} \tag{20}$$

Thus along a streamline $\psi = \text{const}$, we have $d\psi = 0$ and this implies that

$$\frac{\partial\psi}{\partial x}\,dx + \frac{\partial\psi}{\partial y}\,dy = v\,dx - u\,dy = 0 \tag{21}$$

In this case, ψ is called a *stream function*, and lines of constant ψ are the streamlines of the flow. The speeds are proportional to the spacing between the streamlines.

If the flow is two-dimensional but not nondivergent, then the solution is not easily found, a mathematical situation that has a rather simple physical explanation. If the flow is nondivergent, then any vector into a region must be matched by a vector out of the region, and so it is easy to draw a streamline through the region. But if convergence or divergence is allowed, then all vectors on the boundary of some region may be inward (or outward) so that a singular point occurs where streamlines intersect. Thus no simple stream function exists, because the stream function would have to be multiple-valued at such a point in order that the lines of constant value may intersect. In this more general situation, it is therefore more convenient to return to the Helmholtz theorem of Sec. 5.6.

Trajectories represent the paths of fluid elements in (x,y,z,t) space, giving the spatial coordinate of the parcel at each time t. In this case we have the ratios $dx:dy:dz:dt$ of the trajectory and the ratios $u:v:w:1$ of the velocity field which must be put in correspondence. The result is the system

$$\frac{dx}{u} = \frac{dy}{v} = \frac{dz}{w} = dt \tag{22}$$

The mathematical theory of the trajectories thus involves higher dimensional problems than streamlines and is even more complex.

But if the flow is steady, then u, v, w do not depend on time, the last term on the right of Eq. (22) becomes superfluous, and the equations of the trajectories are identical to the equations of the streamlines. Hence we have shown again that streamlines and trajectories coincide in steady flow.

The construction of trajectories is obviously a complicated task, and it is clear that if we could predict the trajectories of all fluid parcels, we would have a result equivalent to the solution of the equations of motion. Furthermore, if we knew the trajectory, we would know the vertical motion of the parcel. Conversely, an accurate trajectory cannot be constructed by assuming two-dimensional motion and neglecting the vertical velocity. We shall consider how trajectories of atmospheric motion are constructed after a discussion of two methods for determining vertical velocities from meteorological data.

9.8 DETERMINATION OF VERTICAL VELOCITIES

The wind vector may always be resolved into a horizontal component and a vertical component, and so far in this chapter we have essentially ignored the presence of the vertical component. From an elementary analysis of orders of magnitude, we would expect that the vertical velocity could usually be neglected because it is so small relative to the horizontal wind. For example, w will typically be of the order of a few centimeters per second for large-scale flow while v_H will be of the order of 10 m/s—a ratio of about 10^{-2}.

But vertical velocities in atmospheric motion present one of the cases in which small causes have large effects. The formation of clouds depends upon vertical motions: the rising parcels of air cool, their water vapor condenses into the droplets that form clouds. The earth's cloud cover is an effective reflector of solar energy and absorber of long-wave radiation, and so the vertical velocities have an important control on the total radiative budget. On the smaller scale, the vertical velocities, necessary for cloud formation, are necessary also for development of precipitation, for the movement of water from sea to continent, and thus necessary for terrestrial life itself.

The magnitude of the vertical velocity depends strongly on scale. A preliminary estimate can be made by assuming that atmospheric motions are nearly nondivergent. Then we use U as a typical amplitude of the horizontal wind and W for the vertical component, we align the x axis with the wind, and so

$$\nabla \cdot \mathbf{v} = \frac{\partial U}{\partial x} + \frac{\partial W}{\partial z} \tag{1}$$

If we take scale lengths L and D in the horizontal and vertical, we have the estimate $W \sim (D/L)U$. For large-scale motion, D will be of order 2 km and L of order 1000 km, and so $W \sim 2 \times 10^{-3} U$. (We shall have occasion to improve the accuracy of this estimate in Chap. 14.) For small-scale motion, such as boundary layer turbulence, $D \sim L$ and so $W \sim U$.

Despite their small size relative to horizontal velocities in large-scale flow, vertical velocities are a significant part of the dynamics of atmospheric motion. They tie the motions together, level by level, and provide a direct connection between the flow patterns at one altitude with those at another. Thus any realistic model of atmospheric motion—whether constructed for theoretical purposes or practical prediction—must take account of the vertical component of the motion.

This is easier said than done, for it is virtually impossible to measure vertical velocities on a synoptic scale. Instruments mounted on towers can measure vertical velocities near the earth's surface, and certain kinds of measurement systems on aircraft can resolve small-scale patterns of vertical motion. But there is no device available that can provide vertical velocity measurements to accompany the synoptic observations of the other variables. Thus meteorologists preparing a weather forecast must always begin without direct information on one of the most significant variables.

The only recourse is the equations of motion. We must in effect determine some of the other variables and their rates of change until we are left with an equation, or several equations, in which the only remaining variable is the vertical velocity. There are various combinations that will work, and we shall investigate two of them in this section, one of them in the next section, and others in the following chapters. The two methods of this section are perhaps the easiest to understand, but they do not give adequate results in most applications.

9.8.1 The Kinematic Method

In Sec. 6.5.3 we pointed out that, assuming the motion is hydrostatic, this forces the vertical velocity field to take a form specified as a function of the other variables. We combined the equation of continuity and the first law of thermodynamics to obtain Richardson's equation (20) in Sec. 6.5.

The equivalent result in isobaric coordinates is much simpler to derive and apply; it is still strongly dependent on the hydrostatic assumption, because as pointed out in Chap. 7 this assumption is crucial in obtaining the simple form of the isobaric equations in general use.

The continuity equation in isobaric coordinates is

$$\nabla_p \cdot \mathbf{v} + \frac{\partial \omega}{\partial p} = 0 \tag{2}$$

and so by integration we find

$$\omega(x,y,p) - \omega(x,y,p^*) = -\int_{p^*}^{p} \nabla_{p_1} \cdot \mathbf{v}\, dp_1 \tag{3}$$

If we take $p^* = 0$, then $\omega(p^*) = 0$ and we have

$$\omega(x,y,p) = -\int_0^p \nabla_{p_1} \cdot \mathbf{v}\, dp_1 \tag{4}$$

But this is not a practical solution, because we must know the divergence of the wind from the level of interest to the top of the atmosphere.

For $p^* = p_0(x,y)$, where p_0 is the surface pressure, we have

$$\omega(x,y,p) = \omega(x,y,p_0) + \int_p^{p_0} \nabla_{p_1} \cdot \mathbf{v}\, dp_1 \tag{5}$$

But

$$\omega(p_0) = \left.\frac{dp}{dt}\right|_{z=0} = \left.\left(\frac{\partial p}{\partial t} + \mathbf{v} \cdot \nabla p\right)\right|_{z=0} \tag{6}$$

The velocities u, v, w must all vanish at the earth's surface if a nonslip, flat boundary is assumed, so then

$$\omega(x,y,p) = \left.\frac{\partial p}{\partial t}\right|_{z=0} + \int_p^{p_0} \nabla_{p_1} \cdot \mathbf{v}\, dp_1 \tag{7}$$

The obvious disadvantage to this kinematic method is that the horizontal divergence must be evaluated, and this is extremely difficult to do with observed data because the computations are sensitive to small variations in wind speed and direction. But the kinematic method is useful in numerical prediction where vertical motion fields are found for time $t + \Delta t$ after the hydrostatic equations have been used to predict the other variables for that time.

PROBLEM 9.8.1 Determine an appropriate form of Eq. (6) for the case of a free-slip boundary condition when the boundary varies only in the x direction and has the equation $z = h(x)$. (Hint: The boundary condition requires that $\eta \cdot \mathbf{v} = 0$, and it is convenient to introduce $\tan \phi = \partial h/\partial x$.)

9.8.2 The Adiabatic Method

Large-scale motions are often nearly isentropic (loosely called adiabatic), and this provides a conceptually simple method of determining the vertical velocity. By definition, we have

$$\frac{d\theta}{dt} = \frac{\partial\theta}{\partial t} + \mathbf{v} \cdot \nabla_z\theta + w\frac{\partial\theta}{\partial z} \tag{8}$$

and so upon solving for w

$$w = \frac{d\theta/dt - \partial\theta/\partial t - \mathbf{v} \cdot \nabla_z\theta}{\partial\theta/\partial z} \tag{9}$$

If we assume that the flow is isentropic, the first term on the right vanishes and thus, denoting the isentropic part of w as w_i, we have

$$w_i = -\frac{\partial\theta/\partial t + \mathbf{v} \cdot \nabla_z\theta}{\partial\theta/\partial z} \tag{10}$$

where

$$w = w_i + w_d = w_i + \left(\frac{\partial\theta}{\partial z}\right)^{-1}\frac{d\theta}{dt} \tag{11}$$

In order to find the isentropic component of w, we need to measure local rates of change of θ and the advection of θ, as well as the vertical stability. In practice the local change must be computed from data 12 hours apart and the advection from station data several hundred kilometers apart, and we usually face the problem of having to take differences of two large numbers.

It is always possible to compute w_i (or ω_i in isobaric coordinates); the question is whether $w \cong w_i$ or, equivalently, whether w_d is small relative to w_i. The most obvious error will occur in regions of condensation or evaporation, and thus isentropic vertical velocities are suspect in the regions of storms and fronts where we would like to know what the patterns of vertical motion actually are.

PROBLEM 9.8.2 Show that in isobaric coordinates

$$\omega_i = -\frac{(\partial T/\partial t_p) + \mathbf{v} \cdot \nabla_p T}{(\partial T/\partial p) + (g/c_p)(\partial h/\partial p)}$$

an expression that involves only variables generally available on standard charts.

9.9 ISENTROPIC TRAJECTORIES

Meteorologists dreaming of ideal circumstances for studying the atmosphere might imagine being able to actually watch air parcels in their motion. For both analysis and prediction, it would be invaluable to be able to see the path the air parcels follow as the patterns of the atmosphere evolve from day to day.

We might see warm, moist parcels being drawn into a cyclone and joining the rising currents that will wring the water from them. We might be able to watch a parcel from the stratosphere coming down a high-altitude frontal surface to mix with tropospheric air. We might watch a parcel, suddenly caught in the jet stream, being whisked halfway around the globe before it escapes again.

Such a capability to watch parcels in motion would obviously improve our understanding and our prediction of atmospheric phenomena; the question is how the presently observed data can be used in practical schemes for constructing accurate trajectories. There are two possibilities:

1 If the equations of motion are solved on a grid with a fine resolution, then trajectories can be computed for any parcel by geometrical methods.
2 If *conservative properties* of the air parcel can be identified, then we can observe their values in a parcel at one time and then find the parcel's new location at a later time by finding where the same combination of values of these variables occurs.

To be more specific about the second alternative, let α, β, γ be variables that are conservative for atmospheric motion, that is variables that do not change as parcels move. Thus

$$\frac{d\alpha}{dt} = \frac{d\beta}{dt} = \frac{d\gamma}{dt} = 0 \tag{1}$$

Now we choose a parcel and find that these variables have values α_0, β_0, and γ_0. At some later time we would analyze our atmospheric data and find somewhere downwind the intersection of the surfaces with value $\alpha = \alpha_0$, $\beta = \beta_0$, $\gamma = \gamma_0$. That point is the new location of the parcel, and by tracing its location at the successive times at which data are available, we would map out its trajectory. The problem, naturally, is what to use for the conservative variables. Various combinations have been proposed, but only one practical method has so far been developed and applied to problems in diagnosis of atmospheric motion.

In the rest of this section, we present the method of trajectory construction developed by Dr. E. F. Danielsen using potential temperature, energy, and a kinematic variable as the conservative variables. In its simplest form, the method assumes parcels are confined to isentropic surfaces, but this is not an essential constraint. Some calculated trajectories for atmospheric motion are shown in Fig. 9.7.

We start with the horizontal equation of motion

$$\frac{d\mathbf{v}}{dt} = -\frac{1}{\rho}\nabla_z p - f\mathbf{k} \times \mathbf{v} \tag{2}$$

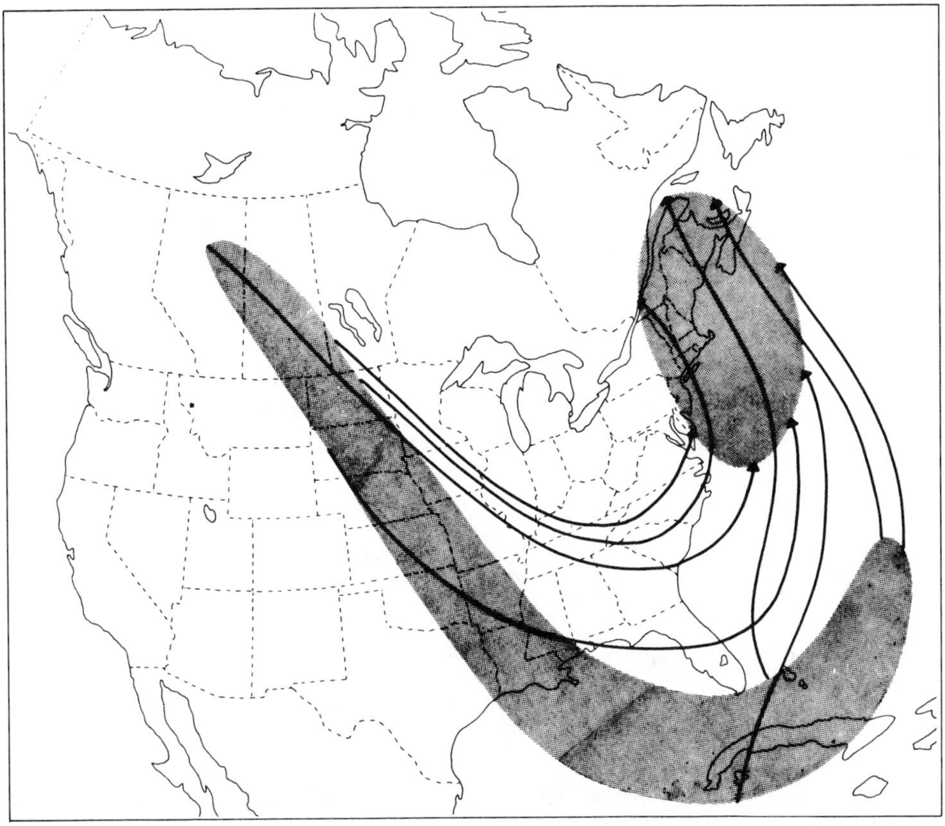

FIGURE 9.7
Trajectories computed over three 12-hour intervals to give the 36-hour path of air parcels that joined to form a cyclone over New England. These trajectories were computed on the 305-K isentropic surface. (*From Sechrist and Dutton, 1970.*)

which gives the energy equation

$$\frac{1}{2}\frac{dv^2}{dt} = -\alpha \mathbf{v} \cdot \nabla_z p \tag{3}$$

The first law of thermodynamics

$$c_p \frac{dT}{dt} - \alpha \frac{dp}{dt} = \frac{c_p T}{\theta} \frac{d\theta}{dt} \tag{4}$$

may be expressed with the aid of the hydrostatic approximation as

$$c_p \frac{dT}{dt} - \alpha \frac{\partial p}{\partial t_z} - \alpha \mathbf{v} \cdot \nabla_z p + g \frac{dz}{dt} = \frac{c_p T}{\theta} \frac{d\theta}{dt} \tag{5}$$

The sum of Eqs. (3) and (5) gives the energy equation

$$\frac{d}{dt}\left(\frac{v^2}{2} + c_pT + gz\right) = \alpha\frac{\partial p}{\partial t_z} + \frac{c_pT}{\theta}\frac{d\theta}{dt} \tag{6}$$

The first term may be modified by noting that the Montgomery stream function

$$\Psi = c_pT + gz = c_pT + gh \tag{7}$$

appears. The next step is to express the first term on the right in isentropic coordinates.

PROBLEM 9.9.1 Show that $\partial\Psi/\partial t_\theta = \alpha\,\partial p/\partial t_z$. [Hint: We know $\partial/\partial t_\theta = d/dt - \mathbf{v}\cdot\nabla_\theta - (d\theta/dt)\partial/\partial\theta$. Use Eq. (42) in Sec. 7.3 and the hydrostatic approximation. Can you find a more direct method using an isentropic version of Eq. (6) in Sec. 7.3?]

Now Eq. (6) may be written

$$\frac{d}{dt}\left(\frac{v^2}{2} + \Psi\right) = \frac{\partial\Psi}{\partial t_\theta} + \frac{c_pT}{\theta}\frac{d\theta}{dt} \tag{8}$$

and with the hydrostatic approximation $\partial\Psi/\partial\theta = c_pT/\theta$, Eq. (8) becomes

$$\frac{d}{dt}\left(\frac{v^2}{2} + \Psi\right) = \frac{\partial\Psi}{\partial t_\theta} + \frac{\partial\Psi}{\partial\theta}\frac{d\theta}{dt} \tag{9}$$

Next we consider a parcel with an initial potential temperature θ_1 and a final value θ_2. We integrate Eq. (9) along the trajectory to obtain

$$\left(\frac{v^2}{2} + \Psi\right)_{x_2,y_2,\theta_2} = \left(\frac{v^2}{2} + \Psi\right)_{x_1,y_1,\theta_1} + \int_{t_1}^{t_2}\left(\frac{\partial\Psi}{\partial t_\theta} + \frac{\partial\Psi}{\partial\theta}\frac{d\theta}{dt}\right)dt \tag{10}$$

In practice, the right side is computed, assuming isentropic motion ($\theta_1 = \theta_2$), from observed data for possible trajectories, and the value obtained at the end of the trial trajectory is compared to observed values of $v^2/2 + \Psi$ at that point. In general a line along which the trajectory might end is found by this method.

Another conservative variable is needed to single out the point on this line which indicates the actual trajectory. The distance the parcel travels from its initial point along the trajectory is

$$S(t_1, t_2) = \int_{t_1}^{t_2} \mathbf{v}\cdot\boldsymbol{\tau}\,dt \tag{11}$$

where $\boldsymbol{\tau}$ is the tangent to the trajectory. Thus the actual distance the parcel moves along the trajectory must correspond to the wind fields through which the parcel passes. With data at t_1 and t_2, this is evaluated as the average

$$S(t_2,t_1) = \frac{(v_2 + v_1)(t_2 - t_1)}{2} \tag{12}$$

where v_2 and v_1 are the actual wind speeds because the trajectory is tangent to the streamlines at each instant of time. The relation (12) thus produces another line along which trajectories may end. In most cases, this line will intersect the line found with Eq. (10) and so the end point is determined.

Cases do occur, however, in which the two lines do not intersect, or in which they become identical over an interval; obviously, then, further analysis is needed. In other cases the trajectory clearly passes into a region in which condensational heating is taking place, and the effects must be accounted for by using the theory of moist thermodynamics. Finally, the use of an additional variable in the method would presumably lead to the end point occurring within a region bounded by three lines in most cases and an estimate of reliability would thus be obtained.

This method of constructing isentropic trajectories is one of the most powerful diagnostic tools available today. It is particularly useful for determining vertical velocities. From the results we know the initial coordinates (x_1,y_1,θ_1) and the final coordinates (x_2,y_2,θ_2). Thus for the vertical velocity at the midpoint of the trajectory we have the estimate

$$w = \frac{h(x_2,y_2,\theta_2) - h(x_1,y_1,\theta_1)}{t_2 - t_1} \tag{13}$$

When trajectories are computed throughout a region of interest, the accompanying vertical velocity patterns are thus easily determined as an additional benefit to the analysis.

PROBLEM 9.9.2 The integral in Eq. (10) is along the trajectory. Develop a scheme for evaluating this integral for isentropic motion assuming the availability of the usual synoptic data.

PROBLEM 9.9.3 Show that the vertical velocity determined (for $\theta_2 = \theta_1$) by Eq. (13) and that of the adiabatic method are theoretically identical. For what reasons might you expect the trajectory method to be superior for obtaining w in practice?

9.10 SOME OTHER TYPES OF WIND

The geostrophic and gradient winds discussed in detail in the beginning of the chapter are defined by requiring a balance of certain terms of the equation of motion. Other model winds can be defined similarly by requiring that other combinations of forces are in balance.

9.10.1 Cyclostrophic Wind

When the flow is sufficiently near the equator so that f is small or when the Coriolis force is negligible compared to the centripetal acceleration, the gradient wind equation becomes

$$\frac{v\mathbf{k} \times \mathbf{v}}{R} = -\frac{1}{\rho}\nabla_z p \tag{1}$$

This equation provides the definition of the *cyclostrophic* wind $\mathbf{v}_c$.

In the gradient wind equation

$$\frac{v\mathbf{k} \times \mathbf{v}}{R} + f\mathbf{k} \times \mathbf{v} + \frac{1}{\rho}\nabla_z p = 0 \tag{2}$$

the ratio of centripetal acceleration to Coriolis force is

$$\frac{|v\mathbf{k} \times \mathbf{v}|}{|Rf\mathbf{k} \times \mathbf{v}|} = \frac{v}{fR} \tag{3}$$

and so the condition for cyclostrophic flow compared to gradient flow is that

$$\frac{v}{fR} \gg 1 \tag{4}$$

It is often suggested that the tropical storms of low latitudes may be in approximate cyclostrophic balance. Let us consider a circular vortex with cyclostrophic flow. Then Eq. (1) may be written

$$\frac{v_c^2}{r} = \frac{1}{\rho}\frac{\partial p}{\partial r} \tag{5}$$

In a region in which the pressure gradient force is given as a power law in r of the form

$$\frac{1}{\rho}\frac{\partial p}{\partial r} = Cr^{\alpha} \tag{6}$$

we have

$$v_c = \sqrt{C}\, r^{(1+\alpha)/2} \tag{7}$$

PROBLEM 9.10.1 Under what conditions will solid rotation with linear speeds $v_c = \omega r$ occur?

PROBLEM 9.10.2 Justify the applicability of the cyclostrophic concept to a tornado and develop a suitable model. Show that the vortex can rotate in either direction.

9.10.2 Antitriptic Wind

The antitriptic wind occurs when the pressure gradient is balanced by the force of friction so that

$$0 = -\frac{1}{\rho}\nabla_z p + \mathbf{F} \tag{8}$$

If we assume that the friction can be represented with an eddy friction coefficient ν_E, in the form

$$\mathbf{F} = \nu_e \frac{\partial^2 \mathbf{v}}{\partial z^2} \tag{9}$$

we have, for the antitriptic wind $\mathbf{v}_a$,

$$\frac{\partial^2 \mathbf{v}_a}{\partial z^2} = \frac{1}{\nu_e \rho} \nabla_z p \tag{10}$$

If an analytic form for the pressure gradient and ν_e is specified, $\mathbf{v}_a$ can be found by integration. Antitriptic winds are thus atmospheric analogs of the Poisseuille flow considered in Sec. 6.10.

9.10.3 Katabatic or Gravity Winds

Some of the most interesting small-scale phenomena among the atmosphere's many motions are the often complex wind structures produced by differential heating on sloping surfaces.

Consider a tall mountain range at the edge of a plain. Heating and cooling by radiation will be most effective near the earth's surface–including the sloping sides of the mountain. Thus at night, air near the mountain will be cooled more than air at the same altitude over the plain. This heavy air will start to slide down the mountain, and as the current accelerates, it may become quite turbulent. In its fall, the air warms, perhaps at the adiabatic rate, and it may become warmer than the ambient air.†

The same processes are present in mountain and valley regions, and along gently sloping seacoasts, with cooled air flowing down the mountain or coastal slopes at night and air warmed near the surface flowing aloft during the day. Complicated, diurnally varying flows develop under the combined effect of the radiational heating and the inertial oscillation induced by the Coriolis force.

†The terms *chinook* and *foehn* are generally applied to a more violent downslope wind than the simple katabatic wind described here. Such chinook winds, which attain speeds of over 50 m/s in the valley, are a mountain lee wave phenomenon that occurs when there is strong flow over a mountain chain and stable stratification. Gravity waves are induced, and an associated downward flux of momentum from aloft produces the strong and often destructive surface wind.

BIBLIOGRAPHIC NOTES

9.1 The geostrophic wind is a concept treated in all textbooks on dynamic meteorology. For an example of earlier views, see:

Shaw, Sir Napier, 1919: *Manual of Meteorology*, vol. IV, *Meteorological Calculus*, Cambridge University Press, Cambridge (revised 1931), 359 pp.

9.2 The treatment of the equations of motion and the geostrophic approximation with analysis of scales is a relatively recent development. The introduction here will be expanded in Chap. 14, and references will be given there.

9.3 An elegant treatment of the Foucault pendulum appears in:

Symon, K. R., 1953: *Mechanics*, Addison-Wesley Publishing Company, Inc., Reading, Mass., 358 pp.

9.4 An extended discussion of inertial stability is given by:

Van Mieghem, Jacques M., 1951: "Hydrodynamic Instability" in T. F. Malone (ed.), *Compendium of Meteorology*, pp. 434–453, American Meteorological Society, Boston.

9.5 The Taylor-Proudman theorem is discussed in some detail, along with photographic illustration of the Taylor column, in the book by:

Greenspan, H. P., 1968: *The Theory of Rotating Fluids*, Cambridge University Press, Cambridge, 327 pp.

The book cited for Sec. 9.1 also contains an early discussion of the thermal wind.

9.6 The gradient wind is another of the classical concepts in meteorology. The construction and theory of wind scales that make application of the concept feasible in synoptic practice are discussed by:

Godske, C. L., T. Bergeron, J. Bjerknes, and R. C. Bungaard, 1957: *Dynamic Meteorology and Weather Forecasting*, American Meteorological Society, Boston, and Carnegie Institution of Washington, Washington, 800 pp.

as well as in most synoptic meteorology textbooks.

9.7 Streamlines and trajectories are discussed in nearly all books on fluid mechanics. Blaton's equation, however, appears to be essentially a meteorological concept.

9.8 The two methods for vertical velocity computation are classical but are not now used extensively in diagnostic work with atmospheric data.

9.9 The material in this section derives from the article by:

Danielsen, E. F., 1961: "Trajectories: Isobaric, Isentropic, and Actual," *J. Meteorol.*, **18**:479–486.

An example of the application of the method is given in the article:

Sechrist, Frank S., and J. A. Dutton, 1970: "Energy Conversions in a Developing Cyclone," *Monthly Weather Rev.*, **98**:354–362.

9.10 Various types of wind are discussed by:

Jeffreys, H., 1922: "On the Dynamics of Wind," *Quart. J. Roy. Meteorol. Soc.*, **48**:29–46.

An extensive discussion of wind phenomena associated with sloping terrain is given by:

Lettau, H., 1967: "Small to Large-Scale Features of Boundary Layer Structure over Mountain Slopes," in *Proceedings, Symposium on Mountain Meteorology* (June 1967), Atmospheric Science Paper 122, Department of Atmospheric Science, Colorado State University.

10

THE ROTATIONAL COMPONENT OF THE WIND: VORTICITY AND CIRCULATION

Any fluid whose thermodynamic equilibrium is disturbed by thermal forcing is set into a pattern of motion that circulates heat from warm to cold regions in the attempt to find a new balance. These currents are modified by the mechanical forces that arise as a result of the motion itself and, in the atmosphere, must also respond to the earth's rotation.

Although the direct, vertical circulations of slow convection are not obvious in the large-scale motion of the atmosphere, the fact that circulation of heat is present is clear, and it leads us to consider to what extent the motions are rotational in nature. We shall reach the conclusion that motion on all scales has a significant amount of spin associated with it, and that by studying the spin we can learn a great deal about atmospheric phenomena.

To illustrate what is meant by spin, consider a stick floating in a stream. As it moves with the current, it will nearly always be spinning to some degree, and the more intense the motions of the stream, the more the stick will spin. In this case, the stick gives evidence of the fact that the parcels of water have a spin imposed on their translational motion.

We see the same spin in the atmosphere. A leaf carried in the breeze will roll and spin, first in one eddy and then in another. Snow falling on a windy day swirls its way from cloud to ground, the white flakes illuminating the spinning chaos of the turbulent winds. At slightly larger scales, we see dust devils, scooping up sand for a brief ride in a whirlwind that will disappear from sight in a few seconds. The same whirlwinds can be seen as snow devils, or as tiny waterspouts. The most intense atmospheric vortices are tornadoes, with tangential wind speeds of several hundred miles per hour.

Satellite photographs provide direct visual evidence of the circulation and large-scale spin in hurricanes and extratropical cyclones (Fig. 10.1). And even in the great, meandering jets of the atmosphere, the air spins its way around the globe.

The spin of the air in motion derives in part from thermodynamic forcing and in part from the spin of the earth itself. Fluid dynamicists and atmospheric scientists have found it useful to study the spin of a fluid as a separate variable derived from the motion field by differentiation. It is called *vorticity*, and vorticity is a variable whose properties reveal much about fluid motion and the thermodynamic motives that it seeks to satisfy.

10.1 THE CONCEPT OF VORTICITY

In this section we shall attempt to show precisely how vorticity quantifies the spin of a parcel and illustrate as succinctly as possible that vorticity is a motion field variable closely entwined with thermodynamic properties.

Vorticity is defined as the curl of the velocity field, so that we have

$$\boldsymbol{\zeta} = \nabla \times \mathbf{v} = \mathbf{i}\left(\frac{\partial w}{\partial y} - \frac{\partial v}{\partial z}\right) + \mathbf{j}\left(\frac{\partial u}{\partial z} - \frac{\partial w}{\partial x}\right) + \mathbf{k}\left(\frac{\partial v}{\partial x} - \frac{\partial u}{\partial y}\right)$$
$$= \mathbf{i}\xi + \mathbf{j}\eta + \mathbf{k}\zeta \qquad (1)$$

The components of the vorticity ξ, η, and ζ are measures of the spin about the x, y, and z axes. Most meteorological texts emphasize the vertical component ζ for reasons we shall discuss in the following sections.

The definition (1) makes it clear that vorticity arises in the variations of a motion field. We shall see how the mathematical curl operation quantifies spin with the following example.

Suppose we have a parcel of air, moving along its trajectory and also spinning about a vertical axis through the center of the parcel. We will assume the parcel completes one revolution about its own axis in time T_p, so that it has an angular velocity $\Omega_p = 2\pi/T_p$. If the parcel's trajectory were circular with radius R, then the tangential speed V of the parcel along the trajectory would give an angular velocity $\Omega = V/R$ with respect to the origin of the circle, and it would take a time $T = 2\pi/\Omega$ for the parcel to complete one revolution on the trajectory.

On a trip around the circle the parcel will spin on its axis and also rotate once because of the revolution, just as the earth does in its orbit around the sun. When we computed the Coriolis force in Sec. 7.1.2, we took both into account in determining the angular velocity of the earth with respect to the stars. We must do the same for our parcel.

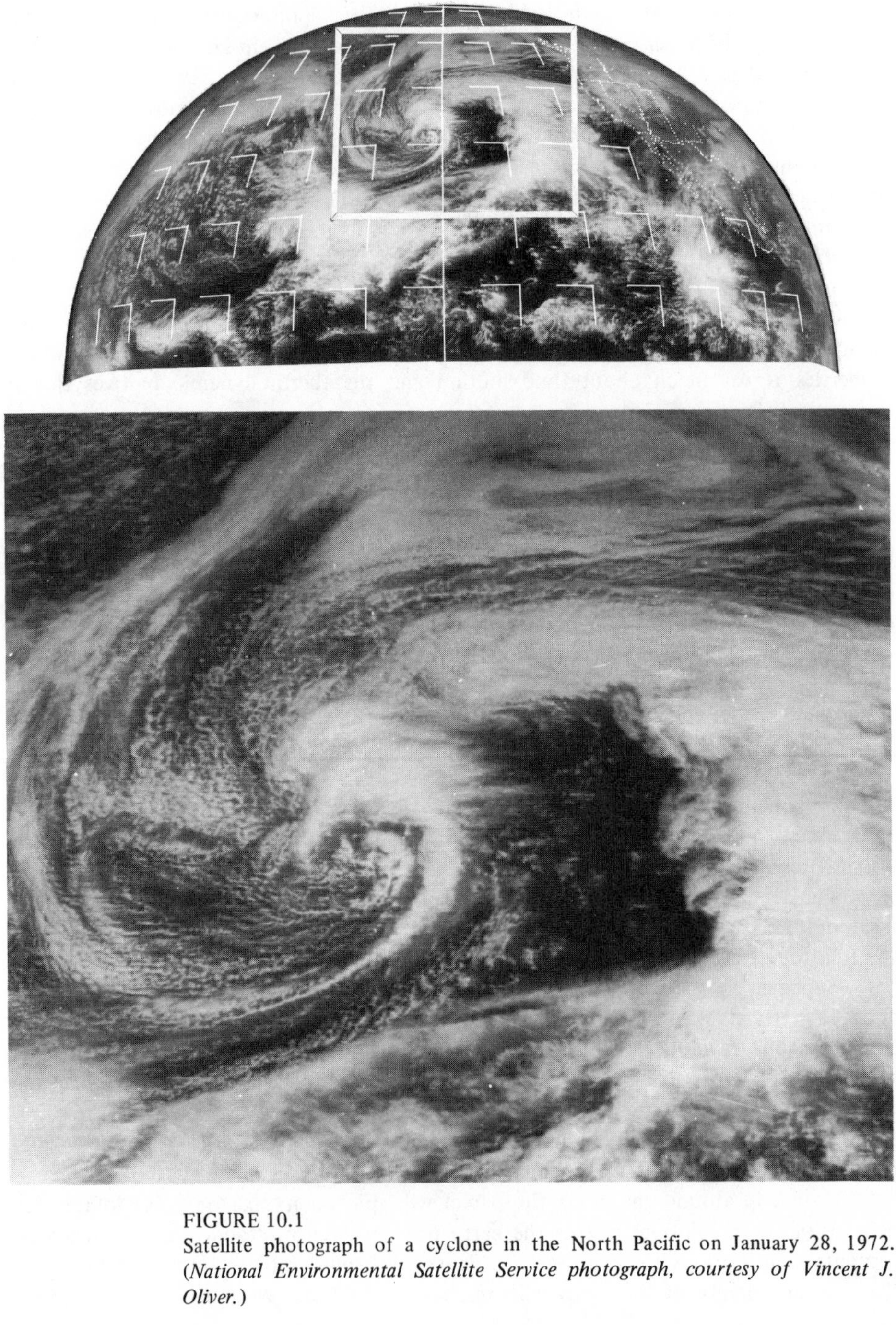

FIGURE 10.1
Satellite photograph of a cyclone in the North Pacific on January 28, 1972. (*National Environmental Satellite Service photograph, courtesy of Vincent J. Oliver.*)

In one revolution, then, around its trajectory, the parcel completes $N = T/T_p = T\Omega_p/2\pi$ rotations about its own axis. The total number of rotations per unit time is therefore given by

$$\frac{N+1}{T} = \frac{\Omega_p}{2\pi} + \frac{1}{T} = \frac{\Omega_p + \Omega}{2\pi} \tag{2}$$

Thus the angular velocity of the parcel with respect to a distant observer is

$$\Omega_{\text{total}} = \Omega_p + \Omega \tag{3}$$

Now we have to see how this compares with the definition of vorticity. For a body in solid rotation, we use cylindrical coordinates with r the radial distance from the axis of rotation $\mathbf{k}$. Then the only component of the vorticity is along this axis and

$$\zeta = \mathbf{k} \cdot \nabla \times \mathbf{V} = \frac{1}{r}\frac{\partial}{\partial r}(rV) \tag{4}$$

where we have used $\nabla \times \mathbf{V}$ in cylindrical coordinates. But because of the solid rotation at angular velocity Ω, we have $V = \Omega r$ and so $\zeta = 2\Omega$. In this case, then, vorticity is exactly twice the angular velocity.

This result allows us to establish the sign conventions used in studying vorticity. The angular velocity Ω is taken to be positive if the rotation is in the counterclockwise direction when viewed from above. Since $\zeta = 2\Omega$, the same is true for vorticity. Thus cyclones have positive vorticity, and anticyclones have negative vorticity (see Fig. 10.2).

Now let us imagine a current of air flowing in a horizontal plane (see Fig. 10.3). We can choose one of the parcels and use a cylindrical coordinate system at the origin of the circle tangent to the trajectory of the parcel. We let the radial distance be r and denote the distance to the center of the parcel by r_0 (the radius of curvature of the trajectory).

Then the air in the neighborhood of the parcel has a tangential velocity component

$$v_\theta = V + \Omega_p(r - r_0) = \Omega r_0 + \Omega_p(r - r_0) \tag{5}$$

The air also has a radial velocity component given by

$$v_r = -\Omega_p r(\theta - \theta_0) \tag{6}$$

where $r\,\Delta\theta = r(\theta - \theta_0)$ is the half width of the parcel as measured by the angle $\Delta\theta$, subtended at the origin of the coordinates. Now the vorticity is

$$\zeta = \mathbf{k} \cdot (\nabla \times \mathbf{v}) = \left\{ \frac{1}{r}\left[\frac{\partial}{\partial r}(rv_\theta) - \frac{\partial v_r}{\partial \theta}\right] \right\}_{\substack{r = r_0 \\ \theta = \theta_0}} = 2(\Omega + \Omega_p) \tag{7}$$

so once again the vorticity is twice the angular velocity.

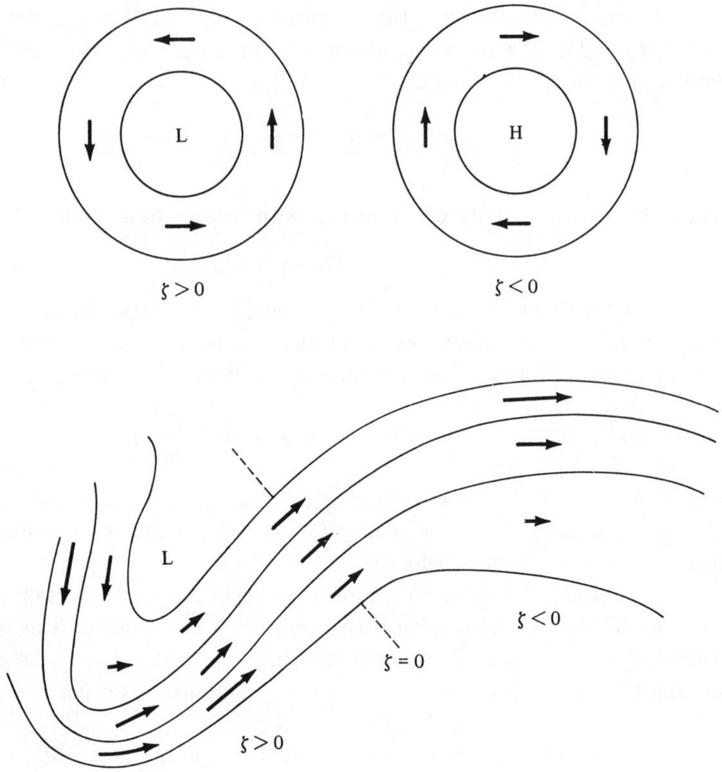

FIGURE 10.2
Sketch of wind patterns associated with surface highs and lows showing the counterclockwise spin and positive vorticity of a cyclone and the negative vorticity of an anticyclone. The lower sketch indicates the vorticity in a trough and a ridge.

Equation (7) thus makes it clear that vorticity is composed of two components–one due to curvature of the trajectory, the other due to wind shear that drives the rotation of the parcel on its own axis, just as a tumbleweed is rolled along the ground. In order for the vorticity to vanish, these two components must both vanish or cancel each other. If there is curvature, the vorticity can vanish only if the parcel rotates slightly in the opposite sense.

For the atmosphere, either we can consider the vorticity ζ of the air relative to the earth, or to obtain the absolute vorticity, we can add to the relative vorticity the contribution 2Ω due to the rotation of the earth. The contribution to the vertical component of the vorticity becomes $2\Omega \sin \phi$, so that $\zeta_a = \zeta + f$.

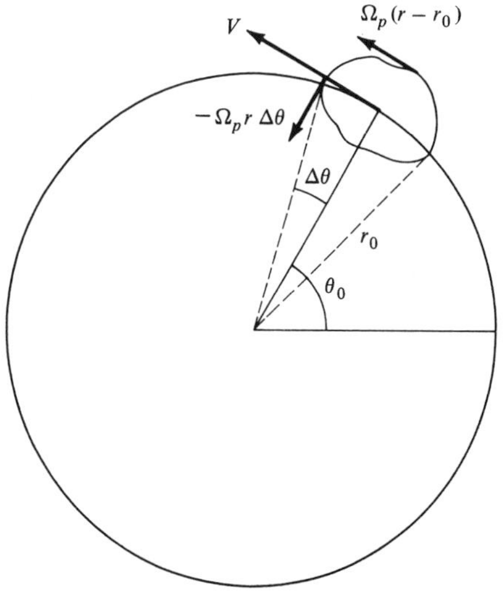

FIGURE 10.3
Illustration of a parcel rotating with angular velocity Ω_p about its own axis while moving around a circular trajectory.

PROBLEM 10.1.1 Show that the vertical component of the vorticity for a horizontal flow depicted in natural coordinates based on the streamlines (Problem 9.7.3) is $\zeta = v/R_S - \partial v/\partial n$. (Hint: Evaluate the curl directly; do not use a determinant. Note that $\partial \tau/\partial n$ need not be calculated explicitly.)

PROBLEM 10.1.2 Consider a flow with concentric, circular streamlines. What velocity pattern is necessary if the vorticity is to vanish? Is the same result valid for cyclones and anticyclones? Is the condition likely to be met in atmospheric systems that are nearly circular?

PROBLEM 10.1.3 Evaluate the absolute vorticity ζ_a for typical values in mid-latitude flow to show that generally $f > |\zeta|$. Find numerical values that would make the absolute vorticity negative.

Now that we have shown what vorticity is, let us see why it is important. To do so, we shall anticipate a general theorem to be studied in Sec. 10.11 and present now a special case. We begin by noting that the equation of motion for an inviscid gas, not in a rotating coordinate system, and free from external forces, is given by Eq. (24) in Sec. 6.5 as

$$\frac{\partial \mathbf{v}}{\partial t} + \nabla \frac{v^2}{2} - \mathbf{v} \times \boldsymbol{\zeta} = T\,\nabla s - c_p\,\nabla T \tag{8}$$

and because $\nabla \times (\nabla\phi)$ vanishes for any scalar ϕ, we have

$$\frac{\partial \boldsymbol{\zeta}}{\partial t} - \nabla \times (\mathbf{v} \times \boldsymbol{\zeta}) = \nabla T \times \nabla s \tag{9}$$

With another vector identity, Eq. (37) in Sec. 5.1, we find

$$\nabla \times (\mathbf{v} \times \boldsymbol{\zeta}) = \mathbf{v}\, \nabla \cdot \boldsymbol{\zeta} + (\boldsymbol{\zeta} \cdot \nabla)\mathbf{v} - \boldsymbol{\zeta}(\nabla \cdot \mathbf{v}) - \mathbf{v} \cdot \nabla \boldsymbol{\zeta} \tag{10}$$

The first term vanishes because $\nabla \cdot (\nabla \times \mathbf{v}) = 0$, and so Eq. (8) becomes

$$\frac{d\boldsymbol{\zeta}}{dt} = -\boldsymbol{\zeta}\, \nabla \cdot \mathbf{v} + (\boldsymbol{\zeta} \cdot \nabla)\mathbf{v} + \nabla T \times \nabla s \tag{11}$$

We have three ways in which the vorticity of a material element can change during inviscid motion:

1 Convergence or divergence will alter vorticity if any is present, with convergence increasing the length of the vorticity vector.

2 Vortex stretching or tilting will alter vorticity as shown by the term $(\boldsymbol{\zeta} \cdot \nabla)\mathbf{v} = |\boldsymbol{\zeta}|\partial \mathbf{v}/\partial l$, in which l is distance in the direction of the vorticity vector, so that vorticity will be increased if the velocity is increasing along the vorticity vector.

3 Baroclinity will alter vorticity, or create vorticity if none is present, since intersections of the isothermal and isentropic surfaces will produce a cross product.

When the divergence in Eq. (11) is eliminated with the continuity equation, we obtain

$$\frac{d\boldsymbol{\zeta}}{dt} + \frac{\boldsymbol{\zeta}}{\alpha}\frac{d\alpha}{dt} = (\boldsymbol{\zeta} \cdot \nabla)\mathbf{v} + \nabla T \times \nabla s \tag{12}$$

which may be written as

$$\frac{d}{dt}(\alpha\boldsymbol{\zeta}) = (\alpha\boldsymbol{\zeta} \cdot \nabla)\mathbf{v} + \alpha\, \nabla T \times \nabla s \tag{13}$$

This result is known as the *Beltrami vorticity equation*. For steady flow (as in Sec. 6.5.4), Eq. (8) becomes

$$\boldsymbol{\zeta} \times \mathbf{v} = T\, \nabla s - \nabla\left(c_p T + \frac{v^2}{2}\right) \tag{14}$$

and shows that the vorticity is closely associated with the thermodynamic properties of the flow.

With this introduction to the concept of vorticity completed, we shall turn now to the study of meteorological vorticity equations and to their application to problems in analysis and prediction of weather patterns. Following that, we shall study the concept of circulation—a form of average vorticity—and investigate the properties of vortices embedded in fluid motion. This will lead us to vector vorticity equations that describe the entire vorticity vector rather than the vertical component considered in synoptic studies. Then we adopt a different viewpoint and look at vorticity theorems for moving parcels, both for incompressible and for isentropic flows. Finally, we end the chapter with a discussion of the effects of friction on vorticity.

PROBLEM 10.1.4 Can the orientation of a vorticity vector be determined in a steady flow without calculating $\nabla \times \mathbf{v}$? (Hint: Refer to Problem 6.5.8.)

PROBLEM 10.1.5 Show that the vortex stretching term $(\boldsymbol{\zeta} \cdot \nabla)\mathbf{v}$ vanishes in two-dimensional flow.

10.2 METEOROLOGICAL VORTICITY EQUATIONS

Vorticity has been an important and relevant concept in analysis of atmospheric flow patterns and the development of simple models to aid in making predictions for a number of reasons. One is that the flows can be decomposed (see Sec. 5.6) into the sum of two vectors: one is nondivergent but has vorticity, the other is divergent but irrotational. For large-scale weather patterns in mid-latitudes, the nondivergent, rotational vector is usually by far the larger of the two, so that by studying vorticity we are studying a representation of most of the flow. A second reason is that the geostrophic vorticity gives a fairly good approximation to the actual vertical component of the vorticity. Thus vorticity is important, and moreover it can be estimated from the height or pressure fields that are more easily observed and analyzed than wind fields.

In this section we shall develop the basic meteorological vorticity equations that are in common use for studying systems that are at least the size of developing cyclones. The equations of motion (18) in Sec. 7.2 are

$$\begin{aligned} \frac{\partial u}{\partial t} + \mathbf{v} \cdot \nabla u &= -\frac{1}{\rho}\frac{\partial p}{\partial x} + fv \\ \frac{\partial v}{\partial t} + \mathbf{v} \cdot \nabla v &= -\frac{1}{\rho}\frac{\partial p}{\partial y} - fu \\ \frac{\partial p}{\partial z} &= -g\rho \end{aligned} \tag{1}$$

in which we are ignoring the terms that depend on the radius of the earth. The three components of the vorticity, Eq. (1) in Sec. 10.1, are

$$\xi = \frac{\partial w}{\partial y} - \frac{\partial v}{\partial z} \qquad \eta = \frac{\partial u}{\partial z} - \frac{\partial w}{\partial x} \qquad \zeta = \frac{\partial v}{\partial x} - \frac{\partial u}{\partial y} \tag{2}$$

so that from the equations (1) we can find directly an equation only for the vertical component ζ. Determination of the other two components necessitates knowing the vertical motion, which is not directly available from Eqs. (1) because of the hydrostatic approximation. Thus meteorologists usually can combine only the first two equations to form a vorticity equation. As made clear by Eq. (2), we want to differentiate the v equation with respect to x and subtract the derivative of the u equation with respect to y. Thus

$$\frac{\partial}{\partial t}\frac{\partial v}{\partial x} + \mathbf{v}\cdot\nabla\frac{\partial v}{\partial x} + \frac{\partial \mathbf{v}}{\partial x}\cdot\nabla v = -\frac{\partial \alpha}{\partial x}\frac{\partial p}{\partial y} - \alpha\frac{\partial^2 p}{\partial x\,\partial y} - f\frac{\partial u}{\partial x} \tag{3}$$

and

$$\frac{\partial}{\partial t}\frac{\partial u}{\partial y} + \mathbf{v}\cdot\nabla\frac{\partial u}{\partial y} + \frac{\partial \mathbf{v}}{\partial y}\cdot\nabla u = -\frac{\partial \alpha}{\partial y}\frac{\partial p}{\partial x} - \alpha\frac{\partial^2 p}{\partial y\,\partial x} + f\frac{\partial v}{\partial y} + \frac{\partial f}{\partial y}v \tag{4}$$

The difference gives, upon use of Eq. (2),

$$\frac{\partial \zeta}{\partial t} + \mathbf{v}\cdot\nabla\zeta + \frac{\partial \mathbf{v}}{\partial x}\cdot\nabla v - \frac{\partial \mathbf{v}}{\partial y}\cdot\nabla u = \frac{\partial \alpha}{\partial y}\frac{\partial p}{\partial x} - \frac{\partial \alpha}{\partial x}\frac{\partial p}{\partial y} - f\left(\frac{\partial u}{\partial x} + \frac{\partial v}{\partial y}\right) - \beta v \tag{5}$$

in which we have used $\beta = \partial f/\partial y$ so that $df/dt = \beta v$.

Now let us observe that

$$\frac{\partial \mathbf{v}}{\partial x}\cdot\nabla v - \frac{\partial \mathbf{v}}{\partial y}\cdot\nabla u = \frac{\partial u}{\partial x}\left(\frac{\partial v}{\partial x} - \frac{\partial u}{\partial y}\right) + \frac{\partial v}{\partial y}\left(\frac{\partial v}{\partial x} - \frac{\partial u}{\partial y}\right) + \frac{\partial w}{\partial x}\frac{\partial v}{\partial z} - \frac{\partial w}{\partial y}\frac{\partial u}{\partial z} \tag{6}$$

and so the vorticity equation becomes

$$\frac{d(\zeta + f)}{dt} = \frac{\partial}{\partial t}(\zeta + f) + \mathbf{v}\cdot\nabla(\zeta + f) = -(\zeta + f)\left(\frac{\partial u}{\partial x} + \frac{\partial v}{\partial y}\right) + \frac{\partial \alpha}{\partial y}\frac{\partial p}{\partial x} - \frac{\partial \alpha}{\partial x}\frac{\partial p}{\partial y} + \frac{\partial w}{\partial y}\frac{\partial u}{\partial z} - \frac{\partial w}{\partial x}\frac{\partial v}{\partial z} \tag{7}$$

We shall make three notational changes to simplify the expression on the right. The horizontal divergence appearing in the first term may be written

$$\nabla_H \cdot \mathbf{v} = \frac{\partial u}{\partial x} + \frac{\partial v}{\partial y} \tag{8}$$

The second term is known as the *solenoidal term* for reasons that will be apparent when we consider the circulation theorem in Sec. 10.9. We may use the Jacobian notation

$$J(p,\alpha) = \frac{\partial(p,\alpha)}{\partial(x,y)} = \begin{vmatrix} \dfrac{\partial p}{\partial x} & \dfrac{\partial p}{\partial y} \\ \dfrac{\partial \alpha}{\partial x} & \dfrac{\partial \alpha}{\partial y} \end{vmatrix} = \frac{\partial p}{\partial x}\frac{\partial \alpha}{\partial y} - \frac{\partial p}{\partial y}\frac{\partial \alpha}{\partial x} = \mathbf{k}\cdot(\nabla p \times \nabla\alpha) \tag{9}$$

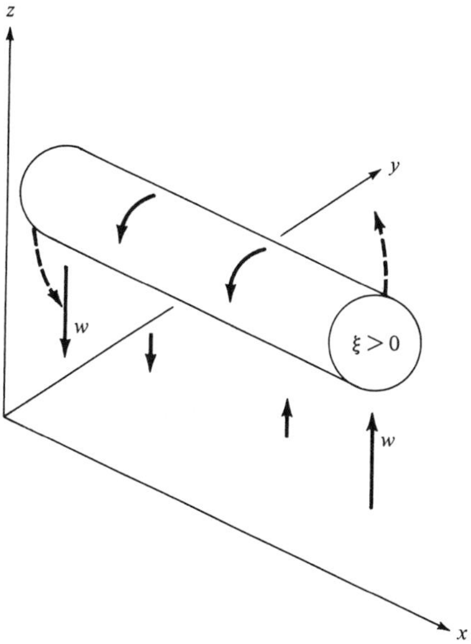

FIGURE 10.4
The tilting of a vortex with vorticity ξ produces a vertical component ζ.

The last term is known as the *tilting term.* From Eq. (2) we see that

$$\frac{\partial w}{\partial y}\frac{\partial u}{\partial z} = \frac{\partial w}{\partial y}\left(\eta + \frac{\partial w}{\partial x}\right) \qquad \frac{\partial w}{\partial x}\frac{\partial v}{\partial z} = \frac{\partial w}{\partial x}\left(\frac{\partial w}{\partial y} - \xi\right) \tag{10}$$

and so

$$\frac{\partial w}{\partial y}\frac{\partial u}{\partial z} - \frac{\partial w}{\partial x}\frac{\partial v}{\partial z} = \eta\frac{\partial w}{\partial y} + \xi\frac{\partial w}{\partial x} \tag{11}$$

Suppose that we have positive vorticity ξ and vertical motion such that $\partial w/\partial x > 0$. Then a material curve, initially in a vertical plane, on which air is circulating around the x axis will be tilted by the vertical motion so that a cyclonic contribution is made to $d\zeta/dt$. (See Fig. 10.4.) This effect is expressed by Eq. (11) and is significant, for it allows the relatively large vorticities ξ and η to be converted into vorticity ζ.

Thus our large-scale vorticity equation is

$$\frac{d}{dt}(\zeta + f) = -(\zeta + f)\nabla_H \cdot \mathbf{v} + J(p, \alpha) + \xi\frac{\partial w}{\partial x} + \eta\frac{\partial w}{\partial y} \tag{12}$$

The absolute vorticity is normally positive in mid-latitudes because of the presence of f. Suppose we have approximately uniform flow at speed V around a curve of radius R. Then from Problem 10.1.1, $\zeta = V/R$, and so for the atmosphere we have the estimate

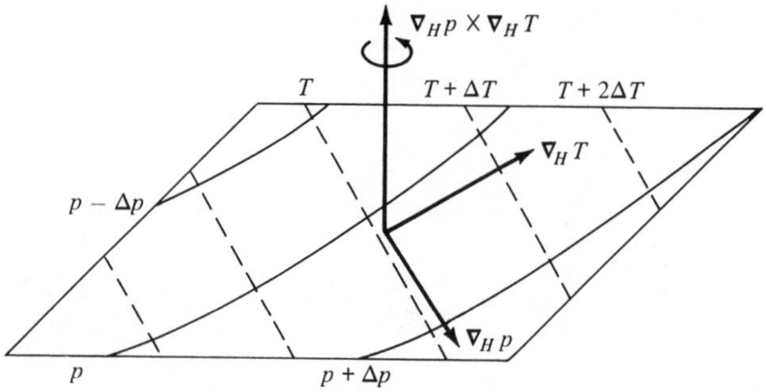

FIGURE 10.5
The vorticity will tend to increase in the direction from the tip of $\nabla_H p$ to the tip of $\nabla_H T$ (through the acute angle) owing to the solenoidal term.

$$\zeta = \pm \frac{10 \text{ m/s}}{1000 \text{ km}} = \pm 10^{-5} \text{ s}^{-1} \tag{13}$$

with the plus sign for cyclonic flow, the negative sign for anticyclonic flow. Because $f \sim 10^{-4}$ in mid-latitudes, it clearly dominates.

The absolute vorticity $\zeta + f$ of a moving parcel can change because:

1 Horizontal convergence ($\nabla_H \cdot \mathbf{v} < 0$) causes the absolute vorticity to become more cyclonic (provided $\zeta + f > 0$) or divergence causes the vorticity to become less cyclonic. Note then that vertical stretching of an air column increases its vorticity.

2 Baroclinity, as expressed in the solenoidal term, causes the spin to accelerate in one direction or the other. This term is evaluated quantitatively by observing that

$$J(p, \alpha) = \mathbf{k} \cdot (\nabla p \times \nabla \alpha) = \mathbf{k} \cdot (\nabla_H p \times \nabla_H \alpha)$$
$$= \mathbf{k} \cdot (\nabla_H p \times \nabla_H T) \frac{R}{p} \tag{14}$$

Thus the vorticity will increase in the direction of the path from the tip of $\nabla_H p$ to the tip of $\nabla_H T$ across the angle (less than π) between them (see Fig. 10.5).

3 Horizontally varying vertical motion tilts vorticity about horizontal axes into vorticity about the vertical axes. Conversely, vertical shear of the horizontal velocity creates vorticity η or ξ by tilting vertical vortices.

Thus the vorticity field is coupled to the field of horizontal convergence, it is changed by baroclinity, and the three vorticity components interact to bring about changes in each other.

The next step is to estimate the relative size of these effects in the large-scale flow we are considering. First we shall need an estimate for the horizontal divergence.

In isobaric coordinates we have

$$\nabla_p \cdot \mathbf{v} = -\frac{\partial \omega}{\partial p} \tag{15}$$

where $\omega = dp/dt \sim -g\rho w$. The vertical velocity will normally be a maximum in the neighborhood of 600 mbar, and we estimate a typical maximum magnitude in a cyclone of about 10 cm/s. Standard density at 600 mbar is about 8×10^{-4} g/cm^3, and so $\omega(600) \sim -8 \times 10^{-3}$ mbar/s. At the surface, we have $\omega = \partial p/\partial t$, and so a pressure fall of 1 mbar/h will give $\omega(1000) = -3 \times 10^{-4}$ mbar/s. Thus $|\partial\omega/\partial p| = 2 \times 10^{-5}$ s^{-1}, and so we have an estimate that $|\nabla_H \cdot \mathbf{v}| \sim |\nabla_p \cdot \mathbf{v}| \sim 2 \times 10^{-5}$ s^{-1}.

In Eq. (14), we use the geostrophic wind and thermal wind to express the solenoidal term as

$$\begin{aligned} J(p,\alpha) &= \mathbf{k} \cdot \left\{ [(\mathbf{v}_g \times \mathbf{k})\rho f] \times \left[\left(\frac{\partial \mathbf{v}_g}{\partial z}\right) \times \mathbf{k}\frac{fT}{g}\right] \right\} \frac{R}{p} \\ &= -\frac{f^2}{g}(\mathbf{v}_g \times \mathbf{k}) \cdot \frac{\partial \mathbf{v}_g}{\partial z} \end{aligned} \tag{16}$$

Thus we can estimate that $J(p,\alpha) \sim f^2 g^{-1}$ (20 m/s) (20 m/s)/10 km $\sim 4 \times 10^{-11}$ s^{-2}.

Finally we must estimate the tilting term. We have

$$\frac{\partial w}{\partial y}\frac{\partial u}{\partial z} = \frac{20 \text{ cm/s}}{500 \text{ km}}\,\frac{20 \text{ m/s}}{10 \text{ km}} \sim \frac{2 \times 10^{-1}}{5 \times 10^{5}} \times \frac{2 \times 10^{1}}{10^{4}} \text{ s}^{-2} = 0.8 \times 10^{-9} \text{ s}^{-2} \tag{17}$$

Note that in the vicinity of an upper-level front we might obtain this variation of w in 50 km, giving an estimate of 10^{-8} s^{-2} for the tilting term.

Therefore we have the estimates

$$\begin{aligned} &\frac{d}{dt}(\zeta + f) = -(\zeta + f)\nabla_H \cdot \mathbf{v} + J(p,\alpha) + \xi\frac{\partial u}{\partial z} + \eta\frac{\partial v}{\partial z} \\ &\qquad\qquad 10^{-4} \quad 10^{-5} \qquad 10^{-11} \qquad 10^{-8} \text{ to } 10^{-9} \qquad \text{s}^{-2} \\ &\qquad\qquad 10^{-9} \end{aligned} \tag{18}$$

The smallness of the terms on the right suggests that conservation of absolute vorticity may be effectively achieved in much of the atmosphere.

10.2.1 Isobaric Coordinates

The vorticity equation becomes slightly simpler in isobaric coordinates because of the use of the hydrostatic equation. The equations (1) are, for isobaric coordinates,

$$\begin{aligned} \frac{\partial u}{\partial t} + \mathbf{v} \cdot \nabla_p u + \omega\frac{\partial u}{\partial p} &= -g\frac{\partial h}{\partial x} + fv \\ \frac{\partial v}{\partial t} + \mathbf{v} \cdot \nabla_p v + \omega\frac{\partial v}{\partial p} &= -g\frac{\partial h}{\partial y} - fu \end{aligned} \tag{19}$$

Now clearly there will be no solenoidal term because the potential gh will be eliminated by the cross-differentiation. Furthermore the tilting term will now involve ω and $\partial(\ )/\partial p$. Thus we find

$$\frac{d}{dt}(\zeta + f)_p = -(\zeta + f)_p \, \nabla_p \cdot \mathbf{v} + \left(\frac{\partial \omega}{\partial y}\frac{\partial u}{\partial p} - \frac{\partial \omega}{\partial x}\frac{\partial v}{\partial p}\right)_p \tag{20}$$

where the subscript p emphasizes that differentiation is to be performed on an isobaric surface.

It should be remembered that the vorticity is measured with respect to an axis of rotation in the direction $\mathbf{k}$; as stressed in Chap. 7, the vertical unit vector does *not* tilt to become the unit normal to the isobaric surface.

PROBLEM 10.2.1 Derive Eq. (20) from Eq. (19).

10.2.2 Isentropic Coordinates

Yet another simplification occurs for isentropic flow. In isentropic coordinates we have the equations ($\dot{\theta} = d\theta/dt$)

$$\begin{aligned} \frac{\partial u}{\partial t} + \mathbf{v} \cdot \nabla_\theta u + \dot{\theta}\frac{\partial u}{\partial \theta} &= -\frac{\partial \Psi}{\partial x} + fv \\ \frac{\partial v}{\partial t} + \mathbf{v} \cdot \nabla_\theta v + \dot{\theta}\frac{\partial v}{\partial \theta} &= -\frac{\partial \Psi}{\partial y} - fu \end{aligned} \tag{21}$$

and so

$$\frac{d}{dt}(\zeta + f)_\theta = -(\zeta + f)_\theta \, \nabla_\theta \cdot \mathbf{v} + \left(\frac{\partial \dot{\theta}}{\partial y}\frac{\partial u}{\partial \theta} - \frac{\partial \dot{\theta}}{\partial x}\frac{\partial v}{\partial \theta}\right)_\theta \tag{22}$$

Thus when the flow is isentropic, we have the simple result

$$\frac{d}{dt}(\zeta + f)_\theta = -(\zeta + f)_\theta \, \nabla_\theta \cdot \mathbf{v} \tag{23}$$

and so in this case the vorticity will change only through horizontal convergence. This is the simplest vorticity equation we have yet encountered, but the remarkable fact is that it can be made even simpler; in the process, we shall find a quantity more conservative than vorticity itself. This new variable will be called *potential vorticity*, and we shall return to it in Sec. 10.11.

We begin by rewriting Eq. (23) in the form

$$\frac{\partial}{\partial t_\theta}(\zeta_\theta + f) + \mathbf{v} \cdot \nabla_\theta(\zeta_\theta + f) = -(\zeta_\theta + f) \, \nabla_\theta \cdot \mathbf{v} \tag{24}$$

From the isentropic continuity equation for isentropic, hydrostatic flow given in Eq.

(53) in Sec. 7.3 we have

$$\frac{\partial}{\partial t_\theta}\left(\frac{\partial p}{\partial \theta}\right) + \mathbf{v} \cdot \nabla_\theta \frac{\partial p}{\partial \theta} + \frac{\partial p}{\partial \theta} \nabla_\theta \cdot \mathbf{v} = 0 \tag{25}$$

But $\partial p/\partial\theta = (\partial\theta/\partial p)^{-1}$, and so Eq. (25) may be written as

$$\frac{\partial}{\partial t_\theta}\left(\frac{\partial \theta}{\partial p}\right) + \mathbf{v} \cdot \nabla_\theta\left(\frac{\partial \theta}{\partial p}\right) - \left(\frac{\partial \theta}{\partial p}\right) \nabla_\theta \cdot \mathbf{v} = 0 \tag{26}$$

Now the sum of Eq. (24) multiplied by $\partial\theta/\partial p$ and Eq. (26) multiplied by $(\zeta_\theta + f)$ yields

$$\frac{\partial}{\partial t_\theta}\left[(\zeta_\theta + f)\frac{\partial \theta}{\partial p}\right] + \mathbf{v} \cdot \nabla_\theta\left[(\zeta_\theta + f)\frac{\partial \theta}{\partial p}\right] = 0 \tag{27}$$

which for the assumed isentropic flow may be written as

$$\frac{d}{dt}\left[(\zeta_\theta + f)\frac{\partial \theta}{\partial p}\right] = 0 \tag{28}$$

Therefore, this new quantity, the potential vorticity,

$$Z_\theta = (\zeta_\theta + f)\frac{\partial \theta}{\partial p} \tag{29}$$

is conserved in inviscid, isentropic, hydrostatic flow. Thus in situations characteristic of large-scale atmospheric flows without condensation or evaporation, the product of the absolute vorticity as measured in isentropic coordinates and the stability factor $\partial\theta/\partial p$ is constant for the moving parcels.

Potential vorticity is one of the few variables that can be used as a tracer to tag parcels of air. Moreover, it allows meteorologists the luxury of vortex stretching as an explanation of observed phenomena.

For example, consider a straight, westerly flow with positive absolute vorticity that encounters a north-south mountain range. If the air flows over the range isentropically, then $\partial\theta/\partial p$ will change from a relatively small negative value to a large negative value at the peak because $\Delta\theta$ does not change but Δp becomes smaller in absolute value as the air layer is compressed over the mountain. But Z_θ must be conserved, and so $\zeta_\theta + f$ must decrease rapidly. If the air flows over the range without changing the horizontal shear present in the original flow, then the flow must acquire anticyclonic curvature to the south. As Δp returns to the original value in the lee of the mountains, f is smaller and the flow will curve back toward the north. Hence a trough is formed in the lee of the range. Note that in this example, however, we have made a crucial assumption: that the shear does not change, so the required vorticity change is allowed to develop only in the anticyclonic curvature initiated at the mountain range.

FIGURE 10.6
The tornado that struck Union City, Oklahoma, on May 24, 1973, as photographed by Dr. J. H. Golden as part of the Tornado Intercept Project of

As another example, meteorologists have long hoped that an explanation for the immense concentrations of vorticity present in tornadoes (Fig. 10.6) might be found with this equation. If we assume that $\zeta_\theta + f$ is positive—as is usually the case—then clearly a sharp decrease in $|\partial\theta/\partial p|$ must occur for the vorticity to be concentrated in a cyclonic tornado. It seems unlikely that a layer of air would undergo violent changes in the potential temperature difference from top to bottom, but it is conceivable that the pressure difference Δp might change rapidly. In a strong updraft, a part of a layer might be pulled suddenly upward so that $|\partial\theta/\partial p|$ would decrease markedly. Or perhaps a large blob of cold air in the thrashing currents of a thunderstorm might be ejected over a layer of warm, lighter air. The sudden sinking and stretching of such air might yield the necessary impetus for tornado formation.

The tilting terms in the vorticity equation (7) give another possible explanation for tornado formation in a thunderstorm. Suppose that in the outdraft ahead of the storm we had a vertical shear of 20 m/s in 100 m in the u component. Suppose also that the vertical velocity reversed from 1 m/s upward to 1 m/s downward in a distance of 100 m. This combination would produce a rate of change of vorticity of 4×10^{-3} s^{-2}. Now the vorticity in the core of a tornado may be estimated from $\zeta = V/R \sim (100 \text{ m/s})/(100 \text{ m}) = 1 \text{ s}^{-1}$. Thus if tilting term effects of the magnitude estimated above could be maintained for 5 minutes, then vorticities of the order of those encountered might be produced.

The point of both these hypotheses is that since tornadoes have immense vorticity, they can form only when conditions are such that one or more of the terms in the vorticity equation become large.

Having thus given brief consideration to the most intense of atmospheric vortices, we shall turn in the next few sections to large-scale flow and use the isobaric vorticity equation to develop some models that are useful in analyzing weather situations and making predictions.

PROBLEM 10.2.2 Derive Eq. (22) from Eq. (21).

PROBLEM 10.2.3 Show that for isentropic flow with $(\zeta_\theta + f) > 0$, $(\zeta_\theta + f)A = (\zeta_\theta + f)_0 A_0$ where A is the horizontal area of a parcel and the subscript denotes initial conditions. What happens when a vortex is stretched under these conditions?

FIGURE 10.6 (*continued*)
the National Severe Storms Laboratory. The movies of the tornado showed that the cloud structure from the top of the visible vortex to the top of the photograph was rotating cyclonically, leading to the interpretation that the vortex made visible by condensation and debris is part of a much larger vortex. Single Doppler radar observations of this storm confirm this interpretation. (*National Severe Storms Laboratory photograph, courtesy of Dr. J. H. Golden.*)

PROBLEM 10.2.4 Show that $\zeta_p = \zeta_\theta - (\nabla_\theta p \times \partial \mathbf{v}/\partial p) \cdot \mathbf{k}$. What does this result imply when the flow is barotropic?

*PROBLEM 10.2.5 Show that the averages over any surface $z = \text{const}$ (not intersecting the earth's surface) of both relative vorticity ζ and the absolute vorticity $\zeta + f$ vanish. Does the result remain true in isobaric and isentropic coordinates? (Hint: Use a corollary of Stokes' theorem.)

*PROBLEM 10.2.6 Show (in accordance with the preceding problem) that the vorticity equations imply that the rate of change of average vorticity on any z, p, or θ surface (not intersecting the ground) is zero. [Hint: The sum of the vertical advection and tilting terms in Eq. (7) is $\nabla_H \cdot [(w\, \partial \mathbf{v}_H/\partial z) \times \mathbf{k}]$. Similarly, $J(p,\alpha) = \nabla_H \cdot (p\, \nabla\alpha \times \mathbf{k})$.]

PROBLEM 10.2.7 (Rossby) Using the assumptions that radii of curvature R_S and R_T of the streamline and trajectory are identical, that the shear normal to the flow is zero, and that the velocity V along the trajectory is constant, find a differential equation for the trajectory along which the absolute vorticity of a parcel will be constant. Let the initial value of the Coriolis parameter be f_0 so that for small displacements $y - y_0$ we have $f = f_0 + \beta(y - y_0)$. Use the small-displacement assumption to show that $y - y_0 = A \sin [(x - x_0)\sqrt{\beta/V}]$ for displacements that are initially without curvature so that $R(0) = \infty$. Express the constant A as a function of the direction of motion at the initial point. (Hint: Use the differential definition of radius of curvature in Sec. 9.7.)

PROBLEM 10.2.8 Derive a potential vorticity equation that includes diabatic effects expressed by $\dot{\theta} \neq 0$.

*PROBLEM 10.2.9 For hydrostatic, isentropic flow, show that for any differentiable function $F(Z_\theta)$,

$$\frac{d}{dt}\int_V \rho J_\theta F(Z_\theta)\, dV_\theta = 0$$

when the integral is taken over any material volume, including the entire atmosphere. In particular, show that both the mean and mean-square potential vorticities are conserved for material volumes. Observe that this result gives an infinite set of integrals that are invariants for isentropic flow.

10.3 PRELIMINARIES TO DIAGNOSIS AND PREDICTION WITH VORTICITY EQUATIONS

Meteorologists usually have the problem of having both too much and too little information at hand, and the complexity of the vast amount of data that assails them is often bewildering. The task for both diagnosis and prediction is to summarize the data that are available in a form that makes them comprehensible and permits the missing information to be inferred.

The concept of vorticity and the accompanying vorticity equation have proved to be most useful aids in both diagnosis and prediction of large-scale atmospheric flow. The efficacy of vorticity techniques derives from two facts. Vorticity and vertical motion are associated, so that the systems of interest to the forecaster can be easily portrayed as centers of strong vorticity with accompanying vertical motion patterns that lead to development or inhibition of precipitation. Second, the atmosphere at some levels behaves to a surprising degree like a barotropic fluid, so that the simple properties of vorticity in barotropic flow provide a quantitative and qualitative approach to short-period prediction.

In the following sections we shall investigate some of these uses of vorticity, considering atmospheric waves, cyclone development, and vertical motion. We begin with an illustration of the simplification that appears in two-dimensional flow, and then we shall apply these ideas to some meteorological phenomena. The reasons for the choices that have been made in developing some of these applications can perhaps be illustrated best by considering a simple case from fluid dynamics.

In an inviscid, incompressible fluid ($\nabla \cdot \mathbf{v} = 0$) with autobarotropic stratification the vorticity equation (11) in Sec. 10.1 becomes

$$\frac{d\boldsymbol{\zeta}}{dt} = (\boldsymbol{\zeta} \cdot \nabla)\mathbf{v} \tag{1}$$

Now if in addition the flow is always two-dimensional so that $\partial\mathbf{v}/\partial z = 0$ and $w = 0$, then $\nabla_H \cdot \mathbf{v} = 0$ and $\boldsymbol{\zeta} = \zeta\mathbf{k}$. Thus Eq. (1) becomes

$$\frac{\partial\zeta}{\partial t} + \mathbf{v}_H \cdot \nabla\zeta = 0 \tag{2}$$

where $\mathbf{v}_H$ denotes the horizontal velocity vector, and therefore the vorticity change at a point is due only to advection of vorticity. Because $\nabla_H \cdot \mathbf{v} = 0$, we may define a stream function $\psi = \psi(x,y)$, so that $\mathbf{v} = \mathbf{k} \times \nabla_H\psi$ and thus $\zeta = \nabla_H{}^2\psi$. Hence Eq. (1) becomes

$$\frac{\partial}{\partial t}\nabla_H{}^2\psi + [(\mathbf{k} \times \nabla_H\psi) \cdot \nabla]\,\nabla_H{}^2\psi = \frac{\partial}{\partial t}\nabla_H{}^2\psi + J(\psi, \nabla_H{}^2\psi) = 0 \tag{3}$$

Upon solving this partial differential equation, we may predict the entire future of the flow.

PROBLEM 10.3.1 Develop in general outline a scheme by which Eq. (3) can be used in numerical integration of the equations for inviscid, two-dimensional flow of a fluid of constant density. State suitable boundary conditions, explain the main steps of the solution method, and ascertain which equation must be solved to find the fields of $\mathbf{v}$ and p. Determine whether your procedure will work also for viscous flow.

Now let us relax the condition of two-dimensionality slightly. The vertical component of Eq. (1) is

$$\frac{\partial \zeta}{\partial t} + \mathbf{v}_H \cdot \nabla \zeta + w \frac{\partial \zeta}{\partial z} = (\boldsymbol{\zeta} \cdot \nabla) w \tag{4}$$

and we have as an additional equation

$$\nabla \cdot \mathbf{v} = \frac{\partial u}{\partial x} + \frac{\partial v}{\partial y} + \frac{\partial w}{\partial z} = 0 \tag{5}$$

The expression on the right of Eq. (4) contains the tilting term; in particular with Eq. (2) in Sec. 10.2 we have

$$(\boldsymbol{\zeta} \cdot \nabla) w = \frac{\partial w}{\partial y}\frac{\partial u}{\partial z} - \frac{\partial w}{\partial x}\frac{\partial v}{\partial z} + \zeta \frac{\partial w}{\partial z} \tag{6}$$

If we allow no vertical shear of the horizontal velocity components, we find that Eq. (4) becomes

$$\frac{\partial \zeta}{\partial t} + \mathbf{v}_H \cdot \nabla \zeta = \zeta \frac{\partial w}{\partial z} \tag{7}$$

because $\partial \zeta / \partial z$ vanishes when there is no shear.

But with no shear present, then $\partial(\nabla_H \cdot \mathbf{v})/\partial z = 0$. Thus if there is a level at which the horizontal divergence vanishes, then it vanishes everywhere and $\nabla_H \cdot \mathbf{v} = -\partial w/\partial z = 0$. In this case Eq. (7) is reduced to the same form as Eq. (2).

These have been severe restrictions, and we would not be very much interested in such a restricted flow. But if we could find a level at which the horizontal winds were a maximum and not changing in direction and at which the vertical velocities were extrema, then at that level we would know that $\partial \mathbf{v}_H/\partial z = \partial w/\partial z = 0$.

Thus the vorticity equation for that level would be, once again, of the simple form given in Eq. (2). But if we assume hydrostatic flow, there is no way we could determine the vertical motion from information only on the level of horizontal nondivergence. Still, we could forecast the movement of vorticity centers on that level, and their local development due to advection, and with this information we might become skilled at inferring what sort of phenomena would accompany intensifying vorticity centers. Because of the similarity of the equations of motion in isobaric coordinates to those of incompressible flow, we shall find these concepts useful near the atmosphere's observed level of horizontal nondivergence. A variety of observational studies have revealed that certain aspects of the average structure of the atmosphere are in accordance with the assumptions presented above. We observe that horizontal convergence at the surface changes with altitude to upper-level divergence, and vice versa. Thus a level at which the horizontal divergence vanishes is generally found in each column, and the average altitude of this level in mid-latitudes is about 600 mbar. At the level of nondivergence, the isobaric vertical velocities ω are either a local maximum or minimum as required by the isobaric continuity equation. At higher altitudes, the core of the polar-front jet occurs somewhere between 400 and 200 mbar and we might view the level of maximum winds in the jet as one in which $\partial \mathbf{v}_H/\partial z = 0$.

In the atmosphere, then, the level of horizontal nondivergence and the level of maximum winds do not coincide. But meteorologists have not been able to resist the temptation to consider models of the atmosphere in which they do, to derive the properties of such model atmospheres, and then to compare the results to observed data to see if they have captured any of the essence of atmospheric flow. The success of this approach has been, perhaps, surprising. In any case, the results have pointed out suitable directions for future work.

10.4 LONG WAVES IN THE WESTERLIES (THE ROSSBY THEORY)

The dominant feature of the upper-air flow in mid-latitudes is a strong, westerly current circling the globe, but meandering slightly. The meanders include waves with lengths so long that there are only two or three waves around the entire hemisphere. These are often called *planetary waves*. They are accompanied by superimposed patterns composed of six to eight waves in a train around the hemisphere. These too are long waves, having lengths of several thousand kilometers, and they may be called *synoptic scale waves*.

To examine such waves, let us coalesce the level of maximum wind and the level of nondivergence and thus construct a model inviscid atmosphere in which, at some isobaric level,

$$\nabla_p \cdot \mathbf{v} = \frac{\partial \omega}{\partial p} = 0 \qquad \text{assumption 1} \tag{1}$$

and

$$\frac{\partial \mathbf{v}_H}{\partial p} = 0 \qquad \text{assumption 2} \tag{2}$$

From Eq. (2) we know that $\partial \zeta/\partial p$ will vanish, and so the vorticity equation, for the special level we are considering, becomes

$$\frac{\partial}{\partial t}(\zeta_p + f) + \mathbf{v} \cdot \nabla_p(\zeta_p + f) = 0 \tag{3}$$

Upon expansion of this equation—which is of the form (2) in Sec. 10.3 again—we find

$$\frac{\partial \zeta_p}{\partial t} + u\frac{\partial \zeta_p}{\partial x} + v\frac{\partial \zeta_p}{\partial y} + v\beta = 0 \tag{4}$$

Now we wish to bring the long wave into consideration. To do so, we assume that

$$u = U = \text{const} > 0 \qquad \text{assumption 3} \tag{5}$$

and

$$v = v(x,t) \qquad \text{assumption 4} \tag{6}$$

Thus the long wave is composed of an invariant eastward zonal velocity U and a varying north-south component v, which gives the total motion a wavelike form. Note that we do not allow the wave to change with latitude and we shall henceforth take

$$\beta = \text{const} \qquad \text{assumption 5} \tag{7}$$

But with assumption 3 we find

$$\zeta_p = \left.\frac{\partial v}{\partial x}\right|_p \qquad \frac{\partial \zeta_p}{\partial y} = 0 \tag{8}$$

and so the vorticity equation (4) now becomes

$$\frac{\partial^2 v}{\partial x\, \partial t} + U\frac{\partial^2 v}{\partial x^2} + v\beta = 0 \tag{9}$$

We give $v(x)$ a wave form depending on wavelength L and wave speed c in the form (see Fig. 10.7)

$$v(x,t) = A \cos\left[\frac{2\pi}{L}(x - ct)\right] \tag{10}$$

and use this in Eq. (9) to determine whether any combination of c and L is possible. We find easily that

$$c = U - \beta\frac{L^2}{4\pi^2} \tag{11}$$

This is the celebrated *Rossby wave formula* and clearly gives the speed of the long waves as a function of zonal velocity U and wavelength L. Now note that for short waves we will have $c < U$, but as L gets very large, the waves will retrogress—move from east to west—a behavior rarely observed. The wavelength L_S of the stationary wave is given by $c = 0$, which yields

$$L_S = 2\pi\sqrt{\frac{U}{\beta}} \tag{12}$$

In the atmosphere we observe about three stationary waves—presumably held in place by the distribution of continents and mountain ranges. There are generally five

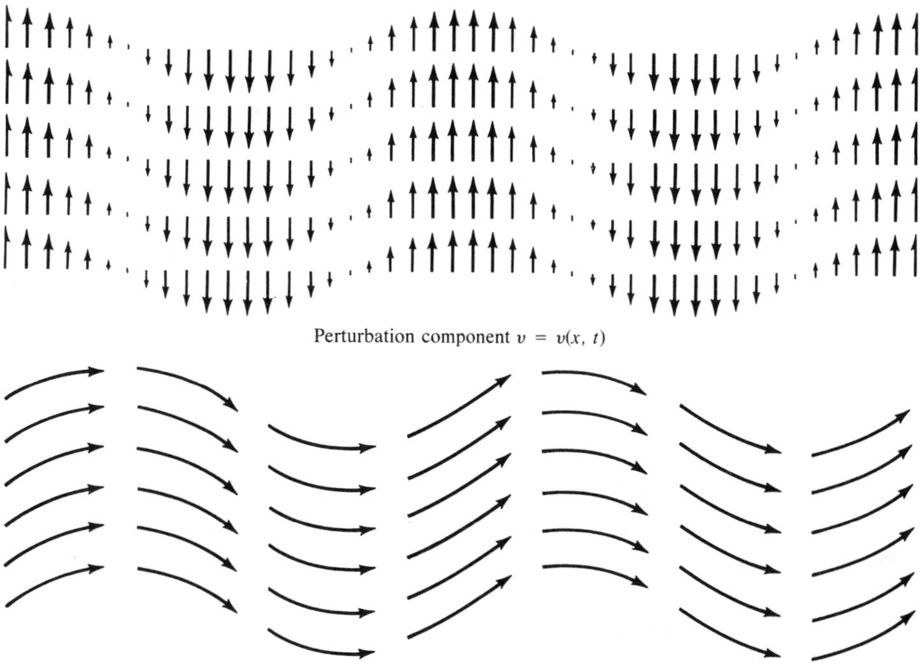

FIGURE 10.7
The flow pattern assumed in the Rossby theory of long waves.

to eight of the waves associated with cyclonic development. Thus the Rossby theory, in the basic form discussed here, is surprisingly successful, considering the severity of the assumptions. Its true significance lies in the fact that a most basic part of atmospheric structure has been represented relatively well by a very simple model. Thus the Rossby theory provided impetus for more searching examination of the mechanisms of long waves and their stability properties—a study that continues today. In historical perspective, it may well appear that the success of the Rossby formula gave meteorologists the courage to attempt numerical weather prediction, because in effect they started out with Eq. (4) as the prognostic equation.

PROBLEM 10.4.1 Show that at 45°, $\beta \cong 1.6 \times 10^{-11}\ \mathrm{s}^{-1}\ \mathrm{m}^{-1}$.

PROBLEM 10.4.2 For waves associated with synoptic disturbances, we observe that $c \sim U/2$. What then are the wavelengths of Rossby waves? How does the number of such waves around the globe compare with observed patterns?

PROBLEM 10.4.3 Suppose that, instead of assumption 5, we took $f = \text{const}$ in Eq. (3). What happens then?

PROBLEM 10.4.4 Convert Eq. (4) into an equation for a stream function and explain in general terms how it may be solved numerically.

PROBLEM 10.4.5 Show that the assumption that $\mathbf{v}$ is geostrophic applied to Eq. (9) will give the same wave formula (11). Hence assume that the geostrophic assumption may be made in Eq. (4), and outline a procedure for a numerical solution that will produce predictions of the height fields.

PROBLEM 10.4.6 Using the model of the previous problem, we have assumptions 1 and 2, and $\mathbf{v} = \mathbf{v}_g$. What then is the pattern of isotherms associated with the Rossby wave at the level of nondivergence?

The vorticity equation also provides a method for determining whether wave patterns in atmospheric flow may be expected to strengthen or to weaken owing to the forcing of changes in circulation patterns by baroclinity. The solenoidal term in the vorticity equation for cartesian coordinates may be written in the form (14) in Sec. 10.2 so that its effect on the vorticity is revealed by

$$\frac{d\zeta}{dt} + \cdots = \mathbf{k} \cdot (\nabla_H p \times \nabla_H T)\frac{R}{p} + \cdots \tag{13}$$

Thus the air moving through the wave pattern may acquire changing vorticity owing to baroclinic stratification.

Figure 10.8 shows two cases. In the first wave, the isotherms are of longer wavelength than the pressure wave. In this case the air moving through the wave will acquire increasing cyclonic vorticity as it approaches the trough and intensification may be expected. In the second case, the isothermal wave is of shorter wavelength and the air moving into the trough will be acquiring increasing anticyclonic vorticity so that the trough may be expected to become less pronounced.

It is of interest to note that in the first case, cold air is being advected equatorward and warm air poleward, so that the expected intensification is associated with motions that will eventually reduce temperature gradients. The opposite is true in the second case, and the expected disappearance of the trough would prevent the advection from occurring, and probably result in the cold pocket shown in the figure being cut off and left to warm.

In general, then, the arrangement of the gradient vectors $\nabla_H p$ and $\nabla_H T$ on a meteorological chart shows what changes baroclinity will contribute to the vorticity of the wind.

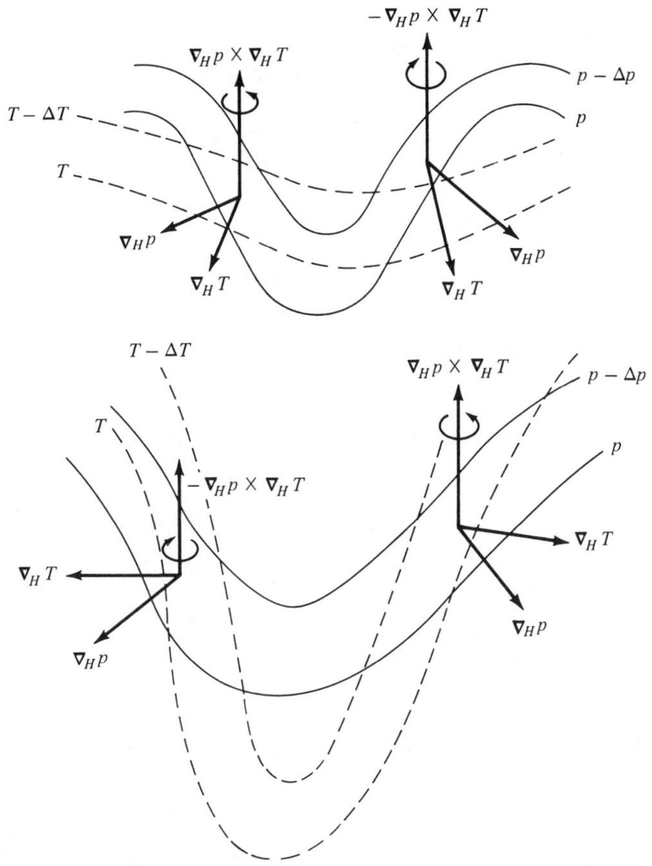

FIGURE 10.8
Illustration of the effect of varying wavelengths in the pressure and temperature fields on the vorticity.

Thus the vorticity equation allows us to predict the existence of long waves from one set of assumptions. By using the vorticity equation in conjunction with the observed patterns of baroclinity, we are then able to infer whether such waves will intensify. The logical chain would be complete if the vorticity equation would tell us, finally, what the vertical motions would be, because then we could predict precipitation patterns.

10.5 VERTICAL MOTION AND VORTICITY

In Chap. 9, we discussed the adiabatic method for determining vertical velocities—a method based on the first law of thermodynamics; we also discussed the kinematic method–based on the equation of continuity. Now we shall develop a method for determining vertical motion based on the vorticity equation.

In isobaric coordinates we have from Eq. (20) in Sec. 10.2

$$\frac{\partial}{\partial t}(\zeta_p + f) + \mathbf{v} \cdot \nabla_p(\zeta_p + f) + \omega \frac{\partial}{\partial p}(\zeta_p + f) = (\zeta_p + f)\frac{\partial \omega}{\partial p} + \frac{\partial \omega}{\partial y}\frac{\partial u}{\partial p} - \frac{\partial \omega}{\partial x}\frac{\partial v}{\partial p} \quad (1)$$

in which we have used the isobaric continuity equation to obtain the first term on the right.

The tilting terms in Eq. (1) are not necessarily small, as we demonstrated in Sec. 10.2, but to try to include them now would lead us far beyond the development of a simple technique. Thus they are dropped at this point solely for computational convenience. Of course, if they are large in a given case, then the vertical velocities produced by this method will be in error. With the further assumption that $\zeta_p + f > 0$ we may multiply Eq. (1) by $(\zeta + f)^{-2}$ and thus obtain

$$\frac{\partial}{\partial p}\left(\frac{\omega}{\zeta_p + f}\right) = \frac{1}{(\zeta_p + f)^2}\left(\frac{\partial}{\partial t} + \mathbf{v} \cdot \nabla_p\right)(\zeta_p + f) \quad (2)$$

Upon integration from level p_0 down to level p we have

$$\omega(p) = (\zeta_p + f)\left[\frac{\omega_0}{(\zeta_p + f)_{p_0}} + \int_0^p \frac{1}{(\zeta_{p'} + f)^2}\left(\frac{\partial}{\partial t} + \mathbf{v} \cdot \nabla_{p'}\right)(\zeta_{p'} + f)\, dp'\right] \quad (3)$$

in which p' is the variable of integration. If we choose $p_0 = 0$, then ω_0 vanishes and so

$$\omega(p) = (\zeta_p + f)\int_0^p \frac{1}{(\zeta_{p'} + f)^2}\left(\frac{\partial}{\partial t} + \mathbf{v} \cdot \nabla_{p'}\right)(\zeta_{p'} + f)\, dp' \quad (4)$$

If we therefore evaluate the local change of vorticity from successive observations and also compute the vorticity advection on many levels from p to the upper part of the atmosphere, we can determine the fields of isobaric vertical velocity on level p. The absence of the tilting terms is a source of error, and so is the fact that the local change and advection of vorticity tend to cancel–as might be expected from the discussion in Secs. 10.3 and 10.4.

The result (4) provides a basis for a concept employed by forecasters. If strong, positive advection of vorticity is observed in the upper part of the atmosphere, say above 500 mbar, then there will be a strong contribution to negative ω, that is, upward vertical motion at lower levels. Thus the identification of regions of positive vorticity advection at upper levels is a valuable guide to prediction of precipitation. As can be shown from Eq. (3), negative vorticity advection at low levels (with $\omega_0 \sim 0$) gives the same result. Hence, forecasters watch for situations in which the advection of vorticity changes sign between about 700 and 500 mbar. Observe, though, that the local derivative and the advection in Eq. (4) tend to cancel each other, so that the argument above must be refined by considering the effect of the wave structure (Prob. 10.5.2) or by using other techniques (Prob. 10.7.5).

PROBLEM 10.5.1 Derive the equivalent of Eq. (4) for $p_0 = 1000$ mbar.

PROBLEM 10.5.2 Let Z (with $Z \ll f$) be a constant and let the flow above level p be described by

$$v = \frac{Z}{\kappa} \sin \kappa\,(x - ct) \qquad u = U + g(x)$$

where U is constant and $U \gg |g(x)|$. Show that the vorticity pattern is wavelike. Justify the approximation

$$\frac{d\zeta_p}{dt} = \frac{\partial \zeta_p}{\partial t} + \mathbf{v} \cdot \nabla_p \zeta_p = (U - c)\frac{\partial \zeta_p}{\partial x}$$

and calculate the isobaric vertical velocity ω on level p. Illustrate with sketches the arrangement of the vertical velocity field relative to the vorticity patterns for the cases $c < c_R$ and $c > c_R$ where c_R is the Rossby wave speed for U and κ. Which case is typical of mid-latitudes and what can you say about the relation between c and c_R for that case? What pattern prevails in the tropics ($U < 0$) where $U \cong c$?

10.6 DEVELOPMENT OF SURFACE CYCLONES (SUTCLIFFE THEORY)

Because the surface cyclone is associated with weather events of interest to both forecasters and the public, meteorologists have devoted much attention to both theoretical and practical study of the development and intensification of cyclones.

If we identify cyclogenesis by falling surface pressure, we are led via the hydrostatic equation to consider

$$\frac{\partial p_0}{\partial t} = \frac{\partial}{\partial t}\int_0^\infty g\rho\, dz = -g\int_0^\infty \nabla \cdot \rho \mathbf{v}\, dz \tag{1}$$

On the assumption that $\rho w \to 0$ as $z \to \infty$ and that $w = 0$ at $z = 0$, we have then

$$\frac{\partial p_0}{\partial t} = -g\int_0^\infty \nabla_z \cdot \rho \mathbf{v}\, dz \tag{2}$$

so that cyclogenesis is possible only if there is upper-level mass divergence, and in fact enough to dominate any mass convergence that may exist in the column. This result is in accordance with the observed structure of the atmosphere in which ridges overlie surface troughs and vice versa. The fact that this alternation of convergence and divergence occurs in a column is known as *Dines' compensation law*, and the level of nondivergence generally occurs about 600 mbar as the divergence changes sign in agreement with Dines' law.

Analysis or prediction of divergence or mass divergence is difficult, however, so we seek an alternative to Eq. (2). Surface cyclones are observed to be centers of cyclonic vorticity, so that $\partial\zeta_0/\partial t > 0$ might be used as an indicator of cyclogenesis. Indeed, with the geostrophic approximation in isobaric coordinates, we have

$$\zeta_0 = \mathbf{k} \cdot \nabla_p \times \mathbf{v} \cong \frac{g}{f} \nabla_p{}^2 h(p_0) \tag{3}$$

where $\zeta_0 = \zeta_p(x,y,p_0)$. Here p_0 is a pressure close to, but less than, surface pressure. Now in the middle of a low-pressure center we will have $\nabla_p h = 0$ and $\nabla_p{}^2 h > 0$ since the height is a minimum.

Thus,

$$\frac{\partial \zeta_0}{\partial t} = \frac{g}{f} \frac{\partial}{\partial t} \nabla_p{}^2 h(p_0) > 0 \tag{4}$$

indicates that the minimum is becoming more pronounced–that the low is deepening and that cyclogenesis is occurring. Thus adoption of Eq. (4) as the criterion of cyclonic development allows us to circumvent the difficulties associated with Eq. (2).

The assumptions we shall make are that:

1 A level of nondivergence exists.

2 The geostrophic vorticity is a sufficiently accurate approximation to the actual vorticity.

3 The thermal wind is a sufficiently accurate approximation to the actual wind shear.

The isobaric vorticity equation may be written from Eq. (20) in Sec. 10.2 as

$$\frac{\partial \zeta_p}{\partial t_p} + \mathbf{v} \cdot \nabla_p(\zeta_p + f) + (\zeta_p + f)\nabla_p \cdot \mathbf{v} = -\omega \frac{\partial}{\partial p}(\zeta_p + f) + \left(\frac{\partial \omega}{\partial y}\frac{\partial u}{\partial p} - \frac{\partial \omega}{\partial x}\frac{\partial v}{\partial p}\right)_p$$

$$= -\mathbf{k} \cdot \nabla_p \times \left(\omega \frac{\partial \mathbf{v}}{\partial p}\right) \tag{5}$$

By the first assumption, we may apply this equation at the level of nondivergence, eliminating the last term on the left. Moreover, the wind $\mathbf{v}$ at the nondivergence level (p_{ND}) is given, according to the third assumption, by $\mathbf{v} = \mathbf{v}_0 + \mathbf{v}_T$, where $\mathbf{v}_T$ is the thermal wind or the difference between geostrophic winds at p_{ND} and p_0. From this it follows that $\zeta = \zeta_0 + \zeta_T$ and so Eq. (5) becomes

$$\frac{\partial \zeta_0}{\partial t_p} + \frac{\partial \zeta_T}{\partial t_p} + [\mathbf{v} \cdot \nabla_p(\zeta_p + f)]_{p_{ND}} = \left[\frac{R}{fp} \nabla_p \cdot (\omega \nabla_p T)\right]_{p_{ND}} \tag{6}$$

In the last term we have used the thermal wind equation $\partial \mathbf{v}_g/\partial p = -(R/fp)\mathbf{k} \times \nabla_p T$ derivable from Problem 9.5.2, the identity (35) in Sec. 5.1, and have taken $\nabla f = 0$.

As a last assumption we use:

4 $|\mathbf{v}_0| \ll |\mathbf{v}_T|$ so that $(\mathbf{v}_0 + \mathbf{v}_T) \cdot \nabla(\ ) = \mathbf{v}_T \cdot \nabla(\ )$.

At this point it is clear that surface cyclogenesis depends on upper-level thermal processes and vorticity advection. Note that positive vorticity advection $-\mathbf{v}_T \cdot \nabla_p \zeta_T \cong -\mathbf{v} \cdot \nabla_p \zeta$ and thermal cyclolysis $(\partial \zeta_T/\partial t_p < 0)$ both contribute to surface intensification.

The crucial step is replacement of the thermal vorticity term with an expression that depends on the temperature or heating field. Thus

$$
\begin{aligned}
\zeta_T &= \mathbf{k} \cdot \nabla_p \times \mathbf{v}_T = -\mathbf{k} \cdot \nabla_p \times \int_{p_{ND}}^{p_0} \frac{\partial \mathbf{v}_g}{\partial p}\, dp \\
&= R\mathbf{k} \cdot \nabla_p \times \int_{p_{ND}}^{p_0} \frac{\mathbf{k}}{f} \times \nabla_p T \, d \ln p
\end{aligned} \tag{7}
$$

We ignore the β effect (∇f) and with Eq. (37) in Sec. 5.1 obtain

$$
\frac{\partial \zeta_T}{\partial t_p} = \frac{R}{f} \nabla_p{}^2 \int_{p_{ND}}^{p_0} \frac{\partial T}{\partial t_p}\, d \ln p \tag{8}
$$

Now we can combine results to obtain

$$
\frac{\partial \zeta_0}{\partial t} = -[\mathbf{v} \cdot \nabla_p(\zeta_p + f)]_{p_{ND}} - \frac{R}{f}\nabla_p{}^2 \int_{p_{ND}}^{p_0} \frac{\partial T}{\partial t_p}\, d \ln p + \left[\frac{R}{fp}\, \nabla_p \cdot (\omega \nabla_p T)\right]_{p_{ND}} \tag{9}
$$

From this we can see that there are two major effects that contribute to cyclogenesis: advection of vorticity aloft and heating between the surface and the level of nondivergence. Forecasters therefore predict surface cyclogenesis by first predicting what will happen to the flow in the middle of the atmosphere.

By using the first law of thermodynamics, the rate of temperature change can be replaced with an expression that depends on observed conditions and we have

$$
\begin{aligned}
\frac{\partial \zeta_0}{\partial t} = &-[\mathbf{v} \cdot \nabla_p(\zeta_p + f)]_{p_{ND}} + \left[\frac{R}{fp}\, \nabla_p \cdot (\omega\, \nabla_p T)\right]_{p_{ND}} \\
&+ \frac{R}{f}\nabla_p{}^2 \int_{p_{ND}}^{p_0} \left[\mathbf{v} \cdot \nabla_p T + \omega\left(\frac{\partial T}{\partial p} - \frac{\alpha}{c_p}\right) - \frac{q}{c_p}\right] d \ln p
\end{aligned} \tag{10}
$$

We comment briefly on each of these terms.

1 The advection term will normally be positive east of an upper-level trough line; it is important to find and predict the location of the maximum of this advection.

2 The term $(R/fp)\ \nabla_p \cdot (\omega \nabla_p T)$ is often omitted, on the grounds that the tilting terms tend to cancel the vertical advection of vorticity. This assumption is not always true, however. In particular, note that upward velocities ($\omega < 0$) in an upper-level temperature maximum ($\nabla_p{}^2 T < 0$) contribute to surface cyclogenesis.
3 The first term in the integral is the negative of the thermal advection. Hence, the region of maximum integrated warm advection contributes most strongly to cyclogenesis. The second term in the integral may be written approximately as $w(\gamma d - \gamma)$ so that maxima of downward velocities contribute to cyclogenesis. Finally, the heating term shows that the region of maximum integrated latent heat release contributes to surface development.

The difficulty with this approach is that the terms and variables on the right of Eq. (9) interact, producing a net pattern that may be quite different from that of any individual term. Still it provides guidance about what processes are important, and permits cyclogenesis to be studied and predicted.

PROBLEM 10.6.1 Sketch the two situations shown in Fig. 10.8. Then use the formula in Problem 10.1.1 (with $\partial v/\partial n = 0$) to sketch in vorticity isopleths, assuming that the wind is geostrophic. Estimate graphically the vorticity advection term and the thermal advection integral of Eq. (10). Where would surface cyclogenesis and cyclolysis be most pronounced?

PROBLEM 10.6.2 Is there justification for ignoring $\mathbf{v}_0$ against $\mathbf{v}_T$, but retaining both ζ_0 and ζ_T so that the advective term in Eq. (9) could be written as $\mathbf{v}_T \cdot \nabla_p(\zeta_0) + \mathbf{v}_T \cdot \nabla_p(\zeta_T + f)$? Explain how the first term can be interpreted as showing that the surface cyclone tends to move in the direction $\mathbf{v}_T$. What does the second term indicate?

PROBLEM 10.6.3 Provide explanations for the following phenomena: (1) In desert regions, a surface low-pressure area often forms during the summer months. (2) Strong cyclogenesis frequently occurs as cold air masses go out to sea over the Middle Atlantic states in North America and over Japan in Asia.

PROBLEM 10.6.4 How can cyclonic disturbances form in a zonal current with $\mathbf{v} = \mathbf{i}u$, $T = T(y,z)$, $\omega = 0$?

10.7 THE OMEGA EQUATION

The use of the vorticity equation to determine vertical velocities and the development theory of the previous sections had serious disadvantages. In the case of the vertical velocities, elimination of the tilting terms and canceling between local and advective changes produce errors. The main defect of the development theory was that the prognostic equation was not part of a closed system; $\partial\zeta_0/\partial t$ depended on various terms for which the theory did not provide a prognosis. To remedy these problems, we turn to a more sophisticated approach, but one that unfortunately results in a process too complicated for use in manual, synoptic techniques.

The geostrophic equation is used to express the horizontal velocity and vorticity in the vorticity equation as functions of the isobaric heights. The hydrostatic equation and the first law are used to obtain a relation between the height fields and the isobaric vertical velocity. The result is a pair of equations linking ω and h, and they have both diagnostic and prognostic features.

We begin with the first law in the form

$$\frac{\partial \theta}{\partial t_p} + \mathbf{v} \cdot \nabla_p \theta + \omega \frac{\partial \theta}{\partial p} = \frac{\theta}{c_p T}(q + f_h) \tag{1}$$

The hydrostatic equation and Poisson's equation give

$$\theta = -\frac{gp}{R}\left(\frac{p_{00}}{p}\right)^{R/c_p} \frac{\partial h}{\partial p} \tag{2}$$

Now the differentiations in the first two terms of Eq. (1) are on isobaric surfaces, and so we may write

$$-\frac{gp}{R}\left(\frac{p_{00}}{p}\right)^{R/c_p}\left(\frac{\partial}{\partial t_p}\frac{\partial h}{\partial p} + \mathbf{v} \cdot \nabla_p \frac{\partial h}{\partial p}\right) + \omega \frac{\partial \theta}{\partial p} = \frac{\theta}{c_p T}(q + f_h) \tag{3}$$

We note that

$$\frac{gp}{R}\left(\frac{p_{00}}{p}\right)^{R/c_p} = g\rho\theta \tag{4}$$

and with the hydrostatic equation we define the stability factor

$$\sigma = -\frac{1}{\theta g \rho}\frac{\partial \theta}{\partial p} = \frac{1}{(g\rho)^2 \theta}\frac{\partial \theta}{\partial z} \tag{5}$$

Thus Eq. (3) becomes

$$\frac{\partial}{\partial t_p}\frac{\partial h}{\partial p} + \mathbf{v} \cdot \nabla_p \frac{\partial h}{\partial p} + \omega\sigma = -\frac{R}{g c_p p}(q + f_h) \tag{6}$$

Now we turn to the vorticity equation and the geostrophic assumption. Thus $\mathbf{v} = (g/f)\mathbf{k} \times \nabla_p h$ gives

$$\zeta_p = \frac{g}{f}\nabla_p{}^2 h + \frac{\beta}{f}u \tag{7}$$

For simplicity we will drop the last term.

Then the vorticity equation (20) in Sec. 10.2 may be written

$$\frac{g}{f}\frac{\partial}{\partial t_p}\nabla_p{}^2 h + \frac{g}{f}(\mathbf{k} \times \nabla_p h) \cdot \nabla_p\left(\frac{g}{f}\nabla_p{}^2 h + f\right) + \frac{g}{f}\omega\frac{\partial}{\partial p}(\nabla_p{}^2 h) = \frac{\partial \omega}{\partial p}\left(\frac{g}{f}\nabla_p{}^2 h + f\right) + \frac{\partial \omega}{\partial y}\frac{\partial u}{\partial p} - \frac{\partial \omega}{\partial x}\frac{\partial v}{\partial p} \tag{8}$$

But the tilting term, upon use of the geostrophic assumption, becomes

$$\frac{\partial \omega}{\partial y}\frac{\partial u}{\partial p} - \frac{\partial \omega}{\partial x}\frac{\partial v}{\partial p} = -\frac{g}{f}\left(\nabla_p \omega \cdot \nabla_p \frac{\partial h}{\partial p}\right) \tag{9}$$

Similarly, in Eq. (6) we use the geostrophic assumption so that

$$\mathbf{v} \cdot \nabla_p \frac{\partial h}{\partial p} = \frac{g}{f} J\left(h, \frac{\partial h}{\partial p}\right) \tag{10}$$

Hence our two equations are

$$\frac{\partial}{\partial t_p}\nabla_p{}^2 h + J\left(h, \frac{g}{f}\nabla_p{}^2 h + f\right) + \omega\frac{\partial}{\partial p}\nabla_p{}^2 h = \frac{\partial \omega}{\partial p}\left(\nabla_p{}^2 h + \frac{f^2}{g}\right) - \nabla_p \omega \cdot \nabla_p \frac{\partial h}{\partial p} \tag{11}$$

and

$$\frac{\partial}{\partial t_p}\frac{\partial h}{\partial p} + \frac{g}{f} J\left(h, \frac{\partial h}{\partial p}\right) + \omega\sigma = -\frac{R}{g c_p p}(q + f_h) \tag{12}$$

But the tendency term in Eq. (11) can be eliminated through differentiation of both Eqs. (11) and (12) to obtain identical first terms. The result is

$$\nabla_p{}^2(\omega\sigma) + \frac{\partial^2 \omega}{\partial p^2}\left(\nabla_p{}^2 h + \frac{f^2}{g}\right) - \omega\,\nabla_p{}^2\frac{\partial^2 h}{\partial p^2} - \frac{\partial}{\partial p}\left(\nabla_p \omega \cdot \nabla_p \frac{\partial h}{\partial p}\right) = -\frac{g}{f}\nabla_p{}^2 J\left(h, \frac{\partial h}{\partial p}\right) + \frac{\partial}{\partial p} J\left(h, \frac{g}{f}\nabla_p{}^2 h + f\right) - \frac{R}{g c_p p}\nabla_p{}^2(q + f_h) \tag{13}$$

and is known as the ω equation.

The pair of equations (12) and (13) form an interesting team. Suppose we know $h(x,y,p,t)$ and $\theta(x,y,p,t)$ for a number of levels. Then we would have approximate values of $\partial h/\partial p$, $\nabla_p{}^2 h$, and $\nabla_p{}^2(\partial h/\partial p)$ as well as σ, as seen from Eqs. (2), (4), and (5). Thus in principle we could solve Eq. (13) for the ω fields. Now we know all the terms except the first of Eq. (12), and so we can use the equation to find $\partial^2 h/\partial t\,\partial p$ and thus we can forecast the height fields. From Eq. (2) we see that these new height fields will yield θ and thus σ, and so we can determine new ω fields from Eq. (13) again. This, then, is a scheme for numerical weather prediction.

These equations are often simplified. Observational evidence suggests that the terms $\omega\,\partial\zeta/\partial p$ and the tilting terms tend to cancel each other, and that their sum is smaller than the other terms of the vorticity equation. If these terms are thus omitted, the vorticity equation becomes

$$\frac{\partial \zeta_p}{\partial t} + \mathbf{v} \cdot \nabla_p(\zeta_p + f) = -(\zeta_p + f)\, \nabla_p \cdot \mathbf{v} = (\zeta_p + f)\, \frac{\partial \omega}{\partial p} \tag{14}$$

and the last two terms on the left of Eq. (13) vanish. The approximation that $f > \zeta$ is made on the right of Eq. (14) and it is assumed that a mean value

$$\bar{\sigma} = \bar{\sigma}(p) \tag{15}$$

can be found such that

$$\nabla_p{}^2 \sigma\omega \cong \bar{\sigma}\, \nabla_p{}^2 \omega \tag{16}$$

Then with an assumption of isentropic motion we have

$$\bar{\sigma}\, \nabla_p{}^2 \omega + \frac{f^2}{g} \frac{\partial^2 \omega}{\partial p^2} = -\frac{g}{f}\, \nabla_p{}^2 J\left(h, \frac{\partial h}{\partial p}\right) + \frac{\partial}{\partial p} J\left(h, \frac{g}{f} \nabla_p{}^2 h + f\right) \tag{17}$$

The benefit of these assumptions is that the equation is now a fairly simple one in the variable ω.

This last equation has been widely used to obtain ω fields from the usually available data. Vertical velocities are needed for many diagnostic studies of atmospheric behavior, including the investigation of energy transformations from one form to another, and they have been provided by this equation. Despite the computational difficulties of working with observed data, even with the simple form of the ω equation, it has given us insight into atmospheric processes that might otherwise remain inscrutable.

PROBLEM 10.7.1 Verify Eqs. (2), (9), (10), and (13).

PROBLEM 10.7.2 Making the same assumptions used to obtain Eq. (17), eliminate ω between Eqs. (11) and (12) to obtain a single prognostic equation for h. [Hint: Use Eq. (14) to obtain a simplified version of Eq. (11).]

PROBLEM 10.7.3 Show that a height field $h = h(y,p)$ is a solution of the prognostic equation of Problem 10.7.2. Retracing the derivation of the equation, let $f_h = 0$ and retain a heating term $q = -\nabla \cdot \mathbf{R}$ in Eq. (12) and obtain the prognostic equation for h with diabatic forcing. Show that longitudinal variations in radiational heating would prevent the simple solution $h = h(y,p)$ from being possible.

PROBLEM 10.7.4 Show that the approximate equation (14) implies that the global mean vorticity on an isobaric surface is constant provided the isobar does not intersect the surface.

PROBLEM 10.7.5 Use the result (17) to discuss the relation of vorticity advection to vertical motion. In particular, show that strong upward motion can be expected on an isobaric surface in a region where geostrophic advection of vorticity changes from negative to positive with increasing height. [Hint: Trace through the derivation of Eq. (17) to determine which of the terms on the right is vorticity advection. What does the other term represent? Let ω be sinusoidal in all three coordinates to obtain the answer easily.]

10.8 KINEMATICS OF CIRCULATION AND VORTICES

Performing the curl operation on a motion field produces a new vector field composed of the vorticity vectors. These vectors reveal by their direction the orientation of the axis about which the parcels are spinning, and by their length they specify the angular velocity with respect to that axis.

Let us consider the field of vorticity vectors at a fixed instant of time. In this vector field we could construct a family of lines so that each line is everywhere tangent to a vorticity vector, just as we constructed streamlines tangent to velocity vectors. Such a line is called a *vortex line*, and obviously the parcels along the vortex line are spinning at various angular velocities with their axis of spin parallel to the line. When water drains from a sink, a visible vortex often forms, and in this case a vortex line would extend from the water's surface down the center of the vortex into the drain.

10.8.1 Circulation

A broader description of the vorticity field is often useful. Let us consider a simple, connected surface S, lying in the fluid, such that the boundary C of the surface forms one closed curve (Fig. 10.9). All the vortex lines passing through C form a surface known as a *vortex tube*. In general, then, a vortex tube includes all the vortex lines passing through the bounding curve of such a surface, and the vortex tube is defined throughout the fluid by the given closed curve.

Once we have found a vortex tube, we might want to compare the strength or intensity of the vortex with some other one. We do this by defining a number

$$\Gamma = \int_S \boldsymbol{\zeta} \cdot \boldsymbol{\eta} \, d\sigma = \int_S \boldsymbol{\eta} \cdot (\nabla \times \mathbf{v}) \, d\sigma \tag{1}$$

where $\boldsymbol{\zeta}$ is the vorticity vector and $\boldsymbol{\eta}$ is the unit normal to the surface S. This quantity is called the *circulation*; to see why, let us apply Stokes' theorem, Eq. (15) in Sec. 5.2, to obtain†

†This transformation from Eq. (1) to Eq. (2) was in fact developed by Lord Kelvin for precisely this use in the study of vorticity. Despite this, it is almost uniformly attributed in the literature to Stokes, who used it on an examination in 1854 after learning of it in a letter from Lord Kelvin. Further details are given in the book by Truesdell (1954) cited in the references.

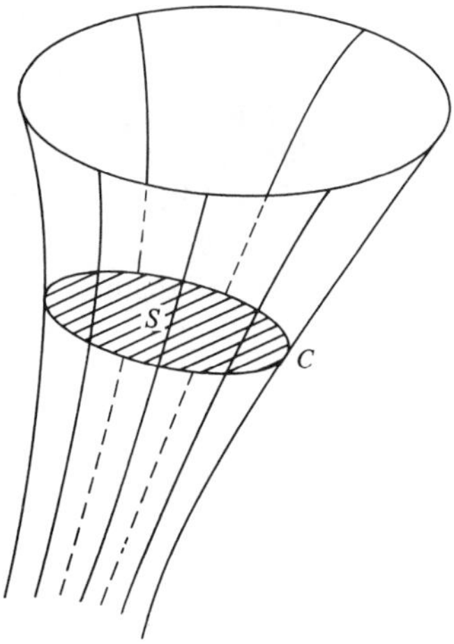

FIGURE 10.9
Illustration of the vortex tube defined by the curve C enclosing the surface S.

$$\Gamma = \int_C \boldsymbol{\tau} \cdot \mathbf{v}\, ds \tag{2}$$

Thus the circulation is the sum of the velocity components tangent to the curve C. By the conventions of Sec. 5.2, Γ is positive when the net flow around C is in the counterclockwise direction looking down the vector $\boldsymbol{\eta}$ (from tip to tail) at the surface element S. Thus on a horizontal surface in the atmosphere, the circulation and the vorticity will be positive in cyclonic flow and negative in anticyclonic flow.

The following problem shows that the existence of circulation around C implies an average vorticity on S, and conversely.

*PROBLEM 10.8.1 Let $\bar{\zeta}_S$ be the average vorticity component normal to S. Show that $\bar{\zeta}_S = \Gamma/S$, where S is the area of the surface S.

We have already proved the fundamental kinematic theorem about vortex tubes in Sec. 5.2 where we showed in Eq. (22) that if Σ is a surface enclosing a volume, then

$$\int_\Sigma \boldsymbol{\eta} \cdot (\nabla \times \mathbf{v})\, d\sigma = 0 \tag{3}$$

Hence we have, for any fluid:

Theorem (Helmholtz, 1858) The circulation on any curve around a vortex tube is constant along the tube.

PROOF Let C_1 and C_2 be two curves around a vortex tube. They enclose surfaces S_1 and S_2 confined to the tube. Let S_3 be the exterior surface of the tube bounded by C_1 and C_2 (see Fig. 10.10). The surfaces S_1, S_2, and S_3 combine to form a surface Σ enclosing a volume in the vortex tube and thus Eq. (3) applies. But by definition, the vorticity vectors are tangent to the sides of the vortex tube, and thus orthogonal to the normal to S_3.

Hence there is no contribution from S_3 and so

$$\int_{\Sigma} \boldsymbol{\eta} \cdot \boldsymbol{\zeta}\, d\sigma = \left(\int_{S_1} + \int_{S_2}\right) \boldsymbol{\eta} \cdot \boldsymbol{\zeta}\, d\sigma = 0 \tag{4}$$

Applying Stokes' theorem to the last two integrals, we have

$$\int_{C_1} \mathbf{v} \cdot \boldsymbol{\tau}_1\, ds + \int_{C_2} \mathbf{v} \cdot \boldsymbol{\tau}_2\, ds = 0 \tag{5}$$

Upon reversing the direction of the tangent vector $\boldsymbol{\tau}_2$, so that C_2 is given the same orientation as C_1, we have $\Gamma_1 = \Gamma_2$ and the proof is complete. ////

A corollary of this theorem is of considerable interest. The theorem asserts that the circulation in a vortex tube cannot change anywhere along the tube. Thus the tube cannot come to an end, except at the boundary of the fluid. Therefore, either vortex tubes are closed or they extend to the boundary. Of course, the tube may take on innumerable changes of shape and may bend and meander through the fluid, but the circulation, and hence the product $S\bar{\zeta}_S$, is constant for any curve encircling the tube, and thus defining a surface S enclosed within the tube. Therefore, if the tube were to end, S would have to vanish, which would imply that $\bar{\zeta}_S$ would become infinite, violating the basic assumption of fluid mechanics that velocities are differentiable.

The surface S and bounding curve C are most often chosen to be material elements; they move with the fluid. According to the theorem Eq. (32) in Sec. 5.3 for differentiation of line integrals we have, assuming C to be material,

$$\frac{\partial \Gamma}{\partial t} = \int_C \frac{d\mathbf{v}}{dt} \cdot \boldsymbol{\tau}\, ds + \int_C \mathbf{v} \cdot [(\boldsymbol{\tau} \cdot \nabla)\mathbf{v}]\, ds \tag{6}$$

The last term may be rewritten as

$$\int_C \mathbf{v} \cdot [(\boldsymbol{\tau} \cdot \nabla)\mathbf{v}]\, ds = \int_C (\boldsymbol{\tau} \cdot \nabla)\frac{\mathbf{v} \cdot \mathbf{v}}{2}\, ds = \int_C d\left(\frac{v^2}{2}\right) = 0 \tag{7}$$

in which we have used the fact that $\boldsymbol{\tau} \cdot \nabla(\;) = \partial(\;)/\partial s$, which derives from the definition of the unit vector $\boldsymbol{\tau}$ as $\boldsymbol{\tau} = \partial \mathbf{x}/\partial s$, where $\mathbf{x}(s)$ is the position of a point at distance s along the curve. Hence

$$\int_C (\boldsymbol{\tau} \cdot \nabla)\frac{v^2}{2}\, ds = \int_C \frac{\partial}{\partial s}\frac{v^2}{2}\, ds = \int_C d\frac{v^2}{2} = 0 \tag{8}$$

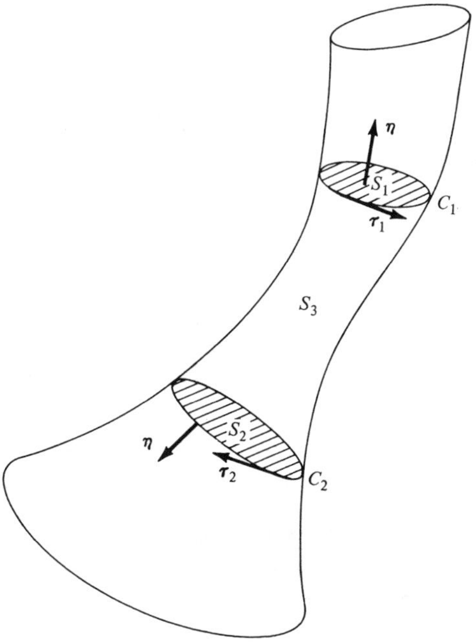

FIGURE 10.10
Illustration for the proof of the first Helmholtz theorem.

The last integral vanishes because $dv^2/2$ is obviously an exact differential and we have for the integral the difference of $v^2/2$ at the same point on the curve.

Therefore we have shown that the *rate of change of circulation following a material curve C* is

$$\frac{\partial \Gamma}{\partial t} = \int_C \frac{d\mathbf{v}}{dt} \cdot \boldsymbol{\tau}\, ds \tag{9}$$

This is the crucial result in investigating the dynamics of circulation.

PROBLEM 10.8.2 Find the rate of change of $\bar{\zeta}_S$ on a material surface S. Show that an increase in the area of S contributes to a decrease in the magnitude of the vorticity.

PROBLEM 10.8.3 What is the rate of change of circulation on a curve fixed in space?

PROBLEM 10.8.4 (Truesdell) What is the necessary condition that the vortex lines be steady?

10.8.2 The Theorems of Kelvin and Helmholtz

The circulation associated with vortex tubes has a number of remarkable properties when we consider cases simpler than those that prevail in the atmosphere. In this section, as an aid to the understanding of the concepts of vorticity and circulation, we investigate the autobarotropic flow of an inviscid fluid. Thus we have $\rho = \rho(p,t)$ or $\alpha = \alpha(p,t)$, and so in principle the indefinite integral

$$A(p,t) = \int \alpha(p,t)\, dp \tag{10}$$

could be determined, and it follows that $\nabla A = (\partial A/\partial p)\,\nabla p = \alpha\,\nabla p$. Thus the equation of motion for an inviscid, autobarotropic flow (in a nonrotating coordinate system) is

$$\frac{d\mathbf{v}}{dt} = -\nabla\left(\int \alpha\, dp\right) - \mathbf{f} \tag{11}$$

We shall assume that the external force $\mathbf{f}$ is *conservative*; that is, it is a function of position (not time) and $\nabla \times \mathbf{f}$ vanishes. Thus there is a potential Φ for the force so that $\mathbf{f} = \nabla\Phi$, and Eq. (11) becomes

$$\frac{d\mathbf{v}}{dt} = -\nabla P \tag{12}$$

where $P = A + \Phi = \int \alpha(p)\, dp + \Phi$. Thus from Eq. (9) we have

$$\frac{\partial \Gamma}{\partial t} = \int_C \frac{d\mathbf{v}}{dt} \cdot \boldsymbol{\tau}\, ds = -\int_C \boldsymbol{\tau} \cdot \nabla P\, ds = -\int_C dP = 0 \tag{13}$$

We have proved:

> **Theorem** (Lord Kelvin, 1869) The circulation around a material curve in the autobarotropic, inviscid flow of a fluid subject to conservative external forces is invariant. ////

This theorem has an obvious corollary: If the vorticity vanishes in any portion of the fluid at an initial time, then the vorticity will vanish in that same portion of the fluid wherever it goes, or however it is deformed, in the subsequent motion. Thus in such a flow, vorticity can be rearranged, but vorticity can never be created or lost in a material element.

This clearly suggests that a vortex line must somehow move with the flow. In particular, we have:

> **Theorem** (Helmholtz, 1858) Vortex lines are material lines (that is, vortex tubes are frozen into the fluid and must move with it) for inviscid, autobarotropic flow with conservative external forces.

PROOF Let C_1 be a closed curve that encloses *no* vortex lines at some time t_0. Then C_1 defines a surface S_1 that is composed of vortex lines (that is, S_1 is everywhere parallel, not orthogonal, to the vortex lines). Because C_1 encloses no vortex lines, $\Gamma_{C_1}(t_0) = 0$. By Kelvin's theorem, $\Gamma_{C_1}(t)$ is also zero for all time t if C_1 is a material curve. Therefore S_1 always consists of vortex lines. The same is true for another material surface S_2, also composed of vortex lines and intersecting S_1, at time t_0.

Because S_1 and S_2 are material surfaces moving with the fluid, their intersection is a material line in the fluid. But both S_1 and S_2 are composed of vortex lines, and thus their intersection is a vortex line that is, therefore, a material line. And if vortex lines are material, then a vortex tube formed from a collection of vortex lines at time t_0 is also a material volume in the fluid, which completes the proof. ////

Finally, by combining the two theorems we have:

Theorem (Helmholtz, 1858) The circulation or strength of a vortex tube is invariant in inviscid, autobarotropic flow with conservative external forces.

PROOF By the preceding theorem of Helmholtz, a vortex tube is material. Thus by Kelvin's theorem the circulation on any curve around the vortex tube is invariant, which completes the proof. ////

Thus we have shown that the vortex tubes and their invariant strengths are conserved throughout the flow of an autobarotropic fluid, and therefore the circulation, or the vorticity in the precise sense $S\bar{\zeta}_S$, is a conservative feature of such flow. This vorticity can be rearranged and S can change in area and orientation, but $S\bar{\zeta}_S$ must remain the same.

These theorems provide a mechanism for creating intense vortices. Suppose we have a vortex tube at time t_0. We let the motion stretch the tube out, pulling it into a thin tube so that the area s_1 of a surface across the thin tube is much less than the area s_0 of the same surface in the original tube. Thus

$$\frac{\bar{\zeta}_{s_1}}{\bar{\zeta}_{s_0}} = \frac{s_0}{s_1} \gg 1 \tag{14}$$

and so the average vorticity rises in proportion to the thickness of the tube. If the tube is compressed so that the areas of internal surfaces increase, then the average vorticity will decrease.

Thus vortex stretching produces a nice analogy to the conservation of angular momentum used by a skater. She starts a slow twirl with arms outstretched, and as they are brought down to her sides, her angular velocity increases in order that her angular momentum remains constant.

Regions of strong internal vorticity are obviously of interest in many applications of fluid dynamics, particularly in problems concerning turbulence, and vortex stretching is a way they are created by the motions.

To modify the results of this section to apply to the atmosphere, we must take account of the baroclinity that generally prevails and the effects of the earth's rotation. In Secs. 10.9 and 10.11 we shall discuss analogs of these theorems for isentropic flow.

*PROBLEM 10.8.5 Use Eq. (12) to obtain a vorticity equation, and specialize it to the case of two-dimensional, incompressible, inviscid flow. Show that $d\zeta^2/dt = 0$, and thus that both the average vorticity and the mean-square vorticity over the fluid are constant for suitable boundary conditions. Can the result be generalized for $F(\zeta)$, where F is an arbitrary, differentiable function?

10.9 ATMOSPHERIC CIRCULATION THEOREMS

Circulation theorems that take account of atmospheric properties ignored in the results of the previous sections have been used to explain meteorological phenomena since the turn of the century. Both the rotation of the earth and the baroclinity typical of atmospheric flows have profound effects on the circulation and vorticity in the atmosphere.

10.9.1 Circulation in a Rotating Coordinate System

We showed in Sec. 10.1 that the vorticity, and thus the circulation as a measure of the average vorticity, are related to a sum of angular velocities. In a rotating system we must take account of the angular velocity of the coordinate system in circulations measured relative to it, just as we did in Chap. 7, where we computed the apparent forces in the rotating system.

Let us consider the absolute velocity $\mathbf{v}_a$, relative to the fixed stars, a material curve C, embedded in the fluid, and define the *absolute circulation* as

$$\Gamma_a = \int_C \mathbf{v}_a \cdot \boldsymbol{\tau}\, ds \tag{1}$$

With the relative velocity $\mathbf{v}$ in the rotating system we can define the *relative circulation* as

$$\Gamma = \int_C \mathbf{v} \cdot \boldsymbol{\tau}\, ds \tag{2}$$

The curve C is a material curve and is embedded in the fluid in either case. But in the rotating system, the curve C may itself rotate and thus contribute to the absolute circulation around C. Obviously, the circulation of C itself will depend upon the angular velocity Ω of the earth's rotation. We let Γ_C be the circulation of C in the absolute system and then we have

$$\Gamma_a = \Gamma + \Gamma_C \tag{3}$$

The velocity $\mathbf{v}_C$ in the absolute system of a point on the curve is $\mathbf{v}_C = \boldsymbol{\Omega} \times \mathbf{r}$, where $\mathbf{r}$ is the vector from the earth's center to the point. Thus in agreement with Eq. (13) in Sec. 7.1 we have $\mathbf{v}_a = \mathbf{v} + \boldsymbol{\Omega} \times \mathbf{r} = \mathbf{v} + \mathbf{v}_C$.

A comparison of Eq. (1) rewritten in the form

$$\Gamma_a = \int_C (\mathbf{v} + \mathbf{v}_C) \cdot \boldsymbol{\tau}\, ds \tag{4}$$

with Eqs. (2) and (3) shows that

$$\Gamma_C = \int_C \mathbf{v}_C \cdot \boldsymbol{\tau}\, ds = \int_C (\boldsymbol{\Omega} \times \mathbf{r}) \cdot \boldsymbol{\tau}\, ds = \int_S \boldsymbol{\eta} \cdot [\nabla \times (\boldsymbol{\Omega} \times \mathbf{r})]\, d\sigma \tag{5}$$

where we have used Stokes' theorem. Now a use of the identity (37) in Sec. 5.1 along with the fact that $\boldsymbol{\Omega}$ is constant shows that

$$\nabla \times (\boldsymbol{\Omega} \times \mathbf{r}) = \boldsymbol{\Omega}\, \nabla \cdot \mathbf{r} - \boldsymbol{\Omega} \cdot \nabla \mathbf{r} = 3\boldsymbol{\Omega} - \boldsymbol{\Omega} = 2\boldsymbol{\Omega} \tag{6}$$

Thus we have

$$\Gamma_C = \int_S 2\boldsymbol{\Omega} \cdot \boldsymbol{\eta}\, d\sigma = 2\Omega \int_S \cos\theta\, d\sigma \tag{7}$$

where θ is the angle between the normal to S and the constant vector $\boldsymbol{\Omega}$. The integral is therefore the area enclosed by the projection of the curve C on the equatorial plane. We denote this area by Σ and thus our result is

$$\Gamma_a = \Gamma + 2\Omega\Sigma \tag{8}$$

and from this we have:

Theorem (Bjerknes, 1902) Let C be a material curve. Then the circulation around C relative to the rotating earth is governed by

$$\frac{d\Gamma}{dt} = \frac{d\Gamma_a}{dt} - 2\Omega \frac{d\Sigma}{dt} \tag{9}$$

where Σ is the area enclosed by the projection of C on the equatorial plane. ////

PROBLEM 10.9.1 Show that an anticyclone moving equatorward and a cyclone moving poleward will both tend to lose intensity of circulation if their absolute circulations are constant and if the area Σ does not change through horizontal divergence or convergence.

PROBLEM 10.9.2 Consider a surface of area S. Under what conditions on $\bar{\zeta}_S$ will the absolute circulation vanish on a curve C bounding S? How do the conditions vary with latitude, if S is horizontal?

It will be useful to have an alternative form of Bjerknes' theorem, in which the dynamic variables rather than the area Σ appear. With the fundamental identity of Eq. (14) in Sec. 7.1 we have

$$\frac{d\Gamma_a}{dt} = \int_C \left[\frac{d\mathbf{v}}{dt} + 2\boldsymbol{\Omega} \times \mathbf{v} + \boldsymbol{\Omega} \times (\boldsymbol{\Omega} \times \mathbf{r})\right] \cdot \boldsymbol{\tau}\, ds \tag{10}$$

The centrifugal forces in Eq. (10) do not contribute to the rate of change of circulation and thus we find that

$$\frac{d\Gamma}{dt} = \frac{d\Gamma_a}{dt} - \int_C (2\boldsymbol{\Omega} \times \mathbf{v}) \cdot \boldsymbol{\tau}\, ds \tag{11}$$

So the two rates of change of circulation differ only in the Coriolis force acting along C. This Coriolian contribution may be balanced, however, by the contributions from $d\Gamma_a/dt$, which we now investigate.

PROBLEM 10.9.3 Show that the last term in Eq. (10) vanishes. (Hint: $\boldsymbol{\tau} = \partial\mathbf{r}/\partial s$.)

10.9.2 Baroclinity and Circulation

In most atmospheric phenomena, the rate of change of Γ_a expresses the effect of thermodynamic forcing on the circulation. In general, we might expect that net acquisition of heat by parcels on a curve C would tend to increase the intensity of the circulation on that curve.

Since we have already included the Coriolis force in Eq. (11), we may use the equation of motion in the absolute form

$$\frac{d\mathbf{v}_a}{dt} = -\alpha\, \nabla p - \nabla\Phi + \mathbf{F} \tag{12}$$

to write

$$\frac{d\Gamma_a}{dt} = -\int_C \alpha\, dp + \int_C \mathbf{F} \cdot \boldsymbol{\tau}\, ds \tag{13}$$

in which we have used the notation $df = \boldsymbol{\tau} \cdot \nabla f\, ds$ and noted that the contribution from the conservative force of gravity thus vanishes.

Two transformations of this equation are possible. First, we may use the equation of state and the fact that dT is an exact differential to replace $-\alpha\, dp$ with $p\, d\alpha$. Second, we may use the fact that $T\,\nabla S = c_p\,\nabla T - \alpha\,\nabla p$, where S is now the specific entropy (we choose the capital letter here to avoid confusion with arc length s). With these transformations, the *relative circulation theorem* for the atmosphere may be written as

$$\begin{aligned}\frac{d\Gamma}{dt} &= \int_C p\,d\alpha + \int_C \mathbf{F}\cdot\boldsymbol{\tau}\,ds - \int_C (2\boldsymbol{\Omega}\times\mathbf{v})\cdot\boldsymbol{\tau}\,ds \\ &= \int_C T\,dS + \int_C \mathbf{F}\cdot\boldsymbol{\tau}\,ds - \int_C (2\boldsymbol{\Omega}\times\mathbf{v})\cdot\boldsymbol{\tau}\,ds \qquad (14)\end{aligned}$$

Now the material curve C is not necessarily a trajectory of a moving parcel, and therefore we must not interpret $d\alpha = \boldsymbol{\tau}\cdot\nabla\alpha\,ds$ as the total differential that appears in the first law of thermodynamics in the work term $p\,d\alpha$. But if we consider a hypothetical parcel that did go around C, then the first term on the right represents the thermodynamic work that would be done by the parcel, as shown by Problem 3.3.1. For such a hypothetical parcel, the condition that $\int p\,d\alpha > 0$ is equivalent via Eq. (20) in Sec. 3.2 to the condition that $\int (q + f_h)\,dt > 0$ because dT is an exact differential. Thus the circulation on C will increase in intensity if the hypothetical parcel gains heat in one circuit around C. It is this energy that is used to do the hypothetical work against the environment required to increase the circulation.

Returning now to the actual situation, it is clear that it is the presence of baroclinity that gives $\int p\,d\alpha$ a nonzero value on the closed curve C. Thus the existence of baroclinity implies that the circulation will be accelerated in a given direction, although the frictional and Coriolian terms may give contributions in the opposite sense.

We have a technique for determining in which direction baroclinity will accelerate the actual circulation. On a (p,α) diagram we could find the area enclosed by the representation of C by counting the number of Δp, $\Delta\alpha$ squares. In the atmosphere, these squares would be transformed into various shapes, but by analyzing the p and α fields within C at the same Δp and $\Delta\alpha$ intervals and counting the number of areas enclosed we would find the same area. The geometrical figure enclosed by the lines p, $p+\Delta p$, α, and $\alpha+\Delta\alpha$ is called a *pressure-volume solenoid*, and the rate of change of circulation is proportional to the number of solenoids inside C. But care must be taken to affix the correct sign to the number of solenoids.

This sign problem can be solved with Stokes' theorem. We have

$$\begin{aligned}\int_C p\,d\alpha &= \int_C p\,\nabla\alpha\cdot\boldsymbol{\tau}\,ds \\ &= \int_S \boldsymbol{\eta}\cdot[\nabla\times(p\,\nabla\alpha)]\,d\sigma \\ &= \int_S \boldsymbol{\eta}\cdot(\nabla p\times\nabla\alpha)\,d\sigma \qquad (15)\end{aligned}$$

Thus when ∇p and $\nabla \alpha$ are known, Eq. (15) reveals immediately what the effect of the solenoidal term will be over a given surface, and thus what it will be around a given curve.

Generally we analyze atmospheric structures for pressure and temperature. With the equation of state we find that

$$\int_C p\, d\alpha = \int_S \boldsymbol{\eta} \cdot \left(\frac{R}{p} \nabla p \times \nabla T \right) d\sigma \tag{16}$$

This result is easily applied. The product $\nabla p \times \nabla T$ defines a direction; in the atmosphere it will usually be nearly parallel to the earth's surface because the largest gradients of p and T are in the vertical. Imagine a plane perpendicular to the vector $\nabla p \times \nabla T$ (Fig. 10.11). The circulation will tend to increase in the same direction in that plane as the path from the tip of ∇p to ∇T across the (smaller) angle between them. In general then, this convention predicts that the warm air will rise and the cool air will sink during the circulation.

To illustrate this remark, we shall consider some applications of these results to the sea breeze and mountain winds, to atmospheric cyclones, and to the general circulation.

PROBLEM 10.9.4 Find the vertical and horizontal components of the baroclinity vector $\mathbf{B} = (R/p) \nabla p \times \nabla T$ and express them (for hydrostatic conditions) as functions of the geostrophic wind and geostrophic wind shear. Use these results to show that

$$\int_C p\, d\alpha = \int_S \boldsymbol{\eta} \cdot \left(\frac{f^2}{g} \mathbf{v}_g \times \frac{\partial \mathbf{v}_g}{\partial z} - f \frac{\partial \mathbf{v}_g}{\partial z} \right) ds$$

PROBLEM 10.9.5 (Eckart's circulation theorem) Define the thermodynamic circulation as $\Gamma_{thm} = -\int \Lambda \nabla S \cdot \boldsymbol{\tau}\, ds$ and show that for a material curve

$$\frac{d}{dt}(\Gamma + \Gamma_{thm}) = -\int_C \Lambda \nabla \frac{dS}{dt} \cdot \boldsymbol{\tau}\, ds - \int_C (2\boldsymbol{\Omega} \times \mathbf{v}) \cdot \boldsymbol{\tau}\, ds + \int_C \mathbf{F} \cdot \boldsymbol{\tau}\, ds$$

if the scalar Λ is defined so that $d\Lambda/dt = T$.

PROBLEM 10.9.6 Define the generalized circulation as $\Gamma_G = \Gamma_a + \Gamma_{thm}$ and show that Γ_G is an invariant for material curves in inviscid, isentropic motion subject to a conservative external force. Develop suitable definitions of generalized vortex lines and vortex tubes and restate the theorems of Helmholtz and Kelvin in Sec. 10.8 so that they apply to the generalized circulation Γ_G.

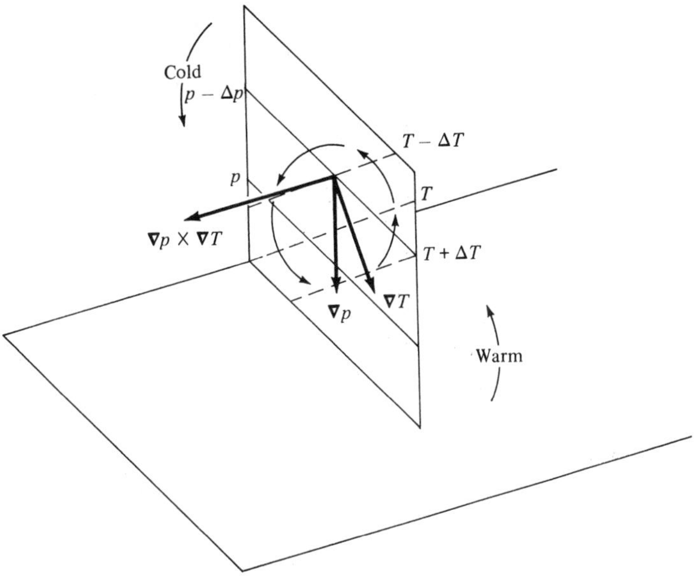

FIGURE 10.11
Illustration of the direction of rate of increase of circulation due to baroclinity.

10.9.3 The Sea Breeze and Mountain Winds

The baroclinity associated with the horizontal temperature gradients that develop along a seacoast on sunny days and clear nights drives a local circulation known as the sea breeze.

Let us start with a barotropic stratification at some time in the early morning. The sunshine heats the land more rapidly than the sea so that ∇T will eventually have a component from sea to land. Thus a circulation will develop so that sea air flows to land at low levels, rises, and returns to sea at higher levels. (See Fig. 10.12.) At night the land cools more rapidly than the sea, ∇T has a seaward component, and the circulation is driven in the opposite direction.

This circulation drives another one parallel to the shoreline. After the circulation has developed during the day, we might imagine a case in which ∇T has no alongshore component. Then $\nabla p \times \nabla T$ vanishes in a plane parallel to the coastline. Thus there is no baroclinic acceleration of the circulation in this plane. But the Coriolian term of Eq. (14) will contribute to the development of a circulation in the counterclockwise direction looking from sea to land.

Thus the sea breeze is a complex phenomenon, in which the plane containing the maximum circulation rotates around an imaginary vertical axis installed at the water's edge.

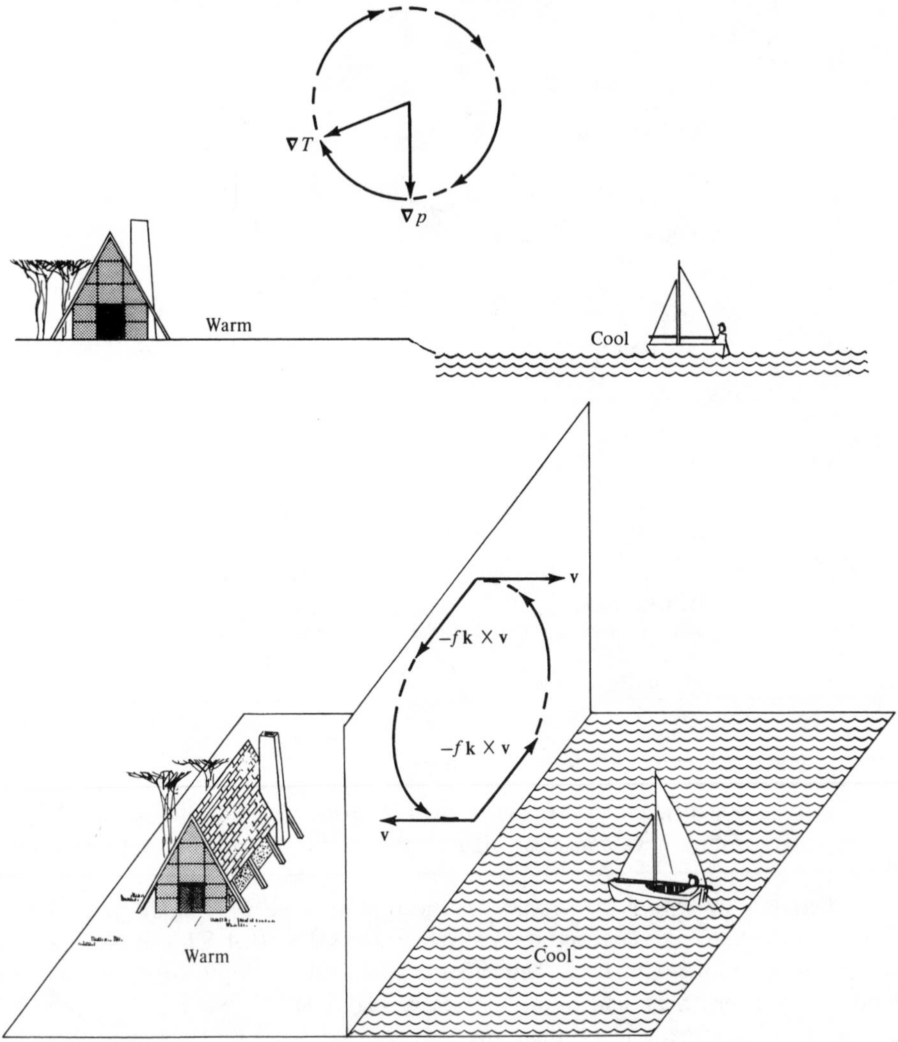

FIGURE 10.12
The top cross section illustrates the origin of the sea breeze. The bottom diagram shows how Coriolis force then creates a circulation in a vertical plane parallel to the shoreline.

Exactly the same reasoning applies to mountain and valley areas; in this case we consider a horizontal surface spanning the valley and ending at the mountainside. The mountain will warm during the day relative to the rest of this surface, and cool at night. Thus an alternating system of winds forms, with upslope flow in the day and downslope flow at night.

PROBLEM 10.9.7 Urban areas are generally warmer than the surrounding countryside. What kind of flow would be expected from the circulation theorem? How will the pattern appear when this flow and a steady, mean flow are combined?

10.9.4 Formation of Cyclones and Anticyclones

The sea breeze problem illustrates that the development of circulation in one plane may, through the Coriolian term of Eq. (14), induce circulations in other planes. We can use this concept to develop models of cyclone formation in response to differential heating.

Suppose we have a pattern of alternating warm and cold regions at the earth's surface created by the different rates of heating that occur on different types of surface. The circulation theorem shows that a vertical circulation will tend to develop with air rising over the warm regions and sinking over the cold. As shown by Fig. 10.13, this will induce circulations at the surface, with cyclonic circulation over the warm regions and anticyclonic circulation over the cold regions.

In a similar manner, a mass of cold air moving into a warm region will induce a vertical circulation with warm air rising and cool air descending over the cold region at the surface. Thus an acceleration of a cyclonic circulation will occur on every horizontal curve encompassing the warm region and passing through the boundary with the cold region. From Fig. 10.13, we see that cold air injection from the northwest will tend to create cyclonic circulations to the southwest of the cold air.

Cyclone formation under such conditions is observed in the United States, where cold air injection from the Great Plains to Texas and the Gulf Coast area initiates cyclone development. It also occurs on east coasts of continents, where cold air moves over the warm, poleward ocean currents so frequently that major cyclogenesis regions are located on the Atlantic coast of North America and off the coast of Japan.

10.10 THE MAINTENANCE OF CIRCULATION

The complete dependence of the motions of the atmosphere on the differential heating that prevents an equilibrium from being reached is a basic concept of large-scale meteorology. With Jeffreys' theorem of Sec. 6.7, we saw that the existence of horizontal temperature gradients necessitated the development of accelerations and motion. But this theorem did not provide information on the character of the motion.

Now we seek the essential character of the atmospheric structure that provides the continual impetus to the flow. How, we ask, must a fluid be arranged so that it has steady circulations?

We start with the atmospheric circulation theorem, Eq. (14) in Sec. 10.9, and to begin we suppose there is no motion and no acceleration. Then the last two terms of this equation vanish, and $d\Gamma/dt$ must also vanish because $d\mathbf{v}/dt$ does. If these conditions prevail throughout the atmosphere, then

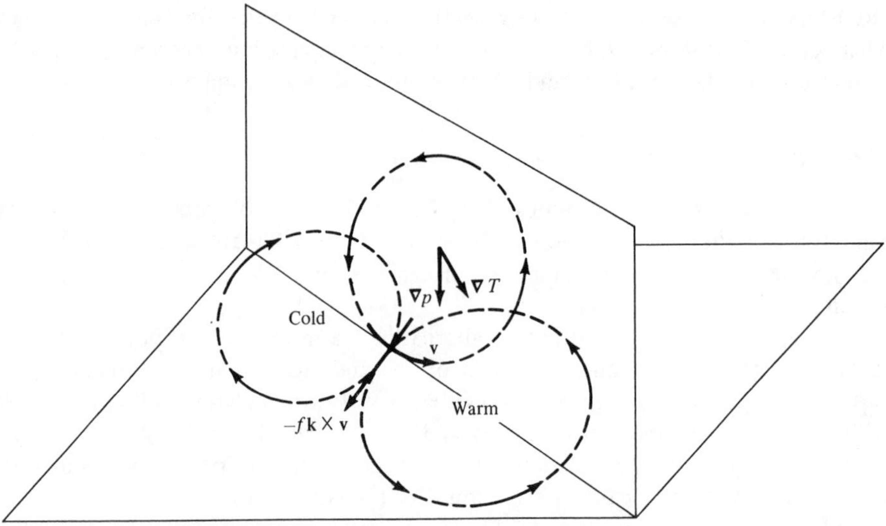

FIGURE 10.13
Model of surface cyclone formation by cold air injection.

$$\int_C T\,dS = 0 \qquad (1)$$

for every closed curve C, and so $T\,dS$ must be an exact differential. But this is possible only if $T = T(S)$ or, turned around, $S = S(T)$. This clearly implies a barotropic stratification.

Note that the force of gravity does not appear explicitly in the dynamics of circulation, and so we are not able to obtain a result as strong as Jeffreys' theorem.

However, we can reach two conclusions from Eq. (1). If the atmosphere is at rest, then circulations can develop only if heating patterns are arranged so that

$$\int_C T\,dS \neq 0 \qquad (2)$$

on some closed curve C. Thus the circulation theorem shows that circulations can develop only if the heating is differential so that temperature gradients are created on isentropic surfaces. Jeffreys' theorem did not specify the structure of the motions that develop, and Problem 6.7.1 demonstrates that uniform heating would cause an expansion of the fluid which would not produce circulations. So we may conclude:

Theorem Circulation around some curve C in a fluid at rest can develop only if there is differential heating that produces baroclinity. ////

The second conclusion concerns steady motion, in which $\partial \mathbf{v}/\partial t$ vanishes throughout the atmosphere. A situation to which the description that follows would apply is a longitudinally invariant circulation with warm, rising currents at the equator, northeastward flow aloft, cool, sinking currents at high latitudes, and then a southwestward return flow at low altitude.

With the aid of Eq. (23) in Sec. 6.5, the steady condition implies that Eq. (9) in Sec. 10.8 is

$$\begin{aligned}\frac{d\Gamma}{dt} &= \int_C (\mathbf{v} \cdot \nabla \mathbf{v}) \cdot \boldsymbol{\tau}\, ds \\ &= \int_C \left(\nabla \frac{v^2}{2} - \mathbf{v} \times \boldsymbol{\zeta}\right) \cdot \boldsymbol{\tau}\, ds = -\int_C (\mathbf{v} \times \boldsymbol{\zeta}) \cdot \boldsymbol{\tau}\, ds \end{aligned} \tag{3}$$

Hence Eq. (14) in Sec. 10.9 becomes

$$\int_C [(\boldsymbol{\zeta} + 2\boldsymbol{\Omega}) \times \mathbf{v}] \cdot \boldsymbol{\tau}\, ds = \int_C T\, dS + \int_C \mathbf{F} \cdot \boldsymbol{\tau}\, ds \tag{4}$$

Now we are free to choose the curve C to be a closed streamline in the steady flow. In this case C is a trajectory, but it does not move relative to the earth because the streamlines and trajectories are fixed in steady flow.

But if C is a streamline, then $\boldsymbol{\tau}$ is parallel to $\mathbf{v}$, and hence the integral on the left of Eq. (4) vanishes. Thus it must be true that

$$\int_C T\, dS = -\int_C \mathbf{F} \cdot \boldsymbol{\tau}\, ds \tag{5}$$

on every closed streamline C. Now it may be possible to find streamlines along which the force of friction is accelerating the velocity because along that streamline the velocities are less than on neighboring streamlines. But over the whole atmosphere, the integral on the right is presumably negative—friction destroys kinetic energy—and thus we might argue that on the average $\mathbf{F} \cdot \boldsymbol{\tau} = \mathbf{F} \cdot \mathbf{v}/|\mathbf{v}| < 0$. If this is the case, then the integral on the left must be positive on the average. Because C is a trajectory, the parcels must take on the values of S and T encountered at each point of C, so they gain entropy where $\nabla S \cdot \boldsymbol{\tau} > 0$. We have:

Theorem If a steady circulation around a closed streamline C is to be maintained against a retarding force of friction, then on the average the moving parcels must gain entropy at high temperature relative to entropy loss at low temperatures.

PROOF Since the force of friction on the streamline is assumed to be retarding, we must have

$$\int_C T\, dS > 0 \tag{6}$$

Let C^+ be the portion of C where $dS > 0$ and C^- the portion where $dS < 0$. Clearly

$$\int_{C^+} dS + \int_{C^-} dS = \int_C dS = 0 \tag{7}$$

We can define

$$T^+ = \left(\int_{C^+} T\,dS\right)\left(\int_{C^+} dS\right)^{-1} \qquad T^- = \left(\int_{C^-} T\,dS\right)\left(\int_{C^-} dS\right)^{-1} \tag{8}$$

Thus Eq. (6) becomes

$$T^+ \int_{C^+} dS + T^- \int_{C^-} dS = (T^+ - T^-) \int_{C^+} dS > 0 \tag{9}$$

so

$$T^+ - T^- > 0 \tag{10}$$

This completes the proof, and gives a precise meaning to the statement of the theorem. ////

It is obviously equivalent to Eq. (6) to write

$$\int_C p\,d\alpha > 0 \tag{11}$$

By an identical argument, we therefore have:

Theorem If a steady circulation is to be maintained on a closed streamline, C, against a retarding force of friction, then the moving parcels must expand at high pressure relative to contraction at low pressure. ////

This result is often given in the literature in a different, but less precise form. The accuracy of the statement generally used depends upon the assumption that the expansion takes place in the region of heating and the contraction in the region of cooling. If this assumption is true, then we may conclude:

Theorem (Sandstrom, 1916) If $d\alpha > 0$ where heat is added and $d\alpha < 0$ where heat is extracted, then a steady circulation can be maintained against a retarding force of friction only if the heating occurs at high pressure relative to cooling at low pressure. ////

This last theorem has been viewed as an explanation of the fact that low-altitude heating in low latitudes and radiational cooling near the tropopause in high latitudes is required for the maintenance of the general circulation of the atmosphere. It has been cited as providing an analogy to a thermodynamic enginė in which heat is added at high pressure and extracted at low pressures. But as pointed out by Jeffreys, its accuracy depends on the assumption that the expansion occurs where the heating is taking place. It is possible to imagine that the flow is arranged so that the expansion occurs on the circuit before the region of heating is reached, thus invalidating the result.

It would thus seem that the previous theorem requiring entropy gain or heating of the moving parcels at high temperature relative to entropy loss or cooling at low temperature is to be preferred, because there is no thermodynamic ambiguity.

PROBLEM 10.10.1 Show that there is no contradiction between the Jeffreys theorem and those of this section. (Hint: There is an important qualification here not present in Sec. 6.7.) Give examples to illustrate the difference between the two cases.

10.11 VORTICITY THEOREMS FOR MOVING PARCELS

In most of this chapter we have pursued understanding of atmospheric flow by studying the vorticity as a property associated in space with velocity fields. But with potential vorticity we found a form of the vorticity that provides considerable insight into how changes in flows develop. In essence, we are able to predict what would happen to the vorticity if we could anticipate the parcel's trajectory.

Here we shall explore more fully the possibilities of a material representation of the vorticity. The results lead to expressions of considerable elegance, and provide an unusual view of the kinematics and dynamics of fluids in motion.

10.11.1 The Vector Vorticity Equation and Ertel's Theorem

The meteorological vorticity equations derived in Sec. 10.2 are a special case of a more general vector vorticity equation. Similarly, the potential vorticity conservation theorem developed in isentropic coordinates is a special case of a more general conservation theorem that applies to a moving parcel. We begin with the equation of motion.

$$\frac{\partial \mathbf{v}}{\partial t} + \nabla \frac{v^2}{2} - \mathbf{v} \times \boldsymbol{\zeta} = -\alpha\, \nabla p - \nabla \Phi - 2\boldsymbol{\Omega} \times \mathbf{v} + \nu\, \nabla^2 \mathbf{v} + \nu\, \nabla(\nabla \cdot \mathbf{v}) \qquad (1)$$

in which we have used the assumption that the dynamic shear and expansion viscosity coefficients are equal and constant (see Sec. 6.8.3).

Upon taking the curl of Eq. (1), we find

$$\frac{\partial \boldsymbol{\zeta}}{\partial t} - \nabla \times (\mathbf{v} \times \boldsymbol{\zeta}) = -\nabla\alpha \times \nabla p - \nabla \times (2\boldsymbol{\Omega} \times \mathbf{v}) + \nu\, \nabla^2 \boldsymbol{\zeta} + \nabla\nu \times [\nabla^2 \mathbf{v} + \nabla(\nabla \cdot \mathbf{v})] \quad (2)$$

The expansion of the second term on the left is given in Eq. (10) in Sec. 10.1; that of the Coriolis force term is obtained from Eq. (37) in Sec. 5.1 as

$$\nabla \times (2\mathbf{\Omega} \times \mathbf{v}) = 2\mathbf{\Omega}\, \nabla \cdot \mathbf{v} + 2\mathbf{v} \cdot \nabla\mathbf{\Omega} - 2\mathbf{\Omega} \cdot \nabla\mathbf{v} \tag{3}$$

in which we have used $\nabla \cdot \mathbf{\Omega} = 0$. Thus these results may be combined as

$$\frac{\partial}{\partial t}(\boldsymbol{\zeta} + 2\mathbf{\Omega}) + \mathbf{v} \cdot \nabla(\boldsymbol{\zeta} + 2\mathbf{\Omega}) - [(\boldsymbol{\zeta} + 2\mathbf{\Omega}) \cdot \nabla]\mathbf{v} = -(\boldsymbol{\zeta} + 2\mathbf{\Omega})\, \nabla \cdot \mathbf{v} - \nabla\alpha \times \nabla p + \nu\, \nabla^2\boldsymbol{\zeta} + \nabla\nu \times [\nabla^2\mathbf{v} + \nabla(\nabla \cdot \mathbf{v})] \tag{4}$$

With the equation of continuity to replace the divergence and the notation $\boldsymbol{\omega} = \boldsymbol{\zeta} + 2\mathbf{\Omega}$, we find exactly as in Eq. (13) in Sec. 10.1 that the general Beltrami formula is

$$\frac{1}{\alpha}\frac{d}{dt}(\alpha\boldsymbol{\omega}) - \boldsymbol{\omega} \cdot \nabla\mathbf{v} = -\nabla\alpha \times \nabla p + \nu\, \nabla^2\boldsymbol{\zeta} + \nabla\nu \times [\nabla^2\mathbf{v} + \nabla(\nabla \cdot \mathbf{v})] \tag{5}$$

Either Eq. (4) or (5) may be viewed as the basic equation for the rate of change of the absolute vector vorticity of atmospheric flow.

For any scalar Θ we have the identity

$$\boldsymbol{\omega} \cdot \frac{d}{dt}\nabla\Theta = \boldsymbol{\omega} \cdot \nabla\frac{d\Theta}{dt} - \nabla\Theta \cdot [(\boldsymbol{\omega} \cdot \nabla)\mathbf{v}] \tag{6}$$

and thus multiplication of Eq. (5) by $\alpha\, \nabla\Theta$ will yield on the left

$$\alpha\, \nabla\Theta \cdot \left[\frac{1}{\alpha}\frac{d}{dt}(\alpha\boldsymbol{\omega}) - \boldsymbol{\omega} \cdot \nabla\mathbf{v}\right] = \frac{d}{dt}(\alpha\, \nabla\Theta \cdot \boldsymbol{\omega}) - \alpha\boldsymbol{\omega} \cdot \nabla\frac{d\Theta}{dt} \tag{7}$$

With these results, Eq. (5) may be rewritten now as

$$\frac{d}{dt}(\alpha\, \nabla\Theta \cdot \boldsymbol{\omega}) - \alpha\boldsymbol{\omega} \cdot \nabla\frac{d\Theta}{dt} = -\alpha\, \nabla\Theta \cdot (\nabla\alpha \times \nabla p) + \alpha\, \nabla\Theta \cdot \left\{\nu\, \nabla^2\boldsymbol{\zeta} + \nabla\nu \times [\nabla^2\mathbf{v} + \nabla(\nabla \cdot \mathbf{v})]\right\} \tag{8}$$

If we choose Θ to be a variable that depends only on the thermodynamic properties of the flow, then $\Theta = \Theta(\alpha,p)$, and so $\nabla\Theta = (\partial\Theta/\partial\alpha)\, \nabla\alpha + (\partial\Theta/\partial p)\, \nabla p$, which implies that $\nabla\Theta \cdot (\nabla\alpha \times \nabla p) = 0$. A further restriction to inviscid motion produces

$$\frac{d}{dt}(\alpha\, \nabla\Theta \cdot \boldsymbol{\omega}) = \alpha\boldsymbol{\omega} \cdot \nabla\frac{d\Theta}{dt} \tag{9}$$

An obvious choice is to take Θ to be potential temperature θ, because of its tendency to be conservative. Thus for isentropic motion

$$\frac{d}{dt}(\alpha\, \nabla\theta \cdot \boldsymbol{\omega}) = 0 \tag{10}$$

and so

$$\alpha \nabla\theta \cdot \boldsymbol{\omega} = \text{const} \qquad \text{along trajectories} \tag{11}$$

If we denote any particular point on a trajectory by subscript zero, we have

$$\boldsymbol{\omega} \cdot \nabla\theta = \frac{\rho}{\rho_0}\boldsymbol{\omega}_0 \cdot (\nabla\theta)_0 \qquad \text{for a parcel} \tag{12}$$

If furthermore the flow is steady, then Eq. (12) holds along streamlines because they coincide with the trajectories.

The quantity in Eq. (12) may be viewed as the projection of the vorticity on the gradient of the potential temperature. Alternatively we may denote differentiation along the vortex line by $d(\ )/dl$ and then Eq. (12) becomes

$$\omega\frac{d\theta}{dl} = \frac{\rho}{\rho_0}\left(\omega\frac{d\theta}{dl}\right)_0 \qquad \text{for a parcel} \tag{13}$$

where ω is the magnitude of the absolute vorticity.

This is a remarkable result and serves to substantiate once again the claim that vorticity and thermodynamic properties are entwined. Both Eqs. (12) and (13) are, in effect, solutions of the equation of motion for isentropic, inviscid flow that tell us how the rotational part of the motion of a parcel must behave as its trajectory passes through regions of different density and potential temperature.

PROBLEM 10.11.1 Verify Eqs. (6) and (7).

PROBLEM 10.11.2 Suppose a parcel starts at high altitude, perhaps in the stratosphere, behind a front in a region where $\nabla\theta = (\partial\theta/\partial z)\mathbf{k}$. Suppose the parcel passes along an isentrope, down through the frontal zone into the warm air ahead of it and into a region where, once again, the isentropes are level.

Show that for the conditions described we might have

$$\left(\frac{\partial\theta}{\partial z}\right)_0 \bigg/ \frac{\partial\theta}{\partial z} \sim \frac{\gamma_d}{T_0}\frac{T}{\frac{1}{3}\gamma_d}\frac{\theta_0}{\theta} \sim 3(p/p_0)^{R/c_p}$$

and assume an initial altitude of 200 mbar and final altitude of 900 mbar.

Assume that only a small change in latitude occurs during the descent and show that this plunging parcel would arrive at 900 mbar with an immense relative vorticity, and severe weather might be expected should the events described be possible.

PROBLEM 10.11.3 Let Z be the potential vorticity $\alpha\boldsymbol{\omega} \cdot \nabla\theta$ in inviscid flow. Show that the integral $\int \rho Z\, dV$ is constant for a material volume when $\dot{\theta}$ vanishes on its boundaries. Assuming that $\dot{\theta}$ vanishes at the top of the atmosphere, show that surface heating can be expected to destroy total potential vorticity in most cases in mid-latitudes. What happens in the tropics?

10.11.2 Basic Definitions

With a change to a distinctly material approach, the analysis will be based on the relation between the coordinates $\mathbf{x}(t)$ of a parcel at time t and its coordinates $\boldsymbol{\xi}$ at time $t = 0$. Some of the aspects of the transformations from material coordinates $\boldsymbol{\xi}$ to the usual spatial coordinates $\mathbf{x}$ were considered in Sec. 5.7. Now we shall be able to exploit some of the deeper implications of the model of fluid motion presented there.

We assume that we have a basic rectangular cartesian coordinate system and that both coordinates $\boldsymbol{\xi}$ and $\mathbf{x}$ are derived from it by the same transformation. Thus in tensor notation we shall use material coordinates ξ^I indexed by capital letters and spatial coordinates x^i indexed by small letters. Both sets of coordinates have the same metric tensor with respect to the cartesian system, g_{IJ} in the material case and g_{ij} in the spatial case.

The relation between the spatial and material systems is specified by the transformation $\mathbf{x} = \mathbf{x}(\boldsymbol{\xi}, t)$, which can be inverted to give $\boldsymbol{\xi} = \boldsymbol{\xi}(\mathbf{x}, t)$ as the initial coordinates of the parcel at location $\mathbf{x}$ at time t.

We define matrices

$$\nabla\boldsymbol{\xi} = \left\{\frac{\partial \xi^I}{\partial x^i}\right\} \qquad \overset{\xi}{\nabla}\mathbf{x} = \left\{\frac{\partial x^i}{\partial \xi^I}\right\} \qquad \begin{matrix} I = 1, 2, 3 \\ i = 1, 2, 3 \end{matrix} \tag{14}$$

and the Jacobian $|J_\xi^x|$ is given by

$$|J_\xi^x| = \det|\overset{\xi}{\nabla}\mathbf{x}| = \epsilon_{ijk} \frac{\partial x^i}{\partial \xi^1} \frac{\partial x^j}{\partial \xi^2} \frac{\partial x^k}{\partial \xi^3} \tag{15}$$

in which we have used Eq. (41) in Sec. 5.5. This Jacobian may be interpreted as the ratio of the volume occupied by a unit mass at time t to the volume occupied by that mass at time $t = 0$, so that

$$J \equiv |J_\xi^x| = \frac{\alpha}{\alpha_0} = \frac{\rho_0}{\rho} \tag{16}$$

In order to express results in vector notation, we need specific conventions on the relation between the del operators and position of the dot representing scalar products. Thus we have three types of expressions:

$$\mathbf{v} \cdot \nabla\boldsymbol{\xi} = v^k \frac{\partial \boldsymbol{\xi}}{\partial x^k} \tag{17}$$

$$\overset{\xi}{\nabla}\mathbf{x} \cdot \mathbf{v} = (\overset{\xi}{\nabla}x^i)v_i = v_i \overset{\xi}{\nabla}x^i \tag{18}$$

and

$$\nabla(\boldsymbol{\xi} \cdot \mathbf{v}) = \nabla(\xi^i v_i) \tag{19}$$

Only Eq. (18) is new, and the notation implies that the scalar product is taken between the two variables, but only the first is differentiated. Finally, we often shall use the notation that an overdot denotes material differentiation, so that $\dot{\mathbf{x}} = \partial\mathbf{x}/\partial t_\xi = d\mathbf{x}/dt = \mathbf{v}$ represents the velocity of the fluid.

Now we are ready to begin.

10.11.3 The Cauchy Formula for Barotropic Flow

For autobarotropic and inviscid flow subject to conservative external forces, the evolution of the vorticity is described by the Beltrami equation (5):

$$\frac{d}{dt}(\alpha\boldsymbol{\omega}) = \alpha\boldsymbol{\omega} \cdot \nabla\mathbf{v} \tag{20}$$

The remarkable fact is that this equation can be integrated with the aid of the material representation. Let us define the vector $\mathbf{c}$ with

$$\boldsymbol{\omega} = \rho\mathbf{c} \cdot \overset{\xi}{\nabla}\mathbf{x} \tag{21}$$

so that upon substitution in Eq. (20) we have

$$\frac{d}{dt}(\mathbf{c} \cdot \overset{\xi}{\nabla}\mathbf{x}) = \frac{\partial\mathbf{c}}{\partial t_\xi} \cdot \overset{\xi}{\nabla}\mathbf{x} + \mathbf{c} \cdot \overset{\xi}{\nabla}\frac{\partial\mathbf{x}}{\partial t_\xi} = (\mathbf{c} \cdot \overset{\xi}{\nabla}\mathbf{x}) \cdot \nabla\mathbf{v} \tag{22}$$

Expansion of the term on the right gives

$$(\mathbf{c} \cdot \overset{\xi}{\nabla}\mathbf{x}) \cdot \nabla\mathbf{v} = c^K \frac{\partial x^j}{\partial \xi^K}\frac{\partial \mathbf{v}}{\partial x^j} = c^K \frac{\partial\mathbf{v}}{\partial\xi^K} = \mathbf{c} \cdot \overset{\xi}{\nabla}\frac{\partial\mathbf{x}}{\partial t_\xi} \tag{23}$$

and thus Eq. (22) becomes

$$\frac{\partial\mathbf{c}}{\partial t_\xi} \cdot \overset{\xi}{\nabla}\mathbf{x} = 0 \tag{24}$$

Therefore it is sufficient to take $\mathbf{c}$ constant on the trajectories to ensure that $\partial\mathbf{c}/\partial t_\xi$ vanishes. Hence we put $\mathbf{c} = \mathbf{c}(\boldsymbol{\xi})$, and evaluation of Eq. (21) at $t = 0$ gives

$$\boldsymbol{\omega}_0 = \rho_0\mathbf{c} \cdot (\overset{\xi}{\nabla}\mathbf{x})_0 = \rho_0\mathbf{c} \cdot (\overset{\xi}{\nabla}\boldsymbol{\xi}) = \rho_0\mathbf{c} \tag{25}$$

and thus we have the solution

$$\boldsymbol{\omega} = \frac{\rho}{\rho_0}\boldsymbol{\omega}_0 \cdot \overset{\xi}{\nabla}\mathbf{x} \tag{26}$$

This is certainly one of the most elegant expressions in fluid mechanics. It was derived by A.-L. Cauchy in 1815.

This compact result allows us to prove the second and third Helmholtz theorems already discussed in Sec. 10.8.

First let us show that vortex lines are material lines for these flows. Let $\boldsymbol{\tau}$ be a tangent vector to a material line; then $\boldsymbol{\tau}$ can be represented as a function s of length along the line in the form

$$\boldsymbol{\tau} = \frac{\partial \mathbf{x}}{\partial s} = \frac{\partial \xi^J}{\partial s}\frac{\partial \mathbf{x}}{\partial \xi^J} = \boldsymbol{\tau}_0 \cdot \overset{\xi}{\nabla}\mathbf{x} \tag{27}$$

At time $t = 0$, we let this material line coincide with a vortex line so that $\boldsymbol{\tau}_0 = \boldsymbol{\omega}_0/|\boldsymbol{\omega}_0|$. Thus with the Cauchy formula

$$\boldsymbol{\tau} = \frac{\boldsymbol{\omega}_0}{|\boldsymbol{\omega}_0|} \cdot \overset{\xi}{\nabla}\mathbf{x} = \frac{\rho_0}{\rho}\frac{\boldsymbol{\omega}}{|\boldsymbol{\omega}_0|} \tag{28}$$

and so $\boldsymbol{\tau}$ is in the direction of the vorticity vector, and thus the vortex line is a material line.

To prove the third Helmholtz theorem, we need to determine how an element of area on a material surface transforms from $\boldsymbol{\xi}$ to $\mathbf{x}$ coordinates. We shall let a material surface be described by an equation $\mathbf{x} = \mathbf{x}(\alpha,\beta,t)$. Then if $\boldsymbol{\eta}$ is the unit normal to the surface and $d\sigma$ the element of area, we have from Sec. 5.3.2 that

$$\boldsymbol{\eta}\, d\sigma = \frac{\partial \mathbf{x}}{\partial \alpha} \times \frac{\partial \mathbf{x}}{\partial \beta}\, d\alpha\, d\beta \tag{29}$$

and also

$$\boldsymbol{\eta}_0\, d\sigma_0 = \frac{\partial \boldsymbol{\xi}}{\partial \alpha} \times \frac{\partial \boldsymbol{\xi}}{\partial \beta}\, d\alpha\, d\beta \tag{30}$$

Upon taking the ith component of Eq. (29), we find with the aid of Eq. (45) in Sec. 5.5 that

$$\begin{aligned} \eta_i\, d\sigma &= \sqrt{g}\,\epsilon_{ijk}\frac{\partial x^j}{\partial \alpha}\frac{\partial x^k}{\partial \beta}\, d\alpha\, d\beta \\ &= \sqrt{g}\,\epsilon_{ijk}\frac{\partial x^j}{\partial \xi^J}\frac{\partial x^k}{\partial \xi^K}\frac{\partial \xi^J}{\partial \alpha}\frac{\partial \xi^K}{\partial \beta}\, d\alpha\, d\beta \end{aligned} \tag{31}$$

With Eq. (42) in Sec. 5.5 we may conclude that

$$\epsilon_{ijk}\frac{\partial x^i}{\partial \xi^I}\frac{\partial x^j}{\partial \xi^J}\frac{\partial x^k}{\partial \xi^K} = |J_\xi^x|\,\epsilon_{IJK} \tag{32}$$

and Eq. (31) leads to

$$\frac{\partial x^i}{\partial \xi^I}\eta_i\, d\sigma = |J_\xi^x|\sqrt{g}\,\epsilon_{IJK}\frac{\partial \xi^J}{\partial \alpha}\frac{\partial \xi^K}{\partial \beta}\, d\alpha\, d\beta = |J_\xi^x|\,\eta_{0I}\, d\sigma_0 \tag{33}$$

Therefore

$$\overset{\xi}{\nabla}\mathbf{x} \cdot \boldsymbol{\eta}\, d\sigma = |J_{\xi}^{x}|\boldsymbol{\eta}_0\, d\sigma_0 = \frac{\rho_0}{\rho}\boldsymbol{\eta}_0\, d\sigma_0 \tag{34}$$

With this result we are able to find from Eq. (26) that

$$\boldsymbol{\omega} \cdot \boldsymbol{\eta}\, d\sigma = \left(\frac{\rho}{\rho_0}\boldsymbol{\omega}_0 \cdot \overset{\xi}{\nabla}\mathbf{x}\right) \cdot \boldsymbol{\eta}\, d\sigma = \boldsymbol{\omega}_0 \cdot \boldsymbol{\eta}_0\, d\sigma_0 \tag{35}$$

Thus for any surface S in a vortex tube (which is material by the second theorem)

$$\Gamma = \int_S \boldsymbol{\omega} \cdot \boldsymbol{\eta}\, d\sigma = \int_{S_0} \boldsymbol{\omega}_0 \cdot \boldsymbol{\eta}_0\, d\sigma_0 \tag{36}$$

and so the circulation is invariant in the vortex tubes moving with the flow.

Thus we see again that the vorticity is a particularly useful characterization of these barotropic flows of inviscid fluids. The vorticity of each parcel is determined only by its initial value and by the subsequent stretching or compression of the vortex tubes that may occur. But the required conditions are not often valid in atmospheric flows, and we must proceed to a more general description.

10.11.4 Vorticity in Isentropic Flows

The assumption of isentropic motion is a realistic one for modeling a large part of atmospheric flow. We have already seen that with it we can transform the equations of motion to yield the conservation of potential vorticity, which, it might now be suspected, is an extension of the Cauchy result to a more general case. We shall show that this supposition is in fact true, and that a result with wider significance can be derived with the material representations.

Let us consider an inviscid fluid in isentropic flow. The relevant equations are (in an absolute coordinate system)

$$\frac{d\mathbf{v}}{dt} = T\,\nabla S - \nabla(h + \Phi) \tag{37}$$

and

$$\frac{dS}{dt} = 0 \tag{38}$$

where S is the specific entropy, h the specific enthalpy, and Φ the potential for external forces such as that of gravity.

We begin with the identity

$$\frac{d}{dt}(\overset{\xi}{\nabla}\mathbf{x} \cdot \mathbf{v}) = \frac{\partial}{\partial t_{\xi}}(\overset{\xi}{\nabla}\mathbf{x} \cdot \mathbf{v}) = \overset{\xi}{\nabla}\mathbf{x} \cdot \dot{\mathbf{v}} + \overset{\xi}{\nabla}\frac{v^2}{2} \tag{39}$$

and with Eq. (37) we have

$$\frac{d}{dt}(\overset{\xi}{\nabla}\mathbf{x}\cdot\mathbf{v}) = \overset{\xi}{\nabla}\mathbf{x}\cdot[(T\,\nabla S) - \nabla(h+\Phi)] + \overset{\xi}{\nabla}\frac{v^2}{2}$$

$$= T\,\overset{\xi}{\nabla}S - \overset{\xi}{\nabla}\left(h + \Phi - \frac{v^2}{2}\right) \tag{40}$$

Now we let $\dot{\Lambda} = T$ and $\dot{\Psi} = h + \Phi - v^2/2$ and, because of Eq. (38), we thus obtain from Eq. (40) the integral

$$\overset{\xi}{\nabla}\mathbf{x}\cdot\mathbf{v} - \mathbf{v}_0 = \Lambda\,\overset{\xi}{\nabla}S - \overset{\xi}{\nabla}\Psi \tag{41}$$

This is an alternative form of the material equations of motion for isentropic, inviscid flow and is *Weber's transformation* as generalized by Serrin. Note that Eq. (41) is composed of three differential equations in the coordinates $\mathbf{x}$ and their rates of change $\dot{\mathbf{x}}$. The function Ψ depends on h and v^2 in addition to Φ, which is a function of position. But h can be related to S and ρ as can Λ, so the addition of the continuity equation (16) completes the system.

But we shall not try to use Eq. (41) in this form. Instead, we take the scalar product with $\nabla\boldsymbol{\xi}$, which produces

$$\mathbf{v} = \Lambda\,\nabla S - \nabla\Psi + \nabla\boldsymbol{\xi}\cdot\mathbf{v}_0 \tag{42}$$

Thus we have an equation specifying the velocity $\mathbf{v}$ that a parcel whose initial velocity was $\mathbf{v}_0$ will have at each point along its trajectory.

Because our present concern is with vorticity, we have (recalling that $\mathbf{v}$ is the absolute velocity here)

$$\nabla\times\mathbf{v} = \boldsymbol{\omega} = \nabla\Lambda\times\nabla S + \nabla\times(\nabla\boldsymbol{\xi}\cdot\mathbf{v}_0) \tag{43}$$

It is desirable to convert the last term into the same form that appears as Cauchy's formula, a possibility that is obviously suggested by the fact that Eq. (43) should reduce to Eq. (26) for autobarotropic flow in which $\nabla\Lambda\times\nabla S$ will vanish.

To derive the result needed to make the necessary transformation, we observe that [see Eq. (41) in Sec. 5.5]

$$|J_\xi^x| = \frac{\partial x^1}{\partial\xi^I}\frac{\partial x^2}{\partial\xi^J}\frac{\partial x^3}{\partial\xi^K}\epsilon^{IJK} \tag{44}$$

and thus

$$\epsilon^{ijk}|J_\xi^x| = \frac{\partial x^i}{\partial\xi^I}\frac{\partial x^j}{\partial\xi^J}\frac{\partial x^k}{\partial\xi^K}\epsilon^{IJK} \tag{45}$$

We multiply this by $\partial\xi^L/\partial x^i$ to obtain

$$\epsilon^{ijk}\frac{\partial\xi^L}{\partial x^i} = \epsilon^{LJK}\frac{\partial x^j}{\partial\xi^J}\frac{\partial x^k}{\partial\xi^K}|J_\xi^x|^{-1} \tag{46}$$

For the last term of Eq. (43) we have

$$\mathbf{i}^i \cdot \nabla \times (\nabla \boldsymbol{\xi} \cdot \mathbf{v}_0) = \frac{\epsilon^{ijk}}{\sqrt{g}} \frac{\partial}{\partial x^j}\left(\frac{\partial \xi^L}{\partial x^k} v_{0_L}\right)$$
$$= \frac{\epsilon^{ijk}}{\sqrt{g}} \frac{\partial \xi^L}{\partial x^k} \frac{\partial v_{0_L}}{\partial x^j} \tag{47}$$

because the terms with $\partial^2(\ )/\partial x^j \partial x^k$ will vanish owing to the asymmetry of ϵ^{ijk}. Now with Eq. (46) we have (note that i and k change roles giving a change of sign)

$$\mathbf{i}^i \cdot \nabla \times (\nabla \boldsymbol{\xi} \cdot \mathbf{v}_0) = -\frac{\epsilon^{LJK}}{\sqrt{g}\,|J_\xi^x|} \frac{\partial x^j}{\partial \xi^J} \frac{\partial x^i}{\partial \xi^K} \frac{\partial v_{0_L}}{\partial x^j}$$
$$= -\frac{\epsilon^{LJK}}{\sqrt{g}\,|J_\xi^x|} \frac{\partial v_{0_L}}{\partial \xi^J} \frac{\partial x^i}{\partial \xi^K}$$
$$= \frac{\epsilon^{KJL}}{\sqrt{g}\,|J_\xi^x|} \frac{\partial v_{0_L}}{\partial \xi^J} \frac{\partial x^i}{\partial \xi^K}$$
$$= \frac{\rho}{\rho_0} \boldsymbol{\omega}_0 \cdot \overset{\xi}{\nabla} x^i \tag{48}$$

Finally then, Eq. (43) now has the desired form

$$\boldsymbol{\omega} = \nabla \Lambda \times \nabla S + \frac{\rho}{\rho_0} \boldsymbol{\omega}_0 \cdot \overset{\xi}{\nabla} \mathbf{x} \tag{49}$$

which is the generalization of the Cauchy formula to isentropic flow. We find

$$\frac{\rho}{\rho_0}(\boldsymbol{\omega}_0 \cdot \overset{\xi}{\nabla} \mathbf{x}) \cdot \nabla S = \frac{\rho}{\rho_0} \omega_0{}^I \frac{\partial x^k}{\partial \xi^I} \frac{\partial S}{\partial x^k} = \frac{\rho}{\rho_0} \omega_0{}^I \frac{\partial S}{\partial \xi^I}$$
$$= \frac{\rho}{\rho_0} \boldsymbol{\omega}_0 \cdot \overset{\xi}{\nabla} S = \frac{\rho}{\rho_0} (\boldsymbol{\omega} \cdot \nabla S)_0 \tag{50}$$

because the entropy S is invariant along the trajectories. Thus we have the potential vorticity formula

$$\boldsymbol{\omega} \cdot \nabla S = \frac{\rho}{\rho_0} (\boldsymbol{\omega} \cdot \nabla S)_0 \tag{51}$$

which is equivalent to Eq. (12) because $\nabla S = c_p\ \nabla \theta / \theta$ and θ/θ_0 is unity for isentropic flow. Thus the conservation of potential vorticity for inviscid isentropic flow is a particular result implied by the more general expression (49).

It is useful to consider the vorticity $\boldsymbol{\omega}$ of the parcel as composed of two components. First we have the barotropic component

$$\boldsymbol{\omega}_{BT} = \frac{\rho}{\rho_0} \boldsymbol{\omega}_0 \cdot \overset{\xi}{\nabla} \mathbf{x} \tag{52}$$

which, by Cauchy's formula, is the vorticity the parcel would have if it were embedded in a barotropic flow. If we choose a material line initially parallel to the vortex line, then this component of the vorticity remains parallel to that material line, and increases and decreases through changes in density or through stretching or compression along the material line.

To see this, we observe that if we let s be distance in $\boldsymbol{\xi}$ space in the direction $\boldsymbol{\omega}_0$, we have $\boldsymbol{\omega}_0 \cdot \overset{\xi}{\nabla}\mathbf{x} = \omega_0\, \partial\mathbf{x}/\partial s$. Hence $\boldsymbol{\omega}_{BT}$ always has the same direction that the material line does as it moves with the flow. The magnitude of $\boldsymbol{\omega}_{BT}$ will increase if $|\partial\mathbf{x}/\partial s|$ exceeds the initial value of $|\partial\boldsymbol{\xi}/\partial s| = 1$ as the material line is stretched.

The second component of $\boldsymbol{\omega}$ is the baroclinic component

$$\boldsymbol{\omega}_{BC} = \nabla\Lambda \times \nabla S \tag{53}$$

It is obvious that we may write

$$\Lambda = \int_0^t T\,dt \tag{54}$$

where the integral follows a parcel. Thus $\boldsymbol{\omega}_{BC}$ depends on the integrated thermal history of the parcel.

More directly, it is clear that

$$\boldsymbol{\omega}_{BC} \cdot \nabla S = 0 \tag{55}$$

so that $\boldsymbol{\omega}_{BC}$ lies in the isentropic surfaces, its orientation determined by $\nabla\Lambda$, and thus does not contribute to the potential vorticity. In the atmosphere, ∇S is a predominantly vertically directed vector, and so the baroclinic component of the vorticity will generally be a vector whose horizontal components are dominant.

Two further points are worth noting. First, the effects of the earth's vorticity are confined to the barotropic component. We may write Eq. (52) as

$$\boldsymbol{\zeta}_{BT} + 2\boldsymbol{\Omega} = \frac{\rho}{\rho_0} 2\boldsymbol{\Omega} \cdot \overset{\xi}{\nabla}\mathbf{x} + \frac{\rho}{\rho_0} \boldsymbol{\zeta}_0 \cdot \overset{\xi}{\nabla}\mathbf{x} \tag{56}$$

to illustrate this fact. Thus changes in the orientation of the material lines, or stretching of the vortex tubes, may cause the relative vorticity to change. Second, because $\boldsymbol{\omega}_{BC}$ is dominantly a horizontal vector, if the vertical component of the vorticity is considered, then the analysis of only the barotropic component may be sufficient. Indeed, barotropic numerical models based on prediction of vorticity capture the dominant effects of changes in $\boldsymbol{\omega}_{BT}$ to forecast the atmospheric patterns that will develop.

PROBLEM 10.11.4 Let the generalized vorticity vector be $\boldsymbol{\omega}_G = \boldsymbol{\omega} - \nabla\Lambda \times \nabla S$. Show that for invisicid, isentropic flow we have the analog of the Cauchy formula in the form $\boldsymbol{\omega}_G = \alpha_0 \rho \boldsymbol{\omega}_{0G} \cdot \overset{\xi}{\nabla}\mathbf{x}$ where Λ is defined by Eq. (54).

PROBLEM 10.11.5 Show that the generalized vortex lines parallel to ω_G are material lines for inviscid isentropic flow.

PROBLEM 10.11.6 Use the result of Problem 10.11.4 to prove the generalized circulation theorems of Problem 10.9.6.

PROBLEM 10.11.7 Use Cauchy's formula to establish Ertel's result that

$$\frac{d}{dt}(J\boldsymbol{\omega} \cdot \nabla\Psi) = J\boldsymbol{\omega} \cdot \nabla\dot{\Psi}$$

for any scalar Ψ. Using Problem 10.11.4, show that this result holds for the generalized vorticity, and use it to obtain an alternative proof of Eq. (10).

10.12 VORTICITY AND VISCOSITY

So far in this chapter we have studied the kinematics and dynamics of vorticity for flows of a scale that could be considered inviscid. The results were in many cases quite elegant, and perhaps surprising, characterizations of important aspects of fluid flow.

Now we turn to the effect of viscosity on vorticity, and we shall find that it has an important role in the creation of vorticity, in diffusing it, and in destroying it. Thus in contrast to the effect of baroclinity, which tilts vortex lines with respect to material lines and changes the strength of vortex tubes, viscosity and the resulting diffusion of vorticity tend to blend vortices together and to bring the entire fluid into a shearless state of uniform motion or a state of rest.

10.12.1 Incompressible Flow

The equations of motion for the incompressible flow of a fluid of constant density may be written

$$\frac{d\mathbf{v}}{dt} = -\nabla\left(\frac{p}{\rho} + \Phi\right) - \frac{\mu}{\rho}\nabla \times \boldsymbol{\zeta} \qquad \nabla \cdot \mathbf{v} = 0 \tag{1}$$

in which we have assumed the dynamic viscosity μ is constant, and used the identity (38) in Sec. 5.1 and $\nabla \cdot \mathbf{v} = 0$ to obtain

$$\nabla \times \boldsymbol{\zeta} = \nabla \times (\nabla \times \mathbf{v}) = -\nabla^2\mathbf{v} \tag{2}$$

The identity (35) in Sec. 5.1 now shows that

$$\nabla \cdot (\mathbf{v} \times \boldsymbol{\zeta}) = \boldsymbol{\zeta} \cdot \boldsymbol{\zeta} - \mathbf{v} \cdot \nabla \times \boldsymbol{\zeta} \tag{3}$$

and thus Eq. (1) yields the energy equation

$$\frac{d}{dt}\int \rho \frac{v^2}{2}\, dV = -\int \mu \zeta^2\, dV - \int_S \boldsymbol{\eta} \cdot [\mathbf{v}(p + \rho\Phi) - \mu \mathbf{v} \times \zeta]\, d\sigma \tag{4}$$

Therefore the presence of vorticity leads to continual destruction of kinetic energy in the interior of the fluid, independently of the last term of Eq. (4) which depends on boundary conditions.

A vorticity equation is easily obtained from Eq. (1) with Eq. (38) in Sec. 5.1, and the identities used in Eqs. (8) to (10) in Sec. 10.1. The result is

$$\frac{d\boldsymbol{\zeta}}{dt} = \boldsymbol{\zeta} \cdot \nabla \mathbf{v} + \frac{\mu}{\rho}\nabla^2 \boldsymbol{\zeta} \tag{5}$$

This equation has a significant implication: Vorticity cannot be created in a region of a viscous, incompressible fluid that is in irrotational motion. This appears to be obvious, because if vorticity $\boldsymbol{\zeta}$ vanishes in a region, then so does each term shown, as well as $\mathbf{v} \cdot \nabla\boldsymbol{\zeta}$, and thus $\partial\boldsymbol{\zeta}/\partial t$ must also vanish. The conclusion must be that the vorticity is diffused into the interior from the region of strong shear, and hence strong vorticity, that will exist at the boundary where the fluid must adhere to the bounding surface.

To explore this further, let us use Eq. (5) to form the equation

$$\frac{1}{2}\frac{d\zeta^2}{dt} = (\zeta_i \zeta_j)\frac{\partial v_i}{\partial x_j} + \frac{\mu}{\rho}\, \zeta_i \frac{\partial^2 \zeta_i}{\partial {x_j}^2} \tag{6}$$

Now if $|\partial v_i/\partial x_j| \leqslant M$ for all t in the interval $[t_0,t_1]$, then through use of the inequality $2ab \leqslant a^2 + b^2$ we will have $\zeta_i\zeta_j\, \partial v_i/\partial x_j \leqslant 3\zeta^2 M$. Suppose a material volume V exists and is bounded by a material surface S on which the vorticity vanishes for all t in $[t_0,t_1]$. Then Eq. (6) gives

$$\frac{1}{2}\overline{\frac{d\zeta^2}{dt}} = \frac{1}{2V}\int_V \frac{d\zeta^2}{dt}\, dV \leqslant 3\overline{\zeta^2} M - \frac{\mu}{\rho}\overline{\left(\frac{\partial \zeta_i}{\partial x_j}\right)^2} \leqslant 3\overline{\zeta^2} M \tag{7}$$

because the integral of $\eta_j\zeta_i\, \partial\zeta_i/\partial x_j$ vanishes on S. The inequality (7) implies that

$$\overline{\zeta^2(t)} \leqslant \overline{\zeta^2(t_0)}\, e^{6M(t-t_0)} \tag{8}$$

So if $\overline{\zeta^2}(t_0)$ is zero and M is finite (which is required for the equations of motion to make sense), then the vorticity remains zero. So we have proved that an irrotational material element will remain irrotational as long as no vorticity reaches its boundary S; for inviscid flow, it will remain irrotational forever. The spread of vorticity in incompressible flows with constant density, then, is due to viscous diffusion, as shown in Problems 10.12.2 and 10.12.3 below.

PROBLEM 10.12.1 Verify the inequality $\zeta_i\zeta_j\, \partial v_i/\partial x_j \leqslant 3\zeta^2 M$, given that $\max|\partial v_i/\partial x_j| = M$.

*PROBLEM 10.12.2 Consider a material volume V in which the vorticity is zero except on the bounding material surface S, which separates V from a region in which the flow is rotational. Show that the viscous term of Eq. (6) will act to diffuse vorticity into the interior of V. (Hint: Show that the term $\eta_j\zeta_i\, \partial\zeta_i/\partial x_j$ must be positive on S.)

PROBLEM 10.12.3 Consider a viscous incompressible flow with constant density that starts from rest over a solid boundary. Explain how vorticity is generated at the boundary and diffused into the fluid. How might you characterize a boundary layer in such a flow?

10.12.2 Vorticity in the General Viscous Flow

The addition of divergence and baroclinity to the flows under consideration produces a more complex situation. We assume for simplicity that the shear and expansion viscosity coefficients are equal, and thus the general vorticity equation (5) in Sec. 10.11 is

$$\frac{d\boldsymbol{\omega}}{dt} + \boldsymbol{\omega}\nabla\cdot\mathbf{v} = \boldsymbol{\omega}\cdot\nabla\mathbf{v} - \nabla\alpha \times \nabla p + \nu\,\nabla^2\boldsymbol{\zeta} + \nabla\nu \times [\nabla^2\mathbf{v} + \nabla(\nabla\cdot\mathbf{v})] \qquad (9)$$

in which $\boldsymbol{\omega} = \boldsymbol{\zeta} + 2\boldsymbol{\Omega}$ is the absolute vorticity vector. In a nonrotating coordinate system, vorticity can be generated in a region of irrotational flow by baroclinity and by gradients of viscosity. In a rotating system such as the atmosphere, the same mechanisms are available for the generation of relative vorticity, as well as those involving interactions of the vorticity $2\boldsymbol{\Omega}$ and the flow velocity.

Thus we are led to the conclusion that in the general case the study of vorticity does not lead to the attractive results that we can obtain with the aid of simplifying conditions. Still these conditions are often approximately met—especially for large-scale flows in which the vorticity behaves in an approximate sense as predicted by the appropriate theorems. Thus we can still gain intuitive understanding, even if precise results are no longer simple.

10.12.3 Viscosity and Circulation

The idea of studying vorticity by examining the properties of its average represented by the circulation has proved fruitful. We add now to our previous results by determining the effects of viscosity on the rate of change of the circulation. For the equation of motion we have Eq. (45) in Sec. 6.8. To write it in vector notation we observe that

$$\frac{\partial}{\partial x_j}(2\mu e_{ij}) = \frac{\partial}{\partial x_j}\mu\left(\frac{\partial v_i}{\partial x_j} + \frac{\partial v_j}{\partial x_i}\right)$$

$$= \nabla\cdot(\mu\,\nabla)v_i + \frac{\partial}{\partial x_j}\left(\frac{\partial \mu v_j}{\partial x_i} - \frac{\partial \mu}{\partial x_i}v_j\right) \qquad (10)$$

$$= \nabla \cdot (\mu \nabla v_i) + \frac{\partial}{\partial x_i} \nabla \cdot \mu \mathbf{v} - \frac{\partial \mu}{\partial x_i} \nabla \cdot \mathbf{v} - \mathbf{v} \cdot \nabla \frac{\partial \mu}{\partial x_i} \tag{10 cont'd}$$

Thus

$$\mathbf{e} = \mathbf{i}_i \frac{\partial}{\partial x_j}(2\mu e_{ij}) = \nabla \cdot (\mu \nabla)\mathbf{v} + \nabla(\nabla \cdot \mu \mathbf{v}) - \nabla\mu \, \nabla \cdot \mathbf{v} - (\mathbf{v} \cdot \nabla) \nabla\mu \tag{11}$$

Expansion of the first and second terms on the right, with the aid of Eq. (36) in Sec. 5.1, produces

$$\mathbf{e} = (\nabla\mu \cdot \nabla)\mathbf{v} + \mu \nabla^2\mathbf{v} + (\mathbf{v} \cdot \nabla) \nabla\mu + (\nabla\mu \cdot \nabla)\mathbf{v} + \nabla\mu \times (\nabla \times \mathbf{v}) + \mu \nabla(\nabla \cdot \mathbf{v}) - (\mathbf{v} \cdot \nabla) \nabla\mu \tag{12}$$

Now application of Eq. (38) in Sec. 5.1 to the second term of Eq. (12) and combination of some other terms gives

$$\mathbf{e} = 2\mu \nabla(\nabla \cdot \mathbf{v}) - \nabla \times (\mu \nabla \times \mathbf{v}) + 2(\nabla\mu \cdot \nabla)\mathbf{v} + 2\nabla\mu \times (\nabla \times \mathbf{v}) \tag{13}$$

Therefore Eq. (45) in Sec. 6.8 may be written, upon adding Coriolis forces and representing the force of gravity with the potential Φ, in the vector form

$$\frac{d\mathbf{v}}{dt} = -\frac{1}{\rho} \nabla p - \nabla\Phi - 2\mathbf{\Omega} \times \mathbf{v} + \alpha \nabla \tfrac{2}{3}(\mu' - \mu) \nabla \cdot \mathbf{v} + 2\mu\alpha \nabla(\nabla \cdot \mathbf{v}) - \alpha \nabla \times (\mu \nabla \times \mathbf{v}) + 2\alpha(\nabla\mu \cdot \nabla)\mathbf{v} + 2\alpha \nabla\mu \times (\nabla \times \mathbf{v}) \tag{14}$$

We let the circulation be

$$\Gamma = \int_C \mathbf{v} \cdot \boldsymbol{\tau} \, ds \tag{15}$$

where C is a material curve, and thus

$$\dot{\Gamma} = \int_C \dot{\mathbf{v}} \cdot \boldsymbol{\tau} \, ds \tag{16}$$

From the first three terms of Eq. (14) we shall have

$$\int_C \alpha \nabla p \cdot \boldsymbol{\tau} \, ds = \int_C \alpha \frac{\partial p}{\partial s} ds = \int_C \alpha \, dp \tag{17}$$

$$\int_C \nabla\Phi \cdot \boldsymbol{\tau} \, ds = \int_C \frac{\partial \Phi}{\partial s} ds = 0 \tag{18}$$

and

$$\int_C (2\mathbf{\Omega} \times \mathbf{v}) \cdot \boldsymbol{\tau} \, ds = \dot{\Gamma}_a - \dot{\Gamma} \tag{19}$$

according to Eq. (11) in Sec. 10.9.

The next three terms give a frictional contribution F_1 of

$$F_1 = \int_C \alpha\, d\left[\tfrac{2}{3}(\mu' - \mu)\,\nabla \cdot \mathbf{v}\right] + \int_C 2\mu\alpha\, d\,(\nabla \cdot \mathbf{v}) - \int_C \alpha\boldsymbol{\tau} \cdot \nabla \times (\mu\boldsymbol{\zeta})\, ds \tag{20}$$

But with the aid of the continuity equation this becomes

$$F_1 = \int_C \alpha\, d\left[\tfrac{2}{3}(\mu' - \mu)\frac{d \ln \alpha}{dt}\right] + \int_C 2\mu\alpha\, d\left(\frac{d \ln \alpha}{dt}\right) - \int_C \alpha\boldsymbol{\tau} \cdot \nabla \times (\mu\boldsymbol{\zeta})\, ds \tag{21}$$

The last two terms of Eq. (14) may be written in tensor form as

$$\begin{aligned}
\mathbf{i}_i \cdot [(\nabla\mu \cdot \nabla)\mathbf{v} + \nabla\mu \times (\nabla \times \mathbf{v})] &= \frac{\partial \mu}{\partial x_j}\frac{\partial v_i}{\partial x_j} + \epsilon_{ijk}\frac{\partial \mu}{\partial x_j}\epsilon_{klm}\frac{\partial v_m}{\partial x_l} \\
&= \frac{\partial \mu}{\partial x_j}\frac{\partial v_i}{\partial x_j} + \epsilon_{ijk}\epsilon_{lmk}\frac{\partial \mu}{\partial x_j}\frac{\partial v_m}{\partial x_l} \\
&= \frac{\partial \mu}{\partial x_j}\frac{\partial v_j}{\partial x_i}
\end{aligned} \tag{22}$$

where

$$\epsilon_{ijk}\epsilon_{lmk} = \delta_{il}\delta_{jm} - \delta_{im}\delta_{jl} \tag{23}$$

Thus for the contribution of this term F_2, we shall have

$$F_2 = \int_C 2\alpha \frac{\partial \mu}{\partial x_j}\frac{\partial v_j}{\partial x_i}\tau_i\, ds = \int_C 2\alpha \frac{\partial \mu}{\partial x_j}\, dv_j = \int_C 2\alpha\, \nabla\mu \cdot d\mathbf{v} \tag{24}$$

Grouping these results together, we find

$$\begin{aligned}
\dot{\Gamma} + \int_C (2\boldsymbol{\Omega} \times \mathbf{v}) \cdot \boldsymbol{\tau}\, ds = -\int_C \alpha\, dp + \int_C \alpha\, d\left[\tfrac{2}{3}(\mu' - \mu)\frac{d \ln \alpha}{dt}\right] + \int_C 2\mu\alpha\, d\left(\frac{d \ln \alpha}{dt}\right) \\
- \int_C \alpha\boldsymbol{\tau} \cdot \nabla \times (\mu\boldsymbol{\zeta})\, ds + \int_C 2\alpha\, \nabla\mu \cdot d\mathbf{v}
\end{aligned} \tag{25}$$

Thus there may be important viscous effects arising from expansion and variations in both vorticity and velocity. Variations in viscosity may cause changes in the circulation provided that, in the fourth integral, $\nabla\mu \times \boldsymbol{\zeta}$ is not parallel to $\boldsymbol{\tau}$ and, in the last, $\mathbf{v}$ varies in the regions in which viscosity is varying. For air, $\nabla\mu$ is proportional to ∇T, so that alterations in the circulation may develop in regions of strong microscale temperature gradients.

For simplicity in application to large-scale flow, we now let the expansion and shear viscosities be equal and constant, and thus

$$\dot{\Gamma} + \int_C (2\boldsymbol{\Omega} \times \mathbf{v}) \cdot \boldsymbol{\tau}\, ds = -\int_C \alpha\, dp + \int_C 2\mu\alpha\, d\left(\frac{d \ln \alpha}{dt}\right) - \int_C \alpha\mu\boldsymbol{\tau} \cdot \nabla \times \boldsymbol{\zeta}\, ds \tag{26}$$

Recalling the argument of Sec. 10.10, we assume steady flow and let C be a stationary closed streamline so that $\boldsymbol{\tau}$ is parallel to $\mathbf{v}$, and hence the necessary condition for $\dot{\Gamma}$ to vanish is that

$$\int_C T\,ds + \int_C 2\mu\alpha\, d\left(\frac{d\ln\alpha}{dt}\right) = \int_C \alpha\mu\boldsymbol{\tau}\cdot\nabla\times\boldsymbol{\zeta}\,ds \tag{27}$$

But with the continuity equation we may rearrange the second term so that

$$\begin{aligned}\int_C 2\mu\alpha\, d\left(\frac{d\ln\alpha}{dt}\right) &= \int_C 2\mu\, d(\alpha\,\nabla\cdot\mathbf{v}) - \int_C 2\mu\,\nabla\cdot\mathbf{v}\,d\alpha \\ &= -\int_C 2\mu\,\nabla\cdot\mathbf{v}\,d\alpha\end{aligned} \tag{28}$$

Now Eq. (27) becomes

$$\int_C T\,dS - \int_C 2\mu\,\nabla\cdot\mathbf{v}\,d\alpha = \int_C \alpha\mu\boldsymbol{\tau}\cdot(\nabla\times\boldsymbol{\zeta})\,ds \tag{29}$$

or

$$\int_C T\,dS + \int_C 2\nu\,\nabla\cdot\mathbf{v}\,d\ln\rho = \int_C \nu\boldsymbol{\tau}\cdot(\nabla\times\boldsymbol{\zeta})\,ds \tag{30}$$

This equation provides explicit representation of the effects of friction considered abstractly in Sec. 10.10; it could have been derived directly from the usual form of the equation of motion for constant viscosity.

We shall end the chapter with this result, a necessary condition for steady circulation on closed streamlines in a steady flow. We have linked in this equation the generation of motion by the baroclinity created through differential heating, the divergence and vorticity that are the consequence of the ceaseless rearrangement of the mass fields by the heating, and the viscosity that soothes the winds into a restless pattern and holds them to their task of seeking equilibrium.

PROBLEM 10.12.4 Obtain Eq. (30) directly for the case in which the shear and expansion viscosities are equal and constant by converting Eq. (14) into the Crocco-Vazsonyi form. Noting the conditions assumed in deriving Eq. (30), show that it is equivalent to

$$\int_C T\,dS + \int_C \frac{\nu}{v}\,\nabla^2\frac{v^2}{2}\,ds = \int_C \nu\left[v(\boldsymbol{\tau}\cdot\nabla\ln\rho)^2 + \frac{1}{v}\left(\frac{\partial v_i}{\partial x_k}\right)^2\right]ds$$

in which $\mathbf{v} = \boldsymbol{\tau}v$, where $\boldsymbol{\tau}$ is the tangent vector to the streamline. What is the necessary and sufficient condition that $\int T\,dS > 0$? Give a simple example in which the condition would be valid. (Hint: What can you say about $\nabla\cdot\mathbf{v}$ for steady flow?)

PROBLEM 10.12.5 Show that, for $\mu = \text{const}$,

$$f_h = \mu\boldsymbol{\zeta} \cdot \boldsymbol{\zeta} + \left(\tfrac{4}{3}\mu + \tfrac{2}{3}\mu'\right)(\nabla \cdot \mathbf{v})^2$$

[Hint: Convert an appropriate equation derived from Eq. (14) into a divergence minus f_h.]

BIBLIOGRAPHIC NOTES

10.1 The interpretation we have given here of the concept of vorticity is due to Stokes. This one and others are discussed by:

Truesdell, C., 1954: *The Kinematics of Vorticity*, Indiana University Publications Science Series no. 19, Indiana University Press, Bloomington, 232 pp.

The close relation between vorticity and thermodynamics is suggested and explored in the monograph:

Truesdell, C., 1952: "Vorticity and the Thermodynamic State in a Gas Flow," *Mémorial des Sciences Mathématiques, fasc. 119*, 1–53, Gauthier-Villars, Paris.

10.2 The discussion of meteorological vorticity equations is standard. The numerical estimates follow the approach used in lectures by Prof. D. O. Staley.

10.4 The Rossby wave equation appears in the article:

Rossby, C.-G., et al., 1938: "Relation between Variations in the Intensity of the Zonal Circulation of the Atmosphere and the Displacements of the Semi-permanent Centers of Action," *J. Marine Research (Sears Foundation)*, 2:38–55.

The properties of Rossby waves will be considered further in Chap. 12.

10.6 The development theory was originated by:

Sutcliffe, R. C., 1947: "A Contribution to the Problem of Development," *Quart. J. Roy. Meteorol. Soc.*, 73:370–383.

The discussion here follows the treatment by:

Petterssen, S., 1956: *Weather Analysis and Forecasting*, vol. 1, McGraw-Hill Book Company, New York, 428 pp.

10.7 The concepts involved in the omega equation can be traced to Sutcliffe's paper cited above, which had an early version of Eq. (11). A pair of equations equivalent to Eqs. (11) and (12) were presented and discussed by:

Eliassen, Arnt, 1948: "The Quasi-static Equations of Motion with Pressure as Independent Variable," *Geofys. Publikasjoner*, 17:5–44.

and

Eliassen, Arnt, 1952: "Simple Dynamic Models of the Atmosphere Designed for the Purpose of Numerical Weather Prediction," *Tellus*, 4:145–156.

A brief search for the original appearance of the omega equation in the form (13) was not successful.

10.8 The material of these sections on the theorems of Kelvin and Helmholtz was compiled from the works:

Lamb, Sir Horace, 1932: *Hydrodynamics*, 6th ed., Cambridge University Press, Cambridge, 738 pp. Reprinted by Dover Publications, Inc., New York, 1954.

Serrin, James, 1959: "Mathematical Principles of Classical Fluid Mechanics," in S. Flügge (ed.), *Handbuch der Physik*, vol. 8, no. 1, pp. 125–263, Springer-Verlag OHG, Berlin.

as well as the book by Truesdell cited first under 10.1.

10.9 The Bjerknes theorem is given in:

Bjerknes, V., 1902: "Zirkulation relativ zu der Erde," *Meteorol. Z.*, **19**:97–108.

and that of Eckart in:

Eckart, Carl, 1960: "Variation Principles of Hydrodynamics," *Phys. Fluids,* 3:421–427.

The other material in the section is standard in meteorological texts. Additional discussion of circulation theorems may be found in:

Godske, C. L., T. Bergeron, J. Bjerknes, and R. C. Bungaard, 1957: *Dynamic Meteorology and Weather Forecasting*, American Meteorological Society, Boston, and Carnegie Institution of Washington, Washington, 800 pp.

10.10 The first two theorems are believed to be new. Sandstrom's theorem is discussed in the book by Godske et al. cited above and in the paper by Jeffreys cited in the Bibliographic Notes for Sec. 6.7.

10.11 Ertel's theorem appears in the article:

Ertel, H., 1942: "Ein neuer hydrodynamischer Wirbelsatz," *Meteorol. Z.*, **59**:277–281, **59**:385.

The discussion in the remainder of this section derives from Truesdell's 1954 book cited under 10.1 and from Serrin's monograph listed under 10.8. The generalization of Weber's transformation is given in the monograph by Serrin. The use of Weber's transformation to derive the generalization of Cauchy's formula for isentropic flow and the potential vorticity equation that follows are believed new, as are the results about the generalized vorticity given in the problems.

10.12 The first topic is standard in books on fluid dynamics. The relation between viscosity and circulation was studied in detail by:

Truesdell, C., 1949: "The Effect of Viscosity on Circulation," *J. Meteorol.*, 6:61–62.

The application of these results to steady flow on closed streamlines is believed new.

11
ATMOSPHERIC ENERGETICS: GLOBAL THERMODYNAMICS TO TURBULENCE

There is much to be learned about atmospheric motion, about the incessant cycle of events involved in the maintenance of the general circulation, by studying averages taken over the entire atmosphere. The integrals of kinetic energy, internal energy, potential energy, a new form known as available potential energy, and entropy—and particularly their rates of change—have been the major quantities of interest in the study of atmospheric thermodynamics and motion on a global scale.

In most of this book, we have been considering the details of atmospheric motion and the accompanying processes by studying the differential relations that apply to individual parcels. Now, by integrating these relations over the entire atmosphere, we obtain averages that represent the net result of all the details. In doing so, we reduce the complexity of the problem sufficiently so that the basic physics of atmospheric motion and the reasons that it never ceases become easier to comprehend.

11.1 BASIC ENERGY AND ENTROPY THEOREMS

The essential idea of atmospheric energetics can be demonstrated with Newton's second law, $m\, d\mathbf{v}/dt = \mathbf{F}$, which may be multiplied by $\mathbf{v}$ to give $\frac{1}{2} m\, d(\mathbf{v} \cdot \mathbf{v})/dt = \mathbf{v} \cdot \mathbf{F}$.

Thus the concept of kinetic energy originates here with this differential equation; the developments of this chapter are an application of this approach to the differential equations governing atmospheric motion.

The energy conservation theorem for the atmosphere has already been considered in Sec. 6.6. Now we take explicit account of the sources and sinks of energy, and examine the effect that the various terms in the equation of motion have on energy conversions. Because certain of the terms require tensor notation, we shall usually express our results with cartesian tensors; this is permitted because the sphericity terms of Eq. (15) in Sec. 7.2 have no net effect on the energy. With the results of Secs. 6.4, 6.8, and 7.2 we write the equations of motion as

$$\rho \frac{dv_i}{dt} = -\frac{\partial p}{\partial x_i} - g\rho\,\delta_{i3} - 2\rho\epsilon_{ijk}\Omega_j v_k + \frac{\partial \Delta_{ij}}{\partial x_j} \tag{1}$$

$$c_V \rho \frac{dT}{dt} = -p\frac{\partial v_i}{\partial x_i} - \frac{\partial R_i}{\partial x_i} + \frac{\partial}{\partial x_i}\left(k\frac{\partial T}{\partial x_i}\right) + C + \frac{\partial v_i}{\partial x_k}\Delta_{ik} \tag{2}$$

and

$$\frac{d\rho}{dt} = -\rho\frac{\partial v_i}{\partial x_i} \tag{3}$$

where the notation is the same as Eqs. (45) to (47) in Sec. 6.8. It was shown in Sec. 6.8.4 that the dissipation term $\epsilon = (\partial v_i/\partial x_k)\,\Delta_{ik}$ is nonnegative, a fact that will be central to our present considerations.

With the aid of the transport theorem Eq. (8) in Sec. 5.3, these equations give the rate of change of the kinetic energy

$$\begin{aligned}\dot{K} = \frac{\partial}{\partial t}\int_V \frac{\rho v_i^2}{2}\,dV &= -\int_V v_i\left(\frac{\partial p}{\partial x_i} + g\rho\,\delta_{i3}\right)dV + \int_V v_i\frac{\partial \Delta_{ij}}{\partial x_j}\,dV \\ &= \int_V\left(p\frac{\partial v_i}{\partial x_i} - g\rho w\right)dV - \int_V \frac{\partial v_i}{\partial x_k}\Delta_{ik}\,dV - \int_B \eta_i(pv_i - v_k\Delta_{ki})\,d\sigma\end{aligned} \tag{4}$$

and the internal energy rate of change

$$\dot{I} = \frac{\partial}{\partial t}\int_V c_V\rho T\,dV = -\int_V\left(p\frac{\partial v_i}{\partial x_i} - \frac{\partial v_i}{\partial x_k}\Delta_{ik}\right)dV + \int_V C\,dV - \int_B \eta_i\left(R_i - k\frac{\partial T}{\partial x_i}\right)d\sigma \tag{5}$$

We also have the rate of change of the potential energy

$$\dot{P} = \frac{\partial}{\partial t}\int_V g\rho z\,dV = \int_V g\rho w\,dV \tag{6}$$

and an integral for the rate of change of latent energy E_L in the form

$$\dot{E}_L = -\int_V C\,dV - \int_B L_i \eta_i \,d\sigma \tag{7}$$

in which L_i is the boundary flux of latent energy in the ith direction.

Thus we have derived the basic energy theorem for the energies $E = K + I + P$ and E_L:

$$\dot{E} + \dot{E}_L = -\int_B \eta_i \left(R_i - k\frac{\partial T}{\partial x_i} - v_k\,\Delta_{ik} + L_i\right) d\sigma \tag{8}$$

which shows that the total energy $E_T = E + E_L$ can change only through the action of processes at the boundary. In writing Eq. (8), we have used the assumption that the boundary integral of $p\mathbf{v} \cdot \boldsymbol{\eta}$ in Eq. (4) vanishes. At the lower boundary, the normal velocity component must vanish, giving $\boldsymbol{\eta} \cdot \mathbf{v} = 0$ as a boundary condition; in the upper regions of the atmosphere we assume that p decreases more rapidly than w increases so that $pw \to 0$ as $z \to \infty$ (see comment at end of Sec. 5.2.1).

The quantity $p\mathbf{v}$ is known as the *energy flux vector*. We shall return to this boundary integral in Sec. 11.1.2.

11.1.1 Energy Conversions

Each of the volume integrals on the right in Eqs. (4) to (7) represents a conversion of energy from one form to another. The usual practice is to find the same term with opposite signs in two of these equations, and then identify it as representing the conversion between two forms of energy.

Notice, though, that such a correspondence between two terms usually must be forced through use of boundary terms. For example, $v_i\,\partial p/\partial x_i$ appears in Eq. (4) and $p\,\partial v_i/\partial x_i$ appears in Eq. (5). In the second line of Eq. (4), we have used an integration by parts to arrange matters so that we have the same term in both Eqs. (4) and (5). In doing so, we may well have altered the physical significance of the conversion statement in one of the integrals.

On the assumption that the boundary integral of $p\mathbf{v} \cdot \boldsymbol{\eta}$ vanishes, the conversion between I and K may be written

$$C(I,K) = -\int_V \mathbf{v} \cdot \nabla p\,dV = \int_V p\,\nabla \cdot \mathbf{v}\,dV \tag{9}$$

where $C(I,K)$ is positive if energy is flowing from I to K, and hence $C(K,I) = -C(I,K)$.

Similarly,

$$C(P,K) = -\int_V g\rho w\,dV \tag{10}$$

and obviously, we have

$$C(I+P,K) = -\int_V \mathbf{v} \cdot (\nabla p + g\rho\mathbf{k})\,dV \tag{11}$$

We will show later that the terms in Eq. (4) that represent frictional loss or dissipation, to be denoted by $D(K)$, are negative. Assuming this result for now, we have proved:

Theorem In order that the total kinetic energy be maintained or increased, it is necessary that $C(I+P,K)>0$. ////

This result has a number of interesting consequences. Let us observe that if the flow is hydrostatic, then we have (again for the whole atmosphere)

$$C(I+P,K) = -\int_V \mathbf{v} \cdot \nabla_z p \, dV = \int_V p \, \nabla_z \cdot \mathbf{v} \, dV \tag{12}$$

Therefore, in order that this term be positive, the angle between $\mathbf{v}$ and $\nabla_z p$ must exceed $\pi/2$, on the average, so there must be cross-isobar flow toward low pressure. Moreover, we have now:

Corollary (Starr, 1951) In order that motion be maintained in a hydrostatic atmosphere, it is necessary that $\int p \, \nabla_z \cdot \mathbf{v} \, dV > 0$. ////

We have used again the fact that $D(K)<0$ if there is motion. The pressure decreases rapidly with altitude, so the integral is probably dominated by processes at lower altitudes. There we find the belt of high pressure associated with the great tropical anticyclones in which we have $p \, \nabla_z \cdot \mathbf{v} > 0$ and in middle latitudes we have the migratory centers of cyclonic activity in which $p \, \nabla_z \cdot \mathbf{v} < 0$. The divergence $\nabla_z \cdot \mathbf{v}$ integrates to zero over any horizontal surface in the atmosphere, so the fact that the pressure is higher where there is divergence and lower where there is convergence allows us to conclude that the net effect of the cyclones and anticyclones results in the integral being positive. This suggests that the subtropical anticyclones are sources of kinetic energy and that the cyclones destroy the kinetic energy thus created.

But if we look instead at the middle term of Eq. (12), we conclude that cross-isobar flow toward lower pressure is the crucial factor, and that therefore the flow toward lower pressure that exists nearly everywhere at low altitudes results in kinetic energy being produced in both cyclones and anticyclones.

So we have arrived at one of the many apparent paradoxes in energetics. What is going on here? The answer is: The integrals in Eq. (12) are equal but their integrands are not. So which one is physically correct? The answer to that depends on what the question is. Equation (4) shows that cross-isobar flow toward lower pressure does indeed create kinetic energy. The integral of $p \, \nabla_z \cdot \mathbf{v}$, when positive, represents a loss of internal energy. Under the assumptions we have made, the two integrals representing the totals over the entire atmosphere are equal, but the patterns formed by the integrands are vastly different. The point is that it is not necessarily valid to give an integrand local significance.

Now we drop the hydrostatic assumption, but assume that the energies K, I, and P are all constant in time. We shall refer to this case as *energetically steady*, and we obviously have (since then $\dot{P}=0$):

Corollary In an energetically steady atmosphere in which $K > 0$, it is necessary that

$$C(I,K) = \int_V p \nabla \cdot \mathbf{v} \, dV = -\int_V \mathbf{v} \cdot \nabla p \, dV > 0 \tag{13}$$

////

Here again we have assumed that $D(K) < 0$ if K does not vanish. The return flow of energy from K to I occurs through frictional loss of kinetic energy and the concomitant heating of the air expressed by the dissipation $\epsilon = (\partial v_i / \partial x_k) \Delta_{ik}$ in Eq. (5).

*PROBLEM 11.1.1 Prove that ageostrophic flow must exist in a hydrostatic, energetically steady atmosphere.

PROBLEM 11.1.2 For hydrostatic flow, derive an estimate of the form $C(I + P, K) = cK$, where c is a constant. From the observational data, $K \sim 15 \times 10^8$ erg/cm^2 and $C(I + P, K) \sim 6$ W/m^2 so that $c \sim 4 \times 10^{-6}$ s^{-1}. What average angle of cross-isobar flow do these estimates imply?

PROBLEM 11.1.3 (Lorenz) The internal energy of the atmosphere is observed to be about 1.7×10^{12} erg/cm^2 so that $K/I \sim 10^{-3}$. What average Mach number does this imply for atmospheric flow? [Hint: The speed of sound is given by $c = (c_p RT/c_V)^{1/2}$.]

PROBLEM 11.1.4 A condition that $p\mathbf{v}$ vanish, for example, at the top of the atmosphere, is commonly referred to as a *radiation condition.* Explain why the implied similarity between the energy flux vector and the vector flux of radiation is a useful concept.

We turn next to the question of how the various terms in the equation of motion (1) affect the energy distribution in an atmospheric flow.

First the pressure gradient term, from which we have the contribution

$$\dot{K}_\alpha^{(p)} = -\int_V v_\alpha \frac{\partial p}{\partial x_\alpha} \, dV = \int_V p \frac{\partial v_\alpha}{\partial x_\alpha} \, dV \qquad K = \sum_{\alpha=1}^{3} K_\alpha \tag{14}$$

in which α denotes a single component so that we *do not sum* on repeated indices α.

Now if the flow were incompressible, then the sum of the terms in Eq. (14) would vanish, so that we would have $\dot{K}_1^{(p)} + \dot{K}_2^{(p)} + \dot{K}_3^{(p)} = 0$, and hence the pressure gradient term would transfer energy from one velocity component to another. In compressible flows, the terms $\partial v_\alpha / \partial x_\alpha$ generally exceed their sum $\nabla \cdot \mathbf{v}$ by an order of magnitude or more, so the main work of the pressure gradient term is still transfer of energy between components with the presumably small residual being converted to or from internal energy.

The gravity term

$$\dot{K}_3^{(g)} = -\int_V g\rho w \, dV \tag{15}$$

transfers energy between kinetic and potential forms; rising air loses K and gains P, but the net result over the atmosphere depends on the association of the density and vertical velocity fields. In a hydrostatic atmosphere, the vertical kinetic energy gained (or lost) this way is equal to the energy lost (or gained) through the vertical pressure gradient by conversion to (or from) the horizontal components and internal energy.

The Coriolis force term gives $\dot{K}_\alpha^{(\Omega)}$ as the integral of $-2\rho\epsilon_{\alpha jk}\Omega_j v_k v_\alpha$ and the sum over α vanishes because of the antisymmetry in α and k of $\epsilon_{\alpha jk}$. Hence the Coriolis force also transfers the energy of one velocity component to another.

The friction term will be studied in the next section; we end this discussion by considering the inertial term that we were able to avoid in Eq. (4) through use of the transport theorem. If we want it to appear, we may write

$$\dot{K}_\alpha = \int_V \frac{\partial}{\partial t}\frac{\rho v_\alpha^2}{2}\, dV = \int_V \left(\frac{v_\alpha^2}{2}\frac{\partial \rho}{\partial t} + \rho v_\alpha \frac{\partial v_\alpha}{\partial t}\right) dV \tag{16}$$

In the first term on the right, we obtain $-\partial \rho v_k/\partial x_k$ from the continuity equation. With the inertial terms $-v_k\, \partial v_\alpha/\partial x_k$ in the equation of motion, we now have

$$\dot{K}_\alpha^{(I)} = -\tfrac{1}{2}\int_V \frac{\partial}{\partial x_k}(\rho v_k v_\alpha^2)\, dV = -\tfrac{1}{2}\int_B \rho \boldsymbol{\eta}\cdot \mathbf{v}\frac{v_\alpha^2}{2}\, d\sigma = 0 \tag{17}$$

in which we have used the fact that $\boldsymbol{\eta}\cdot\mathbf{v}$ vanishes at the surface and in which we have assumed that the density vanishes more rapidly than wv_α^2 might increase as z approaches infinity.

The contribution from the inertial term vanishes separately for each component, and so its sole effect is to alter the distribution of energy within each velocity component. The redistribution may be viewed as transport of energy spatially or as a change in the manner in which energy is distributed by scale within the flow. The nonlinearity of the flow resulting from the inertial term, then, does not alter the total energy or the energy of a velocity component, but by redistributing the energy, it creates the infinite variety of fluid flows and leads to mathematical and physical questions that so far have not been fully answered.

PROBLEM 11.1.5 Analyze the effect of the advective terms in the first law on the energetics of atmospheric flow. Can the term that appears be expressed in a form we have already seen?

PROBLEM 11.1.6 Develop an energy integral for a volume fixed in space. Discuss the significance of the boundary terms that appear in addition to those of Eq. (8).

11.1.2 The Boundary Integral

At the beginning of the section, we showed that the rate of change of energies $E + E_L$ reduced to the boundary integral given in Eq. (8) as the sum of the energy transfers due to radiation, heat conduction, friction, and flux of latent heat.

The evaluation of this integral along an upper free surface at infinite height is, quite frankly, bothersome from a mathematical standpoint. It is easiest to take refuge in an assumption that the only energy flux at infinity is due to solar and infrared radiation, so that the last three terms vanish at the top of the atmosphere.

Each of the terms is important at the lower boundary, however. Since the water vapor in the atmosphere arises from evaporation at the surface, we may assume that the latent energy term is positive in recognition of the latent heat carried upward by the vapor.

For the friction term we write

$$\int_B \eta_i v_k \, \Delta_{ik} \, d\sigma = \int_B \eta_i v_k \left[\mu \left(\frac{\partial v_i}{\partial x_k} + \frac{\partial v_k}{\partial x_i} \right) - \tfrac{2}{3}(\mu - \mu') \frac{\partial v_j}{\partial x_j} \, \delta_{ik} \right] d\sigma$$

$$= \int_B \boldsymbol{\eta} \cdot \left[\mu \left(\mathbf{v} \cdot \nabla \mathbf{v} + \frac{\nabla v^2}{2} \right) - \tfrac{2}{3}(\mu - \mu') \, \mathbf{v} \, \nabla \cdot \mathbf{v} \right] d\sigma \qquad (18)$$

The vector velocity $\mathbf{v}$ in the first and third terms may be expressed as the sum of the components $\mathbf{v}_\parallel$ and $\mathbf{v}_\perp$, parallel and normal to the boundary. Taking account of the boundary condition that both $\mathbf{v}_\parallel$ and $\mathbf{v}_\perp$ must vanish, we have (without making assumptions about derivatives parallel to η),

$$\int_B \eta_i v_k \, \Delta_{ik} \, d\sigma = \int_B \boldsymbol{\eta} \cdot \left[\frac{\mu}{2} \nabla v^2 + \tfrac{1}{3}\left(\mu' + \frac{\mu}{2} \right) \nabla {v_\perp}^2 \right] d\sigma \leqslant 0 \qquad (19)$$

in which the inequality is derived from the fact that v^2 and ${v_\perp}^2$ both increase upward from the boundary while η is an exterior unit normal vector. We have equality in Eq. (19) only if there is no motion in the neighborhood of the boundary. Thus frictional drag at the surface extracts energy from the atmosphere.

This justifies the contention of Sec. 11.1.1 that $D(K)$ is negative when there is motion. The term $D(K)$ is the sum of the integral of Eq. (18) and the integral of $-(\partial v_i/\partial x_k)\,\Delta_{ik}$, which we proved in Sec. 6.8.4 is nonpositive. If the motion extends to the surface, then clearly $D(K)$ is negative by virtue of Eq. (19). If the motion is confined within a region V^* with $\mathbf{v}$ and $\mathbf{v}_\perp$ vanishing on the boundary B^* of V^*, but not vanishing identically within V^* in the neighborhood of B^*, then Eq. (19) will hold on B^* and so $D(K)$ is again the sum of two negative integrals, and thus we find again that $D(K)$ is negative. Hence $D(K)$ is negative whenever there is motion subject to a nonslip boundary condition.

The boundary heat conduction term represented by $\eta \cdot \nabla T$ is undoubtedly positive on the average over the globe. The term appearing in Eq. (8) is the molecular conduction across the air-surface interface, and the resulting heating is concentrated in a very thin layer. But small-scale convection in the boundary layer distributes the heat thus supplied through a much deeper layer. This eddy flux of heat, known as sensible heat transfer, will appear explicitly when we consider the equations for disturbed flow in Sec. 11.6. The heating due to sensible heat flux, shown in Fig. 11.1, provides evidence that the boundary conduction term must be strongly positive, on the average, except in polar regions.

Finally, the radiation term represents the difference between boundary fluxes of radiation at the top and the bottom of the atmosphere. The most important aspect of this term is that the incoming solar radiation is proportional to the cosine of the latitude so that there is a marked decrease in the energy received with latitude. The resulting temperature gradients cause motion that redistributes the heat. The consequence is that the outgoing long-wave radiation is nearly uniform with latitude, and equal in sum to the incoming energy. Hence examination of the net heat budget of the atmosphere reveals strong heating in lower latitudes and strong cooling in higher latitudes, as shown by Fig. 11.1.

PROBLEM 11.1.7 Use the boundary integral Eq. (8) to derive the energy budget at the surface of the earth. Give physical examples that illustrate the processes represented by each term.

11.1.3 Atmospheric Entropy

The processes that appeared in the boundary integral Eq. (8) for energy also have important effects on the total entropy of the atmosphere. Since by definition we have

$$\rho \frac{ds}{dt} = \frac{c_V \rho}{T} \frac{dT}{dt} + \frac{p}{T} \nabla \cdot \mathbf{v} \tag{20}$$

FIGURE 11.1
The components of the winter heat budget of the Northern Hemisphere expressed as $d\theta/dt$ (°C/day): (*a*) net heating rate, (*b*) infrared radiation, (*c*) solar radiation, (*d*) net radiation, (*e*) latent heat release, and (*f*) sensible heat transfer. (*From Johnson and Dutton, 1976*). The net heating graph has been adjusted so that $\rho J_\theta \, d\theta/dt$ summed to zero on each isentrope, thus eliminating mean meridional flow across the equator. (*Data sources: Solar and infrared radiation: Rodgers, C. D., 1967: "The Radiative Budget of the Troposphere and Lower Stratosphere," Report A2, Planetary Circulations Project, Department of Meteorology, Massachusetts Institute of Technology, NSF Grant GA-400. Sensible heat transfer: Davis, P. A., 1963: "An Analysis of the Atmospheric Heat Budget," J. Atmospheric Sci.,* **20**:*5-22. Latent heat release: Brooks, C. E. P., and Theresa M. Hunt, 1930: "The Zonal Distribution of Rainfall over the Earth," Mem. Roy. Meteorol. Soc.,* **3***(28):139-157.*)

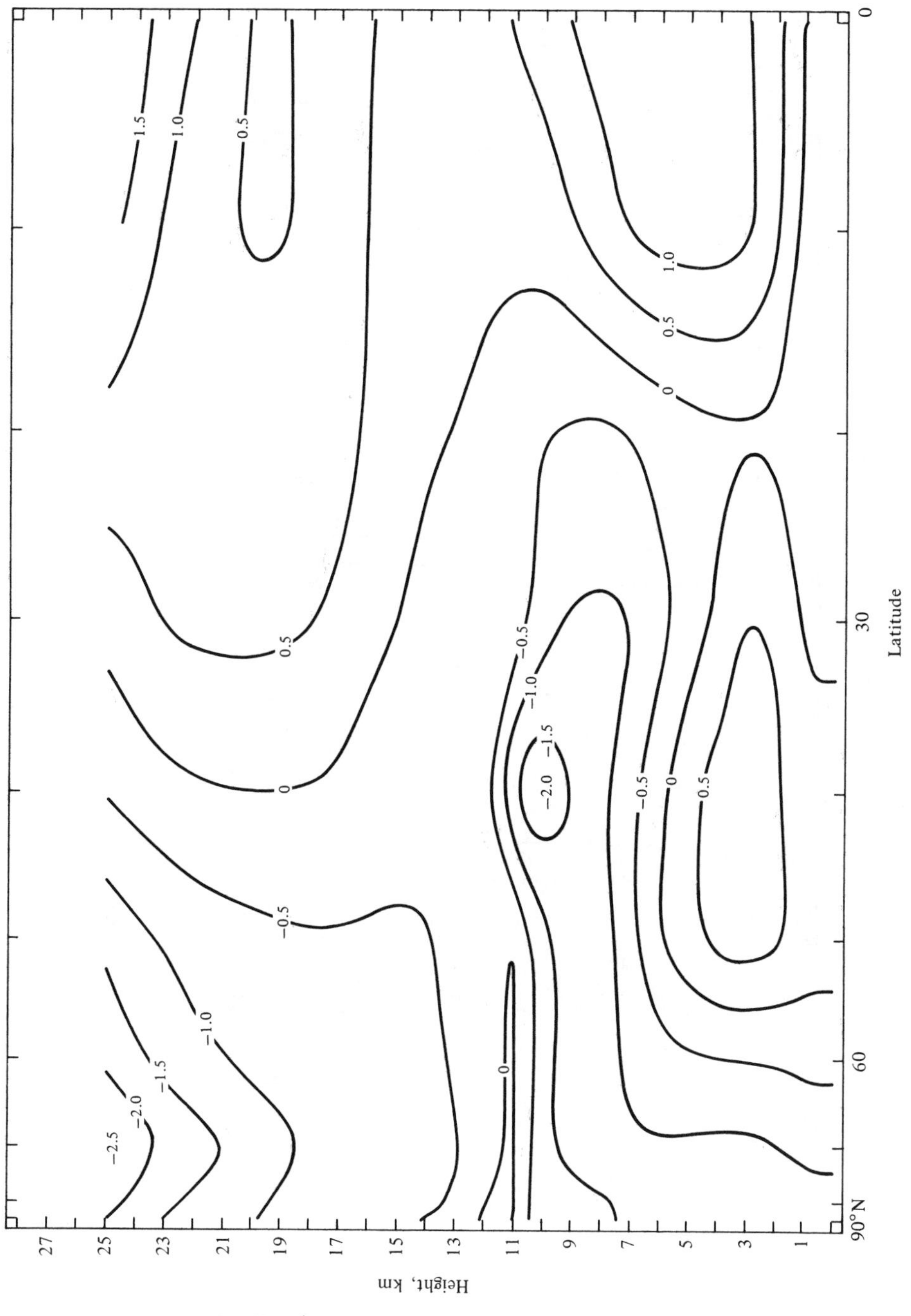

FIGURE 11.1
(*a*) Net heating rate.

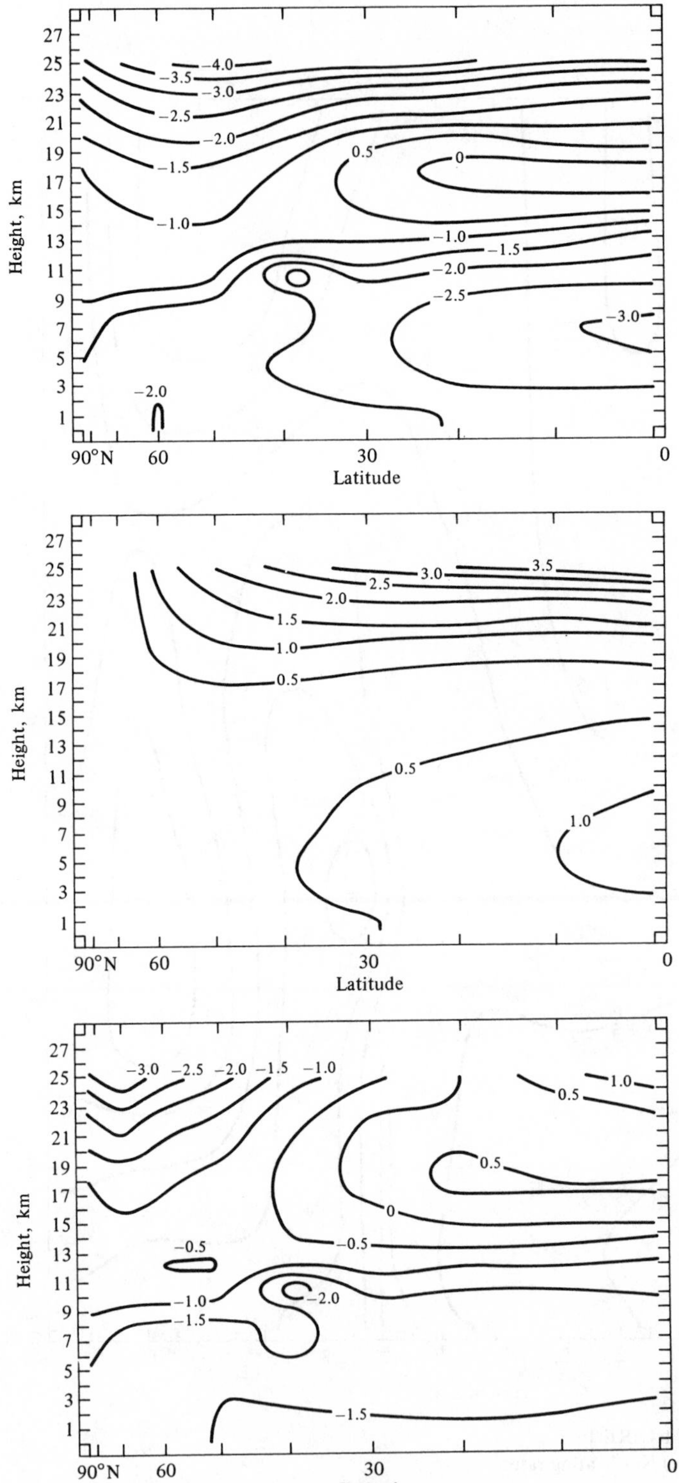

FIGURE 11.1
(*b*) Infrared radiation; (*c*) solar radiation; (*d*) net radiation.

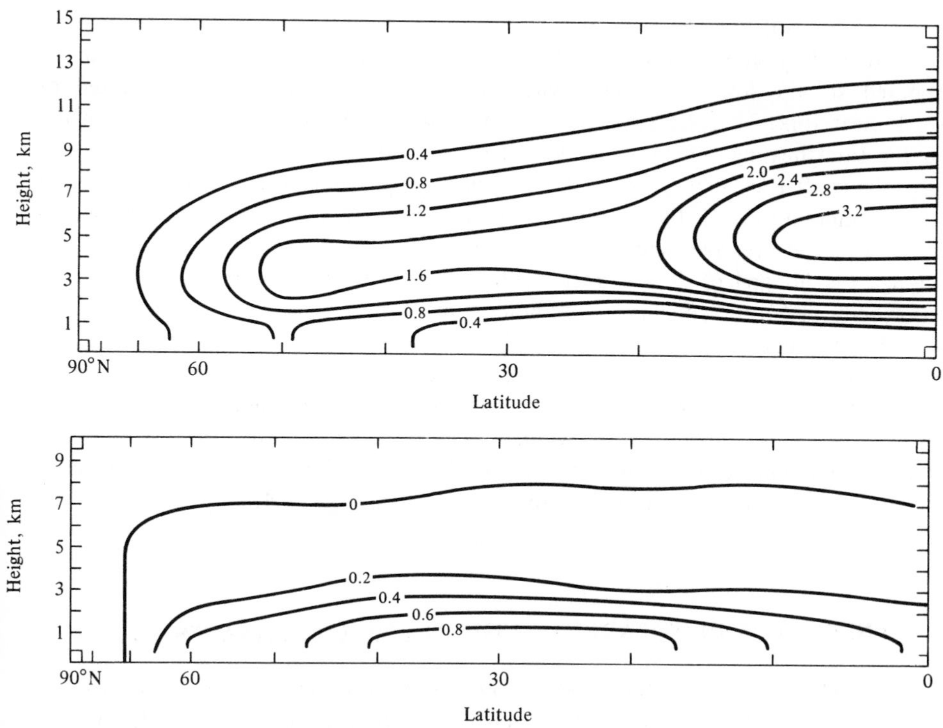

FIGURE 11.1
(*e*) Latent heat release; (*f*) sensible heat transfer.

application of the transport theorem to Eqs. (2) and (20) gives

$$\dot{S} = \frac{\partial}{\partial t}\int_V \rho s\, dV = \int_V \frac{1}{T}\left[-\frac{\partial R_i}{\partial x_i} + \frac{\partial}{\partial x_i}\left(k\frac{\partial T}{\partial x_i}\right) + \frac{\partial v_i}{\partial x_k}\Delta_{ik} + C\right] dV \tag{21}$$

in which S represents the entropy of the air, not the sum of the entropy of the air and the water vapor.

Upon integrating by parts in the conduction term of Eq. (21), we may write

$$\dot{S} = \int_V \left[-\frac{1}{T}\frac{\partial R_i}{\partial x_i} + \frac{\partial}{\partial x_i}\left(\frac{k}{T}\frac{\partial T}{\partial x_i}\right) + \frac{C}{T}\right] dV + \int_V \left[\frac{k}{T^2}\left(\frac{\partial T}{\partial x_i}\right)^2 + \frac{1}{T}\frac{\partial v_i}{\partial x_i}\Delta_{ik}\right] dV \tag{22}$$

If the atmosphere were isolated from the external influences of radiation, water vapor flux, and boundary heat conduction, then only the second term would remain. Experimental evidence shows that the coefficients μ, μ', and k are all positive, so that $\dot{S} \geqslant 0$ in isolation, verifying the applicability of the second law to an isolated atmosphere. Conversely, because the second law is valid for an isolated atmosphere, these three coefficients must be positive.

The effect of the first term of Eq. (22) on the entropy can be analyzed from the net heating field shown in Fig. 11.1, if we note that the conduction term can be integrated to the lower boundary where $\eta \cdot \nabla T$ supplies the heat transported by sensible heat flux. Thus we take the first term of Eq. (22) to be $\int (Q/T)\,dV$, in which Q is the total heating function that would be derived from the values of $d\theta/dt$ shown in the figure. It is clear that the heating occurs at high temperatures relative to cooling at low temperatures. If the integral of Q vanishes, we could surely conclude that $\int (Q/T)\,dV$ were negative since T^{-1} is small where $Q > 0$ and large where $Q < 0$. Although the integral of Q is slightly positive on the long-term average (to balance the friction term in $\dot{E}$), the conclusion still holds, since S is constant on the average.

Thus the net heating of the atmosphere decreases the entropy, which we might expect would cause motion toward the equilibrium represented by a state of maximum entropy. But if we consider only solar radiation, then we find that the entropy increases; thus the initial heating would appear to generate, not destroy, entropy. Apparently we will need a more thorough consideration of this situation to resolve the apparent paradox, and we shall return to this subject in Sec. 11.4.

11.2 ENERGETICS IN ISOBARIC AND ISENTROPIC COORDINATES

Energy transformation equations are easily derived for the hydrostatic isobaric and isentropic coordinates in which observational data are generally represented. To do so, we must take account of the Jacobian of the transformation, which acts to preserve the significance of $\rho\,dV$ as an element of mass. Thus we have

$$K = \int_V \rho \frac{v^2}{2}\,dV = \frac{1}{g}\int_A \left(\int_0^{p_0} \frac{v^2}{2}\,dp \right) d\sigma = -\frac{1}{g}\int_A \left(\int_{\theta_0}^{\theta_T} \frac{v^2}{2}\frac{\partial p}{\partial \theta}\,d\theta \right) d\sigma \tag{1}$$

in which we have effected the transformation by using the hydrostatic equation and the chain rule. The quantities p_0 and θ_0 are the surface values of pressure and potential temperature while $\theta_T\,(=\infty)$ is the potential temperature at the top of the atmosphere. Both p_0 and θ_0 are in general functions of both horizontal position and time.

Similarly,

$$I = \int_V c_V \rho T\,dV = \frac{1}{g}\int_A \left(\int_0^{p_0} c_V T\,dp \right) d\sigma = -\frac{1}{g}\int_A \left(\int_{\theta_0}^{\theta_T} c_V T \frac{\partial p}{\partial \theta}\,d\theta \right) d\sigma \tag{2}$$

and

$$P = \int_V g\rho z\,dV = \int_A \left(\int_0^{p_0} h_p\,dp \right) d\sigma = -\int_A \left(\int_{\theta_0}^{\theta_T} h_\theta \frac{\partial p}{\partial \theta}\,d\theta \right) d\sigma \tag{3}$$

Now from the hydrostatic equation in isobaric coordinates, $\partial h_p/\partial p = -\alpha/g$, we find with an integration by parts that

$$\int_0^{p_0} h_p \, dp = \frac{1}{g}\int_0^{p_0} RT \, dp \tag{4}$$

so we may define the sum of the internal and potential energies as the *total potential energy* and for hydrostatic conditions we have

$$\Pi = I + P = \frac{1}{g}\int_A\left(\int_0^{p_0} c_p T \, dp\right) d\sigma = -\frac{1}{g}\int_A\left(\int_{\theta_0}^{\theta_T} c_p T \frac{\partial p}{\partial \theta}\, d\theta\right) d\sigma \tag{5}$$

PROBLEM 11.2.1 Show in cartesian coordinates that $P = (R/c_V)I$ and that $\Pi = (c_p/c_V)I$ when hydrostatic conditions prevail.

11.2.1 Energy Transformations in Isobaric Coordinates

Because of the hydrostatic assumption, we consider only the kinetic energy K_H of the horizontal component, which we denote as $\mathbf{v}_H$. Thus from Eq. (1)

$$\dot{K}_H = \frac{1}{g}\int_{V_p} \mathbf{v}_H \cdot \frac{\partial \mathbf{v}_H}{\partial t}\, dV_p + \frac{1}{g}\int_A\left(\frac{v_H^2}{2}\frac{\partial p_0}{\partial t}\right)_{p=p_0} d\sigma \tag{6}$$

But $\mathbf{v}_H$ vanishes on the lower boundary owing to the nonslip boundary condition, and so use of Eq. (31) in Sec. 7.3 gives

$$\dot{K}_H = -\frac{1}{g}\int_{V_p} \mathbf{v}_H \cdot \nabla_p \phi_p \, dV_p + D(K_H) \qquad \phi_p = gh_p \tag{7}$$

The contribution of the inertial term to Eq. (7) vanishes because

$$\begin{aligned}
\frac{1}{g}\int_{V_p} \mathbf{v}_H \cdot \left(\mathbf{v}_H \cdot \nabla_p \mathbf{v}_H + \omega\frac{\partial \mathbf{v}_H}{\partial p}\right) dV_p &= \frac{1}{2g}\int_{V_p}\left(\mathbf{v}_H \cdot \nabla_p v_H^2 + \omega\frac{\partial v_H^2}{\partial p}\right) dV_p \\
&= \frac{1}{2g}\int_{V_p}\left(\mathbf{v}_H \cdot \nabla_p v_H^2 - v_H^2\frac{\partial \omega}{\partial p}\right) dV_p \\
&= \frac{1}{2g}\int_A\int_0^{p_0} \nabla_p \cdot \mathbf{v}_H v_H^2 \, dp\, d\sigma \\
&= \frac{1}{2g}\int_A\left[\nabla_p \cdot \left(\int_0^{p_0} \mathbf{v}_H v_H^2 \, dp\right)\right.
\end{aligned} \tag{8}$$

$$\left. - (v_H^2 \mathbf{v} \cdot \nabla_p p_0)_{p_0} \right] d\sigma$$

$$= \frac{1}{2g} \int_A \nabla_z \cdot \left(\int_0^{p_0} \mathbf{v} v_H^2 \, dp \right) d\sigma = 0 \tag{8}$$

(cont'd)

In this sequence of equations we have used the fact that ωv_H^2 vanishes at both $p = 0$ and $p = p_0$ and the isobaric continuity equation to obtain $\nabla_p \cdot \mathbf{v}_H v_H^2$. The term evaluated at p_0 vanishes because $\mathbf{v}_H$ vanishes at the surface where $p = p_0$ and the last integral, in which we have used $\nabla_p \cdot \phi(x,y) = \nabla_z \cdot \phi(x,y)$, vanishes on application of the divergence theorem.

From Eq. (5) in this section and Eq. (29) in Sec. 7.3 we find that

$$\dot{\Pi} = \frac{1}{g} \int_{V_p} \alpha\omega \, dV_p + \frac{1}{g} \int_{V_p} (q + f_h) \, dV_p \tag{9}$$

But two quite different terms have appeared in Eqs. (7) and (9), and the question is whether we can force them to be the same in order to close the energy budget. From the hydrostatic equation, we have through a series of integrations by parts

$$\frac{1}{g} \int_{V_p} \alpha\omega \, dV_p = - \int_{V_p} \frac{\partial h_p}{\partial p} \omega \, dV_p = \int_{V_p} \frac{\partial \omega}{\partial p} h_p \, dV_p = - \int_{V_p} (\nabla_p \cdot \mathbf{v}_H) h_p \, dV_p$$

$$= \frac{1}{g} \int_{V_p} \mathbf{v}_H \cdot \nabla_p \phi_p \, dV_p \tag{10}$$

so that we are indeed justified in writing

$$C(\Pi, K_H) = -\frac{1}{g} \int_{V_p} \mathbf{v}_H \cdot \nabla_p \phi_p \, dV_p = -\frac{1}{g} \int_{V_p} \alpha\omega \, dV_p = -\frac{R}{g} \int_{V_p} \frac{T\omega}{p} \, dV_p \tag{11}$$

This last result has been cited often as showing that kinetic energy is produced through ascent of warm air relative to the descent of cool air on an isobaric surface, a process that lowers the center of mass. This is similar to the paradox already discussed—kinetic energy is produced by flow across height contours and the integrand $\mathbf{v}_H \cdot \nabla_p \phi_p$ is largest at very low altitudes (Fig. 11.2). The integrand $\alpha\omega$ is largest in the middle atmosphere and gives the locally correct picture for loss of potential energy but not for gain of kinetic energy. The point here is the same as it was before: equality of integrals does not require or guarantee equality of integrands.

PROBLEM 11.2.2 Derive the result (9) for $\dot{\Pi}$.

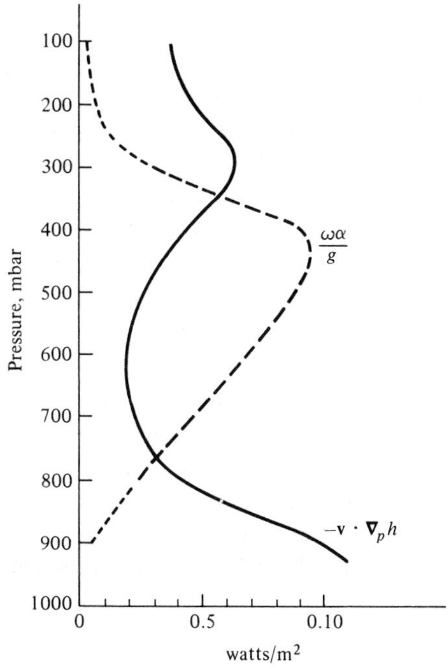

FIGURE 11.2
Comparison of the conversion terms $-\mathbf{v} \cdot \Delta_p h_p$ and $(\omega\alpha)/g$ as integrated over 50-mbar layers. The curve for $-\mathbf{v} \cdot \Delta_p h_p$ is the annual mean over North America (Kung). The shape of the $(\omega\alpha)/g$ curve was derived from measurements over North America for a 3-day period (Eddy) and then adjusted so that the integral equals that of the other curve. *(Kung, E. C., 1969: "Further Study on the Kinetic Energy Balance," Monthly Weather Rev.,* **97**:*573-581. Eddy, A., 1965: "Kinetic Energy Production in a Mid-Latitude Storm," J. Appl. Meteorol.,* **4**:*569-575.)*

PROBLEM 11.2.3 Show that $C(\Pi,K_H)$ as derived in isobaric coordinates is identical to $C(\Pi,K_H)$ in cartesian coordinates under the hydrostatic assumption.

PROBLEM 11.2.4 Show that the integrands $-\alpha\omega$ and $-\mathbf{v}_H \cdot \nabla_p \phi_p$ can be the same everywhere in a hydrostatic atmosphere if and only if the horizontal momentum is nondivergent, that is, if $\nabla_H \cdot \rho\mathbf{v}$ vanishes everywhere.

PROBLEM 11.2.5 Explain how the omega equation could be used to determine $C(\Pi,K_H)$ in isobaric coordinates, from the height fields alone. What assumptions does this involve and what errors might you anticipate?

11.2.2 Energy Transformations in Isentropic Coordinates

A major difficulty with both theoretical and practical applications of isentropic coordinates is that isentropes in the lower atmosphere often intersect the ground, so that we may view the isentropes as continuing underground and reemerging at another location. The problems that arise from this situation are eliminated by defining the pressure at each point on an underground isentrope to be the surface pressure above that point (as suggested by Prof. Edward N. Lorenz). Thus $\partial p/\partial\theta$ vanishes on underground isentropes, and so the integrands in Eqs. (1) to (3) vanish on any underground portions of isentropes. Therefore we may take θ_0 to be any constant value $\theta_0 < \min\theta$ where $\min\theta$ is the lowest value of θ observed in the atmosphere in the period of time in which we are interested.

With the first and third lines of Eq. (53) in Sec. 7.3 and using Eq. (1) for the horizontal velocity components again, we have

$$\dot{K}_H = -\frac{1}{g}\int_{V_\theta}\frac{\partial}{\partial t_\theta}\left(\frac{v_H^2}{2}\frac{\partial p}{\partial\theta}\right)dV_\theta = \frac{1}{g}\int_{V_\theta}\frac{\partial p}{\partial\theta}(\mathbf{v}_H\cdot\nabla_\theta\Psi)\,dV_\theta + D(K_H) \qquad (12)$$

This follows because the inertial terms and the use of the continuity equation give a boundary integral that may be assumed to vanish since $\partial p/\partial\theta$ vanishes on the underground isentrope θ_0.

Before attempting to determine the rate of change of the total potential energy (5), it is useful to express it in a form particularly suited to hydrostatic, isentropic coordinates. With Poisson's equation we find

$$c_pT\frac{\partial p}{\partial\theta} = \frac{c_p\theta}{(1+\kappa)(p_{00})^\kappa}\frac{\partial}{\partial\theta}p^{1+\kappa} \qquad (13)$$

where $\kappa = R/c_p$. Thus with an integration by parts in θ, Eq. (5) becomes

$$\Pi = \frac{c_p}{g[(1+\kappa)(p_{00})^\kappa]}\left[\int_{V_\theta}p^{1+\kappa}\,dV_\theta + \int_A\theta_0 p^{1+\kappa}(\theta_0)\,d\sigma\right] \qquad (14)$$

in which we have used the justifiable assumption that $\theta p^{1+\kappa}$ vanishes as p tends to zero.

For the rate of change of Π, we use the second equation of (53) in Sec. 7.3 and integration by parts to obtain

$$\begin{aligned}\dot{\Pi} &= \frac{1}{g}\int_{V_\theta}\frac{\partial\Psi}{\partial\theta}\frac{\partial p}{\partial t_\theta}\,dV_\theta + \frac{c_p}{gp_{00}{}^\kappa}\int_A\theta_0 p^\kappa(\theta_0)\frac{\partial p(\theta_0)}{\partial t}\,d\sigma \\ &= -\frac{1}{g}\int_{V_\theta}\Psi\frac{\partial^2 p}{\partial t_\theta\,\partial\theta}\,dV_\theta - \frac{1}{g}\int_A\left(\Psi\frac{\partial p}{\partial t_\theta} - \frac{c_p}{p_{00}{}^\kappa}\theta_0 p^\kappa\frac{\partial p}{\partial t}\right)_{\theta=\theta_0}d\sigma \\ &= -\frac{1}{g}\int_{V_\theta}\frac{\partial p}{\partial\theta}\mathbf{v}_H\cdot\nabla_\theta\Psi\,dV_\theta - \frac{c_p}{g}\int_{V_\theta}\left(\frac{p}{p_{00}}\right)^\kappa\frac{d\theta}{dt}\frac{\partial p}{\partial\theta}\,dV_\theta \\ &\qquad - \frac{c_p}{g}\int_A\left[\theta\left(\frac{p}{p_{00}}\right)^\kappa\frac{d\theta}{dt}\frac{\partial p}{\partial\theta}\right]_{\theta=\theta_0}d\sigma \end{aligned} \qquad (15)$$

in which we have employed the isentropic continuity equation in Eq. (53) of Sec. 7.3, and eliminated the second term in the second equation with $\Psi = c_p\theta(p/p_{00})^\kappa + gh$ and the fact that $h(\theta_0) = 0$. The last term vanishes because $\partial p/\partial\theta$ is identically zero on θ_0. The remaining term with $d\theta/dt$ may be denoted as $G(\Pi)$, the notation implying that it represents the generation due to heating.

Here again, the closure of the energy budget has been forced through use of integration by parts.

PROBLEM 11.2.6 Show that the two terms of Eq. (14) may be combined to give

$$\Pi = \frac{c_p}{g[(1+\kappa)(p_{00})^\kappa]} \int_A \int_0^{\theta_T} p^{1+\kappa}\, d\theta\, d\sigma$$

and that $G(\Pi)$ in Eq. (15) may be expressed as

$$G(\Pi) = -\frac{c_p}{g} \int_A \int_0^{\theta_T} \left(\frac{p}{p_{00}}\right)^\kappa \frac{d\theta}{dt} \frac{\partial p}{\partial\theta}\, d\theta\, d\sigma$$

PROBLEM 11.2.7 Show that $C(\Pi, K_H)$ in isentropic coordinates is a sum of two terms, one involving the advection of enthalpy on an isentropic surface, the other related to conversions of potential energy. (Hint: The relation to potential energy conversion may be established by proving that the expression is exactly equal to potential energy conversion when the isentropes are steady in isentropic flow.)

*PROBLEM 11.2.8 Show that $C(\Pi, K_H)$ as obtained in cartesian, isobaric, and isentropic coordinates with the hydrostatic assumption represents the same physical process. Use the geostrophic concept to obtain an analytical expression for this conversion rate that may be applied in any coordinate system.

11.3 AVAILABLE POTENTIAL ENERGY

Observations show that the atmospheric flow in mid-latitudes is composed of an essentially westerly current disturbed by the superimposed alternating pattern of cyclonic and anticyclonic systems. The part of the flow that is parallel to the latitude circles—the zonal component—cannot by itself transport heat poleward. Thus the necessary transport of heat must be effected by either the eddy component arising from the disturbances or by an average meridional and cellular flow pattern, or by a combination of both possibilities.

The question of which form of transport is the dominant one has been a central problem in meteorological research in the last quarter-century. An associated question is whether the disturbances feed on the energy of the zonal current, or whether the eddies actually drive the larger-scale motion. We shall be concerned with these questions, seeking to develop answers here and in Chap. 14.

11.3.1 Zonal and Eddy Components of the Flow

In order to separate the zonal and eddy components in a mathematically deterministic manner, we use zonal averages. Thus we define the *zonal average* of a variable f by

$$\bar{f} = \frac{1}{2\pi} \int_0^{2\pi} f(\lambda, \phi, z, t)\, d\lambda \tag{1}$$

an average over longitude λ. The eddy component is then $f^* = f - \bar{f}$. The energy integrals we have been considering involve mass weighting with the density. For a coordinate system with ϑ as the vertical coordinate, we may introduce this weighting with the Jacobian and define averages and deviations as

$$\hat{f} = \frac{\overline{\rho J_\vartheta f}}{\overline{\rho J_\vartheta}} \qquad f' = f - \hat{f} \tag{2}$$

in which the average is taken along surfaces of constant ϑ. It is clear that $\overline{\rho J_\vartheta f'}$ vanishes.

Thus upon setting $\mathbf{v} = \hat{\mathbf{v}} + \mathbf{v}'$, the kinetic energy becomes

$$K = \int_{V_\vartheta} \tfrac{1}{2}(\overline{\rho J_\vartheta}\, \hat{\mathbf{v}} \cdot \hat{\mathbf{v}} + \overline{\rho J_\vartheta \mathbf{v}' \cdot \mathbf{v}'})\, dV_\vartheta = \hat{K} + K' \tag{3}$$

because the product term $\rho J_\vartheta\, \hat{\mathbf{v}} \cdot \mathbf{v}'$ averages to zero. The kinetic energy is represented now as the sum of the kinetic energy $\hat{K}$ of the zonal velocity component and the kinetic energy K' of the eddy component.

For hydrostatic conditions, we may combine the internal and potential energy into the total potential energy, as shown in Eq. (5) of Sec. 11.2. Applying the same zonal averaging technique gives

$$\Pi = \int_{V_\vartheta} c_p \rho T J_\vartheta\, dV_\vartheta = \int_{V_\vartheta} c_p \overline{\rho J_\vartheta} \hat{T}\, dV_\vartheta = \hat{\Pi} \tag{4}$$

so that there is no eddy form of Π to accompany the decomposition of the kinetic energy K into $\hat{K}$ and K'.

The solution to this problem lies in the theory of *available potential energy*, a concept reintroduced into meteorological thinking by Prof. Edward N. Lorenz in 1955. We shall develop some analytical results first, and then show why the name available potential energy is appropriate.

In isentropic coordinates with hydrostatic conditions, we have the expression in Eq. (14) in Sec. 11.2, as modified in Problem 11.2.6, for Π. Let us define

$$A' = \frac{c_p}{g(1+\kappa)p_{00}{}^{\kappa}} \int_{V_\theta} (p^{1+\kappa} - \bar{p}^{1+\kappa})\, dV_\theta \tag{5}$$

where the range of θ in V_θ is taken to be $(0,\infty)$. The crucial point is that $A' \geqslant 0$. To see this, we use a Taylor series with remainder to write exactly

$$p^{1+\kappa} = \bar{p}^{1+\kappa} + (p-\bar{p})\left(\frac{\partial p^{1+\kappa}}{\partial p}\right)_{\bar{p}} + \frac{(p-\bar{p})^2}{2}\left(\frac{\partial^2 p^{1+\kappa}}{\partial p^2}\right)_{p_*} \tag{6}$$

where p_* is the particular value, between p and $\bar{p}$, that gives equality. The third term of Eq. (6) is always nonnegative and the second vanishes upon averaging, so that we have proved that

$$\overline{p^{1+\kappa}} \geqslant \bar{p}^{1+\kappa} \tag{7}$$

Now if $\bar{\bar{p}}$ denotes the average of p over an entire isentropic surface, then the same reasoning gives $\overline{\overline{(\bar{p})^{1+\kappa}}} - (\bar{\bar{p}})^{1+\kappa} \geqslant 0$ so that

$$\hat{A} = \frac{c_p}{g(1+\kappa)p_{00}{}^{\kappa}} \int_{V_\theta} [(\bar{p})^{1+\kappa} - (\bar{\bar{p}})^{1+\kappa}]\, dV_\theta \tag{8}$$

is also nonnegative.

The *available potential energy* is defined by

$$A = \hat{A} + A' = \frac{c_p}{g(1+\kappa)p_{00}{}^{\kappa}} \int_{V_\theta} (p^{1+\kappa} - \bar{\bar{p}}^{1+\kappa})\, dV_\theta \tag{9}$$

and has a number of significant features. This expression can be converted by various means to approximate forms that give expressions for A in other coordinate systems. Alternatively, perturbation equations (embodying the quasi-geostrophic assumption, for example) often lead to energy forms (see Chaps. 13 and 14) that may be viewed as available potential energies. These expressions thus provide exact versions of A appropriate to approximate sets of equations; we shall encounter some of them in the chapters cited. For now, we shall use the definitions (5), (8), and (9) in isentropic coordinates, in which they lead to exact results for hydrostatic conditions.

We begin our analysis of available potential energy by considering its rate of change. The derivative of the first term in Eq. (9) is obviously $\dot{\Pi} = -C(\Pi, K) + G(\Pi)$. To compute the second term, we use

$$p = -\int_\theta^\infty \frac{\partial p}{\partial \theta'}\, d\theta' \tag{10}$$

along with the continuity equation in isentropic coordinates to find the product of $\partial\bar{\bar{p}}/\partial t_\theta$ and the area of the isentrope as

$$\begin{aligned} \frac{\partial}{\partial t_\theta}\int_A p\, d\sigma_\theta &= \int_\theta^\infty \left\{ \int_A \left[\nabla_\theta \cdot \left(\frac{\partial p}{\partial \theta}\mathbf{v}\right) + \frac{\partial}{\partial \theta}\left(\frac{\partial p}{\partial \theta}\frac{d\theta}{dt}\right)\right] d\sigma_\theta \right\} d\theta \\ &= -\overline{\overline{\left(\frac{\partial p}{\partial \theta}\frac{d\theta}{dt}\right)}} \int_A d\sigma_\theta \end{aligned} \tag{11}$$

The combination of Eq. (11), Eq. (15) in Sec. 11.2, and $G(\Pi)$ from Problem 11.2.6 gives

$$G(A) = -\frac{c_p}{g}\int_{V_\theta}\left[\left(\frac{p}{p_{00}}\right)^\kappa - \left(\frac{\bar{\bar{p}}}{p_{00}}\right)^\kappa\right]\frac{\partial p}{\partial \theta}\frac{d\theta}{dt}\,dV_\theta \tag{12}$$

We have therefore proved:

Theorem (Lorenz, 1955) When hydrostatic conditions prevail, then the available potential energy defined by Eq. (9) obeys

$$\dot{A} = -C(\Pi, K) + G(A) \tag{13}$$

and

$$\dot{K} + \dot{A} = G(A) + D(K) \tag{14}$$

in which $G(A)$ is given by Eq. (12) and vanishes if the entire flow is isentropic. Maintenance of circulation requires that $G(A) > 0$. ////

11.3.2 Properties of Available Potential Energy

The usefulness of the preceding theorem is that the available potential energy has a number of conceptual advantages as a replacement for the total potential energy. The first is that it can be decomposed into zonal and eddy components, as already pointed out.

The second is that in frictionless, isentropic flow we have from Eq. (14) the result that

$$K + A = \text{const} \tag{15}$$

Thus the maximum amount of kinetic energy that could be obtained under these conditions is $K_{\text{max}} = K + A$. Moreover, we may define the potential energy of the average (or reference) state as Π_r, where $-\Pi_r$ is the second term of Eq. (9). Thus we may write $\Pi = A + \Pi_r$. Following through the derivation in Eqs. (10) to (12), we see that $\dot{\Pi}_r = G(\Pi) - G(A)$ so that changes in Π_r are not associated with $C(\Pi,K)$. Put the other way around, any kinetic energy that is realized by the flow to offset $D(K)$ must have come from A. Therefore we have succeeded in partitioning Π into a portion A that is available for conversion to K and a portion Π_r that is not available.

This partitioning provides the solution to the problem of what an appropriate decomposition of Π might be. We have now that $\Pi = \Pi_r + \hat{A} + A'$ so that $\Pi_r + \hat{A}$ is a zonal form and A' is an eddy form. But as we have just seen, Π_r does not participate in $C(\Pi,K)$ so that we may use the active part, $\hat{A} + A'$, in studying energy conversions.

As pointed out in the theorem, the maintenance of the general circulation can now be viewed as a question of how available potential energy is generated. As made clear by Eq. (12), generation requires an association between the mass field and the heating field. The expression in brackets in Eq. (12) may be written as

$(p/p_{00})^{\kappa}[1-(\bar{\bar{p}}/p)^{\kappa}]$ and is positive for $p > \bar{\bar{p}}$ and negative otherwise. Thus heating at high pressure or cooling at lower pressure on an isentrope generates A and permits maintenance of the circulation. Again, then, we see that the pattern of heating at low altitude in low latitudes and the higher altitude cooling in higher latitudes is responsible for the continuation of atmospheric flow.

PROBLEM 11.3.1 Show that over the entire atmosphere, the amount of mass between two isentropes can change only if there are regions in which $d\theta/dt$ does not vanish. Show that changes in Π_r are related to such changes in mass distribution with respect to potential temperature.

PROBLEM 11.3.2 Show that the outbreaks of cold air over warm currents in the cyclogenetic regions on the east coast of continents can be expected to lead to strong local contributions to $G(A)$.

PROBLEM 11.3.3 Describe completely the atmospheric state in which $A = 0$. In particular, show that the statement "if the atmosphere is hydrostatic and barotropic, then A vanishes" is not correct in the sense that the hypothesis is not strong enough to support the conclusion.

PROBLEM 11.3.4 Let a typical (not extreme) value of $p - \bar{\bar{p}} = \pm 0.1\bar{\bar{p}}$. Show that A should be about $1.5 \times 10^{-3}\Pi$.

PROBLEM 11.3.5 Using the estimates obtained in earlier problems, find the ratio between K and A. Assume that present conditions are energetically steady to estimate $D(K)$. How long would it take the circulation to run down if $G(A) = 0$ and if $D(K)$ were steady at this rate? How long would it take if $D(K) = -\lambda K$, where λ is to be determined from present conditions? [Hint: You will have to assume that $A > 0$ implies that $C(A,K) > 0$.]

11.3.3 Interactions between A and K

The available potential energy is an integral that summarizes some aspects of the mass field just as the kinetic energy is a measure of certain properties of the velocity field. We have seen that kinetic energy in the atmosphere is created from the reservoir of available energy. Now we wish to consider the details of the interaction between these two quantities.

The first question we might ask is whether we can have A without K, and the second might be whether the presence of A implies that $C(A,K)$ is positive, as we assumed in the last problem.

In answer to the first we have:

Theorem If stable hydrostatic conditions prevail and if $A > 0$, then there is motion in the sense that $\mathbf{v}$ and $\dot{\mathbf{v}}$ cannot both vanish identically.

PROOF If $A > 0$, then there must be pressure gradients on isentropic surfaces (including, perhaps, those which go underground, implying surface pressure gradients) so that $p \not\equiv \bar{p}$, and hence $\nabla_\theta p \not\equiv 0$. But

$$\nabla_\theta p = \nabla_z p - \frac{\partial p}{\partial \theta} \nabla_z \theta \tag{16}$$

so that if $\nabla_\theta p \not\equiv 0$, then $\nabla_z p$ and $\nabla_z \theta$ cannot both vanish identically. Referring to the discussion of Jeffreys' theorem in Sec. 6.7, we see that if $\nabla_z p \not\equiv 0$, then either $\mathbf{v}$ or $\dot{\mathbf{v}}$ must be nonzero somewhere. If $\nabla_z p \equiv 0$, then $\nabla_z \theta \not\equiv 0$, which implies that $\nabla_z T \not\equiv 0$, which gives the same conclusion. ////

Corollary If stable hydrostatic conditions prevail and if $A > 0$ in an interval $[t_1, t_2]$, then K cannot vanish identically in that interval.

PROOF Assume the contrary so that $K \equiv 0$ in the interval, which implies that $\mathbf{v} \equiv 0$. But then $\dot{K} = \ddot{K} = 0$ in at least (t_1, t_2), so that $\ddot{K} = \int \rho(\dot{\mathbf{v}}^2 + \ddot{\mathbf{v}} \cdot \mathbf{v})\, dV = 0$ implies that $\dot{\mathbf{v}} \equiv 0$ since the last term vanishes. Thus the contrary assumption leads to $\mathbf{v} \equiv \dot{\mathbf{v}} \equiv 0$, which contradicts the theorem, and so the corollary is established. ////

The answer to the second question is presumably that $A > 0$ does not imply that conversion to K must occur, because we can at least imagine counterexamples (for example, purely geostrophic flow). Still, by virtue of the corollary we have the assurance that if differential solar radiation creates and maintains the supply of available energy (despite the destruction by conduction and thermal radiation, for example), then motion must inevitably follow.

Next we turn to the transformations between $\hat{K}, K', \hat{A}$, and A'. To establish the kinetic energy part of the budget, we need only to compute the rate of change of $\hat{K}$. Thus

$$\frac{\partial \hat{K}}{\partial t} = \frac{\partial}{\partial t} \int_{V_\vartheta} \frac{\overline{\rho J_\vartheta}}{2} \hat{\mathbf{v}} \cdot \hat{\mathbf{v}}\, dV_\vartheta = \int_{V_\vartheta} \left(\overline{\rho J_\vartheta}\, \hat{\mathbf{v}} \cdot \frac{\partial \hat{\mathbf{v}}}{\partial t} + \frac{\hat{v}^2}{2} \frac{\partial}{\partial t} \overline{\rho J_\vartheta} \right) dV_\vartheta \tag{17}$$

The second term will contribute to a transformation between $\hat{K}$ and K'.

To aid in the analysis of the effect of choosing the various coordinate systems, we write the equation of motion as

$$\frac{\partial \mathbf{v}}{\partial t_\vartheta} + (\mathbf{v} \cdot \nabla \mathbf{v})_\vartheta = -\mathbf{P}_\vartheta - g\mathbf{k} - f\mathbf{k} \times \mathbf{v} + \mathbf{F}_\vartheta \tag{18}$$

so that with the continuity equation

$$\frac{\partial}{\partial t}(\rho J_\vartheta) + (\nabla \cdot \rho J_\vartheta \mathbf{v})_\vartheta = 0 \tag{19}$$

we find immediately that

$$\overline{\rho J_\vartheta}\,\frac{\partial \hat{\mathbf{v}}}{\partial t} = -\overline{\rho J_\vartheta(\mathbf{P}_\vartheta + g\mathbf{k})} - f\,\overline{\rho J_\vartheta}\,\mathbf{k} \times \hat{\mathbf{v}} + \overline{\rho J_\vartheta}\,\hat{\mathbf{F}}_\vartheta - \overline{\rho J_\vartheta(\mathbf{v} \cdot \nabla \mathbf{v})_\vartheta}$$
$$- \overline{\mathbf{v}(\nabla \cdot \rho J_\vartheta \mathbf{v})_\vartheta} - \hat{\mathbf{v}}\frac{\partial}{\partial t}\overline{\rho J_\vartheta} \tag{20}$$

From this we see that

$$C_\vartheta(A,\hat{K}) = -\int_{V_\vartheta} \hat{\mathbf{v}} \cdot \overline{[\rho J_\vartheta(\mathbf{P}_\vartheta + g\mathbf{k})]}\, dV_\vartheta = -\int_{V_\vartheta} \hat{\mathbf{v}} \cdot [\hat{\mathbf{P}}_\vartheta + g\mathbf{k}]\,\overline{\rho J_\vartheta}\, dV_\vartheta \tag{21}$$

Because the averages on the two terms in brackets may be removed, and because $C(A,K) = C(A,\hat{K}) + C(A,K')$, we obtain

$$C_\vartheta(A,K') = -\int_{V_\vartheta} \mathbf{v}' \cdot (\mathbf{P}_\vartheta + g\mathbf{k})\rho J_\vartheta\, dV_\vartheta = -\int_{V_\vartheta} \mathbf{v}' \cdot \mathbf{P}'_\vartheta \rho J_\vartheta\, dV_\vartheta \tag{22}$$

The next step is to compute the rate of change of $\hat{A}$ from Eq. (8), and this must be done in isentropic coordinates. We assume hydrostatic conditions for simplicity so that use of Eq. (19) in the manner of Eqs. (10) and (11) gives

$$\frac{\partial \bar{p}}{\partial t_\theta} = \int_\theta^{\theta_T} \left[\overline{\frac{\partial}{\partial y}\left(\frac{\partial p}{\partial \theta_1} v\right)} + \overline{\frac{\partial}{\partial \theta_1}\left(\frac{\partial p}{\partial \theta_1}\frac{d\theta_1}{dt}\right)} \right] d\theta_1 \tag{23}$$

in which, for convenience, we use y instead of latitude. The term in Eq. (23) with $d\theta/dt$ combines with Eq. (11) to give

$$G(\hat{A}) = -\frac{c_p}{g}\int_{V_\theta} \left[\left(\frac{\bar{p}}{p_{00}}\right)^\kappa - \left(\frac{\bar{\bar{p}}}{p_{00}}\right)^\kappa\right] \frac{\partial p}{\partial \theta}\frac{d\theta}{dt}\, dV_\theta \tag{24}$$

and the remaining term in Eq. (23) yields

$$\frac{\partial \hat{A}}{\partial t} - G(\hat{A}) = \frac{c_p}{g}\int_{V_\theta} \left\{ \left(\frac{\bar{p}}{p_{00}}\right)^\kappa \cdot \int_\theta^{\theta_T} \frac{\partial}{\partial y}\overline{\left[\left(\frac{\partial p}{\partial \theta_1}\right) v\right]}\, d\theta_1 \right\} dV_\theta$$
$$= \frac{c_p}{g}\int_{V_\theta} \left\{ \left[\int_\theta^{\theta} \left(\frac{\bar{p}}{p_{00}}\right)^\kappa d\theta_1 \right] \frac{\partial}{\partial y}\overline{\left(\frac{\partial p}{\partial \theta} v\right)} \right\} dV_\theta \tag{25}$$

$$= -\frac{c_p}{g}\int_{V_\theta}\left[\hat{v}\frac{\partial \bar{p}}{\partial \theta}\frac{\partial}{\partial y}\int_0^\theta\left(\frac{\bar{p}}{p_{00}}\right)^\kappa d\theta_1\right]dV_\theta \tag{25 (cont'd)}$$

It is convenient to define a zonally symmetric version of the Montgomery stream function by

$$\Psi_s = c_p\int_{\theta_0}^{\theta}\left(\frac{\bar{p}}{p_{00}}\right)^\kappa d\theta_1 + c_p\bar{T}_0 = c_p\int_0^\theta\left(\frac{\bar{p}}{p_{00}}\right)^\kappa d\theta_1 \tag{26}$$

so that finally we obtain

$$\frac{\partial \hat{A}}{\partial t} - G(\hat{A}) = -\frac{1}{g}\int_{V_\theta}\frac{\partial p}{\partial \theta}\hat{\mathbf{v}}\cdot\nabla_\theta\Psi_s\,dV_\theta \tag{27}$$

For hydrostatic conditions Eq. (21) becomes

$$C_\theta(A,\hat{K}) = \frac{1}{g}\int_{V_\theta}\hat{\mathbf{v}}\cdot\overline{\frac{\partial p}{\partial \theta}\nabla_\theta\Psi}\,dV_\theta \tag{28}$$

So, as we have found before, the two conversion terms in Eqs. (27) and (28) do not match. There are a number of alternatives, with varying implications; we shall choose the conventional one now, even though it has disadvantages in later developments that we cannot consider here. Taking account of all the possibilities, we have

$$\begin{aligned}\frac{\partial \hat{A}}{\partial t} - G(\hat{A}) &= -C(\hat{A},\hat{K}) - C(\hat{A},K') - C(\hat{A},A')\\ C(A,\hat{K}) &= C(\hat{A},\hat{K}) + C(A',\hat{K})\end{aligned} \tag{29}$$

If the values $C(\hat{A},K') = C(A',\hat{K}) = 0$ are arbitrarily assigned, then Eqs. (27) and (28) along with Eq. (29) give the result that

$$C_\theta(\hat{A},A') = \frac{1}{g}\int_{V_\theta}\hat{\mathbf{v}}\cdot\left[\overline{\frac{\partial p}{\partial \theta}\nabla_\theta(\Psi_s - \Psi)}\right]dV_\theta \tag{30}$$

The budget we have derived is shown in Fig. 11.3.

PROBLEM 11.3.6 Show that

$$C(\hat{K},K') = -\int_{V_\vartheta}\rho J_\vartheta \mathbf{v}'\cdot(\mathbf{v}\cdot\nabla\hat{\mathbf{v}})_\vartheta\,dV_\vartheta$$

Explain what mechanisms or flow structures lead to conversions.

PROBLEM 11.3.7 Study the transformation used in Eq. (25) and verify the result. (Comment: This technique is a useful trick to remember.)

PROBLEM 11.3.8 Ascertain the relation between Ψ_S and $\overline{\Psi}$ for hydrostatic conditions. With the binomial theorem, find an approximate expression for $\Psi_S - \Psi$ that allows you to explain how the conversion $C(\hat{A},A')$ is effected.

PROBLEM 11.3.9 Taking all possible conversions into account, write out a complete energy budget for the system $\hat{K}$, K', $\hat{A}$, A' in symbolic form [using $C(\ ,\ )$ notation]. How many conversion terms must be set arbitrarily in order to close the budget? How many energy budgets are possible?

PROBLEM 11.3.10 Construct an energy budget for isentropic coordinates in which $C(\hat{A},A')$ and $C(\hat{A},K')$ vanish. Discuss the relation of this budget with the one given in the text and Fig. 11.3.

11.3.4 Comparison of Modes of Response

At the beginning of this section, we raised the question of whether atmospheric response to differential heating involved a significant meridional circulation. The notion that the basic mode involves ascent of warm air in the tropics, poleward and westerly flow at high altitudes, descent of cooled air at high latitudes, and equatorward and easterly flow near the surface was proposed by Hadley in 1735 in an attempt to explain the trade winds. Such a direct thermal circulation is now known as a *Hadley cell.*

An alternative to a direct Hadley circulation as a mechanism for accomplishing the required transport of heat is a series of circular or asymmetric quasi-horizontal disturbances; such a flow is called a *Rossby mode* of response. Certainly such disturbances are observed in middle latitudes, and the conclusion of two decades of work with observational data has been that the atmospheric circulation is basically a Hadley response in the tropics and a Rossby regime in higher latitudes.

So even though we do not have a Hadley type of response in middle and higher latitudes, we might ask whether the zonal averages of the disturbances would produce an average direct circulation of Hadley type. Here again, the present answer is that the mean meridional circulation in extratropical regions is very weak.

But these are coordinate dependent results, for all of the observational data that has been used is in isobaric coordinates. Now surely the physics of the situation is not dependent on the coordinate system, but what we obtain for averages does depend on what variables are held constant while the averaging operation is performed. Furthermore, there may be a coordinate system in which the mathematical description of the physics takes a simpler form.

The point of the present subsection is that the view we obtain of the general circulation is quite different in the cartesian and isobaric coordinates than it is in isentropic coordinates. The main reason is that the vertical velocities in the quasi-horizontal z and p systems have physical meaning and implication significantly different from those of the vertical velocity $\dot{\theta}$ in isentropic coordinates. The

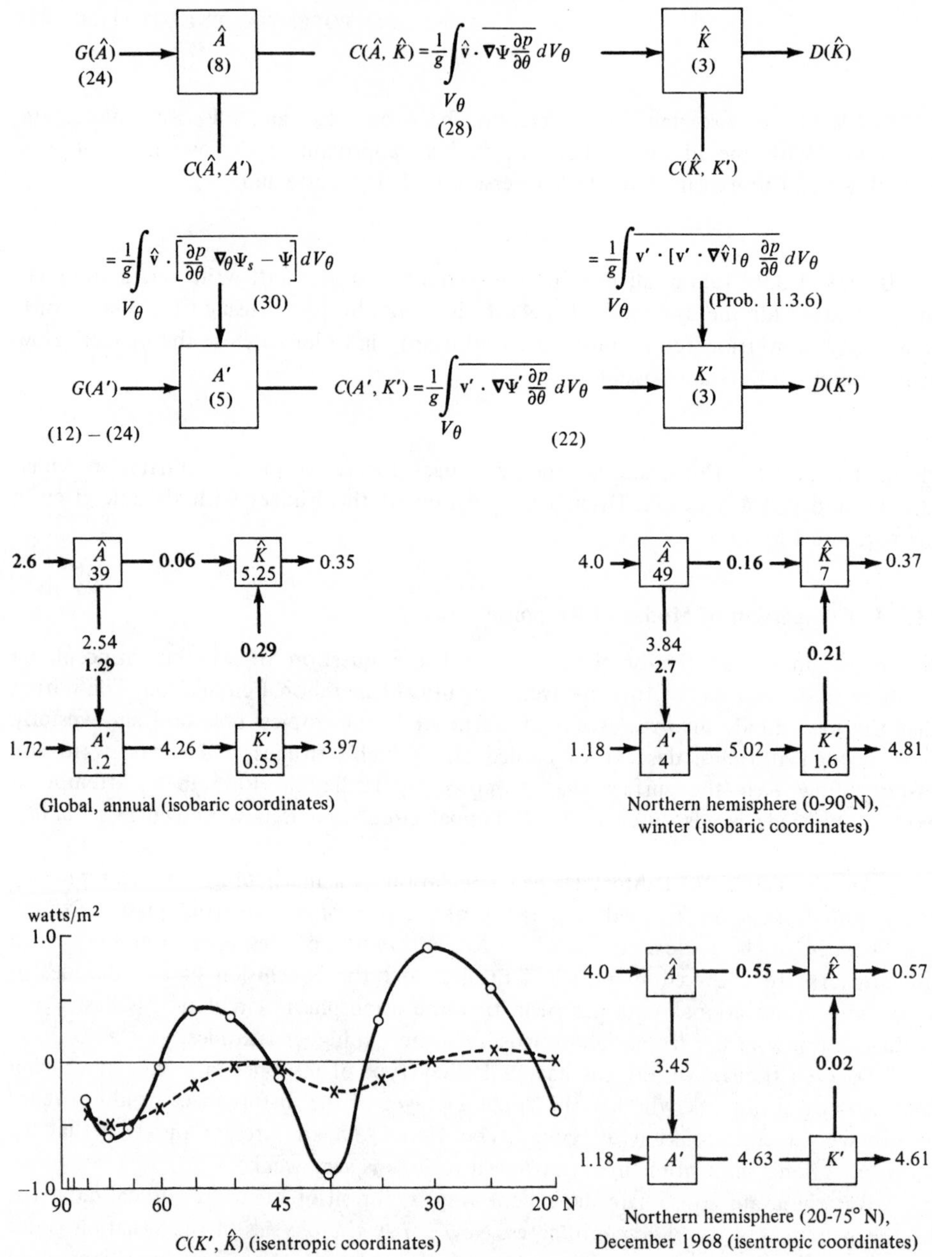

FIGURE 11.3
Atmospheric energy budgets. The top figure shows the budget derived in the text for isentropic coordinates, with the relevant equations indicated in parentheses. The three budgets giving numerical values to the conversion rates were developed from reported results and were forced to balance by assuming energetically steady conditions. The transformation rates computed from observed data by the original authors are given in boldface type.

In these budgets, the multiannual dissipation rates of 4.32 W/m^2 for the year and 5.18 W/m^2 for the winter computed (Kung) for North America were consequences of this fact for atmospheric energetics were first perceived by Prof. Donald R. Johnson.

Let us begin with the continuity equation (19). Upon taking a zonal average, we find that

$$\hat{v}(y,\vartheta,t) = (\overline{\rho J_\vartheta})^{-1} \int_y^{y_P} \left[\frac{\partial}{\partial t_\vartheta} \overline{\rho J_\vartheta} + \frac{\partial}{\partial \vartheta} \left(\overline{\rho J_\vartheta \frac{d\vartheta}{dt}} \right) \right] dy_\vartheta \tag{31}$$

in which y_P denotes the North Pole. If the zonally averaged mass distribution does not change with time, then the first term vanishes. For the statistically quasi-steady circulations we observe in the atmosphere, this is a reasonable assumption and permits the exposition to concentrate on the significant essentials; we therefore adopt this assumption along with the hydrostatic restriction throughout this subsection.

In the three coordinate systems, then

$$\hat{v} = \begin{cases} (\bar{\rho})^{-1} \displaystyle\int_y^{y_P} \frac{\partial}{\partial z} (\overline{\rho w})\, dy_1 & \text{cartesian} \\ \displaystyle\int_y^{y_P} \frac{\partial \bar{\omega}}{\partial p}\, dy_1 & \text{isobaric} \\ \left(\overline{\frac{\partial p}{\partial \theta}}\right)^{-1} \displaystyle\int_y^{y_P} \frac{\partial}{\partial \theta} \left(\overline{\frac{\partial p}{\partial \theta} \frac{d\theta}{dt}} \right) dy_1 & \text{isentropic} \end{cases} \tag{32}$$

FIGURE 11.3 (*continued*)

used to obtain a balance. The two budgets in isobaric coordinates (Newell) represent the global, annual mean and the Northern Hemisphere (0–90°N) winter. The results in isentropic coordinates (Henderson) were computed from seven evenly spaced days in December 1968 from National Meteorological Center data. In this isentropic budget, the $G(\hat{A})$ value of 4.0 W/m² was adopted from the isobaric computations and the winter dissipation rate of 5.18 W/m² was used.

The remaining graph gives the values of $C(K', \hat{K})$ as a function of latitude from the computations (Henderson) in isentropic coordinates (the dotted line gives at latitude ϕ the integral from 75°N to latitude ϕ). The isobaric Northern Hemisphere budget (Newell) had an equivalent integrated value at $\phi = 30°$N of $C(K', \hat{K}) = -0.20$ W/m².

The balance imposed on the two Northern Hemisphere budgets assumes that there were no energy fluxes through the boundaries of the computational region. The amounts of the four forms of energy are shown in the isobaric budgets in units of 10^5 J/m². (*Kung, E. C., 1969: "Further Study on the Kinetic Energy Balance," Monthly Weather Rev.,* **97**:*573–581. Newell, R. E., D. G. Vincent, T. G. Dopplick, D. Ferruza, and J. W. Kidson, 1970: "The Energy Balance of the Global Atmosphere," in G. A. Corby (ed.), The Global Circulation of the Atmosphere, Royal Meteorological Society, London. Henderson, H. W., 1971: "The Atmospheric Energy Budget in Isentropic Coordinates," master of science thesis, The Pennsylvania State University, University Park.*)

These integrals are taken along the coordinate surfaces in each case. Observational data generally show that average values of ρw and ω are small, so that the zonally averaged meridional momentum would be small. But in isentropic coordinates, $\hat{v}$ is directly coupled to the heating fields that drive the motion, suggesting that a direct response of Hadley type may be observed in the average meridional momentum field.

The next step is to determine the effect of zonal averaging on the geostrophic velocity. We use x for longitude and find that

$$\hat{v}_g = \frac{\overline{\rho J_\vartheta v_g}}{\overline{\rho J_\vartheta}} = \begin{cases} (\bar{\rho} f)^{-1} \dfrac{\overline{\partial p}}{\partial x} = 0 & \text{cartesian} \\[2ex] g f^{-1} \dfrac{\overline{\partial h_p}}{\partial x_p} = 0 & \text{isobaric} \\[2ex] \left(f \dfrac{\overline{\partial p}}{\partial \theta} \right)^{-1} \overline{\dfrac{\partial p}{\partial \theta} \dfrac{\partial \Psi}{\partial x_\theta}} & \text{isentropic} \end{cases} \tag{33}$$

so that $\hat{v}_g$ vanishes in the quasi-horizontal z and p coordinates for hydrostatic conditions. Thus $\hat{v} = \hat{v}_{ag}$, in which v_{ag} denotes the ageostrophic velocity component. The average meridional flux F_h of any quantity $\rho J_\vartheta h$ becomes

$$F_h = \overline{\rho J_\vartheta v h} = \hat{v}\hat{h}\, \overline{\rho J_\vartheta} + \overline{\rho J_\vartheta v' h'} \tag{34}$$

and, consequently, the first term may be expected to be small in the quasi-horizontal coordinates. Therefore, eddy motions giving rise to v' and h' must be a significant and dominant feature of the statistics of the quasi-geostrophic flow observed in isobaric coordinates in middle and high latitudes.

But in isentropic coordinates, we have a strong relation between $\hat{v}$ and the heating fields, and we may expect that the first term of Eq. (34) will give a significant contribution.

Applying these results to the energy budget, we see that $C(\hat{A},\hat{K})$ given in Eq. (21) may also be expected to be small in the quasi-horizontal systems compared to $C(A',K')$ in Eq. (22). The generation of available energy is dominated by the zonal term, so the main cycle of energy involves creation of $\hat{A}$ through $G(\hat{A})$, conversion to A' through $C(\hat{A},A')$, realization as K', most of which is lost to $D(K')$, the remainder being converted through $-C(\hat{K},K')$ to $\hat{K}$ and then lost to $D(\hat{K})$. Thus the energy cycle in quasi-horizontal coordinates, shown in Fig. 11.3 is an indirect one with kinetic energy flowing from shorter wavelengths to longer ones. Further discussion of this process will be presented in Chap. 14 when we study geostrophic turbulence.

In isentropic coordinates, we can expect that $C(\hat{A},\hat{K})$ will be larger relative to $C(A',K')$ than it is in quasi-horizontal systems, and that the conversions will be more direct, as is illustrated by the observational data in Fig. 11.3.

The implication of these results is not that the disturbances are absent in isentropic coordinates so that only a meridional flow is evident. Rather, the disturbances are arranged so that the covariance $\overline{\rho J_\theta v}$ is significant and directly related to the heating fields as expressed by $\dot{\theta}$.

As a final illustration of these ideas, we shall consider the global entropy budget, with steadiness conditions imposed for simplicity. If we could separate the region of heating from the region of cooling by a vertical plane at constant latitude, then we could let P and E denote the polar and tropical regions separated by the vertical plane, and the flux of entropy through the plane would be obtained in the expressions [see Eq. (7) in Sec. 5.3]

$$\begin{aligned} \dot{S}_P &= \int_{V_P} \rho J_\vartheta \frac{ds}{dt}\, dV + \int_B \rho J_\vartheta \mathbf{v} \cdot \boldsymbol{\eta} s\, d\sigma \\ \dot{S}_E &= \int_{V_E} \rho J_\vartheta \frac{ds}{dt}\, dV - \int_B \rho J_\vartheta \mathbf{v} \cdot \boldsymbol{\eta} s\, d\sigma \end{aligned} \tag{35}$$

where η is a poleward unit vector exterior to V_E and B is the plane.

It is reasonable to assume that both S_P and S_E are constant, so that the flux F_s that balances the local gain of entropy in V_E and the loss in V_P is

$$F_s = \int_B \rho J_\vartheta \mathbf{v} \cdot \boldsymbol{\eta} s\, d\sigma = \int_{V_E} \rho J_\vartheta \frac{ds}{dt}\, dV > 0 \tag{36}$$

This flux of entropy is given for hydrostatic conditions by

$$\bar{F}_s = \begin{cases} \displaystyle\int_0^\infty (\bar{\rho}\, \hat{v}\hat{s} + \overline{\rho v' s'})\, dz & \text{cartesian} \\ \displaystyle\frac{1}{g}\int_0^{p_0} (\bar{v}\bar{s} + \overline{v^* s^*})\, dp & \text{isobaric} \\ \displaystyle -\frac{1}{g}\int_{\theta_0}^{\theta_T} \frac{\partial \bar{p}}{\partial \theta}\, \hat{v} s\, d\theta & \text{isentropic} \end{cases} \tag{37}$$

because s is constant along the θ surfaces. The first term can be expected to be small in the first two integrals (and vanishes exactly for purely geostrophic modes), so that the entropy budget is balanced mainly by the eddy motions. But in isentropic coordinates, the eddy motions do not even appear in the mean entropy flux and the balance is effected solely by the mean meridional momentum.

This integral even allows us to determine the orientation of the circulation. A requirement that the mass in V_E and V_P be constant shows that the integral of $(\partial \bar{p}/\partial \theta)\hat{v}$ over all θ vanishes. Hence $F_s > 0$ requires that $\hat{v}$ be positive at upper altitudes where s is large and negative at lower altitudes where s is small.

Having reached the conclusion that such a direct circulation must be a significant part of the atmosphere's response as seen in isentropic coordinates, we can use the last equation in Eq. (32) to produce the meridional velocities that would be associated with a given heating field. Figure 11.4 shows streamlines obtained from the net heating data of Fig. 11.1. A comparison of this meridional circulation with that shown in Fig. 4.9 makes it clear that they are radically different.

In isentropic coordinates, then, a significant, direct thermal circulation in response to the differential heating is evident. The fact that the atmosphere chooses to maintain the meridional distributions of mass and entropy at quasi-steady values ordains the circulations that must effect the balance.

PROBLEM 11.3.11 What pattern of heating must occur for $\hat{v}$ to vanish when the mass distribution is steady?

PROBLEM 11.3.12 Prove that if steady conditions prevail with $\partial h/\partial\theta > 0$ everywhere, and if heating exists on an isentrope, then cooling must also be present on that isentrope.

11.4 GLOBAL THERMODYNAMICS

Of the global quantities such as the mass, the various energy forms, and the entropy that we have considered in this book and in this chapter, only the concept of entropy contains a prediction about the direction a thermodynamic process must take. In the absence of external forcing, energies balance but entropy increases. We shall combine the concepts of energy and entropy in this section, and thus obtain a new view of the global thermodynamics of the atmosphere that will help us understand more clearly both the subtleties of the balances and the causes of the unrest.

We begin with a natural, dry atmospheric state that has total energy E and total mass M. Corresponding to this state, we define an *associated equilibrium state* that is motionless and hydrostatic and has the constant temperature

$$T_0 = \frac{\bar{E}}{c_p \bar{M}} \tag{1}$$

and the surface pressure

$$p_0(0) = g\bar{M} \tag{2}$$

in which overbars now denote horizontal averages over the entire atmosphere. From these definitions and the hydrostatic condition it follows that

$$\rho_0(z) = \frac{g\bar{M}}{RT_0}\, e^{-gz/RT_0} \tag{3}$$

Thus there are only vertical variations in the associated equilibrium state. This state has the maximum entropy found in any state with the same mass and energy. It will take

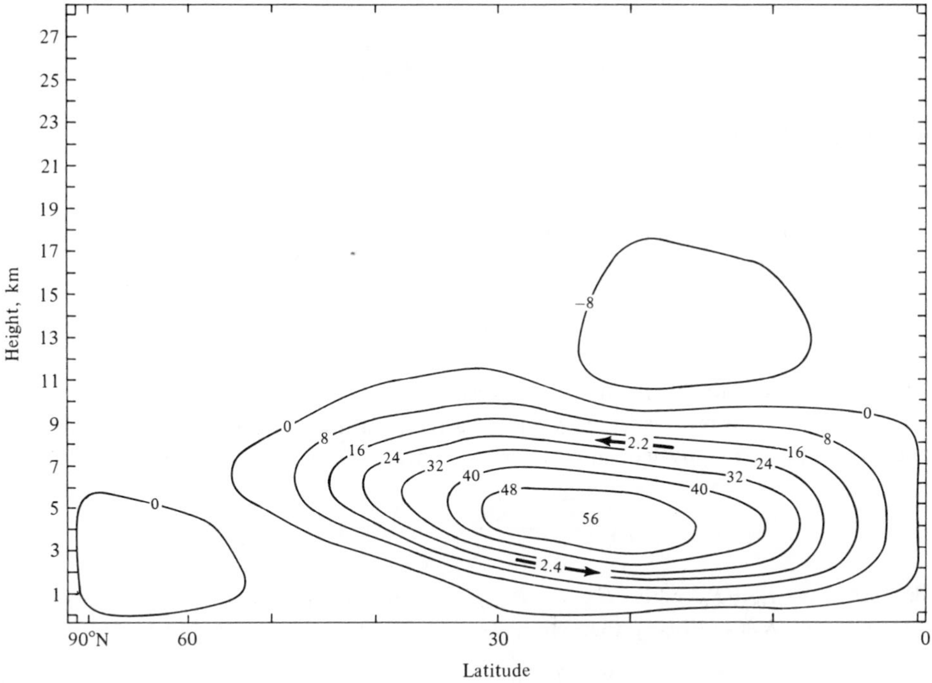

FIGURE 11.4
Streamlines of the mean meridional circulation induced in isentropic coordinates by the net heating field of Fig. 11.1. The stream function shown has the properties that $\partial\psi/\partial\theta = -\rho J_\theta \hat{v}$ and $\partial\psi/\partial y = \rho J_\theta \, d\theta/dt$, and is given in units of 10^2 kg/m · s. The velocities implied by the gradients shown are indicated by the arrows with the speed given in meters per second. (*From Johnson and Dutton, 1976.*)

a few pages to prove this statement, but the consequences of doing so are of considerable significance.

In measuring entropy, we must take account of a reference value s_c with constant values T_c, ρ_c, and p_c. Thus we have

$$s = s_c + c_V \ln \frac{T}{T_c} + R \ln \frac{\alpha}{\alpha_c} \tag{4}$$

and we shall use these constants for all determinations of entropy. In computing the differences of total entropy, the constants cancel, and we have

$$\begin{aligned} S_0 - S &= \int_V (\rho_0 s_0 - \rho s)\, dV \\ &= \int_V \rho(s_0 - s)\, dV + \int_V (\rho_0 - \rho) s_0 \, dV \end{aligned} \tag{5}$$

In the second integral, one term is

$$\int_V (\rho_0 - \rho)\left(c_V \ln \frac{T_0}{T_c} + s_c\right) dV = 0 \tag{6}$$

because T_0, T_c, and s_c are all constant and the total masses M_0 and M are identical. The other term is

$$R \int_V (\rho_0 - \rho) \ln \frac{\alpha_0}{\alpha_c}\, dV = R \int_V (\rho_0 - \rho)\left(\ln \frac{\alpha_0(0)}{\alpha_c} + \frac{gz}{RT_0}\right) dV$$
$$= \frac{P_0 - P}{T_0} \tag{7}$$

because the first term in the integral also vanishes.

Now we shall rewrite Eq. (5), adding a temperature difference in the first integral and subtracting an equivalent internal energy difference in the second. The result is

$$S_0 - S = \int_V \rho\left(c_V \frac{T - T_0}{T_0} + s_0 - s\right) dV + \frac{I_0 + P_0 - I - P}{T_0} \tag{8}$$

Because $E_0 = I_0 + P_0$ while $E = I + P + K$ and we are requiring $E_0 = E$, the second term becomes K/T_0. Finally then we have

$$N = T_0(S_0 - S) = K + T_0 \Sigma \tag{9}$$

where N is called the *entropic energy* and $T_0\Sigma$ is the *static entropic energy* with

$$\Sigma = \int_V \rho\left(c_V \frac{T - T_0}{T_0} + s_0 - s\right) dV = \int_V \rho\left[c_V\left(\frac{T - T_0}{T_0} - \ln \frac{T}{T_0}\right) - R \ln \frac{\alpha}{\alpha_0}\right] dV \tag{10}$$

In Eq. (9), each quantity may vary with time, the changes in T_0 and S_0 arising from variations in the total energy E of the natural state (assuming that M is constant).

The crucial point is that Σ is nonnegative. To see this, we use Taylor's theorem with Lagrange's form of the remainder to write

$$\ln \frac{T_0}{T} = (T - T_0)\left(\frac{\partial \ln T_0/T}{\partial T}\right)_{T_0} + \frac{(T - T_0)^2}{2} [T_0 + \lambda(T - T_0)]^{-2}$$
$$= \frac{T_0 - T}{T_0} + \frac{1}{2}\left[\frac{(T - T_0)/T_0}{1 + \lambda(T - T_0)/T_0}\right]^2 \qquad 0 \leqslant \lambda \leqslant 1 \tag{11}$$

where $\lambda = \lambda(T, T_0)$ is precisely that value which makes the equation exact. For the specific volume, we have the same result using a parameter κ in the remainder. With the observation that

$$\int_V \rho \frac{\alpha_0 - \alpha}{\alpha_0}\, dV = \int_V (\rho - \rho_0)\, dV = 0 \tag{12}$$

we may write Eq. (10) as

$$\Sigma = \int_V \frac{\rho}{2} \left\{ c_V \left[\frac{(T - T_0)/T_0}{1 + \lambda(T - T_0)/T_0} \right]^2 + R \left[\frac{(\alpha - \alpha_0)/\alpha_0}{1 + \kappa(\alpha - \alpha_0)/\alpha_0} \right]^2 \right\} dV \tag{13}$$

and we have:

Theorem The associated equilibrium state has the maximum entropy of any state with the same total mass and energy.

PROOF Referring to Eq. (9), we see that $N \geqslant 0$, with equality achieved only if $\mathbf{v} \equiv 0$, $T \equiv T_0$, and $\alpha \equiv \alpha_0$. ////

A century ago, J. W. Gibbs recognized the thermodynamic importance of such a situation and defined an associated equilibrium state to be *thermodynamically stable* if every other state with the same mass and energy has smaller entropy. This is precisely the case for the atmosphere, and we may say that it is stable in Gibbs' sense, a fact that has important dynamical consequences.

We shall refer to an atmosphere as *energetically isolated* if [see Eq. (8) in Sec. 11.1]

$$\dot{E} = -\int_B \eta_i \left(R_i - k \frac{\partial T}{\partial x_i} - v_k \Delta_{ik} \right) d\sigma = 0 \tag{14}$$

and as *entropically isolated* if the first term of Eq. (22) in Sec. 11.1 vanishes so that

$$\dot{S} = \int_V \left[\frac{k}{T^2} \left(\frac{\partial T}{\partial x_i} \right)^2 + \frac{1}{T} \frac{\partial v_i}{\partial x_k} \Delta_{ik} \right] dV \geqslant 0 \tag{15}$$

Thus we may be sure that the second law applies to entropically isolated atmospheres, since the total entropy is a nondecreasing function of time. With these definitions, we can prove the fundamental stability theorem for ideal gases:

Theorem Let an atmosphere be energetically and entropically isolated for all time $t \geqslant \tau$. Then the deviations of natural states from the associated equilibrium state are for all $t \geqslant \tau$ restricted by

$$N = K + T_0 \Sigma \leqslant T_0 [S_0 - S(\tau)] \tag{16}$$

PROOF Under energetic isolation T_0 and S_0 in Eq. (9) are constant. Because of entropic isolation, $\dot{S} \geqslant 0$ so that $S(t) \geqslant S(\tau)$ and hence $S_0 - S \leqslant S_0 - S(\tau)$, completing the proof. ////

PROBLEM 11.4.1 Prove that as long as energetic and entropic isolation prevail, we have

$$\dot{N} = \dot{K} + T_0\dot{\Sigma} \leqslant 0 \tag{17}$$

For what conditions will we have $\dot{N} \geqslant 0$?

These results characterize the flows of isolated atmospheres; they tend inexorably toward equilibrium. The remarkable property of Eq. (16) is that the departures from equilibrium of the motion fields, the temperature fields, and the mass fields are bounded for all time by the entropy difference at the time isolation occurred or was imposed.

It is of interest to ascertain the relative size of the various quantities involved in this version of the global thermodynamics of atmospheric motion. Some estimates are given in Table 11.1. Of particular note is the estimate that the net departure of the atmosphere from equilibrium is only 1.6 percent as measured by $(S_0 - S)/S_0$, which indicates that the kinetic energy and the intensity of the circulation are probably quite sensitive to the difference $S - S_0$ and the mechanisms by which it is maintained at a relatively constant value.

For the atmosphere, we are now faced once again with the question of how its circulation is maintained. Clearly it is not an isolated system, and we have already observed that the net heating field provides the entropy destruction that permits N to avoid the monotonic evolution specified by Eq. (17).

Let us look into this more closely. We define $H = -\nabla \cdot \mathbf{R} + \nabla \cdot \mathrm{k}\, \nabla T$, and observe from Eq. (14) that if $\dot{E} = 0$, then with the same assumptions made in Sec. 11.1, the integral of H must be nonnegative. But from Eq. (21) in Sec. 11.1, we see that $\dot{S} \leqslant 0$ requires

Table 11.1 ESTIMATES OF GLOBAL THERMODYNAMIC VARIABLES COMPUTED ALONG 75°W IN THE NORTHERN HEMISPHERE

Quantity		Winter (Dec.–Feb.)	Summer (July–Sept.)
Total energy	E J/m^2	2.52×10^9	2.58×10^9
Equilibrium temperature	T deg K	243.1	248.9
Kinetic energy*	K J/m^2	8.6×10^5	2.4×10^5
Total entropy†	S J/m$^2\cdot$K	5.89×10^7	5.91×10^7
Equilibrium entropy†	S_0 J/m$^2\cdot$K	5.98×10^7	6.01×10^7
Entropy difference	$S_0 - S$	9.844×10^5	9.604×10^5
Static entropic energy	$T_0\Sigma$ J/m^2	2.38×10^8	2.39×10^8
Ratio	$K/T_0\Sigma$	3.61×10^{-3}	1.00×10^{-3}
Departure from equilibrium	$(S_0 - S)/S_0$	0.0164	0.0160

*The kinetic energy values are from R. Newell et al., 1970, in G. A. Corby (ed.), *The Global Circulation of the Atmosphere*, Royal Meteorological Society, London; the values are for the entire Northern Hemisphere, with summer being the period June–August.

†Entropies are computed with the constant potential temperature $\theta_c = 1$ K.

SOURCE: After Dutton (1973).

$$\int_V \frac{H}{T}\, dV \leqslant -\int_V \frac{\epsilon}{T}\, dV \tag{18}$$

where ϵ/T is the frictional dissipation term appearing in Eq. (21) in Sec. 11.1. It is convenient to divide H into its positive and negative parts by defining

$$H^+ = \begin{cases} H & \text{if } H > 0 \\ 0 & \text{if } H \leqslant 0 \end{cases} \qquad H^- = \begin{cases} -H & \text{if } H \leqslant 0 \\ 0 & \text{if } H > 0 \end{cases} \tag{19}$$

so that $H = H^+ - H^-$. Then with the definitions

$$T^\pm = \left(\int_V H^\pm\, dV\right)\left[\int_V \frac{H^\pm}{T}\, dV\right]^{-1} \tag{20}$$

Eq. (18) may be written as

$$\frac{1}{T^+}\int_V H^+\, dV - \frac{1}{T^-}\int_V H^-\, dV \leqslant -\int_V \frac{\epsilon}{T}\, dV \tag{21}$$

If this is rearranged, it becomes

$$\begin{aligned} T^- &\leqslant T^+\left[\int_V H^-\, dV\left(\int_V H^+\, dV + T^+\int_V \frac{\epsilon}{T}\, dV\right)^{-1}\right] \\ &\leqslant T^+\left(\int_V H^-\, dV\right)\left(\int_V H^+\, dV\right)^{-1} \end{aligned} \tag{22}$$

But since the integral of H is nonnegative, we have

$$\int_V H^+\, dV \geqslant \int_V H^-\, dV \tag{23}$$

and so $T^- \leqslant T^+$.

Therefore we have proved that if H^- does not vanish identically, then subject to the assumptions made, $\dot{E} = 0$ and $\dot{S} \leqslant 0$ simultaneously require that the average temperature T^+ in the region of heating as defined by Eq. (20) must be greater than T^-. This leads to the result:

Theorem For maintenance or intensification of circulation in the sense that $\dot{N} \geqslant 0$ in an energetically isolated atmosphere in which H^- does not vanish identically, it is necessary and sufficient that the weighted average temperature T^- in the region of cooling be less than T^+ in the region of heating as specified by the first line of Eq. (22).

PROOF Because $\dot{E} = 0$, $\dot{N} \geqslant 0$ requires only that $\dot{S} \leqslant 0$ since T_0 and S_0 are constant. For this to be true, the first line of Eq. (22) has been shown to be necessary and sufficient. ////

In this sense, the atmosphere resembles a heat engine. Its circulation continues because heat is received at high temperatures and given up at cool temperatures, a heating pattern that maintains the temperature gradients that drive the motion.

But the question posed at the end of Sec. 11.1 still remains: Why does the entropy-producing solar radiation field cause motion?

For simplicity we consider only this solar component of the radiation field now and upon letting $R_s = -\nabla \cdot \mathbf{R}^{(s)}$ we have

$$\dot{S}^{(s)} = \int_V \frac{R_s}{T}\, dV > 0 \tag{24}$$

and

$$\dot{E}^{(s)} = \int_V R_s\, dV > 0 \tag{25}$$

The next task is to determine the effect of this solar component on S_0. Applying Eq. (4) to S_0 gives

$$S_0 = \int_V \rho_0 s_0\, dV = Ms_c + c_V M \ln \frac{T_0}{T_c} + RM \ln \frac{\alpha_0(0)}{\alpha_c} + RM \tag{26}$$

in which the last term is obtained by an integration using Eq. (3). Thus for $\dot{M} = 0$, we have $\dot{p}_0(0) = 0$ so that $\dot{\alpha}_0(0) = R\dot{T}_0/p_0(0)$ and hence

$$\dot{S}_0 = c_p M \frac{\dot{T}_0}{T_0} \tag{27}$$

But from Eqs. (1) and (25) we have now

$$\dot{T}_0^{(s)} = (c_p M)^{-1} \int_V R_s\, dV \tag{28}$$

and therefore

$$(\dot{S}_0 - \dot{S})^{(s)} = \int_V \left(\frac{1}{T_0} - \frac{1}{T}\right) R_s\, dV \tag{29}$$

For a hydrostatic atmosphere in which $K \ll I$, we know from Problem 11.2.1 that $T_0 \cong (c_p/c_V)I/c_p M = \tilde{T}$, where $\tilde{T}$ is the mass-weighted average temperature of the atmosphere. Thus

$$(\dot{S}_0 - \dot{S})^{(s)} = \int_V \left(\frac{1}{\tilde{T}} - \frac{1}{T}\right) R_s\, dV \tag{30}$$

We may conclude that this integral is positive because the solar radiation is primarily received where $T > \tilde{T}$. So at last we have the result that

$$\dot{N}^{(s)} = \left[\frac{d}{dt}(K + T_0\Sigma)\right]^{(s)} = \dot{T}_0^{(s)}(S_0 - S) + T_0(\dot{S}_0 - \dot{S})^{(s)} > 0 \tag{31}$$

and we see that the solar radiation increases the total entropic energy both by increasing T_0 and by increasing S_0 more than S. The increasing entropic energy gives the atmosphere a range of choices from which to select a response that, presumably, will involve the motions that attempt to reach the ever-distant equilibrium state. And so the sunshine falling on the round earth creates entropic energy, and the atmosphere then destroys it by using the winds to transport the heat that eases the gradients in the thermal field.

*PROBLEM 11.4.2 Show that if $\dot{N} \not\equiv 0$ in an interval $[t_1, t_2]$, then $K \not\equiv 0$ in that interval. (Hint: Show first that $\dot{N} \not\equiv 0$ implies that both $\mathbf{v}$ and $\dot{\mathbf{v}}$ cannot vanish identically.)

PROBLEM 11.4.3 Show that

$$\dot{K} + T_0\dot{\Sigma} = T_0(\dot{S}_0 - \dot{S}) + \dot{T}_0 \frac{K}{T_0} = \left(1 + \frac{K}{E}\right)\dot{E} - T_0\dot{S}$$

so that

$$\dot{\Sigma} = \frac{-C(\Pi, K)}{T_0} + G(\Sigma)$$

where

$$T_0 G(\Sigma) = -D(K) + \left(1 + \frac{K}{E}\right)\dot{E} - T_0\dot{S}$$

Explain which processes produce Σ and which destroy it.

11.5 VARIATIONAL DERIVATION OF THE EQUATION OF MOTION

The equations of motion, upon conversion to the energy integrals being discussed in this chapter, yield predictions about the temporal behavior of the global quantities of atmospheric energetics. It is noteworthy that this process can be reversed, in the sense that hypotheses about trends in time of global quantities will allow us to derive the equations of motion for inviscid, isentropic flow.

The basic motivation of the analysis we shall present might be said to be the concept that fluid systems are efficient in a thermodynamic sense. The mechanical or thermodynamic forcing imposed upon them sets them in motion transferring heat, momentum, and other quantities from one boundary surface to another. The concept of efficiency involves the realization that these transfers are accomplished without waste motion. In other words, the kinetic energy is only as large as is necessary to accomplish the task mandated by the external forcing. Thus the quantity $K - \Pi$ might be expected to be large and negative in fluid motion and it certainly is in the atmosphere. A successful mathematical formulation of this idea constitutes the application of Hamilton's principle to fluid dynamics. The discussion here is based upon a method developed by Prof. Francis P. Bretherton.

The crucial quantity is the action

$$A = \int_0^\infty (K - \Pi)\, dt = \int_0^\infty \int_V \rho \left[\tfrac{1}{2}\mathbf{v}^2 - e(\rho,s) - \Phi\right] dV\, dt \qquad (1)$$

and because we shall minimize A, the domain of ideas under consideration is often denoted as the *least-action principle.*

11.5.1 Kinematic Analysis

Certain constraints must be imposed on the action during the minimization to ensure that all the properties of fluids are correctly modeled. First, the motion must obey the equation of continuity. Next, it must obey a suitable first law or entropy relation. Finally, it must have the properties discussed in Sec. 5.7, so the identification of each parcel with its initial coordinate $\boldsymbol{\xi}$ is expressed by the constraint

$$\frac{d\boldsymbol{\xi}}{dt} = \frac{\partial \boldsymbol{\xi}}{\partial t_{\xi}} = 0 \qquad (2)$$

The precise sense in which we minimize the action is also important. We may view the motion, in the spirit of Sec. 5.7, as a collection of trajectories. The idea of Hamilton's principle is that the natural motions will actually be the particular trajectories in the interval $t_0 \leqslant t \leqslant t_1$ that minimize A over that interval. We consider, then, trajectories fixed for $t < t_0$ and $t > t_1$. There are many possible collections of trajectories between these two points, all of them giving actions A greater than the set giving the minimum action supposed to correspond to the actual motion.

To analyze the behavior of A for different trajectories, we must consider two kinds of variations. In the first we displace the parcel trajectory holding $\boldsymbol{\xi}$ fixed; these will be denoted by δ; in the second we have a variation Δ at a fixed point in space. [Variations δ correspond to the material derivative $d(\ )/dt$; Δ *corresponds to* $\partial(\ )/\partial t$.]

We let ϕ represent a property of a parcel, and for two different trajectories $\mathbf{x}(\boldsymbol{\xi},t)$ and $\mathbf{x}_1(\boldsymbol{\xi},t)$ we have values

$$\phi = \phi(\mathbf{x}(\boldsymbol{\xi},t),t) \qquad \phi_1 = \phi_1(\mathbf{x}_1(\boldsymbol{\xi},t),t) \qquad (3)$$

and therefore $\delta\phi = \phi_1(\mathbf{x}_1, t) - \phi(\mathbf{x}, t)$ while $\Delta\phi = \phi_1(\mathbf{x}, t) - \phi(\mathbf{x}, t)$. Thus, we may calculate

$$\begin{aligned}\delta\phi &= \phi_1(\mathbf{x}, t) - \phi(\mathbf{x}, t) + \phi_1(\mathbf{x}_1, t) - \phi_1(\mathbf{x}, t) \\ &= \Delta\phi + \phi(\mathbf{x}_1, t) - \phi(\mathbf{x}, t) + [\phi_1(\mathbf{x}_1, t) - \phi_1(\mathbf{x}, t) - \phi(\mathbf{x}_1, t) + \phi(\mathbf{x}, t)] \\ &= \Delta\phi + \delta\mathbf{x} \cdot \nabla\phi + [\delta\mathbf{x} \cdot \nabla(\Delta\phi)] + o(\delta x)^2 \end{aligned} \tag{4}$$

in which we have used Taylor's theorem to express the differences. Thus to first order we find

$$\delta\phi = \Delta\phi + \delta\mathbf{x} \cdot \nabla\phi \tag{5}$$

for the relation between the two types of variation. This is obviously a variational version of Euler's relation, as anticipated by the choice of notation.

Applying Eq. (5) to the scalar components of the velocities, we find

$$\delta\mathbf{v} = \Delta\mathbf{v} + \delta\mathbf{x} \cdot \nabla\mathbf{v} \tag{6}$$

where

$$\delta\mathbf{v} = \mathbf{v}_1 - \mathbf{v} = \frac{\partial}{\partial t_\xi}(\mathbf{x}_1 - \mathbf{x}) = \frac{\partial \delta\mathbf{x}}{\partial t_\xi} \tag{7}$$

Then the continuity equation applied to the two trajectories gives

$$\frac{1}{\rho_1}\frac{\partial \rho_1}{\partial t_\xi} - \frac{1}{\rho}\frac{\partial \rho}{\partial t_\xi} = -\nabla \cdot (\mathbf{v}_1 - \mathbf{v}) = -\nabla \cdot \left(\frac{\partial \delta\mathbf{x}}{\partial t_\xi}\right) \tag{8}$$

Since $\delta\rho = \delta\mathbf{v} = 0$ at $t < t_0$, an integration of Eq. (8) using logarithmic derivatives on the left produces

$$\ln \frac{\rho_1}{\rho} = \ln\left(1 + \frac{\delta\rho}{\rho}\right) = -\nabla \cdot \delta\mathbf{x} \tag{9}$$

so that to first order

$$\frac{\delta\rho}{\rho} = -\nabla \cdot \delta\mathbf{x} \tag{10}$$

11.5.2 Derivation of the Equation of Motion

The action in Eq. (1) is an integration over the physical space. We define $A + \epsilon\,\Delta A$ to be the value of Eq. (1) obtained by substituting $\rho + \epsilon\,\Delta\rho$, $s + \epsilon\,\Delta s$, and $\mathbf{v} + \epsilon\,\Delta\mathbf{v}$ for each variable. Here ρ, s, and $\mathbf{v}$ are assumed to be the minimizing fields. Then we have

$$\begin{aligned}\Delta A &= \lim_{\epsilon \to 0} \frac{\partial}{\partial \epsilon}(A + \epsilon\,\Delta A) \\ &= \iint \Delta\rho\left(\tfrac{1}{2}\mathbf{v}^2 - e - \Phi\right) dV\,dt + \iint \rho\left(\mathbf{v} \cdot \Delta\mathbf{v} - \frac{\partial e}{\partial \rho}\,\Delta\rho - \frac{\partial e}{\partial s}\,\Delta s\right) dV\,dt \end{aligned} \tag{11}$$

The suppressed limits on the integrals are identical to those in Eq. (1), although the integrands are nonzero only for $t_0 \leqslant t \leqslant t_1$.

From Eqs. (5) and (10) we conclude that

$$\Delta\rho = \delta\rho - \delta\mathbf{x} \cdot \nabla\rho = -\nabla \cdot \rho\delta\mathbf{x} \tag{12}$$

and so with Eq. (7) and with Eq. (5) applied to entropy s, we may convert Eq. (11) to an expression involving the variations $\delta\rho$, δs, and $\delta\mathbf{x}$ due to trajectory displacements. The result is

$$\Delta A = \iint -\nabla \cdot (\rho\,\delta\mathbf{x})(\tfrac{1}{2}\mathbf{v}^2 - e - \Phi)\,dV\,dt + \iint \rho \left\{ \mathbf{v} \cdot \left[\frac{\partial \delta\mathbf{x}}{\partial t_\xi} - (\delta\mathbf{x} \cdot \nabla)\mathbf{v}\right] + \frac{\partial e}{\partial \rho} \nabla \cdot (\rho\,\delta\mathbf{x}) - \frac{\partial e}{\partial s}(\delta s - \delta\mathbf{x} \cdot \nabla s) \right\} dV\,dt \tag{13}$$

Three steps are necessary to complete the derivation. First, the variations are assumed to vanish at the boundary of V and so the first term of Eq. (13) may be integrated by parts. Second, in the second term we have

$$\rho\mathbf{v} \cdot \left[\frac{\partial \delta\mathbf{x}}{\partial t} + (\mathbf{v} \cdot \nabla)\,\delta\mathbf{x} - (\delta\mathbf{x} \cdot \nabla)\mathbf{v}\right] = \rho\mathbf{v} \cdot \frac{\partial \delta\mathbf{x}}{\partial t} + \frac{\partial}{\partial x_j}(\rho v_k v_j\,\delta x_k) - \delta x_k \frac{\partial}{\partial x_j}(\rho v_k v_j) - \rho\,\delta\mathbf{x} \cdot \nabla \frac{\mathbf{v}^2}{2} \tag{14}$$

The first term may be integrated by parts in time, taking account of the fact that $\delta\mathbf{x}$ vanishes outside of $[t_0, t_1]$. The second is a divergence and vanishes. The last cancels the kinetic energy term obtained in the first step.

Finally, we observe from Table 3.1 that $(\partial e/\partial s)_\rho = T$ and that $(\partial e/\partial\rho)_s = p/\rho^2$; then we integrate by parts on the $\partial e/\partial\rho$ term. These three steps give

$$\Delta A = -\iint \delta\mathbf{x} \cdot \left[\frac{\partial}{\partial t}\rho\mathbf{v} + \frac{\partial}{\partial x_j}(\rho\mathbf{v}v_j) + \rho\,\nabla\Phi + \rho\,\nabla\left(e + \frac{p}{\rho}\right) - \rho T\,\nabla s\right] dV\,dt - \iint \rho T\,\delta s\,dV\,dt \tag{15}$$

The last two terms of the first integral combine to give ∇p (see Problem 11.5.1), and upon expanding the first two terms and using the continuity equation, we find at last that

$$\Delta A = -\iint \rho\,\delta\mathbf{x} \cdot \left(\frac{\partial \mathbf{v}}{\partial t} + \mathbf{v} \cdot \nabla\mathbf{v} + \frac{1}{\rho}\,\nabla p + \nabla\Phi\right) dV\,dt - \iint \rho T\,\delta s\,dV\,dt \tag{16}$$

This result now can be applied to the isentropic flow of an inviscid fluid. Then $s(\mathbf{x}(\xi,t),t) = s(\xi,0)$ for every parcel so that $\delta s = 0$. If A is to be a minimum, then ΔA

must vanish for arbitrary variation $\delta \mathbf{x}$ vanishing outside $[t_0, t_1]$ and on the boundaries. This is possible only if the expression in parentheses in Eq. (16) vanishes, and thus we have the equation of motion. In this case, the isentropic condition represents an energy constraint.

It is not known whether this approach can be extended to a more general case. Hamilton's principle generally applies only to conservative systems, and the modification necessary in order to develop a successful version of the least-action principle for dissipative systems is not obvious.

PROBLEM 11.5.1 Show that

$$\rho \nabla \left(e + \frac{p}{\rho} \right) - \rho T \nabla s = \nabla p$$

[Hint: Apply the chain rule to $e = e(\rho, s)$.]

PROBLEM 11.5.2 Show that $\Delta I = \int \rho \, \delta e \, dV$ when boundary values are fixed. Is this relation true for $F = \int \rho f \, dV$?

PROBLEM 11.5.3 Show that the term $\int_0^\infty \Delta K \, dt$ leads to the $d\mathbf{v}/dt$ term in the equation of motion.

11.6 THE ENERGETICS OF DISTURBED OR TURBULENT MOTION

Motion patterns in the atmosphere range in size from the jet stream structure of planetary scale to small eddies millimeters or less in diameter. The energy source is the global arrangement of differential heating, and the energy is carried from one scale to another until it is lost to friction in the smallest eddies. As we have seen in Eq. (17) in Sec. 11.1, it basically is the nonlinear inertial term $(\mathbf{v} \cdot \nabla \mathbf{v})$ that is responsible for moving energy from one wavelength to another.

These interactions of motions of vastly different scale pose serious problems for atmospheric scientists, as pointed out in Chap. 1. The student of the general circulation and synoptic systems is quickly overwhelmed by the details of turbulence; the investigator of turbulence cannot take into account all the details of the larger-scale flow in which it is embedded.

But there is no simple way to separate a given flow into components by scale or size, because, as a result of the inertial term, a motion pattern of any size interacts to some degree with patterns of every other size. Thus we use an abstract model, a generalization of the technique used to decompose the general circulation into zonal and eddy components.

To develop the idea, let us restrict ourselves to a certain geometry of the flow domain D. Here D might represent the atmospheric volume or a small container in a laboratory. Next, let us impose a set $\mathcal{C}$ of conditions on the flows we will consider, including the condition that they satisfy the equations of motion. All flows on D satisfying the conditions $\mathcal{C}$ form an *ensemble* $\mathcal{F}$ of flows; each particular flow in the ensemble is called a *realization*. Depending on the conditions in $\mathcal{C}$, there may be many or few flows in $\mathcal{F}$.

If we were to specify boundary and initial conditions exactly and let them be conditions in $\mathcal{C}$, then we would assume the resulting flow would be unique so that in this case $\mathcal{F}$ would have only one flow. Since they cannot be specified exactly in many cases, we might let $\mathcal{F}$ be all flows whose boundary and initial conditions are the same within some tolerance of measurement. Now $\mathcal{F}$ is an infinite family.

Let a particular flow in $\mathcal{F}$ be denoted by a real number α and let $p(\alpha)$ be the probability that the conditions producing the αth flow might occur. A flow variable ϕ is now a function

$$\phi = \phi(\mathbf{x}, t; \alpha) \tag{1}$$

The *expectation* or *ensemble average* of ϕ is defined by

$$\bar{\phi} = \int_{-\infty}^{\infty} \phi(\mathbf{x}, t; \alpha) p(\alpha)\, d\alpha = \bar{\phi}(\mathbf{x}, t) \tag{2}$$

and the variance is

$$\overline{(\phi - \bar{\phi})^2} = \int_{-\infty}^{\infty} (\phi - \bar{\phi})^2 p(\alpha)\, d\alpha \tag{3}$$

Every flow may now be represented as

$$\phi = \bar{\phi} + \phi^* \tag{4}$$

where obviously $\overline{\phi^*} = 0$. In applying this concept, we often suppose that we have arranged the conditions in $\mathcal{C}$ so that $\bar{\phi}$ is typical of large-scale flows in which we are interested, and that ϕ^* represents a particular small-scale disturbance that might be superimposed on $\bar{\phi}$. In doing this, we tacitly assume that the spatial mean-square value of ϕ^* is small, in a suitable sense, for every α.

This model is particularly suited for, and motivated by, the study of turbulence. Everybody knows what turbulence is, and yet no one can define it. A turbulent flow is easily recognized if its patterns can be seen visually, but no specification that provides a sharp, mathematical distinction between turbulent and nonturbulent flow is yet available.

True turbulence is three-dimensional and nonlinear, vortices are stretched, energy is dissipated, passive quantities in the flow are diffused.

In terms of the equations of motion, the inertial term $\mathbf{v} \cdot \nabla \mathbf{v}$ is vital to the structure of turbulent flow. As we have seen on several occasions, simple solutions are constructed with assumptions that make this term vanish. The ratio of the inertial to viscous terms (the Reynolds number, see Sec. 9.2) must be large for turbulence to occur. Thus it is the fully nonlinear character of the flow that makes theoretical analysis of turbulent flow a continuing, formidable challenge.

In turbulent flows, we cannot predict what the velocities will be at a point, although we may be able to predict averages at a point or over the entire flow. Thus turbulence appears to us to be random, and so we use statistical techniques to describe it. To some degree, then, turbulence is related to the scale of the observer.

11.6.1 Equations for the Mean and Disturbance Components

We are free to choose ϕ' to be a turbulent component of a flow variable ϕ, although it may also represent a nonturbulent disturbance. In either case, we use the Reynolds convention to define the ensemble average and deviation

$$\hat{\mathbf{v}} = \frac{\overline{\rho \mathbf{v}}}{\bar{\rho}} \qquad \mathbf{u} = \mathbf{v} - \hat{\mathbf{v}}$$

$$\hat{\phi} = \frac{\overline{\rho \phi}}{\bar{\rho}} \qquad \phi' = \phi - \hat{\phi} \tag{5}$$

so that as in Eq. (3) in Sec. 11.3,

$$\bar{K} = \hat{K} + K' = \tfrac{1}{2} \int_V (\bar{\rho}\hat{v}^2 + \overline{\rho u^2})\, dV \tag{6}$$

where V is the volume of the domain D.

To proceed to energetics, we must determine the equations of motion for the mean and disturbance components. We substitute Eq. (5) into the momentum form

$$\frac{\partial}{\partial t}(\rho v_i) + \frac{\partial}{\partial x_k}(\rho v_i v_k) = -\frac{\partial p}{\partial x_i} - g\rho\delta_{i3} - f\epsilon_{i3j}\rho v_j + F_i \tag{7}$$

in which F_i is the viscous force and the summation convention applies. (We are making the usual meteorological approximation that only the vertical component of 2Ω is necessary and that the sphericity terms may be ignored.) Thus by averaging we obtain

$$\frac{\partial}{\partial t}(\bar{\rho}\hat{v}_i) + \frac{\partial}{\partial x_k}(\bar{\rho}\hat{v}_i\hat{v}_k + \overline{\rho u_i u_k}) = -\frac{\partial \bar{p}}{\partial x_i} - g\bar{\rho}\delta_{i3} - f\epsilon_{i3j}\bar{\rho}\hat{v}_j + \bar{F}_i \tag{8}$$

The continuity equation, by the same procedure, becomes

$$\frac{\partial \bar{\rho}}{\partial t} + \frac{\partial}{\partial x_k}(\bar{\rho}\hat{v}_k) = 0 \tag{9}$$

The first law of thermodynamics in the form

$$c_V\left[\frac{\partial}{\partial t}(\rho T) + \frac{\partial}{\partial x_k}(\rho v_k T)\right] + p\frac{\partial v_j}{\partial x_j} = Q \tag{10}$$

gives

$$c_V\left[\frac{\partial}{\partial t}(\bar{\rho}\hat{T}) + \frac{\partial}{\partial x_k}(\bar{\rho}\hat{v}_k\hat{T} + \overline{\rho u_k T'})\right] + \bar{p}\frac{\partial \hat{v}_j}{\partial x_j} + \overline{p\frac{\partial u_j}{\partial x_j}} = \bar{Q} \tag{11}$$

The continuity equation (9) along with Eqs. (8) and (11) then produces

$$\frac{\partial \hat{v}_i}{\partial t} + \hat{v}_k\frac{\partial \hat{v}_i}{\partial x_k} + \frac{1}{\bar{\rho}}\frac{\partial}{\partial x_k}(\overline{\rho u_i u_k}) = -\frac{1}{\bar{\rho}}\frac{\partial \bar{p}}{\partial x_i} - g\delta_{i3} - f\epsilon_{i3j}\hat{v}_j + \frac{\bar{F}_i}{\bar{\rho}} \tag{12}$$

and

$$c_V\left[\frac{\partial \hat{T}}{\partial t} + \hat{v}_k\frac{\partial \hat{T}}{\partial x_k} + \frac{1}{\bar{\rho}}\frac{\partial}{\partial x_k}(\overline{\rho u_k T'})\right] + \frac{\bar{p}}{\bar{\rho}}\frac{\partial \hat{v}_j}{\partial x_j} + \frac{\overline{p\,\partial u_j/\partial x_j}}{\bar{\rho}} = \frac{\bar{Q}}{\bar{\rho}} \tag{13}$$

By subtracting these equations for the ensemble mean from the original equations, we obtain the equations governing the disturbance flow:

$$\frac{\partial u_i}{\partial t} + \hat{v}_k\frac{\partial u_i}{\partial x_k} + u_k\frac{\partial \hat{v}_i}{\partial x_k} + u_k\frac{\partial u_i}{\partial x_k} - \frac{1}{\bar{\rho}}\frac{\partial}{\partial x_k}(\overline{\rho u_i u_k}) = \frac{1}{\bar{\rho}}\frac{\partial \bar{p}}{\partial x_i} - \frac{1}{\rho}\frac{\partial p}{\partial x_i} - f\epsilon_{i3j}u_j + \frac{F_i}{\rho} - \frac{\bar{F}_i}{\bar{\rho}} \tag{14}$$

and

$$c_V\left[\frac{\partial T'}{\partial t} + \hat{v}_k\frac{\partial T'}{\partial x_k} + u_k\frac{\partial \hat{T}}{\partial x_k} + u_k\frac{\partial T'}{\partial x_k} - \frac{1}{\bar{\rho}}\frac{\partial}{\partial x_k}(\overline{\rho u_k T'})\right] + \frac{p}{\rho}\frac{\partial v_j}{\partial x_j} - \frac{\bar{p}}{\bar{\rho}}\frac{\partial \hat{v}_j}{\partial x_j} - \frac{\overline{p\,\partial u_j/\partial x_j}}{\bar{\rho}} = \frac{Q}{\rho} - \frac{\bar{Q}}{\bar{\rho}} \tag{15}$$

and, for the special definition $\rho' = \rho - \bar{\rho} = \rho^*$,

$$\frac{\partial \rho'}{\partial t} + \frac{\partial}{\partial x_k}(\rho'\hat{v}_k + \rho u_k) = 0 \tag{16}$$

Two types of interactions between the mean and the disturbance flow appear in these equations. First, the divergences of $\overline{\rho u_i u_k}$ and $\overline{\rho u_k T'}$ appear in both sets of equations. The divergence of the *Reynolds stress*, $\overline{\rho u_i u_k}$, represents a loss (gain) of momentum by the mean flow and a gain (loss) of momentum by the disturbances. The *eddy heat flux*, $\partial(\overline{\rho u_k T'})/\partial x_k$, is a similar connection between the components. Second, there are direct interactions of the form $\hat{v}_k\,\partial u_i/\partial x_k$ and $u_k\,\partial \hat{v}_i/\partial x_k$ in the eddy equation, these representing mean advection of eddy velocity and eddy advection of the mean.

Clearly then, because of nonlinearity, the mean flow will evolve differently from a deterministic flow $(\mathbf{v},\rho,T) = (\hat{\mathbf{v}},\bar{\rho},\hat{T})$ for which the Reynolds stress and eddy heat flux vanish. Moreover, to predict the evolution of the mean flow, we must know the evolution of the Reynolds stress. But if we use Eq. (14) to form an equation for

$\overline{\partial(\rho u_i u_k)/\partial t}$, then the result will have averages of triple products, one of them being $\overline{\rho u_j\, \partial(u_i u_k/\partial x_j)}$, for example.

Thus the system cannot be closed, for the mean equation involves the mean of a second-order product of u_i; the equation for that product involves third-order products, and so on. The decomposition we thought might help has led to an infinite series. To make the technique successful, some hypothesis that relates higher-order statistics to those of lower order must be invoked. This is more simply said than done, and it forms a book-length subject by itself.

So although we do not have a completely successful technique for analyzing the mean flow, let us see whether taking the mean flow as known allows us to study the deviation flows. From Eqs. (14) to (16) we see that indeed it does, even though we cannot actually know the mean flow without knowing the eddy flow because of the eddy momentum and heat fluxes. This logical inconsistency is often ignored, however, and Eqs. (14) to (16) are used in the theory of turbulence by assuming simple forms (which may not really be solutions to the full mean equation) for $\hat{v}_i$.

Now let us turn to energetics. If we multiply Eq. (12) by $\bar{\rho}\hat{v}_i$ and Eq. (9) by $\hat{v}_i^2/2$, then the sum is

$$\frac{\partial}{\partial t}\left(\bar{\rho}\frac{\hat{v}_i^2}{2}\right)+\frac{\partial}{\partial x_k}\left(\bar{\rho}\frac{\hat{v}_k\hat{v}_i^2}{2}\right)+\hat{v}_i\frac{\partial}{\partial x_k}(\overline{\rho u_i u_k}) = -\hat{v}_i\frac{\partial \bar{p}}{\partial x_i} - g\hat{v}_3\bar{\rho} + \hat{v}_i\bar{F}_i \tag{17}$$

For internal energy, we have Eq. (11) as

$$c_V\left[\frac{\partial}{\partial t}(\bar{\rho}\hat{T}) + \frac{\partial}{\partial x_k}(\bar{\rho}\hat{v}_k\hat{T}) + \frac{\partial}{\partial x_k}(\overline{\rho u_k T'})\right] + \bar{p}\frac{\partial \hat{v}_j}{\partial x_j} + \overline{p\frac{\partial u_j}{\partial x_j}} = \bar{Q} \tag{18}$$

Thus for

$$\frac{\partial \bar{P}}{\partial t} = \int_V gz\,\frac{\partial \bar{\rho}}{\partial t}\,dV = \int_V g\bar{\rho}\hat{v}_3\,dV \tag{19}$$

which is obtained with the aid of Eq. (9), we have

$$\frac{\partial}{\partial t}(\hat{K}+\bar{I}+\bar{P}) = -\int_B\left(\bar{p}\hat{v}_k + \bar{\rho}\hat{v}_k\frac{\hat{v}_i^2}{2} + \hat{v}_j\overline{\rho u_j u_k} + c_V\bar{\rho}\hat{v}_k\hat{T} + \overline{c_V\rho u_k T'}\right)\eta_k\,dB$$
$$+\int_V\left(\overline{\rho u_i u_j}\frac{\partial \hat{v}_i}{\partial x_j} - \overline{p\frac{\partial u_j}{\partial x_j}} + \hat{v}_i\bar{F}_i + \bar{Q}\right)dV \tag{20}$$

The boundary integral vanishes when the usual conditions are applied.

The kinetic energy equation for the disturbance motion is developed analogously and becomes

$$\frac{\partial}{\partial t}\left(\overline{\frac{\rho u_i^2}{2}}\right) + \frac{\partial}{\partial x_k}\overline{(\rho\hat{v}_k+\rho u_k)\frac{u_i^2}{2}} + \overline{\rho u_i u_k}\frac{\partial \hat{v}_i}{\partial x_k} = -\overline{u_i\frac{\partial p}{\partial x_i}} + \overline{u_i F_i} \tag{21}$$

so that

$$\frac{\partial K'}{\partial t} = -\int_B \left[\overline{(\rho\hat{v}_k + \rho u_k)\frac{u_i^2}{2}} + \overline{u_k p} \right] \eta_k \, dB + \int_V \left(-\overline{\rho u_i u_k} \frac{\partial \hat{v}_i}{\partial x_k} + \overline{p \frac{\partial u_i}{\partial x_i}} + \overline{u_i F_i} \right) dV \quad (22)$$

Thus the two terms involving eddy quantities appear with opposite sign in Eqs. (20) and (22), showing how energy is interchanged between the two components. Of these, the product $\overline{\rho u_i u_k}\, \partial\hat{v}_i/\partial x_k$ is usually the most important and shows how energy is transferred from the mean to the disturbance flow through the Reynolds stress acting on the shear of the mean flow. To see that this is generally the direction of transfer, we consider $\overline{\rho u_1 u_3}\, \partial\hat{v}_1/\partial z$, where $\partial\hat{v}_1/\partial z > 0$. Now a parcel at level $z + \Delta z$ will have momentum $\rho v_1(z + \Delta z) = \rho[\hat{v}_1(z) + \Delta z\, \partial\hat{v}_1/\partial z + u_1(z + \Delta z)]$. If such a parcel is carried a distance Δz downward ($u_3 < 0$) without significant change of momentum, then we have for that parcel at level $z, u_1(z) = \Delta z\, \partial\hat{v}_1/\partial z + u_1(z + \Delta z)$. Thus if $|\partial\hat{v}_1/\partial z| > |u_1(z + \Delta z)/\Delta z|$, then $u_1(z) > 0$. On the average then, $\overline{\rho u_1 u_3}$ will have the opposite sign of $\partial\hat{v}_1/\partial z$ so that we may generally expect that $\overline{\rho u_i u_j}\, \partial\hat{v}_i/\partial x_k$ is negative.

In the atmosphere, the momentum flux divergence in Eq. (12) is usually much larger than $\overline{F}_i$. Hence the viscous effect on the mean motion is ignored and we write Eq. (12) as

$$\frac{\partial \hat{v}_i}{\partial t} + \hat{v}_k \frac{\partial \hat{v}_i}{\partial x_k} = -\frac{1}{\bar{\rho}} \frac{\partial \bar{p}}{\partial x_i} - g\,\delta_{i3} - f\epsilon_{i3j}\hat{v}_j + \frac{1}{\bar{\rho}} \frac{\partial}{\partial x_k} \tau_{ik} \quad (23)$$

where $\tau_{ik} = -\overline{\rho u_i u_k}$. It is often assumed that a function $K_{iklm}(\mathbf{x}) > 0$ may be found such that

$$\tau_{ik} = K_{iklm} \frac{\partial \hat{v}_l}{\partial x_m} \quad (24)$$

This is a closure hypothesis that uses the *eddy mixing coefficient* K to relate an eddy average to mean flow properties. It is important to note that an inertial term effect has been converted to a linear form that resembles the viscous term. The so-called eddy friction $\partial\tau_{ik}/\partial x_k$ is initially and in reality a nonlinear contribution from $\mathbf{v} \cdot \nabla\mathbf{v}$, no matter how it is subsequently modified by hypotheses.

PROBLEM 11.6.1 Show that the disturbance equation of state for an ideal gas is, to first order,

$$\frac{p^*}{\bar{p}} = \frac{T'}{\hat{T}} + \frac{\rho'}{\bar{\rho}}$$

PROBLEM 11.6.2 Show that $\overline{\rho u F(\hat{\phi})} = 0$ for any function F, a result frequently used in the derivations in the text.

PROBLEM 11.6.3 Show that $\overline{\rho u_3 T'}$ may be expected to be opposite in sign to $\partial\hat{\theta}/\partial z$. Use this fact to define an eddy mixing coefficient so that the heat flux may be replaced by average variables.

PROBLEM 11.6.4 Show that the eddy mixing coefficient closure scheme cannot be valid in general by considering an asymmetric jet. Let $\hat{v}_1(z)$ be the maximum and let $\hat{v}_1(z+\Delta z) > \hat{v}_1(z-\Delta z)$. Show that a negative eddy stress may be expected at z while the eddy theory predicts no stress.

PROBLEM 11.6.5 Consider a set of synoptic measurements made at time $t = 0$ that are contaminated by random observational errors. Let the ensemble (v, ρ, T) represent all conceivable sets of observations of the true value plus the error. If the errors are of zero mean, then $\hat{v}$, $\bar{\rho}$, $\bar{T}$ would be the true values. What system of equations describes the evolution of the true flow? Let the total error be measured by

$$\epsilon = \tfrac{1}{2}\int_V \overline{\rho\left[u^2 + \frac{c_V(T')^2}{T_0} + \frac{RT_0(\rho')^2}{\rho_0{}^2}\right]}\, dV$$

where T_0 and ρ_0 are constants. How does ϵ evolve in time under the assumption that the boundary conditions are known exactly? What processes may be expected to lead to increases in the error?

PROBLEM 11.6.6 (Robinson) In numerical prediction, we write a system of equations in the form (23), in which $\hat{\mathbf{v}}$, $\bar{\rho}$, $\bar{p}$ are assumed to be averages representative of the variables in a neighborhood of the observation point. In other words, we might let

$$\bar{\phi}(\mathbf{x}, t) = \int_V K(\mathbf{x} - \boldsymbol{\xi})\phi(\boldsymbol{\xi}, t)\, d\boldsymbol{\xi}$$

where K is a smoothing function that vanishes for $|\mathbf{x} - \boldsymbol{\xi}| > L$. Show that such a procedure applied to the actual equations of motion will not give a set like Eq. (23). Explain why it fails.

PROBLEM 11.6.7 Suppose the overbar denotes a spatial or temporal average. Are the equations of this section then valid? [Hint: Considei the implication of a definition for which $\overline{(\bar{\phi})} \neq \bar{\phi}$. What happens then to $\overline{\phi'}$?] Are the equations valid for a zonal average? For an average over all time? Will the operator in Problem 11.6.6 give back the mean and disturbance equations?

11.6.2 Simplified Energy Equations for Turbulence

The energy equation (21) is often simplified in studying atmospheric turbulence by taking advantage of certain properties of the flow. For example, if we assume that the mean flow is hydrostatic and that horizontal pressure gradients are balanced by the Coriolis force, then

$$\frac{\partial \bar{p}}{\partial x_i} + \bar{\rho} f \epsilon_{i3j} \hat{v}_j = \frac{\partial \bar{p}}{\partial x_i} \delta_{i3} = -g\bar{\rho}\delta_{i3} \tag{25}$$

This gives

$$\overline{u_i \frac{\partial p}{\partial x_i}} = \overline{u_i \frac{\partial p^*}{\partial x_i}} - g\bar{\rho}\bar{u}_i\delta_{i3} - f\epsilon_{i3j}\bar{\rho}\bar{u}_i\hat{v}_j \tag{26}$$

But $\bar{u}_i = -\overline{\rho' u_i}/\bar{\rho}$ because $\overline{\rho u_i} = 0$, and so now with some manipulations involving the rule for differentiation of products we arrive at

$$\frac{\partial}{\partial t}\frac{\overline{\rho u_i^2}}{2} + \frac{\partial}{\partial x_k}\left[\hat{v}_k \frac{\overline{\rho u_i^2}}{2} + \left(\overline{\rho u_k \frac{u_i^2}{2}} + \overline{p' u_k}\right)\right]$$
$$= \overline{p^* \frac{\partial u_k}{\partial x_k}} - \overline{\rho u_i u_k}\frac{\partial \hat{v}_i}{\partial x_k} - \overline{\rho' u_i}\,[g\delta_{i3} + f\epsilon_{i3j}\hat{v}_j] + \overline{u_i F_i} \tag{27}$$

At this point, the equation is exact for hydrostatic, geostrophic mean flow. Several approximations valid for small-scale flow lead to further simplification. These assumptions, known as the Boussinesq approximation, are discussed in detail in Chap. 13. For now we observe that if fractional pressure fluctuations $p^*/\bar{p}$ are much smaller than fractional temperature fluctuations $T'/\hat{T}$, then from Problem 11.6.1 we have $\rho'/\bar{\rho} = -T'/\hat{T}$. We also take the eddy flow to be nondivergent. Finally, $f\hat{v}_j/g \ll 1$.

Application of these assumptions and the additional simplification that $\hat{v}_k = \hat{u}(z)\,\delta_{1k}$ gives

$$\frac{\partial}{\partial t}\frac{\overline{\rho u_i^2}}{2} + \hat{u}\frac{\partial}{\partial x}\frac{\overline{\rho u_i^2}}{2} + \frac{\partial}{\partial x_k}\left[\left(\overline{\rho u_k \frac{u_i^2}{2}}\right) + \overline{p' u_k}\right] = -\overline{\rho u_1 u_3}\frac{\partial \hat{u}}{\partial z} + g\overline{\frac{T'}{\hat{T}} u_3 \bar{\rho}} + \overline{u_i F_i} \tag{28}$$

The next assumption is that the turbulence is *horizontally homogeneous*; that is, all averages depend only on z. Now Eq. (28) reduces to

$$\frac{\partial}{\partial t}\frac{\overline{\rho u_i^2}}{2} + \frac{\partial}{\partial z}\left(\overline{\rho u_3 \frac{u_i^2}{2}}\right) + \frac{\partial}{\partial z}(\overline{p' u_3}) = -\overline{\rho u_1 u_3}\frac{\partial \hat{u}}{\partial z} + \frac{g}{\hat{T}}\overline{T' u_3}\,\bar{\rho} + \overline{u_i F_i} \tag{29}$$

The flux Richardson number

$$Rf = \frac{\overline{g(T'/\hat{T}) u_3 \bar{\rho}}}{\overline{\rho u_1 u_3}(\partial \hat{u}/\partial z)} \tag{30}$$

is the ratio of turbulent energy lost to buoyant forces to the energy gained by eddy stress acting on the mean shear. From Eq. (29), then, the necessary condition for turbulent energy to be maintained or increased is that

$$-\overline{\rho u_1 u_3}\frac{\partial \hat{u}}{\partial z}(1 - Rf) \geqslant \frac{\partial}{\partial z}\left(\rho u_3 \frac{\overline{u_i^2}}{2}\right) + \frac{\partial}{\partial z}\overline{p' u_3} - \overline{u_i F_i} \tag{31}$$

and because $\overline{u_i F_i}$ is assumed negative and the fluxes of turbulent energy may be assumed to be out of the turbulent region, the right side is positive and so the necessary condition becomes

$$Rf < 1 \tag{32}$$

Thus the flux Richardson number is the crucial nondimensional number for turbulence in stratified, shearing flow.

PROBLEM 11.6.8 Using eddy mixing theory, find a relation between Rf and Ri.

PROBLEM 11.6.9 Are there mechanisms through which turbulence could decay when unstable conditions prevail?

11.6.3 The Planetary Boundary Layer

At heights far enough above the earth's surface, the winds are often quite close to the geostrophic wind. But near the surface, large departures are found, due in the final analysis to the effects of friction at the earth's surface. The region in which the boundary effects are clearly reflected in the flow—through both ageostrophic motion and turbulence—is called the *planetary boundary layer* and is a kilometer or two in depth.

Over relatively homogeneous terrain, conditions in the boundary layer are often horizontally homogeneous and we can apply the results of the last subsection. Three subregions of the planetary boundary layer are revealed by observational data:

1 The *viscous* or *laminar sublayer* very near the surface in which viscous forces dominate, eddy stress is nearly absent, and the mean wind speed has large positive shear

2 The *constant stress layer*, in which the eddy stress is approximately constant to a height of 100 m or so above the surface and the wind profile is approximately logarithmic

3 The *Ekman layer*, in which the actual wind spirals with height toward the geostrophic wind at the top of the boundary layer

We shall present examples of the simplest cases in the constant stress and Ekman layers, assuming neutral stratification so that the heat flux may be ignored. The two derivations here are prototypes for the extensive theory of the planetary boundary layer, which has probably received more attention and filled more books than all the rest of dynamic meteorology. Despite this attention, the theory is still quasi-empirical and to some degree unsatisfying. The problems are complex ones, and no clear direction for future research is evident.

The two cases we analyze represent simple but basic applications of the equations for the mean and turbulent flow. Starting with the constant stress layer, we use Eq. (29) and assume steady conditions, neutral stratification, and that the energy fluxes $\overline{u_3 u_1{}^2/2}$ and $\overline{p' u_3}$ are constant. It is convenient to define the *friction velocity* as $u_*{}^2 = -\overline{\rho u_1 u_3}/\bar{\rho}$ and to write $\epsilon = -\overline{u_i F_i}$. Hence Eq. (29) becomes

$$\bar{\rho} u_*{}^2 \frac{\partial \hat{u}}{\partial z} = \epsilon \tag{33}$$

or in nondimensional form

$$\frac{z}{u_*} \frac{\partial \hat{u}}{\partial z} = \frac{1}{\bar{\rho}} \frac{\epsilon z}{u_*{}^3} \tag{34}$$

where by assumption u_* is constant.

The basic assumption is that nondimensional quantities depend only on z/z_0 where z_0 is the *roughness length* and is related to the depth of the laminar sublayer. Hence

$$\frac{z}{u_*} \frac{\partial \hat{u}}{\partial z} = \frac{\epsilon z}{\bar{\rho} u_*{}^3} = f\left(\frac{z}{z_0}\right) \tag{35}$$

where the function f is arbitrary so far. For $z \gg z_0$, we assume no explicit dependence on z_0 so that $f(z/z_0) \to$ const. Thus Eq. (35) yields

$$\frac{\hat{u}}{u_*} = (\text{const}) \ln z + \text{const} = \frac{1}{k} \ln \frac{z}{z_0} + \frac{\hat{u}(z_0)}{u_*} \tag{36}$$

The von Kármán constant in the second version of Eq. (36) is $k \sim 0.4$ as determined by measurements. This relation is verified by observations when a sufficient number of individual profiles are averaged to give $\hat{u}$.

Now let us turn to the Ekman layer and apply Eq. (12), assuming steady, horizontally homogeneous, horizontal mean flow. Thus

$$f\mathbf{k} \times (\hat{\mathbf{v}}_g - \hat{\mathbf{v}}) + \frac{1}{\bar{\rho}} \frac{\partial \boldsymbol{\tau}}{\partial z} + \frac{\overline{\mathbf{F}}}{\bar{\rho}} = 0 \tag{37}$$

Eddy mixing theory for $\boldsymbol{\tau} = K\, \partial \hat{\mathbf{v}}/\partial z$ and the assumption that eddy friction dominates viscous friction give

$$f\mathbf{k} \times (\hat{\mathbf{v}}_g - \hat{\mathbf{v}}) + \frac{1}{\bar{\rho}} \frac{\partial}{\partial z}\left(K \frac{\partial \hat{\mathbf{v}}}{\partial z}\right) = 0 \tag{38}$$

The simplest case here is when $K =$ const. Then we recognize that $\bar{\rho}$ varies little in the planetary boundary layer and set $K_m = K/\bar{\rho}$. Next we use complex notation to write $\mathbf{v} = u + iv$ so that the equation becomes ($i\mathbf{v} \sim \mathbf{k} \times \mathbf{v}$)

$$\frac{\partial^2 \hat{\mathbf{v}}}{\partial z^2} + i\frac{f}{K_m}(\hat{\mathbf{v}}_g - \hat{\mathbf{v}}) = 0 \tag{39}$$

and this is easily solved for $\hat{\mathbf{v}}_g = \text{const}$. The usual exponential substitution gives the result

$$\hat{\mathbf{v}} - \hat{\mathbf{v}}_g = \mathbf{A} \exp\,[(1+i)\lambda z] + \mathbf{B} \exp\,[-(1+i)\lambda z] \tag{40}$$

where $\lambda = (f/2K_m)^{1/2}$.

For boundary conditions we require that $\hat{\mathbf{v}} = 0$ at $z = 0$ and that $\hat{\mathbf{v}} = \hat{\mathbf{v}}_g$ as $z \to \infty$, which gives

$$\hat{\mathbf{v}} - \hat{\mathbf{v}}_g = -\hat{\mathbf{v}}_g \exp\,[-(1+i)\lambda z] \tag{41}$$

so that for $\hat{v}_g = 0$ and $\hat{u}_g = |\hat{\mathbf{v}}_g|$, we have

$$\hat{u} = \hat{u}_g(1 - e^{-\lambda z} \cos \lambda z) \qquad \hat{v} = \hat{u}_g \sin \lambda z e^{-\lambda z} \tag{42}$$

which show that the wind makes an angle of 45° to the left of $\hat{\mathbf{v}}_g$ at the surface. This value is clearly too large, a failure that may be attributed both to the assumption that K_M is constant and to the asymptotic boundary conditions.

PROBLEM 11.6.10 Let z_g be the height at which the wind first has the direction of $\hat{\mathbf{v}}_g$. If $z_g \sim 1$ km, show that $K_m \sim 5\ \text{m}^2/\text{s}$.

PROBLEM 11.6.11 Let $\hat{\mathbf{v}} = \hat{\mathbf{v}}_g$ at a height z_g and let the wind have speed V_0 and cross-isobar angle α_0 at height z_0. For constant K_m, what then is the behavior of the wind between z_0 and z_g?

PROBLEM 11.6.12 Define the so-called Monin-Obukov-Lettau length L by $L^{-1} = -gk\overline{T'u_3}/u_*^3\hat{T}$ in which k is the von Kármán constant. Show that L may be expected to be positive in stable conditions and negative in unstable conditions, and that $L \to \infty$ for neutral stratification. Find an approximate version of Eq. (29) that generalizes Eq. (34), and assume that $z\,\partial\hat{u}/u_*\,\partial z = f(z/z_0, z/L)$. Show that if $\hat{u}(z_0) = 0$, then

$$\frac{\hat{u}}{u_*} = \frac{1}{k} \ln \frac{z}{z_0} + \frac{cz}{L}$$

for conditions near neutral, where c is a constant. [Hint: For $z \sim L$, $z \gg z_0$ so that $f(z/z_0, z/L) \sim g(z/L) = g(0) + g'(0)z/L + \ldots$.]

PROBLEM 11.6.13 Let the turbulence be isotropic in the sense that scales and amplitudes are independent of direction. If $E = \rho(u^2 + v^2 + w^2)/2$ is constant with height, show that we may take $\epsilon \cong a/z$. Integrate the differential equation of the preceding problem for the constant stress layer with $L = \text{const}$. (Hint: Express ϵ as simply as possible as a function of E and a scale l.)

BIBLIOGRAPHIC NOTES

11.1 The basic results given in this section are standard material. A general discussion is given in:

Lorenz, E. N., 1967: *The Nature and Theory of the General Circulation of the Atmosphere*, no. 218TD115 World Meteorological Organization, Geneva, 161 pp.

Maintenance of the circulation was discussed by:

Starr, V. P., 1951: "Application of Energy Principles to the General Circulation," in T. F. Malone (ed.), *Compendium of Meteorology*, American Meteorological Society, Boston.

The effect of each of the terms in the equation of motion as resolved in wavenumber space is illustrated in:

Dutton, J. A., 1962: "The Rate of Change of the Kinetic Energy Spectrum of Flow in a Compressible Fluid," *J. Atmospheric Sci.*, **20**:107–114.

11.2 These results are also standard and can be found in various versions in the meteorological literature.

11.3 Further material on available potential energy is presented in:

Dutton, J. A., and D. R. Johnson, 1967: "The Theory of Available Potential Energy and a Variational Approach to Atmospheric Energetics," *Advances in Geophys.*, **12**:333–436.

Johnson, D. R., 1970: "The Available Potential Energy of Storms," *J. Atmospheric Sci.*, **27**:727–741.

Johnson, D. R., and J. A. Dutton, 1976: "Atmospheric Energetics and the General Circulation Viewed from Isentropic Coordinates." In preparation.

Lorenz, E. N., 1955: "Available Potential Energy and the Maintenance of the General Circulation," *Tellus*, **7**:157–167.

11.4 This section presents a summary of:

Dutton, J. A., 1973: "The Global Thermodynamics of Atmospheric Motion," *Tellus*, **25**:89–110.

11.5 The most recent reference here is:

Bretherton, F. P., 1970: "A Note on Hamilton's Principle for Perfect Fluids," *J. Fluid Mech.*, **44**:19–31.

This article contains a bibliography of earlier work.

11.6 Bibliographies devoted to turbulence can be book-length lists by themselves. The ideas discussed here are treated more fully in:

Batchelor, G. K., 1956: *The Theory of Homogeneous Turbulence*, Cambridge University Press, Cambridge, 197 pp.

Lumley, J. L., and H. A. Panofsky, 1964: *The Structure of Atmospheric Turbulence*, John Wiley & Sons, Inc. (Interscience), New York, 239 pp.

Tennekes, H., and J. L. Lumley, 1972: *A First Course in Turbulence*, The M.I.T. Press, Cambridge, Mass., 300 pp.

Vinnichenko, N., N. Pinus, S. Shmeter, and G. Shur, 1973: *Turbulence in the Free Atmosphere*, Plenum Press, Plenum Publishing Corporation, New York, 259 pp.

The last reference includes a review of clear air turbulence that contains some of the ideas presented here.

12

ATMOSPHERIC WAVE MOTION

Even casual observers of the sky will often see cloud patterns that remind them of the waves on a lake or ocean. A common example is the arrangement of cirrus clouds in long bands, resembling perhaps the whitecaps on a water surface on a windy day. At lower altitudes, cumulus clouds often show similar patterns, especially over the oceans.

In these cases, the clouds reveal regions of upward motion while the clear spaces between are regions of downward motion. The patterns in the clouds show that the ascending and descending motions are arranged in alternating sequence. The high-resolution sounding techniques that became available in the 1960s produced the surprising result that the horizontal components of the wind often have vertical oscillations in both speed and direction, again with a regularity suggesting that wave motion is present.

On the synoptic scale, too, we see on weather charts the wavelike oscillations in the westerlies of mid-latitudes and the easterlies of the tropics. In this case, the wavelengths may be thousands of kilometers, and again the synoptic patterns of cloudiness show that there is an associated pattern of vertical motion.

These observations suggest that many of the motions of the atmosphere have a wavelike behavior. Although the existence of wave motion in the atmosphere has been studied since Newton considered sound waves, the investigation of wave phenomena has intensified in recent years for a number of reasons.

First, the various types of wave motion found in the atmosphere represent solutions of a simplified form of the equations of motion in which nonlinearities are eliminated. Thus they can be studied by standard mathematical techniques. The recently acquired observations that appear to show wavelike motions have provided impetus for these investigations.

Second, because these waves are solutions of the equations, they will appear when the equations are integrated numerically. Some of these waves cause difficulty in attempts to produce numerical predictions, and so ways must be found to modify the equations in order to filter them out of the solutions.

Third, certain of the waves may have important consequences for atmospheric dynamics because they may be propagating significant amounts of energy from one place to another in a manner analogous to the radiative propagation of energy by electromagnetic waves.

Fourth, under some conditions waves may grow rapidly in amplitude, become nonlinear phenomena, and perhaps break or roll up. Thus an important part of the study of wave motion, considered in Chap. 13, is concerned with stability problems, and an attempt is made to determine in which conditions small-amplitude disturbances will be unstable and grow in amplitude. Such instability of wave motion is now believed to be the cause of at least part of the clear air turbulence that has been both nuisance and hazard in the jet age.

Fifth, in the quasi-geostrophic theory (to be discussed in Chap. 14), the system of partial differential equations relevant to large-scale atmospheric flow is reduced to a single partial differential equation. When linearized, its solutions are generalized Rossby waves; in nonlinear form it gives a mathematical model of large-scale flow more compact and presumably more tractable than the original set of equations. The equation allows us to begin the study of nonlinear phenomena and predictability problems with a relatively simple case.

The various modes of wave motion thus have varying significance for atmospheric dynamics. They are all solutions of the linearized equations of motion, but it would be confusing to try to study them all at once. Thus an important aspect of the study of wave motion involves development of rational and consistent methods of modifying the equations of motion to apply to particular temporal and spatial scales of motion so that the various modes of wave motion may be studied separately. In this chapter, we shall consider various types of pure waves.

12.1 BASIC CONCEPTS OF WAVE MOTION

In the study of differential equations, we usually try to discover functions that will provide solutions. In the study of wave motion, we often turn this around and try to find out under what conditions the wavelike shapes that are suggested by observation will be solutions.

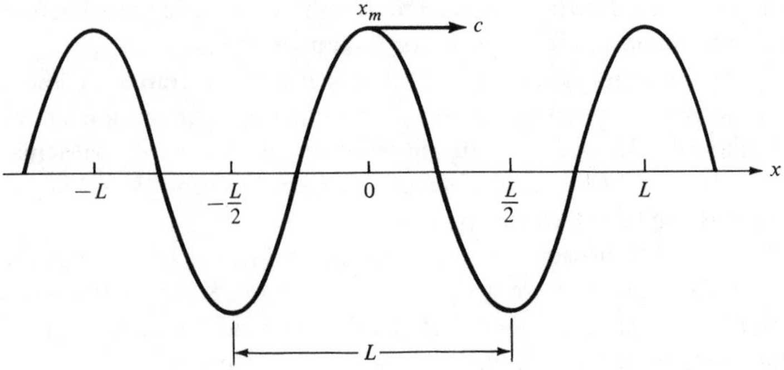

FIGURE 12.1
The basic wave form.

The most common choice of wave is a sine or cosine function that is constrained to propagate in one spatial direction. Hence we assume that a variable $\phi(x,t)$ can be represented as

$$\phi(x,t) = \phi_0 \cos [\kappa (x - ct)] \tag{1}$$

in which ϕ_0 is the amplitude and the wavenumber $\kappa = 2\pi/L$, L being the wavelength. At $t = 0$, the wave has a maximum at $x = 0$ (and at $x = \pm L, \pm 2L, \ldots$). As t increases, the maximum will move in the positive direction along the x axis, with the position x_m of the maximum controlled by $x_m - ct = 0$ so that the wave speed is $dx_m/dt = c$ (Fig. 12.1). We generally assume that the x axis extends from $-\infty$ to $+\infty$ so that we do not have to worry about wave reflections at a solid boundary and do not have to introduce some form of spatial damping to make the waves disappear at $\pm\infty$.

More complex phenomena may be represented by letting the amplitude ϕ_0 vary in the other spatial directions so that we might have $\phi_0 = \phi_0(y,z)$. Thus the variable ϕ will be represented by

$$\phi(x,y,z,t) = \phi_0(y,z) \cos [\kappa(x - ct)] \tag{2}$$

But in analyzing the physics of wave motion, rather than using Eq. (2) it is often more convenient to use frequency $\omega = 2\pi/T$, where T is the period, and represent the wave as

$$\phi = \phi_0 \cos (\omega t + \kappa x) \tag{3}$$

This form is substituted in the dynamic equations, which contain parameters $\pi_1, \ldots, \pi_n$ describing the structure of the atmosphere (for example, the acceleration of gravity is often such a parameter), and we find that the assumed wave solutions are possible only if the frequency obeys an equation of the form

$$\omega = \omega(\kappa, \pi_1, \ldots, \pi_n) \tag{4}$$

This is the *characteristic equation*, and all wave motion problems involve this determination of permitted combinations of frequency and wavenumber.

A comparison of Eqs. (2) and (3) shows that we have $c = -\omega/\kappa$, so that an equivalent characteristic equation is $c = c(\kappa, \pi_1, \ldots, \pi_n)$. Recognizing that the dynamic equations impose this restriction on the wave speed c, we observe that either

$$\frac{\partial \phi}{\partial t} + c\frac{\partial \phi}{\partial x} = 0 \tag{5}$$

or

$$\frac{\partial^2 \phi}{\partial t^2} - c^2 \frac{\partial^2 \phi}{\partial x^2} = 0 \tag{6}$$

is a differential equation describing wave motion of the type specified by Eq. (2).

*PROBLEM 12.1.1 Show that $\phi = \phi(x - ct)$ is a general solution to the wave equations (5) and (6).

Most atmospheric wave problems are more complicated than the type considered so far, and we use complex notation. Thus if we want waves to be able to propagate in the horizontal plane, we use

$$\phi(x,y,z,t) = \phi_0(z)e^{i(\omega t + \kappa_1 x + \kappa_2 y)} \tag{7}$$

in order to find the characteristic values (ω, κ_1, κ_2) for which wave solutions can actually exist. In application of Eq. (7), it is always understood that $\phi_0(z)$ is complex, and that the physically significant part of the solution is $\phi = \mathrm{Re}(\phi_0 e^{i(\omega t + \kappa_1 x + \kappa_2 y)})$.

In these more complicated problems, we find that the characteristic equation relating ω, κ_1, and κ_2 is more difficult too, often depending on variables that describe the structure of the atmosphere. It often turns out that ω is itself complex so that $\omega = \omega_R + i\omega_I$ and then

$$\phi = \phi_0 e^{-\omega_I t} e^{i(\omega_R t + \kappa_1 x + \kappa_2 y)} \tag{8}$$

Thus if ω_I is positive, the wave will be damped in time and the motion is *stable*; if ω_I is negative, the wave will grow exponentially and the motion is *unstable*. Usually we shall find that

$$\omega = \pm(\omega_R + i\omega_I) \tag{9}$$

because the characteristic equations are generally quadratic or of higher order, so that the solutions must then be written

$$\phi = (\phi_1 e^{i(\omega_R + i\omega_I)t} + \phi_2 e^{-i(\omega_R + i\omega_I)t})e^{i(\kappa_1 x + \kappa_2 y)} \tag{10}$$

Thus one of the terms grows exponentially and we have instability whenever $\omega_I \neq 0$.

Now let us determine the direction of propagation of a wave of the form (7). Let $\boldsymbol{\kappa} = \mathbf{i}\kappa_1 + \mathbf{j}\kappa_2$. Then we can express the argument of the exponential in Eq. (7) as

$$\omega t + \kappa_1 x + \kappa_2 y = \boldsymbol{\kappa} \cdot \left(\frac{\omega \boldsymbol{\kappa}}{\kappa^2} t + \mathbf{x}\right) \tag{11}$$

We define a *vector propagation velocity* as†

$$\mathbf{c} = -\frac{\omega \boldsymbol{\kappa}}{\kappa^2} \tag{12}$$

so that the wave will propagate in the x direction with speed $-\omega\kappa_1/\kappa^2$ and in the y direction with speed $-\omega\kappa_2/\kappa^2$. Thus the direction of propagation will be given by the angle θ measured from the positive x axis by

$$\theta = \arctan \frac{c_y}{c_x} = \arctan \frac{-\omega\kappa_2}{-\omega\kappa_1} \tag{13}$$

in agreement with the previous results. The signs determine the correct quadrant.

The representation (7) is the most convenient form for substitution into dynamical equations because it allows the direction of propagation and the wavelength to be varied conveniently. But it is an assumption that the waves are moving with parallel crests. The direction of propagation and the speed in that direction can be determined from the characteristic equation once the wavenumbers κ_1 and κ_2 are specified.

PROBLEM 12.1.2 Find a coordinate system (X,Y) such that the wave form specified in Eq. (7) propagates in the X direction with crests parallel to the Y axis.

Now we shall turn to the questions associated with wave interference—a process that occurs when two waves are propagating in the same medium. Because we have linearized equations, if $\phi_1 = A_1 \cos(\kappa x + \omega t)$ and $\phi_2 = A_2 \cos[(\kappa + \delta\kappa)x + (\omega + \delta\omega)t]$ are both solutions, then $\phi_1 + \phi_2$ will also be a solution. Thus we may determine what the propagation characteristics of such a sum of two basic wave forms will be. With the aid of a common trigonometric identity, we have

$$\begin{aligned}\phi_1 + \phi_2 = {} & [A_1 + A_2 \cos(x\,\delta\kappa + t\,\delta\omega)] \cos(\kappa x + \omega t) \\ & - A_2 \sin(x\,\delta\kappa + t\,\delta\omega) \sin(\kappa x + \omega t)\end{aligned} \tag{14}$$

†The reader will probably wonder why we did not choose to use exp $[i(\kappa x - \omega t)]$ rather than exp $[i(\kappa x + \omega t)]$ and thus avoid the minus sign in Eq. (12). There is a good reason. Anyone who blunders into the wave motion business will very soon find himself bedeviled trying to keep the plus and minus signs correct. This trick helps, because both temporal and spatial derivatives of the term exp $[i(\kappa x + \omega t)]$ have their original signs. Thus when the representation (7) is substituted into the dynamic equations the signs remain the same, and at least one small battle in a horrendous war has been won.

Now we put

$$A \cos \psi = A_1 + A_2 \cos (x\, \delta\kappa + t\, \delta\omega)$$
$$A \sin \psi = A_2 \sin (x\, \delta\kappa + t\, \delta\omega) \tag{15}$$

so that Eq. (14) becomes

$$\phi_1 + \phi_2 = A \cos (\kappa x + \omega t + \psi) \tag{16}$$

in which the phase angle ψ is found from

$$\tan \psi = \frac{A_2 \sin (x\, \delta\kappa + t\, \delta\omega)}{A_1 + A_2 \cos (x\, \delta\kappa + t\, \delta\omega)} \tag{17}$$

and we have

$$A^2 = (A_1{}^2 + A_2{}^2)\left[1 + 2\frac{A_1 A_2}{A_1{}^2 + A_2{}^2} \cos (x\, \delta\kappa + t\, \delta\omega)\right] \tag{18}$$

In the special case in which $A_1 = A_2$, we find with trigonometric identities that

$$A = 2A_1 \cos\left[\tfrac{1}{2}(x\, \delta\kappa + t\, \delta\omega)\right] \qquad \psi = \tfrac{1}{2}(x\, \delta\kappa + t\, \delta\omega) \tag{19}$$

Combination of this with Eq. (16) yields

$$\phi_1 + \phi_2 = 2A_1 \cos\left[\tfrac{1}{2}(x\, \delta\kappa + t\, \delta\omega)\right] \cos (\kappa x + \omega t + \psi) \tag{20}$$

The combined wave formed from ϕ_1 and ϕ_2 thus has a more complicated structure. We could say that it is a wave moving at speed $c = -(\omega + \frac{1}{2}\delta\omega)/(\kappa + \frac{1}{2}\delta\kappa)$ with the first cosine in Eq. (20) providing an amplitude modulation of the wave form. This modulation also travels like a wave form, in this case with speed

$$c_g = -\frac{\delta\omega}{\delta\kappa} \tag{21}$$

This wave speed is called the *group velocity*, and gives the propagation speed of the maximum in the pattern formed by two waves with nearly the same frequency and wavenumber. The speed $c = -\omega/\kappa$ of a single wave is known as a *phase speed*. Properties of wave groups are shown in Fig. 12.2.

In general, we solve the characteristic equation for $\omega = \omega(\kappa)$ and then write Eq. (21) in the explicit differential form

$$c_g = -\frac{d\omega}{d\kappa} = \frac{d}{d\kappa}\,[\kappa c(\kappa)] \tag{22}$$

With waves on deep water, we can often distinguish the individual wave forms and the group pattern. In this case $c = 2c_g$, so that the individual wave forms moving faster than the group can be seen catching up to the group and passing through it.

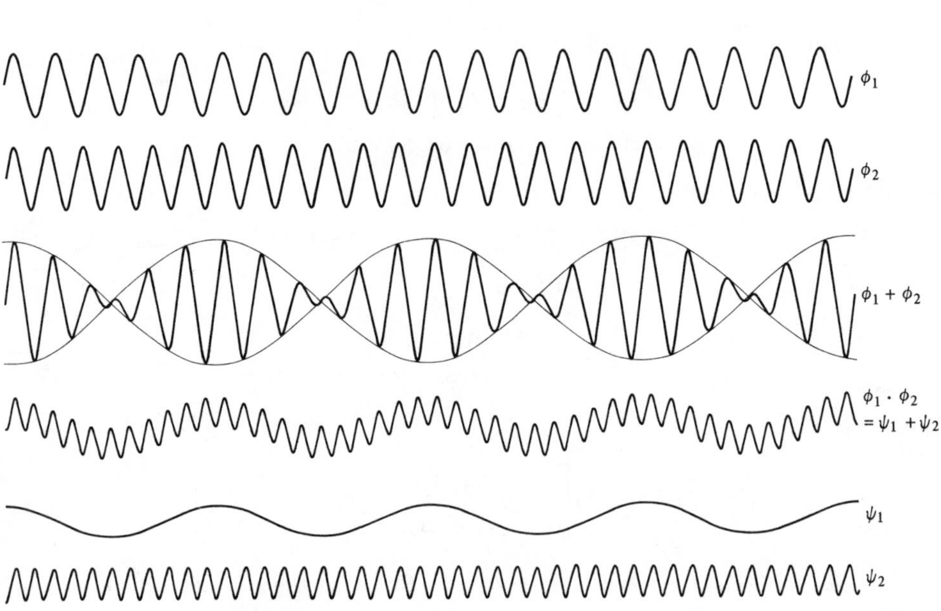

FIGURE 12.2
Combinations of two wave forms ϕ_1 and ϕ_2 with slightly different wavenumbers and frequencies. The sum $\phi_1 + \phi_2$ shows how the individual waves are amplitude-modulated into groups that propagate at the group velocity. The lower part of the figure shows how the product of ϕ_1 and ϕ_2 can be represented as a sum of waves that have both longer and shorter wavelengths than the original waves—a phenomenon that is crucial in nonlinear interactions.

PROBLEM 12.1.3 Show that the group concept is valid even if A_1 and A_2 are not identical. [Hint: Use Eq. (18) to show that A satisfies a wave equation with speed c_g.]

For wave motion in two or more dimensions

$$\phi = \phi_0 e^{i(\omega t + \boldsymbol{\kappa} \cdot \mathbf{x})} = \phi_0 e^{i\boldsymbol{\kappa} \cdot (\mathbf{x} - \mathbf{c}t)} \tag{23}$$

satisfies the equation

$$\frac{\partial \phi}{\partial t} + \mathbf{c} \cdot \nabla \phi = 0 \qquad \mathbf{c} = \mathbf{c}(\kappa_1, \kappa_2, \kappa_3, \pi_1, \ldots, \pi_n) \tag{24}$$

Two solutions of this equation (for different $\mathbf{c}$) of the form

$$\begin{aligned} \phi_1 &= A_1 e^{i(\omega t + \boldsymbol{\kappa} \cdot \mathbf{x})} \\ \phi_2 &= A_2 e^{i[(\omega + \delta\omega)t + (\boldsymbol{\kappa} + \delta\boldsymbol{\kappa}) \cdot \mathbf{x}]} \end{aligned} \tag{25}$$

will give the solution

$$\phi_1 + \phi_2 = (A_1 + A_2 e^{i(t\,\delta\omega + \mathbf{x}\cdot\delta\boldsymbol{\kappa})})e^{i(\omega t + \boldsymbol{\kappa}\cdot\mathbf{x})}$$
$$= A(t\delta\omega + \mathbf{x}\cdot\delta\boldsymbol{\kappa})e^{i(\omega t + \boldsymbol{\kappa}\cdot\mathbf{x})} \tag{26}$$

so that again the sum of solutions is a combination of the original harmonic motion and a varying amplitude.

Because of the characteristic equation, we may treat $\delta\omega$ as an infinitesimal and use the chain rule to write $\delta\omega = \delta\boldsymbol{\kappa}\cdot\nabla_\kappa\omega$ where

$$\nabla_\kappa = \mathbf{i}\frac{\partial}{\partial\kappa_1} + \mathbf{j}\frac{\partial}{\partial\kappa_2} + \mathbf{k}\frac{\partial}{\partial\kappa_3} \tag{27}$$

Thus the amplitude may be written as

$$A = A_1 + A_2 e^{i\,\delta\boldsymbol{\kappa}\cdot(\mathbf{x} - \mathbf{c}_g t)} \tag{28}$$

in which we have defined $\mathbf{c}_g = -\nabla_\kappa\omega$. But now we see that

$$\frac{\partial A}{\partial t} + \mathbf{c}_g\cdot\nabla A = 0 \tag{29}$$

so that by comparison with Eq. (24) we may interpret $\mathbf{c}_g$ as the vector group velocity giving the direction and speed of the group motion resulting from the superposition of waves with different phase speeds.

It is the group propagation of patterns created by individual waves interfering with each other that makes the observational study of wave motion quite difficult. As many waves combine to form a group, the pattern propagating at the group velocity may have little resemblance to a simple harmonic wave. Thus observational results must be tested against the phase version and the group version of the characteristic equation to determine whether they do indeed represent the kind of wave motion predicted by theory. It must be kept in mind that wave motion in the atmosphere is a mode of motion that obeys a characteristic equation relating speed of propagation to wavelength.

PROBLEM 12.1.4 Suppose that a variable ϕ obeys the equation

$$\left(\frac{\partial^2}{\partial t^2} - \alpha^2\,\nabla^2 - \beta^2\,\nabla^4\right)\phi = 0$$

Find the characteristic equation and determine for what wavelengths the solution is unstable. What is the group velocity? What change in the equation would make the wave stable for all wavelengths?

PROBLEM 12.1.5 For wave motion in three dimensions, find the necessary and sufficient conditions that the group velocity be in the same direction as the vector phase velocity (assumed constant in space). (Hint: What is $\nabla_\kappa\kappa$?)

PROBLEM 12.1.6 Suppose that a wave equation

$$\frac{\partial^2 \phi}{\partial t^2} - c^2 \nabla^2 \phi = 0$$

is valid with $c = c(x,y,z)$ rather than being constant. Assume solutions of the form $\phi = \phi_0 \exp i\Phi$ where both the amplitude ϕ_0 and the phase Φ may be functions of (x,y,z,t).

(*a*) Let $\Phi = \omega t + \boldsymbol{\kappa} \cdot \mathbf{x}$ where $\boldsymbol{\kappa} = \mathbf{i}\kappa_1 + \mathbf{j}\kappa_2 + \mathbf{k}\kappa_3$. What equation must ϕ_0 satisfy?

(*b*) Let ϕ_0 be constant (this a form of the "WKB approximation"). What equation must Φ satisfy? By analogy with the simple case $c =$ const, develop suitable definitions of the propagation vector $\boldsymbol{\kappa}$ and frequency ω. Show that when the differential equation for Φ is expressed with these definitions, it reduces for $c =$ const to the characteristic equation for the simple case. (Hint: Using the simple case as a guide, determine how $\boldsymbol{\kappa}$ and ω can be found by differentiation.)

PROBLEM 12.1.7 (Eckart) In the preceding problem, let $c = c(z)$, and apply the WKB approximation with $\Phi = \omega t + \boldsymbol{\kappa} \cdot \mathbf{x} + F(\omega,\kappa_1,\kappa_2,z)$ where $\boldsymbol{\kappa} = \mathbf{i}\kappa_1 + \mathbf{j}\kappa_2$. Add the two solutions specified by the frequency and wavenumber sets $(\omega,\kappa_1,\kappa_2)$ and $(\omega + \delta\omega, \kappa_1 + \delta\kappa_1, \kappa_2 + \delta\kappa_2)$. Find an expression for the components of the group velocity. [Hint: Use a Taylor series to find a linear approximation for $F(\omega + \delta\omega, \kappa_1 + \delta\kappa_1, \kappa_2 + \delta\kappa_2, z) - F(\omega,\kappa_1,\kappa_2,z)$. Find the values of (x,y,z,t) for which the modulation will have a constant amplitude, and use these equations to determine the velocity of the point of constant amplitude.]

12.2 EXTERNAL GRAVITY WAVES

Although our main interest is with internal atmospheric waves, we shall first consider external waves of the kind that form on the surface of lakes or oceans. These external gravity waves have a relatively simple theory that illustrates the methods to be applied in the remainder of the chapter. Furthermore, external waves are a complicating factor in certain kinds of numerical experiments so that the theory has been put to good use in meteorology as well as oceanography.

We consider a fluid of constant density ρ_0 with an equilibrium free surface at height $z = 0$ and a flat, rigid bottom at $z = -h$. We assume that the density of the fluid above $z = 0$ is negligible compared to ρ_0 so that the presence of any upper fluid may be ignored. For water waves, we have $\rho/\rho_0 \sim 10^{-3}$, and in studying the simplest properties of these waves we assume that they are unaffected by the air above them despite the obvious fact that they are generated by the wind blowing over the water surface.

In the fluid at rest, hydrostatic equilibrium would prevail and so $\partial p_0/\partial z = -g\rho_0$ We assume that the waves are of small amplitude so that we may neglect the inertia term, and we consider only two-dimensional motion. With the definition $p = p_0(z) + p'(x,z,t)$, the equations of motion for the incompressible, inviscid fluid become

$$\frac{\partial u}{\partial t} + \frac{1}{\rho_0}\frac{\partial p'}{\partial x} = 0 \qquad \frac{\partial w}{\partial t} + \frac{1}{\rho_0}\frac{\partial p'}{\partial z} = 0$$
$$\frac{\partial u}{\partial x} + \frac{\partial w}{\partial z} = 0 \tag{1}$$

The boundary conditions require that w vanish on the rigid surface at $z = -h$, and that dp/dt vanish on the upper free surface. Hence the upper condition becomes, again ignoring products of terms representing the small-amplitude wave motion,

$$\frac{\partial p'}{\partial t} + w\frac{\partial p_0}{\partial z} = \frac{\partial p'}{\partial t} - g\rho_0 w = 0 \qquad \text{free surface} \tag{2}$$

Because we consider small-amplitude waves with small surface displacement, we may apply this condition at $z = 0$ with only negligible error.

The two equations of motion in Eq. (1) show immediately that

$$\frac{\partial}{\partial t}\left(\frac{\partial w}{\partial x} - \frac{\partial u}{\partial z}\right) = 0 \tag{3}$$

and if we assume that the motion started from a state of rest without vorticity, no vorticity can develop in this inviscid, autobarotropic fluid. Thus Eq. (3) may be written as

$$\frac{\partial w}{\partial x} - \frac{\partial u}{\partial z} = 0 \tag{4}$$

Because of the incompressibility condition, we may define a stream function by $u = -\psi_z$, $w = \psi_x$, in which the subscript denotes partial differentiation. Hence Eq. (4) yields

$$\psi_{xx} + \psi_{zz} = 0 \tag{5}$$

To obtain wavelike solutions, we now put $\psi = \Psi(z)\cos[\kappa(x - ct)]$ and thus Eq. (5) gives

$$\Psi_{zz} - \kappa^2\Psi = 0 \tag{6}$$

which has the solution

$$\Psi = Ae^{\kappa z} + Be^{-\kappa z} \tag{7}$$

where A and B are constants.

According to the boundary condition that w vanish at $z = -h$, we must have $\psi_x = 0$ at $z = -h$. But this can be true only if $\Psi(-h) = 0$, which implies that $B = -A\exp(-2\kappa h)$. Thus we have

$$\psi = D \sinh [\kappa(z+h)] \cos [\kappa(x-ct)] \tag{8}$$

in which $D = 2A \exp(-\kappa h)$ (see Fig. 12.3).

This form of the stream function gives

$$w = \psi_x = -\kappa D \sinh [\kappa(z+h)] \sin [\kappa(x-ct)] \tag{9}$$

and so

$$\frac{\partial w}{\partial t} = \kappa^2 c D \sinh [\kappa(z+h)] \cos [\kappa(x-ct)] \tag{10}$$

But the equation of motion shows that

$$\frac{\partial p'}{\partial z} = -\rho_0 \frac{\partial w}{\partial t} \tag{11}$$

and so by an indefinite integration we have

$$p' = -\kappa \rho_0 c D \cosh [\kappa(z+h)] \cos [\kappa(x-ct)] \tag{12}$$

We have used the condition that p' vanishes when the wave motions vanish ($D = 0$) to eliminate the arbitrary function of (x,t) that appears in the integration.

In order to find the characteristic equation, we may apply the upper boundary condition of Eq. (2) to Eqs. (12) and (9) with the result

$$\kappa \rho_0 D \sin [\kappa(x-ct)](-c^2 \kappa \cosh \kappa h + g \sinh \kappa h) = 0 \qquad z = 0 \tag{13}$$

or

$$c^2 = g\kappa^{-1} \tanh \kappa h = \frac{gL}{2\pi} \tanh \frac{2\pi h}{L} \tag{14}$$

*PROBLEM 12.2.1 Show that $c^2 = g\kappa^{-1} = gL/2\pi$ for deep water ($h \gg L$) and that $c^2 = gh$ for shallow water ($h \ll L$). Show that the deep-water waves are *dispersive* because the phase speed depends on wavelength and yields the group velocity $c_g = c/2$. Show that the shallow-water waves are not dispersive because $c = c_g$.

It is of interest to determine the trajectories of the water parcels, in addition to knowing the characteristics of the propagating wave form. We use Eq. (9) and the equivalent result for the x component to write the pair of equations

$$\begin{aligned} w &= -\kappa D \sinh [\kappa(z+h)] \sin [\kappa(x-ct)] \\ u &= -\psi_z = -\kappa D \cosh [\kappa(z+h)] \cos [\kappa(x-ct)] \end{aligned} \tag{15}$$

Now let us choose a parcel of fluid and denote its average position by (x^*, z^*). Then we denote its actual position by writing $x = x^* + \xi$ and $z = z^* + \zeta$ as a function of the displacements ξ and ζ. Thus we have $u = d\xi/dt$ and $w = d\zeta/dt$.

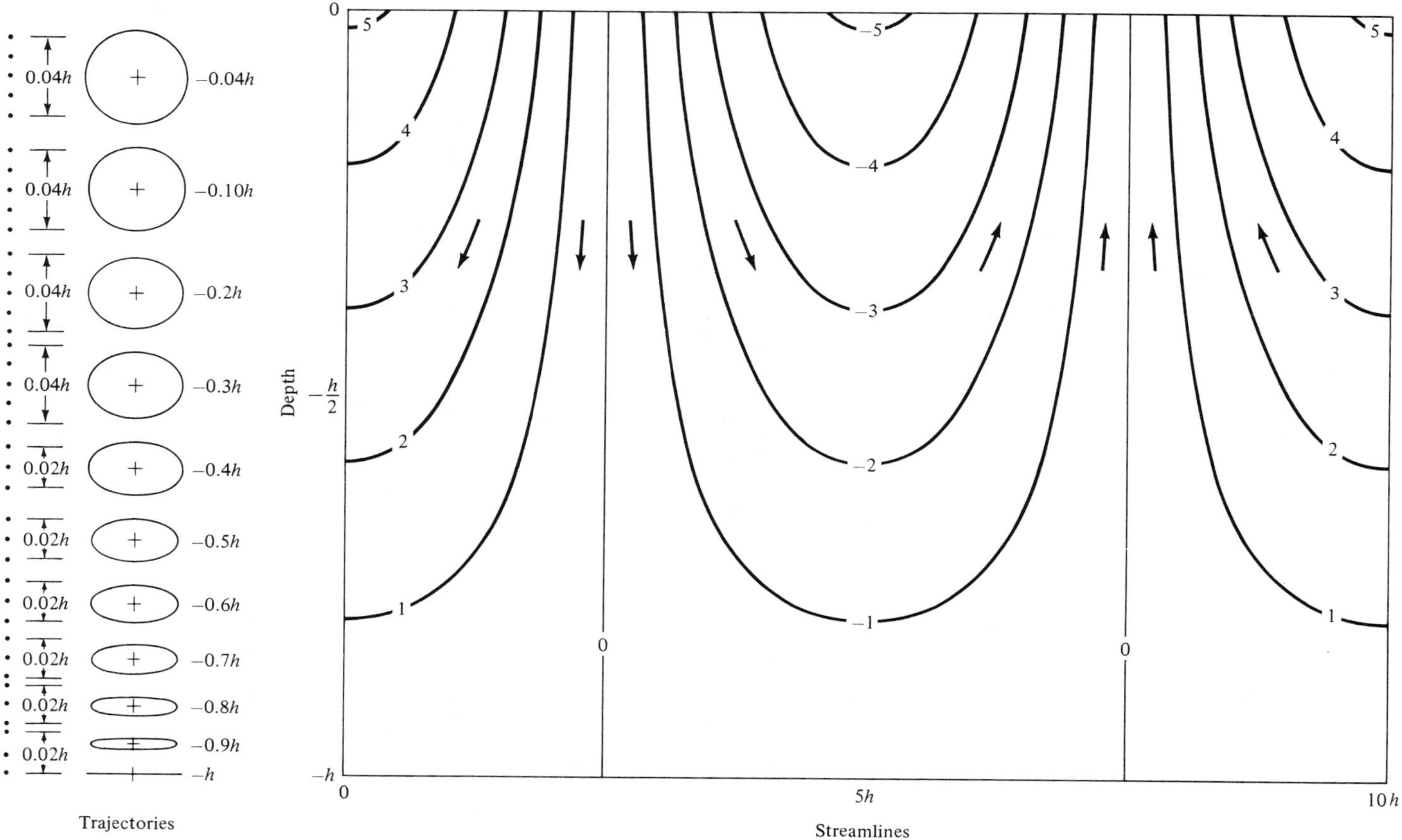

FIGURE 12.3
An example of streamlines and trajectories for external gravity waves. Values of the stream function are given in relative units. The ellipses are the trajectories at the depths shown; note that the ellipses are plotted to the same scale even though they are not centered on a linear depth scale.

Next we assume that the local velocity at $(x^* + \xi, z^* + \zeta)$ is approximately equal to the velocity at (x^*, z^*). With this assumption, Eqs. (15) may be integrated immediately to show that

$$\xi = \frac{D}{c} \cosh\,[\kappa(z^* + h)]\, \sin\,[\kappa(x^* - ct)]$$
$$\zeta = -\frac{D}{c} \sinh\,[\kappa(z^* + h)]\, \cos\,[\kappa(x^* - ct)] \tag{16}$$

This result shows that the parcel trajectory is an ellipse, because with

$$a = \frac{D}{c} \cosh\,[\kappa(z^* + h)] \qquad b = -\frac{D}{c} \sinh\,[\kappa(z^* + h)] \tag{17}$$

we find

$$\frac{\xi^2}{a^2} + \frac{\zeta^2}{b^2} = 1 \tag{18}$$

Near the free surface in deep water when h is large, $z^* \ll h$ so that $\sinh(\kappa h) \cong \cosh(\kappa h)$ and thus the trajectories are nearly circular. Near the bottom, where $z^* \cong -h$, $\sinh\,[\kappa(z^* + h)] \cong 0$ and $\cosh\,[\kappa(z^* + h)] \cong 1$ so that the trajectories are flattened ellipses and the majority of the displacement is horizontal.

PROBLEM 12.2.2 In what direction do the parcels move at a crest and trough near the surface? (Hint: A cork bobbing in the waves will only oscillate back and forth with the parcels; it will not move with the wave form.)

PROBLEM 12.2.3 From the equations of motion, determine the energy equation for external gravity waves and find an explicit expression for the energy propagation vector. Calculate this vector, and upon integrating over one wavelength, find the average values of the vertical and horizontal components.

12.3 PURE TYPES OF ATMOSPHERIC WAVE MOTION

There are four basic modes of atmospheric wave propagation that are of interest in meteorology. In this section we attempt to illustrate them in their simplest form by reducing the equations of motion, perhaps somewhat arbitrarily, to the point that they will yield only one mode of wave motion in each case. The four types are:

1 Acoustic waves, speed controlled by temperature
2 Gravity waves, speed controlled by stability
3 Inertial waves, speed controlled by Coriolis force

4 Rossby waves, speed controlled by latitudinal variation in Coriolis force

Acoustic waves Sound propagates in the atmosphere in a compressional, isentropic wave motion. We consider only acoustic waves propagating along the x axis, so that the equations of motion are

$$\frac{\partial u}{\partial t} + u\frac{\partial u}{\partial x} + \frac{1}{\rho}\frac{\partial p}{\partial x} = 0 \tag{1}$$

$$\frac{\partial p}{\partial t} + u\frac{\partial p}{\partial x} - \frac{c_p}{c_V}\frac{p}{\rho}\left(\frac{\partial \rho}{\partial t} + u\frac{\partial \rho}{\partial x}\right) = 0 \tag{2}$$

$$\frac{\partial \rho}{\partial t} + \frac{\partial}{\partial x}(\rho u) = 0 \tag{3}$$

The second equation is obtained from the isentropic assumption.

We denote a motionless state with a subscript zero and put $p = p_0(z) + p'(x,t)$ and $\rho = \rho_0(z) + \rho'(x,t)$. The substitution of these two expressions in Eqs. (1) to (3) and subsequent linearization by neglecting all products of u, p', and ρ' with each other gives the set of equations

$$\frac{\partial u}{\partial t} + \frac{1}{\rho_0}\frac{\partial p'}{\partial x} = 0 \tag{4}$$

$$\frac{\partial p'}{\partial t} - \frac{c_p}{c_V}\frac{p_0}{\rho_0}\frac{\partial \rho'}{\partial t} = 0 \tag{5}$$

$$\frac{\partial \rho'}{\partial t} + \rho_0\frac{\partial u}{\partial x} = 0 \tag{6}$$

Combination of Eqs. (5) and (6) gives

$$\frac{\partial p'}{\partial t} + \frac{c_p p_0}{c_V}\frac{\partial u}{\partial x} = 0 \tag{7}$$

and upon differentiation of Eq. (4) with respect to time and use of Eq. (7) we have the wave equation

$$\frac{\partial^2 u}{\partial t^2} - \frac{c_p}{c_V}\frac{p_0}{\rho_0}\frac{\partial^2 u}{\partial x^2} = 0 \tag{8}$$

Thus the assumption of a solution $u = U\cos[\kappa(x - ct)]$ yields the characteristic equation

$$c^2 = \frac{c_p}{c_V}\frac{p_0}{\rho_0} = \frac{c_p}{c_V}RT_0 \tag{9}$$

for the (Laplacian) speed of sound. Clearly, acoustic waves are nondispersive waves and have a group velocity equal to their phase velocity.

PROBLEM 12.3.1 When Newton attempted to find the speed of sound waves, he erroneously assumed that acoustic waves propagate isothermally. What speed of sound does this assumption imply?

Internal gravity waves To exhibit the basic characteristics of internal atmospheric gravity waves, we shall consider isentropic, incompressible motion with an x and z component in a hydrostatic atmosphere with constant stability.

For this analysis we need to observe that we may write the equation of state as

$$\frac{p_0 + p'}{p_0} = \frac{(\rho_0 + \rho')(T_0 + T')R}{R\rho_0 T_0} \tag{10}$$

The equilibrium variables (subscript zero) also obey the equation of state, and so neglecting the product $\rho' T'$ gives

$$\frac{\rho'}{\rho_0} + \frac{T'}{T_0} = 0 \tag{11}$$

when we assume for simplicity that the fractional changes in temperature are much greater than those in pressure.

From Poisson's equation we find

$$\frac{\theta'}{\theta_0} = \frac{T'}{T_0} - \frac{R}{c_p}\frac{p'}{p_0} \tag{12}$$

and this same assumption gives $\theta'/\theta_0 = T'/T_0$.

Now the hydrostatic relation may be written as

$$\frac{1}{\rho}\frac{\partial p}{\partial z} + g = \frac{1}{\rho_0 + \rho'}\frac{\partial}{\partial z}(p_0 + p') + g = 0 \tag{13}$$

But the relations $(\rho_0 + \rho')^{-1} \cong [1 - (\rho'/\rho_0)]/\rho_0$ and $\partial p_0/\partial z = -g\rho_0$ give

$$\frac{1}{\rho_0}\frac{\partial p'}{\partial z} + g\frac{\rho'}{\rho_0} - \frac{\rho'}{\rho_0^2}\frac{\partial p'}{\partial z} = 0 \tag{14}$$

With Eq. (11) and the simplified form of (12) the linearized hydrostatic condition is

$$\frac{1}{\rho_0}\frac{\partial p'}{\partial z} - g\frac{\theta'}{\theta_0} = 0 \tag{15}$$

A further approximation is useful. We may write Eq. (15) as

$$\frac{\partial}{\partial z}\left(\frac{p'}{\rho_0}\right) + \frac{p'}{\rho_0}\frac{\partial}{\partial z}(\ln \rho_0) - g\frac{\theta'}{\theta_0} = 0 \tag{16}$$

Now we wish to compare the magnitudes of the first and second terms. To do so, we shall use a common method of estimating the size of the derivative in the first term. We suppose that the variations of p'/ρ_0 in the vertical direction may be represented locally in the neighborhood of a point z^* as a function of wavelength L in the form

$$\frac{p'}{\rho_0} = B \cos\left[\frac{2\pi}{L}(z - z^*)\right] \tag{17}$$

so that

$$\frac{\partial}{\partial z}\left(\frac{p'}{\rho_0}\right) = -\frac{2\pi}{L} B \sin\left[\frac{2\pi}{L}(z - z^*)\right] \tag{18}$$

The derivative at a point midway between the points at which it has the values 0 and $-2\pi B/L$ may be estimated by evaluating at $(z - z^*)/L = \frac{1}{8}$ so that (Fig. 12.4)

$$\left|\frac{\partial(p'/\rho_0)}{\partial z}\right| \sim \frac{2\pi}{L} B \frac{\sqrt{2}}{2} \sim \frac{4}{L}\left|\frac{p'}{\rho_0}\right| \sim \frac{|p'/\rho_0|}{L_{p'}} \tag{19}$$

in which we use the symbol $\sim$ to denote order of magnitude and where the scale $L_{p'} = L/4$. In Chap. 4 we defined the scale height $H^{-1} = -\partial \ln \rho_0/\partial z$, which has a value of about 8 km for the atmosphere. Thus the ratio of the first two terms of Eq. (16) is

$$\left|\frac{\partial(p'/\rho_0)/\partial z}{p'/\rho_0 H}\right| \sim \frac{4H}{L} \tag{20}$$

Therefore if we apply our results only to waves whose vertical oscillations have wavelengths of less than, say, $0.4H$, we may neglect the second term of Eq. (16) and for an approximate hydrostatic equation use the result

$$\frac{\partial}{\partial z}\left(\frac{p'}{\rho_0}\right) - g\frac{\theta'}{\theta_0} = 0 \tag{21}$$

The remaining equations, linearized as before, are

$$\frac{\partial u}{\partial t} + \frac{1}{\rho_0}\frac{\partial p'}{\partial x} = 0 \tag{22}$$

$$\frac{\partial \theta'/\theta_0}{\partial t} + \frac{w}{\theta_0}\frac{\partial \theta_0}{\partial z} = 0 \tag{23}$$

and

$$\frac{\partial u}{\partial x} + \frac{\partial w}{\partial z} = 0 \tag{24}$$

We shall let $p'/\rho_0 = P$ and $\theta'/\theta_0 = \Theta$, and we shall denote the amplitude of a variable ϕ by $\hat{\phi}$. Then we give all variables a trial wave form defined by

$$\frac{u}{\hat{u}} = \frac{w}{\hat{w}} = \frac{\Theta}{\hat{\Theta}} = \frac{P}{\hat{P}} = Ae^{i(\omega t + \kappa_1 x + \kappa_3 z)} \tag{25}$$

in which A is a constant, possibly complex. Thus the equations become

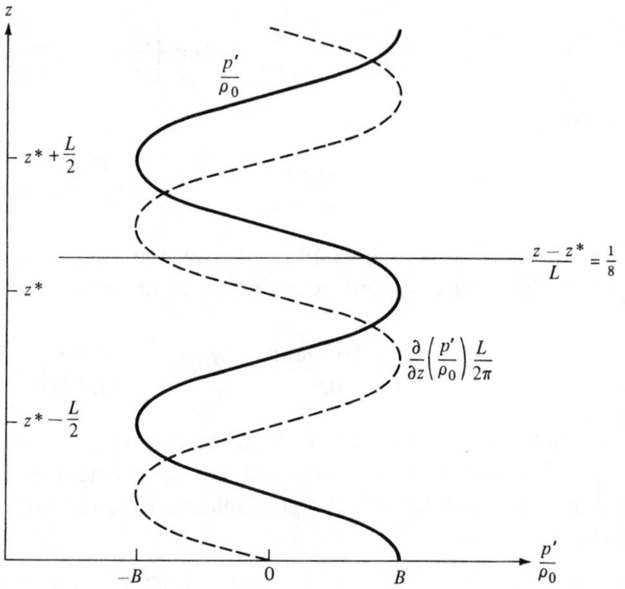

FIGURE 12.4
Estimation of the derivative midway between its values 0 and $-2\pi B/L$.

$$\begin{aligned} i\omega\hat{u} + i\kappa_1\hat{P} &= 0 \\ i\kappa_3\hat{P} - g\hat{\Theta} &= 0 \\ i\omega\hat{\Theta} + \frac{\hat{w}\omega_g^{\,2}}{g} &= 0 \\ i\kappa_1\hat{u} + i\kappa_3\hat{w} &= 0 \end{aligned} \tag{26}$$

in which we have defined the Brunt-Väisälä frequency as $\omega_g^{\,2} = (g/\theta_0)\,\partial\theta_0/\partial z$; by assumption, it is constant, thus permitting the solution (26) with constant amplitudes.

Thus we have four homogeneous, linear equations with the four amplitudes as unknowns. In order for a solution to these equations to exist, it is necessary that the determinant of the coefficients vanish. This condition, which we may write explicitly as

$$\begin{array}{cccc} \hat{u} & \hat{w} & \hat{P} & \hat{\Theta} \end{array}$$
$$\begin{vmatrix} i\omega & 0 & i\kappa_1 & 0 \\ 0 & 0 & i\kappa_3 & -g \\ 0 & \dfrac{\omega_g^{\,2}}{g} & 0 & i\omega \\ i\kappa_1 & i\kappa_3 & 0 & 0 \end{vmatrix} = 0 \tag{27}$$

will produce a characteristic equation. The result of evaluating the determinant is

$$\omega^2 = \frac{\omega_g^2 \kappa_1^2}{\kappa_3^2} = \frac{\omega_g^2 L_3^2}{L_1^2} \tag{28}$$

where L_3 and L_1 are the wavelengths in the vertical and horizontal respectively. We shall see in the next section that this result is valid in a more general case for low frequencies. For now we will assume that $L_3 \ll L_1$, and the phase velocity is in the direction above the horizontal specified by Eq. (13) in Sec. 12.1 as

$$\theta = \tan^{-1} \frac{-\omega\kappa_3}{-\omega\kappa_1} \tag{29}$$

With L_1 much greater than L_3, the phase propagation is nearly vertical. The group velocity is

$$\mathbf{c}_g = -\nabla_\kappa \omega = -\omega_g \nabla_\kappa \frac{\kappa_1}{\kappa_3} = \frac{\omega_g}{\kappa_3^2}(-\mathbf{i}\kappa_3 + \mathbf{k}\kappa_1) \tag{30}$$

The significance of this result is that the group velocity is predominantly horizontal when the phase propagation is nearly vertical (see Fig. 12.5). As we shall show later, the energy propagation due to gravity wave motion is in the direction of the group velocity.

PROBLEM 12.3.2 Find a necessary and sufficient condition for the gravity wave motion described here to be unstable.

PROBLEM 12.3.3 Show that the trajectories in gravity wave motion are straight lines inclined from the horizontal.

PROBLEM 12.3.4 Using Eqs. (21) to (24), insert a stream function for the velocities, and reduce the system to one equation in the stream function. Taking the real part of the solution for the stream function, show that $\psi = \psi_0 \cos(\omega t + \kappa_1 x + \kappa_3 z + \gamma)$, where ψ_0 is the amplitude and γ is the phase angle. Sketch the pattern made by this stream function in the xz plane and with the sketch illustrate that the pattern is consistent with the result of the preceding problem.

Inertial waves We have already considered, in discussing the geostrophic wind in Chap. 9, the properties of local inertial oscillations of the wind. Now we show that the Coriolis force permits a pure form of wave motion. Neglecting all other forces, we have for the equations of motion

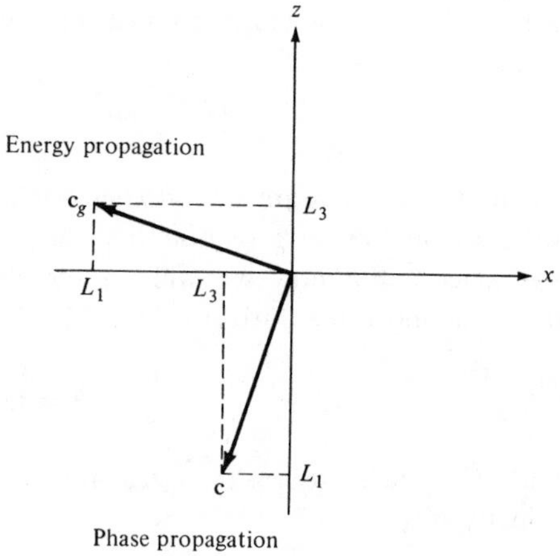

FIGURE 12.5
Relation between the phase velocity and group velocity for internal gravity waves.

$$\begin{aligned}\frac{\partial u}{\partial t} - 2\Omega(v \sin\phi - w\cos\phi) &= 0\\ \frac{\partial v}{\partial t} + 2\Omega u \sin\phi &= 0\\ \frac{\partial w}{\partial t} - 2\Omega u \cos\phi &= 0\end{aligned} \tag{31}$$

We try three-dimensional wave solutions of the form

$$\frac{u}{\hat{u}} = \frac{v}{\hat{v}} = \frac{w}{\hat{w}} = Ae^{i(\omega t + \boldsymbol{\kappa}\cdot\mathbf{x})}$$

and from the determinant of the equations for the amplitudes, the characteristic equation is

$$\begin{array}{ccc}\hat{u} & \hat{v} & \hat{w}\end{array}$$
$$\begin{vmatrix} i\omega & -2\Omega\sin\phi & 2\Omega\cos\phi \\ 2\Omega\sin\phi & i\omega & 0 \\ -2\Omega\cos\phi & 0 & i\omega \end{vmatrix} = -i\omega^3 + i\omega 4\Omega^2(\cos^2\phi + \sin^2\phi) = 0 \tag{32}$$

Hence

$$\omega^2 = 4\Omega^2 \tag{33}$$

From the theory of equations we know that the amplitudes are given by a constant B times a set of cofactors of the determinant in Eq. (32). We choose the cofactors of the first row and find that

$$\begin{aligned}\hat{u} &= -B\omega^2 = -B4\Omega^2\\ \hat{v} &= -iB\omega 2\Omega\sin\phi = -i4B\Omega^2\sin\phi\end{aligned} \tag{34}$$

$$\hat{w} = i\omega B 2\Omega \cos\phi = i4B\Omega^2 \cos\phi \tag{34}$$
(*cont'd*)

Inertial waves, then, have amplitudes varying as a function of latitude, with the vertical component vanishing at the poles and the meridional component vanishing at the equator.

The phase speed of the waves is

$$c = -\frac{\omega}{\kappa} = \mp\frac{2\Omega}{\kappa} = \mp\frac{2\Omega}{(\kappa_1^2 + \kappa_2^2 + \kappa_3^2)^{1/2}} \tag{35}$$

so that the speed increases linearly with increasing wavelength. The group velocity vanishes, and thus inertial waves do not propagate energy.

PROBLEM 12.3.5 Compare the phase speeds of inertial and gravity waves. For what wavelengths are they identical?

Rossby-Haurwitz waves We have already considered Rossby waves in chap. 10, where we derived the characteristic equation with the aid of a vorticity equation. Here we study a more general case, applying wave analysis methods to a rotating fluid with density varying only in the vertical. We shall restrict the analysis to horizontal and incompressible motion and consider the wave motion as superimposed on a constant current U in the x direction. Thus the linearized equations of motion are (with p a perturbation pressure)

$$\begin{aligned}
&\frac{\partial u}{\partial t} + U\frac{\partial u}{\partial x} - fv + \frac{\partial}{\partial x}\frac{p}{\rho} = 0 \\
&\frac{\partial v}{\partial t} + U\frac{\partial v}{\partial x} + fu + \frac{\partial}{\partial y}\frac{p}{\rho} = 0 \\
&\frac{\partial u}{\partial x} + \frac{\partial v}{\partial y} = 0
\end{aligned} \tag{36}$$

in which f is the Coriolis parameter.

On the assumption that the waves propagate in the x direction only, we try solutions of the form

$$\frac{u}{\hat{u}(y)} = \frac{v}{\hat{v}(y)} = \frac{P}{\hat{P}(y)} = Ae^{i(\omega t + \kappa x)} \tag{37}$$

where $P = p/\rho$. Thus the equations become

$$\begin{aligned}
&i(\omega + \kappa U)\hat{u} - f\hat{v} + i\kappa\hat{P} = 0 \\
&i(\omega + \kappa U)\hat{v} + f\hat{u} + \frac{\partial \hat{P}}{\partial y} = 0 \\
&i\kappa\hat{u} + \frac{\partial \hat{v}}{\partial y} = 0
\end{aligned} \tag{38}$$

Elimination of P between the first two equations and the use of the incompressibility condition give

$$(\omega + \kappa U)\left(-\frac{\partial^2 \hat{v}}{\partial y^2} + \kappa^2 \hat{v}\right) - \kappa\beta\hat{v} = 0 \tag{39}$$

A condition that $c = -\omega/\kappa$ not be equal to U is required to prevent the first factor from vanishing. With this restriction, we have

$$\frac{\partial^2 \hat{v}}{\partial y^2} + \left(\frac{\beta}{U - c} - \kappa^2\right)\hat{v} = 0 \tag{40}$$

Provided that the factor in parentheses is positive, the solutions to this equation will be harmonic functions. Thus

$$\hat{v} = C_1 \cos\left[\left(\frac{\beta}{U - c} - \kappa^2\right)^{1/2} y\right] + C_2 \sin\left[\left(\frac{\beta}{U - c} - \kappa^2\right)^{1/2} y\right] \tag{41}$$

We choose a channel of width D and require that $|\hat{v}|$ be maximum at $y = 0$ and that $\hat{v}$ vanish at $y = \pm D/2$.

PROBLEM 12.3.6 Show that these boundary conditions require that

$$C_2 = 0$$

$$\left(\frac{\beta}{U - c} - \kappa^2\right)^{1/2} \frac{D}{2} = \frac{\pi n}{2} \qquad n = \pm 1, \pm 3, \ldots$$

From the second relation we find the characteristic equation first derived by Prof. Bernhard Haurwitz:

$$c = U - \frac{\beta}{\kappa^2 + (n\pi/D)^2} = U - \frac{\beta L^2/4\pi^2}{1 + (nL/2D)^2} \tag{42}$$

We note that as D approaches infinity, the Rossby wave formula is given as a special case. These Haurwitz waves are illustrated in Fig. 12.6.

PROBLEM 12.3.7 Compare the predictions of the Rossby-Haurwitz wave theory for the cases in which $D = \infty$ and in which D represents the distance between 40 and 55° latitude. Consider both phase speeds and group speeds. [Hint: Recall that observations show that the wave speed is about one-half the wind speed. A good estimate is $U = 20$ m/s at long waves, and you may find the simplification that results when $(L/2D)^2 \gg 1$ is useful.]

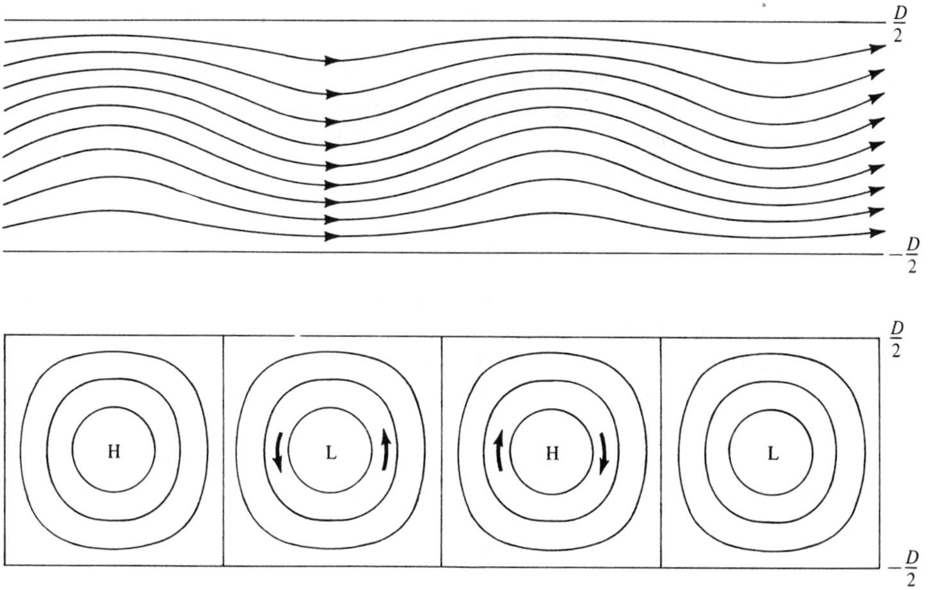

FIGURE 12.6
Streamlines for Haurwitz waves. The lower figure shows the pattern obtained with $U = 0$. The upper pattern is a superposition of the lower pattern and a zonal velocity $U = \text{const}$. As the speed U increases relative to the amplitude of the perturbation field, the streamlines in the upper pattern become less disturbed from a purely straight flow.

*PROBLEM 12.3.8 The westward winds in the tropics are retarded by friction at the earth's surface, thus giving up westward momentum or, equivalently, gaining eastward momentum. The westerlies of mid-latitudes are similarly losing eastward momentum. Thus there must be a transfer of eastward momentum from the tropics to mid-latitudes. Let the momentum transfer rate be $\overline{\rho_0 uv}$ where $\rho_0 = \rho_0(z)$ and the overbar denotes an average around a latitude circle at a fixed time and altitude. Show that $\overline{\rho_0 uv} = 0$ for Rossby-Haurwitz waves by integration over one wavelength $L = 2\pi/\kappa$. Thus we can conclude that sinusoidal waves of this type will not accomplish the necessary transport. Sketch an asymmetric wave form that will.

12.4 SOUND AND GRAVITY WAVES IN AN ISOTHERMAL ATMOSPHERE

In the previous section we considered acoustic and gravity waves separately, obtaining them from different reduced forms of the equations of motion. Now we shall see how the equations of motion provide a separation between these two modes of wave propagation in an isothermal, inviscid atmosphere in which the waves are the only motion. Waves in the atmosphere are strongly affected by both variations in stability and wind speed, as we shall see in Sec. 13.4. Still the analysis here will reveal some of the important properties of atmospheric wave motion, and perhaps more importantly, it will allow us to discover some of the main principles involved in filtering unwanted waves from numerical solutions.

12.4.1 Derivation of the Characteristic Equation

We shall denote the values of variables in the quiet, hydrostatic atmosphere with a subscript of zero. We define the scale height $H^{-1} = -\partial \ln \rho_0/\partial z$ and because of the isothermal condition we have $H^{-1} = g/RT_0 \sim (8 \text{ km})^{-1}$.

For the quiet isothermal state, we have $\rho_0(z) = \rho_0(0) \exp(-z/H)$, $p_0(z) = p_0(0) \exp(-z/H)$, and $\theta_0(z) = T_0 \exp(Rz/c_pH)$.

It is convenient to define the ratio of specific heats as $\gamma = c_p/c_V$ and from the previous section we know that the speed of sound is given by $c_0{}^2 = \gamma RT_0$ and hence

$$H = \frac{c_0{}^2}{g\gamma} \tag{1}$$

The equation of motion for inviscid flow is

$$\frac{\partial \mathbf{u}}{\partial t} + \mathbf{u} \cdot \nabla \mathbf{u} = -\alpha \nabla p - g\mathbf{k} - 2\boldsymbol{\Omega} \times \mathbf{u} \tag{2}$$

Exactly as in Eq. (19) in Sec. 12.3, we estimate the local derivative by

$$\left|\frac{\partial \mathbf{u}}{\partial t}\right| \sim \frac{4}{T} |\mathbf{u}| \tag{3}$$

in which T is the period of the motion.

The ratio of the Coriolis force and the temporal derivative is

$$\left|\frac{2\boldsymbol{\Omega} \times \mathbf{u}}{\partial \mathbf{u}/\partial t}\right| = \frac{2\Omega T}{4} = \frac{4\pi}{\text{day}} \frac{T}{4} \tag{4}$$

From Eq. (28) in Sec. 12.3 we find for gravity waves that

$$T = \frac{2\pi}{\omega} = \frac{2\pi L_1}{L_3 \omega_g} = \frac{2\pi L_1}{L_3(g\, \partial\theta_0/\theta_0\, \partial z)^{1/2}} \tag{5}$$

It is easily shown that $g\, \partial \ln \theta_0/\partial z = gR/c_pH$ in isothermal conditions and thus $T = 330(L_1/L_3)$ s. From this it follows that we may ignore the Coriolis force term provided that $L_1 \ll 10^2 L_3$.

PROBLEM 12.4.1 Show that Coriolis force effects on horizontally propagating acoustic waves may be neglected provided that $L \ll 8000$ km.

In both cases then, we can ignore the rotation of the earth provided that the wavelengths of the motion are sufficiently small. With this restriction, the equation of motion becomes

$$\frac{\partial \mathbf{u}}{\partial t} + \mathbf{u} \cdot \nabla \mathbf{u} = -\alpha \nabla p - g\mathbf{k} \tag{6}$$

and we define perturbation quantities associated with the wave motion as before by $p = p_0(z) + p'$ and $\alpha = \alpha_0(z) + \alpha'$.

Now we may use the hydrostatic relation in the basic state to obtain

$$\begin{aligned} \alpha \nabla p + g\mathbf{k} &= \alpha_0\left(1 + \frac{\alpha'}{\alpha_0}\right)\nabla p' - g\frac{\alpha'}{\alpha_0}\mathbf{k} \\ &= \frac{c_0^2}{\gamma}\nabla\left(\frac{p'}{p_0}\right) + g\left(\frac{\rho'}{\rho_0} - \frac{p'}{p_0}\right)\mathbf{k} \end{aligned} \tag{7}$$

in which we have linearized by neglecting products of perturbations and used the fact that to first order

$$\frac{\rho'}{\rho_0} = -\frac{\alpha'}{\alpha_0} \tag{8}$$

PROBLEM 12.4.2 Establish Eq. (8).

Therefore the linearized equation of motion is

$$\frac{\partial \mathbf{u}}{\partial t} + \frac{c_0^2}{\gamma}\nabla\frac{p'}{p_0} + g\left(\frac{\rho'}{\rho_0} - \frac{p'}{p_0}\right)\mathbf{k} = 0 \tag{9}$$

The continuity equation may be written upon linearization as

$$\frac{\partial}{\partial t}\left(\frac{\rho'}{\rho_0}\right) - \frac{w}{H} + \nabla \cdot \mathbf{u} = 0 \tag{10}$$

and the linearized first law is

$$\frac{\partial}{\partial t}\left(\frac{T'}{T_0}\right) - \frac{R}{c_V}\left[\frac{\partial}{\partial t}\left(\frac{\rho'}{\rho_0}\right) - \frac{w}{H}\right] = 0 \tag{11}$$

and so with the perturbation equation of state,

$$\frac{p'}{p_0} = \frac{\rho'}{\rho_0} + \frac{T'}{T_0} \tag{12}$$

which follows from Eq. (10) in Sec. 12.3 upon linearization, we have

$$\frac{\partial}{\partial t}\left(\frac{p'}{p_0}\right) - \gamma\frac{\partial}{\partial t}\left(\frac{\rho'}{\rho_0}\right) + (\gamma - 1)\frac{w}{H} = 0 \tag{13}$$

as an alternative form of the linearized first law. Thus Eqs. (9), (10), and (13) form a complete set in the variables $\mathbf{u}$, p'/p_0, and ρ'/ρ_0.

Upon denoting u by u_1, v by u_2, w by u_3, p'/p_0 by P, and ρ'/ρ_0 by Q, we may use amplitudes identified for variable ϕ by $\hat{\phi}$ to form trial solutions:

$$\frac{u_1}{\hat{u}_1} = \frac{u_2}{\hat{u}_2} = \frac{u_3}{\hat{u}_3} = \frac{P}{\hat{P}} = \frac{Q}{\hat{Q}} = Ae^{i(\boldsymbol{\kappa}\cdot\mathbf{x}+\omega t)} \tag{14}$$

Here A is a possibly complex constant and $\boldsymbol{\kappa} = \kappa_1\mathbf{i} + \kappa_2\mathbf{j} + \kappa_3\mathbf{k}$ is a three-dimensional vector wavenumber.

Upon substitution of Eq. (14) into the equations, we obtain the matrix of coefficients

$$D = \begin{array}{c} \\ \\ \\ \\ \\ \\ \end{array}\begin{array}{ccccc} \hat{u}_1 & \hat{u}_2 & \hat{u}_3 & \hat{P} & \hat{Q} \end{array}$$

$$D = \left\{\begin{array}{ccccc} i\omega & 0 & 0 & \dfrac{i\kappa_1 c_0^2}{\gamma} & 0 \\ 0 & i\omega & 0 & \dfrac{i\kappa_2 c_0^2}{\gamma} & 0 \\ 0 & 0 & i\omega_h & \dfrac{i\kappa_3 c_0^2}{\gamma} - g & g \\ 0 & 0 & (\gamma-1)H^{-1} & i\omega & -i\gamma\omega \\ i\kappa_1 & i\kappa_2 & i\kappa_3 - H_c^{-1} & 0 & i\omega_c \end{array}\right\} \begin{array}{l} u \text{ component} \\ v \text{ component} \\ w \text{ component} \\ \text{Eq. (13)} \\ \text{Eq. (10)} \end{array} \tag{15}$$

(columns: $\hat{u}_1$, $\hat{u}_2$, $\hat{u}_3$, $\hat{P}$, $\hat{Q}$)

In Eq. (15) we have specifically identified the frequency arising in the equation for the vertical component by subscript h; for hydrostatic conditions we will have $\omega_h = 0$. Similarly for incompressible conditions, $\nabla \cdot \mathbf{u} = 0$ and so the frequency identified as ω_c and the scale height H_c would not appear in the last row of the matrix. Thus we can ascertain the effect of assuming either hydrostatic equilibrium or incompressible flow by eliminating these specially identified quantities from the matrix of coefficients.

In order for a solution to exist, the determinant of the matrix must vanish, and this condition produces the characteristic equation. To evaluate it, we use the method of diagonalization, multiplying the first row by $-\kappa_1/\omega$ and adding it to the last row and multiplying the second row by $-\kappa_2/\omega$ and also adding it to the last row. Then the definition $\kappa^2 = \kappa_1{}^2 + \kappa_2{}^2$, use of Eq. (1) to replace H, and identification of the term with H_c by γ_c after the substitution, give the characteristic equation

$$\omega\omega_h\omega_c - \gamma\omega\left(\frac{\kappa_3 c_0^2}{\gamma} + ig\right)\left(\kappa_3 + \frac{ig\gamma_c}{c_0^2}\right) + \frac{g^2(\gamma-1)\kappa^2}{\omega} + ig\omega\left(\kappa_3 + \frac{ig\gamma_c}{c_0^2}\right) - \omega_h\kappa^2 c_0^2$$
$$+ \frac{i\omega_c(\gamma-1)g\gamma}{c_0^2}\left(\frac{\kappa_3 c_0^2}{\gamma} + ig\right) = 0 \tag{16}$$

In the most general case, in which we do not make the hydrostatic or incompressibility assumptions, we drop the subscripts c and h, and the characteristic equation becomes

$$\omega^4 - c_0^2\omega^2(\kappa^2 + \kappa_3^2) + g^2(\gamma - 1)\kappa^2 - ig\gamma\omega^2\kappa_3 = 0 \tag{17}$$

An obvious difficulty with this equation is that it has both real and imaginary parts. We want κ_1 and κ_2 to be real so that the waves are not changing in amplitude in the horizontal plane, and we also want ω to be real so that the waves are not growing or decaying in amplitude in the case we are now studying, in which we have provided no wind shear or hydrostatic instability as an energy source. The only remaining way to satisfy Eq. (17) is for the vertical wavenumber κ_3 to be complex.

PROBLEM 12.4.3 Put $\kappa_3 = \kappa_3{}^* + iI$ and show that in order for the imaginary part of Eq. (17) to vanish, we must have $I = -g\gamma/2c_0{}^2 = -1/2H$ and that the solutions are now of the form

$$\phi = \hat{\phi}Ae^{z/2H}e^{i(\kappa_1 x + \kappa_2 y + \kappa_3 z + \omega t)}$$

in which κ_3 now identifies the real part $\kappa_3{}^*$.

As shown by the problem, these neutrally stable waves have an amplitude that increases exponentially with altitude. But any mass-weighted quantity will be of the form

$$\rho_0\phi = \rho_0(0)\hat{\phi}Ae^{-z/2H}e^{i(\boldsymbol{\kappa}\cdot\mathbf{x} + \omega t)} \tag{18}$$

and will vanish as z approaches infinity. For example, if we take the real part of the solution and form the kinetic energy $\rho_0(z)\mathbf{u}\cdot\mathbf{u}$, then we will find that the average over the vertical wavelength L_3 is constant with height for each component of the wave motion.

Returning to the characteristic equation (17), we use Problem 12.4.3 to rewrite the real part in a revealing form obtained by judicious factoring. Thus

$$\omega^2\left[\omega^2 - \frac{g^2\gamma^2}{4c_0{}^2}\right] + c_0{}^2(\kappa^2 + \kappa_3{}^2)\left[\frac{g^2(\gamma - 1)\kappa^2}{c_0{}^2(\kappa^2 + \kappa_3{}^2)} - \omega^2\right] = 0 \tag{19}$$

The two factors in brackets must have opposite signs, and so the crucial terms are the two appearing in the brackets with ω^2. Their ratio is

$$\begin{aligned}\frac{g^2\gamma^2/4c_0{}^2}{g^2(\gamma - 1)\kappa^2/c_0{}^2(\kappa^2 + \kappa_3{}^2)} &= \frac{\gamma^2}{4(\gamma - 1)}\left(1 + \frac{\kappa_3{}^2}{\kappa^2}\right) \\ &= \frac{(1.4)^2}{1.6}\left(1 + \frac{\kappa_3{}^2}{\kappa^2}\right) > 1\end{aligned} \tag{20}$$

Thus there are only two possibilities. Either

$$\omega^2 > \frac{g^2\gamma^2}{4c_0{}^2} = \omega_a{}^2 \tag{21}$$

so that the first term is positive and the second negative, or

$$\omega^2 < \frac{g^2(\gamma - 1)\kappa^2}{c_0^{\,2}(\kappa^2 + \kappa_3^{\,2})} = \omega_g^{\,2}\frac{\kappa^2}{\kappa^2 + \kappa_3^{\,2}}$$ (22

so that the second term is positive and first is negative. In Eq. (22) we have use $\omega_g^{\,2} = g^2(\gamma - 1)/c_0^{\,2}$, in which ω_g is the Brunt-Väisälä frequency in the undisturbe atmosphere.

Thus Eqs. (21) and (22) show that the possible waves are separated into high-frequency mode specified by Eq. (21) and a low-frequency mode specified by E (22). The first are *acoustic waves* with frequencies greater than the bound ω_a; th second are *gravity waves* with frequencies less than ω_g. We may now write th characteristic equation as

$$(\omega^2 - \omega_a^{\,2})\frac{\omega^2}{c_0^{\,2}} - \omega^2(\kappa^2 + \kappa_3^{\,2}) + \omega_g^{\,2}\kappa^2 = 0$$ (2

The solutions to this equation are illustrated in the diagnostic diagram shown in Fi 12.7. Another representation is given in Fig. 12.8, where the direction of the grou velocity is also indicated.

PROBLEM 12.4.4 Show that for very-high-frequency waves $(\omega^2 \gg \omega_a^{\,2})$ th characteristic equation for acoustic waves is approximately $\omega^2 = c_0^{\,2}(\kappa^2 + \kappa_3^{\,2})$. [Hin Use Eq. (20) to show that $\omega_a^{\,2} > \omega_g^{\,2}\kappa^2(\kappa^2 + \kappa_3^{\,2})^{-1}$.]

PROBLEM 12.4.5 Show that for low-frequency waves $(\omega^2 \ll \omega_a^{\,2})$ the characterist equation is approximately $\omega^2 = \omega_g^{\,2}\kappa^2(\kappa^2 + \kappa_3^{\,2} + \omega_a^{\,2}/c_0^{\,2})^{-1}$. Simplify it f $L_3 \leqslant 10\text{ km} \ll 4\pi H$. Show that for either very low frequencies specified by $\omega^2 \ll \omega_g$ or for $L_3 \ll L = 2\pi/\kappa$, $\omega^2 = \omega_g^{\,2}\kappa^2/\kappa_3^{\,2} = \omega_g^{\,2}L_3^{\,2}/L^2$.

PROBLEM 12.4.6 Explain with suitable equations and sketch diagrams how t directions of phase and group propagation in Fig. 12.8 are determined.

PROBLEM 12.4.7 Express the characteristic equation as a function of phase speed and show that for fixed κ and κ_3 the speeds of acoustic waves are greater than tho of gravity waves. Construct a diagnostic (c,κ) diagram modeled after Fig. 12.7.

12.4.2 Soundproof Equations

The propagation of acoustic waves is not generally of interest in meteorologic problems, and in fact their presence is detrimental to practical methods of numeric

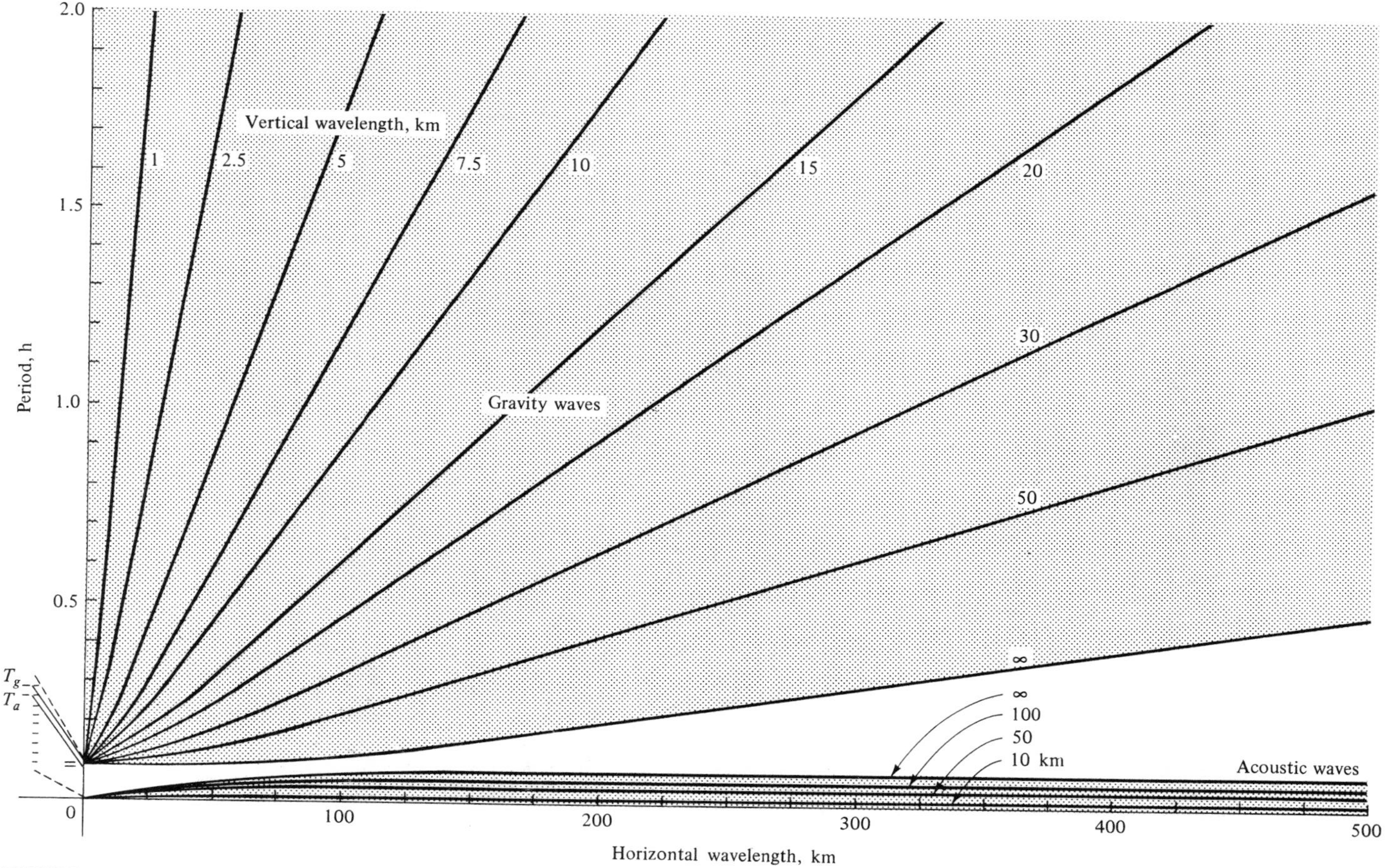

FIGURE 12.7
Diagnostic diagram showing the separation of gravity and acoustic waves. The computations were made for an isothermal atmosphere with $T_0 = 250$ K.

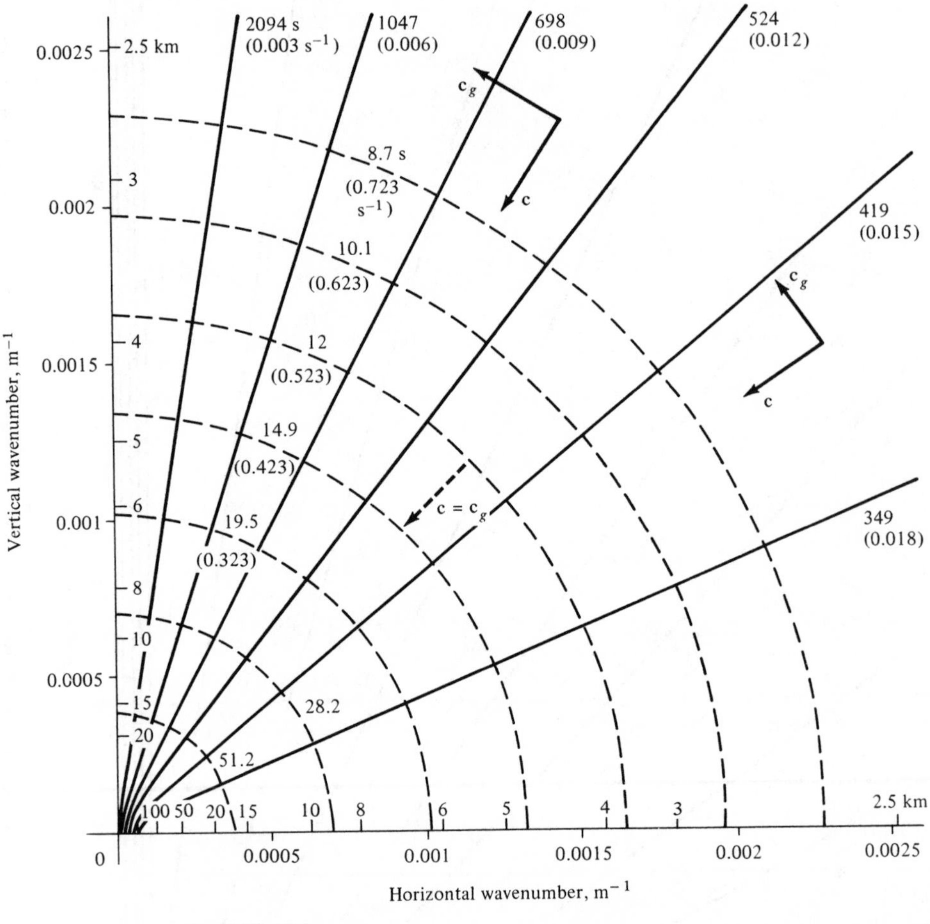

FIGURE 12.8
The period of gravity waves (solid hyperbolic lines) and acoustic waves (dashed circles). The direction of phase propagation for both waves at any point on the diagram is parallel to radial lines. For acoustic waves (dashed arrows), the group velocity is identical to the phase velocity; for gravity waves, it is orthogonal to the phase propagation vector, as shown. The angular frequencies ω are shown in parentheses under the period in seconds, and wavelengths are shown in kilometers on the inside of the axes. The diagnostic diagram is shown in the first quadrant; the entire diagram in all four quadrants is obtained by reflecting this diagram across the axes into the second, third, and fourth quadrants. The computations were made for an isothermal atmosphere with $T_0 = 250$ K.

weather prediction. Either the hydrostatic or incompressibility assumptions are sufficient to eliminate acoustic waves as solutions, and thus either assumption will produce a set of soundproof equations of motion.

In Sec. 12.3 we saw that sound propagates as longitudinal, compressional waves. Thus we would certainly expect that invoking incompressibility would prevent them from occurring. To see precisely what wave motion is permitted with this assumption, we set ω_c and γ_c both equal to zero in Eq. (16) and thus obtain the characteristic equation

$$-\omega^2[(\kappa^2 + \kappa_3^2)c_0^2 + i(\gamma - 1)g\kappa_3] + g^2(\gamma - 1)\kappa^2 = 0 \tag{24}$$

Now we find that $I = -(\gamma - 1)g/2c_0^2$ will ensure that the characteristic equation is real, and we have

$$\omega^2 = \omega_g^2 \frac{\kappa^2}{\kappa^2 + \kappa_3^2 + [(\gamma - 1)/2\gamma H]^2} \tag{25}$$

We are again using κ_3 for $\kappa_3{}^*$ in Eq. (25) once the imaginary part of the equation is eliminated.

The last term in the denominator may be neglected if $L_3 \ll 4\pi\gamma H/(\gamma - 1) \sim 40H$ so that $L_3 \leqslant 4H \sim 40$ km is certainly sufficient to give the equation

$$\omega^2 = \omega_g^2 \frac{\kappa^2}{\kappa^2 + \kappa_3^2} \tag{26}$$

that controls low-frequency gravity wave propagation. Thus we have shown that when the flow is incompressible, then the linear equations of motion do not permit acoustic waves to exist in an isothermal, nonrotating atmosphere.

If we retain the compressibility but let the motion be hydrostatic, then ω_h vanishes and Eq. (16) reduces to

$$\omega^2\left(\kappa_3^2 + \frac{ig\gamma\kappa_3}{c_0^2}\right) - \omega_g^2\kappa^2 = 0 \tag{27}$$

Now we have $I = -(2H)^{-1}$ again, and Eq. (27) becomes (denoting $\kappa_3{}^*$ by κ_3)

$$\omega^2 = \omega_g^2 \frac{\kappa^2}{\kappa_3^2 + (1/2H)^2} \tag{28}$$

This again is a characteristic equation for motion of the gravity wave type. If $L_3 \ll 4\pi H \sim 100$ km or $L_3 \leqslant 10$ km, we have the simple form

$$\omega^2 = \omega_g^2\left(\frac{L_3}{L_1}\right)^2 \tag{29}$$

In summary, we see that the propagation of sound requires that the density be able to adjust to the compression and expansion of the wave. In the incompressible case this is clearly prohibited. In the hydrostatic case, the pressure and density are bound together by both the first law and the hydrostatic relation, and the freedom necessary for acoustic waves is eliminated.

These results have another implication that is often quoted. For gravity waves, we obtained the characteristic equation (26), which will reduce to Eq. (29) provided that $(L_3/L_1) \ll 1$. Thus for linear solutions to the equations, this condition that the vertical wavelengths are small compared to horizontal wavelengths is equivalent to the hydrostatic assumption. Therefore, because this assumption will produce the same characteristic equation as the hydrostatic assumption, the condition $L_3 \ll L_1$ is often used to justify making a hydrostatic approximation. We shall see later that for equations describing the motions of synoptic scale, this condition is indeed sufficient.

PROBLEM 12.4.8 Do the results of this section justify an assertion that the consequences of the hydrostatic and incompressible assumptions are the same either in general or for a restricted class of wavelengths?

12.4.3 The Polarization Relations, Energy Flux, and Group Velocity

In addition to the characteristic equation describing the relation between wavenumbe[r] and frequency, the matrix of coefficients allows us to deduce information about th[e] relative size of the amplitudes of the variables involved in the wave motion. Fo[r] simplicity, we assume that the x axis is aligned with the projection of the direction o[f] wave propagation on the horizontal plane. Then in the matrix of coefficients, Eq. (15[)] we may eliminate the second row and second column. Upon replacing the comple[x] vertical wavenumber with its real and imaginary parts, and using κ_3 for the real par[t] we have as the reduced matrix:

$$D_2 = \begin{pmatrix} \omega & 0 & \dfrac{\kappa_1 c_0^2}{\gamma} & 0 \\ 0 & \omega & \dfrac{\kappa_3 c_0^2}{\gamma} + \dfrac{ig}{2} & -ig \\ 0 & -i(\gamma-1)H^{-1} & \omega & -\gamma\omega \\ \kappa_1 & \kappa_3 + \dfrac{i}{2H} & 0 & \omega \end{pmatrix} \quad \begin{matrix} \hat{u}_1 & \hat{u}_3 & \hat{P} & \hat{Q} \end{matrix} \qquad (3[0])$$

We may denote the elements of this matrix by A_{ij} (first index = row second = column). The characteristic equation is derived from the necessary conditio[n] that $\det\{A_{ij}\} = 0$ for solutions to exist. From the theory of equations we know th[at] the solutions $\hat{\phi}_i$ are given for any appropriate j by

$$\hat{\phi}_i = KA^{ji} \qquad i = 1, 2, 3, 4; j \text{ fixed} \qquad (3$$

in which K is a constant and A^{ji} is the cofactor of the element A_{ji}. We shall take $j = 2$, so that the cofactor of the variable $\hat{\phi}_i$ is obtained from Eq. (30) by striking both the second row and the column headed $\hat{\phi}_i$, and then attaching the appropriate sign to the resulting determinant. The sign is determined exactly as it is in expanding a determinant. Thus we have the relations

$$\begin{aligned}\frac{\hat{u}_1}{K} &= \omega\kappa_1 c_0^2\left[\kappa_3 + \frac{ig}{c_0^2}\left(1 - \frac{\gamma}{2}\right)\right] \\ \frac{\hat{u}_3}{K} &= \omega(\omega^2 - \kappa_1^2 c_0^2) \\ \frac{\hat{P}}{K} &= -\omega^2\gamma\left[\kappa_3 + i\frac{g}{c_0^2}\left(1 - \frac{\gamma}{2}\right)\right] \\ \frac{\hat{Q}}{K} &= -\omega^2\left(\kappa_3 + i\frac{g\gamma}{2c_0^2}\right) + i\kappa_1^2 g(\gamma - 1)\end{aligned} \tag{32}$$

PROBLEM 12.4.9 Verify the relations given in Eq. (32). Show that (*a*) the magnitude of $\hat{u}_1$ is dominated by the real part in Eq. (32), if $L_3 \leqslant 3H$ so that $\hat{u}_1 = K\omega\kappa_1\kappa_3 c_0^2$; (*b*) for the sound waves, $\hat{u}_3 = K\omega c_0^2\kappa_3^2$; (*c*) for gravity waves, $\hat{u}_3 = -K\omega c_0^2\kappa_1^2$ if $L_3 \leqslant 4H$ and $L_3 \ll L_1$.

*PROBLEM 12.4.10 Show that for acoustic waves the parcels oscillate along a straight line in the direction of phase propagation, whereas for gravity waves the line along which the parcels oscillate is orthogonal to the direction of phase propagation.

PROBLEM 12.4.11 Devise a ground-based observational network and procedure using rawinsondes that would produce enough data so that in principle all the characteristics of gravity wave motion could be determined if a case of phase propagation at a single wavenumber vector were encountered in an isothermal, shear-free layer.

It is the oscillation of the parcels that makes the transport of energy possible, so that we expect sound and gravity waves to have quite different energy propagation characteristics.

As shown in Chap. 11, the equation of motion and the first law of thermodynamics for isentropic motion combine to yield the energy equation

$$\rho\frac{d}{dt}\left(\frac{\mathbf{u}\cdot\mathbf{u}}{2} + c_V T + gz\right) = -\nabla\cdot p\mathbf{u} \tag{33}$$

so that the quantity $p\mathbf{u}$ is an *energy flux vector*, and it determines in which direction and at what rate energy is being transported by the motion.

Upon using the results of Problem 12.4.9 to simplify the polarization relations for $\hat{u}_1$ and $\hat{P}$, we have $\hat{P} = -K\omega^2\gamma\kappa_3$ and $\hat{u}_1 = K\omega\kappa_1\kappa_3 c_0{}^2$, and we have $\hat{u}_3 = K\omega c_0{}^2 k^2$ in which $k^2 = \kappa_3{}^2$ for acoustic waves and $-\kappa_1{}^2$ for gravity waves.

Now we have

$$\begin{Bmatrix} u_1 \\ u_3 \end{Bmatrix} = \begin{Bmatrix} \kappa_1\kappa_3 \\ k^2 \end{Bmatrix} \mathrm{Re}(AK\omega c_0{}^2 e^{i(\omega t + \kappa_1 x + \kappa_3 z)}) e^{z/2H} \tag{34}$$

We let $AK = U_R + iU_I$, where R and I denote real and imaginary parts and we set $\chi = \omega t + \kappa_1 x + \kappa_3 z$. Thus

$$\begin{Bmatrix} u_1 \\ u_3 \end{Bmatrix} = \begin{Bmatrix} \kappa_1\kappa_3 \\ k^2 \end{Bmatrix} \omega c_0{}^2 (U_R \cos\chi - U_I \sin\chi) e^{z/2H} \tag{35}$$

and for the pressure $p = p_0 + p'$ we have

$$\begin{aligned} p &= p_0(0)e^{-z/2H}[e^{-z/2H} + \mathrm{Re}(\hat{P}Ae^{i\chi})] \\ &= p_0(0)e^{-z/2H}[e^{-z/2H} - \omega^2\gamma\kappa_3(U_R\cos\chi - U_I\sin\chi)] \end{aligned} \tag{36}$$

We may calculate the energy flux $\mathbf{F}$ at a point (x,z) by integrating over one cycle of the wave of period $T = 2\pi/\omega$. Thus

$$\mathbf{F} = \langle p\mathbf{u}\rangle = \frac{1}{T}\int_0^T p\mathbf{u}\, dt \tag{37}$$

We use the easily derived results that $\langle\cos\chi \sin\chi\rangle = 0$ and $\langle\cos^2\chi\rangle = \langle\sin^2\chi\rangle = \frac{1}{2}$ to obtain

$$\begin{Bmatrix} F_1 \\ F_3 \end{Bmatrix} = \begin{Bmatrix} \langle pu_1\rangle \\ \langle pu_3\rangle \end{Bmatrix} = -\begin{Bmatrix} \kappa_1\kappa_3 \\ k^2 \end{Bmatrix} \omega^3\gamma\kappa_3 c_0{}^2 \frac{U_R{}^2 + U_I{}^2}{2} p_0(0) \tag{38}$$

Thus for $\omega > 0$

$$\frac{F_3}{F_1} = \frac{-\kappa_3 k^2}{-\kappa_1\kappa_3{}^2} = \begin{cases} \dfrac{-\kappa_3}{-\kappa_1} & \text{for acoustic waves} \\[2ex] \dfrac{\kappa_3\kappa_1{}^2}{-\kappa_1\kappa_3{}^2} & \text{for gravity waves} \end{cases} \tag{39}$$

If $\omega < 0$, the signs in both numerator and denominator change.

For sound waves the direction of energy propagation is the same as the direction of phase propagation. But for gravity waves, these directions are orthogonal. Inspection of the ratios in Eq. (29) in Sec. 12.3 and Eq. (39) show that if $\omega > 0$ and $\kappa_3 < 0$, then the phase propagation will have an upward component. If $\kappa_3 > 0$, the phase propagates with a downward component and energy with an upward component (see Figs. 12.5 and 12.8).

Finally, we consider the group velocity. For the high-frequency acoustic waves we have the characteristic equation specified in Problem 12.4.4, and by differentiation

$$\omega \nabla_\kappa \omega = c_0^2(\mathbf{i}\kappa_1 + \mathbf{k}\kappa_3) \tag{40}$$

We find that

$$\mathbf{c}_g = -\frac{c_0^2}{\omega}(\mathbf{i}\kappa_1 + \mathbf{k}\kappa_3) \tag{41}$$

Thus the energy propagates in the direction of the group velocity.

For low-frequency gravity waves with $L_3 \ll L_1$ we use the characteristic equation $\omega^2 = \omega_g^2\kappa^2/\kappa_3^2$ and so

$$\mathbf{c}_g = -\frac{\omega_g^2}{\omega}\left(\mathbf{i}\frac{\kappa_1}{\kappa_3^2} - \mathbf{k}\frac{\kappa_1^2}{\kappa_3^3}\right) \tag{42}$$

The direction of group propagation becomes (for $\omega > 0$)

$$\theta = \arctan\left(\frac{\mathbf{c}_g \cdot \mathbf{k}}{\mathbf{c}_g \cdot \mathbf{i}}\right) = \arctan\left(\frac{\kappa_1^2/\kappa_3^3}{-\kappa_1/\kappa_3^2}\right) = \arctan\left(\frac{\kappa_1^2\kappa_3}{-\kappa_1\kappa_3^2}\right) \tag{43}$$

in which we have multiplied both numerator and denominator by κ_3^4, which does not affect the quadrant in which the angle falls. For gravity waves, too, the direction of group propagation and energy flux coincide.

12.4.4 Comment on Boundary Conditions

All the analysis presented in this section has proceeded unhampered by the reality forced upon the solutions by boundary conditions. There are two possible approaches.

In the first, we let the amplitudes appearing in Eq. (14) be functions of z and we set $\kappa_3 = 0$ in the exponential factor. Thus we do not presume initially that only harmonic solutions are possible. Then Eqs. (9), (10), and (13) can be applied to the new trial solutions and a second-order equation in $\hat{u}_3(z)$ will result. This equation will have as solutions functions that permit us to have harmonic variations in the vertical that satisfy the boundary condition $\hat{u}_3(0) = 0$. We would then find that $\hat{u}_3 = We^{z/2H} \sin \kappa_3 z$, and so

$$w = \mathrm{Re}(We^{z/2H} \sin \kappa_3 z \, e^{i(\kappa_1 x + \kappa_2 y + \omega t)}) \tag{44}$$

where W is a complex constant.

Upon returning to the equations we can find explicit expressions for $\hat{P}$, $\hat{u}_1$, and $\hat{u}_2$. But we note that according to Eq. (44) the wave forms propagate only in the horizontal. Moreover, the analysis of the trajectories shows that they are ellipses that have a complicated dependence on height. And finally, to compute vertical energy flux, we must take into account the combined solutions that arise from adding the basic solutions resulting from the multiple roots of the characteristic equation.

An equivalent approach, discussed by Eckart (1960), is to combine solutions of the form (14) as

$$\phi = \hat{\phi}(e^{i(\omega t + \kappa_1 x + \kappa_2 y + \kappa_3 z)} - e^{i(\omega t + \kappa_1 x + \kappa_2 y - \kappa_3 z)}) \tag{45}$$

so that $\phi = 0$ when $z = 0$. Now we have a downward propagating wave in effect reflecting at the earth's surface into an upward propagating wave. Thus the vertical energy flux due to the basic form (45) vanishes, and we must rely again on the multiple roots of the characteristic equation to produce solutions with vertical fluxes.

The point here is that the assumptions of an isothermal atmosphere without wind shear render the problem sufficiently different from the actual case that this sort of complicated analysis does not seem warranted. Actual gravity waves in the atmosphere are affected strongly by variations in stability and wind shear. Moreover they can form on free surfaces at inversions, thus requiring a boundary condition other than $\hat{u}_3(0) = 0$.

Thus we may view the analysis presented in this section as revealing how acoustic and gravity waves differ, and as revealing the main properties of the phase and energy propagation associated with the basic wave forms. The details of what will happen in an individual case depend on both boundary and initial conditions, and it will probably turn out that we are most interested in how wind shear and stability variations affect the waves anyway.

PROBLEM 12.4.12 Carry out the analysis described as the first alternative in this section. In particular, verify Eq. (44) as the solution, determine the characteristic equation, and show that the trajectories are height-dependent ellipses.

BIBLIOGRAPHIC NOTES

12.1 This section contains standard material.

12.2–12.3 Further information about pure wave types and external gravity waves may be found in:

Eckart, Carl, 1960: *Hydrodynamics of Oceans and Atmospheres*, Pergamon Press, New York, 290 pp.

Haltiner, G. J., 1971: *Numerical Weather Prediction*, John Wiley & Sons, Inc., New York, 317 pp.

Thompson, P. D., 1961: *Numerical Weather Analysis and Prediction*, The Macmillan Company, New York, 170 pp.

Water waves are discussed in considerable detail by:

Kinsman, Blair, 1965: *Wind Waves, Their Generation and Propagation on the Ocean Surface*, Prentice-Hall, Inc., Englewood Cliffs, N.J., 676 pp. Reprinted by Dover Publications, Inc., New York, 1984.

Phillips, O. M., 1966: *The Dynamics of the Upper Ocean*, Cambridge University Press, Cambridge, 261 pp.

12.4 The material in this section derives from the article:

Hines, C. O., 1960: "Internal Atmospheric Waves at Ionospheric Heights," *Can. J. Phys.*, **38**:1441–1481.

Complications introduced by wind shear and stability variations are discussed by:

Booker, J. R., and F. P. Bretherton, 1967; "The Critical Layer for Internal Gravity Waves in a Shear Flow," *J. Fluid Mech.*, **27**:513–539.

13
APPROXIMATE EQUATIONS OF MOTION FOR SMALL-SCALE FLOW

As has been said many times in many ways, a difficulty with the meteorological equations of motion is that they are too complete and contain as solutions phenomena that may not be of interest in the study of a particular problem. For example, the linearized equations for small-scale motion include gravity and acoustic waves, even though we do not believe that acoustic phenomena are significant in the study of either gravity waves or convection. Still this is a small problem compared to the so far insurmountable difficulties caused by the nonlinearity of the equations. Despite the fact that all motions are nonlinear to some degree, certain phenomena may have scales and amplitudes such that the nonlinearity is not crucial and so that wavelike patterns are observed. Thus for nearly a century, students of fluid mechanics and atmospheric science have sought to reduce the complexity of the equations of motion and yet retain a reasonably accurate description of motion within certain limits of temporal and spatial scales.

In attempting to find such equations, we must do three things: First, we must ascertain what are the essential physical aspects of the motion we hope to study. Second, we must find ways of making approximations that are consistent in all the equations by using quantitative estimates based upon our physical perception of th problem. Third, we must determine for what scales the estimates accurately reflect physical reality.

In this chapter, we shall investigate approximate equations in which buoyancy forces are a significant part of the total forcing, either providing energy to the flow or providing an energy sink through stable stratification. We shall find that we must use slightly different equations for cases described as shallow and deep convection. Then, we shall go on to investigate the stability properties of the equations.

13.1 SCALE ANALYSIS OF THE EQUATIONS

In seeking approximate equations for small-scale flow, we must ascertain which terms in the equations are small compared to others. To do so, we shall compare both the amplitudes of the variations in the physical variables and the sizes of the derivatives that appear. As we have seen in the previous chapter, derivatives can be estimated as a ratio of the amplitude to a scale of variation.

The idea is to determine the relative sizes of the various terms in the equations by comparing amplitudes and scales, and then to neglect the smaller terms. We are thus invoking an assumption that small causes have small effects—a postulate that is not always valid in either fluid mechanics or meteorology.

Nevertheless, the systematic use of the analysis technique produces a consistent set of approximate equations for small-scale flow. Their notable feature is that they are linear in the thermodynamic variables, and that they change form depending on the vertical wavelength of the perturbation motion.

13.1.1 The Reference State and the Equation of State

We intend to consider the flows to be studied as composed of a reference flow whose properties vary only in the vertical and perturbation motions superimposed upon the reference flow. Because we wish now to emphasize the perturbation motions, we shall make the specification of the reference state as simple as possible. Hence we assume that the reference motion is steady, and that the effects of viscosity, heat conduction, and radiation are not important. Moreover, we do not consider the effects of the earth's rotation on the reference state. Thus it is possible to choose an arbitrary velocity profile $\mathbf{U} = \mathbf{i}U(z)$ and to consider the x axis as aligned with the direction of reference flow.

The equations of motion for such a flow reduce to $\nabla_z p_0 = 0$ and $\partial p_0/\partial z = -g\rho_0$ in which the subscript denotes values of the thermodynamic variables in the reference flow. As shown in Chap. 6, these conditions imply that $\nabla_z \alpha_0 = \nabla_z T_0 = 0$, and so the reference state is described by $U(z)$, $p_0(z)$, $\rho_0(z)$, and $T_0(z)$. These are arbitrary functions, provided they are sufficiently differentiable and provided they satisfy the hydrostatic condition and the equation of state. They may be chosen to represent typical ambient conditions in the atmosphere in a region of interest. The restriction to wind profiles with changes in speed only is not an essential one, but does simplify the calculations slightly.

In the vertical, the scale of variation of the wind field and the thermodynamic variables may be of the same order as the vertical scale of the perturbation motion, and interactions may be expected.

PROBLEM 13.1.1 Show that in a reference state in which viscous effects are important, the conditions $U = U(z)$ and $p_0 = p_0(z)$ are incompatible.

PROBLEM 13.1.2 Show that if the effects of rotation are retained, then the conditions $U = U(z)$ and $\nabla_z p_0 = 0$ are incompatible. What conditions must such a state satisfy?

We describe the sum of the reference and perturbation quantities with the variables

$$\mathbf{v} = \mathbf{i}U(z) + \mathbf{u}(x,y,z,t)$$
$$\psi = \psi_0(z) + \psi'(x,y,z,t) \qquad \psi = p, T, \alpha \tag{1}$$

in which the prime denotes a perturbation.

Substitution of the thermodynamic variables in the equation of state and use of the assumption that $p_0\alpha_0 = RT_0$ along with a restriction on the size of the fractional perturbation quantities so that the product term is much smaller than the others gives

$$\frac{p'}{p_0} + \frac{\alpha'}{\alpha_0} = \frac{T'}{T_0} \tag{2}$$

which we may make as accurate as we please by limiting the excursions we permit from reference values of the thermodynamic variables.

13.1.2 The Equations of Motion

In order to have a consistent method of estimating the size of various quantities that will appear, we shall use a Fourier representation for the variable ϕ of the form

$$\phi(x,y,z,t) = \hat{\phi}(\omega, \kappa_1, \kappa_2, z)e^{i(\omega t + \kappa_1 x + \kappa_2 y)} \tag{3}$$

and we shall denote equality of order of magnitude with the symbol $\sim$. Thus we have, for example,

$$\phi \sim |\hat{\phi}| \qquad \frac{\partial \phi}{\partial t} \sim |\omega\hat{\phi}| \qquad \frac{\partial \phi}{\partial x} \sim \kappa_1|\hat{\phi}| \tag{4}$$

Needless to say, we do not consider signs in estimating orders of magnitudes. For vertical derivatives we shall define a scale L_ϕ by

$$\frac{\partial \phi}{\partial z} \sim \left|\frac{\partial \phi}{\partial z}\right| = \frac{|\hat{\phi}|}{L_\phi} \tag{5}$$

The quantity L_ϕ may be related to a wavelength in the vertical as in Eq. (19) in Sec. 12.3 so that for $\kappa_3 = 2\pi/L_3$

$$\left|\frac{\partial \hat{\phi}}{\partial z}\right| \sim \frac{4|\hat{\phi}|}{L_3} \sim \frac{|\hat{\phi}|}{L_\phi} \tag{6}$$

where we have defined $L_\phi \sim L_3/4$. In comparing estimates, we shall always assume in this chapter that $L_{p'} = L_{\alpha'} = L_w$.

PROBLEM 13.1.3 Show that essentially the same estimate of the scale L_ϕ is obtained by using the root-mean-square of the derivative to estimate $|\partial\hat{\phi}/\partial z|$.

For convenience we assume inviscid flow now; the equations of motion are

$$\frac{\partial \mathbf{v}}{\partial t} + \mathbf{v} \cdot \nabla \mathbf{v} = -\alpha \nabla p - g\mathbf{k} - 2\mathbf{\Omega} \times \mathbf{v} \tag{7}$$

If there is no reference flow $U(z)$, then we have

$$\frac{\partial \mathbf{v}}{\partial t} \sim \omega|\mathbf{v}| \qquad 2\mathbf{\Omega} \times \mathbf{v} \sim 2\Omega|\mathbf{v}| \tag{8}$$

Thus if $\omega \gg 2\Omega$, we may ignore Coriolis force on the perturbation motion. Because $\omega = 2\pi/T$, the condition on the period T is that $T \ll 12$ h. In general, the symbol $\ll$ may be interpreted as requiring a factor of 10 or more. Thus we might say $T \leqslant 1.2$ h is sufficient for an accuracy within 10 percent. Henceforth we ignore Coriolis force.

Now we find as before that

$$-\alpha \nabla p - g\mathbf{k} = -\alpha_0\left(1 + \frac{\alpha'}{\alpha_0}\right) \nabla p' + g\frac{\alpha'}{\alpha_0}\mathbf{k} \tag{9}$$

In order to linearize this expression, we thus require that $|\alpha'/\alpha_0| \ll 1$. Because of the equation of state (2), this is a further restriction on p'/p_0 and T'/T_0; both quantities must be small enough to give Eq. (2) and also sufficiently small so that $|\alpha'/\alpha_0| \ll 1$ is satisfied. Thus the equation of motion is

$$\frac{\partial \mathbf{v}}{\partial t} + \mathbf{v} \cdot \nabla \mathbf{v} = -\alpha_0 \nabla p' + g\frac{\alpha'}{\alpha_0}\mathbf{k} \tag{10}$$

This equation demonstrates the logic behind what is known as the *Boussinesq approximation*: the multiplication of α'/α_0 by the acceleration of gravity creates a term that is relatively large. It is the combination of the force of gravity and the small variations in density represented by α'/α_0 that drives the motion in convection or stabilizes it in gravity wave motion that can feed on shear of the reference wind field.

The use of Eq. (3) on the linearized vertical component of Eq. (10) produces

$$i\omega\hat{w} = -\alpha_0 \frac{\partial \hat{p}'}{\partial z} + g\frac{\hat{\alpha}'}{\alpha_0} \qquad (11$$

Various modifications of this equation are possible in different physical circumstance In hydrostatic perturbation motion we would keep only the right side. In convectio the buoyancy term creates an acceleration, but the pressure term extracts vertic kinetic energy and converts it into the energy of horizontal motion. Thus we ma conceive of a situation in which the two terms on the right of Eq. (11) will tend to b out of phase. In this case then the acceleration is governed in magnitude by th buoyancy force and we have the estimate

$$|\omega\hat{w}| \sim \left|\frac{g\hat{\alpha}'}{\alpha_0}\right| \qquad (12$$

But if the pressure term is reasonably effective at converting vertical kinetic energy t horizontal energy, thereby controlling to some degree the rate of growth of th disturbance, then

$$\alpha_0 \frac{\partial p'}{\partial z} \sim g\frac{\alpha'}{\alpha_0} \qquad (13$$

or

$$\frac{p_0}{g\rho_0}\left|\frac{p'}{p_0}\right| \sim L_{p'}\frac{\alpha'}{\alpha_0} \qquad (14$$

The estimates (12) and (14) are the basic ones for determining consisten approximations in the other equations.

PROBLEM 13.1.4 Define the scale height $H_i = p_0/g\rho_0$. Show that if $L_{p'} \ll H_i$, the Eq. (2) becomes $T'/T_0 = \alpha'/\alpha_0 = -\rho'/\rho_0$.

PROBLEM 13.1.5 (Gough) Consider a case in which vertical kinetic energy is create by buoyant forces and then converted to horizontal kinetic energy by the action c the pressure fluctuations.

(*a*) On the assumption that the vertical kinetic energy depends on a vertic scale, D, and the buoyancy force, obtain an estimate of w^2 and show th $\omega^2 \sim g|\alpha'/\alpha_0|D^{-1}$.

(*b*) Compute the gain in horizontal kinetic energy due to pressure fluctuations a parcel moves through one horizontal wavelength. Show that if this energy extracted from the vertical kinetic energy, thus balancing buoyant production, the $|p'| \sim \rho_0 w^2$.

(*c*) From these results recover the estimate (12) and discuss the implications o ω^2 and w^2 of the two cases $D \ll p_0/g\rho_0 = H_i$ and $D \sim H_i$.

13.1.3 The Continuity Equation

We may expand the continuity equation

$$\frac{d\alpha}{dt} = \alpha \nabla \cdot \mathbf{v} \tag{15}$$

in the form

$$\frac{\partial \alpha'}{\partial t} + \mathbf{v} \cdot \nabla_H \alpha' + w \frac{\partial \alpha'}{\partial z} + w \frac{\partial \alpha_0}{\partial z} = \alpha_0 \left(1 + \frac{\alpha'}{\alpha_0}\right) \nabla \cdot \mathbf{v} = \alpha_0 \nabla_H \cdot \mathbf{v} + \alpha_0 \frac{\partial w}{\partial z} \tag{16}$$

We have used $|\alpha'/\alpha_0| \ll 1$ on the right.

We shall denote the horizontal components with a single amplitude identified by subscript H and estimate derivatives with a single wavenumber $\kappa_H = 2\pi/L_H$. To determine whether any terms can be ignored, we estimate the ratio of each term to the last term on the left. Thus for $H_\alpha{}^{-1} = \partial \ln \alpha_0 / \partial z$ we have

$$\frac{\partial \alpha'}{\partial t} \div w \frac{\partial \alpha_0}{\partial z} \sim \frac{|\omega \hat{\alpha}'|}{|\hat{w}||\partial \alpha_0/\partial z|} \sim \frac{|\omega|^2 \, |\hat{\alpha}'/\alpha_0|}{|\omega \hat{w}| H_\alpha{}^{-1}} = \frac{|\omega|^2 H_\alpha}{g} \tag{17}$$

$$\mathbf{v} \cdot \nabla_H \alpha' \div w \frac{\partial \alpha_0}{\partial z} \sim \frac{|\hat{v}_H \kappa_H|}{|\hat{w}|} \left|\frac{\hat{\alpha}'}{\alpha_0}\right| H_\alpha = \left|\frac{\hat{v}_H}{\hat{w}}\right| \left(\frac{2\pi H_\alpha}{L_H}\right) \left|\frac{\hat{\alpha}'}{\alpha_0}\right| \tag{18}$$

$$w \frac{\partial \alpha'}{\partial z} \div w \frac{\partial \alpha_0}{\partial z} \sim \frac{H_\alpha}{L_{\alpha'}} \left|\frac{\hat{\alpha}'}{\alpha_0}\right| \tag{19}$$

$$\alpha_0 \nabla_H \cdot \mathbf{v} \div w \frac{\partial \alpha_0}{\partial z} \sim \left|\frac{\hat{v}_H}{\hat{w}}\right| \frac{2\pi H_\alpha}{L_H} \tag{20}$$

and

$$\alpha_0 \frac{\partial w}{\partial z} \div w \frac{\partial \alpha_0}{\partial z} \sim \frac{H_\alpha}{L_w} \tag{21}$$

It is easily seen that the first three terms on the left of Eq. (16) are small compared to the fourth if

$$|\omega|^2 \ll \frac{g}{H_\alpha} \qquad \left|\frac{\hat{v}_H}{\hat{w}}\right| \leqslant \frac{L_H}{2\pi H_\alpha} \qquad L_{\alpha'} \sim H_\alpha \tag{22}$$

Because of Eq. (22), the ratio in Eq. (20) may approach unity. With the condition that $L_w \sim L_{\alpha'}$ the ratio in Eq. (21) may also be of order unity. Thus the terms that remain are

$$\alpha_0 \nabla_H \cdot \mathbf{v} + \alpha_0 \frac{\partial w}{\partial z} - w \frac{\partial \alpha_0}{\partial z} = 0 \tag{23}$$

and this may be written

$$\nabla \cdot (\rho_0 \mathbf{v}) = \rho_0 \nabla_H \cdot \mathbf{v} + \frac{\partial}{\partial z}(\rho_0 w) = 0 \tag{24}$$

We shall refer to this case as *deep convection* because the scales $L_{\alpha'} \sim L_W$ are allowed to be of the order of the scale height H_α. Because of Eq. (6), we thus have $L_3 \sim 4L_W \sim 4H_\alpha$ so that with $H_\alpha \sim 8$ km in the atmosphere, the vertical wavelength may be several tens of kilometers. Thus motions whose vertical scale is quite large are permitted provided that the various other conditions are met.

PROBLEM 13.1.6 Show that Eq. (24) may not be valid for very small scales $L_{\alpha'}$.

To find an approximate form for motions with small vertical scales, we divide each term instead by $\alpha_0\, \partial w/\partial z$. But because of Eq. (21), to complete this calculation we need only to multiply each ratio in Eqs. (17) to (20) by L_W/H_α.

Thus we have

$$\frac{\partial \alpha'}{\partial t} \div \alpha_0 \frac{\partial w}{\partial z} \sim \frac{|\omega|^2 L_W}{g} \tag{25}$$

$$\mathbf{v} \cdot \nabla_H \alpha' \div \alpha_0 \frac{\partial w}{\partial z} \sim \left|\frac{\hat{v}_H}{\hat{w}}\right| \frac{|\hat{\alpha}'|}{|\alpha_0|} \left(\frac{2\pi L_W}{L_H}\right) \tag{26}$$

$$w \frac{\partial \alpha'}{\partial z} \div \alpha_0 \frac{\partial w}{\partial z} \sim \left|\frac{\hat{\alpha}'}{\alpha_0}\right| \frac{L_W}{L_{\alpha'}} \tag{27}$$

$$\alpha_0 \nabla_H \cdot \mathbf{v} \div \alpha_0 \frac{\partial w}{\partial z} \sim \left|\frac{\hat{v}_H}{\hat{w}}\right| \frac{2\pi L_W}{L_H} \tag{28}$$

So in this case if

$$|\omega|^2 \ll \frac{g}{L_W} \qquad \left|\frac{\hat{v}_H}{\hat{w}}\right| \leqslant \frac{L_H}{2\pi L_W} \qquad L_W \sim L_{\alpha'} \qquad L_W \ll H_\alpha \tag{29}$$

then all the ratios except Eq. (28) are small and so we need to keep only the terms

$$\nabla_H \cdot \mathbf{v} + \frac{\partial w}{\partial z} = \nabla \cdot \mathbf{v} = 0 \tag{30}$$

Thus we are assuming, as shown by Eq. (28), that

$$\frac{|\hat{v}_H|}{L_H} \sim \frac{4}{2\pi} \frac{|\hat{w}|}{L_3} \sim \frac{|\hat{w}|}{L_3} \tag{31}$$

Because vertical scales are now restricted to $L_3 = 4L_W \ll 4H_\alpha$, we are assuming that L_3 is a few kilometers at most. Thus this case is called *shallow convection.*

It is important to consider the restrictions on the frequency further. In deep convection, the frequency of oscillation must be small compared to a frequency derived from the scale height H_α. But it is easily shown that $\omega_g{}^2 = g\, \partial \ln \theta_0/\partial z$

$< g/H_\alpha$. Thus the condition is similar to that we found for gravity waves in an isothermal atmosphere in which $\omega^2 \ll {\omega_g}^2$. Later on, we shall show that ω_g is indeed the limiting frequency for the wave motion allowed by the approximate equations.

However, in shallow convection, we have a restriction on frequency depending on vertical scale, so that higher frequencies are permitted for the components of the flow with smaller vertical scale. The associated fact in Eq. (31) that the magnitudes of the horizontal and vertical velocity components may become identical as the vertical scale approaches the horizontal scale thus shows that these equations may be used for the study of atmospheric turbulence (see Problem 13.1.7 below), provided that viscous terms are added to the equation of motion.

In shallow convection, then, the frequency is limited by the vertical scale, which is small enough that expansion and contraction of the moving parcels does not affect their density significantly. Thus shallow convection is effectively incompressible. We shall show in the next section that $H_i \sim H_\alpha$, so Prob. 13.1.4 shows that density variations in shallow convection are associated directly with variations in the temperature, and relative pressure fluctuations are negligible in comparison. In deep convection, however, the rising parcels must expand in proportion to the rate of decrease of reference density and the momentum $\rho_0 \mathbf{v}$ is nondivergent.

PROBLEM 13.1.7 An important feature in atmospheric turbulence and other small-scale motion is advection by the mean wind, U. Show that ω should be replaced by $\omega + \kappa_H U = \kappa_H (U - c)$. Rederive the frequency condition in Eq. (29) and show that it is now replaced by $(U-c)^2 \ll \frac{1}{9} g L_H$ for motions for which $L_3 \sim L_H$.

13.1.4 The First Law of Thermodynamics

We turn now to the first law of thermodynamics to find an approximate version consistent with the other equations we have established. We write it in the form

$$c_p \frac{dT}{dt} - \alpha \frac{dp}{dt} = T \frac{ds}{dt} \tag{32}$$

and upon substituting Eq. (1) we have

$$c_p\left(\frac{dT'}{dt} + w\frac{\partial T_0}{\partial z}\right) - \alpha_0\left(1 + \frac{\alpha'}{\alpha_0}\right)\frac{d}{dt}\left[p_0\left(1 + \frac{p'}{p_0}\right)\right] = c_p\left(\frac{dT'}{dt} + w\frac{\partial T_0}{\partial z}\right)$$
$$- \alpha_0\left(1 + \frac{\alpha'}{\alpha_0}\right)\left[-g\rho_0 w + p_0 \frac{d}{dt}\left(\frac{p'}{p_0}\right)\right] = T\frac{ds}{dt} \tag{33}$$

We neglect α'/α_0 in comparison with unity, and upon dividing by T_0 we have

$$c_p \frac{d}{dt}\left(\frac{T'}{T_0}\right) + c_p w\left[\frac{1}{T_0}\left(\gamma_d + \frac{\partial T_0}{\partial z}\right) - T'\frac{\partial}{\partial z}\frac{1}{T_0}\right] - R\frac{d}{dt}\left(\frac{p'}{p_0}\right) = \frac{ds}{dt} \tag{34}$$

in which we have used the fact that $T'/T_0 \ll 1$ on the right. But in the brackets we have

$$-T'\frac{\partial}{\partial z}\frac{1}{T_0} = \frac{T'}{T_0^2}\frac{\partial T_0}{\partial z} \ll \frac{1}{T_0}\frac{\partial T_0}{\partial z} \tag{35}$$

so that our final approximate form becomes

$$c_p\frac{d}{dt}\left(\frac{T'}{T_0}\right) + \frac{c_p w}{T_0}\left(\gamma_d + \frac{\partial T_0}{\partial z}\right) - R\frac{d}{dt}\left(\frac{p'}{p_0}\right) = \frac{ds}{dt} \tag{36}$$

To determine how the equation might vary between the cases of deep and shallow convection, we substitute the equation of state (2) and collect terms, which gives

$$c_p\frac{d}{dt}\left(\frac{\alpha'}{\alpha_0} + \frac{c_V}{c_p}\frac{p'}{p_0}\right) + \frac{c_p w}{T_0}\left(\gamma_d + \frac{\partial T_0}{\partial z}\right) = \frac{ds}{dt} \tag{37}$$

For shallow convection we may ignore the pressure term; for deep convection it must be retained.

We may now make a change in the equation of motion for the vertical component to correspond to the differing forms of the first law. The vertical component of the term on the right in Eq. (10) is

$$-\alpha_0\frac{\partial p'}{\partial z} + g\frac{\alpha'}{\alpha_0} = -\alpha_0\frac{\partial p'}{\partial z} + g\left[\frac{T'}{T_0} - (1-\delta_s)\frac{p'}{p_0}\right] \tag{38}$$

Here we have introduced the symbol δ_s, which we will take to be 1 for shallow convection and 0 for deep convection.

*PROBLEM 13.1.8 Show that Eq. (37) may be written as

$$\frac{d}{dt}\left(\frac{\theta'}{\theta_0}\right) + \frac{w}{\theta_0}\frac{\partial \theta_0}{\partial z} = \frac{1}{c_p}\frac{ds}{dt}$$

PROBLEM 13.1.9 Show that the ratio of the second to the first pressure terms in Eq (38) for deep convection is $L_{p'}/H_i$.

13.2 THE EQUATIONS, BOUNDARY CONDITIONS, AND ENERGY FORMS

The analysis of this chapter has now proceeded to the point at which we may collec our set of approximate equations and investigate the mechanisms that supply an transform the energy of the perturbation motion. In order to obtain an energy balance we must be sure for deep convection that the pressure term from the first law will match that from the equation of motion. In Eq. (37) in Sec. 13.1 we have

$$\frac{c_V}{c_p}\frac{p'}{p_0} = \frac{c_V}{c_p}\frac{p'}{g\rho_0 H_i} \tag{1}$$

where $H_i = p_0/g\rho_0$. As we shall see, we shall need this term to be $p'/g\rho_0 H_\alpha$ instead. We have (compare Prob. 4.3.4)

$$\frac{1}{\theta_0}\frac{\partial\theta_0}{\partial z} = \frac{1}{T_0}\left(\gamma_d + \frac{\partial T_0}{\partial z}\right) = \frac{1}{H_\alpha} - \frac{c_V}{c_p}\frac{1}{H_i} \tag{2}$$

With the estimate that $\partial T_0/\partial z = -\gamma_d/2$, we have $\gamma_d/2T_0 \sim \frac{1}{50}\ \text{km}^{-1}$, and by virtue of the fact that $H_\alpha \sim 8$ km we see that $H_\alpha{}^{-1} \cong c_V/c_p H_i$. To obtain a precise estimate, we write $(c_V/c_p)(H_\alpha/H_i) = 1 - \frac{8}{50}$ and thus this is an error of about 16 percent, which is tolerable in view of the other approximations. The requirement that $\theta_0(z) = \text{const}$, which makes the substitution exact, can be used, but it seems unnecessarily restrictive. As can be shown, the need for the substitution arises because α'/α_0 is eliminated from the equation of continuity—a step that filters sound waves. When this term is retained, energy equations can be obtained directly, as shown by Eckart (1960). With this change we may now use

$$\delta_d = \begin{cases} 1 & \text{deep convection} \\ 0 & \text{shallow convection} \end{cases} \qquad \delta_s = \begin{cases} 1 & \text{shallow convection} \\ 0 & \text{deep convection} \end{cases} \tag{3}$$

and rewrite our results as

$$\frac{\partial \mathbf{v}}{\partial t} + \mathbf{v}\cdot\nabla\mathbf{v} = -\alpha_0\,\nabla p' + g\frac{\alpha'}{\alpha_0}\mathbf{k} = -\alpha_0\,\nabla p' + g\left(\delta_s\frac{T'}{T_0} + \delta_d\frac{\alpha'}{\alpha_0}\right)\mathbf{k} \tag{4}$$

$$\frac{d}{dt}\left(\frac{\alpha'}{\alpha_0} + \delta_d\frac{p'}{g\rho_0 H_\alpha}\right) + \frac{w}{\theta_0}\frac{\partial\theta_0}{\partial z} = \frac{1}{c_p}\frac{ds}{dt} \tag{5}$$

$$\nabla\cdot\mathbf{v} + \delta_d\frac{w}{\rho_0}\frac{\partial\rho_0}{\partial z} = 0 \tag{6}$$

$$\frac{\alpha'}{\alpha_0} = \frac{T'}{T_0} - \delta_d\frac{p'}{p_0} \tag{7}$$

For shallow convection, it is most convenient to use $\mathbf{v}$, T'/T_0, and p' as variables; for deep, $\mathbf{v}$, α'/α_0, and p'.

In Eqs. (4) to (7), we have a set of approximate equations that are useful for the study of a variety of small-scale atmospheric phenomena—gravity waves, convection, turbulence. But these equations are valid only if the solutions obey the set of conditions we have used in justifying them. It is possible to have either initial conditions or variations of the reference state that will lead the perturbation motion to violate one or more of these conditions. At the instant one of the conditions is violated, then the motion cannot be studied with these equations. Thus if a solution to these equations is obtained either analytically or numerically, it must be tested to be sure that all the conditions necessary to justify the validity of the approximate equations are indeed met.

Next we consider boundary conditions. We shall use the cyclic continuity assumption that the solutions are periodic on each horizontal plane, and following Eckart (1960) we admit only *regular solutions*, that is, those for which the variables are differentiable and bounded throughout the part of the atmosphere affected by the motion and on the boundaries. If there is only a finite distance between the vertical boundaries, we require that w vanish on these boundaries, whether they are free or rigid, thus filtering external gravity waves. This condition may be substituted in Eqs. (4) to (7) to derive additional boundary conditions.

If we consider the flow to extend throughout an infinitely deep atmosphere, then we shall require a rigid lower boundary and an upper diffuse region in which

$$\lim_{z \to \infty} p_0 = \lim_{z \to \infty} \rho_0 = \lim_{z \to \infty} \frac{\partial \rho_0}{\partial z} = 0 \tag{8}$$

Moreover, we shall insist that

$$\lim_{z \to \infty} p' = 0 \tag{9}$$

Finally, for shallow convection, it is often convenient to use the boundary condition (sometimes called the dynamic-kinematic condition) that

$$\lim_{z \to Z_T} \frac{dp}{dt} = 0 \tag{10}$$

where Z_T is the height of the upper surface.

We intend to apply the approximate equations to linear stability problems, and so we shall restrict the present analysis of energetics to that case. Therefore, the linearized versions of Eqs. (4) and (5) are

$$\left(\frac{\partial}{\partial t} + U\frac{\partial}{\partial x}\right)\mathbf{u} + w\frac{\partial U}{\partial z}\mathbf{i} = -\alpha_0\,\nabla p' + \frac{g\alpha'}{\alpha_0}\mathbf{k} \tag{11}$$

and

$$\left(\frac{\partial}{\partial t} + U\frac{\partial}{\partial x}\right)\left(\frac{\alpha'}{\alpha_0} + \delta_d\frac{p'}{g\rho_0 H_\alpha}\right) + \frac{w}{g}\omega_g^2(z) = \frac{1}{c_p}\frac{ds}{dt} \tag{12}$$

From Eq. (11) we have the kinetic energy equation

$$\left(\frac{\partial}{\partial t} + U\frac{\partial}{\partial x}\right)\rho_0\frac{\mathbf{u}\cdot\mathbf{u}}{2} + \rho_0 uw\frac{\partial U}{\partial z} = -\nabla\cdot p'\mathbf{u} + \left(\frac{\delta_d p'}{\rho_0 H_\alpha} + g\frac{\alpha'}{\alpha_0}\right)\rho_0 w \tag{13}$$

in which we have used Eq. (6) to eliminate a divergence on the right side. To obtain an energy balance, we must make the term in Eq. (12) with w match that in Eq. (13).

Thus we have

$$\left(\frac{\partial}{\partial t} + U\frac{\partial}{\partial x}\right)\left[\frac{\rho_0}{2\omega_g^2(z)}\left(g\frac{\alpha'}{\alpha_0} + \delta_d\frac{p'}{\rho_0 H_\alpha}\right)^2\right] + \rho_0 w\left(\frac{g\alpha'}{\alpha_0} + \delta_d\frac{p'}{\rho_0 H_\alpha}\right)$$
$$= \frac{\rho_0\gamma_d}{\omega_g^2(z)}\left(g\frac{\alpha'}{\alpha_0} + \delta_d\frac{p'}{\rho_0 H_\alpha}\right)\frac{ds}{dt} \quad (14)$$

Upon integrating over the entire atmosphere, using the assumption of cyclic continuity, we have

$$\frac{\partial}{\partial t}\int_V \frac{\rho_0}{2}\left[u^2 + \frac{1}{\omega_g^2(z)}\left(g\frac{\alpha'}{\alpha_0} + \delta_d\frac{p'}{\rho_0 H_\alpha}\right)^2\right] dV$$
$$= -\int_V \left[\rho_0 uw\frac{\partial U}{\partial z} - \frac{\rho_0\gamma_d}{\omega_g^2(z)}\left(g\frac{\alpha'}{\alpha_0} + \frac{\delta_d p'}{\rho_0 H_\alpha}\right)\frac{ds}{dt}\right] dV \quad (15)$$

When the basic flow is constant with height and when the motions are isentropic, then the energy integral on the left is constant. It is the sum of the perturbation kinetic energy and a perturbation available potential energy. The first term on the right shows that the Reynolds stress factor uw may interact with the shear to extract perturbation kinetic energy from the basic flow. The last term shows that heat sources and sinks may either excite or damp the motion depending on the phasing of the entropy changes and the perturbations themselves. It is interesting to note that the energy integral on the left is finite for regular solutions if the atmosphere has finite mass and is statically stable everywhere. Furthermore, the integral shows that large amounts of energy may become available to the perturbation motion as the static stability tends toward neutral conditions.

It is obvious from Eqs. (13) and (14) that the rate of transformation of available energy to kinetic energy is

$$C(A,K) = \int_V \left(\delta_d\frac{p'}{\rho_0 H_\alpha} + g\frac{\alpha'}{\alpha_0}\right)\rho_0 w\, dV \cong \int_V g\frac{\theta'}{\theta_0}\,\rho_0 w\, dV \quad (16)$$

Thus kinetic energy is realized by the motion when, on the average, ascending parcels have higher potential temperature than descending parcels. This is the essential mechanism by which convection proceeds. But if the parcels in a wavelike motion are rising isentropically, then under stable conditions θ' will be negative and the buoyancy will extract the energy and store it in the reservoir of available energy. But θ' will be positive for a parcel descending from its initial position, and so it will also contribute to the available energy. Of course, this energy is used as the parcels return to their original position.

Thus we have shown that it is possible to express the approximate equations in such a way that they yield energy forms with suitable properties. It is of particular interest that we were able to derive one equation for deep and shallow convectio simultaneously. The next task is to show that wave equations can be derived that als allow us to treat these cases at the same time.

PROBLEM 13.2.1 Include the nonlinear perturbation terms on the left of Eqs. (4 and (5) and analyze their effect on the energy equations for deep convection. Explai the physical significance of the new term that appears.

PROBLEM 13.2.2 Draw an energy diagram that displays reservoirs of available energy kinetic energy of the horizontal components, and kinetic energy of the vertic components. Compute the energy transfers between the three components and displa them along with the energy sources. Is there a fundamental difference between dee and shallow convection?

13.3 THE LINEAR WAVE EQUATIONS

The assumption that the perturbation motion is isentropic and harmonic in both spac and time allows us to reduce the system of approximate equations to a sing differential equation describing the vertical variation of one of the amplitudes. We sha put

$$\phi(x,y,z,t) = \hat{\phi}(\omega,\kappa_1,\kappa_2,z)e^{i(\omega t + \kappa_1 x + \kappa_2 y)} \tag{1}$$

with the understanding that the actual solutions are to be constructed from a sum (the basic solutions of form (1). This may be expressed as

$$\phi(x,y,z,t) = \int_{-\infty}^{\infty}\int_{-\infty}^{\infty} \hat{\phi}(\omega,\kappa_1,\kappa_2,z)e^{i(\omega t + \kappa_1 x + \kappa_2 y)}\, d\kappa_1\, d\kappa_2 \tag{2}$$

so that by formal inverse Fourier transformation

$$\hat{\phi}(\omega,\kappa_1,\kappa_2,z) = \frac{1}{4\pi^2}\int_{-\infty}^{\infty}\int_{-\infty}^{\infty} \phi(x,y,z,0)e^{-i(\kappa_1 x + \kappa_2 y)}\, dx\, dy \tag{3}$$

We are neglecting here questions of convergence of the integral, which can be force by standard techniques from the theory of spectral analysis. It is usually convenient t apply the boundary conditions to the amplitude $\hat{\phi}$ for every value of κ_1 and κ_2.

By defining a new set of canonical variables that have a different meaning fo deep and shallow convection, we can obtain a linear equation that applies to bot cases simultaneously. Hence we define

Variable	Deep	Shallow
$\boldsymbol{\xi}$	$\rho_0 \mathbf{u}$	$\mathbf{u}$
π	p'	$p'\alpha_0$
σ	$\dfrac{\alpha'}{{\alpha_0}^2}$	$\dfrac{\alpha'}{\alpha_0}$

Then the equation of motion (11) in Sec. 13.2 may be written

$$\left(\frac{\partial}{\partial t} + U\frac{\partial}{\partial x}\right)\boldsymbol{\xi} + \xi_3 \frac{\partial U}{\partial z}\mathbf{i} = -\nabla_H \pi - \left(\frac{\partial}{\partial z} - \frac{\delta_s}{H_\alpha}\right)\pi \mathbf{k} + g\sigma \mathbf{k} \tag{4}$$

The first law, Eq. (12) in Sec. 13.2, for isentropic motion becomes

$$\left(\frac{\partial}{\partial t} + U\frac{\partial}{\partial x}\right)\left(\sigma + \delta_d \frac{\pi}{gH_\alpha}\right) + \frac{\xi_3}{g}\,{\omega_g}^2(z) = 0 \tag{5}$$

and the equation of continuity (6) in Sec. 13.2 is

$$\nabla \cdot \boldsymbol{\xi} = 0 \tag{6}$$

Upon substitution of Eq. (1) into these equations along with use of the definitions $\Omega = \omega + \kappa_1 U$ and $D = \partial/\partial z$, we obtain

$$i\Omega\hat{\xi}_j + \hat{\xi}_3 DU\,\delta_{j1} + i\kappa_j\hat{\pi} = 0 \qquad j = 1, 2 \tag{7}$$

$$i\Omega\hat{\xi}_3 + \left(D - \frac{\delta_s}{H_\alpha}\right)\hat{\pi} - g\hat{\sigma} = 0 \tag{8}$$

$$i\Omega\left(\hat{\sigma} + \delta_d \frac{\hat{\pi}}{gH_\alpha}\right) + \frac{\hat{\xi}_3}{g}\,{\omega_g}^2(z) = 0 \tag{9}$$

$$i\kappa_1\hat{\xi}_1 + i\kappa_2\hat{\xi}_2 + D\hat{\xi}_3 = 0 \tag{10}$$

We shall reduce this system to a single equation in the amplitude $\hat{\xi}_3$. Equations (7) and (10) combine to give

$$\Omega D\hat{\xi}_3 - \kappa_1\hat{\xi}_3\,DU - i\kappa^2\hat{\pi} = 0 \qquad \kappa^2 = {\kappa_1}^2 + {\kappa_2}^2 \tag{11}$$

Elimination of σ between Eqs. (8) and (9) gives

$$\Omega^2\hat{\xi}_3 - i\Omega\left(D + \frac{\delta_d - \delta_s}{H_\alpha}\right)\hat{\pi} - {\omega_g}^2(z)\hat{\xi}_3 = 0 \tag{12}$$

and then we eliminate $\hat{\pi}$ between Eqs. (11) and (12) to obtain

$$\Omega^2 D^2\hat{\xi}_3 - \Omega\kappa_1\hat{\xi}_3\,D^2U + \Omega\frac{\delta_d - \delta_s}{H_\alpha}\left[\Omega\,D\hat{\xi}_3 - \kappa_1(DU)\,\hat{\xi}_3\right] + \kappa^2({\omega_g}^2 - \Omega^2)\hat{\xi}_3 = 0 \tag{13}$$

The final equation is obtained by putting Eq. (13) in the classical self-adjoin form

$$\Omega^2 D(q_0 D\hat{\xi}_3) - \kappa_1 \Omega \hat{\xi}_3 D(q_0 DU) + \kappa^2(\omega_g{}^2 - \Omega^2) q_0 \hat{\xi}_3 = 0 \qquad q_0 = \delta_s \rho_0 + \delta_d \alpha_0 \quad (14$$

where q_0 is α_0 for deep convection and ρ_0 for shallow.

Equation (14) is a quite general result describing the vertical variation of th amplitude of the vertical component of motion for small-scale flow that is periodic i time and on horizontal planes. It presents formidable difficulties, however. The first i that the functions $U(z)$, q_0, $\omega_g{}^2$, and Ω (through U) are arbitrary functions of height and so it is not possible to find a general analytic solution. The second is that if ω i real, then Ω may vanish at some point where $c = U$, and thus the equation has second-order singularity. Still, Eq. (14) is mathematically equivalent to the equatio studied in a variety of investigations of fluid instabilities, and thus we shall be able t apply known results to the problem of the stability of deep or shallow convection o wave motion in the atmosphere.

PROBLEM 13.3.1 Verify Eq. (14).

PROBLEM 13.3.2 Show that Eq. (13) contains the gravity wave motion of Chap. 1 as a special case. [Hint: Apply the assumption about $U(z)$ used in that analysis.]

Before taking up the stability problem, let us show that the approximat equations are soundproof when $DU = 0$. We multiply Eq. (14) by the comple conjugate $\bar{\xi}_3$ and integrate by parts in the vertical using the boundary conditions an the fact that Ω is constant to find

$$\int_0^{Z_T} \Omega^2 q_0 |D\hat{\xi}_3|^2 \, dz = \int_0^{Z_T} \kappa^2(\omega_g{}^2 - \Omega^2) q_0 |\hat{\xi}_3|^2 \, dz \qquad (15$$

PROBLEM 13.3.3 Show that the term that appears in the integration by parts

$$\Omega^2 \bar{\hat{\xi}}_3 q_0 \, D\hat{\xi}_3 \Big|_0^{Z_T}$$

vanishes in deep and shallow convection, even if Z_T is infinite. [Hint: Use Eq. (11).

We might suppose that the frequency ω has an imaginary part so tha $\Omega = \Omega_R + i\omega_I$. But then the imaginary part of Eq. (15) becomes

$$2i\omega_I \int_0^{Z_T} \Omega_R q_0 (|D\hat{\xi}_3|^2 + \kappa^2 |\hat{\xi}_3|^2) \, dz = 0 \qquad (16$$

Thus for $U = \text{const}$ either $\omega_R = -\kappa_1 U$ or $\omega_I = 0$ provided that κ is real so that the motion is harmonic on horizontal planes. The first case would correspond to perturbations being carried along exactly at the wind speed so that the motions would be standing waves in a coordinate system moving at speed U. We conclude that if $\omega_R \neq -\kappa_1 U$ and κ is real, then ω_I vanishes and the wave motion neither amplifies nor decays. Clearly, then, some variation of U is necessary for amplification.

With the result then that $\omega_I = 0$, we may write Eq. (15) as

$$\Omega^2 \int_0^{z_T} q_0 \, |D\hat{\xi}_3|^2 \, dz = \kappa^2(\omega_g{}^2 - \Omega^2) \int_0^{z_T} q_0 \, |\hat{\xi}_3|^2 \, dz \tag{17}$$

in which all factors are positive except for the first one on the right. Thus it must be positive also, and so

$$(\omega + \kappa_1 U)^2 \leqslant \omega_g{}^2 \tag{18}$$

Thus when U vanishes, the frequencies are limited by the Brunt-Väisälä frequency and acoustic waves are not permitted. As is obvious from the analysis of the continuity equation, this results from the frequency limitation that allowed us to neglect α'/α_0.

13.4 INTEGRAL STABILITY THEOREMS AND THE CRITICAL RICHARDSON NUMBER

The question we seek to answer now is whether conditions on the profiles of wind and stability can be found that will allow us to conclude that the motion will be either stable or unstable. As pointed out in Sec. 12.1, we need to ascertain whether the frequency ω has an imaginary part and whether its sign is positive or negative.

The technique we shall use provides a wealth of information and has been applied to many problems in fluid dynamics. The success of the method for deep and shallow convection in the atmosphere derives from the fact that we were able to express the basic differential equation in the self-adjoint form of Eq. (14) in Sec. 13.3.

Because we are interested now in stability questions, we assume that the imaginary part ω_I is not zero. Therefore $\Omega = (\omega_R + \kappa_1 U) + i\omega_I$ cannot vanish and we may define the functions $F_n(z)$ by

$$\hat{\xi}_3 = \Omega^n F_n(z) \tag{1}$$

Substitution of this quantity for $\hat{\xi}_3$ in Eq. (14) in Sec. 13.3 gives

$$\Omega^2 D[q_0 D(\Omega^n F_n)] - \kappa_1 \Omega^{n+1} F_n D(q_0 DU) + \kappa^2(\omega_g{}^2 - \Omega^2) q_0 \Omega^n F_n = 0 \tag{2}$$

We denote a complex conjugate with an overbar, multiply Eq. (2) by $\Omega^{n-2}\,\overline{F_n}$, and integrate over the layer in which the motion is occurring to obtain

$$\int_0^{z_T} \Omega^n \bar{F}_n D[q_0 D(\Omega^n F_n)] \, dz - \int_0^{z_T} \kappa_1 \Omega^{2n-1} D(q_0 DU) |F_n|^2 \, dz \tag{3}$$

$$+\int_0^{Z_T} \kappa^2(\omega_g{}^2 - \Omega^2) q_0 \Omega^{2n-2} |F_n|^2 \, dz = 0 \qquad (3) \; (cont'd)$$

We emphasize now that if $\omega_g{}^2 \equiv 0$, then we lose a first power of Ω in Eq. (14) in Sec. 13.3 and a second power in Eq. (2), which would necessitate multiplication by Ω^{n-1} rather than Ω^{n-2}. Hence, in all that follows we assume that $\omega_g{}^2$ cannot vanish identically. If it does, then Eq. (14) in Sec. 13.3 becomes equivalent to the inviscid Orr-Sommerfeld equation and different conclusions would result.

We denote the first term of Eq. (3) on the left as T_1. Integration by parts produces (we drop the subscript on F for convenience)

$$T_1 = \left[\Omega^n q_0 \bar{F} D(\Omega^n F)\right]_0^{Z_T} - \int_0^{Z_T} q_0 D(\Omega^n \bar{F}) D(\Omega^n F) \, dz \tag{4}$$

We use

$$D(\Omega^n F) = \Omega^n \, DF + n\Omega^{n-1}\kappa_1 \, DU \, F \tag{5}$$

in the integral to find that

$$T_1 = B_1 - \int_0^{Z_T} q_0 [\Omega^{2n}|DF|^2 + n^2\Omega^{2n-2}\kappa_1{}^2(DU)^2 \, |F|^2] \, dz$$
$$- \int_0^{Z_T} q_0 n\Omega^{2n-1}\kappa_1 \, DU(F \, D\bar{F} + \bar{F} \, DF) \, dz \tag{6}$$

in which B_1 is the boundary term of Eq. (4). But

$$F \, D\bar{F} + \bar{F} \, DF = D(F\bar{F}) = D|F|^2 \tag{7}$$

With Eq. (7) substituted in the last integral of Eq. (6), we can employ integration by parts again to obtain finally

$$T_1 = B_1 - \left[nq_0\Omega^{2n-1}\kappa_1 \, DU|F|^2\right]_0^{Z_T} - \int_0^{Z_T} q_0 [\Omega^{2n} \, |DF|^2 + n^2\Omega^{2n-2}\kappa_1{}^2(DU)^2|F|^2] dz$$
$$+ \int_0^{Z_T} n\Omega^{2n-1}\kappa_1 D(q_0 DU)|F|^2 \, dz + \int_0^{Z_T} n(2n-1)\Omega^{2n-2}\kappa_1{}^2(DU)^2 q_0|F|^2 \, dz \tag{8}$$

Before proceeding, let us analyze the boundary term. The condition that $\hat{\xi}_3$ vanish on the rigid boundary at $z = 0$ means that $\Omega^n F_n(0)$ must vanish, and so the two boundary terms in Eq. (8) vanish at $z = 0$. The same condition applies if Z_T is finite. If Z_T is infinite, then a condition analogous to Eq. (9) in Sec. 13.2 is that $\hat{\pi}$ must vanish at the top of the upper diffuse region. With this condition, Eq. (11) in Sec. 13.3 with the definition (1) implies that

$$\Omega^n q_0 \bar{F}_n \, D(\Omega^n F_n) = \kappa_1 \Omega^{2n-1} q_0 DU|F_n|^2 \tag{9}$$

Therefore, the sum B of the two boundary terms in Eq. (8) is

$$B = -[(n-1)\Omega^{2n-1}\kappa_1\, DU\, q_0\, |F_n|^2]_{Z_T=\infty} \tag{10}$$

and appears in the equation only when $Z_T = \infty$.

Now we collect terms in powers of Ω after substituting Eq. (8) back into Eq. (3) to obtain

$$\int_0^{Z_T} \Omega^{2n}(|DF_n|^2 + \kappa^2|F_n|^2)q_0\, dz - \int_0^{Z_T} \Omega^{2n-2}[n(n-1)\kappa_1^2(DU)^2 + \kappa^2\omega_g^2]|F_n|^2 q_0\, dz$$
$$+ \int_0^{Z_T} [\Omega^{2n-1}(1-n)\kappa_1\, D(q_0\, DU)\,|F_n|^2]\, dz$$
$$+ [(n-1)\Omega^{2n-1}\kappa_1\, DU\, q_0|F_n|^2]_{Z_T=\infty} = 0 \tag{11}$$

This complicated equation turns out to be a very powerful result, for with it we can deduce some important theorems about fluid instabilities by choosing the proper values of n.

The sign of ω_I The first result we obtain from Eq. (11) shows that damped motion is impossible for inviscid, isentropic perturbation flow. We take $n = 1$ and define $|G|^2 = |DF|^2 + \kappa^2|F|^2$, dropping the subscript on F as we shall throughout this section. Thus Eq. (11) becomes

$$\int_0^{Z_T} \Omega^2|G|^2 q_0\, dz - \int_0^{Z_T} \kappa^2\omega_g^2|F|^2 q_0\, dz = 0 \tag{12}$$

But with $\Omega^2 = \omega^2 + 2\kappa_1\omega U + (\kappa_1 U)^2$ we have

$$\omega^2 I_2 + 2\omega I_1 + I_0 = 0 \tag{13}$$

where

$$I_2 = \int_0^{Z_T} |G|^2 q_0\, dz \qquad I_1 = \int_0^{Z_T} \kappa_1 U|G|^2 q_0\, dz$$

$$I_0 = \int_0^{Z_T} [(\kappa_1 U)^2|G|^2 - \kappa^2\omega_g^2|F|^2]q_0\, dz \tag{14}$$

From Eq. (13) we have then the solution

$$\omega = -\frac{I_1}{I_2} \pm \left(\frac{I_1^2}{I_2^2} - \frac{I_0}{I_2}\right)^{1/2} \tag{15}$$

Therefore if ω is complex, it appears in conjugate pairs and one of the terms in the solution will amplify and the other will decay whenever $\omega_I \neq 0$, as we are now assuming in order for Eq. (11) to be valid. Thus $\omega_I \neq 0$ is a sufficient condition for instability.

From Eq. (15), the necessary and sufficient condition for the assumed instability is $I_1{}^2 < I_2 I_0$. With the aid of the Schwarz inequality we see that

$$\begin{aligned} I_1{}^2 &= \left(\int_0^{Z_T} \kappa_1 U |G|^2 q_0 \, dz\right)^2 \\ &\leqslant \int_0^{Z_T} |G|^2 q_0 \, dz \int_0^{Z_T} (\kappa_1 U)^2 |G|^2 q_0 \, dz \\ &= I_2\left(I_0 + \int_0^{Z_T} \kappa^2 \omega_g{}^2 |F|^2 q_0 \, dz\right) \end{aligned} \tag{16}$$

We have equality (on the middle line) only if U is constant. If $\omega_g{}^2 < 0$ because $D\theta_0 < 0$ everywhere in the layer in motion, then Eq. (16) shows that for any distribution of U the motion is unstable. If $\omega_g{}^2 \leqslant 0$ (but not zero everywhere), then a variation of U, however slight, will also lead to instability.

The result (15) shows that the value of ω_I, and thus the rate of growth, depends on interaction between the basic flow and the perturbation motion. Because the solutions are dependent on initial conditions, it thus seems unlikely that sufficient conditions for instability that depend only on the basic flow can be found when the atmosphere is statically stable.

The critical Richardson number In Chap. 4 we showed by a heuristic argument that the critical Richardson number should be $\frac{1}{4}$; now we can obtain a proof of this result. In Eq. (11) we set $n = \frac{1}{2}$ and in the second integral we write $\Omega^{-1} = \bar{\Omega}/|\Omega|^2$ so that we have (dropping the subscript $\frac{1}{2}$ on F and using $|G|^2 = |DF|^2 + \kappa^2 |F|^2$)

$$\int_0^{Z_T} \Omega(|G|^2) q_0 \, dz + \int_0^{Z_T} \bar{\Omega}\left[\frac{\kappa_1{}^2}{4}(DU)^2 - \kappa^2 \omega_g{}^2\right]\left|\frac{F}{\Omega}\right|^2 q_0 \, dz$$
$$+ \int_0^{Z_T} \tfrac{1}{2}\kappa_1 D(q_0 DU)|F|^2 \, dz - (\tfrac{1}{2}\kappa_1 DU q_0 |F|^2)_{Z_T=\infty} = 0 \tag{17}$$

The imaginary part of this expression is

$$\omega_I \int_0^{Z_T} \left\{ |G|^2 + \left[\kappa^2 \omega_g{}^2 - \frac{\kappa_1{}^2}{4}(DU)^2\right]\left|\frac{F}{\Omega}\right|^2 \right\} q_0 \, dz = 0 \tag{18}$$

Thus there are only two possibilities. If

$$\kappa^2 \omega_g{}^2 - \frac{\kappa_1{}^2}{4}(DU)^2 \geqslant 0 \tag{19}$$

then ω_I must vanish. Conversely, we can have $\omega_I \neq 0$ only if

$$\kappa^2\omega_g{}^2 - \frac{\kappa_1{}^2}{4}(DU)^2 < 0 \tag{20}$$

somewhere. Hence we have:

Theorem (Miles-Howard, 1961) If the Richardson number $Ri = \omega_g{}^2/(DU)^2$ obeys the inequality $Ri \geqslant \kappa_1{}^2/4\kappa^2$ everywhere, then the motion must be stable. If the motion is unstable, then somewhere in the layer in motion it must be true that $Ri < \kappa_1{}^2/4\kappa^2$. ////

This result shows that the gradient Richardson number characterizes instability of wave motion or convection. But we have only a sufficient condition for stability and a necessary condition for instability; there is not a sufficient condition for instability except when $\omega_g{}^2 < 0$ everywhere. That a sufficient condition could exist seems unlikely because we cannot be sure that the motion will arrange itself to accomplish the necessary energy transformations.

PROBLEM 13.4.1 Express the theorem in a form that is valid if DU vanishes somewhere in the layer.

PROBLEM 13.4.2 Show that two-dimensional motion is more unstable than the equivalent three-dimensional motion.

Howard's semicircle theorem We begin by defining $C_R = -\omega_R/\kappa_1$ and $C_I = -\omega_I/\kappa_1$ so that

$$\Omega^2 = \kappa_1{}^2[(C_R{}^2 - C_I^2 + U^2 - 2UC_R) - 2iC_I(U - C_R)] \tag{21}$$

Thus the imaginary part of Eq. (12) may be written for $C_I \neq 0$ as

$$C_R = \left(\int_0^{Z_T} U|G|^2 q_0\,dz\right)\left(\int_0^{Z_T} |G|^2 q_0\,dz\right)^{-1} \tag{22}$$

and so

$$U_{\min} \leqslant C_R \leqslant U_{\max} \tag{23}$$

where min and max refer to the minimum and maximum values of U in the entire layer $(0, Z_T)$. Obviously, the speed of phase propagation in the direction of U must equal some value of U in the profile because U is assumed differentiable and therefore is continuous.

With Eq. (21) we may write the real part of Eq. (12) as

$$\int_0^{Z_T} (C_R{}^2 - C_I{}^2 + U^2 - 2C_R U)|G|^2 q_0\,dz - \int_0^{Z_T}\left(\frac{\kappa}{\kappa_1}\right)^2 \omega_g{}^2|F|^2 q_0\,dz = 0 \tag{24}$$

The crucial step is to observe that it is always true that

$$(U - U_{\min})(U - U_{\max}) \leqslant 0 \tag{25}$$

and therefore

$$-U^2 + U(U_{\min} + U_{\max}) - U_{\min} U_{\max} \geqslant 0 \tag{26}$$

We multiply Eq. (26) by $|G|^2 q_0$, integrate, and then add the result to Eq. (24) using Eq. (22) to eliminate U. Hence we have

$$-(C_R^2 + C_I^2) + C_R(U_{\min} + U_{\max}) - U_{\min} U_{\max} - \left[\int_0^{z_T} \left(\frac{\kappa}{\kappa_1}\right)^2 \omega_g^2 |F|^2 q_0 \, dz\right] \left(\int_0^{z_T} |G|^2 q_0 \, dz\right)^{-1} \geqslant 0 \tag{27}$$

We denote the last term by β^2 and obtain

$$C_R^2 + C_I^2 \leqslant C_R(U_{\min} + U_{\max}) - U_{\min} U_{\max} - \beta^2 \tag{28}$$

But this may be rewritten as

$$\left(C_R - \frac{U_{\min} + U_{\max}}{2}\right)^2 + C_I^2 \leqslant \left(\frac{U_{\max} - U_{\min}}{2}\right)^2 - \beta^2 \tag{29}$$

The depiction of this relation in the diagram shown in Figure 13.1 makes it clear why it is known as the semicircle theorem. The eigenvalues C_R and C_I are constrained to lie in the semicircle shown, so that the growth rate is bounded (assuming $\omega_g^2 > 0$), and as C_I approaches its maximum value, the phase propagation must adopt the speed $(U_{\min} + U_{\max})/2$. In this sense, the most unstable waves are moving with the wind.

PROBLEM 13.4.3 Show that

$$\beta^2 \leqslant \max \frac{\omega_g^2}{\kappa_1^2}$$

where β^2 is defined in Eqs. (27) and (28). Propose definitions of $\overline{\omega_g^2}$ and k_3^2 such that (here the overbar denotes a suitable average)

$$\beta^2 = \left(\frac{\kappa}{\kappa_1}\right)^2 \overline{\omega_g^2} (\kappa^2 + k_3^2)^{-1}$$

The growth rate inequality The fact that the growth rate is bounded suggests that further information concerning it may be discovered. We use again the assumption that $\omega_I \neq 0$ and rewrite Eq. (18) as

$$\int_0^{z_T} \kappa^2 |F|^2 q_0 \, dz \leqslant \int_0^{z_T} \left[\tfrac{1}{4} \kappa_1^2 (DU)^2 - \kappa^2 \omega_g^2\right] q_0 \frac{|F|^2}{\omega_I^2} \, dz - \int_0^{z_T} |DF|^2 q_0 \, dz \tag{30}$$

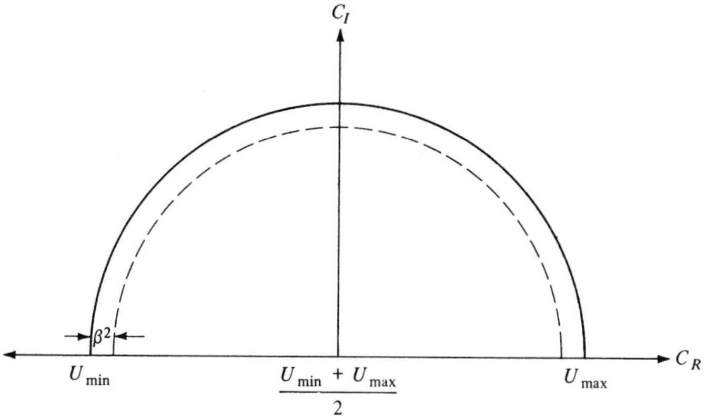

FIGURE 13.1
The range of permitted combinations of C_R and C_I is the interior of the semicircle with radius $[(U_{max} - U_{min})/2]^2 - \beta^2$.

To obtain the inequality, we have used the fact that $|\Omega|^2 \geqslant \omega_I^2$. Clearly then

$$\kappa^2\omega_I^2 \leqslant \max\left[\tfrac{1}{4}\kappa_1^2(DU)^2 - \kappa^2\omega_g^2\right] \tag{31}$$

in which max refers again to the layer $(0,Z_T)$. Thus

$$\omega_I^2 \leqslant \max\left[\left(\frac{1}{4}\frac{\kappa_1^2}{\kappa^2}\frac{1}{Ri} - 1\right)\omega_g^2\right] \tag{32}$$

We let the maximum value of ω_g in the layer be ω_{gM} and then may write

$$\left(\frac{\omega_I}{\omega_{gM}}\right)^2 \leqslant \frac{1}{4}\left(\frac{\kappa_1}{\kappa}\right)^2 \frac{1}{\min Ri} - 1 \qquad Ri > 0 \text{ in } (0, Z_T) \tag{33}$$

We define a time period by $T_g = 2\pi/\omega_{gM}$, and as shown by the limitation on the frequency this period is less than the time for one oscillation of the motion. Then the ratio of amplitudes after time T_g to the amplitude at time $t = 0$ is

$$\left|\frac{\hat{\phi}(z, T_g)}{\hat{\phi}(z, 0)}\right| = e^{T_g\omega_I} = e^{2\pi\omega_I/\omega_{gM}} \tag{34}$$

Thus

$$\left|\frac{\hat{\phi}(z, T_g)}{\hat{\phi}(z, 0)}\right| \leqslant \exp\left(2\pi\left\{\frac{1}{4[1 + (L_1/L_2)^2]\min Ri} - 1\right\}^{1/2}\right) \tag{35}$$

For waves propagating parallel to the wind, $L_2 = \infty$, and we have amplitude ratio bounds of 1 for min $Ri = 0.25$, 23 for 0.2, and 170 for 0.15. For waves propagating nearly perpendicular to the wind direction, $L_1 \gg L_2$, and very small Richardson numbers are needed to permit rapid growth.

These results suggest the manner in which turbulence may form from wave instability. If the growth rate is small in one period T_g, then in the actual case we might expect that viscous forces would control the instability. But if the growth is very rapid in a period T_g, then the nonlinear phenomena become important and we may expect that the unstable motion would become turbulent.

The growth of perturbations in a shear flow on an interface between fluids of different densities is known as the Kelvin-Helmholtz instability. It has been studied in the laboratory by Thorpe (1968); some photographs from his experiment are shown in his article. Similar patterns of motion are sometimes seen in the atmosphere. A significant fraction of clear air turbulence in the atmosphere is undoubtedly due to the instability of gravity waves and formation of turbulence by the Kelvin-Helmholtz process. However, definite experimental verification of this hypothesis has not yet been obtained.

BIBLIOGRAPHIC NOTES

13.1-13.3 Approximate equations of motion for small-scale flow have been studied recently by:

Dutton, J. A., and G. H. Fichtl, 1969: "Approximate Equations of Motion for Gases and Liquids," *J. Atmospheric Sci.*, **26**:241-254.

Gough, D. O., 1969: "The Anelastic Approximation for Thermal Convection," *J. Atmospheric Sci.*, **26**:448-456.

Ogura, Y., and N. Phillips, 1962: "Scale Analysis of Deep and Shallow Convection in the Atmosphere," *J. Atmospheric Sci.*, **19**:173-179.

A comparison of early approximations is available in:

Eckart, C., and H. G. Ferris, 1956: "Equations of Motion of the Ocean and Atmosphere," *Rev. Mod. Phys.*, **28**:48-52.

The method for deriving energy equations for approximate equations is outlined in this reference. The complete reference for Eckart's 1960 book will be found in the notes to Chap. 12.

13.4 Integral stability theorems are a useful technique of some significance in fluid mechanics. A basic reference is:

Lin, C. C., 1955: *The Theory of Hydrodynamic Stability*, Cambridge University Press, Cambridge, 155 pp.

The results given here were obtained in the paper by Dutton and Fichtl cited above and derive from the article:

Howard, L. N., 1961: "Note on a Paper of John W. Miles," *J. Fluid Mech.*, **10**:509-512.

Further information about the Kelvin-Helmholtz instability and about its possible relation to clear air turbulence may be found in:

Dutton, J. A., 1971: "CAT, Aviation, and Atmospheric Science," *Rev. Geophys. and Space Sci.*, **9**:613–657.

Dutton, J. A., and H. A. Panofsky, 1970: "Clear Air Turbulence: A Mystery May Be Unfolding," *Science*, **167**:937–944.

Thorpe, S. A., 1968: "A Method of Producing a Shear Flow in a Stratified Fluid," *J. Fluid Mech.*, **32**:693–707.

14

THE QUASI-GEOSTROPHIC THEORY OF LARGE-SCALE FLOW

The large-scale flow of the atmosphere has the essential features of being nearly hydrostatic, nearly geostrophic, and wavelike in appearance. The systems that produce the clouds and weather of mid-latitudes seem to develop, propagate eastward, and dissipate in regular progression.

These observations certainly suggest that it should be possible to use the concepts of wave motion to develop a theory that will portray adequately the major features of large-scale flow. The Rossby wave theory represents such an attempt to abstract the characteristics of these waves in a simple model. In the late 1940s, the first attempts at numerical weather prediction with the primitive computers then available necessitated further development of as simple as possible a set of equations that would describe the large-scale flow.

A major contribution was a 1947 article by Prof. Jule Charney that applied the scale analysis techniques of fluid dynamics to the meteorological equations as a rational approach to finding a consistent method to use the geostrophic approximations. Research in this area has continued to the present, with the newest development being the theory of quasi-geostrophic turbulence.

The major task of this chapter is to incorporate the assumption that the motions are nearly geostrophic, and yet retain freedom for the energy transformations that provide the impetus for the cyclonic development. An associated problem is to provide a mechanism by which the ageostrophic component of the wind brings about the change in the largest-scale geostrophic components of the flow.

14.1 QUASI-GEOSTROPHIC SCALE ANALYSIS

It will be convenient to replace the pressure by Exner's function $\pi = c_p(p/p_{00})^{R/c_p}$ and use the system [see Eq. (9) in Sec. 6.3]:

$$\frac{\partial \mathbf{v}}{\partial t} + \mathbf{v} \cdot \nabla \mathbf{v} = -\theta\, \nabla \pi - g\mathbf{k} - f\mathbf{k} \times \mathbf{v} \tag{1}$$

$$\frac{\partial \theta}{\partial t} + \mathbf{v} \cdot \nabla \theta = \frac{\theta}{c_p T}(q + f_h) = H \tag{2}$$

$$\frac{\partial \pi}{\partial t} + \mathbf{v} \cdot \nabla \pi + \frac{R\pi}{c_V}\, \nabla \cdot \mathbf{v} = \frac{R}{c_V}\frac{\pi}{\theta} H \tag{3}$$

These equations form a complete system in $\mathbf{v}$, θ, and π if the heating rate H is specified. We shall generally take $H = 0$, which is clearly an assumption of isentropic motion.

14.1.1 Basic Scaling Assumptions

The orders of magnitude of the variables and their derivatives in Eqs. (1) to (3) are determined by the fact that the motions we shall consider are essentially geostrophic and hydrostatic. We again denote equality in magnitude by $\sim$, and we again ignore signs. From the vertical component of Eq. (1) we introduce the hydrostatic assumption by requiring that

$$\theta \frac{\partial \pi}{\partial z} \sim g \tag{4}$$

and we use the horizontal components to introduce the geostrophic scaling by defining the velocity magnitude V_g as

$$V_g \sim \frac{\theta}{f}\frac{\partial \pi}{\partial x} \sim \frac{\theta}{f}\frac{\partial \pi}{\partial y} \tag{5}$$

These definitions have several significant consequences. First,

$$c_p T = \pi\theta = \frac{\theta\, \partial\pi/\partial z}{\partial\pi/\pi\, \partial z} \sim gH_\pi \tag{6}$$

in which we use the definition that $H_\pi^{-1} = -\partial \ln \pi/\partial z$. From Eq. (6) then $H_\pi = c_p T/g \sim 25$ km.

Second, we have

$$\frac{1}{\pi}\frac{\partial \pi}{\partial x} = \frac{\theta\, \partial\pi/\partial x}{\theta\pi} \sim \frac{f V_g}{g H_\pi} \tag{7}$$

so that the magnitudes of the horizontal derivatives $\partial\pi/\partial x$ and $\partial\pi/\partial y$ are determined.

Third, because the winds are assumed nearly geostrophic, the thermal wind relation should be a good approximation. From Eqs. (4) and (5) we have

$$\frac{\partial V_g}{\partial z} \sim \frac{g\theta}{f}\frac{\partial}{\partial x}\frac{1}{\theta} + \frac{1}{\theta}\frac{\partial\theta}{\partial z}V_g \sim \frac{g}{f\theta}\frac{\partial\theta}{\partial x} + \frac{1}{\theta}\frac{\partial\theta}{\partial z}V_g \tag{8}$$

But for hydrostatic conditions and $\gamma \sim 2\gamma_d/3$

$$\frac{1}{\theta}\frac{\partial\theta}{\partial z} = \frac{1}{T}(\gamma_d - \gamma) \sim \frac{g}{3c_pT} \sim \frac{1}{3H_\pi} \tag{9}$$

Therefore, if L_H is the horizontal wavelength, then

$$\frac{g}{f\theta}\frac{\partial\theta}{\partial x} \div \frac{1}{\theta}\frac{\partial\theta}{\partial z}V_g \sim \frac{4g}{fL_H}\frac{3H_\pi}{V_g} \tag{10}$$

As an estimate we use $V_g \sim 10$ m/s, $f \sim 10^{-4}$ s^{-1} so that $g/fV_g \sim 10^4$. For horizontal wavelengths of order 4000 km, we have $12H_\pi/L_H \sim 7.5 \times 10^{-2}$ so that we need to keep only the first term of Eq. (8) and we have the scale relation

$$\frac{1}{\theta}\frac{\partial\theta}{\partial x} \sim \frac{f}{g}\frac{\partial V_g}{\partial z} \sim \frac{fV_g}{gD} \tag{11}$$

in which D is a new scale derived from $\partial V_g/\partial z \sim V_g/D$. The average speed of polar-front jet winds is about 20 m/s so that the net shear below the jet might be $\partial V_g/\partial z \sim (20 \text{ m/s})/10$ km. Taking this to be the typical shear in developing situations and $V_g \sim 10$ m/s, we have $D \sim 5$ km, or about half the depth of the troposphere.

We may define $\theta_0(z)$ and $\pi_0(z)$ to represent atmospheric conditions that would prevail in the absence of motion. They must obey the hydrostatic condition

$$\theta_0\frac{\partial\pi_0}{\partial z} = -g \tag{12}$$

We shall represent the actual variables in the form

$$\theta = \theta_0(z)(1 + \Theta\theta') \qquad \pi = \pi_0(z)(1 + \Pi\pi') \tag{13}$$

in which θ' and π' are dimensionless quantities of order 1 and Θ and Π are nondimensional quantities giving the approximate fractional size of the perturbations associated with the large-scale motion.† For example, if we were to write $\theta = \theta_0 + \theta^*$ in which θ^* is the actual perturbation, then $\theta^*/\theta_0 \sim \Theta$.

We estimate the quantities Θ and Π from the basic scaling relations. From Eq. (11) we find

$$\frac{fV_g}{gD} \sim \frac{1}{\theta}\frac{\partial\theta}{\partial x} \sim \Theta\frac{\partial\theta'}{\partial x} \sim \frac{\Theta}{L} \tag{14}$$

where L is a length given by $4L \sim L_H$. Thus

$$\Theta \sim \frac{fLV_g}{gD} \sim \frac{10^{-4}\ \text{s}^{-1}\ 10^6\ \text{m}\ 10\ \text{m/s}}{10\ \text{m/s}^2\ 5 \times 10^3\ \text{m}} \sim 2 \times 10^{-2} \tag{15}$$

so that $\theta^* \sim \pm 7$ K. From Eq. (7) we find similarly that $\Pi \sim fLV_g/gH_\pi \sim D\Theta/H_\pi$.

Now we turn to the basic temporal scaling. The equation of motion may be written as

$$\frac{\partial \mathbf{v}}{\partial t} + \mathbf{v} \cdot \nabla \mathbf{v} = f\mathbf{k} \times (\mathbf{v}_g - \mathbf{v}) - \mathbf{k}\left(\theta\frac{\partial \pi}{\partial z} + g\right) \tag{16}$$

in which we shall denote geostrophic and ageostrophic components as

$$\mathbf{v} = \mathbf{v}_g + \mathbf{v}_a + \mathbf{k}w \tag{17}$$

where we are assuming, say, that $|\mathbf{v}_a| \div |\mathbf{v}_g| \ll 1$. With the aid of a time scale T we obtain from Eq. (16) the linear estimate for the horizontal component $V_g/T \sim f|\mathbf{v}_a| = fv_a$.

It is worth pointing out here that we are often forced in our analysis to make linear estimates of nonlinear terms. For example, we use $\mathbf{v} \cdot \nabla_H\phi \sim V\phi/L$, although in

†The usual approach in nonlinear problems with a small parameter is illustrated by the equation

$$y'' + \lambda y + \epsilon y^2 = 0$$

The expansion $y = y_0 + \epsilon y_1 + \epsilon^2 y_2 + \cdots$ produces the series of equations

$$\begin{aligned} &y_0'' + \lambda y_0 = 0 \\ &y_1'' + \lambda y_1'' + y_0^2 = 0 \\ &y_2'' + \lambda y_2'' + 2y_1 y_0 = 0 \\ &\cdots \end{aligned}$$

which can be solved in succession.

In the development here, we are considering only the first two terms in each variable (for example, $\theta = \theta_0 + \Theta\theta_0\theta'$) and the resulting equations will apply only to the large-scale quasi-geostrophic and associated large-scale ageostrophic components of the motions. Implicit in the analysis is the assumption that we have only one scale of motion; quite different equations would result if we let the ageostrophic component be of much smaller scale than the quasi-geostrophic component.

many cases $\mathbf{v}$ and $\nabla\phi$ may be orthogonal. We must employ some physical intuition for these estimates to provide a realistic appraisal of the orders of magnitude.

In synoptic systems, local changes are of the order of advective changes so that

$$\frac{\partial}{\partial t}(\ ) \sim \mathbf{v}_g \cdot \nabla(\ ) \tag{18}$$

or $T^{-1} \sim V_g/L$. But then we have $V_g{}^2/L \sim fv_a$ and we define the *Rossby number* as

$$Ro = \frac{V_g}{fL} \sim \frac{v_a}{V_g} \tag{19}$$

Because

$$Ro \sim \frac{10 \text{ m/s}}{10^{-4} \text{ s}^{-1}\ 10^{6} \text{ m}} \sim 0.1 \tag{20}$$

we are permitting ageostrophic components to be about 10 percent of the geostrophic wind.

Furthermore, we now have

$$\frac{1}{T} \sim \frac{fv_a}{V_g} \sim fRo \sim 10^{-5} \text{ s}^{-1} \sim 1 \text{ day}^{-1} \tag{21}$$

so that the actual period T_p is $T_p \sim 4T \sim 4$ day.

14.1.2 Scaling Results from the First Law and Continuity Equation

Upon substituting Eq. (17) in Eq. (2) and using $v_a \sim 0.1\,V_g$ we obtain

$$\frac{1}{\theta}\frac{\partial\theta}{\partial t} + \mathbf{v}_g \cdot \frac{\nabla_H\theta}{\theta} + \frac{w}{\theta}\frac{\partial\theta}{\partial z} = 0 \tag{22}$$

Thus the use of the estimates (9), (14), and (18) along with (13) gives

$$\frac{\Theta V_g}{L} + \frac{V_g\Theta}{L} + \frac{W}{3H_\pi} = 0 \tag{23}$$

in which W is the characteristic vertical velocity. Thus we have as an estimate of order of magnitude

$$\frac{W}{3H_\pi} \sim \frac{fV_g{}^2}{gD} \tag{24}$$

With the definition of the Richardson number

$$Ri = \frac{(g/\theta)(\partial\theta/\partial z)}{|\partial\mathbf{v}/\partial z|^2} = \frac{g/3H_\pi}{V_g{}^2/D^2} \tag{25}$$

we find that $W \sim fD/Ri$. From Eq. (25) we see that $Ri \sim 30$ and thus the characteristic vertical velocity is $W \sim 2$ cm/s.

By definition we must have $\mathbf{v}_g$ orthogonal to $\nabla_H \pi$ so that the equation of continuity (3) may be written as

$$\frac{1}{\pi}\frac{\partial \pi}{\partial t} + \mathbf{v}_a \cdot \frac{\nabla_H \pi}{\pi} + \frac{w}{\pi}\frac{\partial \pi}{\partial z} + \frac{R}{c_V}\nabla \cdot \mathbf{v} = \frac{R}{c_V}\frac{H}{\theta} \tag{26}$$

The divergence may be expressed as

$$\nabla \cdot \mathbf{v} = \nabla \cdot \mathbf{v}_g + \nabla \cdot \mathbf{v}_a + \frac{\partial w}{\partial z} \sim \frac{\beta V_g}{f} + \frac{v_a}{L} + \frac{W}{D} \tag{27}$$

in which we have now adopted D as the characteristic scale of the oscillations in the vertical velocity. We have defined $\beta = \partial f/\partial y = \partial f/a\,\partial\phi$, and so at latitude $\phi = 45°$, we have $\beta/f \sim 1/a$ where a is the earth's radius; furthermore $v_a/L \sim Ro V_g/L$ so that the magnitudes in Eq. (26) are given by

$$\frac{V_g \Pi}{L} + \frac{Ro V_g}{L}\Pi + \frac{W}{H_\pi} + \frac{R}{c_V}\left(\frac{V_g}{a} + \frac{Ro V_g}{L} + \frac{W}{D}\right) \sim \frac{R}{c_V}\frac{H}{\theta_0} \tag{28}$$

But

$$\frac{V_g \Pi}{L} \sim \frac{f V_g^2}{g H_\pi} \sim \frac{WD}{3H_\pi^2} \tag{29}$$

so that upon collecting the dominant terms we have

$$W\left(\frac{1}{H_\pi} + \frac{R}{c_V}\frac{1}{D} + \frac{D}{3H_\pi^2}\right) \sim \frac{R}{c_V}\frac{V_g}{L}\left(\frac{L}{a} + Ro\right) + \frac{R}{c_V}\frac{H}{\theta_0} \tag{30}$$

Therefore, because the middle term on the left dominates, we arrive at the estimate

$$\frac{W}{D} \sim \frac{V_g}{L}\left(\frac{L}{a} + Ro\right) + \frac{H}{\theta_0} \tag{31}$$

This relation was first derived by Burger (1958) for the isentropic case with the vorticity equation.

The point here is that the important mechanisms in waves with $L/a \ll 1$ and those with $L \sim a$ may be different. The analysis shows that for small waves the vertical velocity is coupled to the ageostrophic divergence fields, while for planetary waves ($Ro \sim 0.01$) the β effect becomes significant. For waves of cyclone size we shall have $L/a \sim 0.15$ and $Ro \sim 0.1$, so both effects will have to be retained in the continuity equation.

PROBLEM 14.1.1 Show that for heating rates ($\partial\theta/\partial t \sim H$) less than ~ 3 deg/day, the second term of Eq. (31) may be neglected.

PROBLEM 14.1.2 Justify the result used in Eq. (27) that $\nabla \cdot \mathbf{v}_g \sim \beta V_g/f$.

*PROBLEM 14.1.3 Show that the vertical velocity scale W increases as stability ($\omega_g{}^2$) decreases. Show that D increases with decreasing stability for waves of very small scale, for cyclonic waves, and for planetary waves.

PROBLEM 14.1.4 The ultralong waves (say three or fewer around the globe) tend to be stationary. It is tempting to argue that therefore we should have $\partial\mathbf{v}/\partial t \sim \partial\mathbf{v}_a/\partial t$ and $\mathbf{v} \cdot \nabla \mathbf{v} \sim \mathbf{v}_g \cdot \nabla \mathbf{v}_a$. Show that this temptation leads to unrealistic estimates. (Hint: For such waves $Ro \sim 0.01$.) Use the basic estimates $T^{-1} \sim V_g/L$ and $V_g{}^2/L \sim f v_a$ to develop a consistent set of numerical estimates for these ultralong waves.

14.1.3 The Approximate Equations

The preceding analysis has produced relations among the scales and magnitudes of the variables that permit us to reduce the equations of motion to an approximate set appropriate to the quasi-geostrophic regime in the atmosphere.

We put

$$\mathbf{v} = V_g(\mathbf{u}_g + Ro\mathbf{u}') + \mathbf{k}Ww \tag{32}$$

where the horizontal velocities $\mathbf{u}_g$ and $\mathbf{u}'$ and the vertical component w' are all of order unity. The relation (19) was used in scaling $\mathbf{u}'$. The horizontal component of Eq. (16) may be written

$$\frac{V_g{}^2}{L}\left[\frac{\partial \mathbf{u}_g}{\partial t'} + \mathbf{u}_g \cdot \nabla'_H \mathbf{u}_g + Ro\left(\frac{\partial \mathbf{u}'}{\partial t'} + \mathbf{u}_g \cdot \nabla'_H \mathbf{u}' + \mathbf{u}' \cdot \nabla'_H \mathbf{u}_g + Ro\mathbf{u}' \cdot \nabla'_H \mathbf{u}'\right)\right]$$
$$+ \frac{w' V_g}{D}\left(\frac{\partial \mathbf{u}_g}{\partial z'} + Ro\frac{\partial \mathbf{u}'}{\partial z'}\right) V_g \frac{D}{L}\left(\frac{L}{a} + Ro\right) = -f V_g Ro \mathbf{k} \times \mathbf{u}' \tag{33}$$

In this equation we have used the nondimensional quantities $t' = t/T$, $z' = z/D$, and $\nabla'_H(\ ) = L\nabla_H(\ )$, and we have substituted Eq. (31) with $H = 0$ for the magnitude W in the last term on the left. We bring the variation of the Coriolis force into effect by defining

$$f = f_0 + \left(\frac{\partial f}{\partial \phi}\right)_{\phi_0}(\phi - \phi_0) \tag{34}$$

$$= f_0\left(1 + \cot\phi_0 \frac{y}{a}\right) = f_0\left(1 + \beta^* \frac{L}{a} y'\right) \tag{34 (cont'd)}$$

Henceforth when f appears in a scale parameter, we use f_0 to estimate its value.

Upon dividing Eq. (33) by $f_0 V_g$ and canceling one power of the Rossby number, we retain the terms of order unity and we have

$$\frac{\partial \mathbf{u}_g}{\partial t'} + \mathbf{u}_g \cdot \nabla'_H \mathbf{u}_g + w' \frac{\partial \mathbf{u}_g}{\partial z'}\left(\frac{L}{a} + Ro\right) = -\mathbf{k} \times \mathbf{u}' \tag{35}$$

We shall investigate the vertical component of Eq. (16) in two steps. The term R on the right, with the use of Eq. (12), becomes

$$\begin{aligned} R &= \theta\pi_z + g \\ &= \theta_0(1 + \Theta\theta')\left[-\frac{g}{\theta_0}(1 + \Pi\pi') + \pi_0 \Pi \frac{\partial \pi'}{\partial z}\right] + g \\ &= -g\Pi\pi' - g\Theta\theta' - g\Theta\Pi\pi'\theta' + \theta_0\pi_0\Pi(1 + \Theta\theta')\frac{\partial \pi'}{\partial z} \end{aligned} \tag{36}$$

But $\Pi/\Theta \sim D/H_\pi$, so that a nondimensional form of Eq. (36) is

$$\frac{R}{g\Theta} = -\left(\frac{D}{H_\pi}\pi' + \theta' + \Pi\pi'\theta'\right) + \frac{\theta_0\pi_0}{gH_\pi}\frac{\partial \pi'}{\partial z'}(1 + \Theta\theta') \tag{37}$$

With Eqs. (6) and (15) and the fact that $D/H_\pi \sim \frac{1}{5}$, the dominant part of this expression is

$$\frac{R}{g\Theta} = \frac{\partial \pi'}{\partial z'} - \theta' \tag{38}$$

For the vertical component of Eq. (16) then we have

$$\frac{V_g W}{L}\left(\frac{\partial w'}{\partial t'} + \mathbf{u}_g \cdot \nabla'_H w' + Ro\,\mathbf{u}' \cdot \nabla'_H w'\right) + \frac{W^2}{D}\frac{\partial w'}{\partial z'} = -g\Theta\left(\frac{\partial \pi'}{\partial z'} - \theta'\right) \tag{39}$$

But with Eqs. (15) and (31) and $H = 0$, we find

$$\frac{V_g W/L}{g\Theta} = Ro\frac{D^2}{L^2}\left(\frac{L}{a} + Ro\right) \qquad \frac{W^2}{gD\Theta} \sim Ro\frac{D^2}{L^2}\left(\frac{L}{a} + Ro\right)^2 \tag{40}$$

so we clearly need to retain only the perturbation hydrostatic equation

$$\frac{\partial \pi'}{\partial z'} - \theta' = 0 \tag{41}$$

It is worth noting here that the smallness of the ratios in Eq. (40) arises from $D/L \ll 1$, so that this condition is now seen to be a sufficient one for validating the hydrostatic approximation.

From the first law, Eq. (2), we obtain with the aid of Eq. (24)

$$\theta_0\Theta\frac{V_g}{L}\left(\frac{\partial\theta'}{\partial t'}+\mathbf{u}_g\cdot\nabla'_H\theta'+Ro\mathbf{u}'\cdot\nabla'_H\theta'\right)+\frac{fV_g{}^2 3H_\pi}{gD^2}w'\left[\frac{\partial\theta_0}{\partial z'}(1+\Theta\theta')+\theta_0\Theta\frac{\partial\theta'}{\partial z'}\right]=H \tag{42}$$

The magnitude of the first term and the leading, and dominant, term in the brackets thus have the ratio

$$\frac{\theta_0\Theta V_g/L}{(fV_g{}^2 3H_\pi/gD^2)\dfrac{\partial\theta_0}{\partial z'}}=\frac{fV_g{}^2/gD}{(fV_g{}^2/gD)\left(\dfrac{3H_\pi}{D}\dfrac{1}{\theta_0}\dfrac{\partial\theta_0}{\partial z'}\right)}\sim 1 \tag{43}$$

in which we have used Eqs. (9) and (15). Thus the approximate first law is

$$\frac{\partial\theta'}{\partial t'}+\mathbf{u}_g\cdot\nabla'_H\theta'+w'\sigma=3\frac{Ri}{gDf_0}(q+f_h)=Q' \tag{44}$$

where the stability factor $\sigma(z)$ is

$$\sigma=\frac{3H_\pi}{D}\frac{1}{\theta_0}\frac{\partial\theta_0}{\partial z'}\sim 1 \tag{45}$$

For isentropic motion, Q' vanishes.

The continuity equation (3) may be similarly written for $H\lesssim 3$ deg/day as

$$\frac{\pi_0\Pi V_g}{L}\left(\frac{\partial\pi'}{\partial t'}+\mathbf{u}_g\cdot\nabla'_H\pi'+Ro\mathbf{u}'\cdot\nabla'_H\pi'\right)+\frac{V_g}{L}\left(\frac{L}{a}+Ro\right)\left[\frac{\partial\pi_0}{\partial z'}(1+\Pi\pi')+\pi_0\Pi\frac{\partial\pi'}{\partial z'}\right]w'$$
$$+\frac{R}{c_V}\left(-V_g\frac{\cot\phi}{a}v_g+\frac{RoV_g}{L}\nabla'_H\cdot\mathbf{u}'+\frac{W}{D}\frac{\partial w'}{\partial z'}\right)\pi_0(1+\Pi\pi')=0 \tag{46}$$

As with the first law, the ratio of the dominant parts of the first two terms is

$$\frac{\pi_0 fLV_g/gH_\pi}{(L/a+Ro)\partial\pi_0/\partial z'}\sim\frac{fLV_g/gH_\pi}{(L/a+Ro)D/H_\pi}=\frac{fLV_g/gD}{(L/a+Ro)}\sim\frac{2\times 10^{-2}}{(L/a+Ro)} \tag{47}$$

so that the first term is small in comparison to the second.

The two remaining terms may be rewritten as

$$\left(\frac{L}{a}+Ro\right)w'\frac{\partial\pi_0}{\partial z'}+\frac{R}{c_V}\pi_0\left[-\frac{L}{a}\cot\phi\, v_g+Ro\,\nabla'_H\cdot\mathbf{u}'+\left(\frac{L}{a}+Ro\right)\frac{\partial w'}{\partial z'}\right]=0 \tag{48}$$

in which we have used Eq. (31) in the last term. But now with Poisson's equation we find

$$\frac{1}{\pi_0}\frac{\partial\pi_0}{\partial z}=\frac{R}{c_p}\frac{1}{p_0}\frac{\partial p_0}{\partial z}=\frac{R}{c_V}\left(\frac{1}{\theta_0}\frac{\partial\theta_0}{\partial z}+\frac{1}{\rho_0}\frac{\partial\rho_0}{\partial z}\right) \tag{49}$$

For the atmosphere, the first term on the right is of order (50 km)$^{-1}$ while the second is of order (10 km)$^{-1}$. Thus we may replace the logarithmic derivative of π_0 in Eq. (48) with that of ρ_0 according to Eq. (49). We use ϕ_0 in place of ϕ and thus

$$\frac{1}{Ro}\left(\frac{L}{a} + Ro\right)w'\frac{\partial \rho_0}{\partial z'} + \rho_0\left[-\frac{L}{aRo}\cot \phi_0 v_g + \nabla'_H \cdot \mathbf{u}' + \frac{1}{Ro}\left(\frac{L}{a} + Ro\right)\frac{\partial w'}{\partial z'}\right] = 0 \quad (50)$$

For cyclone scale motion, we might have $L_H = 2000 - 4000$ km so that $L \sim 500 - 1000$ km and hence $L/a \sim 0.07 \sim 0.14$. Thus for $Ro \sim 0.1$ we must keep each of these terms if we wish accuracy to, say, 10 percent. We define $\lambda = L/aRo$ and thus we have

$$\rho_0 \nabla'_H \cdot \mathrm{u}' + (1 + \lambda)\frac{\partial}{\partial z'}(\rho_0 w') - \rho_0 \beta' v_g = 0 \tag{51}$$

where $\beta' = \lambda \cot \phi_0$.

We note that for $L/a \ll 1$ and $\beta' = 0$, the continuity equation reduces, appropriately, to that obtained for deep convection.

PROBLEM 14.1.5 Verify the form given in Eq. (44) for the heating term. Show that for rates $1 \lesssim H \lesssim 3$ deg/day this term should be in Eq. (44) although it may be neglected in the continuity equation. What total rainfall/day out of a layer 400 mbar thick would correspond to these bounds on H?

14.2 DISCUSSION OF THE QUASI-GEOSTROPHIC EQUATIONS

We have derived a set of quasi-geostrophic equations for large-scale, moderately heated flow that should be useful in the study of the evolution of synoptic systems in the atmosphere. For convenience we assemble the entire set here, dropping the prime on the independent variables that indicates they are nondimensional. We have

$$\frac{\partial \mathbf{u}_g}{\partial t} + \mathbf{u}_g \cdot \nabla_H \mathbf{u}_g + Ro(1 + \lambda)w'\frac{\partial \mathbf{u}_g}{\partial z} = -\mathbf{k} \times \mathbf{u}' \tag{1}$$

$$\frac{\partial \pi'}{\partial z} - \theta' = 0 \tag{2}$$

$$\frac{\partial \theta'}{\partial t} + \mathbf{u}_g \cdot \nabla_H \theta' + w'\sigma = Q' \tag{3}$$

$$\nabla_H \cdot \mathbf{u}' + \frac{1 + \lambda}{\rho_0}\frac{\partial}{\partial z}(\rho_0 w') - \beta' v_g = 0 \tag{4}$$

From the basic definition of the geostrophic velocity we have

$$\mathbf{v}_g = V_g \mathbf{u}_g = \frac{\theta_0(1 + \Theta\theta')\pi_0 \Pi L^{-1}\mathbf{k} \times \nabla_H \pi'}{f_0[1 + \beta^*(Ly/a)]} \tag{5}$$

But with the definition of Π and with Eq. (6) of Sec. 14.1, we find upon retention of dominant terms that

$$\mathbf{u}_g = \mathbf{k} \times \nabla_H \pi' \tag{6}$$

Thus the geostrophic velocity in the set of equations can be replaced with Eq. (6), and we have a complete set of equations in π', θ', $\mathbf{u}'$, and w'.

We shall restrict attention now to the case of interest in the study of development of weather systems so that $Ro \sim L/a \sim 0.1$. With this assumption the vertical velocity term in Eq. (1) may be neglected and the system will reduce to a relatively simple nonlinear partial differential equation in π'. From Eqs. (1) and (6) we have

$$\left[\frac{\partial}{\partial t} + (\mathbf{k} \times \nabla_H \pi') \cdot \nabla_H\right] \nabla_H \pi' = -\mathbf{u}' \tag{7}$$

and Eqs. (2) and (3) combine upon invoking the isentropic assumption to give

$$\left[\frac{\partial}{\partial t} + (\mathbf{k} \times \nabla_H \pi') \cdot \nabla_H\right] \frac{\partial \pi'}{\partial z} + w'\sigma = 0 \tag{8}$$

Now we eliminate $\mathbf{u}'$ between Eqs. (7) and (4). For the nonlinear term in Eq. (7) we will have

$$\nabla_H \cdot \{[(\mathbf{k} \times \nabla_H \pi') \cdot \nabla_H] \nabla_H \pi'\} = \frac{\partial}{\partial x_i}\left(\epsilon_{j3m} \frac{\partial \pi'}{\partial x_m} \frac{\partial^2 \pi'}{\partial x_j \, \partial x_i}\right)$$

$$= \epsilon_{j3m} \frac{\partial \pi'}{\partial x_m} \frac{\partial}{\partial x_j} \frac{\partial^2 \pi'}{\partial x_i{}^2} + \epsilon_{j3m} \frac{\partial^2 \pi'}{\partial x_m \, \partial x_i} \frac{\partial^2 \pi'}{\partial x_j \, \partial x_i} \tag{9}$$

The second term vanishes owing to the asymmetry of ϵ_{j3m} and thus Eqs. (7) and (4) give

$$\left[\frac{\partial}{\partial t} + (\mathbf{k} \times \nabla_H \pi') \cdot \nabla_H\right] \nabla_H{}^2 \pi' = \frac{1+\lambda}{\rho_0} \frac{\partial}{\partial z}(\rho_0 w') - \beta' v_g \tag{10}$$

We note that the last term on the right is $\beta' v_g = \beta' \mathbf{u}_g \cdot \nabla y$. We multiply Eq. (8) by ρ_0 and observe that one of the terms that develops in the differentiation of $\rho_0 w'$ required by Eq. (10) vanishes. Thus finally

$$\left[\frac{\partial}{\partial t} + (\mathbf{k} \times \nabla_H \pi') \cdot \nabla_H\right]\left[\nabla_H{}^2 \pi' + \frac{1+\lambda}{\rho_0} \frac{\partial}{\partial z}\left(\frac{\rho_0}{\sigma} \frac{\partial \pi'}{\partial z}\right) + \beta' y\right] = 0 \tag{11}$$

This single equation controls the evolution of the synoptic scale disturbances as modeled by the set of quasi-geostrophic equations. It has a relatively straightforward physical interpretation. From Eq. (6) we see that the geostrophic vorticity $\mathbf{k} \cdot (\nabla \times \mathbf{u}_g)$ is $\zeta_g = \nabla_H{}^2 \pi'$ and we define

$$\frac{\delta}{\delta t} = \frac{\partial}{\partial t} + (\mathbf{k} \times \nabla_H \pi') \cdot \nabla_H \tag{12}$$

so that Eq. (11) becomes

$$\frac{\delta}{\delta t}\left[\zeta_g + \frac{1+\lambda}{\rho_0}\frac{\partial}{\partial z}\left(\frac{\rho_0}{\sigma}\theta'\right) + \beta' y\right] = \frac{\delta Z_g}{\delta t} \tag{13}$$

The vorticity form in the brackets is conserved following the motion in the sense of Eq. (12), and it involves the vorticity as well as a measure of static stability and the variation of the Coriolis parameter. We are therefore justified in viewing it as the *quasi-geostrophic potential vorticity*, and so we denote it, as indicated in Eq. (13), by Z_g.

Equation (11) also permits us to develop an energy integral for the system of equations. We assume cyclic continuity on horizontal surfaces, and we see from Eq. (8) that $w' = 0$ on a horizontal, rigid lower surface will yield the boundary condition

$$\left[\frac{\partial}{\partial t} + (\mathbf{k} \times \nabla_H \pi') \cdot \nabla_H\right]\frac{\partial \pi'}{\partial z} = 0 \qquad z = 0 \tag{14}$$

Moreover, we consider only solutions that are regular in the sense that the energy forms $\rho_0(\mathbf{u}_g \cdot \mathbf{u}_g)$, $\rho_0(w')^2$, $\rho_0(\theta')^2$, and $\rho_0(\pi')^2$ are integrable over the entire atmosphere. This condition thus implies that these quantities must vanish as $z \to \infty$.

We shall multiply Eq. (11) by $\rho_0 \pi'$ and integrate over the entire atmosphere. With Eq. (9) the first term becomes

$$\int_V \rho_0 \pi' \nabla_H \cdot \left[\left(\frac{\partial}{\partial t} + \mathbf{u}_g \cdot \nabla_H\right)\nabla_H \pi'\right] dV = \int_V \rho_0 \nabla_H \cdot \left[\pi'\left(\frac{\partial}{\partial t} + \mathbf{u}_g \cdot \nabla_H\right)\nabla_H \pi'\right] dV$$
$$- \int_V \frac{\rho_0}{2}\left(\frac{\partial}{\partial t} + \mathbf{u}_g \cdot \nabla_H\right)|\nabla_H \pi'|^2 \, dV \tag{15}$$

The first integral here vanishes because of cyclic continuity. The next term in Eq. (11) gives

$$\int_V \rho_0 \pi'\left(\frac{\partial}{\partial t} + \mathbf{u}_g \cdot \nabla_H\right)\left[\frac{1+\lambda}{\rho_0}\frac{\partial}{\partial z}\left(\frac{\rho_0}{\sigma}\frac{\partial}{\partial z}\pi'\right)\right] dV$$

$$= \int_V \frac{\partial}{\partial z}\left[\pi'\left(\frac{\partial}{\partial t} + \mathbf{u}_g \cdot \nabla_H\right)\left(\frac{1+\lambda}{\sigma}\rho_0\frac{\partial \pi'}{\partial z}\right)\right] dV$$
$$- \int_V \frac{\rho_0}{2}\left(\frac{1+\lambda}{\sigma}\right)\left(\frac{\partial}{\partial t} + \mathbf{u}_g \cdot \nabla_H\right)\left(\frac{\partial \pi'}{\partial z}\right)^2 dV - \int_V \frac{\rho_0(1+\lambda)}{\sigma}\pi'\frac{\partial \mathbf{u}_g}{\partial z} \cdot \nabla_H \frac{\partial \pi'}{\partial z}\, dV \tag{16}$$

The first term on the right vanishes as $z \to \infty$ because of the regularity condition (see Problem 14.2.2 below) and vanishes at $z = 0$ because of the boundary condition at the lower surfaces. The last vanishes because

$$\frac{\partial \mathbf{u}_g}{\partial z} \cdot \nabla_H \frac{\partial \pi'}{\partial z} = \left(\mathbf{k} \times \nabla_H \frac{\partial \pi'}{\partial z}\right) \cdot \nabla_H \frac{\partial \pi'}{\partial z} = 0 \tag{17}$$

The third term in Eq. (11) gives

$$\int_V \rho_0 \beta' \pi' \frac{\partial \pi'}{\partial x} \, dV = \int_V \rho_0 \beta' \frac{\partial}{\partial x}\left(\frac{\pi'^2}{2}\right) dV = 0 \tag{18}$$

Thus we have

$$\int_V \frac{\rho_0}{2}\left(\frac{\partial}{\partial t} + \mathbf{u}_g \cdot \nabla_H\right)\left[|\nabla_H \pi'|^2 + \left(\frac{1+\lambda}{\sigma}\right)\left(\frac{\partial \pi'}{\partial z}\right)^2\right] dV = 0 \tag{19}$$

But for any cyclically continuous scalar ϕ

$$\int_V \mathbf{u}_g \cdot \nabla_H \phi \, dV = \int_V [\nabla_H \cdot \phi \mathbf{u}_g - \phi \nabla_H \cdot (\mathbf{k} \times \nabla_H \pi')] \, dV = 0 \tag{20}$$

which vanishes because the first term is eliminated by cyclic continuity and the second because $\nabla_H \times (\nabla_H \pi')$ vanishes. Therefore upon using Eq. (2) the energy equation is

$$\frac{\partial}{\partial t} \int_V \frac{\rho_0}{2}\left(|\nabla_H \pi'|^2 + \frac{1+\lambda}{\sigma} |\theta'|^2\right) dV = 0 \tag{21}$$

The first term is the geostrophic kinetic energy, the second another perturbation form of the available potential energy. Thus growth of disturbances can occur because the kinetic energy in the first term is rearranged to produce local increases of kinetic energy, or they can grow because the available energy is converted into geostrophic kinetic energy. More significantly, the quasi-geostrophic equations show again that the motions of the atmosphere arise from the differential heating that creates variations in potential temperature and thus maintains the supply of available energy.

*PROBLEM 14.2.1 Show that Rossby waves are included as a special case in Eq. (11).

PROBLEM 14.2.2 Show that the boundary term in Eq. (16) is

$$-\int_V \frac{\partial}{\partial z} [(1+\lambda)\rho_0 \pi' w'] \, dV$$

so that it vanishes at infinity because of the regularity condition. (Hint: Apply the Schwarz inequality to the integral over the upper surface.)

PROBLEM 14.2.3 Assuming Q' does not vanish in Eq. (3), what modifications must be made in Eq. (11)? Show that the rate of change of energies in Eq. (21) is equal to $\int_V (1+\lambda)\sigma^{-1}\rho_0\theta' Q'\,dV$ if Q' vanishes at $z=0$ and is regular in the sense that $\rho_0 Q'^2$ is integrable.

*PROBLEM 14.2.4 Let Q' vanish. Show that for the globe $\partial/\partial t \int_V \rho_0 Z_g{}^2\,dV = 0$ and that if $F(x)$ is any differentiable scalar function then $\partial/\partial t \int_V \rho_0 F(Z_g)\,dV = 0$. Show that cyclic continuity in x and a condition that $v_g = \pi'_x$ vanish on lateral boundaries at $y=0$ and $y=1$ will give the same result. How do these results change if Q' does not vanish in Eq. (3)?

PROBLEM 14.2.5 The answer to Problem 10.7.2 is the quasi-geostrophic equation in isobaric coordinates. Make a list of the assumptions used to obtain the equation via the argument culminating in that problem and compare it to the assumptions here. What information did we obtain with the lengthy derivation of this chapter that we did not find in the approach of Problem 10.7.2?

PROBLEM 14.2.6 Find a new vertical coordinate ξ such that the $1+\lambda$ in Eq. (11) does not appear explicitly. If σ is constant, show that the same procedure will allow the middle term of Eq. (11) to be written $\alpha_0\,\partial(\rho_0\,\partial\pi'/\partial\xi)/\partial\xi$.

14.3 LINEAR STABILITY OF QUASI-GEOSTROPHIC MOTIONS

The partial differential equation derived in the previous section can be linearized and used to study the stability of zonal flow to small perturbations. We shall derive an energy equation for disturbances, illustrate instability with a simple model, and then investigate stability with integral theorems.

14.3.1 The Linear Equation and Its Energetics

To begin, we observe that $\mathbf{u}_g = \mathbf{k} \times \nabla_H \pi'$ is nondivergent so that we may use a stream function $\psi = \pi'$ so that $u_g = -\psi_y$, $v_g = \psi_x$, in which subscripts x and y denote differentiation with respect to the nondimensional variables. But with the aid of the hydrostatic equation (2) in Sec. 14.2 we have

$$\frac{\partial \mathbf{u}_g}{\partial z} = -\psi_{yz}\mathbf{i} + \psi_{xz}\mathbf{j} = \mathbf{k} \times \nabla_H \frac{\partial \pi'}{\partial z} = \mathbf{k} \times \nabla_H \theta' \qquad (1)$$

and so we can use the stream function also to represent potential temperature as $\psi_z = \theta'$.

We shall use Λ to denote $(1+\lambda)/\sigma$, now assumed constant, and the partial differential equation (11) in Sec. 14.2 becomes (upon dropping the prime on β)

$$\left(\frac{\partial}{\partial t} - \psi_y \frac{\partial}{\partial x} + \psi_x \frac{\partial}{\partial y}\right)\left[\psi_{xx} + \psi_{yy} + \frac{\Lambda}{\rho_0}(\rho_0 \psi_z)_z + \beta y\right] = 0 \tag{2}$$

A stream function $\Psi = \Psi(y,z)$ that varies only with y and z is a solution of Eq. (2), and so we may take a zonal current varying only with latitude and height as a basic state upon which to superimpose perturbations. From this definition, we have $U = -\Psi_y$, $U_z = -\Psi_{yz}$ and we add to this basic flow a small disturbance by setting

$$\psi = \Psi(y,z) + \phi(x,y,z,t) \tag{3}$$

in which we assume that $|\phi| \ll |\Psi|$. Upon substituting Eq. (3) in Eq. (2) and linearizing, we have

$$\left(\frac{\partial}{\partial t} - \Psi_y \frac{\partial}{\partial x}\right)\left[\phi_{xx} + \phi_{yy} + \frac{\Lambda}{\rho_0}(\rho_0 \phi_z)_z\right] + \phi_x\left[\beta + \Psi_{yyy} + \frac{\Lambda}{\rho_0}(\rho_0 \Psi_{yz})_z\right]$$

$$= \left(\frac{\partial}{\partial t} + U\frac{\partial}{\partial x}\right)\left[\phi_{xx} + \phi_{yy} + \frac{\Lambda}{\rho_0}(\rho_0 \phi_z)_z\right] + \phi_x\left[\beta - U_{yy} - \frac{\Lambda}{\rho_0}(\rho_0 U_z)_z\right]$$

$$= 0 \tag{4}$$

It is easily shown that the assumptions that $U = \text{const}$ and that $\phi = \hat{\phi} e^{i\kappa(x - Ct)}$ along with Eq. (4) will yield the Rossby wave speed formula when $\hat{\phi}$ is constant. Thus we have allowed in Eq. (4) for the modification of the Rossby mode of wave propagation by horizontal and vertical shear of U and by latitudinal and vertical variations of the perturbation.

In order to proceed further with the linear equation, we must set appropriate boundary conditions. We assume cyclic continuity in the x direction, and we assume that the flow is confined by vertical walls at $y = \pm l$. Thus from the requirement that $v = \psi_x = 0$ at $y = \pm l$ we have $\phi_x = 0$ at $y = \pm l$. But for harmonic variation ($\kappa \neq 0$) in the x direction, this condition requires that $\phi = 0$ at $y = \pm l$. We adopt this stronger condition, so that perturbations in both velocity and potential temperature are not present at the lateral wall.

At the lower rigid boundary, we let $Z_B(x,y)$ denote the height of the surface. A moving parcel at the lower boundary must follow this contour and so in dimensional variables

$$w = \frac{dZ_B}{dt} = u\frac{\partial Z_B}{\partial x} + v\frac{\partial Z_B}{\partial y} \qquad \text{lower surface} \tag{5}$$

We let the characteristic amplitude of variations in Z_B be B, so that with the nondimensional variable $h(x,y)$ we have

$$Z_B = Bh(x,y) \tag{6}$$

But $w = Ww'$, where according to Eq. (31) in Sec. 14.1 we have $W/D \sim (Ro V_g/L)(1+\lambda)$ and thus Eq. (5) in nondimensional variables becomes

$$w' = \frac{B}{DRo(1+\lambda)}(\psi_x h_y - \psi_y h_x) \tag{7}$$

To eliminate w', we use Eq. (3) in Sec. 14.2 in the form

$$\left(\frac{\partial}{\partial t} - \psi_y \frac{\partial}{\partial x} + \psi_x \frac{\partial}{\partial y}\right)\psi_z + \sigma w' = 0 \tag{8}$$

and so

$$(\psi_{zt} - \psi_y \psi_{xz} + \psi_x \psi_{yz}) + \frac{B}{DRo\Lambda}(\psi_x h_y - \psi_y h_x) = 0 \tag{9}$$

The linearized form of Eq. (9) is

$$\phi_{zt} - \Psi_y \phi_{xz} - U_z \phi_x + \frac{B}{DRo\Lambda}[\phi_x h_y + (U - \phi_y)h_x] = 0 \tag{10}$$

If the upper boundary at Z_T is a rigid surface, we may use the same condition with $B = 0$. But if the upper boundary is at infinity, we shall want the condition to be based upon energy considerations. We shall determine an appropriate form after computing an energy integral with Eq. (4).

We multiply Eq. (4) by $\rho_0 \phi$ and integrate over the volume V. A rearrangement of the terms gives

$$\int_V \left\{\rho_0 \phi \frac{\partial}{\partial t}\left[\phi_{xx} + \phi_{yy} + \frac{\Lambda}{\rho_0}(\rho_0 \phi_z)_z\right]\right\} dV + \int_V (\rho_0 \phi U \phi_{xxx})\, dV$$

$$+ \int_V \rho_0 \phi (U \phi_{yyx} - \phi_x U_{yy})\, dV + \int_V \rho_0 \phi \left[U \frac{\partial}{\partial x}\frac{\Lambda}{\rho_0}(\rho_0 \phi_z)_z - \frac{\phi_x \Lambda}{\rho_0}(\rho_0 U_z)_z\right] dV = 0 \tag{11}$$

The term involving β vanished immediately because of cyclic continuity and the fact that $(\rho_0)_x = 0$. We shall analyze each of the four terms in Eq. (11) separately.

With integration by parts and use of cyclic continuity and the lateral boundary condition, the first term T_1 becomes

$$T_1 = -\frac{\partial}{\partial t}\int_V \frac{\rho_0}{2}[(\phi_x)^2 + (\phi_y)^2 + \Lambda(\phi_z)^2]\, dV + \int_x \int_y \left[\Lambda \rho_0 \phi \phi_{zt}\right]_{Z_B}^{Z_T} dx\, dy \tag{12}$$

PROBLEM 14.3.1 Show that the second term of Eq. (11) vanishes.

The third term T_3 may be integrated by parts to become

$$T_3 = -\int_V \rho_0 \left(U \phi_x \phi_{yy} - \frac{\phi^2}{2} U_{yyx}\right) dV \tag{13}$$

and the second term of this vanishes because U_x does. The remaining term, again upon integration by parts, is

$$T_3 = -\int_V \left[(\rho_0 U \phi_x \phi_y)_y - \rho_0 \phi_x \phi_y U_y - \frac{\rho_0 U}{2} (\phi_y{}^2)_x \right] dV \tag{14}$$

The first term vanishes at the lateral boundaries, the last by virtue of cyclic continuity, and so

$$T_3 = \int_V \rho_0 \phi_x \phi_y U_y \, dV \tag{15}$$

The second term of the last integral in Eq. (11) vanishes because of cyclic continuity. Thus we have

$$\begin{aligned} T_4 &= \int_V \Lambda \phi U \frac{\partial}{\partial x} (\rho_0 \phi_z)_z \, dV \\ &= \int_V \Lambda \left[(U \phi \rho_0 \phi_{xz})_z + U_z \phi_x \rho_0 \phi_z - \frac{\rho_0}{2} U \frac{\partial}{\partial x} (\phi_z)^2 \right] dV \end{aligned} \tag{16}$$

The last term vanishes by cyclic continuity and so integration of the boundary term gives

$$T_4 = \int_V \rho_0 \Lambda \phi_x \phi_z U_z \, dV + \int_x \int_y \left[\Lambda U \rho_0 \phi \phi_{xz} \right]_{Z_B}^{Z_T} dx \, dy \tag{17}$$

The sum, T_B, of the two boundary terms from T_1 and T_4 gives

$$T_B = \int_x \int_y \left[\Lambda \rho_0 \phi (\phi_{zt} + U \phi_{xz}) \right]_{Z_B}^{Z_T} dx \, dy \tag{18}$$

The linear Eq. (10) may be written

$$\phi_{zt} + U \phi_{xz} - \phi_x U_z + \sigma w' = 0 \tag{19}$$

so that Eq. (18) becomes

$$T_B = \int_x \int_y \left[\rho_0 (1 + \lambda) \phi (-w' + \sigma^{-1} \phi_x U_z) \right]_{Z_B}^{Z_T} dx \, dy \tag{20}$$

The last term of this integral vanishes by cyclic continuity.

Upon collecting results, we have the energy equation

$$\begin{aligned} \frac{\partial}{\partial t} \int_V \frac{\rho_0}{2} [(\phi_x)^2 + (\phi_y)^2 + \Lambda (\phi_z)^2] \, dV &= \int_V (\rho_0 \phi_x \phi_y U_y + \rho_0 \Lambda \phi_x \phi_z U_z) \, dV \\ &\quad - \int_x \int_y \left[\rho_0 (1 + \lambda) \phi w' \right]_{Z_B}^{Z_T} dx \, dy \end{aligned} \tag{21}$$

Each of the terms of this energy equation has a straightforward physical interpretation. The first two on the left are the perturbation kinetic energy and the third is the available potential energy created by the presence of the disturbance. The first term on the right involves a Reynolds stress factor $\phi_x \phi_y$ acting on the horizontal shear of the basic wind. This energy source is the one responsible for *barotropic instability*. The second term on the right may be interpreted as a heat flux ($v\phi_z$), due to the disturbance, that interacts with the latitudinal gradient of potential temperature ($U_z \cong \theta'_y$). Thus this is the mechanism of *baroclinic instability*.

We left open until this point the question of a suitable boundary condition if Z_T were infinity. Certainly we do not want to allow energy to flux into or out of the atmosphere at its top, and so we require that

$$\int_x \int_y (\rho_0 \phi w')_{Z_T}\, dx\, dy = 0 \tag{22}$$

On a flat rigid boundary at Z_B, the boundary integral will vanish.

PROBLEM 14.3.2 Find the boundary term in Eq. (21) for the case in which the lower boundary is not flat. Show that it vanishes if $h_x = 0$.

PROBLEM 14.3.3 Show that the nonlinear terms, involving products of ϕ and its derivatives, that were dropped in going from Eq. (2) to Eq. (4), make no contribution to the energy budget. [Hint: Let $L(\phi) = \phi_{xx} + \phi_{yy} + (\Lambda/\rho_0)(\rho_0 \phi_z)_z$.]

14.3.2 A Simple Model of Baroclinic Instability

The baroclinic instability mechanism contained in Eq. (4) is revealed by the energy equation and is illustrated in its simplest form in a model due to Eady (1949). We let $\beta \to 0$, let $H_\alpha \to \infty$ so that $\rho_0 = \rho_0(0)$, and take a basic flow $U = Sz$. We use z for the vertical coordinate in which $1 + \lambda$ does not appear (Prob. 14.2.6), and then Eq. (4) reduces to

$$\left(\frac{\partial}{\partial t} + U\frac{\partial}{\partial x}\right)(\phi_{zz} + \sigma\, \nabla_H{}^2 \phi) = 0 \tag{23}$$

and from Eq. (8) we have the boundary condition on upper and lower rigid surfaces at $z = 0$ and $z = 1$

$$\left(\frac{\partial}{\partial t} + U\frac{\partial}{\partial x}\right)\phi_z - \phi_x S = 0 \tag{24}$$

We put $\phi = \hat{\phi}(z) e^{i(\kappa_1 x + \kappa_2 y - \kappa_1 C t)}$ where $\kappa = (\kappa_1{}^2 + \kappa_2{}^2)^{1/2}$ and in which we assume that $Sz - C \neq 0$. Thus we have from Eqs. (23) and (24)

$$\begin{aligned} &\hat{\phi}_{zz} - \sigma\kappa^2 \hat{\phi} = 0 \\ &(U - C)\hat{\phi}_z - \hat{\phi} S = 0 \qquad z = 0, 1 \end{aligned} \tag{25}$$

Thus $\hat{\phi} = Ae^{\Sigma z} + Be^{-\Sigma z}$ where $\Sigma = (\sigma\kappa^2)^{1/2}$. Application of the boundary conditions to this solution yields two homogeneous equations for A and B, and in order for there to be a solution, it is necessary that

$$\begin{vmatrix} C\Sigma + S & S - C\Sigma \\ [-C\Sigma + S(\Sigma - 1)]e^{\Sigma} & [C\Sigma - S(\Sigma + 1)]e^{-\Sigma} \end{vmatrix} = 0 \tag{26}$$

When this determinant is expanded, it leads to the characteristic equation

$$(C\Sigma)^2 - CS\Sigma^2 + S^2(\Sigma \coth \Sigma - 1) = 0 \tag{27}$$

and hence

$$C = \frac{S}{2} \pm \frac{S}{2\Sigma}(\Sigma^2 - 4\Sigma \coth \Sigma + 4)^{1/2} \tag{28}$$

but with $\mu = \Sigma/2$ this becomes

$$C = \frac{S}{2} \pm \frac{S}{2\mu}[(\mu - \tanh \mu)(\mu - \coth \mu)]^{1/2} \tag{29}$$

The first factor in the brackets is always positive, the second changes sign from negative to positive at $\mu = 1.1997$. Hence the waves are unstable for $\mu < 1.1997$.

With Eq. (45) in Sec. 14.1 we can apply this result to the atmosphere by writing

$$\mu^2 = \frac{\sigma\kappa^2}{4} = \frac{f^2L^2 3H_\pi \kappa^2}{4gD^2} Ri Ro^2 \tag{30}$$

where κ is a nondimensional wavenumber. For $L \sim 1000$ km, $D \sim 5$ km, the condition on μ gives $\kappa^2 Ri Ro^2 < 1.9$. But for "square" waves of length l in both x and y, we have $\kappa^2 = 2(L/l)^2(2\pi)^2$ so that the condition for instability is

$$Ri Ro^2 < \left(\frac{l}{L}\right)^2 2.4 \times 10^{-2} \tag{31}$$

and thus for wavelength $l = 1.1(4L)$, instability will occur provided $Ri Ro^2 < 0.5$ so that for $Ro \sim 0.1$ we must have $Ri < 50$. Summarizing the physics of the situation, we say that $Ri Ro^2 < 0.5$ is the condition for baroclinic energy conversion near the basic scale $L \sim 1000$ km. Still, for any value of $Ri Ro^2$, there is a limiting wavelength such that all longer waves are unstable.

*PROBLEM 14.3.4 Verify that the maximum growth rate occurs for $\mu = 0.8$. Obtain a relation between Ro, Ri, and l that gives conditions for maximum growth of square waves. Show that for the estimates in the text of $Ro \sim 0.1$, $Ri \sim 50$, we have $l_{max} \sim 7000$ km.

PROBLEM 14.3.5 Write out the complete solution for the stream function ϕ in the Eady model, assuming κ_1 and κ_2 are positive. Show that the waves propagate at the average speed of the basic current $U(z)$. Of the two components, one amplifying in time and one decaying in time, show that the trough line in the growing component tilts poleward and westward with height. [Hint: Because of the two signs in Eq. (29), you must have four constants A^+, A^-, B^+, B^- where the superscripted signs show the relation with the signs in Eq. (29). Moreover, be sure to take the presence of the zonal current U into account in determining the complete pressure field.]

14.3.3 Integral Theorems on Stability and Growth Rate

Inspection of Eq. (4) reveals that all the factors of the derivatives of the perturbation stream function ϕ are independent of x and t so that harmonic functions in one dimension may be used to simplify the equation. Thus we put $\phi = \hat{\phi}(y,z)e^{i\kappa(x-Ct)}$ in which the real part is the physically significant portion of the solution. With this substitution, Eq. (4) becomes

$$(U-C)\left[-\kappa^2\hat{\phi} + \hat{\phi}_{yy} + \frac{\Lambda}{\rho_0}(\rho_0\hat{\phi}_z)_z\right] + \hat{\phi}\left(\frac{\partial Z_G}{\partial y}\right) = 0 \tag{32}$$

We have used the definition (13) in Sec. 14.2 in the last term and denoted the potential vorticity of the zonal flow by Z_G.

For neutrally stable waves with $\mathrm{Im}\,(C) = C_I = 0$ the equation has a singularity at any point at which $U = C$. The equation is a form of the inviscid Orr-Sommerfeld equation; the complications introduced by this singularity are discussed in detail for this equation by Burger (1962).

Rather than consider the singularity, we use the same approach we applied to small-scale instability. We assume that C_I is not zero so that $U - C$ does not then vanish anywhere, and hence we may put $\hat{\phi} = (U-C)^n F_n$, so that Eq. (32) is now

$$(U-C)\left\{-\kappa^2(U-C)^n F_n + \frac{\partial^2}{\partial y^2}[(U-C)^n F_n] + \frac{\Lambda}{\rho_0}\frac{\partial}{\partial z}\left[\rho_0\frac{\partial}{\partial z}(U-C)^n F_n\right]\right\}$$
$$+ (U-C)^n F_n \frac{\partial Z_G}{\partial y} = 0 \tag{33}$$

We let F_n^* be the complex conjugate of F_n, multiply Eq. (33) by $\rho_0(U-C)^{n-1}F_n^*$, and integrate to obtain

$$-\int_y\int_z \rho_0\kappa^2(U-C)^{2n}|F_n|^2\,dy\,dz - \int_y\int_z \rho_0\frac{\partial}{\partial y}[(U-C)^n F_n^*]\frac{\partial}{\partial y}[(U-C)^n F_n]\,dy\,dz$$
$$+ \int_{Z_B}^{Z_T}\left[\rho_0(U-C)^n F_n^*\frac{\partial}{\partial y}(U-C)^n F_n\right]_{-l}^{l} dz$$
$$- \int_y\int_z \rho_0\Lambda\frac{\partial}{\partial z}[(U-C)^n F_n^*]\frac{\partial}{\partial z}[(U-C)^n F_n]\,dy\,dz \tag{34}$$

$$+ \int_{-l}^{l} \left\{ \rho_0 \Lambda (U-C)^n F_n^* \frac{\partial}{\partial z} [(U-C)^n F_n] \right\}_{Z_B}^{Z_T} dy$$

$$+ \int_y \int_z \rho_0 (U-C)^{2n-1} |F_n|^2 \frac{\partial Z_G}{\partial y} \, dy \, dz = 0 \qquad (34) \; (cont'd)$$

The boundary condition that ϕ vanishes at $y = \pm l$ implies, because $U - C$ is not zero, that

$$\hat{\phi} = F_n = F_n^* = 0 \quad \text{at } y = \pm l \qquad (35)$$

Thus the boundary term in the second line vanishes. We expand the second term L_2 in the first line and find

$$L_2 = -\int_y \int_z \rho_0 \left[n^2 (U-C)^{2n-2} (U_y)^2 |F_n|^2 + (U-C)^{2n} \left| \frac{\partial F_n}{\partial y} \right|^2 \right] dy \, dz$$

$$- \int_y \int_z \rho_0 \left[(U-C)^{2n-1} n U_y \left(F_n \frac{\partial F_n^*}{\partial y} + F_n^* \frac{\partial F_n}{\partial y} \right) \right] dy \, dz \qquad (36)$$

But $f \, \partial f^*/\partial y + f^* \, \partial f/\partial y = \partial |f|^2/\partial y$, so that upon denoting the first integral of L_2 by M_2, we have

$$L_2 = -M_2 + \int_y \int_z \rho_0 [(U-C)^{2n-1} n U_{yy} + n(2n-1)(U-C)^{2n-2} U_y^2] |F_n|^2 \, dy \, dz \qquad (37)$$

in which we have again used the boundary condition (35).

The expansion of the third line of Eq. (34) proceeds in precisely the same way and we find, upon adding the boundary term of the fourth line, that

$$L_3 = -\int_y \int_z \rho_0 \Lambda \left[(U-C)^{2n} \left| \frac{\partial F_n}{\partial z} \right|^2 + n^2 (U-C)^{2n-2} (U_z)^2 |F_n|^2 \right] dy \, dz$$

$$+ \int_y \int_z \Lambda \left[n(U-C)^{2n-1} (\rho_0 U_z)_z + \rho_0 n(2n-1)(U-C)^{2n-2} (U_z)^2 \right] |F_n|^2 \, dy \, dz$$

$$+ \int_{-l}^{l} \left(\rho_0 \Lambda \left\{ (U-C)^n F_n^* \frac{\partial}{\partial z} [(U-C)^n F_n] - n(U-C)^{2n-1} U_z |F_n|^2 \right\} \right)_{Z_B}^{Z_T} dy \qquad (38)$$

The boundary term B_3 in the last expression requires further analysis. With $h_x = 0$, the harmonic functions give Eq. (10) the form

$$(U-C)\hat{\phi}_z - U_z \hat{\phi} + \frac{B}{DRo\Lambda} \hat{\phi} h_y = 0 \qquad (39)$$

Thus we have at the lower boundary

$$B_3|_{Z_B} = -\int_{-l}^{l} \rho_0 \Lambda \left\{ (U-C)^{2n-1} |F_n|^2 \left[(1-n)U_z - \frac{B}{DRo\Lambda} h_y \right] \right\}_{Z_B} dy \tag{40}$$

Moreover the boundary term of Eq. (38) vanishes when evaluated at $Z_T = \infty$ with the condition on the basic, zonal flow that the wind shear is constant above some arbitrarily large, but finite, height, and the regularity conditions that $\rho_0 |\hat{\phi}_z|^2$ and $\rho_0 |\hat{\phi}|^2$ vanish at infinity.

PROBLEM 14.3.6 Show that the boundary term in Eq. (38) vanishes at infinity with the condition on U_z stated above. (Hint: Use the Schwarz inequality on one term; for the other, use $|U-C| \geqslant |C_I| > 0$.)

Upon collecting the various terms we have

$$\begin{aligned}
&\int_y \int_z \rho_0 (U-C)^{2n} \left(\kappa^2 |F_n|^2 + \left| \frac{\partial F_n}{\partial y} \right|^2 + \Lambda \left| \frac{\partial F_n}{\partial z} \right|^2 \right) dy\, dz \\
&+ \int_y \int_z \rho_0 |F_n|^2 (U-C)^{2n-1} \left\{ (1-n) \left[U_{yy} + \frac{\Lambda}{\rho_0} (\rho_0 U_z)_z \right] - \beta \right\} dy\, dz \\
&- \int_y \int_z \rho_0 (U-C)^{2n-2} |F_n|^2 n(n-1) [(U_y)^2 + \Lambda (U_z)^2]\, dy\, dz \\
&+ \int_{-l}^{l} \rho_0 \Lambda \left\{ (U-C)^{2n-1} |F_n|^2 \left[(1-n)U_z - \frac{B}{DRo\Lambda} h_y \right] \right\}_{z=Z_B} dy = 0
\end{aligned} \tag{41}$$

As with the small-scale stability problem, we can reach various conclusions for different choices of the parameter n.

The potential vorticity theorem In the study of an unstratified, shearing flow, the Rayleigh theorem asserts that the basic flow must have an inflection point ($D^2 U$ changes sign) if there is instability. In the large-scale flow, we obtain a similar condition through use of the potential vorticity.

For $n = 0$, Eq. (41) becomes

$$\begin{aligned}
&\int_y \int_z \rho_0 \left\{ |G|^2 + |F|^2 \frac{(U - C^\dagger)}{|U-C|^2} \left[U_{yy} + \frac{\Lambda}{\rho_0} (\rho_0 U_z)_z - \beta \right] \right\} dy\, dz \\
&+ \int_{-l}^{l} \left[\rho_0 \Lambda \frac{(U - C^\dagger)}{|U-C|^2} |F|^2 \left(U_z - \frac{B}{DRo\Lambda} h_y \right) \right]_{z=Z_B} dy = 0
\end{aligned} \tag{42}$$

in which

$$|G|^2 = \kappa^2 |F|^2 + \left|\frac{\partial F}{\partial y}\right|^2 + \Lambda\left|\frac{\partial F}{\partial z}\right|^2 \tag{43}$$

and in which we have dropped the subscript and denoted complex conjugates by a dagger.

The imaginary part of this equation is

$$C_I\left\{\int_y\int_z \rho_0\left\{U_{yy} + \frac{\Lambda}{\rho_0}(\rho_0 U_z)_z - \beta\right\}\frac{|F|^2}{|U-C|^2}\,dy\,dz\right.$$
$$\left.+\int_{-l}^{l}\left[\rho_0\Lambda\left\{U_z - \frac{B}{DRo\Lambda}h_y\right\}\frac{|F|^2}{|U-C|^2}\right]_{z=Z_B} dy\right\} = 0 \tag{44}$$

The quantity in braces in the first integral is the negative of the latitudinal rate of change of the potential vorticity Z_G of the zonal flow. Thus if the shear of the zonal current vanishes at a flat lower boundary, then $\partial Z_G/\partial y$ must change sign if there is instability. If the gradient $\partial Z_G/\partial y$ is of one sign, then stability depends on the boundary conditions.

To obtain an interpretation of the boundary term in Eq. (44), we use Eqs. (8) and (10) in Sec. 14.1 to write (the asterisk denotes dimensional variables)

$$\frac{\partial u_g^*}{\partial z^*} = -\frac{g}{f}\frac{\partial \ln \theta^*}{\partial y^*} = \frac{g}{f}\frac{\partial h_\theta^*}{\partial y^*}\frac{\partial \ln \theta^*}{\partial z^*} \tag{45}$$

so that

$$U_z = \frac{D}{V_g}\frac{g}{f}\frac{\partial \ln \theta^*}{\partial z^*}\frac{\partial h_\theta^*}{\partial y^*} = \frac{Dg}{V_g f}\frac{\sigma}{3H_\pi}\frac{\partial h_\theta^*}{\partial y^*} \tag{46}$$

in which we have used Eq. (45) in Sec. 14.1. Now the boundary term in Eq. (44) becomes

$$\Lambda\left(U_z - \frac{B}{DRo\Lambda}h_y\right) = \frac{fL^2}{DV_g}\left[\frac{(1+\lambda)}{2}\frac{2gD^2}{3H_\pi f^2L^2}\frac{\partial h_\theta^*}{\partial y^*} - B\frac{\partial h}{\partial y^*}\right] \tag{47}$$

With the numerical estimates we have been using, we have

$$\frac{2gD^2}{3H_\pi f^2L^2} = \frac{2\times 10\ \text{m/s}^2\ 25\times 10^6\ \text{m}^2}{10^{-8}\ \text{s}^{-2}\ 7.5\times 10^4\ \text{m}\ 10^{12}\ \text{m}^2} = 0.67 \tag{48}$$

But $L/aRo \sim 1$ so that $1+\lambda \sim 2$, and so we have for the boundary term B_0 in Eq. (44) the approximation

$$B_0 = \int_{-l}^{l} \left[\rho_0 \left(\frac{\partial h_\theta^*}{\partial y^*} - \frac{\partial Z_B^*}{\partial y^*} \right) \frac{fL^2}{DV_g} \frac{|F|^2}{|U-C|^2} \right]_{z=Z_B} dy \tag{49}$$

Thus the boundary term depends on the slope of the isentropes at the surface relative to the slope of the surface itself. In general, the isentropes slope upward toward the poles, so that the difference in Eq. (49) is positive. When the boundary term is positive, then we must have

$$U_{yy} + \frac{\Lambda}{\rho_0} \frac{\partial}{\partial z}(\rho_0 U_z) - \beta = -\frac{\partial Z_G}{\partial y} < 0 \tag{50}$$

somewhere in the flow for instability to be possible.

For the barotropic case, U_z vanishes and so stability depends on $U_{yy} - \beta$ and the boundary term. If $h_y = 0$, then $U_{yy} - \beta$ must change sign at least once, and so a current U varying linearly with y is neutrally stable if $\beta \neq 0$.

More information can be gained from an analysis of the real part of Eq. (42). Because of the assumption that $C_I \neq 0$, the term multiplying C_R will vanish according to Eq. (44) so that we have

$$\int_y \int_z \rho_0 \left(|G|^2 - U \frac{\partial Z_G}{\partial y} \frac{|F|^2}{|U-C|^2} \right) dy\, dz + \int_{-l}^{l} \left[\rho_0 U \left(\frac{\partial h_\theta^*}{\partial y^*} - \frac{\partial Z_B^*}{\partial y^*} \right) \frac{fL^2}{DV_g} \frac{|F|^2}{|U-C|^2} \right]_{z=Z_B} dy = 0 \tag{51}$$

Thus a necessary condition for instability is that

$$\int_y \int_z \rho_0 U \frac{\partial Z_G}{\partial y} \frac{|F|^2}{|U-C|^2} dy\, dz - \int_{-l}^{l} \left[\rho_0 U \left(\frac{\partial h_\theta^*}{\partial y^*} - \frac{\partial Z_B^*}{\partial y^*} \right) \frac{fL^2}{DV_g} \frac{|F|^2}{|U-C|^2} \right]_{z=Z_B} dy > 0 \tag{52}$$

As an illustration, we consider the baroclinic case with $U = sz$. If the isentrope coincides with the lower boundary, then

$$\int_y \int_z \rho_0 U \frac{\partial Z_G}{\partial y} \frac{|F|^2}{|U-C|^2} dy\, dz > 0 \tag{53}$$

for instability. But

$$U \frac{\partial Z_G}{\partial y} = sz \left(\beta - \frac{\Lambda}{\rho_0} s \frac{\partial \rho_0}{\partial z} \right) = sz \left(\beta + \frac{\Lambda s D}{H} \right) \tag{54}$$

Thus instability at some wavelength may be possible for any positive value of the shear. But if Eq. (53) does not hold, then $C_I = 0$ so that a sufficient condition for neutral stability is that

$$sz\left(\beta + \frac{\Lambda sD}{H}\right) \leqslant 0 \tag{55}$$

everywhere. But there is no value of $s > 0$ that will satisfy the inequality.

PROBLEM 14.3.7 Show that if the constant shear of an easterly current is small enough, then the flow will be stable.

In the general case in which the boundary integral is positive, instability is possible only if $U\,\partial Z_g/\partial y$ is positive somewhere. Because

$$U\frac{\partial Z_g}{\partial y} = U\left[\beta - \frac{\Lambda}{\rho_0}(\rho_0 U_z)_z - U_{yy}\right] = U\left(\beta + \frac{\Lambda U_z D}{H} - \Lambda U_{zz} - U_{yy}\right) \tag{56}$$

we see that the maximum value of U that would occur in a westerly jet stream would give a large value of $U\,\partial Z_g/\partial y$ and thus might provide the necessary destabilizing effect.

In summary of the results so far, we have:

Theorem (Charney and Stern, 1962; Pedlosky, 1964) If the isentropes intersect the surface so that $U_z > Bh_y/D\,Ro\Lambda$, then a necessary condition for instability is that the gradient of potential vorticity $\partial Z_G/\partial y$ be such that both $\partial Z_G/\partial y$ and $U\,\partial Z_G/\partial y$ are positive somewhere in the flow (although not necessarily at the same point). If the gradient is uniformly negative, then the flow is neutrally stable. ////

The semicircle theorem A result confining C_R and C_I to a semicircle may be developed for the large-scale stability problem in exact analogy to the method used for small-scale motion. For $n = 1$ and $h_y = 0$, the basic result (41) becomes

$$\int_y\int_z \rho_0[(U-C)^2|G|^2 - \beta(U-C)|F|^2]\,dy\,dz = 0 \tag{57}$$

in which we have again dropped the subscripts and used $|G|^2$ to denote the sum of three terms in the first integral.

Therefore $C_R \leqslant U_{\max}$ if $C_I \neq 0$, but C_R may be less than $U_{\min}$ owing to the β term. To obtain a lower bound on C_R, we use a theorem (see Problem 14.3.8 below) that says that if the function f is suitably behaved with $f(l) = f(-l) = 0$, then

$$\int_{-l}^{l}\left|\frac{\partial f}{\partial y}\right|^2 dy \geqslant \frac{\pi^2}{4l^2}\int_{-l}^{l}|f|^2\,dy \tag{58}$$

Thus

$$\int_{-l}^{l}\rho_0\left(\kappa^2|F|^2 + \left|\frac{\partial F}{\partial y}\right|^2 + \Lambda\left|\frac{\partial F}{\partial z}\right|^2\right)dy \geqslant \left(\kappa^2 + \frac{\pi^2}{4l^2}\right)\int_{-l}^{l}\rho_0|F|^2\,dy \tag{59}$$

Hence Eq. (57) implies that

$$(U_{\min} - C_R)\left(\kappa^2 + \frac{\pi^2}{4l^2}\right) \leqslant \frac{\beta}{2} \tag{60}$$

and so

$$U_{\min} - \frac{\beta}{2}\left(\kappa^2 + \frac{\pi^2}{4l^2}\right)^{-1} \leqslant C_R \leqslant U_{\max} \tag{61}$$

Now we use an overbar to denote the integration over the meridional plane so that the real part of Eq. (57) is

$$\overline{(U^2 - 2UC_R + C_R{}^2 - C_I{}^2)\rho_0|G|^2} = \overline{\beta(U - C_R)\rho_0|F|^2} \tag{62}$$

and the imaginary part yields

$$\overline{\rho_0 U|G|^2} = \frac{\beta}{2}\overline{\rho_0|F|^2} + C_R\overline{\rho_0|G|^2} \tag{63}$$

Combination of these two to eliminate $U\rho_0 |G|^2$ in Eq. (62) gives

$$\overline{[U^2 - (C_R{}^2 + C_I{}^2)]\rho_0|G|^2} = \beta\,\overline{\rho_0 U|F|^2} \tag{64}$$

As before, we use the inequality

$$0 \geqslant \overline{(U - U_{\max})(U - U_{\min})\rho_0|G|^2} = \overline{[U^2 - U(U_{\max} + U_{\min}) + U_{\max}U_{\min}]\rho_0|G|^2} \tag{65}$$

With Eqs. (63) and (64) we may rearrange this expression to read

$$0 \geqslant \left[\left(C_R - \frac{U_{\max} + U_{\min}}{2}\right)^2 + C_I{}^2\right]\overline{\rho_0|G|^2} + \beta\,\overline{\left(U - \frac{U_{\max} + U_{\min}}{2}\right)\rho_0|F|^2} - \left(\frac{U_{\max} - U_{\min}}{2}\right)^2\overline{\rho_0|G|^2} \tag{66}$$

Now we use

$$U - \frac{U_{\max} + U_{\min}}{2} \geqslant \frac{U_{\min} - U_{\max}}{2} \tag{67}$$

in order to put Eq. (66) in the form

$$\left(\frac{U_{\max} - U_{\min}}{2}\right)^2 + \frac{\beta}{2}(U_{\max} - U_{\min})\frac{\overline{\rho_0|F|^2}}{\overline{\rho_0|G|^2}} \geqslant \left(C_R - \frac{U_{\max} + U_{\min}}{2}\right)^2 + C_I^2 \tag{68}$$

But in the present situation, Eq. (59) is

$$\overline{\rho_0|G|^2} \geqslant \left(\kappa^2 + \frac{\pi^2}{4l^2}\right)\overline{\rho_0|F|^2} \tag{69}$$

and so finally we have the quasi-geostrophic semicircle theorem (Pedlosky, 1964)

$$\left(\frac{U_{\max}-U_{\min}}{2}\right)^2+\frac{\beta(U_{\max}-U_{\min})}{2[\kappa^2+(\pi^2/4l^2)]}\geqslant\left(C_R-\frac{U_{\max}+U_{\min}}{2}\right)^2+C_I^2 \quad (70)$$

As shown in Fig. 14.1, the modulus of the complex wave speed must lie in the semicircle with radius

$$R^2=\left(\frac{U_{\max}-U_{\min}}{2}\right)^2+\frac{\beta(U_{\max}-U_{\min})}{2[\kappa^2+(\pi^2/4l^2)]} \quad (71)$$

but because $C_R \leqslant u_{\max}$, a portion of the entire circle is in fact cut off.

Two important conclusions are to be drawn. First, instability is possible only if there are variations in speed of the zonal current. Second, the maximum possible growth rate is a direct function of the variations of the zonal current. We turn now to a more precise characterization of this relation.

The growth rate inequality Upon setting $n=\frac{1}{2}$ in Eq. (41) we find, in the same notation we have been using,

$$\int_y\int_z \rho_0\left\{(U-C)|G|^2+\tfrac{1}{2}\left[U_{yy}+\frac{\Lambda}{\rho_0}(\rho_0U_z)_z-2\beta\right]|F|^2\right.$$

$$\left.+\tfrac{1}{4}[(U_y)^2+\Lambda(U_z)^2]\frac{|F|^2}{|U-C|^2}(U-C^\dagger)\right\}dy\,dz$$

$$+\int_{-l}^{l}\left[\rho_0\Lambda|F_n|^2\left(\tfrac{1}{2}U_z-\frac{B}{DRo\Lambda}h_y\right)\right]_{z=Z_B}dy \quad (72)$$

so that the imaginary part requires that

$$C_I\left(\int_y\int_z\left\{\rho_0|G|^2-\tfrac{1}{4}[(U_y)^2+\Lambda(U_z)^2]\frac{\rho_0|F|^2}{|U-C|^2}\right\}dy\,dz\right)=0 \quad (73)$$

Thus when the flow is unstable, the integral must vanish.

But using the fact that $[U-C]^2\geqslant C_I^2$ and the inequality (59), we find that

$$\left(\kappa^2+\frac{\pi^2}{4l^2}\right)\overline{\rho_0|F|^2}\leqslant\overline{\rho_0|G|^2}\leqslant\tfrac{1}{4}\overline{[(U_y)^2+\Lambda U_z{}^2]\frac{\rho_0|F|^2}{C_I^2}} \quad (74)$$

Thus

$$(C_I\kappa)^2\leqslant\frac{\max[(U_y)^2+\Lambda(U_z)^2]}{4[1+(\pi^2/4l^2\kappa^2)]} \quad (75)$$

The term in the numerator represents the energy available in the horizontal shear of

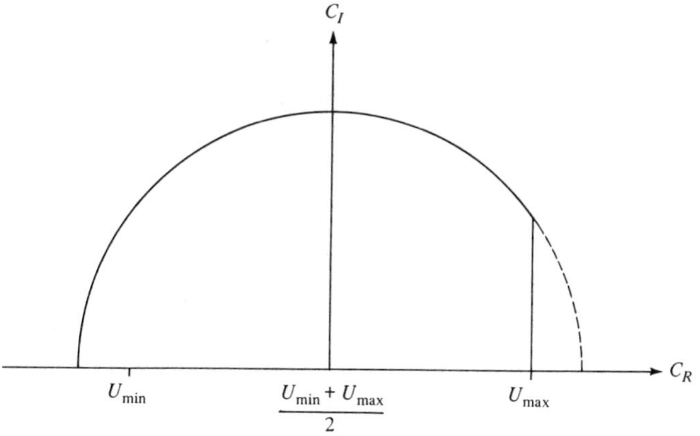

FIGURE 14.1
The range of permitted combinations of C_R and C_I for quasi-geostrophic motion is the interior of the region bounded by the solid line. The remainder of the semicircle is cut off by the requirement that $C_R \leqslant U_{max}$.

the basic flow $U_y{}^2/2$, plus the available potential energy, $(\Lambda/2)(U_z)^2 \cong (\Lambda/2)(\partial\theta'/\partial y)^2$. We denote the sum by A, put $\kappa = 2\pi/L$, and find (Pedlosky, 1964)

$$\omega_I{}^2 \leqslant \frac{\max A}{2[1 + (L^2/16l^2)]} \tag{76}$$

Thus the bound on the growth rate decreases with increasing horizontal scale of the perturbations.

PROBLEM 14.3.8 Prove the theorem stated in Eq. (58). [Hints: Let $f(x)$, defined on $(-L,L)$, be sufficiently well-behaved that Parseval's relation between the integral of $|f|^2$ and the sum of the squared Fourier coefficients applies to both f and its derivative f'. Show that then

$$\int_{-L}^{L} |f'(x)|^2 \, dx \geqslant \frac{\pi}{L^2} \int_{-L}^{L} |f - \bar{f}|^2 \, dx$$

where $\bar{f}$ is the average. Now extend f into $(-2L,2L)$ by reflecting the part in $(0,L)$ into $(L,2L)$ with opposite sign and similarly in $(-2L,-L)$. Take $f(L) = f(-L) = 0$ and apply the previous result.]

PROBLEM 14.3.9 Use Eq. (57) to show that the necessary and sufficient condition for instability is that

$$\left(\frac{\beta}{2}\overline{\rho_0|F|^2} - \overline{U\rho_0|G|^2}\right)^2 + \beta\,\overline{U\rho_0|F|^2}\;\overline{\rho_0|G|^2} < \overline{U^2\rho_0|G|^2}\;\overline{\rho_0|G|^2}$$

Show that the motion is always unstable for $\beta = 0$ if there is any variation in U.

14.4 THE NONLINEAR QUASI-GEOSTROPHIC EQUATION

The quasi-geostrophic equation is surely one of the great accomplishments of dynamic meteorology, for it presents a model of synoptic-scale flow in which the equations of motion reduce to a single nonlinear partial differential equation in the stream function ψ. It offers a remarkable opportunity for studying nonlinear aspects of atmospheric motion and the associated questions of predictability with a model that presumably resembles the actual flow at synoptic scales, but is much less complicated than the full equations of motion.

Nonlinearity is responsible for the mechanisms by which energy is transferred from one scale of the flow to others, a transfer that is impossible in a linear flow. Thus explanations of the observed distribution of energy by wavelength and the transfer rates may be sought with the quasi-geostrophic equation. Of particular interest is the phenomenon, discussed in Chap. 11, that atmospheric flow as represented in quasi-horizontal coordinates exhibits a transfer of energy from smaller to larger scales.

It is only through study of these nonlinear phenomena in atmospheric flow that we can approach the predictability problem, which is of great interest as we attempt to achieve significant extensions in the range for which numerical predictions are useful. The problem of atmospheric predictability may be formulated as two questions:

1 If two flows evolve from identical initial conditions under identical boundary conditions, will they remain the same?

2 If two flows evolve from slightly different initial conditions under identical boundary conditions, how will their initial difference evolve in time?

The issue raised in the first question is whether two or more different atmospheric flows could exist for a given set of initial and boundary conditions and whether a given flow could switch spontaneously from one such solution to another. The second issue is how rapidly observational errors in the initial data will grow to produce significant prediction errors.

Thus we conclude this chapter with a brief examination of some of these aspects of the nonlinearity of atmospheric flow as revealed analytically in the quasi-geostrophic model. In order to simplify the discussion and make exact statements possible, we shall investigate only a simple problem on a bounded domain.

With the elliptic operator

$$L(\psi) = \nabla_H{}^2\psi + \frac{1}{\rho_0}\left(\frac{\rho_0}{\sigma}\psi_z\right)_z \tag{1}$$

the quasi-geostrophic material derivative Eq. (12) in Sec. 14.2, and the Jacobian operator $J(f,g) = f_x g_y - f_y g_x$, we may write the equation as

$$\frac{\delta}{\delta t}[L(\psi) + \beta y] = \frac{\partial}{\partial t}L(\psi) + J(\psi, L(\psi)) + \beta\psi_x = 0 \tag{2}$$

where $\mathbf{v}_g = \mathbf{k} \times \nabla_H \psi$. All variables, including ρ_0, are nondimensional.

We apply the equations on a domain V with lateral boundaries at $y = 0$ and $y = 1$, and in which $0 \leqslant z \leqslant Z_T < \infty$. The solutions are required to be cyclically continuous on horizontal planes, to have $\psi_x = 0$ at $y = 0$ and $y = 1$, and to satisfy the further boundary conditions that $\psi_z = 0$ at $z = 0$ and $z = Z_T$. Admissible solutions are subject to the regularity conditions that the integrals over V of ρ_0, $\rho_0\psi^2$, $\rho_0|\mathbf{v}_g|^2$, $\rho_0 w^2$, $\rho_0|\psi_z|^2$, and $\rho_0|L(\psi)|^2$ all be finite. This class of solutions we shall call *simple quasi-geostrophic flows.*

Two comments about these boundary conditions are in order. First, the condition on the surfaces at $z = 0$ and $z = Z_T$ is the simplest way to satisfy the first law, Eq. (8) in Sec. 14.2, once we have set w to zero on these boundaries. It is admittedly somewhat unrealistic, since both the general mid-latitude structure with isentropes sloping to intersect the surface and special features such as fronts require that $\theta' = \psi_z$ not vanish at $z = 0$. The spectral techniques used in this section have not been generalized to the point that they can cope with time-dependent boundary conditions, however. Second, the condition of cyclic continuity in both x and y along with the requirement that $v_g = \psi_x$ vanish on the lateral boundaries is intended to give a model that portrays the conditions present in the spherical geometry of the globe. It leads to the same conservation properties that prevail in the spherical case.

If ϕ is continuous and periodic in x with summable, continuous derivatives, then for simple quasi-geostrophic stream functions ψ

$$\int \rho_0 J(\psi, \phi)\, dV = 0 \tag{3}$$

and so we have the *quasi-geostrophic transport theorem*

$$\begin{aligned}\frac{\partial}{\partial t}\int \rho_0\phi\, dV = \int \rho_0 \frac{\partial \phi}{\partial t}\, dV &= \int \rho_0\left[\frac{\delta\phi}{\delta t} - J(\psi, \phi)\right] dV \\ &= \int \rho_0 \frac{\delta\phi}{\delta t}\, dV\end{aligned} \tag{4}$$

All integrals without limits are over the domain V.

The energy constraint Eq. (21) in Sec. 14.2 now follows immediately from integration by parts, which gives

$$\begin{aligned}2\dot{E} &= -2\int \rho_0\psi \frac{\delta}{\delta t}[L(\psi) + \beta y]\, dV \\ &= \frac{\partial}{\partial t}\int \rho_0\left(|\nabla_H\psi|^2 + \frac{1}{\sigma}|\psi_z|^2\right) dV - 2\int_A \frac{\rho_0\psi}{\sigma}\frac{\partial}{\partial t}\psi_z\bigg|_{z=0}^{z=Z_T} dA = 0\end{aligned} \tag{5}$$

and the last term vanishes because of the boundary condition. Similarly, we find a mean-square vorticity constraint

$$2\dot{F} = 2\int \rho_0 L(\psi)\frac{\delta}{\delta t}[L(\psi) + \beta y]\, dV$$

$$= \frac{\partial}{\partial t}\int \rho_0 |L(\psi)|^2\, dV + 2\beta \int_A \frac{\rho_0}{\sigma}\psi_z \psi_x \Big|_{z=0}^{z=Z_T} dA = 0 \tag{6}$$

and the potential vorticity constraint

$$\dot{G} = \frac{\partial}{\partial t}\int \rho_0 H(Z_g)\, dV = 0 \tag{7}$$

follows from Problem 14.2.4 for every differentiable function H. Here $Z_g = L(\psi) + \beta y$.

These relations can be converted into an infinite family of constraints. We take $H(x) = x^2$ and then we have

$$G_2(t) = \int \rho_0 Z_g^{\,2}\, dV = \int \rho_0 [L(\psi) + \beta y]^2\, dV$$

$$\leqslant \left[\left(\int \rho_0 |L(\psi)|^2\, dV\right)^{1/2} + \left(\int \rho_0 \beta^2 y^2\, dV\right)^{1/2}\right]^2$$

$$= \left[(2F)^{1/2} + \left(\int \rho_0 \beta^2 y^2\, dV\right)^{1/2}\right]^2 \tag{8}$$

in which we have used Minkowski's inequality. Hence G_2 is finite when F is. Application of Hölder's inequality [$p = 2/\alpha$, $q = 2/(2-\alpha)$, $1/p + 1/q = 1$] gives, for $\alpha < 2$, the result

$$G_\alpha = \int \rho_0 |Z_g|^\alpha\, dV = \int \rho_0^{\alpha/2} |Z_g|^\alpha \rho_0^{(2-\alpha)/2}\, dV \leqslant \left(\int \rho_0 |Z_g|^2\, dV\right)^{\alpha/2} \left(\int \rho_0\, dV\right)^{(2-\alpha)/2} \tag{9}$$

Hence G_α is finite for every $0 < \alpha \leqslant 2$ when F is finite. Our conclusion is:

Theorem Simple quasi-geostrophic motion on a finite domain V is subject to the infinite family of constraints

$$E(t) = E(0) \qquad F(t) = F(0) \qquad G_\alpha(t) = G_\alpha(0) \qquad 0 < \alpha \leqslant 2 \tag{10}$$

////

This family (or any other infinite family) is central to the analysis of the quasi-geostrophic equation, for the theorem suggests that the evolution of the flow is subject to restrictions not generally found in fluid dynamics.

PROBLEM 14.4.1 Verify Eq. (3) and then prove the quasi-geostrophic transport theorem in Eq. (4).

PROBLEM 14.4.2 Does the fact that L is a linear operator imply that $\delta L(\psi)/\delta t = L(\delta\psi/\delta t)$?

PROBLEM 14.4.3 Use the Schwarz inequality to obtain the inequality in the middle line of Eq. (8).

PROBLEM 14.4.4 Suppose a quasi-geostrophic flow is represented by data at N points in V. How many of the infinite family of constraints can be satisfied?

14.4.1 The Spectral Theory and Quasi-Geostrophic Turbulence

There is a set of functions, analogous to the sine and cosine functions of Fourier series, that appears in quasi-geostrophic theory and has revealing consequences, as shown by Charney (1971).

We consider the problem of finding solutions to

$$L(\psi) = -\lambda\psi \tag{11}$$

and it turns out that only for certain eigenvalues λ_n are there solutions, known as the associated eigenfunctions ψ_n, for the boundary conditions imposed.

Proceeding as we did to find the energy constraint, we may multiply Eq. (11) by $\rho_0\psi$ and integrate, applying the boundary conditions, to obtain

$$\int \rho_0\left(|\nabla_H\psi|^2 + \frac{1}{\sigma}|\psi_z|^2\right) dV - \lambda \int \rho_0\psi^2\, dV = 0 \tag{12}$$

This relation enables us to find the eigenvalues and eigenfunctions by a variational procedure. First we observe that $\psi = \text{const}$ satisfies the boundary conditions and Eq. (11) for $\lambda = 0$. Hence $\lambda_0 = 0$, and we take $\psi_0 = (\int\rho_0\, dV)^{-1/2}$. To find λ_1 and ψ_1 we add the constraints $\int\rho_0\psi_1{}^2 = 1$ and $\int\rho_0\psi_0\psi_1\, dV = 0$ to Eq. (12) and find the minimizing values. As the additional constraints $\int\rho_0\psi_n{}^2\, dV = 1$ and $\int\rho_0\psi_n\psi_k\, dV = 0$ for $k < n$ are added to Eq. (12), then eigenvalues $0 < \lambda_1 < \lambda_2 < \cdots \lambda_n$ and associated eigenfunctions are found. Thus the eigenfunctions are an orthonormal sequence ($\int\rho_0\psi_n\psi_m\, dV = \delta_{nm}$), and for a finite domain they may be used to represent ψ. It can be shown that if V is bounded, and if

$$a_n = \int_V \rho_0\psi_n\psi\, dV \tag{13}$$

then

$$\sum_{n=0}^{\infty} a_n^2 = \int_V \rho_0 \psi^2 \, dV \tag{14}$$

so that

$$\lim_{N \to \infty} \int_V \rho_0 \left| \psi - \sum_{n=0}^{N} a_n \psi_n \right|^2 dV = 0 \tag{15}$$

This is a statement that the series $\sum_{n=0}^{N} a_n \psi_n$ *converges in mean square* to ψ. Thus with this exact interpretation in mind we shall write

$$\psi(\mathbf{x}, t) = \sum_{n=0}^{\infty} a_n(t) \psi_n(\mathbf{x}) \tag{16}$$

and because $L(\psi_n) = -\lambda_n \psi_n$, we have the formal result

$$L(\psi) = L\left(\sum_{n=0}^{\infty} a_n \psi_n \right) = -\sum_{n=1}^{\infty} \lambda_n a_n \psi_n \tag{17}$$

We say "formal" here because we are differentiating an infinite series, but not investigating whether the resulting series really converges.

The functions ψ_n are the analog of the trigonometric wave shapes that add together to make a light beam. When we plot the energy in a beam of light as a function of wavelength, we call the graph a spectrum. Here $a_n{}^2$, as a function of n, shows how the mass-weighted mean square of ψ is distributed among the functions ψ_n, and is also called a spectrum. Thus Eq. (13) gives a *spectral representation* of ψ.

The constraints on E, F, and G_α can now be expressed as

$$\begin{aligned} 2E(t) &= -\int \rho_0 \psi L(\psi) \, dV = \sum_{n=1}^{\infty} \lambda_n |a_n(t)|^2 \\ &= \sum_{n=1}^{\infty} \lambda_n |a_n(0)|^2 \end{aligned} \tag{18}$$

$$2F(t) = \sum_{n=1}^{\infty} \lambda_n{}^2 |a_n(t)|^2 = \sum_{n=1}^{\infty} \lambda_n{}^2 |a_n(0)|^2 \tag{19}$$

and

$$G_\alpha(t) = \int \rho_0 |L(\psi) + \beta y|^\alpha \, dV = G_\alpha(\lambda_1 a_1(t), \lambda_2 a_2(t), \ldots)$$
$$= G_\alpha(\lambda_1 a_1(0), \lambda_2 a_2(0), \ldots) \tag{20}$$

The first two constraints here have a remarkable implication for geostrophic flow. Let us define

$$2E_{M,N} = \sum_{n=M}^{N} \lambda_n a_n{}^2 \qquad 2F_{M,N} = \sum_{n=M}^{N} \lambda_n{}^2 a_n{}^2 \tag{21}$$

Then because $\lambda_N < \lambda_{N+1}$, we may write that

$$2E_{M,\infty} \leqslant \frac{1}{\lambda_M} \sum_{n=M}^{\infty} \lambda_n{}^2 a_n{}^2 = \frac{2F_{M,\infty}}{\lambda_M} \leqslant \frac{2F_{1,\infty}(0)}{\lambda_M}$$

$$2F_{1,M-1} = \sum_{n=1}^{M-1} \lambda_n{}^2 a_n{}^2 \leqslant \lambda_{M-1} \sum_{n=1}^{M-1} \lambda_n a_n{}^2 = 2\lambda_{M-1} E_{1,M-1} \tag{22}$$

The quantity $F_{1,\infty}(0)/\lambda_M$ vanishes as $M \to \infty$, so that we may conclude that energy may not accumulate at large "wavenumbers" M. If we combine the two inequalities in Eq. (22), we have

$$\frac{E_{M,\infty}}{E_{1,M-1}} \leqslant \frac{\lambda_{M-1}}{\lambda_M} \frac{F_{M,\infty}}{F_{1,M-1}} \tag{23}$$

This relation prohibits a flow structure in which energy is perpetually passing from wavenumbers less than M to those greater than M ($E_{1,M-1} \to E_{M,\infty}$) at the same time enstrophy (mean-square vorticity) is transferred the opposite direction ($F_{M,\infty} \to F_{1,M-1}$). Thus the only possible flows that have fluxes of energy and enstrophy each in a single direction for all time must have energy moving toward low M ($E_{M,\infty} \to E_{1,M-1}$) and enstrophy toward large M ($F_{1,M-1} \to F_{M,\infty}$).

An identical result is obtained in the theory of inviscid, incompressible two-dimensional turbulence, where it follows from the conservation of vorticity law $d\zeta/dt = 0$, which implies that $d\zeta^n/dt = 0$ for every parcel. In quasi-geostrophic turbulence, the infinite family of constraints based on energy and potential vorticity provides the equivalent restriction on the evolution of the flow.

The point here for dynamics is that the transfer of energy from eddies to the mean flow observed in isobaric coordinates now has a theoretical analog in this result from quasi-geostrophic theory.

Suppose that now we change from an eigenfunction expansion to a Fourier series. Then we may represent the specific energy spectrum of the flow by the square moduli of the Fourier coefficients. If we have a wavelength L in the argument of a trigonometric function, then the wavenumber is $\kappa = 2\pi/L$ and the (specific energy) spectral density $E(\kappa)$ is defined so that

$$E = \int_0^\infty E(\kappa)\, d\kappa = \frac{1}{V}\int_V \frac{\mathbf{v}_g \cdot \mathbf{v}_g}{2}\, dV \tag{24}$$

so that $[E(\kappa)] = (L^2/T^2)L$. Next, we assume that the energy spectrum $E(\kappa)$ depends only on the wavenumber and the rate η at which enstrophy is transferred across wavenumber κ. This mean-square vorticity transfer rate will thus have dimensions $[\eta] = T^{-3}$. The simplest dimensionally correct relation then is $E(\kappa) = C\eta^{2/3}\kappa^{-3}$ where C is a constant. Observational evidence appears to support this conclusion, because a range in which $E(\kappa)$ is proportional to κ^{-3} is indeed found in mid-latitudes.

PROBLEM 14.4.5 Find an intuitive argument that proves that if $\lambda_n \neq \lambda_m$, then $\lambda_n > \lambda_m$ for $n > m$. [Hint: What effect do the constraints have on the minimum in Eq. (12)?]

PROBLEM 14.4.6 Let ψ_n and ψ_m be solutions to Eq. (11) corresponding to distinct eigenvalues $\lambda_n \neq \lambda_m$. Prove that ψ_n and ψ_m are orthogonal with respect to ρ_0, that is, $\int \rho_0 \psi_n \psi_m \, dV = \delta_{nm} \int \rho_0 \psi_n^2 \, dV$. [Hint: Show first that $\int \rho_0 \psi_n L(\psi_m)\, dV = \int \rho_0 \psi_m L(\psi_n)\, dV$.]

PROBLEM 14.4.7 Assume that Eq. (16) is valid in the usual sense. Show that Eqs. (14) and (15) follow formally.

PROBLEM 14.4.8 Show that $\lambda_n |a_n| \to 0$ as $n \to \infty$ for simple quasi-geostrophic flow.

PROBLEM 14.4.9 Establish the spectral forms (18) to (20) of the constraints. What can be said about the relative size of E and F?

PROBLEM 14.4.10 Derive a spectral version of the quasi-geostrophic equation by substituting Eqs. (16) and (17) in Eq. (2) and multiplying by $\rho_0 \psi_k$ to obtain $\lambda_k \, \partial a_k/\partial t$ as the leading term. Explain how the resulting system of equations could be integrated numerically if only n coefficients a_n were retained. What spectral constraints and how many apply to this system? What is to be done about detail in the initial data represented by coefficients $a_n(0)$ for $n > N$?

14.4.2 Predictability and Uniqueness

The two questions about atmospheric predictability raised in the introduction to this section may be investigated with the quasi-geostrophic equation. We shall take them up in order, first proving uniqueness of simple flows.

Thus we assume that simple solutions exist in V for $0 \leqslant t \leqslant T$. We suppose that two solutions ψ and $\widetilde{\psi}$ can be found for the same initial data and boundary conditions. Letting $\Phi(\mathbf{x},t) = L(\psi)$, we have then

$$\frac{\delta}{\delta t}(\Phi + \beta y) = 0 \qquad \frac{\widetilde{\delta}}{\delta t}(\widetilde{\Phi} + \beta y) = 0 \tag{25}$$

Our aim is to prove that Φ and $\widetilde{\Phi}$ must be identical. The work is greatly simplified by the following lemmas:

Lemma 1 For cyclically continuous functions with continuous derivatives:

$$\begin{aligned}
&\text{(i)} && J(f,g) = -J(g,f) \\
&\text{(ii)} && J(f+g,h) = J(f,h) + J(g,h) \\
&\text{(iii)} && \textstyle\int \rho_0 h J(f,g)\, dV = \int \rho_0 f J(g,h)\, dV \\
&\text{(iv)} && \textstyle\int \rho_0 f J(f,g)\, dV = 0
\end{aligned}$$

Lemma 2 (Serrin, 1959) Let $\phi' = \widetilde{\phi} - \phi$ and note that $\widetilde{\delta}/\delta t = \partial/\partial t + \widetilde{\mathbf{v}}_g \cdot \nabla_H$. Then

$$\frac{\widetilde{\delta}\widetilde{\phi}}{\delta t} - \frac{\delta \phi}{\delta t} = \frac{\delta \phi'}{\delta t} + J(\psi', \widetilde{\phi})$$

The proof of these lemmas follows from direct computations and is left as a problem. ////

Upon subtracting the two equations in Eq. (25), we arrive at

$$\frac{\delta \Phi'}{\delta t} + J(\psi', \widetilde{\Phi}) + \beta v_g' = 0 \tag{26}$$

which is the *quasi-geostrophic difference equation* and controls the evolution of the difference between two solutions. Note that it is nonlinear, containing both $J(\psi,\Phi')$ and $J(\psi',\widetilde{\Phi})$.

We are going to develop a series of inequalities that relate the mean square of Φ' to $\Phi'(0)$. Thus from Eq. (26) and the transport theorem we find

$$\begin{aligned}
\frac{\partial}{\partial t}\int \rho_0 \frac{|\Phi'|^2}{2}\, dV &= \int \rho_0 \Phi' \frac{\delta \Phi'}{\delta t}\, dV \\
&= -\int \rho_0 \Phi' [J(\psi', \widetilde{\Phi}) + \beta v_g']\, dV
\end{aligned} \tag{27}$$

The β term here vanishes, as it did in Eq. (6), because both ψ_z' and ψ_x' vanish at $z = 0$ and Z_T. For the remaining term we have

$$\left| \int \rho_0 \Phi' J(\psi', \widetilde{\Phi})\, dV \right| \leqslant \int \rho_0 |\Phi'|\, |\mathbf{v}_g'|\, |\nabla_H \widetilde{\Phi}|\, dV \tag{28}$$

$$\leqslant M\left(\int \rho_0 |\Phi'|^2\, dV \int \rho_0 |\mathbf{v}_g'|^2\, dV\right)^{1/2}$$
$$\leqslant 2M(F'E')^{1/2} \tag{28 (cont'd)}$$

Here we know that $\nabla_H \widetilde{\Phi}$ must be finite because it appears in the quasi-geostrophic equation, in which every term must be finite for a solution $\widetilde{\psi}$. Hence $M = \max |\nabla_H \widetilde{\Phi}| < \infty$.

We need to convert the last expression in Eq. (28) to one involving F' only. To do so, we use an indirect argument.

The requirement on ψ and $\widetilde{\psi}$ is that they satisfy the same boundary conditions, so that ψ' must also satisfy the conditions set on the problem. Then the eigenvalue problem $L(\psi') = -\mu\psi'$ will have an eigenvalue $\mu = 0$ only for constant functions ψ'. Thus we let $\mu_1 > 0$ be the smallest eigenvalue associated with a varying function ψ' that minimizes $\mu > 0$ in $L(\psi') = -\mu\psi'$. From this equation, by the technique leading to Eq. (12), we obtain

$$\mu = \frac{\int \rho_0 \left(|\nabla_H \psi'|^2 + \frac{1}{\sigma}|\psi_z'|^2\right) dV}{\int \rho_0 |\psi'|^2\, dV} \tag{29}$$

If ψ' is not the first nonconstant eigenfunction, then $\mu > \mu_1$, so for every ψ' and $\Phi' = L(\psi')$ satisfying the boundary conditions we have with the Schwarz inequality the result (known as Rayleigh's principle) that

$$0 < \mu_1 \leqslant -\frac{\int \rho_0 \psi' \Phi'\, dV}{\int \rho_0 |\psi'|^2\, dV} \leqslant \left(\frac{\int \rho_0 |\Phi'|^2\, dV}{\int \rho_0 |\psi'|^2\, dV}\right)^{1/2} = \left(\frac{2F'}{D'}\right)^{1/2} \tag{30}$$

where we have set $\int \rho_0\, |\psi'|^2\, dV = D$. But Eq. (30) also implies that

$$E' = -\tfrac{1}{2}\int \rho_0 \psi' \Phi'\, dV \leqslant \tfrac{1}{2}(2F'D')^{1/2} \leqslant \frac{F'}{\mu_1} \tag{31}$$

Upon combining Eqs. (27), (28), and (31), we arrive at

$$\frac{\partial F'}{\partial t} \leqslant \frac{2MF'}{\sqrt{\mu_1}} \tag{32}$$

This differential inequality gives

$$F'(t) \leqslant F'(0) \exp\left(\frac{2M}{\sqrt{\mu_1}}\, t\right) \tag{33}$$

According to this result, derived from a series of inequalities of rather wide latitude, mean-square vorticity errors are at least bounded exponentially in growth.

We have now proved uniqueness of the vorticity fields, for if $\Phi'(\mathbf{x},0) = 0$, then $F'(0) = 0$ and so $F'(t) = 0$ and hence

$$\Phi'(\mathbf{x},t) = 0 \qquad 0 \leqslant t \leqslant T \tag{34}$$

Moreover, it now follows from Eq. (31) that the velocity fields are identical, and finally from Eq. (30) [in its original form $D' \leqslant (2F'D')^{1/2}/\mu_1$] it follows that the stream functions are unique. Thus we have:

Theorem If simple quasi-geostrophic flows exist as classical solutions of the quasi-geostrophic equation (2) in the interval $0 \leqslant t \leqslant T$, then they are unique.

////

This has been a lengthy development, but we have now answered the first of our predictability questions in the context of the quasi-geostrophic equation: the solutions are unique for fixed boundary conditions and parameter values when the initial conditions are known exactly.

Proceeding to the second question, we wish to examine how the difference fields for solutions starting from different initial conditions evolve. An error energy equation is obtained from Eq. (26) upon multiplying by $\rho_0 \psi'$ and integrating, which gives

$$\frac{\partial E'}{\partial t} = \frac{\partial}{\partial t} \int \frac{\rho_0}{2} \left(|\nabla_H \psi'|^2 + \frac{1}{\sigma} |\psi_z'|^2 \right) dV$$

$$= \int \rho_0 \psi' J(\psi, \Phi') \, dV = \int \rho_0 \Phi' J(\psi', \psi) \, dV \tag{35}$$

The integral of $\rho_0 \psi' J(\psi', \widetilde{\Phi})$ vanishes by virtue of Lemma 1 (iv).

Let us now consider ψ to represent the solution obtained from the actual initial data $\psi(\mathbf{x},0)$ and $\widetilde{\psi}$ to represent the solution evolving from initial data $\widetilde{\psi}(\mathbf{x},0) = \psi(\mathbf{x},0) + \psi'(\mathbf{x},0)$, which contains the error ψ'. Thus Eq. (35) gives the rate of error growth as measured by the error energy E'. The equation makes it clear that error growth is a nonlinear phenomenon involving both the true solution ψ and the error ψ'.

The integrals in Eq. (35) reveal that error growth has baroclinic and barotropic components that are exactly analogous to the equivalent energy sources for the linear equation (4) in Sec. 14.3. There the energy source terms in Eq. (21) in Sec. 14.3 arise from $\rho_0 \phi J(\Psi, L(\phi))$, which may be compared with Eq. (35). We use the second integral, and for the first part of it we obtain, after a use of Lemma 1 (iv), the result

$$\int \rho_0 (\nabla_H{}^2 \psi') J(\psi', \psi) \, dV = - \int \rho_0 \, \nabla_H \psi' \cdot (\mathbf{v}_g{}' \cdot \nabla_H) \, \nabla_H \psi \, dV$$

$$= - \int \mathbf{v}_g' \cdot (\mathbf{v}_g' \cdot \nabla_H) \mathbf{v}_g \, dV \tag{36}$$

where we have used $(\mathbf{v}_g{}' \times \mathbf{k}) \cdot (\mathbf{v}_g \times \mathbf{k}) = \mathbf{v}_g{}' \cdot \mathbf{v}_g$. This is the *barotropic error growth rate.*

From the second part of the integral, for $\sigma = \sigma(z)$ we have

$$\int \left(\frac{\rho_0}{\sigma}\psi'_z\right)_z J(\psi', \psi)\, dV = -\int \frac{\rho_0}{\sigma}\psi'_z \mathbf{v}'_g \cdot \nabla_H \psi_z\, dV$$

$$= -\int \frac{\rho_0}{\sigma}\psi'_z \mathbf{k} \cdot \left(\mathbf{v}'_g \times \frac{\partial \mathbf{v}_g}{\partial z}\right) dV \tag{37}$$

which gives the *baroclinic error growth rate*.

These two terms are too complicated for further analysis here, but they demonstrate that the study of error growth and propagation is a difficult one, and that it has strong analogies to the study of barotropic and baroclinic instability. From this analogy, we can infer that the same processes that provide the energy for the growth of disturbances will stimulate error growth.

There are two processes by which growth of error can be controlled or reduced. The first is irreversible dissipation of error, exactly as kinetic energy and temperature variance are dissipated. The other is for the flow to be strongly forced, and in the atmosphere we have the differential heating driving the motion.

If the analysis of heating developed in Secs. 14.1 and 14.2 and in the problems there is carried through, we arrive at the quasi-geostrophic equation for moderately heated flow in the form

$$\frac{\delta}{\delta t}[L(\psi) + \beta y] = \frac{1}{\rho_0}\frac{\partial}{\partial z}\left(\frac{\rho_0}{\sigma}Q'\right) \tag{38}$$

where Q' is defined in Eq. (44) in Sec. 14.1 as $Q' = K(q + f_h)$. Here we use a simple model for the radiative heating that forces the motion. We set

$$\rho_0 q = c_p \frac{\rho_0(T_e - T)}{\tau} \tag{39}$$

in which T_e is the temperature of a motionless atmosphere in radiative equilibrium and τ is a relaxation time. The horizontal gradients of T_e would be much stronger than those observed in the atmosphere owing to the absence of advection in the motionless state. We define the potential temperature $\theta_e = T_e {\pi_0}^{-1} c_p$ and use $c_p T = \pi_0 \theta = \theta_0 \pi_0 (1 + \Theta\theta^*)$ where θ^* is the θ' of Sec. 14.1. Now with $\theta^* = \partial\psi/\partial z$ we may express the heating as

$$\frac{1}{\rho_0}\frac{\partial}{\partial z}\left(\frac{\rho_0}{\sigma}Q'\right) = \frac{K}{\tau\rho_0}\frac{\partial}{\partial z}\left\{\frac{\rho_0\pi_0}{\sigma}\left[\theta_e - \theta_0\left(1 + \Theta\frac{\partial\psi}{\partial z}\right)\right]\right\} \tag{40}$$

where $\theta_0\pi_0 = gH_\pi$, which we take to be constant.

With this heating expression, the error equation becomes

$$\frac{\delta}{\delta t}\Phi' + J(\psi', \widetilde{\Phi}) + \beta v'_g = -\frac{T_A}{\rho_0 \tau}\frac{\partial}{\partial z}\left(\frac{\rho_0}{\sigma}\frac{\partial \psi'}{\partial z}\right) \tag{41}$$

where $T_A = L/V_g = KgH_\pi\Theta$ is the advective time scale.

Now it follows immediately that the error energy growth rate is

$$\frac{\partial E'}{\partial t} = \int \rho_0 \psi' J(\psi, \Phi')\, dV - \frac{T_A}{\tau} \int \frac{\rho_0}{\sigma} \left(\frac{\partial \psi'}{\partial z}\right)^2 dV \tag{42}$$

Clearly then, the heating supplies a term that leads to error reduction, that helps to stabilize the solutions by adding a tendency for them to converge. The problem in Eq. (42) is that analysis of the nonlinear term is important, because it may lead to greater error growth in the heated case than the unheated one because both ψ' and ψ may be quite different in the two cases.

It is clear, however, that at the initial time $\partial E'/\partial t$ is less with heating than without, because the first term of Eq. (42) is identical if it contains the same initial error and the same actual flow represented by ψ. Thus at least initially, the heating reduces the error. It is reasonable to expect that this situation will continue, but no proof has yet been developed. But we can show that the bound on F' is reduced by heating.

To do so, let us return to the uniqueness proof based on the mean-square vorticity error. The heating function will contribute (to $\dot{F}'$) the term

$$\begin{aligned} H &= -\frac{T_A}{\tau} \int \Phi' \frac{\partial}{\partial z}\left(\frac{\rho_0}{\sigma}\frac{\partial \psi'}{\partial z}\right) dV \\ &= -\frac{T_A}{\tau} \int \left[\nabla_H{}^2 \psi' + \frac{1}{\rho_0}\frac{\partial}{\partial z}\left(\frac{\rho_0}{\sigma}\frac{\partial \psi'}{\partial z}\right)\right] \frac{\partial}{\partial z}\left(\frac{\rho_0}{\sigma}\frac{\partial \psi'}{\partial z}\right) dV \\ &= \frac{T_A}{\tau} \int \left[\nabla_H{}^2 \frac{\partial \psi'}{\partial z} + \frac{\partial}{\partial z}\alpha_0 \frac{\partial}{\partial z}\left(\frac{\rho_0}{\sigma}\frac{\partial \psi'}{\partial z}\right)\right] \frac{\rho_0}{\sigma}\frac{\partial \psi'}{\partial z}\, dV \\ &= -\frac{T_A}{\tau} \int \frac{\rho_0}{\sigma}\left[\left|\nabla_H \frac{\partial \psi'}{\partial z}\right|^2 + \alpha_0{}^2 \sigma \left|\frac{\partial}{\partial z}\left(\frac{\rho_0}{\sigma}\frac{\partial \psi'}{\partial z}\right)\right|^2\right] dV \end{aligned} \tag{43}$$

Thus we have shown that heating will reduce the bound on the rate of error growth given in Eq. (33). If we write the last line of Eq. (43) as $-K[(\psi'_z)^2, t]$, then Eq. (32) becomes

$$\frac{\partial F'}{\partial t} \leqslant \frac{2M}{\sqrt{\mu_1}} F' - K[(\psi'_z)^2, t] \tag{44}$$

and so the equivalent of Eq. (33) is

$$F' \leqslant e^{(2M/\sqrt{\mu_1})t} \left\{ F'(0) - \int_0^t e^{-(2M/\sqrt{\mu_1})\tau} K[(\psi'_z)^2, \tau]\, d\tau \right\} \tag{45}$$

Hence the error bound is reduced at every time t, and errors may even be eliminated. But we shall not know for sure until Eq. (43) can be expressed as a function of F', and that result may be hard to find.

So we face severe difficulties at this point in quasi-geostrophic theory, but it is undoubtedly true that we shall overcome them before we succeed with the full set of equations of motion.

We have continued at some length in this chapter developing necessary conditions on quasi-geostrophic solutions. Such results depend for their relevance on the existence of solutions. We cannot go into this subject here, but the existence of solutions for the simple quasi-geostrophic model has been established. Thus we may proceed to study the equation with confidence in its validity as a mathematical abstraction of synoptic-scale flow. The quasi-geostrophic theory provides us with a model atmosphere that, despite any lapses it may have in accuracy of representation of the real atmosphere, gives us considerable insight into the processes of atmospheric circulation and allows us to develop mathematical approaches that someday can be extended to more realistic and comprehensive models.

PROBLEM 14.4.11 Establish Lemmas 1 and 2.

PROBLEM 14.4.12 Show that the barotropic and baroclinic error growth rates in Eqs. (36) and (37) are analogous to the terms in Eq. (21) of Sec. 14.3.

PROBLEM 14.4.13 Show that $T_A = L/V_g = KgH_\pi\Theta$ where $K = 3R_i/gDf_0$.

PROBLEM 14.4.14 Show that

$$\frac{\partial E'}{\partial t} \leqslant \frac{N}{\sqrt{\mu_1}} F' - \frac{T_A}{2\tau}\int \frac{\rho_0}{\sigma}\left(\frac{\partial \psi'}{\partial z}\right)^2 dV$$

PROBLEM 14.4.15 In Eq. (31) we have $E' \leqslant F'/\mu_1$. Is it true that $E \leqslant F/\lambda_1$? Can you establish your answer both with the eigenfunction expansion and by an argument that does not assume the validity of the expansion (16)?

PROBLEM 14.4.16 Show, for H as defined in Eq. (43), that

$$H = \frac{T_A}{\tau}\int \alpha_0 L^*(\phi)\phi \, dV$$

by finding suitable definitions of ϕ and $L^*(\phi)$. From this show that

$$H \geqslant -\frac{\gamma_1 T_A}{\tau} \int \frac{\rho_0}{\sigma^2} \left| \frac{\partial \psi'}{\partial z} \right|^2 dV$$

(Hint: L^* is similar in form to L, and γ_1 is a suitable eigenvalue.)

BIBLIOGRAPHIC NOTES

14.1–14.3 Among the many articles on quasi-geostrophic motion, the following are noteworthy and were used in preparing the material here:

Burger, A. P., 1958: "Scale Consideration of Planetary Motions of the Atmosphere," *Tellus,* **10**:195–205.

Burger, A. P., 1962: "On the Non-Existence of Critical Wavelengths in a Continuous Baroclinic Stability Problem," *J. Atmospheric Sci.,* **19**:30–38.

Charney, J. G., 1947: "The Dynamics of Long Waves in a Baroclinic Westerly Current," *J. Meteorol.,* **4**:135–162.

Charney, J. G., and M. E. Stern, 1962: "On the Stability of Internal Baroclinic Jets in a Rotating Atmosphere," *J. Atmospheric Sci.,* **19**:159–172.

Eady, E. T., 1949: "Long Waves and Cyclone Waves," *Tellus,* **1**:33–52.

Pedlosky, J., 1964: "The Stability of Currents in the Atmosphere and Ocean, Part I," *J. Atmospheric Sci.,* **21**:201–219.

Phillips, N. A., 1963: "Geostrophic Motion," *Rev. Geophys.,* **2**:123–176.

Further discussion of some of the topics considered here may be found in:

Monin, A. S., 1972: *Weather Forecasting as a Problem in Physics,* (translation of the 1969 Russian edition by Paul Superak), The M.I.T. Press, Cambridge, Mass., 199 pp.

This book also gives references to the articles in which Soviet scientists developed a quasi-geostrophic theory in parallel with that of the West.

14.4 Additional discussion of the nonlinear quasi-geostrophic equation and further information on predictability may be found in:

Charney, J. G., 1971: "Geostrophic Turbulence," *J. Atmospheric Sci.,* **28**:1087–1095.

Dutton, J. A., 1974: "The Non-linear Quasi-geostrophic Equation: Existence and Uniqueness of Solutions on a Bounded Domain," *J. Atmospheric Sci.,* **31**:427–433.

Lorenz, E. N., 1969: "The Predictability of a Flow Which Possesses Many Scales of Motion," *Tellus,* **21**:289–307.

The model of radiative heating was proposed by:

Spiegel, E. A., 1959: "The Smoothing of Temperature Fluctuations by Radiative Transfer," *Astrophys. J.,* **126**:202–210.

and Serrin's lemma appears in the article cited in Chap. 6.

15
ATMOSPHERIC MODELING AND METAMODELING

The study of the atmosphere proceeds on a number of levels simultaneously. On the most fundamental level, observations are made of atmospheric phenomena and then examined to detect patterns and structures, or sequences that reveal causes and effects. On another level, theories are suggested by available knowledge and their implications are tested by comparison with the observations. Numerical experiments are performed by evaluating the equations of motion with computational methods in order to simulate or predict atmospheric processes. Finally, we attempt to isolate the essential physical relations in simple models that reveal the qualitative behavior we see in certain regimes of atmospheric flow. Obviously, the endeavors at each level interact with the others to produce enhanced understanding.

Dynamic meteorology is central to the study of the atmosphere, for it is concerned with the motions that tie atmospheric processes together, the motions that create the changing and complex atmospheric system that so far has defied a comprehensive description and explanation.

In its own evolution as the science of atmospheric motion, dynamic meteorology has involved four interacting phases:

1 Derivation of the equations of motion. This phase began in the seventeenth century with Newton's laws of motion, which were transformed by Euler in the eighteenth century into the equations we use today. The first law of thermodynamics was codified near the end of the nineteenth century. This phase of dynamic meteorology continues with the expression of the equations in meteorological coordinates or in approximate forms suitable for various purposes.

2 *Deduction and comparison.* The deduction of the implications of the equations and the comparison with observed phenomena form a second phase of dynamic meteorology. Throughout this book, we have seen how this process proceeds. The geostrophic approximation and the Rossby wave formula are two meteorological and oceanographic abstractions developed by deduction and comparison.

3 *Simulation and prediction.* Although the equations of motion were used to guide weather prediction long before the modern era of digital computers, the success of numerical integration on computers of increasing power has led to an emphasis on simulation and prediction in the second half of this century. These simulations will become increasingly important as we attempt to ascertain whether future climate variations and the effects of human activities may threaten the continuation of life on the planet.

4 *Topological dynamics and metamodeling.* Deceptively simple nonlinear models containing only a few degrees of freedom have been developed recently to represent specific atmospheric phenomena. These models possess an unexpected mathematical richness that promotes interest in the topology of solution spaces. The new directions and new results obtained from these models have stimulated interest in what we may call metamodeling—the study of the modeling process itself.

Throughout these four interacting phases of dynamic meteorology, the goals remain the same: derive the equations of atmospheric motion and utilize them to understand and predict phenomena involving atmospheric flow. Despite the invariance of the goals, new challenges for dynamic meteorology continue to emerge.

The traditional separation of interest between synoptic scale and turbulent scale flow has now been fractured by the necessity of understanding the intermediate mesoscale flows that are responsible for many of the important impacts of weather on human activities. The time scales of interest have expanded as a result of a new emphasis on the dynamics, the simulation, and the prediction of climatic processes.

The existence of regimes of motion within distinct ranges of temporal and spatial scales has long been recognized, and has required that dynamic meteorologists seek appropriate approximate systems of equations. The traditional approach has been to emphasize processes within an identifiable regime and neglect at least some of the interactions between regimes. While valuable for examining dynamics within a flow regime, this approach tends to minimize questions about the processes and principles that determine when a particular regime will appear as a response to the forcing that drives atmospheric motions. Thus a broader view is needed for us to understand the entire atmospheric system.

In the next section of this chapter, we develop a unified scaling procedure that produces approximate equations for the large-scale, the mesoscale, and the turbulent- or convective-scale. As in Chaps. 13 and 14, we emphasize physical estimates rather than formal expansion procedures.

In the second section, we turn to the question of determining which regimes of circulation will prevail when various external conditions are specified. Instead of attempting to isolate regimes, we use scaling procedures to develop a canonical set of equations whose solutions will depend on external parameters rather than the more familiar internal ones such as Rossby and Richardson numbers. Converting these equations to spectral form will lead us to a remarkable result about the asymptotic properties of their solutions.

Finally, we examine some of the issues raised by the diversity and success of atmospheric modeling by stepping back to consider the process of modeling. The final section, then, is a preliminary look at metamodeling.

15.1 ATMOSPHERIC FLOW REGIMES

Dynamic meteorology generally identifies three regimes of atmospheric motion: the large-scale or synoptic-scale component, the mesoscale component, and the smaller-scale convective or turbulent components (Fig. 15.1). Studies of the larger scales have traditionally considered cyclones, but now problems related to blocking, to global relationships between flow patterns, and to climate evolution are all under active investigation.

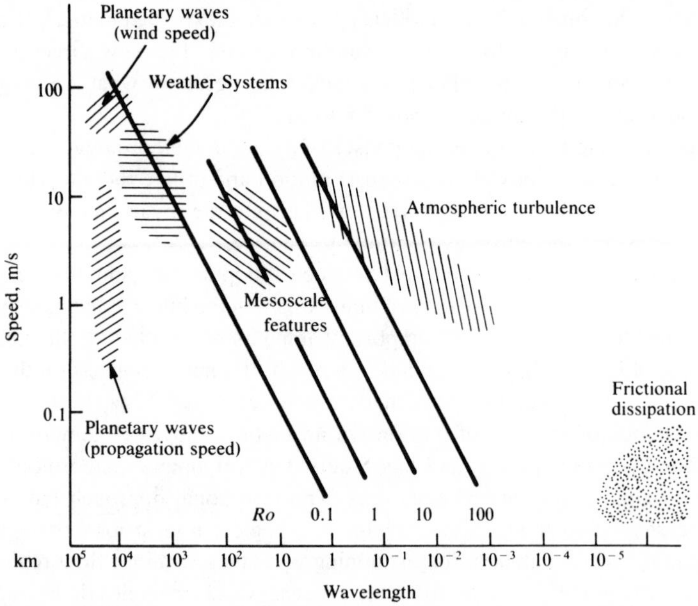

FIGURE 15.1
Scales of atmospheric phenomena and the relation of regimes to the Rossby number *Ro*, redrawn from Fig. 1.1.

Interest in the mesoscale as a separate regime is relatively new and is motivated and focused by the fact that many of the most significant weather effects on human activities arise from mesoscale systems, some of which can be quite intense. Smaller-scale convection and turbulent flow have long been of concern, especially in engineering and in efforts to protect environmental quality; now these flows are also studied because of their role in shaping the interaction of the atmosphere with processes at the surface of the earth.

Although it is widely recognized that the various regimes of motion interact with one another in important ways, it is of value and interest to attempt to isolate each regime to some degree in order to identify and study its features and to understand the physical processes and principles that control its evolution.

The objective of this section is to develop a scaling procedure that will produce a set of equations that can be specialized to each of these regimes by the proper choice of dimensionless parameters. In Chaps. 13 and 14 we developed separately approximate forms of the equations that are applicable to small-scale and to large-scale or quasi-geostrophic flow. In this chapter we use an approach that provides a consistent set of estimates across the spectrum of atmospheric flow regimes. The present method provides less detailed information (in its present form), but has broader applicability. As mentioned before, we emphasize physical reasoning rather than formal mathematical expansions.

15.1.1 Preliminary Analysis

Two basic dimensionless parameters that appear in the analysis of the equations of atmospheric motion are the Rossby number

$$Ro = \frac{U}{fL} \sim \frac{|\mathbf{v}\cdot\nabla\mathbf{v}|}{|f\mathbf{k}\times\mathbf{v}|} \tag{1}$$

specifying the relative importance of the inertial to Coriolis forces and the Richardson number

$$Ri = \frac{g}{\theta}\frac{\partial\theta}{\partial z} \Big/ \left|\frac{\partial\mathbf{v}}{\partial z}\right|^2 \tag{2}$$

expressing the squared ratio of the shear to stratification time scales.

Alternatively, we can consider the aspect ratio D/L of vertical to horizontal length scales and the squared ratio of rotational to stratification time scales

$$\omega_g^2/f^2 = \frac{RiU^2/D^2}{f^2} = RiRo^2(D/L)^{-2} \tag{3}$$

as basic dimensionless parameters.

The development of approximate equations requires selection of variables that permit us to obtain estimates of the relevant variables. To simplify specification of the thermodynamic state variables, we choose again the maximum entropy state of Sec. 11.4 with

$$T_0 = \text{const}, \qquad p_0/p_0(0) = \rho_0/\rho_0(0) = e^{-gz/RT_0}$$

$$\theta_0/\theta_0(0) = e^{gz/c_pT_0} \tag{4}$$

The state is specified by the height scales

$$\frac{1}{H_\theta} = \frac{1}{\theta_0}\frac{\partial \theta_0}{\partial z} = \frac{g}{c_p T_0} = \frac{R}{c_p}\frac{1}{H}, \qquad \frac{1}{H} = -\frac{1}{\rho_0}\frac{\partial \rho_0}{\partial z} \tag{5}$$

and the hydrostatic relation between p_0 and ρ_0.

As we saw in Chap. 14, it is advantageous to use Exner's function $\pi = c_p T/\theta$ to write the pressure gradient and gravitational forces in the form

$$\alpha \nabla p + g\mathbf{k} = \theta \nabla \pi + g\mathbf{k} \tag{6}$$

and then expand

$$\begin{aligned} \theta \nabla \pi + g\mathbf{k} &= \theta_0 \left(1 + \frac{\theta'}{\theta_0}\right) \nabla(\pi_0 + \pi') + g\mathbf{k} \\ &= \theta_0 \nabla \pi' - g\frac{\theta'}{\theta_0}\mathbf{k} \end{aligned} \tag{7}$$

in which we have used the hydrostatic relation $\theta_0 \partial \pi_0/\partial z = -g$ and neglected (*a priori*) the product $\theta' \nabla \pi'$ relative to $\theta_0 \nabla \pi'$. Now we are interested in

$$\theta_0 \nabla \pi' = \nabla \theta_0 \pi' - \pi' \frac{\partial \theta_0}{\partial z}\mathbf{k} \tag{8}$$

and using estimates of the form (13.1.5–6) we find that

$$\left(\pi' \frac{\partial \theta_0}{\partial z}\right) \Big/ \left(\frac{\partial \theta_0 \pi'}{\partial z}\right) = \left(\frac{\theta_0 \pi'}{H_\theta}\right) \Big/ \left(\frac{\partial \theta_0 \pi'}{\partial z}\right) \sim D/H_\theta \tag{9}$$

This ratio is of order 5/25 and we shall neglect the second term of (8).

15.1.2 Scaling Estimates

The basic strategy for developing a set of approximate equations valid across the spectrum of flow regimes is presented in Table 15.1. For larger-scale flow we use geostrophic scaling to estimate π'/π_0 and for smaller scales we assume that the horizontal acceleration is of the order of the pressure gradient force. The ratio θ'/θ_0 is determined by hydrostatic scaling

Table 15.1 VARIABLES AND SOURCES OF ESTIMATES

	Large-scale	**Mesoscale and smaller**
π'/π_0	Geostrophic scaling	Horizontal acceleration
θ'/θ_0	Hydrostatic scaling or thermal wind	Energy transfer (out-of-phase hydrostatic scaling)
w	Continuity equation	Continuity equation
Dimensionless parameters	First law	First law

and the vertical velocity is estimated with the continuity equation. When these scaled variables are substituted in the first law of thermodynamics we can expect the dimensionless parameters to combine in a manner that ensures consistency of the scaling.

We shall find that different estimates appear for large-scale (quasi-geostrophic) flow and for those of smaller scales, but that we can find scaling relations that are valid across the entire spectrum.

The geostrophic scaling

$$\theta_0 \nabla_H \pi' \sim f\mathbf{k} \times \mathbf{v} \tag{10}$$

implies that

$$\theta_0 \pi' \sim fLU \tag{11}$$

while for smaller scales the assumption that

$$\frac{\partial \mathbf{v}_H}{\partial t} \sim \theta_0 \nabla_H \pi' \tag{12}$$

suggests that

$$\theta_0 \pi' \sim LU/T \sim U^2 \tag{13}$$

because of the choice of the advective time scale $1/T = U/L$ obtained from $\partial(\ )/\partial t \sim \mathbf{v}\cdot\nabla(\ )$.

From these estimates we may write that

$$\frac{\pi'}{\pi_0} \sim \left[\frac{fLU}{gH_\theta}, \frac{U^2}{gH_\theta}\right] \tag{14}$$

in which the first component is for large-scale flow and the second for the smaller scales. Here we have used (5) to conclude that $\theta_0 \pi_0 = c_p T_0 = gH_\theta$. Continuing with (14) we write

$$\begin{aligned} \frac{\pi'}{\pi_0} &= \frac{fLU}{gH_\theta}[1, Ro] \\ &= \frac{fLU}{gH_\theta}\left(\frac{1 + Ro^2}{1 + Ro}\right) \end{aligned} \tag{15}$$

in which the function chosen in (15) interpolates adequately at the values $Ro = 0.1$ for larger scales, $Ro = 1$ for mesoscale and $Ro \geq 10$ for turbulent or convective scales. This form shows that the geostrophic and acceleration scaling are equivalent on the mesoscale.

For large-scale flow, direct application of the hydrostatic approximation to (7) gives

$$\theta_0 \frac{\partial \pi'}{\partial z} \sim g \frac{\theta'}{\theta_0} \tag{16}$$

or

$$\frac{\theta'}{\theta_0} \sim \frac{\theta_0 \pi'}{gD} \sim \frac{H_\theta}{D} \frac{\pi'}{\pi_0} \tag{17}$$

For smaller-scale or convective motion, we apply the argument given on page 492: hydrostatic scaling of amplitudes is justified by the energy cycle in which vertical kinetic energy is created by buoyancy forces $g\theta'/\theta_0$ and then converted to horizontal kinetic energy through the action of the pressure gradient forces. Thus we may expect approximate equality of the amplitudes of the vertical pressure gradient and buoyancy terms, as in (16), even though they may be out of phase. With this reasoning, (17) becomes, for all scales,

$$\frac{\theta'}{\theta_0} \sim \frac{fLU}{gD}\left(\frac{1 + Ro^2}{1 + Ro}\right) \tag{18}$$

It is easy to verify that this scaling satisfies the thermal wind requirement for the larger scales for which $Ro \leq 0.1$. Then we have from (18) for hydrostatic and geostrophic conditions

$$f\frac{\partial u}{\partial z} \sim g\frac{\partial}{\partial y}\frac{\theta'}{\theta_0} \sim \frac{fU}{D} \tag{19}$$

Turning to the mass conservation equation, we write $\rho = \rho_0(1 + \rho'/\rho_0)$ and thus

$$\frac{d}{dt}\frac{\rho'}{\rho_0} + \frac{w}{\rho_0}\frac{\partial \rho_0}{\partial z} + \nabla\cdot\mathbf{v} = 0 \tag{20}$$

Now the first law of thermodynamics

$$\frac{c_p}{\theta}\frac{d\theta}{dt} = \frac{q}{T} \tag{21}$$

can be written in the form

$$\frac{d}{dt}\left(\frac{\theta'}{\theta_0}\right) + \frac{w}{H_\theta} = h \qquad h = q/c_pT \tag{22}$$

Then from $\theta = \rho^{-1}(\pi/c_p)^{c_V/R}(p_{00}/R)$ we deduce that

$$\frac{\rho'}{\rho_0} = \frac{c_V}{R}\frac{\pi'}{\pi_0} - \frac{\theta'}{\theta_0} \tag{23}$$

and thus (20) may be written as

$$\frac{c_V}{R}\frac{d}{dt}\frac{\pi'}{\pi_0} + w\left(\frac{1}{H_\theta} - \frac{1}{H}\right) + \nabla\cdot\mathbf{v} = h \tag{24}$$

To estimate the terms in (24) we note that $H/H_\theta \sim 7/25 \sim 0.3$. Moreover, for geostrophic scaling we have $\nabla_H\cdot\mathbf{v} \sim \nabla_H\cdot\mathbf{v}_a$. Now the large-scale acceleration is proportional to $f\mathbf{v}_a$, where $\mathbf{v}_a$ is the ageostrophic component, and so (page 516)

$$\frac{\partial \mathbf{v}}{\partial t} \sim f\mathbf{k} \times \mathbf{v}_a \tag{25}$$

implies that for larger-scales

$$v_a \sim \frac{U}{fT} \sim \frac{U^2}{fL} \sim RoU \tag{26}$$

Hence we have

$$\nabla_H \cdot \mathbf{v} = \frac{U}{L}[Ro, 1] \tag{27}$$

The ratio of the first term of (24) to $\nabla_H \cdot \mathbf{v}$ thus can be estimated as

$$\frac{\dfrac{c_V}{R}\dfrac{U}{L}\dfrac{fLU}{gH_\theta}[1, Ro]}{\dfrac{U}{L}[Ro, 1]} = \frac{c_V}{R}\frac{f^2L^2}{gH_\theta}\frac{Ro[1, Ro]}{[Ro, 1]}$$

$$= \frac{c_V}{R}\frac{f^2L^2}{gH_\theta}[1, Ro^2] \tag{28}$$

The estimates in Table 15.2 show that the first term of (24) may thus be neglected and so may the heating term if the rates

$$\frac{1}{\theta}\frac{d\theta}{dt} \sim h \ll \frac{U}{L}[Ro, 1] \sim fRo[Ro, 1]$$

$$\sim 10Ro[Ro, 1]\,\mathrm{day}^{-1} \tag{29}$$

Large-scale flow is the most restrictive case, and the heating term may be neglected if $d\theta/dt \leq 2.5$ deg/day.

Now the continuity equation is

$$\nabla_H \cdot \mathbf{v} + \frac{\partial w}{\partial z} + \frac{w}{\rho_0}\frac{\partial \rho_0}{\partial z} = 0 \tag{30}$$

Table 15.2 BASIC ESTIMATES

Quantity	Large-scale	Mesoscale	Small-scale
L (length)	$\geqslant 10^3$ km	100 km	$\leqslant 1$ km
D (depth)	5 km[1]	$\leqslant 5$ km	1 km
U (horizontal velocity)	10 m/s	10 m/s	1 – 10 m/s
T ($\sim L/U$)	1 day	0.1 day[2]	100 s[2]
$H_\theta \left(\left[\dfrac{1}{\theta_0}\dfrac{\partial\theta}{\partial z}\right]^{-1}\right)$	25 km	25 km – ∞[3]	25 km – ∞[3]
Ro (U/fL)	$\leqslant 0.1$	1	$\geqslant 10$
Ri ($gD^2/H_\theta U^2$)	100	10 – 100[3]	1 – 10[3]
$(fL)^2/gH_\theta$	0.04	$< 4\ 10^{-4}$	$< 10^{-7}$

Notes: (1) Determined from shear or quarter wavelength of w variation.
(2) Advective time scale; internal time scales may be much smaller.
(3) Negative values occur for unstable stratification.

For small-scale flow with $D/H \ll 1$ we have the estimate

$$W \sim DU/L \tag{31}$$

For large-scale flow we assume $D \sim H$ and so

$$\frac{\partial w}{\partial z} \sim \nabla_H \cdot \mathbf{v}_a + \nabla_H \cdot \mathbf{v}_g \tag{32}$$

or (see page 292)

$$\frac{W}{D} \sim \frac{URo}{L} + \frac{U}{a} \operatorname{cotan} \phi$$

$$\sim \frac{U}{L}\left(Ro + \beta \frac{L}{a}\right) = \frac{URo}{L}(1 + \beta')$$

$$\beta' = (L/a\,Ro) \operatorname{cotan} \phi \tag{33}$$

We assume for simplicity that $L/a \sim Ro$ and so finally

$$W \sim \frac{DU}{L}[Ro, 1] \sim \frac{DURo}{L}\left(\frac{1 + Ro}{1 + Ro^2}\right) \tag{34}$$

15.1.3 Reconciliation of Estimates

Now that we have estimates for each variable we can substitute in the first law of thermodynamics, essentially unused so far in the scaling, to determine the dimensionless parameters that describe the characteristics of the flow regime.

Thus from (18) we set

$$\frac{\theta'}{\theta_0} = \left(\frac{fLU}{gD}\right)\left(\frac{1 + Ro^2}{1 + Ro}\right)\tau^* \tag{35}$$

in which τ^* is the new scaled variable. We also use

$$w = \frac{DURo}{L}\left(\frac{1 + Ro}{1 + Ro^2}\right) w^* \tag{36}$$

and $t = Tt^*$; then the first law (22) gives

$$\frac{d\tau^*}{dt^*} + Ro^2 Ri \left(\frac{1 + Ro}{1 + Ro^2}\right)^2 w^* = h^* \tag{37}$$

The factor of w^* now contains the dimensionless parameters that allow us to distinguish the flow regimes.

15.1.4 The Approximate Equations

With the scaling complete and reconciled in the first law, we may proceed to develop the approximate equations for the various flow regimes. The definitions of the scaled variables are shown in Table 15.3.

Table 15.3 DIMENSIONLESS SCALED VARIABLES

Original form	(Scales) (Dimensionless variables)
(x,y)	$L\ (x,y)$
z	Dz
t	$Tt = (L/U)t$
(u,v)	$U(u,v)$
w	$\frac{DU}{L} Ro \left(\frac{1 + Ro}{1 + Ro^2}\right) w$
π'/π_0	$\frac{f_0 LU}{gH_\theta} \left(\frac{1 + Ro^2}{1 + Ro}\right) \frac{\pi'}{\pi_0}$
θ'/θ_0	$\frac{f_0 LU}{gD} \left(\frac{1 + Ro^2}{1 + Ro}\right) \tau$
$\theta_0 \pi'$	$f_0 LU \left(\frac{1 + Ro^2}{1 + Ro}\right) \vartheta$
Definitions	
$Ro = U/f_0 L$	$f_0 = 2\Omega \sin \phi_0$
$Ri = \frac{g}{H_\theta}\left(\frac{\partial U}{\partial z}\right)^2 \sim \frac{gD^2}{H_\theta U^2}$	$\frac{1 + Ro^2}{1 + Ro} \cong [1, Ro]$
$T = L/U \sim 1/f_0 Ro$	

First we assess the relative sizes of the components of the material derivative by writing

$$\begin{array}{ccccc} \frac{d}{dt} & = \frac{\partial}{\partial t} & + \mathbf{v}_H \cdot \nabla & + w \frac{\partial}{\partial z} \\ \frac{1}{T} & \frac{1}{T} & \frac{U}{L} = \frac{1}{T} & \frac{1}{T} Ro \left(\frac{1 + Ro}{1 + Ro^2}\right) \end{array} \tag{38}$$

The vertical advection term will be smaller than the others for large-scale flow, but of the same size for smaller scales.

To proceed we set $f_0 = 2\Omega \sin \phi_0$, where ϕ_0 is a central latitude, and use the inviscid equations (7.2.15) of motion. Then the horizontal equation of motion, in dimensionless variables (with asterisks now eliminated) obtained from the substitutions in Table 15.3, is

$$Ro \frac{d\mathbf{v}_H}{dt} + \frac{LRo}{a} \tan \phi \, (u^2 \mathbf{j} - uv\mathbf{i}) + \frac{D}{a} Ro^2 \left[\frac{1 + Ro}{1 + Ro^2}\right] (u\mathbf{i} + v\mathbf{j}) w =$$
$$- \left(\frac{1 + Ro^2}{1 + Ro}\right) \nabla_H \vartheta - (f/f_0) \mathbf{k} \times \mathbf{v} - \left(\frac{\cos \phi}{\sin \phi_0}\right) \frac{D}{L} Ro \left[\frac{1 + Ro}{1 + Ro^2}\right] w\mathbf{i} \tag{39}$$

while the vertical equation of motion becomes

$$Ro^2\left(\frac{1+Ro}{1+Ro^2}\right)^2\left(\frac{D}{L}\right)^2\frac{dw}{dt}-\frac{D}{a}Ro\left(\frac{1+Ro}{1+Ro^2}\right)(u^2+v^2)$$
$$=-\frac{\partial\vartheta}{\partial z}+\tau+\left(\frac{\cos\phi}{\sin\phi_0}\right)\left(\frac{1+Ro}{1+Ro^2}\right)\frac{D}{L}u \quad (40)$$

As before, the first law is

$$\frac{d\tau}{dt}+\left[Ro^2Ri\left(\frac{1+Ro}{1+Ro^2}\right)^2\right]w=h \quad (41)$$

and the continuity equation is replaced by

$$\nabla_H\cdot\mathbf{v}+Ro\left(\frac{1+Ro}{1+Ro^2}\right)\left(\frac{\partial w}{\partial z}-\frac{D}{H}w\right)=0 \quad (42)$$

The approximate equations for the three regimes of motion may now be obtained by specifying the parameters Ro, Ri, and D/L appropriately.

For turbulent flow or small-scale convection, we choose $Ro = 10$, $L/a \ll 1$, $D/H \ll 1$, and $D/L = 1$, and thus obtain

$$\frac{d\mathbf{v}}{dt}=-\nabla\vartheta+\tau\mathbf{k}+\cdots$$

$$\frac{d\tau}{dt}+Ri\,w=h+\cdots$$

$$\nabla\cdot\mathbf{v}=0 \quad (43)$$

which are the usual Boussinesq equations with the missing dissipation terms indicated by the ellipses. An analysis of the energetics of these equations and a discussion of their application to turbulent flow is available in Panofsky and Dutton (1984).

For mesoscale flow we choose $Ro = 1$, $L/a \ll 1$, and $D/L \ll 1$ to produce

$$\frac{d\mathbf{v}_H}{dt}=-\nabla_H\vartheta-(f/f_0)\mathbf{k}\times\mathbf{v}+\cdots$$

$$\frac{\partial\vartheta}{\partial z}=\tau$$

$$\frac{d\tau}{dt}+Ri\,w=h+\cdots$$

$$\nabla\cdot\rho_0\mathbf{v}=0 \quad (44)$$

in which the term $-Dw/H$ in the last equation may be neglected for shallow structures in which D/H is small.

For large-scale quasi-geostrophic flow we have $Ro \sim 0.1$ and $D/L \ll 1$. We choose a central latitude ϕ_0 to evaluate f_0 and then define for the dimensionless velocity

$$\mathbf{v}_H=\mathbf{v}_g+Ro\mathbf{v}_a$$

$$\mathbf{v}_g=(f/f_0)\mathbf{k}\times\nabla_H\vartheta \quad (45)$$

and thus obtain, to first order in Ro, the system

$$\frac{d\mathbf{v}_g}{dt} = -(f/f_0)\mathbf{k} \times \mathbf{v}_a + \cdots$$

$$\frac{\partial \vartheta}{\partial z} = \tau$$

$$\frac{d\tau}{dt} + Ro^2 Ri\, w = h + \cdots$$

$$\nabla_H \cdot \mathbf{v}_a + \frac{1}{\rho_0}\frac{\partial}{\partial z}\rho_0 w - \beta v_g = 0, \quad \beta = \frac{L}{a\, Ro} \operatorname{cotan} \phi \tag{46}$$

which for $f = f_0$ are the quasi-geostrophic equations of Sec. 14.2, where now (38) is

$$\frac{d}{dt} = \frac{\partial}{\partial t} + \mathbf{v}_g \cdot \nabla \tag{47}$$

The preceding analysis has been restricted to the individual flow regimes. In many cases, we shall be interested in interactions between regimes, especially when we consider mesoscale or turbulent regimes imbedded in larger-scale flows. In these cases, we would consider velocities $\mathbf{v} = \mathbf{U} + \mathbf{u}$ and temperature fields $\tau = \Theta + \tau'$. The advective time scale would then be given by $1/T = U/L$ and we would have for material derivatives

$$\frac{d\mathbf{u}}{dt} = \frac{\partial \mathbf{u}}{\partial t} + \mathbf{U} \cdot \nabla \mathbf{u} + \mathbf{u} \cdot \nabla \mathbf{U} + \mathbf{u} \cdot \nabla \mathbf{u}$$

$$\frac{d\tau'}{dt} = \frac{\partial \tau'}{\partial t} + \mathbf{U} \cdot \nabla \tau' + \mathbf{u} \cdot \nabla \Theta + \mathbf{u} \cdot \nabla \tau' \tag{48}$$

As we have seen in Chap. 11, the terms $\mathbf{u} \cdot (\mathbf{u} \cdot \nabla \mathbf{U})$ and $\tau'(\mathbf{u} \cdot \nabla \Theta)$ are central to the energetics and possible instabilities of the flow. Moreover, the effect of the larger-scale flow in modifying the hydrostatic stability of the maximum entropy state now appears in the $w\partial\Theta/\partial z$ term.

15.2 CANONICAL EQUATIONS AND THE MODE OF RESPONSE PROBLEM

In the previous section, we developed equations appropriate for some of the regimes of motion that we observe in the atmosphere. Now we turn to the broader question of elucidating which regimes might appear as we vary the external characteristics that force and shape atmospheric motion. This is the mode of response problem and we may state it broadly in the form:

> Determine the characteristics of the flow and the climate of an atmosphere with well-defined physical properties on a planet whose relevant characteristics are specified along with the solar constant and the rotation rate.

The problem can be approached on different levels of complexity. In its simplest form, we can consider that boundary conditions are fixed and that internal heating fields are specified so that we are concerned only with the atmospheric portion of the response. In more complex forms of the problem, we would have to take account of the interactions of the atmosphere with the oceans, surface features, vegetation, and chemical cycles that might exist on the planet.

As a specific example of this problem, we might be interested in determining the conditions that may have been responsible for the variations in climate that have occurred on earth, the most notable examples being the well-known oscillations between glacial and interglacial periods. Are these oscillations caused by external factors such as changes in the intensity of the solar radiation or in variations of the earth's orbit around the sun, or are they caused by long-term internal oscillations between quasi-steady states?

The first and crucial task in attacking the mode of response problem is to determine an appropriate model—a suitable set of equations of motion. Clearly, the approximate equations of the previous section are not appropriate because they contain *a priori* assumptions about the characteristics of the response. We must avoid specification of such internal characteristics as the values of D/L, Ro, or Ri. We must express our equations in a form in which the variations in solutions—the different regimes—will appear as a function of external parameters such as the solar constant and the rotation rate of the planet. Thus we need a canonical set of equations that contain dimensionless variables with full freedom to respond to the external parameters.

For the present exposition, we will examine the most basic form of the mode of response problem for the atmosphere, concentrating on internal dynamics by using the usual temporally independent boundary conditions. To simplify the dynamics even further, we will not treat fluxes or changes of phase of water explicitly, although suitable equations could be added to the model.

15.2.1 The Canonical Equations

There are both physical and mathematical advantages to beginning with a general form of the Boussinesq equations applicable to both larger- and smaller-scale flows. Of course, if the solutions obtained do not satisfy the requirements for the validity of the approximation, then those solutions must be abandoned.

To begin, we assume that we can estimate the total mass and the mean temperature that we need to specify the maximum entropy state of Sec. 11.4. We use this state to obtain the basic state variables (15.1.4). Then we use (15.1.8)–(15.1.9) to modify the equation of motion. The first law is taken to be (15.1.22), with appropriate forms of the irreversible terms appended, and the equation of continuity is replaced by the form (15.1.30). The equations in dimensional form are then

$$\frac{\partial \mathbf{v}}{\partial t} + \mathbf{v}\cdot\nabla\mathbf{v} = -\nabla\theta_0\pi' + g\frac{\theta'}{\theta_0}\mathbf{k} - 2\mathbf{\Omega}\times\mathbf{v} + \alpha_0\nabla\cdot\rho_0\nu\nabla\mathbf{v}$$

$$\frac{\partial}{\partial t}\frac{\theta'}{\theta_0} + \mathbf{v}\cdot\nabla\frac{\theta'}{\theta_0} + \frac{w}{H_\theta} = \frac{Q}{c_p\rho_0 T_0} + \alpha_0\nabla\cdot\rho_0\kappa\nabla\frac{\theta'}{\theta_0}$$

$$\nabla\cdot\rho_0\mathbf{v} = 0 \tag{1}$$

It is important to note that we have not made the hydrostatic approximation and have not introduced the notion of geostrophic balance; it is conceivable that modes of response could appear in which neither would be appropriate. The equations are clearly general enough to permit hydrostatic and geostrophic modes to develop if required. The forms of the viscous force in the equation of motion and the heat conduction term in the first law are chosen in part for mathematical convenience. We have neglected frictional heating in the first law on the grounds that it is negligible compared to radiative heating.

The use of the maximum entropy state has introduced the scale height into the equations (1), and, as indicated earlier, this depends on the average temperature or total energy of the atmosphere and is thus one of the response characteristics in which we are interested. This introduces the necessity of an iteration procedure in which we choose a value of the total energy, perhaps by the procedure given later in this section, and then see if the energy of the response state is sufficiently close; if not, we must choose a new value of the total energy and try again. Presumably, this procedure would converge.

In order to isolate the relevant external parameters, we put the equations in dimensionless form, being careful not to introduce any of the common meteorological assumptions based on the present modes of response. In particular, we eschew introducing the shallowness of the atmosphere into the scaling. Thus we shall introduce a single length scale l (the planet's radius, for example) and shall scale time with the rotation rate. The definitions of the dimensionless variables are shown in Table 15.4 and lead to the system of equations

$$\frac{\partial\mathbf{v}}{\partial t} + \mathbf{v}\cdot\nabla\mathbf{v} = -\nabla\vartheta + \Gamma\tau\mathbf{k} - \boldsymbol{\omega}\times\mathbf{v} + \alpha_0\nabla\cdot\nu\rho_0\nabla\mathbf{v}$$

$$\frac{\partial\tau}{\partial t} + \mathbf{v}\cdot\nabla\tau = -\Gamma w + \Lambda q + P^{-1}\alpha_0\nabla\cdot\nu\rho_0\nabla\tau$$

$$\nabla\cdot\rho_0\mathbf{v} = 0 \tag{2}$$

For the present discussion we shall use the standard boundary conditions that

$$\mathbf{v} = 0 \text{ at } z = 0 \text{ and } z = Z$$

$$\frac{\partial\tau}{\partial z} = 0 \text{ at } z = 0$$

$$\tau = 0 \text{ at } z = Z \tag{3}$$

where Z is the top of the atmosphere and can be any finite value. With the choice $\partial\tau/\partial z = 0$ at $z = 0$ we have eliminated conductive heating and so all of the thermal forcing is contained in the term q. Alternatively, we might require that $\tau = \tau_B(x,y)$ at $z = 0$ as a means of introducing the effect of surface heating and the consequent effects of the distribution of land, ocean, and ice. To do this, it would probably be most convenient to set

Table 15.4 RELATIONS BETWEEN THE DIMENSIONAL AND DIMENSIONLESS VARIABLES

Dimensional variables	(Scales)·(Dimensionless variables)
(x,y)	$l(x,y)$
z	lz
t	$\Omega^{-1}t$
$\mathbf{v}_H = \mathbf{i}u + \mathbf{j}v$	$\Omega l \mathbf{v}_H$
w	$\Omega l w$
$\theta_0 \pi'$	$l^2\Omega^2\vartheta$
$\dfrac{\theta'}{\theta_0}$	$\Gamma\dfrac{\Omega^2 l}{g}\tau$
ν	$l^2\Omega\nu$
κ	$l^2\Omega\kappa = P^{-1}\nu$
Q	$c_p\rho_0 T_0\Omega q$

Definitions

$\Gamma = \dfrac{N}{\Omega}$	$\omega = \dfrac{2\Omega}{\Omega}$
$\Lambda = \dfrac{g}{N\Omega l}$	$N^2 = \dfrac{g}{\theta_0}\dfrac{\partial\theta_0}{\partial z} = \dfrac{g}{H_\theta}$
$\nabla_H = \nabla - \mathbf{k}\dfrac{\partial}{\partial z}$	

$\tau = \tau_B + \tau'$ and then require that $\tau' = 0$ at $z = 0$; the function τ_B and its spatial derivatives would then appear in the equations as forcing terms.

The next task is to identify the parameters whose values will control the structure of the solutions to the equations. The explicit dimensionless forms are Γ and Λ, depending on g, l, N and Ω. Additional contributions may be made by the dimensionless heating function q.

The effects of heating can be examined on several levels. On the first, we presumably would let the intensity of the heating field be proportional to the solar constant and specify the variation with latitude and altitude. The next level of sophistication would be to allow the evolving circulations to affect the heating patterns, and a convenient approach might be to use the Newtonian formulation discussed on page 550. In either case, we shall have to take account of the fact that the maximum entropy structure depends, as mentioned earlier, on the intensity of the solar heating.

In order to obtain a first estimate, we might use the familiar argument that the radiation emitted over the surface of the earth must equal that absorbed on the disc facing the sun. Thus for a solar constant I_0, the equilibrium is specified by

$$4\pi a^2\sigma T_0^4 = \pi a^2(1-\alpha)I_0 \tag{4}$$

where a is the earth's radius, α the albedo, and σ the Stefan-Boltzmann constant. For a solar constant of 1400 watts/m and an albedo of 35 percent, this equation gives an average planetary temperature of about 250 K, and thus Eq. (4) can be used as a first estimate. In developing a more sophisticated model, it would be necessary to model the albedo as a function of the response, for the albedo depends on both cloudiness and the amount of snow cover, and to include an estimate of the effect of atmospheric absorption of the outgoing infrared radiation.

For our present purposes, the balance is written as

$$T_0 = \left[\frac{1}{4}\left(\frac{1-\alpha}{\sigma}\right) I_0\right]^{1/4} \tag{5}$$

and since

$$N^2 = g/H_\theta = g^2/c_p T_0 \tag{6}$$

we can find $N = N(I_0)$. If we put

$$Q = I_0 f(y,z)/H_\theta \tag{7}$$

where f is a dimensionless function specifying the spatial distribution of the net heating, then we have

$$\Lambda q = \frac{f(y,z)\rho_0(0)}{\rho_0(z)}\left[\frac{I_0 N^3}{g^2\Omega^2 l \rho_0(0)}\right] = G(y,z)\,\Delta \tag{8}$$

where now the parameter $\Delta = \Delta_0(I_0, \Omega)$ and the spatial variation is specified by the function G. The same arguments can be used in the Newtonian heating formulation, since the equilibrium temperatures T_e and θ_e that appear there will also depend on the solar constant.

Now we can proceed to consider the solutions to this boundary value problem. We observe first that Jeffreys' theorem of Sec. 6.7 applies, and thus requires that motion be present if $\nabla_H q$ does not vanish. Because of the global implications of this result, we may refer to it as the *first fundamental theorem of atmospheric science.*

We would be interested next in whether an axisymmetric solution would appear for weak axisymmetric heating fields. Dutton and Kloeden (1983) have shown that such a solution exists and that it will be stable for weak-enough heating. Even though we do not know the exact character of such a Hadley regime, we can presume that it will become unstable as the intensity of the heating increases and that more complicated solutions will appear.

We would also be interested in the processes by which circulation is maintained, and with the techniques of Chap. 10 we find with the aid of the boundary conditions and the incompressibility condition that the rates of change of the energy integrals are

$$\dot{K}_H = \frac{\partial}{\partial t}\int \frac{\rho_0 v_H^2}{2}\,d\mathbf{x} = \int \rho_0 w \frac{\partial \vartheta}{\partial z}\,d\mathbf{x} - \int \rho_0 \nu\,|\nabla \mathbf{v}_H|^2\,d\mathbf{x} - B_H + C$$

$$\dot{K}_3 = \frac{\partial}{\partial t}\int \frac{\rho_0 w^2}{2}\,d\mathbf{x} = -\int \rho_0 w \frac{\partial \vartheta}{\partial z}\,d\mathbf{x} + \Gamma \int \rho_0 w \tau\,d\mathbf{x} - \int \rho_0 \nu\,|\nabla w|^2\,d\mathbf{x} - B_3 - C$$

$$\dot{A} = \frac{\partial}{\partial t}\int \frac{\rho_0 \tau^2}{2}\, d\mathbf{x} = -\Gamma \int \rho_0 w\tau\, d\mathbf{x} + \Lambda \int \rho_0 \tau q\, d\mathbf{x} - P^{-1}\int \rho_0 \nu |\nabla\tau|^2\, d\mathbf{x} - B_T \tag{9}$$

Here the integration is over the entire model atmosphere, the terms B_H, B_3, and B_T are all nonnegative boundary dissipation integrals, and C represents conversions between horizontal and vertical kinetic energy induced by the Coriolis forces.

These energy integrals demonstrate that the available energy A is created by covariance resulting from heating of warm regions and cooling of cold regions. As expected in Boussinesq models, the available energy is transferred to vertical kinetic energy through heat flux $w\tau$, and the horizontal and vertical kinetic energies are exchanged through the pressure gradients, expressed as integrals of $\rho_0 w\partial\vartheta/\partial z$ with the aid of incompressibility.

The problem in which we are interested is to determine the modes of response of this model atmosphere as a function of the solar constant I_0, the rotation rate Ω, and the other constants appropriate to the planet and the atmosphere. We would be interested in seeing how the flow regimes change as the parameters are varied, and we would be particularly interested in variations in climate that might appear with variations in heating rate.

In the rest of this section, we look at some of the mathematical techniques that are available to attack this problem analytically.

15.2.2 Spectral Equations and Their Energetics

The spectral technique introduced in Sec. 14.4 allows us to convert the partial differential equations of the model into a set of ordinary differential equations governing the coefficients of a Fourier representation of the variables. Our attention will then turn from the physical space in which the flow of the model atmosphere occurs to the behavior of solution trajectories in the phase space of Fourier coefficients.

As we did with the quasi-geostrophic equation, it is convenient to determine the orthonormal functions to be used in the spectral model by extracting an eigenvalue problem from the partial differential equations themselves. Thus we take the linear, unforced equations, uncouple them by setting $\Gamma = 0$, and let the variables $\mathbf{v}$ and ϑ vary with respect to $\exp(\gamma t)$ to obtain

$$\nabla\cdot\nu\rho_0\nabla\boldsymbol{\psi} = -\gamma\rho_0\boldsymbol{\psi} + \rho_0\nabla\vartheta, \qquad \nabla\cdot\rho_0\boldsymbol{\psi} = 0 \tag{10}$$

from the equation of motion and the incompressibility condition and, similarly,

$$\nabla\cdot\nu\rho_0\nabla\varphi = -\eta\rho_0\varphi \tag{11}$$

from the first law. With the boundary conditions stated earlier, these equations produce discrete eigenvalues $\gamma_1, \gamma_2, \ldots$, and $\eta_1, \eta_2, \ldots$, increasing without bound, and produce associated orthonormal eigenfunctions $\boldsymbol{\psi}_i$ and φ_i, $i = 1, 2, \ldots$. With these eigenfunctions we may write the expansions

$$\mathbf{v}(\mathbf{x},t) = \sum_m a_m(t)\boldsymbol{\psi}_m(\mathbf{x})$$

$$\tau(\mathbf{x},t) = \sum_m b_m(t)\varphi_m(\mathbf{x}) \tag{12}$$

We substitute these expansions in the equations, multiply by the appropriate eigenfunctions and integrate to obtain the spectral equations

$$\dot{a}_n + D_{lmn}a_l a_m = \Gamma E_{mn}b_m - C_{mn}a_m - \nu\lambda_n a_n$$

$$\dot{b}_n + F_{lmn}a_l b_m = -\Gamma E_{nm}a_m + \Lambda\hat{q}_n - P^{-1}\nu\mu_n b_n, \qquad l, m, n = 1, 2, \cdots \tag{13}$$

In these equations we sum on the repeated indices l and m, and with ν taken to be constant we have used the definitions $\gamma_n = \nu\lambda_n$ and $\eta_n = \nu\mu_n$. The coefficients in these equations are defined by

$$D_{lmn} = \int \rho_0 \boldsymbol{\psi}_n\cdot(\boldsymbol{\psi}_l\cdot\nabla)\boldsymbol{\psi}_m \, d\mathbf{x} = -D_{lnm}$$

$$F_{lmn} = \int \rho_0 \varphi_n(\boldsymbol{\psi}_l\cdot\nabla)\varphi_m \, d\mathbf{x} = -F_{lnm}$$

$$C_{mn} = \int \rho_0 \boldsymbol{\omega}\cdot(\boldsymbol{\psi}_m \times \boldsymbol{\psi}_n) \, d\mathbf{x} = -C_{nm}$$

$$E_{mn} = \int \rho_0(\boldsymbol{\psi}_n\cdot\mathbf{k})\varphi_m \, d\mathbf{x}$$

$$\hat{q}_n = \int \rho_0 \varphi_n q \, d\mathbf{x} \tag{14}$$

As we saw with the quasi-geostrophic equation in spectral form, the antisymmetry of the coefficients displayed in the first three definitions creates energy relations for the spectral equations analogous to those for the partial differential equations.

In particular, if we truncate the system (13) with $l, m, n, = 1, 2, \ldots, N$, then we have a finite set of ordinary differential equations whose solutions obey the energy relation

$$\dot{E}_N = \frac{1}{2}\frac{d}{dt}\sum_{n=1}^{N}(a_n^2 + b_n^2)$$

$$= \Lambda\sum_{n=1}^{N} b_n\hat{q}_n - \nu\sum_{n=1}^{N}(\lambda_n a_n^2 + P^{-1}\mu_n b_n^2) \tag{15}$$

This energy relation allows us to show that the energy of the system is always finite and that solutions to the truncated systems exist.

With the Schwarz inequality (Problem 2.2.4) for sums applied to the first term and with the fact that the eigenvalues increase with n, we may convert (15) into the inequality

$$\dot{E}_N \leq \Lambda\left(\sum_{n=1}^{N} b_n^2 \sum_{n=1}^{N} \hat{q}_n^2\right)^{1/2} - \nu\,\mathrm{Min}(\lambda_1, \mu_1)\sum_{n=1}^{N}(a_n^2 + P^{-1}b_n^2)$$

$$\leq \sqrt{2}\Lambda E_N^{1/2} Q_N - 2\nu\eta E_N, \qquad \eta = \frac{\mathrm{Min}(\lambda_1, \mu_1)}{\mathrm{Max}(1, P)} \tag{16}$$

in which

$$Q_N = \left(\sum_{n=1}^{N} \hat{q}_n^2\right)^{1/2} \tag{17}$$

This differential inequality implies that

$$E_N^{1/2}(t) \leq e^{-\nu\eta t} E_N^{1/2}(0) + \frac{\Lambda Q_N}{\sqrt{2}\nu\eta}[1 - e^{-\nu\eta t}] \tag{18}$$

and this shows that $E_N(t)$ is uniformly bounded in t when $E_N(0)$ and Q_N are finite. Moreover, it is evident that if

$$E_N(0) > \frac{1}{2}\left(\frac{\Lambda Q_N}{\nu\eta}\right)^2 \tag{19}$$

then

$$E_N(t) < E_N(0), \qquad t > 0 \tag{20}$$

Finally, these relations and inequalities are valid for all values of the truncation limit N when both Q_∞ and $E_\infty(0)$ are finite, as we may assume for applications to the atmosphere.

The truncated equations are analytic in the coefficients **a** and **b** and we have shown above that the solutions will have finite energy for all finite values of the truncation limit N. These facts are sufficient to ensure that for all $N < \infty$ the equations (13) truncated at N have solutions that exist for all time and have nice properties; the details are given in Dutton (1982).

15.2.3 The Flow in Phase Space

Study of the spectral equations is facilitated by turning our attention from the physical space in which the flow occurs to the behavior of the Fourier coefficients in phase space. We replace the original physical flows with the equivalent trajectories in phase space; the properties of the original flows are revealed by the study of the topological properties of the flows in phase space. In its abstract form, this approach is known as topological dynamics.

We observe, then, that the system (13) is equivalent to the autonomous ordinary differential system

$$\dot{\mathbf{y}} = \mathbf{F}(\mathbf{y}) \tag{21}$$

in which the $2N$-vector $\mathbf{y} = (a_1, \ldots, a_N, b_1, \ldots, b_N)$ represents a point in the phase space R^{2N}. The temporal evolution of the velocity and temperature fields

$$\mathbf{v}^N(\mathbf{x},t) = \sum_{n=1}^{N} a_n(t)\boldsymbol{\psi}_n(\mathbf{x})$$

$$\tau^N(\mathbf{x},t) = \sum_{n=1}^{N} b_n(t)\varphi_n(\mathbf{x}) \tag{22}$$

is thus equivalent to the movement of the phase point $\mathbf{y}(t)$ along a trajectory in R^{2N}. The power of this method appears when we let $\phi_t(\mathbf{y})$ be the solution of (21) starting from the initial point $\mathbf{y}$ at $t = 0$. Then the map or flow

$$\phi_t : R^{2N} \to R^{2N} \tag{23}$$

describes a transformation of R^{2N} into itself that represents the evolution of the velocity and temperature fields for all initial conditions in an appropriate space. Conceptually, we now observe the deformation of the phase space and attempt to prove theorems about its limiting behavior.

15.2.4 The Trapping Theorem and Its Consequences

The energy inequality (18) shows that the flow $\phi_t(\mathbf{y})$ maps any bounded set in R^{2N} into the ball

$$B_N = \left\{ \mathbf{y} \,\middle|\, |\mathbf{y}| \le \frac{\Lambda Q_N}{\nu\eta} \right\} \tag{24}$$

as $t \to \infty$. Using the technique introduced by Lorenz (1963), we can obtain a remarkable theorem.

We complete the square in (15) to obtain

$$\dot{E}_N = -\nu \sum_{n=1}^{N} \lambda_n a_n^2 - P^{-1} \sum_{n=1}^{N} \mu_n \left(b_n - \frac{\Lambda \hat{q}_n P}{2\mu_n} \right)^2 + P\Lambda^2 \sum_{n=1}^{N} \frac{\hat{q}_n^2}{4\mu_n} \tag{25}$$

The hyperellipse C_N specified by $\dot{E}_N = 0$ divides the phase space R^{2N} into an interior domain in which $\dot{E}_N > 0$ (let $a_n = 0$, $b_n = \Lambda \hat{q}_n P/2\mu_n$) and an exterior domain in which $\dot{E}_N \le 0$ (when the magnitudes of **a** and **b** become large enough, the first two negative terms dominate).

The consequence is that trajectories starting from initial conditions with large values of E_N must continuously cross surfaces $E_N =$ const toward smaller values of E_N until they reach the hyperellipse $\dot{E}_N = 0$. Now we let E_N^* be the surface of constant energy that intersects $\dot{E}_N = 0$ tangentially at just one point and thus encloses it completely. Clearly, trajectories starting at large E_N are eventually trapped inside E_N^*, and those starting from small values of E_N can never pass outside of E_N^*.

From (25) we see that the hyperellipse C_N passes through the origin $\mathbf{a} = \mathbf{b} = 0$ in phase space and through the point $a_n = 0$, $b_n = \Lambda \hat{q}_n P/\mu_n$, $n = 1, 2, \ldots, N$. The conclusions we have reached are illustrated in Fig. 15.2.

To obtain a more formal statement of our result, let A_N be the closed ball in R^{2N} with exterior boundary E_N^*. The set A_N is invariant, because with the surface $\dot{E}_N = 0$ contained inside A_N, no trajectory in A_N can cross its boundary toward larger values of E_N. All trajectories with initial points outside of A_N have distances from its surface E_N^* that decrease monotonically with time since the trajectory must move toward smaller values of energy as long as it is outside of the surface C_N on which $\dot{E}_N = 0$. Thus because all trajectories are eventually confined inside A_N, we refer to it as a global attractor, and we may state the trapping theorem:

Theorem (Lorenz, 1963) Let Q be finite. Then for $N < \infty$, any trajectory $\phi_t(\mathbf{y})$ in phase space starting at a point $\mathbf{y}$ with finite energy proceeds asymptotically to the

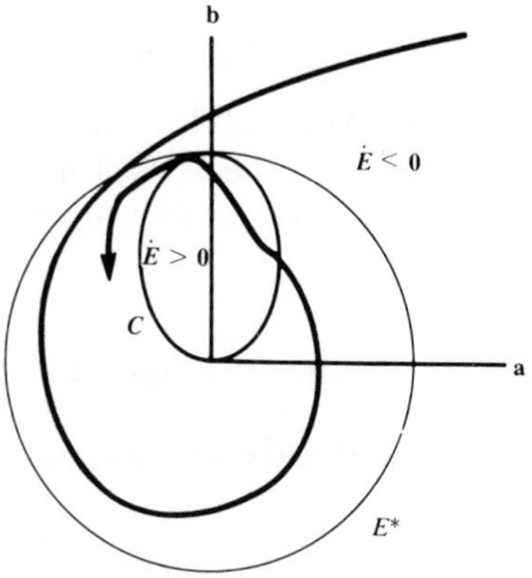

FIGURE 15.2
Schematic illustration of the geometry in phase space, with the ball contained inside the sphere E^* being a global attractor.

closed ball A_N with exterior boundary E_N^* and so A_N is a global attractor as $t \to \infty$ for any bounded part of phase space. ////

We shall refer to this result as the *second fundamental theorem of atmospheric science*. It asserts that dissipation and heating combine to collect all trajectories in a restricted region of phase space. The mathematical and physical aspects of interest will thus be revealed by studying the behavior of the trajectories inside the attractor A_N; we have reduced the problem considerably by bounding the region of interest in phase space.

Corollary 1 The system of equations (13) has at least one steady solution, or equivalently, the transformation $\phi_t(\mathbf{y})$ has at least one fixed point in A_N.

PROOF Any ball B in R^{2N} with boundary E_N = const $> E_N^*$ is a closed set. The continuous transformation $\phi_t(\mathbf{y})$ maps B into itself and so by Brouwer's fixed point theorem, there exists for every fixed $t > 0$ at least one point such that $\phi_t(\mathbf{x}) = \mathbf{x}$. The finite intersection property applied to sets of these fixed points can then be used to show that they have at least one common member $\mathbf{x}_0$ for which $\phi_t(\mathbf{x}_0) = \mathbf{x}_0$ for all t; details are given in Dutton (1982). ////

At such a fixed point, the energy is constant and so all of the fixed points lie on the hyperellipse C_N and thus are in the attractor A_N.

Corollary 2 Every trajectory $\phi_t(\mathbf{y})$ starting at an initial point with finite energy has at least one limit point in A_N, and hence there is at least one point for which the trajectory is recurrent.

PROOF The sequence $\phi_{t_n}(\mathbf{y})$, $n = 1, 2, \ldots$, is bounded since the energy $E_N(t)$ is either less than the initial energy or is bounded by E_N^*. The Bolzano-Weierstrass theorem that every bounded infinite sequence contains a convergent subsequence thus implies the existence of a subsequential limit point $\mathbf{y}^L$ in A. By the definition of such a subsequential limit point, for every $\varepsilon > 0$ there exists an M such that

$$|\phi_{t_{nk}}(\mathbf{y}) - \mathbf{y}^L| < \varepsilon, \quad n_k > M \tag{26}$$

Thus the trajectory comes arbitrarily close to $\mathbf{y}^L$ arbitrarily often and hence is recurrent in any neighborhood containing $\mathbf{y}^L$. ////

This is again a remarkable result for the meteorologist or climatologist, since it says that the finite-dimensional flow in physical space will repeatedly pass through patterns that are indistinguishable from the limiting pattern. Furthermore, it is well known that the trajectory $\phi_t(\mathbf{y}^L)$ starting from $\mathbf{y}$ is composed entirely of limit points of $\phi_t(\mathbf{y})$ when ϕ_t is the flow of an autonomous equation. Hence there is in A_N a collection of trajectories composed entirely of limit points.

It can be shown that the set S of all limit points in A_N is a closed and invariant (but not necessarily connected) set. Since all limit points generate a set of trajectories composed entirely of limit points, the set S is a collection of limiting trajectories that govern the asymptotic behavior of the finite-dimensional flow. These trajectories determine the climate of the mathematical model.

Being able to confine our attention to a specific set of trajectories offers some hope that we may be able someday to answer the basic questions underlying a theory of climate:

- What is the precise location of S inside A?
- Is S connected when the heating is strong enough?
- Are there periodic orbits in S and are they dense in S?
- Is the flow ϕ_t transitive? (Is there a dense orbit?)
- Is ϕ_t ergodic in the sense that we can find a probability measure such that averages along trajectories are equal to integrals over S?
- Do the properties of the flow ϕ_t and of the limit set change with slight changes in the approximate equations?
- How do the properties of ϕ_t and of S change as I_0 and Ω change?

The answers to these questions would furnish a mathematical theory of the topological dynamics of climate. The answer to the last question would help us to understand flow transitions and climatic change.

It is possible, with a simple result, to outline what we would expect to find. The energy equation (15) shows that the origin $(\mathbf{a},\mathbf{b}) = 0$ is a stable fixed point when q vanishes, since the perturbation energies will decrease monotonically. As heating is introduced, the fixed point will move away from the origin, and because the eigenvalues determining linear stability about the fixed point depend continuously on the $\hat{q}_n$, we know that the fixed point will be stable for weak-enough heating. Then when the fixed point becomes unstable, there will be a bifurcation with more fixed points appearing or perhaps a bifurcation to a stable periodic solution. As the heating is increased further, the limit set S will become increasingly complex. The attractor may become "strange," in the sense that it contains no stable fixed points or periodic trajectories, and the flow may begin to resemble turbulence at some scales.

The properties of spectral models such as these have turned out to be quite complex, even when severe truncation is applied. Nevertheless, these finite-dimensional systems are presumably easier to work with than the partial differential equations and there is hope that a fairly complete theory can be obtained. An open question is how large the truncation limit N must be for the spectral models to have solutions that adequately represent those of the partial differential equations.

As an indication that the discrete and continuous models may share important features, we prove a trapping theorem for the partial differential equations.

The two eigenvalue problems establish the Poisson inequalities that

$$\int \nu\rho_0 \mid \nabla \mathbf{v} \mid^2 d\mathbf{x} \geq \gamma_1 \int \rho_0 \mid \mathbf{v} \mid^2 d\mathbf{x} \tag{27}$$

and

$$\int \nu\rho_0 \mid \nabla \tau \mid^2 d\mathbf{x} \geq \eta_1 \int \rho_0 \mid \tau \mid^2 d\mathbf{x} \tag{28}$$

because the minimum values γ_1 and η_1 are attained only for the first eigenfunctions. With these inequalities and the Schwarz inequality (Problem 2.2.4), the energy equation (15) can be converted into the inequality

$$\begin{aligned} \dot{E} &= \dot{K}_H + \dot{K}_3 + \dot{A} \\ &= \Lambda \int \rho_0 \tau q \, d\mathbf{x} - \int \rho_0 \nu [\mid \nabla \mathbf{v} \mid^2 + P^{-1} \mid \nabla \tau \mid^2] \, d\mathbf{x} - B \\ &\leq \sqrt{2} \Lambda E^{1/2} Q - 2\nu\eta E, \quad \eta = \frac{\text{Min } (\gamma_1, \eta_1)}{\text{Max } (1, P)} \end{aligned} \tag{29}$$

and thus we obtain

$$E^{1/2}(t) \leq e^{-\nu\eta t} E^{1/2}(0) + \frac{\Lambda Q}{\sqrt{2}\nu\eta} [1 - e^{-\nu\eta t}] \tag{30}$$

as the trapping theorem for the solutions of the partial differential equations.

We might continue from this point to examine the question of the existence of solutions for the boundary value problem (2). An attractive approach is to let $N \to \infty$ in the spectral model. With the results of Ladyzhenskaya (1969) as a prototype, we would expect to be able to show that the solutions $\mathbf{v}^N(\mathbf{x},t)$ and $\tau^N(\mathbf{x},t)$ converge weakly to functions $\mathbf{v}(\mathbf{x},t)$ and $\tau(\mathbf{x},t)$ that are weak solutions of the boundary value problem in finite intervals $0 \leq t \leq T$. In this case, weak solutions are defined by

$$\int_0^T \int \mathbf{\Phi} \cdot \left[\frac{\partial \mathbf{u}}{\partial t} + F(\mathbf{u}) \right] d\mathbf{x} \, dt = 0, \qquad \mathbf{u} = (\mathbf{v}, \tau) \tag{31}$$

with the original differential system represented by the expression in brackets. Here $\mathbf{\Phi}$ is a member of a suitable space of smooth functions. The notion of weak solution is invaluable here, and permits specific results to be obtained. In some cases, it is possible to go further and show that such weak solutions satisfy the boundary value problem in the usual, classical sense. To date, there is no proof that solutions exist for all time, even though they do for the spectral model for all truncation limits $N < \infty$.

15.3 METAMODELING IN ATMOSPHERIC SCIENCE

Throughout this book we have been concerned with constructing and examining mathematical models of atmospheric flow patterns and flow regimes. The equations of motion, it is often said, govern the flow of the atmosphere, but that statement is not correct. The equations of motion do not govern, but rather attempt to represent atmospheric flow, to portray in mathematical form the interactions that shape its motions.

Despite the success of atmospheric science in representing various aspects of atmospheric motion with partial differential equations and parameterizations, such forms will never be exact. The full spectrum of atmospheric phenomena includes a range of processes too complex for us ever to measure or represent in detail, either analytically or numerically. We choose to turn to models to describe, comprehend, and predict atmospheric phenomena. Many of us, then, devote ourselves to the study of models, perhaps because they represent atmospheric events, or perhaps because the models are of interest themselves and present intellectual challenges that engage our curiosity.

We might well improve our abilities to devise models and to determine their implications by studying the process of modeling. In this section then, we turn to a preliminary examination of modeling in atmospheric science. Because we are taking a broader view and looking at modeling as a process, we shall refer to the study of modeling as metamodeling. The antecedent for this terminology appears in the metamathematical studies of the processes of mathematics.

15.3.1 Purposes of Models

It is useful to start with a definition:

> A model is a reduced and parsimonious representation of a physical, chemical, or biological system in a mathematical, abstract, numerical, or experimental form.

The notions of reduction and parsimony introduced in the definition are both important. One value of a model lies in the simplification achieved by reduction of complexity; we use models to represent physical or natural systems in simplified forms of manageable proportions—either in size or complexity. Parsimony is also crucial, for it is the efficiency or elegance of representation that makes a model useful as a reduced version of the prototype system.

The definition is intended to cover a wide range of representations—empirically based descriptions or simplifications, systems of equations whose solutions resemble the behavior of the prototype, numerical versions of such solutions obtained with computers, and finally the physical analogues developed for laboratory experimentation. Theories, while intended to codify knowledge about the prototype system, are usually constructed relative to a model of the system.

Models have a number of related purposes:

1 Organize and document knowledge. The sciences are replete with examples in which models serve to encapsulate the understanding obtained from observational or experimental efforts. Obvious examples in atmospheric science include the Norwegian cyclone model, the logarithmic wind-profile model of the mean wind in the surface layer, the Rossby wave formula model of the oscillations in the westerlies, and of course the equations of motion as a mathematical model of the dynamics and thermodynamics of atmospheric flow.

2 Reveal links between cause and effect. Scientists are curious about the causes of the events or processes they see around them, and often represent their conclusions in models. Newton's second law tells us to look for forces as the causes of accelerations. The concept of hydrostatic stability leads us to expect strong convection when the atmosphere is conditionally unstable.

3 Test understanding; reveal issues for research and observation. If a predictive model is successful, then presumably we have achieved an understanding of the process we are studying; if the model fails to predict correctly, then we must examine our assumptions and perhaps seek additional observations. Sometimes in modeling, we seek to simplify models to the point that they are qualitatively correct even though some quantitative detail is lost. In this attempt at parsimony, we often achieve some understanding of the major interactions of the physical system.

4 Predict changes in structure as forcing or essential parameters vary. How does the period of a pendulum depend on its length or the acceleration of gravity? What controls whether flow in a pipe will be laminar or turbulent? When will a mountain snowfield begin to move and become a roaring avalanche? How will the climate change if we increase or decrease the solar constant?

5 Provide forecasts of evolutionary behavior. The concern for the future is widespread, and many disciplines use models of various kinds to predict or simulate future events. Meteorologists use both empirical models and mathematical models, in approximate forms or in numerical forms on computers, in the attempt to predict the weather for periods up to ten days in advance. What kinds of models might be developed that could predict climate on the time scale of hundreds of years?

15.3.2 Types of Models

The definition of models stated earlier emphasizes the representation of a natural system without specifying the many levels or forms of models that are possible. Despite the difficulties of categorizing models, we enumerate some of the most common types used in science.

The first type we refer to as *organizational models.* These summarize observation or knowledge and often are used to reduce complex fields or patterns into conceptuall simple forms. The classical meteorological example is the model of fronts associated wit cyclones. The analyst of a weather chart uses this model to organize the synoptic dat and emphasize the regions of interest. A meteorologist can look at such a frontal patter and infer the associated weather even though the detailed observations are not presented.

Other forms of organizational models exist, too. The computer programs that are used to create numerical models exist at several levels: the overall plan for developing such a program is itself an organizational model, for complex physical and dynamical relations are organized in a way that provides understanding and suggests areas of research.

The second type contains the *dynamical models* in which rates of change are inferred from current conditions. We can divide this group of models into three kinds:

Predictive models are created with the specific aim of anticipating future events. The numerical weather-prediction models are an obvious example. The value of predictive models is to be assessed solely by their ability to make accurate and useful predictions, although considerations such as cost and speed may govern their availability and applicability.

Simulation models are developed with the goal of simulating processes and in numerical form are often used to perform experiments that would be impossible in the atmosphere. Such efforts can be expected to contribute to the discussion of questions about the future viability of life on this planet. For example, computer models are used to simulate the effects on the global carbon cycle of fossil-fuel utilization.

The numerical models of the general circulation are intended to simulate the statistics of large-scale atmospheric flow and are judged by their success in doing so, not by their success in weather prediction. Similarly, we may find that successful models of climatic processes will simulate the range of variation that has occurred, but not be useful for prediction.

Abstract mathematical models are created to encapsulate important aspects of a physical process in a manageable form while ignoring aspects that may be of secondary importance or relevant in broader contexts than those for which the model is intended. Abstract mathematical models are parsimonious in nature, trading detailed accuracy for tractability. Examples include the representation in Chap. 4 of a vertically moving parcel as a harmonic oscillator, a bulk model of climate processes that includes mean atmospheric and oceanic temperatures and an ice boundary position as its three variables, and severely truncated spectral models.

As mentioned earlier, simple spectral models can reveal important aspects of the transition from steady to periodic flows as forcing is increased, but would not be expected to model correctly a fully turbulent flow. Thus abstract mathematical models are to be judged by their contribution to the understanding of physical or mathematical structure. In this sense the simple spectral models and the associated topological dynamics help us to comprehend some of the processes of nonlinear fluid dynamics. Whether or not such models become truly representative of actual processes as the truncation limit is increased is a valid and important concern of metamodeling.

The third type of model includes *laboratory or experimental devices* designed to simulate processes in a controlled, geometrically and dynamically similar environment. Wind tunnels have long been used to investigate flow around miniature versions of bodies such as aircraft or buildings. Experiments of some relevance to atmospheric flow are performed with differentially heated liquids confined to rotating containers.

15.3.3 Principles for Constructing Effective Models

Most scientists who work with models have well-developed intuitions about what const tutes an effective model. They know that they must always compromise accuracy fc tractability or economical use of computers, and yet they will only be satisfied with model that deals with those features of the physical system that appear to be essential i the context in which they are studying it. In the spirit of a preliminary look at metamodeling it is useful to try to list some characteristics of successful models.

A model should:

1 Possess adequate complexity. The point in modeling is to reduce the represen tation to the essentials, but not to strip away phenomena or processes that are contextuall essential. The model must be simple, but not simplistic. Thus we may model the transitio between conduction and steady convection in a fluid heated below by systems with onl a few degrees of freedom, but nonlinearity must be retained in the model for it is essentia to the physics of the transition. Geostrophic equilibrium is a useful concept for estimatin wind speeds, but does not portray the processes responsible for maintaining the kineti energy against friction; the quasi-geostrophic theory necessarily adds a divergent com ponent of the wind.

2 Possess conservation properties. The basic conservation laws of physics con cerning mass, energy, and momentum should not be violated in reduced or approximat systems. We have seen in Part III and in Sec. 15.2 that these laws sometimes take nev forms in approximate systems of equations, but nevertheless they are present in a forn appropriate to the context. Similarly, in reducing partial differential equations to spectra or grid point models, it is essential that conservation laws be available for the new form of the variables.

3 Emphasize the energy-containing features. Although interactions among al scales of motion are present in fluid flow, we generally will be interested in the shapes patterns, or processes associated with the temporal or spatial scales that contain the bul of the energy. No model can be completely successful unless it accounts for the observe maxima of energy at certain locations in either physical or phase space. The great succes of Planck's model of the energy spectrum of black-body radiation was that it successfull represented the shape of the spectral density at the maximum between the two spectra tails modeled by the previously available asymptotic formulas.

4 Maximize the information content per component or degree of freedom. Re gardless of its actual orientation in (x,y,z) space, we always represent the motion of simple pendulum in a two-dimensional plane rotated to coincide with the plane of motion and we usually further simplify matters by recognizing that the radius is constant and tha only the angle θ from the vertical need be considered. Similarly, in fluid dynamics it i essential to try to maximize the information represented by each degree of freedom. Spec tral models are often constructed with basis functions that reflect the geometry of th domain, but as we have seen in Chap. 14 and Sec. 15.2, it may be advantageous to us the equations themselves to generate basis functions for spectral expansions (see Dutto and Wells, 1984). No formal theory of maximizing information content in fluid dynami models now exists (to my knowledge), but such a theory would be an important contribution to the theory of modeling.

5 *Preserve overall structure under sufficiently small perturbations of the form of the model.* We would be suspicious of a model in which quite different solutions were obtained with slight alterations in the constants or with an arbitrarily small term added to an equation. Thus we require that successful models be reasonably robust in their formulation and not be sensitive to variations in the values of parameters that are smaller than those we can measure. In mathematics, this property is called structural stability.

6 *Represent configurations that occur with finite probability.* A model representing a situation that we would expect to encounter with only zero probability is evidently of limited practicality. Many models are restricted to zero-probability cases in an attempt at simplicity, but such models usually need dramatic restructuring to represent reality. As an example, note that a model with viscous forces, however small the viscosity, will be quite different from an inviscid companion. A flow over a horizontal surface is to be expected to occur with zero probability compared to the large probability of a flow occurring over a surface with small slope. We are concerned, then, with whether our models are mathematically generic, with whether they represent the usual case.

7 *Preserve structure under extension to more degrees of freedom or to more complex forms.* We often work with models that are reduced or truncated to be as simple as possible, but we do not want our view of the phenomena we study to be radically altered as we make the models more realistic. Thus we know that a simple spectral model will not represent a turbulent flow, but we should be sure if we truncate the original model to examine a situation characterized by weak forcing that we obtain the same components that would be active in a fully general model for the same intensity of forcing. In a sense, we have returned to our first statement that models should possess adequate complexity. Given that complexity, a model can be simplified for special cases with the assurance that it can be extended again as the scope of our interest widens.

15.3.4 Metamodeling

Modeling, as we have seen, is concerned with representing physical systems or phenomena in forms that allow us to understand them and to predict their changes in structure or their evolution. Metamodeling is concerned with the process of modeling and with the properties of models. The goal of metamodeling is to develop a theory of modeling.

Metamodeling will evolve, we would hope, from descriptions of types of models and desirable properties to a discipline in which theorems about models could be proved, just as theorems about mathematics are proved in metamathematics. Models are always an abstraction of an actual physical system; metamodeling seeks to abstract the properties of models. Many of our models, such as those composed of systems of partial differential equations, provide us with a language that we can use to examine or to discuss a physical process. Metamodeling seeks to provide a language with which to discuss modeling and to understand the process of modeling more thoroughly.

We see, then, a progression from the actual physical system to higher levels of abstraction. In general, higher-level abstractions or higher-level languages provide structures that are intellectually more effective and conceptually more revealing. Consider the great progress that has been made in scientific computing because scientists can think and write programs in higher-level languages rather than within the stultifying confines of machine languages. In a similar sense, the transformation of the dynamics represented by the physical equations into the topological dynamics of the flow of a phase space provides new insight and powerful new mathematical concepts with which to examine the evolution of the system.

So we seek in metamodeling to arrive at higher-level concepts and perhaps higher-level linguistic structures with which to examine models of the physical systems in which we are interested. As we saw in the previous subsection, metamodeling leads us to consider what generic properties models should possess. It leads us to issues of structural stability. As we progress in metamodeling, we shall be increasingly concerned with the generic properties of hydrodynamic attractors, and we shall attempt to take advantage of the fact that the flows in phase space are often, perhaps generically, confined to rather restricted portions of those phase spaces. Precise formulations of these concepts so that theorems can be proved will certainly provide new insight into such issues as hydrodynamic predictability and the ranges of variation that can be expected in both the internal free oscillations and the externally forced oscillations of climate systems.

The concept of metamodeling and higher-level approaches also leads to the notion of a metamodel—a mathematical or computational structure that generates models and provides information about their properties. An effective metamodel would again move the work of the scientist to a higher level, a level at which the interactions of the human mind and the metamodel would be expected to produce even more profound results than have been obtained from the contemplation of models.

The history of science is clearly tied to progress in constructing and verifying models of natural systems; the success of science and mathematics lies in the steady progression to higher levels of structured thought. Metamodeling may well provide important new advances in atmospheric sciences, in the understanding of the ceaseless motions of the atmosphere in which we live.

ACKNOWLEDGMENTS

Some of the material presented in this chapter was developed under the sponsorship of the National Science Foundation through grant ATM-7908354 and the National Aeronautics and Space Administration through grant NAS 8-33794 to The Pennsylvania State University.

The ideas discussed in Sec. 15.3 have been shaped through discussion with many colleagues, most notably Hampton N. Shirer and Robert Wells of Penn State, and with my associates on the NASA Earth System Science Committee and its Modeling Working Group. Comments by Alistair B. Fraser stimulated improvements in the manuscript.

The Society for Industrial and Applied Mathematics has kindly consented to liberal quotations from my *SIAM Review* article cited below.

BIBLIOGRAPHIC NOTES

15.1 Additional discussion of this topic can be found in Chaps. 13 and 14 and in the references cited there, as well as in the following:

Panofsky, H. A., and J. A. Dutton, 1984: *Atmospheric Turbulence, Models and Methods for Engineering Applications,* John Wiley & Sons, New York, 397 pp.

Pielke, Roger A., 1984: *Mesoscale Meteorological Modeling,* Academic Press, Orlando, Fla., 612 pp.

Different and more advanced approaches to large-scale approximate equations for quasi-geostrophic flow regimes are given in:

Hoskins, Brian J., 1975: "The Geostrophic Momentum Approximation and the Semi-Geostrophic Equations," *J. Atmosph. Sci.,* **32**:233–242.

Pedlosky, Joseph, 1984: "The Equations for Geostrophic Motion in the Ocean," *J. Phys. Oceanogr.,* **14**:448–455.

15.2 The section derives entirely from:

Dutton, J. A., 1982: "Fundamental Theorems of Climate Theory—Some Proved, Some Conjectured," *SIAM Review,* **24**:1–33.

Related discussion may be found in:

Lorenz, E. N., 1963: "Deterministic Non-Periodic Flow," *J. Atmosph. Sci.,* **20**:130–141.

Serrin, James, 1962: "The Initial Value Problem for the Navier-Stokes Equations," in R. Langer (ed.), *Nonlinear Problems,* University of Wisconsin Press, Madison.

Ladyzhenskaya, O. A., 1969: *The Mathematical Theory of Viscous Incompressible Flow* (trans. by R. A. Silverman), 2d ed., Gordon and Breach, Science Publishers, Inc., New York, 184 pp.

Ladyzhenskaya, O. A., 1975: "Mathematical Analysis of Navier-Stokes Equations for Incompressible Fluids," *Ann. Rev. Fluid Mech.,* **7**:249–272.

Dutton, J. A., and P. E. Kloeden, 1983: "The Existence of Hadley Convective Regimes of Atmospheric Motion," *J. Austral. Math. Soc. Ser. B,* **24**:318–338.

An application of the ideas of this section to the predictability problem is given in:

Dutton, J. A., and R. Wells, 1984: "Topological Issues in Hydrodynamic Predictability," in Greg Holloway and Bruce J. West (eds.), *Predictability of Fluid Motions* (La Jolla Institute—1983), American Institute of Physics.

15.3 An exposition of the mathematical aspects of topological dynamics relevant to metamodeling may be found in:

Chillingworth, D. R. J., 1976: *Differential Topology with a View to Applications,* Pitman Publishing Ltd., London, 291 pp.

Metamodeling issues associated with the unfolding of singularities are examined in detail by:

Shirer, H. N., and R. Wells, 1983: "Mathematical Structure of the Singularities at the Transitions Between Steady States in Hydrodynamic Systems," *Lecture Notes in Physics 185,* Springer-Verlag, Berlin, 276 pp.

A recent review of bulk climate models is:

Saltzman, Barry, 1983: "Climatic System Analysis," *Advances in Geophysics,* **25**:173–233.

Some metamodeling issues relative to the dimension of hydrodynamic attractors and the roles of stable and unstable manifolds are examined in the article by Dutton and Wells (1984), cited above.

APPENDIX 1

THE FIRST LAW OF THERMODYNAMICS

This brief appendix provides a discussion of the first law of thermodynamics at the level usually adopted in beginning dynamics courses. It is an alternative to Sec. 3.1, but it provides considerably less information than the totality of information developed there.

Conservation of energy is a basic principle of physics; the first law is an expression of this principle applicable to thermodynamic systems. It is an axiom that codifies our experience with thermodynamic substances.

The basic energy budget, then, says that the change in the energy of a thermodynamic system (macroscopically at rest) must equal the sum of the mechanical work done on (or by) the system and the thermal work done on (or by) the system. Thus

$$\dot{I} = W + Q \tag{A.1}$$

where I is the thermodynamic or internal energy and $\dot{I}$ is its rate of change, W is the rate at which mechanical work is done on the system, and Q is the thermal power applied to the system. If $W < 0$, then the system is performing work on its environment; if $Q < 0$, then thermal energy is flowing from the system to the environment. The existence of the internal energy follows from the observation that if

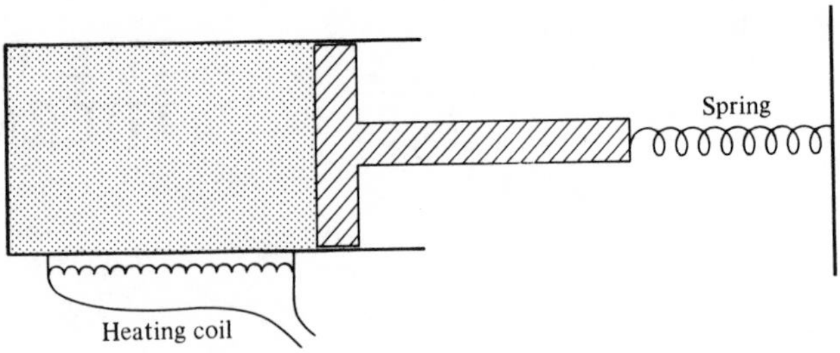

FIGURE A.1
A system illustrating the first law of thermodynamics.

work is done on a system whose macroscopic kinetic energy and potential energy do not change, then there must be another form of energy that did increase.

These concepts are illustrated by the system shown in Fig. A.1. The gas is confined in a cylinder by a piston. If the piston compresses the gas further, then $W > 0$; if the gas expands, pushing the piston against the spring, then $W < 0$. If we heat the gas, we would expect that its energy would increase and that it would expand, in agreement with Eq. (A.1).

Let the cylinder in Fig. A.1 be of cross-sectional area A. The piston moves inward a distance ds compressing the gas. Then, since the pressure p of the gas must be overcome by the piston, the work rate (force · velocity) is

$$W = pA\frac{ds}{dt} = -p\frac{dV}{dt} \tag{A.2}$$

where V is the volume occupied by the gas and $dV/dt < 0$ for compression of the gas so that $W > 0$. If the gas expands, then $dV/dt > 0$ and $W < 0$.

In the figure, thermal power is applied by heat conduction through the bottom of the cylinder. Other forms of thermal power include radiation and heating due to latent heat released by condensation of water vapor.

We assume that the internal energy is a function of the variables p, V, and T describing the state of the system. Because of the equation of state, we can always reduce it to a function of two of these variables, say,

$$I = I(T, V) \tag{A.3}$$

By the chain rule, we have

$$\dot{I} = \left(\frac{\partial I}{\partial T}\right)_V \dot{T} + \left(\frac{\partial I}{\partial V}\right)_T \dot{V} \tag{A.4}$$

Now suppose the heater is turned off, and the piston is pulled out of the cylinder at a speed faster than the speed of the fastest molecule of gas. Then the gas will expand into the surrounding vacuum. But in this expansion, it does no work because there is no compressive force. Moreover, the mass of the gas does not change, nor does its kinetic energy. But from Problem 2.3.1, we see that the temperature of the gas is proportional to the molecular kinetic energy and is thus constant. From Eqs. (A.1) and (A.4) we therefore have the result that

$$\dot{I} = \left(\frac{\partial I}{\partial V}\right)_T \dot{V} = 0 \tag{A.5}$$

which implies that $(\partial I/\partial V)_T$ vanishes since $\dot{V} > 0$. Hence the internal energy is a function $I = I(T)$ of temperature alone for an ideal gas. We define the specific heat

$$C_V = \left(\frac{\partial I}{\partial T}\right)_V$$

and so the first law for an ideal gas now is

$$C_V \frac{dT}{dt} + p\frac{dV}{dt} = Q \tag{A.6}$$

From this equation we find that

$$\frac{C_V}{T}\frac{dT}{dt} + \frac{p}{T}\frac{dV}{dt} = \frac{C_V}{T}\frac{dT}{dt} + \frac{RM}{V}\frac{dV}{dt} = \frac{Q}{T} \tag{A.7}$$

Thus by dividing by T, we have converted the left side of Eq. (A.6) into an exact derivative

$$\frac{dS}{dt} = \frac{C_V d\ln T}{dt} + \frac{RM\,d\ln V}{dt} = \frac{Q}{T} \tag{A.8}$$

The quantity S is known as the entropy of the system.

The reader may now return to the main text on page 42, to the paragraph containing Eq. (5) in Sec. 3.2, although the definitions of thermodynamic terms in Table 3.1 should be noted.

APPENDIX 2

INTERNATIONAL SYSTEM OF UNITS / MATHEMATICAL SYMBOLS

As this book went to press, the American Meteorological Society adopted the International System of Units (known as SI, from its name Le Système International) for use in its scientific journals. SI is basically a meter-kilogram-second metric system, and thus differs from the centimeter-gram-second system generally used here mainly in the powers of 10 that occur in numerical expressions.

The basic units of the SI system are shown in Table A.1. Derived SI units for various quantities that appear in the book are shown in Table A.2.

The most significant impact of SI on meteorology would be changing to use of the pascal as the unit for pressure. Since $1 \text{ Pa} = 1 \text{ kg/m}\cdot\text{s}^2 = 10 \text{ g/cm} \cdot \text{s}^2 = 10^{-2}$ mbar, the usual 1000 mbar $= 10^5$ Pa. Thus 1013 mbar = 101.3 kPa. At present, the millibar is still the unit accepted by international meteorological agreement for observations and data compilations.

The advantage of any uniform and consistent system of units (be it SI or cgs) is that if all units in an equation are expressed in the system, then the computed numerical values will be expressed in the system. Regardless of the system chosen, some conversion of quantities perceived or observed on scales other than those in the system will have to be made. For example, the sizes of many meteorological phenomena are measured as thousands of kilometers on synoptic charts.

Table A.1 BASIC SI UNITS

Quantity	Name	Symbol
Length	meter	m
Mass	kilogram	kg
Time	second	s
Electric current	ampere	A
Thermodynamic temperature	kelvin	K
Amount of substance	mole	mol
Luminous intensity	candela	cd

Prefixes denoting decimal multiples and fractions that allow these conversions to be made easily are shown in Table A.3. It is noteworthy (and perhaps a bit inconsistent) that the basic unit of mass in SI is 1000 g, and that prefixes are to be attached to grams, not to kilograms.

In any case, today's meteorology student will have to be familiar with SI, the cgs system, and, in some applications, British engineering units. Facility in conversion is thus more essential than slavish adherence to any particular system.

Table A.2 SOME SI DERIVED UNITS

		Symbol		
Quantity	Name	Powers of base units	Special name	Expressed in other units
Area	square meter	m^2		
Volume	cubic meter	m^3		
Speed, velocity	meter per second	m/s		
Acceleration	meter per second squared	m/s^2		
Divergence	per second	s^{-1}		
Vorticity	per second	s^{-1}		
Geopotential	meter squared per second squared	m^2/s^2		
Density	kilogram per cubic meter	kg/m^3		
Specific volume	cubic meter per kilogram	m^3/kg		
Force	newton	$m \cdot kg/s^2$	N	
Pressure	pascal	$kg/m \cdot s^2$	Pa	N/m^2
Energy	joule	$m^2 \cdot kg/s^2$	J	$N \cdot m$
Power	watt	$m^2 \cdot kg/s^3$	W	J/s
Dynamic viscosity	pascal second	$m^{-1} \cdot kg/s$	. . .	$Pa \cdot s$
Moment of force	newton meter	$m^2 \cdot kg/s^2$	. . .	$N \cdot m$
Heat flux density	watt per square meter	kg/s^3	. . .	W/m^2
Entropy	joule per kelvin	$m^2 \cdot kg/s^2 \cdot K$	. . .	J/K
Gas constant, universal	joule per kelvin	$m^2 \cdot kg/s^2 \cdot K$	. . .	J/K
Specific heat capacity	joule per kilogram kelvin	$m^2/s^2 \cdot K$	. . .	$J/kg \cdot K$
Specific energy	joule per kilogram	m^2/s^2	. . .	J/kg
Thermal conductivity	watt per meter kelvin	$m \cdot kg/s^3 \cdot K$	. . .	$W/m \cdot K$
Energy density	joule per cubic meter	$m^{-1} \cdot kg/s^2$	. . .	J/m^3

A fuller account of the details of SI in meteorology may be found in the *Bulletin of the American Meteorology Society*, 55:926–930, August 1974.

Table A.3 PREFIXES FOR DECIMAL MULTIPLES AND FRACTIONS

Multiple	Prefix	Symbol	Fraction	Prefix	Symbol
10^{12}	tera	T	10^{-1}	deci	d
10^{9}	giga	G	10^{-2}	centi	c
10^{6}	mega	M	10^{-3}	milli	m
10^{3}	kilo	k	10^{-6}	micro	μ
10^{2}	hecto	h	10^{-9}	nano	n
10^{1}	deka	da	10^{-12}	pico	p
			10^{-15}	femto	f
			10^{-18}	atto	a

Table A.4 MATHEMATICAL SYMBOLS

$\cong$	approximately equal
$\sim$	equal in magnitude or approximately equal in numerical estimates
$\ll$	much less than
$f(z) \to a$	$\lim f(z) = a$
(a, b) $[a, b]$	open and closed intervals
const	constant
max, min	maximum and minimum
ln	logarithm to base e
$\overline{(\ )}$	average (416, 440) or complex conjugate
$(\hat{\ })$	mass weighted average (416, 441) or Fourier coefficient (467, 490)
Re, Im	real and imaginary parts of a complex number
$[a_{ij}]$	matrix with elements a_{ij}; i = row, j = column
$[a_{ij}]^T = [a_{ji}]$	transpose of a matrix
$\lvert a_{ij}\rvert$	determinant of $[a_{ij}]$; expansion of, 143
δ_{ij}	Kronecker delta, 126
ϵ_{ijk}	permutation symbol, 135
g_{jk}, g^{jk}	metric tensors, 137
$g = \lvert g_{ij}\rvert = \lvert J_x^x\rvert$	magnitude of metric tensor, 138
$f_t = \partial f/\partial t$ $f_x = \partial f/\partial x$	partial derivatives, 19
$(\partial/\partial x)^n f = \partial^n f/\partial x^n$	nth order derivative, 29
$d(\ )/dt = \dot{(\ )} = \partial(\ )/\partial t_\xi$	material derivative, 31, 162
∇	gradient operator, 103; $\nabla_H = \nabla_z$, ∇_p, ∇_θ, ∇_ϑ denote ∇ with subscripted variable constant, but for ∇_κ see 459

Table A.4 MATHEMATICAL SYMBOLS (*Continued*)

$\nabla \cdot \mathbf{A}$ $\nabla \times \mathbf{A}$	divergence and curl, 103, 104
$(\mathbf{v} \cdot \nabla\phi)_\vartheta$, $(\nabla \cdot \mathbf{v})_\vartheta$	operations in ϑ coordinates, 240, 242
$x_{,\xi} = \partial x/\partial \xi$	partial derivative, 118
$A_{i;k}$ $A^i_{;k}$	covariant derivatives, 140
$\Gamma_{ik}{}^t$	Christoffel symbol, 140
$\lvert J^x_\xi \rvert = \lvert \partial x_i/\partial \xi_j \rvert$	Jacobian determinant, 123, 125
$J(f,g) = f_x g_y - f_y g_x$	Jacobian operation, 340
$\int(\ )dV = \int_V (\ )dV = \iiint_{\text{atm}}(\ )dV$	integral over the atmosphere, 20
$\int_{x_o}^{x} f(x')dx'$	x' (sometimes x_1) is a dummy variable of integration
$\int_C \mathbf{A} \cdot \tau\, ds$ $\int_S \eta \cdot \mathbf{A}\, d\sigma$	line and surface integrals, 108
$\oint(\ )dt = \int_t^{t+\tau} (\ )dt'$	τ is the time to complete a circuit
$\iint_{xy}(\ )dx\,dy$	integration over the xy plane
$\sum_{i=1}^{N} b_i = b_1 + b_2 + \ldots b_N$	summation
$a_i b_i = \sum_{i=1}^{3} a_i b_i$	summation convention

APPENDIX 3

USEFUL FORMULAS, IDENTITIES, AND CONSTANTS

Meteorological Coordinate Transformations

$$(x,y,z) \to (x,y,\phi(x,y,z,t)) \qquad (\partial\phi/\partial z \neq 0)$$

$$f(x,y,z,t) \to f(x,y,\phi(x,y,z,t),t)$$

$$\frac{\partial f}{\partial t_z} = \frac{\partial f}{\partial t_\phi} + \frac{\partial f}{\partial \phi}\frac{\partial \phi}{\partial t_z}$$

$$\frac{\partial f}{\partial z} = \frac{\partial f}{\partial \phi}\frac{\partial \phi}{\partial z}$$

$$\nabla_z f = \nabla_\phi f + \frac{\partial f}{\partial \phi}\nabla_z \phi$$

$$= \nabla_\phi f - \frac{\partial f}{\partial z}\nabla_\phi h_\phi$$

$$\frac{df}{dt} = \frac{\partial f}{\partial t_\phi} + \mathbf{v}\cdot\nabla_\phi f + \frac{d\phi}{dt}\frac{\partial f}{\partial \phi}$$

$$(\nabla\cdot\mathbf{v})_\phi = \nabla_\phi\cdot\mathbf{v} + \frac{\partial}{\partial\phi}\left(\frac{d\phi}{dt}\right) \qquad (\mathbf{v}\cdot\nabla f)_\phi = \mathbf{v}\cdot\nabla_\phi f + \frac{d\phi}{dt}\frac{\partial f}{\partial\phi}$$

	CARTESIAN OR SPHERICAL	ISOBARIC	ISENTROPIC
Complete equations of motion	Eqs. 15–19, Sec. 7.2	Eq. 31, Sec. 7.3	Eq. 53, Sec. 7.3
Horizontal pressure gradient force	$-\alpha\,\nabla_z p$	$-g\,\nabla_p h_p$	$-\nabla_\theta(c_p T + g h_\theta) = -\nabla_\theta \Psi$
Hydrostatic equation	$\frac{\partial p}{\partial z} = -g\rho$	$g\frac{\partial h_p}{\partial p} = -\alpha$	$\frac{\partial \Psi}{\partial \theta} = c_p(p/p_{00})^{R/c_p} = c_p T/\theta$
Geostrophic wind velocity	$\mathbf{v}_g = \frac{1}{\rho f}\mathbf{k}\times\nabla_z p$	$\frac{g}{f}\mathbf{k}\times\nabla_p h_p$	$\frac{1}{f}\mathbf{k}\times\nabla_\theta\Psi$

Integral Theorems

The Divergence Theorem

$$\iiint_V \nabla \circ \mathbf{A}\, dV = \iint_S \eta \circ \mathbf{A}\, d\sigma$$

Stokes' Theorem

$$\iint_S \boldsymbol{\eta} \cdot (\nabla \times \mathbf{A})\, d\sigma = \int_C \boldsymbol{\tau} \cdot \mathbf{A}\, ds$$

$$\iint_S (\eta \times \nabla) \circ \mathbf{A}\, d\sigma = \int_C \boldsymbol{\tau} \circ \mathbf{A}\, ds$$

The Transport Theorem

$$\frac{d}{dt}\int_V f\, dV = \int_V \left(\frac{\partial f}{\partial t} + \nabla \cdot f\mathbf{w}\right) dV$$

$$\frac{d}{dt}\int_V \rho f\, dV = \int_V \left\{\rho \frac{df}{dt} + \nabla \cdot [\rho f(\mathbf{w} - \mathbf{v})]\right\} dV$$

$\mathbf{w}$ = *velocity of points in volume V*
$\mathbf{v}$ = *fluid velocity*
$\mathbf{w} = \mathbf{v}$ *for a material volume*

Cylindrical Coordinates

Radius r, azimuth λ, height z

$h_r = 1, h_\lambda = r, h_z = 1, J = r$

$$\nabla f = \mathbf{i}_r \frac{\partial f}{\partial r} + \frac{\mathbf{i}_\lambda}{r}\frac{\partial f}{\partial \lambda} + \mathbf{i}_z \frac{\partial f}{\partial z}$$

$$\nabla \cdot \mathbf{A} = \frac{1}{r}\frac{\partial}{\partial r}(rA_r) + \frac{1}{r}\frac{\partial A_\lambda}{\partial \lambda} + \frac{\partial A_z}{\partial z}$$

$$\nabla \times \mathbf{A} = \mathbf{i}_r\left(\frac{1}{r}\frac{\partial A_z}{\partial \lambda} - \frac{\partial A_\lambda}{\partial z}\right) + \mathbf{i}_\lambda\left(\frac{\partial A_r}{\partial z} - \frac{\partial A_z}{\partial r}\right) + \mathbf{i}_z\left[\frac{1}{r}\frac{\partial}{\partial r}(rA_\lambda) - \frac{1}{r}\frac{\partial A_r}{\partial \lambda}\right]$$

$$\nabla^2 f = \frac{1}{r}\frac{\partial}{\partial r}\left(r\frac{\partial f}{\partial r}\right) + \frac{1}{r^2}\frac{\partial^2 f}{\partial \lambda^2} + \frac{\partial^2 f}{\partial z^2}$$

Physical Constants

Mean radius of the earth	$a = 6371$ km
Acceleration due to gravity	$g = 9.80616$ m/s^2 (sea level, latitude 45°)
Earth's angular velocity	$\Omega = 7.292 \cdot 10^{-5}$/s
Coriolis parameter	$f = 2\Omega \sin\phi = 10^{-4}$/s at $\phi = 43°17'$
Rate of change of f	$\beta = \partial f/a\partial\phi = 1.618 \cdot 10^{-11}$/m · s at $\phi = 45°$
Mechanical equivalent of heat	1 cal = 4.186 J
Latent heat of vaporization of water	$L_{lv} = [597 - 0.57\,(T - 273)]$ cal/g
Latent heat of fusion of water	$L_{il} = 80$ cal/g

Dry Air

Gas constant	$R = 0.28704$ J/g·K
Specific heats	$c_p = 1.005$ J/g·K (at 273 K)
	$= 1.006$ J/g·K (288 K)
	$c_V = c_p - R = 0.718$ J/g·K (at 273 K)
	$= 0.719$ J/g·K (at 288 K)
Viscosity	$\mu = 1.72 \cdot 10^{-4}$ g/cm · s (at 273 K)
Thermal conductivity	$k = 2.43 \cdot 10^{3}$ erg/cm · s · K (at 273 K)
Speed of sound	$c = 331$ m/s (at 273 K)

Liquid Water (*at 288 K, 1 atm*)

Specific heat	$c_p = 4.186$ J/g·K ($c_V = 0.997\, c_p$)
Expansion coefficients	$\epsilon = 1.5 \cdot 10^{-4}$/K (isobaric expansion)
	$\eta = 5 \cdot 10^{-5}$/atm (isothermal expansion)
Viscosity	$\mu = 1.14 \cdot 10^{-2}$ g/cm·s
Thermal conductivity	$k = 5.9 \cdot 10^{-3}$ J/cm · s·K

Note: K denotes degrees Kelvin = degrees C + 273 *1 atm = 1013 mbar*

Vector Differentiation Identities (Sec. 5.1)

Eq. (29) $\nabla \times \nabla\phi = 0$

Eq. (30) $\nabla \cdot (\nabla \times \mathbf{A}) = 0$

Eq. (31) $\nabla \cdot \mathbf{x} = 3$

Eq. (32) $\nabla \times \mathbf{x} = 0$

Eq. (33) $\nabla \cdot (\phi\mathbf{A}) = \phi\nabla \cdot \mathbf{A} + (\mathbf{A} \cdot \nabla)\phi$

Eq. (34) $\nabla \times (\phi\mathbf{A}) = \phi\nabla \times \mathbf{A} + (\nabla\phi) \times \mathbf{A} = \phi\nabla \times \mathbf{A} - \mathbf{A} \times (\nabla\phi)$

Eq. (35) $\nabla \cdot (\mathbf{A} \times \mathbf{B}) = \mathbf{B} \cdot (\nabla \times \mathbf{A}) - \mathbf{A} \cdot (\nabla \times \mathbf{B})$

Eq. (36) $\nabla(\mathbf{A} \cdot \mathbf{B}) = (\mathbf{A} \cdot \nabla)\mathbf{B} + (\mathbf{B} \cdot \nabla)\mathbf{A} + \mathbf{A} \times (\nabla \times \mathbf{B}) + \mathbf{B} \times (\nabla \times \mathbf{A})$

Eq. (37) $\nabla \times (\mathbf{A} \times \mathbf{B}) = \mathbf{A}\nabla \cdot \mathbf{B} + (\mathbf{B} \cdot \nabla)\mathbf{A} - \mathbf{B}\,\nabla \cdot \mathbf{A} - (\mathbf{A} \cdot \nabla)\mathbf{B}$

Eq. (38) $\nabla \times (\nabla \times \mathbf{A}) = \nabla(\nabla \cdot \mathbf{A}) - \nabla^2\mathbf{A}$

Eq. (39) $\mathbf{A} \times (\nabla \times \mathbf{B}) - (\mathbf{A} \times \nabla) \times \mathbf{B} = \mathbf{A}\nabla \cdot \mathbf{B} - (\mathbf{A} \cdot \nabla)\mathbf{B}$

Spherical Coordinates

Radius r, longitude λ, latitude φ

$h_r = 1, h_\lambda = r\cos\phi, h_\phi = r, J = r^2\cos\phi$

$$\nabla f = \mathbf{i}_r\frac{\partial f}{\partial r} + \frac{\mathbf{i}_\lambda}{r\cos\phi}\frac{\partial f}{\partial \lambda} + \frac{\mathbf{i}_\phi}{r}\frac{\partial f}{\partial \phi}$$

$$\nabla \cdot \mathbf{A} = \frac{1}{r^2}\frac{\partial}{\partial r}(r^2A_r) + \frac{1}{r\cos\phi}\frac{\partial A_\lambda}{\partial \lambda} + \frac{1}{r\cos\phi}\frac{\partial}{\partial \phi}(\cos\phi\, A_\phi)$$

$$\nabla \times \mathbf{A} = (r^2\cos\phi)^{-1}\left\{\mathbf{i}_r\left[\frac{\partial rA_\phi}{\partial \lambda} - \frac{\partial}{\partial \phi}(r\cos\phi\, A_\lambda)\right] + r\cos\phi\,\mathbf{i}_\lambda\left[\frac{\partial A_r}{\partial \phi} - \frac{\partial rA_\phi}{\partial r}\right] + r\mathbf{i}_\phi\left[\frac{\partial}{\partial r}(r\cos\phi\, A_\lambda) - \frac{\partial A_r}{\partial \lambda}\right]\right\}$$

$$\nabla^2 f = \frac{1}{r^2}\frac{\partial}{\partial r}\left(r^2\frac{\partial f}{\partial r}\right) + \frac{1}{r^2\cos^2\phi}\frac{\partial^2 f}{\partial \lambda^2} + \frac{1}{r^2\cos\phi}\frac{\partial}{\partial \phi}\left(\cos\phi\frac{\partial f}{\partial \phi}\right)$$

INDEXES

INDEX

(Separate Index to Chapter 15 follows.)

INDEX TO CHAPTER 15